Lecture Notes in Computer Science   9582

Commenced Publication in 1973
Founding and Former Series Editors:
Gerhard Goos, Juris Hartmanis, and Jan van Leeuwen

Editorial Board

David Hutchison
  Lancaster University, Lancaster, UK
Takeo Kanade
  Carnegie Mellon University, Pittsburgh, PA, USA
Josef Kittler
  University of Surrey, Guildford, UK
Jon M. Kleinberg
  Cornell University, Ithaca, NY, USA
Friedemann Mattern
  ETH Zurich, Zürich, Switzerland
John C. Mitchell
  Stanford University, Stanford, CA, USA
Moni Naor
  Weizmann Institute of Science, Rehovot, Israel
C. Pandu Rangan
  Indian Institute of Technology, Madras, India
Bernhard Steffen
  TU Dortmund University, Dortmund, Germany
Demetri Terzopoulos
  University of California, Los Angeles, CA, USA
Doug Tygar
  University of California, Berkeley, CA, USA
Gerhard Weikum
  Max Planck Institute for Informatics, Saarbrücken, Germany

Ilias S. Kotsireas · Siegfried M. Rump
Chee K. Yap (Eds.)

Mathematical Aspects of Computer and Information Sciences

6th International Conference, MACIS 2015
Berlin, Germany, November 11–13, 2015
Revised Selected Papers

 Springer

Editors
Ilias S. Kotsireas
Wilfrid Laurier University
Waterloo, ON
Canada

Chee K. Yap
New York University
New York, NY
USA

Siegfried M. Rump
Hamburg University of Technology
Hamburg
Germany

ISSN 0302-9743   ISSN 1611-3349   (electronic)
Lecture Notes in Computer Science
ISBN 978-3-319-32858-4   ISBN 978-3-319-32859-1   (eBook)
DOI 10.1007/978-3-319-32859-1

Library of Congress Control Number: 2016935965

LNCS Sublibrary: SL1 – Theoretical Computer Science and General Issues

Printed on acid-free paper

This Springer imprint is published by Springer Nature
The registered company is Springer International Publishing AG Switzerland

Preface

Mathematical Aspects of Computer and Information Sciences (MACIS) is a series of biennial conferences focusing on research in mathematical and computational aspects of computing and information science. It is broadly concerned with algorithms, their complexity, and their embedding in larger logical systems. At the algorithmic level, there is a rich interplay along the numerical/algebraic/geometric/topological axes. At the logical level, there are issues of data organization, interpretation, and associated tools. These issues often arise in scientific and engineering computation where we need experimental and case studies to validate or enrich the theory. MACIS is interested in outstanding and emerging problems in all these areas. Previous MACIS conferences have been held in Beijing (2006, 2011), Paris (2007), Fukuoka (2009), and Nanning (2013). MACIS 2015 was held at the Zuse Institute Berlin (ZIB) located in the capital of Germany, in the vicinity of the Freie Universität Berlin. Named after Konrad Zuse, the inventor of the first programmable computer, ZIB is an interdisciplinary research institute for applied mathematics and data-intensive high-performance computing. Its research areas in modeling, simulation, and optimization in partnership with academia and industry are exemplary of the goals of MACIS.

We are grateful to the session organizers (and their referees) for their critical role in putting together the successful technical program. We also wish to extend our gratitude to all MACIS 2015 conference participants—all of them contributed in making the conference a success. The conference would not have been possible without the hard work of the local organizers from ZIB, Winfried Neun, and Benedikt Bodendorf, and the generous support of our sponsors, namely, Maplesoft and Zuse Institute Berlin (ZIB).

This volume contains 55 refereed papers, i.e., seven invited papers and 48 submitted papers, all of which were presented at MACIS. The papers are organized in sections corresponding to 12 special sessions featured in the MACIS 2015 conference. The topics of the MACIS 2015 sessions cover a wide array of research areas as follows:

SS1: Vikram Sharma: Curves and Surfaces
SS2: Jon Hauenstein: Applied Algebraic Geometry
SS3: Johannes Blömer: Implementations of Cryptography
SS4: Takeshi Ogita: Verified Numerical Computation
SS5: Johannes Blömer and Jan Camenisch: Cryptography and Privacy
SS6: Chengi Mou and Eric Schost: Polynomial System Solving
SS7: Maxime Crochemore and Costas Iliopoulos: Managing Massive Data
SS8: Viktor Levandovskyy, Alexey Ovchinnikov, Michael Wibmer: Computational Theory of Differential and Difference Equations
SS9: Xiaoyu Chen and Jie Luo: Data and Knowledge Exploration
SS10: Rudolf Fleischer and Stefan Schirra: Algorithm Engineering in Geometric Computing

SS11: Akitoshi Kawamura and Martin Ziegler: Real Complexity: Theory and Practice

SS12: Jordan Ninin: Global Optimization

We wish to thank all the session organizers for their hard work in putting together these sessions.

February 2016

Ilias S. Kotsireas
Siegfried M. Rump
Chee K. Yap

Organization

General Chair

Ilias S. Kotsireas — Wilfrid Laurier University, Canada

Local Organization

Winfried Neun — Zuse Institute Berlin, Germany
Benedikt Bodendorf — Zuse Institute Berlin, Germany

Program Chairs

Siegfried Rump — Hamburg University of Technology, Germany
Chee Yap — Courant Institute, NYU, USA

Program Committee

Johannes Blömer — University of Paderborn, Germany
Jan Camenisch — IBM Zurich, Switzerland
Xiaoyu Chen — Beihang University, Beijing, China
Maxime Crochemore — Kings College London, UK
Rudolf Fleischer — German University of Technology in Oman
Mark Giesbrecht — University of Waterloo, Canada
Jonathan Hauenstein — Notre Dame University, USA
Costas Iliopoulous — Kings College London, UK
Akitoshi Kawamura — University of Tokyo, Japan
R. Baker Kearfott — University of Louisiana, Lafayette, USA
Viktor Levandovskyy — RWTH Aachen University, Germany
Jie Luo — Beihang University, Beijing, China
Chenqi Mou — Beihang University, China
Jordan Ninin — ENSTA Bretagne, France
Takeshi Ogita — Tokyo Woman's Christian University, Japan
Alexey Ovchinnikov — CUNY Graduate Center, USA
Mohab Safey El-Din — University of Pierre and Marie Curie, Paris, France
Michael Sagraloff — Max Planck Institute, Saarbrücken, Germany
Stefan Schirra — Otto von Guericke University Magdeburg, Germany
Éric Schost — Western University, London, Ontario, Canada
Vikram Sharma — Institute of Math Sciences, Chennai, India

Thomas Sturm	Max Planck Institute, Saarbrücken, Germany
Michael Wibmer	RWTH Aachen, Germany; University of Pennsylvania, USA
Martin Ziegler	Technical University Darmstadt, Germany

MACIS Steering Committee

Thomas Sturm (Chair)	Universitat Autonoma de Barcelona, Spain
Ilias Kotsireas	Wilfrid Laurier University, Canada
Stefan Ratschan	Institute of Computer Science, Academy of Sciences of the Czech Republic
Dongming Wang	CNRS, Paris, France
Jinzhao Wu	Guangxi University for Nationalities, China
Zhiming Zheng	Peking University, China

Abstracts of Invited Papers

Current Challenges in Developing Open Source Computer Algebra Systems

Janko Böhm[1(✉)], Wolfram Decker[1], Simon Keicher[2] and Yue Ren[1]

[1] University of Kaiserslautern, 67663 Kaiserslautern, Germany
{boehm, decker, ren}@mathematik.uni-kl.de
[2] Universidad de Concepción, Casilla 160-C, Concepción, Chile
simonkeicher@googlemail.com

Abstract. This note is based on the plenary talk given by the second author at MACIS 2015, the Sixth International Conference on Mathematical Aspects of Computer and Information Sciences. Motivated by some of the work done within the Priority Programme SPP 1489 of the German Research Council DFG, we discuss a number of current challenges in the development of Open Source computer algebra systems. The main focus is on algebraic geometry and the system SINGULAR.

The first author acknowledges support from the DFG projects DE 410/8-1 and -2, DE 410/9-1 and -2, and from the OpenDreamKit Horizon 2020 European Research Infrastructures project (#676541). The third author was supported partially by the DFG project HA 3094/8-1 and by proyecto FONDECYT postdoctorado no 3160016.

Modeling Side-Channel Leakage

Stefan Dziembowski

University of Warsaw

Abstract. Physical side-channel attacks that exploit leakage emitted from devices (see, e.g., [8]) are an important threat to cryptographic implementations. A recent trend in cryptography [9, 10] is to construct cryptographic algorithms that are secure in a given leakage model. Over the past 15 years several such models have been proposed in the literature, starting with the probing model of [9], where the computation is modeled as a Boolean circuit, and the adversary can learn a limited number of them. Other models studied in the theory community include the *bounded-leakage paradigm* [1, 5], the *only computation leaks model* [10], the *independent leakage model* [7], the *auxiliary input model* [3], and many others.

Some of these models have been received with skepticism by the practitioners, who often argued that it is much more realistic to model leakage as a noisy function of the secret data. The first model for noisy leakage was proposed in [2], and fully formalized in [11]. Recently in [4] it has been shown that in fact the noisy leakage model of [11] can be reduced the probing model (i.e.: every noisy leakage function can be simulated be a probing function), which, in particular, greatly simplifies several proofs in the noisy leakage model, and can be viewed as establishing a bridge between theory and practice in this area.

In this talk we give an overview of the leakage models used in the literature. We then present the reduction from [4], and talk about some follow-up work [6].

References

1. Akavia, A., Goldwasser, S., Vaikuntanathan, V.: Simultaneous hardcore bits and cryptography against memory attacks. In: Reingold, O. (ed.) TCC 2009. LNCS, vol. 5444, pp. 474–495. Springer, Heidelberg (2009)
2. Chari, S., Jutla, C.S., Rao, J.R., Rohatgi, P.: Towards sound approaches to counteract power-analysis attacks. In: Wiener, M.J. (ed.) CRYPTO 1999. LNCS, vol. 1666, pp. 398–412. Springer, Heidelberg (1999)
3. Dodis, Y., Goldwasser, S., Kalai, Y.T., Peikert, C., Vaikuntanathan, V.: Public-key encryption schemes with auxiliary inputs. In: Micciancio, D., (ed.) TCC 2010. LNCS, vol. 5978, pp. 361–381. Springer, Heidelberg (2010)
4. Duc, A., Dziembowski, S., Faust, S.: Unifying leakage models: from probing attacks to noisy leakage. In: Nguyen, P.Q., Oswald, E. (eds.) EUROCRYPT 2014. LNCS, vol. 8441, pp. 423–440. Springer, Heidelberg (2014)

Partly supported by the WELCOME/2010-4/2 grant founded within the framework of the EU Innovative Economy (National Cohesion Strategy) Operational Programme.

Solving Structured Polynomial Systems
with Gröbner Bases

Jean-Charles Faugère

Inria, Equipe POLSYS, Centre Paris Rocquencourt, F-75005, Paris, France
Sorbonne Universits, UPMC Univ Paris 06, Equipe POLSYS,
LIP6, F-75005, Paris, France
CNRS, UMR 7606, LIP6, F-75005, Paris, France

Abstract. In most cases, the number of solutions of a polynomial system is exponential, and in finite fields, solving polynomial systems is NP-hard. However, problems coming from applications usually have additional structures. Consequently, a fundamental issue is to design a new generation of algorithms exploiting the special structures that appear ubiquitously in the applications.

At first glance, multi-homogeneity, weighted homogeneity overdeterminedness, sparseness and symmetries seem to be unrelated structures. Indeed, until recently we have obtained specific results for each type of structure: we obtain dedicated algorithm and sharp complexity results too handle a particular structure. For instance, we handle bilinear systems by reducing the problem to determinantal ideals; we also propose ad-hoc techniques to handle symmetries.

All these results have been obtained separately by studying each structure one by one. Recently we found a new unified way to analyze these problems based on monomial sparsity. To this end, we introduce a new notion of sparse Gröbner bases, an analog of classical Gröbner bases for semigroup algebras. We propose sparse variants of the F4/F5 and FGLM algorithms to compute them and we obtain new and sharp estimates on the complexity of solving them (for zero-dimensional systems where all polynomials share the same Newton polytope). As a by product, we can generalize to the multihomogeneous case the already useful bounds obtained in the bilinear case. We can now handle in a uniform way several type of structured systems (at least when the type of structure is the same for every polynomial). From a practical point of view, all these results lead to a striking improvement in the execution time.

We also investigate the non convex case when only a small subset of monomials appear in the equations: the fewnomial case. We can relate the complexity of solving the corresponding algebraic system with some combinatorial property of a graph associated with the support of the polynomials. We show that, in some cases, the systems can be solved in polynomial time.

Joint work with Jules Svartz and Pierre-Jean Spaenlehauer.

Exploiting Structure in Floating-Point Arithmetic

Claude-Pierre Jeannerod

Inria
Laboratoire LIP (CNRS, ENSL, Inria, UCBL), Université de Lyon

Abstract. The analysis of algorithms in IEEE floating-point arithmetic is most often carried out via repeated applications of the so-called standard model, which bounds the relative error of each basic operation by a common epsilon depending only on the format. While this approach has been eminently useful for establishing many accuracy and stability results, it fails to capture most of the low-level features that make floating-point arithmetic so highly structured. In this paper, we survey some of those properties and how to exploit them in rounding error analysis. In particular, we review some recent improvements of several classical, Wilkinson-style error bounds from linear algebra and complex arithmetic that all rely on such structure properties.

Keywords: Floating-point arithmetic · IEEE standard 754-2008 · Rounding error analysis · High relative accuracy

Symbolic Geometric Reasoning with Advanced Invariant Algebras

Hongbo Li[✉]

Key Laboratory of Mathematics Mechanization, Academy of Mathematics
and Systems Science, Chinese Academy of Sciences, Beijing 100190, China
hli@mmrc.iss.ac.cn

Abstract. In symbolic geometric reasoning, the output of an algebraic method is expected to be geometrically interpretable, and the size of the middle steps is expected to be sufficiently small for computational efficiency. Invariant algebras often perform well in meeting the two expectations for relatively simple geometric problems. For example in classical geometry, symbolic manipulations based on basic invariants such as squared distances, areas and volumes often have great performance in generating readable proofs. For more complicated geometric problems, the basic invariants are still insufficient and may not generate geometrically meaningful results.

An advanced invariant is a monomial in an "advanced algebra", and can be expanded into a polynomial of basic invariants that are also included in the algebra. In projective incidence geometry, Grassmann-Cayley algebra and Cayley bracket algebra are an advanced algebra in which the basic invariants are determinants of homogeneous coordinates of points, and the advanced invariants are Cayley brackets. In Euclidean conformal geometry, Conformal Geometric Algebra and null bracket algebra are an advanced algebra where the basic invariants are squared distances between points and and signed volumes of simplexes, and the advanced invariants are Clifford brackets.

This paper introduces the above advanced invariant algebras together with their applications in automated geometric theorem proving. These algebras are capable of generating extremely short and readable proofs. For projective incidence theorems, the proofs generated are usually two-termed in that the conclusion expression maintains two-termed during symbolic manipulations. For Euclidean geometry, the proofs generated are mostly one-termed or two-termed.

Keywords: Grassmann-Cayley algebra · Cayley bracket algebra · Conformal Geometric Algebra · Null bracket algebra · Automated geometric theorem proving

Decidability from a Numerical Point of View

Stefan Ratschan

Institute of Computer Science
Czech Academy of Sciences

Abstract. An important application of computation is the automatic analysis of mathematical models of real-world systems, for example by simulation or formal verification. Here, the systems to be automatically analyzed can be physical systems (e.g., the wing of an airplane) or computational systems (e.g., computer software). In the past, research in this direction has happened largely independently for those two types of systems: Algorithms for automatically analyzing models of physical systems have been developed mainly by engineers and numerical mathematicians, resulting in notions such as "well-posed problem", and "condition number", and algorithms for automatically analyzing models of computational systems have been developed mainly by computer scientists based on logic, and notions such as "decision procedure", "decidability", and "computational complexity".

Nowadays, the boundary between physical and computational systems is vanishing, since computation is more and more intertwined with our everyday physical world (cf. the notion of cyber-physical system). This makes it necessary for the boundary between the two research strands mentioned above to be overcome as well. In the talk, we discussed some examples of results obtained by the speaker that point into this direction, especially results, where inspiration from numerical analysis helps to solve problems that are considered undecidable by computer scientists [1–3].

References

1. Franek, P., Ratschan, S., Zgliczynski, P.: Quasi-decidability of a fragment of the first-order theory of real numbers. J. Autom. Reason. (2015). http://dx.doi.org/10.1007/s10817-015-9351-3
2. Ratschan, S.: Continuous first-order constraint satisfaction. In: Calmet, J., Benhamou, B., Caprotti, O., Henocque, L., Sorge, V. (eds.) Artificial Intelligence, Automated Reasoning, and Symbolic Computation. LNCS, vol. 2385, pp. 181–195. Springer, Berlin (2002)
3. Ratschan, S.: Safety verification of non-linear hybrid systems is quasi-decidable. Formal Methods Syst. Des. **44**(1), 71–90 (2014)

The research published in this paper was supported by GAČR grant 15-14484S and with institutional support RVO:67985807.

ORCID: 0000-0003-1710-1513

Congruence Testing of Point Sets in Three and Four Dimensions Results and Techniques

Günter Rote[✉]

Institut für Informatik, Freie Universität Berlin
rote@inf.fu-berlin.de

Abstract. I will survey algorithms for testing whether two point sets are congruent, that is, equal up to an Euclidean isometry. I will introduce the important techniques for congruence testing, namely dimension reduction and pruning, or more generally, condensation. I will illustrate these techniques on the three-dimensional version of the problem, and indicate how they lead for the first time to an algorithm for four dimensions with near-linear running time (joint work with Heuna Kim). On the way, we will encounter some beautiful and symmetric mathematical structures, like the regular polytopes, and Hopf-fibrations of the three-dimensional sphere in four dimensions.

Contents

Cryptography

Verified Numerical Computation

Polynomial System Solving

Data and Knowledge Exploration

Algorithm Engineering in Geometric Computing

Real Complexity: Theory and Practice

Global Optimization

General Session

Invited Papers

Current Challenges in Developing Open Source Computer Algebra Systems

Janko Böhm[1]([⊠]), Wolfram Decker[1], Simon Keicher[2], and Yue Ren[1]

[1] University of Kaiserslautern, 67663 Kaiserslautern, Germany
{boehm,decker,ren}@mathematik.uni-kl.de
[2] Universidad de Concepción, Casilla 160-C, Concepción, Chile
keicher@mail.mathematik.uni-tuebingen.de

Abstract. This note is based on the plenary talk given by the second author at MACIS 2015, the Sixth International Conference on Mathematical Aspects of Computer and Information Sciences. Motivated by some of the work done within the Priority Programme SPP 1489 of the German Research Council DFG, we discuss a number of current challenges in the development of Open Source computer algebra systems. The main focus is on algebraic geometry and the system SINGULAR.

1 Introduction

The goal of the nationwide Priority Programme SPP 1489 of the German Research Council DFG is to considerably further the algorithmic and experimental methods in algebraic geometry, number theory, and group theory, to combine the different methods where needed, and to apply them to central questions in theory and practice. In particular, the programme is meant to support the further development of Open Source computer algebra systems which are (co-)based in Germany, and which in the framework of different projects may require crosslinking on different levels. The cornerstones of the latter are the well-established systems GAP [34] (group and representation theory), POLYMAKE [35] (polyhedral geometry), and SINGULAR [25] (algebraic geometry, singularity theory, commutative and non-commutative algebra), together with the newly evolving system ANTIC [41] (number theory), but there are many more systems, libraries, and packages involved (see Sect. 2.4 for some examples).

In this note, having the main focus on SINGULAR, we report on some of the challenges which we see in this context. These range from reconsidering the efficiency of the basic algorithms through parallelization and making abstract concepts constructive to facilitating the access to Open Source computer algebra systems. In illustrating the challenges, which are discussed in Sect. 2, we take examples from algebraic geometry. In Sects. 3 and 4, two of the examples are

The second author acknowledges support from the DFG projects DE 410/8-1 and -2, DE 410/9-1 and -2, and from the OpenDreamKit Horizon 2020 European Research Infrastructures project (#676541). The third author was supported partially by the DFG project HA 3094/8-1 and by proyecto FONDECYT postdoctorado no 3160016.

I.S. Kotsireas et al. (Eds.): MACIS 2015, LNCS 9582, pp. 3–24, 2016.
DOI: 10.1007/978-3-319-32859-1_1

highlighted in more detail. These are the parallelization of the classical Grauert-Remmert type algorithms for normalization and the computation of GIT-fans. The latter is a show-case application of bringing SINGULAR, POLYMAKE, and GAP together.

2   Seven Challenges

2.1   Reconsidering the Efficiency of the Basic Algorithms

Motivated by an increasing number of success stories in applying algorithmic and experimental methods to algebraic geometry (and other areas of mathematics), research projects in this direction become more and more ambitious. This applies both to the theoretical level of abstraction and to the practical complexity. On the computer algebra side, this not only requires innovative ideas to design high-level algorithms, but also to revise the basic algorithms on which the high-level algorithms are built. The latter concerns efficiency and applicability.

Example 1 (The NEMO *Project).* NEMO is a new computer algebra package written in the JULIA[1] programming language which, in particular, aims at highly efficient implementations of basic arithmetic and algorithms for number theory and is connected to the ANTIC project. See http://nemocas.org/index.html for some benchmarks.

In computational algebraic geometry, aside from polynomial factorization, the basic work horse is Buchberger's algorithm for computing Gröbner bases [22] and, as remarked by Schreyer [46] and others, syzygies. While Gröbner bases are specific sets of generators for ideals and modules which are well-suited for computational purposes, the name syzygies refers to the relations on a given set of generators. Syzygies carry important geometric information (see [28]) and are crucial ingredients in many basic and high-level algorithms. Taking syzygies on the syzygies and so forth, we arrive at what is called a free resolution. Here is a particular simple example.

Example 2 (The Koszul Complex of Three Variables). In the SINGULAR session below, we first construct the polynomial ring $R = \mathbb{Q}[x, y, z]$, endowed with the degree reverse lexicographical order dp. Then we compute the successive syzygies on the variables x, y, z.

```
> ring R = 0, (x,y,z), dp;
> ideal I = x,y,z;
> resolution FI = nres(I,0);
> print(FI[2]);
 0,-y,-z,
-z, x, 0,
 y, 0, x
```

---

[1] See http://julialang.org.

```
> print(FI[3]);
 x,
 z,
-y
```

In the following example, we show how Gröbner basis and syzygy computations fit together to build a more advanced algorithm.

Example 3 (Parametrizing Rational Curves). We study a degree-5 curve C in the projective plane which is visualized as the red curve in Figs. 1 and 2. To begin with, after constructing the polynomial ring $R = \mathbb{Q}[x, y, z]$, we enter the homogeneous degree-5 polynomial $f \in \mathbb{Q}[x, y, z]$ which defines C:

```
> ring R = 0, (x,y,z), dp;
> poly f = x5+10x4y+20x3y2+130x2y3-20xy4+20y5-2x4z-40x3yz-150x2y2z
          -90xy3z-40y4z+x3z2+30x2yz2+110xy2z2+20y3z2;
```

Our goal is to check whether C is rational, and if so, to compute a rational parametrization. For the first task, recall that an algebraic curve is rational if and only if its geometric genus is zero. In the example here, this can be easily read off from the genus formula for plane curves, taking into account that the degree-5 curve has three ordinary double points and one ordinary triple point (see the aforementioned visualization). An algorithm for computing the genus in general, together with an algorithm for computing rational parametrizations, is implemented in the SINGULAR library **paraplanecurves.lib** [15]:

```
> LIB "paraplanecurves.lib";
> genus(f);
  0
> paraPlaneCurve(f);
```

Rather than displaying the result, we will now show the key steps of the algorithm at work. The first step is to compute the ideal generated by the adjoint curves of C which, roughly speaking, are curves which pass with sufficiently high multiplicity through the singular points of C. The algorithm for computing the adjoint ideal (see [11]) builds on algorithms for computing normalization (see Sect. 3) or, equivalently, integral bases (see [10]). In all these algorithms, Gröbner bases are used as a fundamental tool.

```
> ideal AI = adjointIdeal(f);
> AI;
  _[1]=y3-y2z
  _[2]=xy2-xyz
  _[3]=x2y-xyz
  _[4]=x3-x2z
```

The resulting four cubic generators of the adjoint ideal define the curves depicted in Fig. 1, where the thickening of a line indicates that the line comes with a double structure. A general adjoint curve, that is, a curve defined by a general linear combination of the four generators, is shown in Fig. 2.

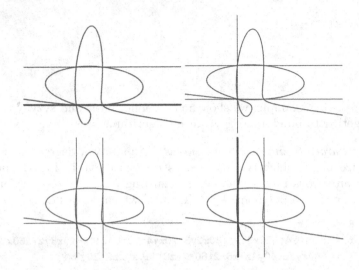

Fig. 1. Cubic curves defined by the generators of the adjoint ideal of a degree-5 curve with three ordinary double points and one ordinary triple point. The degree-5 curve is shown in red (Color figure online).

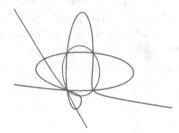

Fig. 2. A general adjoint curve of C of degree 3 (Color figure online).

The four generators give a birational map from C to a curve $\widetilde{C}$ in projective 3-space $\mathbb{P}^3$. We obtain $\widetilde{C}$ via elimination, a typical application of Gröbner bases:

```
> def Rn = mapToRatNormCurve(f,AI);
> setring(Rn);
> RNC;
RNC[1]=y(2)*y(3)-y(1)*y(4)
RNC[2]=20*y(1)*y(2)-20*y(2)^2+130*y(1)*y(4)
       +20*y(2)*y(4)+10*y(3)*y(4)+y(4)^2
RNC[3]=20*y(1)^2-20*y(1)*y(2)+130*y(1)*y(3)
       +10*y(3)^2+20*y(1)*y(4)+y(3)*y(4)
```

Note that $\widetilde{C}$ is a variant of the projective twisted cubic curve, the rational normal curve in $\mathbb{P}^3$ (for a picture see Fig. 10). This non-singular curve is mapped isomorphically onto the projective line $\mathbb{P}^1$ by the anticanonical linear system, which can be computed using syzygies:

```
> rncAntiCanonicalMap(RNC);
 _[1]=2*y(2)+13*y(4)
 _[2]=y(4)
```

Composing all maps in this construction, and inverting the resulting birational map, we get the desired parametrization. In general, depending on the number of generators of the adjoint ideal, the rational normal curve computed by the algorithm is embedded into a projective space of odd or even dimension. In the latter case, successive applications of the canonical linear system map the normal curve onto a plane conic. Computing a rational parametrization of the conic is equivalent to finding a point on the conic. It can be algorithmically decided whether we can find such a point with rational coordinates or not. In the latter case, we have to pass to a quadratic field extension of $\mathbb{Q}$.

Remark 1. The need of passing to a field extension occurs in many geometric constructions. Often, repeated field extensions are needed. The effective computation of Gröbner bases over (towers of) number fields is therefore of utmost importance. One general way of achieving higher speed is the parallelization of algorithms. This will be addressed in the next section, where we will, in particular, discuss a parallel version of the Gröbner basis (syzygy) algorithm which is specific to number fields [18]. New ideas for enhancing syzygy computations in general are presented in [31]. Combining the two approaches in the case of number fields is a topic of future research.

2.2 Parallelization

Parallelizing computer algebra systems allows for the efficient use of multicore computers and high-performance clusters. To achieve parallelization is a tremendous challenge both from a computer science and a mathematical point of view.

From a computer science point of view, there are two possible approaches:

- Distributed and multi-process systems work by using different processes that do not share memory and communicate by message passing. These systems only allow for *coarse-grained parallelism*, which limits their ability to work on large shared data structures, but can in principle scale up indefinitely.
- Shared memory systems work by using multiple threads of control in a single process operating on shared data. They allow for more *fine-grained parallelism* and more sophisticated concurrency control, down to the level of individual CPU instructions, but are limited in their scalability by how many processors can share efficient access to the same memory on current hardware.

For best performance, typically hybrid models are used, which exploit the strengths of both shared memory and distributed systems, while mitigating their respective downsides.

From its version 3.1.4 on, SINGULAR has been offering a framework for coarse-grained parallelization, with a convenient user access provided by the library **parallel.lib** [48]. The example below illustrates the use of this framework:

Example 4 (Coarse Grained Parallelization in SINGULAR*).* We implement a SINGULAR procedure which computes a Gröbner basis for a given ideal with respect to a given monomial ordering. The procedure returns the size of the Gröbner basis. We apply it in two parallel runs to a specific ideal in $\mathbb{Q}[x_1, \ldots, x_4]$, choosing for one run the lexicographical monomial ordering lp and for the other run the degree reverse lexicographical ordering dp:

```
> LIB "parallel.lib"; LIB "random.lib";
> proc sizeGb(ideal I, string monord){
       def R = basering; list RL = ringlist(R);
       RL[3][1][1] = monord; def S = ring(RL); setring(S);
       return(size(groebner(imap(R,I))));}
> ring R = 0,x(1..4),dp;
> ideal I = randomid(maxideal(3),3,100);
> list commands = "sizeGb","sizeGb";
> list args = list(I,"lp"),list(I,"dp");
> parallelWaitFirst(commands, args);
  [1] empty list
  [2] 11
> parallelWaitAll(commands, args);
  [1] 55
  [2] 11
```

As expected, the computation with respect to dp is much faster and leads to a Gröbner basis with less elements.

Using ideas from the successful parallelization of GAP within the HPC-GAP project (see [4–6]), a multi-threaded prototype of SINGULAR has been implemented. Considerable further efforts are needed, however, to make this accessible to users without a deep background in parallel programming.

From a mathematical point of view, there are algorithms whose basic strategy is inherently parallel, whereas others are sequential in nature. A prominent example of the former type is Villamayor's constructive version of Hironaka's desingularization theorem, which will be briefly discussed in Sect. 2.3. A prominent example of the latter type is the classical Grauert-Remmert type algorithm for normalization, which will be addressed at some length in Sect. 3.

The systematic design of parallel algorithms for applications which so far can only be handled by sequential algorithms is a major task for the years to come. For normalization, this problem has recently been solved [14]. Over the field of rational numbers, the new algorithm becomes particularly powerful by combining it with modular methods, see again Sect. 3.

Modular methods are well-known for providing a way of parallelizing algorithms over $\mathbb{Q}$ (more generally, over number fields). For the fundamental task of computing Gröbner bases, a modular version of Buchberger's algorithm is due to Arnold [1]. More recently, Boku, Fieker, Steenpaß and the second author [18] have designed a modular Gröbner bases algorithm which is specific to number fields. In addition to using the approach from Arnold's paper, which is to compute Gröbner bases modulo several primes and then use Chinese remaindering together with rational reconstruction, the new approach provides a second level of parallelization as depicted in Fig. 3: If the number field is presented as $K = \mathbb{Q}(\alpha) = \mathbb{Q}[t]/\langle f \rangle$, where $f \in \mathbb{Q}[t]$ is the minimal polynomial of α, and if generators $g_1(X, \alpha), \ldots, g_s(X, \alpha)$ for the ideal under consideration are given, represented by polynomials $g_1(X, t), \ldots, g_s(X, t) \in \mathbb{Q}[X, t] = \mathbb{Q}[x_1, \ldots, x_n, t]$, we wish to compute a Gröbner basis for the ideal $\tilde{I} = \langle g_1(X, t), \ldots, g_s(X, t), f \rangle \subset \mathbb{Q}[X, t]$. The idea then is to reduce $\tilde{I}$ modulo a suitable number of primes $p_1, \ldots, p_k$ (level 1 of the algorithm), get the second level of parallelization by factorizing the reductions of f modulo the p_i, and use, for each i, polynomial Chinese remaindering to put the results modulo p_i together (level 3 of the algorithm).

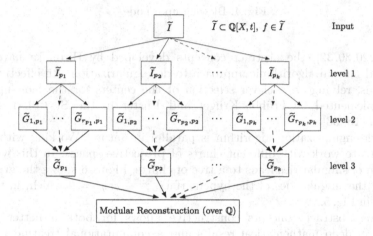

Fig. 3. Two-fold parallel modular approach to Gröbner bases over number fields.

2.3   Make More and More of the Abstract Concepts of Algebraic Geometry Constructive

The following groundbreaking theorem proved by Hironaka in 1964 shows the existence of resolutions of singularites in characteristic zero. It is worth mentioning that, on his way, Hironaka introduced the idea of standard bases, the power series analogue of Gröbner bases.

Theorem 1 (Hironaka, 1964). *For every algebraic variety over a field K of characteristic zero, a desingularization can be obtained by a finite sequence of blow-ups along smooth centers.*

We illustrate the blow-up process by a simple example:

Example 5. As shown in Fig. 4, a node can be resolved by a single blow-up: we replace the node by a line and separate, thus, the two branches of the curve intersecting in the singularity.

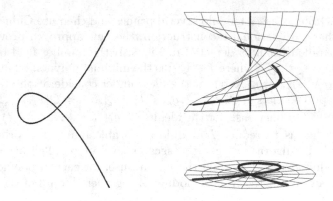

Fig. 4. Blowing up a node.

In [9,20,30,32], the abstract concepts developed by Hironaka have been translated into an algorithmic approach to desingularization. An effective variant of this, relying on a clever selection of the centers for the blow-ups, has been implemented by Frühbis-Krüger and Pfister in the SINGULAR library `resolve.lib` [33].

The desingularization algorithm is parallel in nature: Working with blow-ups means to work with different charts of projective spaces. In this way, the resolution of singularities leads to a tree of charts. Figure 6 shows the graph for resolving the singularities of the hypersurface $z^2 - x^2y^2 = 0$ which, in turn, is depicted in Fig. 5.

Making abstract concepts constructive allows for both a better understanding of deep mathematical results and a computational treatment of the concepts. A further preeminent example for this is the constructive version of the Bernstein-Gel'fand-Gel'fand correspondence (BGG-correspondence) by Eisenbud, Fløystad, and Schreyer [29]. This allows one to express properties of sheaves over projective spaces in terms of exterior algebras. More precisely, if $\mathbb{P}(V)$ is the projective space of lines in a vector space V, and E is the exterior algebra $E = \Lambda V$, then the BGG-correspondence relates coherent sheaves over $\mathbb{P}(V)$ to free resolutions over E. Since E contains only finitely many monomials, (non-commutative) Gröbner basis and syzygy computations over E are often preferable to (commutative) Gröbner basis and syzygy computations over the homogeneous coordinate ring of $\mathbb{P}(V)$. One striking application of this, which is implemented in MACAULAY2 [37] and SINGULAR, gives a fast way of computing sheaf cohomology. Providing computational access to cohomology in all its disguises is a long-term goal of computational algebraic geometry.

The BGG-correspondence is an example of an equivalence of derived categories. As we can see from the above discussion, such equivalences are not only interesting from a theoretical point of view, but may also allow for creating more effective algorithms – provided they can be accessed computationally.

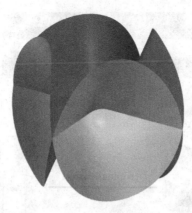

Fig. 5. The surface $z^2 - x^2 y^2 = 0$.

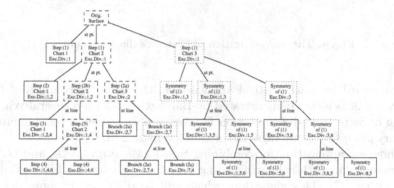

Fig. 6. The tree of charts.

2.4 Interaction and Integration of Computer Algebra Systems and Libraries from Different Areas of Research

On the theoretical side, mathematical breakthroughs are often obtained by combining methods from different areas of mathematics. Making such connections accessible to computational methods is another major challenge. Handling this challenge requires, in particular, that computer algebra systems specializing in different areas are connected in a suitable way. One goal of the Priority Programme SPP 1489, which was already mentioned in the introduction, is to interconnect GAP, POLYMAKE, SINGULAR, and ANTIC. So far, this has lead to directed interfaces as indicated in Fig. 7, with further directions and a much tighter integration of the systems subject to future development.

In fact, the picture is much more complicated: The four systems rely on further systems and libraries such as NORMALIZ [21] (affine monoids) and FLINT [42] (number theory), and there are other packages which use at least one of the four systems, for example homalg [49] (homological algebra) and a-tint [40] (tropical intersection theory) (Fig. 8).

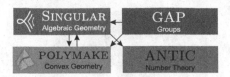

Fig. 7. Directed interfaces.

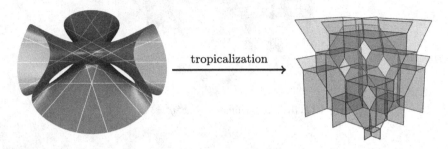

Fig. 8. The Tropicalization of Clebsch's diagonal cubic.

With regard to mathematical applications, the value of connecting GAP and SINGULAR is nicely demonstrated by Barakat's work on a several years old question of Serre to find a prediction for the number of connected components of unitary groups of group algebras in characteristic 2 [3,47].

A showcase application for combining SINGULAR, POLYMAKE, and GAP is the symmetric algorithm for computing GIT-fans [17] by the first, third and fourth author [17]. This algorithm, which will be discussed in more detail in Sect. 4, combines Gröbner basis and convex hull computations, and can make use of actions of finite symmetry groups.

2.5 A Convenient Hierarchy of Languages

Most modern computer algebra systems consist of two major components, a kernel which is typically written in C/C++ and a high level language for direct user interaction, which in particular provides a convenient way for users to extend the system. While the kernel code is precompiled and, thus, performant, the user language is interpreted, which means that it operates at a significantly slower speed. In addition to the differences in speed, the languages involved provide different levels of abstraction with regard to modeling mathematical concepts. In view of the integration of different systems, a number of languages has to be considered, leading to an even more complicated situation. To achieve the required level of performance and abstraction in this context, we need to set up a convenient hierarchy of languages. Here, we propose in particular to examine the use of just-in-time compiled languages such as JULIA.

2.6   Create and Integrate Electronic Libraries and Databases Relevant to Research

Electronic libraries and databases of certain classes of mathematical objects provide extremely useful tools for research in their respective fields. An example from group theory is the *SmallGroups* library, which is distributed as a GAP package. An example from algebraic geometry is the *Graded Ring Database*,[2] written by Gavin Brown and Alexander Kasprzyk, with contributions by several other authors. The creation of such databases often depends on several computer algebra systems. On the other hand, a researcher using the data may wish to access the database within a system with which he is already familiar. This illustrates the benefits of a standardized approach to connect computer algebra systems and mathematical databases.

2.7   Facilitating the Access to Computer Algebra Systems

Computational algebraic geometry (and computer algebra in general) has a rapidly increasing amount of applications outside its original core areas, for example to computational biology, algebraic vision, and physics. As more and more non-specialists wish to use computer algebra systems, the question of how to considerably ease the access to the systems arises also in the Open Source community. Virtual research environments such as the one developed within the *OpenDreamKit* project[3] may provide an answer to this question. Creating Jupyter notebooks[4] for systems such as GAP and SINGULAR is one of the many goals of this project. A SINGULAR prototype has been written by Sebastian Gutsche, see Fig. 9.

3   A Parallel Approach to Normalization

In this section, focusing on the normalization of rings, we give an example of how ideas from commutative algebra can be used to turn a sequential algorithm into a parallel algorithm.

The normalization of rings is an important concept in commutative algebra, with applications in algebraic geometry and singularity theory. Geometrically, normalization removes singularities in codimension one and "improves" singularities in higher codimension. In particular, for curves, normalization yields a desingularization (see Examples 6 and 7 below). From a computer algebra point of view, normalization is fundamental to quite a number of algorithms with applications in algebra, geometry, and number theory. In Example 3, for instance, we have used normalization to compute adjoint curves and, thus, parametrizations of rational curves.

---

[2] See http://www.grdb.co.uk.
[3] See http://opendreamkit.org.
[4] See http://jupyter.org.

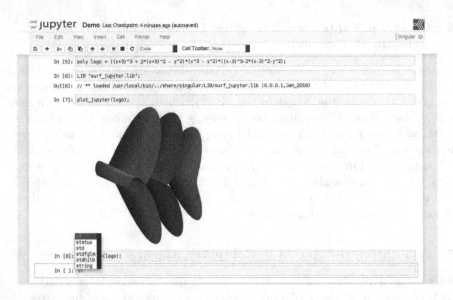

Fig. 9. Jupyter notebook for SINGULAR.

The by now classical Grauert-Remmert type approach [23,24,38] to compute normalization proceeds by successively enlarging the given ring until the Grauert-Remmert normality criterion [36] tells us that the normalization has been reached. Obviously, this approach is completely sequential in nature. As already pointed out, it is a major challenge to systematically design parallel alternatives to basic and high-level algorithms which are sequential in nature. For normalization, this problem has recently been solved in [14] by using the technique of localization and proving a local version of the Grauert-Remmert normality criterion.

To explain this in more detail, we suppose for simplicity that the ring under consideration is an affine domain over a field K. That is, we consider a quotient ring of type $A = K[x_1, \ldots, x_n]/I$, where I is a prime ideal. We require that K is a perfect field.

We begin by recalling some basic definitions and results.

Definition 1. *The **normalization** of A is the integral closure $\overline{A}$ of A in its quotient field $Q(A)$,*

$$\overline{A} = \{a \in Q(A) \mid \text{there exists } f \in A[t] \text{ monic with } f(a) = 0\}.$$

*We call A **normal** if $A = \overline{A}$.*

By Emmy Noether's finiteness theorem (see [43]), we may represent $\overline{A}$ as the set of A-linear combinations of a finite set of elements of $\overline{A}$. That is:

Theorem 2 (Emmy Noether). *$\overline{A}$ is a finitely generated A-module.*

We also say that the ring extension $A \subset \overline{A}$ is finite. In particular, $\overline{A}$ is again an affine domain over K.

Example 6. For the coordinate ring $A = K[x,y]/I$ of the nodal plane curve $C = V(I)$ defined by the prime ideal $I = \langle x^3 + x^2 - y^2 \rangle \subset K[x,y]$, we have

$$A = K[x,y]/I \cong K[t^2 - 1, t^3 - t] \subset K[t] \cong \overline{A}.$$
$$\overline{x} \mapsto t^2 - 1$$
$$\overline{y} \mapsto t^3 - t$$

In particular, $\overline{A}$ is generated as an A-module by 1 and $\frac{y}{x}$.

Geometrically, the inclusion map $A \hookrightarrow \overline{A}$ corresponds to the parametrization

$$\mathbb{A}^1(K) \to C \subset \mathbb{A}^2(K), \qquad t \mapsto (t^2 - 1, t^3 - t).$$

In other words, the parametrization is the normalization (desingularization) map of the rational curve C.

Historically, the first Grauert-Remmert-type algorithm for normalization is due to de Jong [23,24]. This algorithm has been implemented in SINGULAR, MACAULAY2, and MAGMA [19]. The algorithm of Greuel, Laplagne, and Seelisch [38] is a more efficient version of de Jong's algorithm. It is implemented in the SINGULAR library `normal.lib` [39].

The starting point of these algorithms is the following lemma:

Lemma 1 ([38]). *If $J \subset A$ is an ideal and $0 \neq g \in J$, then there are natural inclusions of rings*

$$A \hookrightarrow Hom_A(J,J) \cong \frac{1}{g}(gJ :_A J) \subseteq \overline{A} \subset Q(A), \quad a \mapsto \varphi_a, \quad \varphi \mapsto \frac{\varphi(g)}{g},$$

where φ_a is the multiplication by a.

Now, starting from $A_0 = A$ and $J_0 = J$, and setting

$$A_{i+1} = \frac{1}{g}(gJ_i :_{A_i} J_i) \text{ and } J_i = \sqrt{JA_i},$$

we get a chain of finite extensions of affine domains which becomes eventually stationary by Theorem 2:

$$A = A_0 \subset \cdots \subset A_i \subset \cdots \subset A_m = A_{m+1} \subseteq \overline{A}.$$

The Grauert-Remmert-criterion for normality tells us that for an appropriate choice of J, the process described above terminates with the normalization $A_m = \overline{A}$. In formulating the criterion, we write $N(A)$ for the **non-normal locus** of A, that is, if

$$\mathrm{Spec}(A) = \{P \subset A \mid P \text{ prime ideal}\}$$

denotes the **spectrum** of A, and A_P the localization of A at P, then

$$N(A) = \{P \in \mathrm{Spec}(A) \mid A_P \text{ is not normal}\}.$$

Theorem 3 (Grauert-Remmert [36]**).** *Let* $\langle 0 \rangle \neq J \subset A$ *be an ideal with* $J = \sqrt{J}$ *and such that*

$$N(A) \subseteq V(J) := \{P \in \mathrm{Spec}(A) \mid P \supseteq J\}.$$

Then A *is normal if and only if* $A \cong \mathrm{Hom}_A(J, J)$ *via the map which sends* a *to multiplication by* a.

The problem now is that we do not know an algorithm for computing $N(A)$, except if the normalization is already known to us. To remedy this situation, we consider the **singular locus** of A,

$$\mathrm{Sing}(A) = \{P \in \mathrm{Spec}(A) \mid A_P \text{ is not regular}\},$$

which contains the non-normal locus: $N(A) \subseteq \mathrm{Sing}(A)$. Since we work over a perfect field K, the Jacobian criterion tells us that $\mathrm{Sing}(A) = V(\mathrm{Jac}(I))$, where $\mathrm{Jac}(I)$ is the Jacobian ideal[5] of A (see [27]). Hence, if we choose $J = \sqrt{\mathrm{Jac}(I)}$, the above process terminates with $A_m = \overline{A}$ by the following lemma.

Lemma 2 ([38]**).** *With notation as above,* $N(A_i) \subseteq V(\sqrt{JA_i})$ *for all* i.

Example 7. For the coordinate ring A of the plane algebraic curve C from Example 6, the normalization algorithm returns the coordinate ring of a variant of the twisted cubic curve $\overline{C}$ in affine 3-space, where the inclusion $A \subset \overline{A}$ corresponds to the projection of $\overline{C}$ to C via $(x, y, z) \mapsto (x, y)$ as shown in Fig. 10. This result fits with the result in Example 6: The curve $\overline{C}$ is rational, with a parametrization given by

$$\mathbb{A}^1(K) \to \overline{C} \subset \mathbb{A}^3(K), \qquad t \mapsto (t^2 - 1, t^3 - t, t).$$

Composing this with the projection, we get the normalization map from Example 6.

Now, following [14], we describe how the normalization algorithm can be redesigned so that it becomes parallel in nature. For simplicity of the presentation, we focus on the case where $\mathrm{Sing}(A)$ is a finite set. This includes the case where A is the coordinate ring of an algebraic curve.

In the example above, the curve under consideration has just one singularity. If there is a larger number of singularities, the normalization algorithm as discussed so far is global in the sense that it "improves" all singularities at the same time. Alternatively, we now aim at "improving" the individual singularities separately, and then put the individual results together. In this local-to-global approach, the local computations can be run in parallel. We make use of the following result.

---

[5] The Jacobian ideal of A is generated by the images of the $c \times c$ minors of the Jacobian matrix $\left(\frac{\partial f_i}{\partial x_j}\right)$, where c is the codimension and $f_1, \ldots, f_r$ are polynomial generators for I.

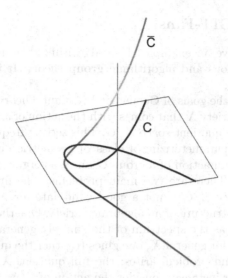

Fig. 10. The normalization of the nodal plane curve $C = V(x^3 + x^2 - y^2)$ is a variant of the twisted cubic curve $\overline{C}$ in 3-space.

Theorem 4 ([14]). *Suppose that* $\mathrm{Sing}(A) = \{P_1, \ldots, P_r\}$ *is finite. Then:*

1. For each i, let

$$A \subseteq B_i \subseteq \overline{A}$$

be the intermediate ring obtained by applying the normalization algorithm with P_i in place of J. Then

$$(B_i)_{P_i} = \overline{A_{P_i}}, \text{ and}$$
$$(B_i)_Q = A_Q \text{ for all } P_i \neq Q \in \mathrm{Spec}(A).$$

*We call B_i the **minimal local contribution** to $\overline{A}$ at P_i.*
2. We have

$$\overline{A} = B_1 + \ldots + B_r.$$

This theorem, together with the local version of the Grauert-Remmert criterion, whose proof is given in [14], yields an algorithm for normalization which is often considerably faster than the global algorithm presented earlier, even if the local-to-global algorithm is not run in parallel. The reason for this is that the cost for "improving" just one singularity is in many cases much less than that for "improving" all singularities at the same time. The new algorithm is implemented in the SINGULAR library `locnormal.lib` [12]. Over the rationals, the algorithm becomes even more powerful by combining it with a modular approach. This version of the algorithm is implemented in the SINGULAR library `modnormal.lib` [13].

4   Computing GIT-Fans

In this section, we give an example of an algorithm that uses Gröbner bases, polyhedral computations and algorithmic group theory. It is also suitable for parallel computations.

Recall that one of the goals of Geometric Invariant Theory (GIT) is to assign to a given algebraic variety X that comes with the action of an algebraic group G in a sensible manner a quotient space $X /\!\!/ G$. This setting frequently occurs when we face a variety X parameterizing a class of geometric objects, for example algebraic curves, and an action of a group G on X emerging from isomorphisms between the objects. There are two main problems. The first problem is that the homogeneous space X/G is not a good candidate for $X /\!\!/ G$ as it does not necessarily carry the structure of an algebraic variety. One then defines for affine X the quotient $X /\!\!/ G$ as the spectrum of the (finitely generated) invariant ring of the functions of X; for general X, one glues together the quotients of an affine covering. Now a second problem arises: the full quotient $X /\!\!/ G$ may not carry much information: For instance, consider the action of $\mathbb{C}^* := \mathbb{C} \setminus \{0\}$ on $X = \mathbb{C}^2$ given by component-wise multiplication

$$\mathbb{C}^* \times X \to X, \qquad (t, (x, y)) \mapsto (tx, ty). \tag{1}$$

Then the quotient $X /\!\!/ \mathbb{C}^*$ is isomorphic to a point. However, considering the open subset $U := X \setminus \{(0,0)\}$ gives us $U /\!\!/ \mathbb{C}^* = \mathbb{P}^1$, the projective line. For general X, there are many choices for these open subsets $U \subseteq X$, where different choices lead to different quotients $U /\!\!/ G$. To describe this behaviour, Dolgachev and Hu [26] introduced the **GIT-fan**, a polyhedral fan describing this variation of GIT-quotients. Recall that a **polyhedral fan** is a finite collection of strongly convex rational polyhedral cones such that their faces are again elements of the fan and the intersection of any two cones is a common face.

Fig. 11. A polyhedral fan in $\mathbb{R}^2$.

Of particular importance is the action of an algebraic torus $G = (\mathbb{C}^*)^k$, on an affine variety $X \subseteq \mathbb{C}^r$. In this case, Berchtold/Hausen and the third author [7,44] have developed a method for computing the GIT-fan, see Algorithm 1. The input of the algorithm consists of

- an ideal $\mathfrak{a} \subseteq \mathbb{C}[T_1, \ldots, T_r]$ which defines X and
- a matrix $Q = (q_1, \ldots, q_r) \in \mathbb{Z}^{k \times r}$ such that $\mathfrak{a}$ is homogeneous with respect to the multigrading defined by setting $\deg(T_i) := q_i \in \mathbb{Z}^k$.

Note that the matrix Q encodes the action of $(\mathbb{C}^*)^k$ on X. For instance, the action (1) is encoded in $Q = (1, 1)$.

Algorithm 1 can be divided into three main steps. For the first step, we decompose $\mathbb{C}^r$ into the 2^r disjoint torus orbits

$$\mathbb{C}^r = \bigcup_{\gamma \subseteq \{1, \ldots, r\}} O(\gamma), \qquad O(\gamma) := \{(z_1, \ldots, z_r) \in \mathbb{C}^r \mid z_i \neq 0 \Leftrightarrow i \in \gamma\}.$$

The algorithm then identifies in line 1 which of the torus orbits $O(\gamma)$ have a non-trivial intersection with X. The corresponding $\gamma \subseteq \{1, \ldots, r\}$ (interpreted as faces of the positive orthant $\mathbb{Q}^r_{\geq 0}$) are referred to as $\mathfrak{a}$-**faces**. Using the equivalence

$$X \cap O(\gamma) \neq \emptyset \quad \Longleftrightarrow \quad (\mathfrak{a}|_{T_i = 0 \text{ for } i \notin \gamma}) : \langle T_1 \cdots T_r \rangle^\infty \neq \langle 1 \rangle,$$

the $\mathfrak{a}$-faces can be determined by computing the saturation through Gröbner basis techniques available in SINGULAR. In the second step (line 2 of the algorithm), the $\mathfrak{a}$-faces are projected to cones in $\mathbb{Q}^k$. For each $\mathfrak{a}$-face γ, defining inequalities and equations of the resulting **orbit cones**

$$Q(\gamma) := \mathrm{cone}(q_i \mid i \in \gamma) \subseteq \Gamma := \mathrm{cone}(q_1, \ldots, q_r) \subseteq \mathbb{Q}^k$$

are determined, where by $\mathrm{cone}(v_1, \ldots, v_k)$ we mean the polyhedral cone obtained by taking all non-negative linear combinations of the v_i. Computationally, this can be done via the double description method available in POLYMAKE. We denote by Ω the set of all orbit cones. In the final step, the GIT-fan is obtained as

$$\Lambda(\mathfrak{a}, Q) := \{\lambda_\Omega(w) \mid w \in \Gamma\} \quad \text{where} \quad \lambda_\Omega(w) := \bigcap_{w \in \eta \in \Omega} \eta.$$

To compute $\Lambda(\mathfrak{a}, Q)$, we perform a fan-traversal in the following way: Starting with a random maximal GIT-cone $\lambda_\Omega(w_0) \in \Lambda(\mathfrak{a}, Q)$, we compute its facets, determine the GIT-cones $\lambda_\Omega(w)$ adjacent to it, and iterate until the support of the fan equals $\mathrm{cone}(q_1, \ldots, q_r)$. Figure 12 illustrates three steps in such a process.

In line 9 of Algorithm 1, we write $\ominus$ for the symmetric difference in the first component. Again, computation of the facets of a given cone is available through the convex hull algorithms in POLYMAKE.

Algorithm 1 is implemented in the SINGULAR library `gitfan.lib` [16]. The SINGULAR to POLYMAKE interface `polymake.so` [45] provides key convex geometry functionality in the SINGULAR interpreter through a kernel level interface written in C++. We illustrate the use of this interface by a simple example.

Example 8. We compute the normal fan F of the Newton polytope P of the polynomial $f = x^3 + y^3 + 1$, see Fig. 11. Note that F is the Gröbner fan of the ideal $\langle f \rangle$ and its codimension one skeleton is the tropical variety of $\langle f \rangle$.

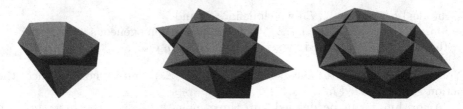

Fig. 12. Fan traversal.

Algorithm 1. GIT-fan

Input: An ideal $\mathfrak{a} \subseteq \mathbb{C}[T_1, \ldots, T_r]$ and a matrix $Q \in \mathbb{Z}^{k \times r}$ of full rank such that $\mathfrak{a}$ is homogeneous with respect to the multigrading given by Q.

Output: The set of maximal cones of $\Lambda(\mathfrak{a}, Q)$.

1: $\mathcal{A} := \{\gamma \subseteq \mathbb{Q}_{\geq 0}^r \text{ face} \mid \gamma \text{ is an } \mathfrak{a}\text{-face}\}$
2: $\Omega := \{Q(\gamma) \mid \gamma \in \mathcal{A}\}$
3: Choose a vector $w_0 \in Q(\gamma)$ such that $\dim(\lambda_\Omega(w_0)) = k$.
4: $\mathcal{C} := \{\lambda_\Omega(w_0)\}$
5: $\mathcal{F} := \{(\tau, \lambda_\Omega(w_0)) \mid \tau \subseteq \lambda_\Omega(w_0) \text{ facet with } \tau \not\subseteq \partial\Gamma\}$.
6: **while** there is $(\eta, \lambda) \in \mathcal{F}$ **do**
7:    Find $w \in Q(\gamma)$ such that $w \notin \lambda$ and $\lambda_\Omega(w) \cap \lambda = \eta$.
8:    $\mathcal{C} := \mathcal{C} \cup \{\lambda_\Omega(w)\}$
9:    $\mathcal{F} := \mathcal{F} \ominus \{(\tau, \lambda_\Omega(w)) \mid \tau \subseteq \lambda_\Omega(w) \text{ facet with } \tau \not\subseteq \partial\Gamma\}$
10: **return** $\mathcal{C}$

```
> LIB "polymake.so";
  Welcome to polymake version 2.14
  Copyright (c) 1997-2015
  Ewgenij Gawrilow, Michael Joswig (TU Berlin)
  http://www.polymake.org
  // ** loaded polymake.so
> ring R = 0,(x,y),dp; poly f = x3+y3+1;
> polytope P = newtonPolytope(f);
> fan F = normalFan(P); F;
 RAYS:
 -1 -1 #0
  0  1 #1
  1  0 #2
 MAXIMAL_CONES:
 {0 1} #Dimension 2
 {0 2}
 {1 2}
```

For many relevant examples, the computation of GIT-fans is challenged not only by the large amount of computations in lines 1 and 6 of Algorithm 1, but also by the complexity of each single computation in some boundary cases.

Making use of symmetries and parallel computations, we can open up the possibility to handle many interesting new cases by considerably simplifying and speeding up the computations. For instance, the computations in line 1 of Algorithm 1 can be executed independently in parallel. Parallel computation techniques can also be applied in the computation of $\lambda_\Omega(w_0)$ and the traversal of the GIT-fan. This step, however, is not trivially parallel.

An example of the use of symmetries is [17]; here, the first, third and fourth authors have applied and extended the technique described above to obtain the cones of the Mori chamber decomposition (the GIT-fan of the action of the characteristic torus on its total coordinate space) of the Deligne-Mumford compactification $\overline{M}_{0,6}$ of the moduli space of 6-pointed stable curves of genus zero that lie within the cone of movable divisor classes. A priori, this requires to consider 2^{40} torus orbits in line 1. Hence, a direct application of Algorithm 1 in its stated form is not feasible. However, moduli spaces in algebraic geometry often have large amounts of symmetry. For example, on $\overline{M}_{0,6}$ there is a natural group action of the symmetric group S_6 which Bernal [8] has extended to the input data $\mathfrak{a}$ and Q required for Algorithm 1. The GIT-fan $\Lambda(\mathfrak{a}, Q)$, and all data that arises in its computation reflect these symmetries. Hence, by computing an orbit decomposition under the action of the group of symmetries of the set of all torus orbits, we can restrict ourselves to a distinct set of representatives. Also the fan-traversal can be done modulo symmetry. To compute the orbit decomposition, we apply the algorithms for symmetric groups implemented in GAP.

Example 9. We apply this technique in the case of the affine cone X over the Grassmannian $\mathbb{G}(2,5)$ of 2-dimensional linear subspaces in a 5-dimensional vector space, see also [17]. By making use of the action of S_5, the number of monomial containment tests in line 1 can be reduced from $2^{10} = 1024$ to 34. A distinct set of representatives of the orbits of the 172 $\mathfrak{a}$-faces consists of 14 elements. The

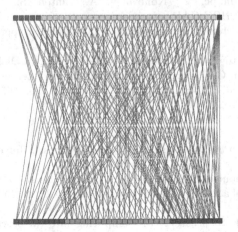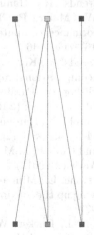

Fig. 13. The adjacency graph of the set of maximal cones of the GIT-fan of $\mathbb{G}(2,5)$ and the adjacency graph of the orbits of these cones under the S_5-action.

GIT-fan has 76 maximal cones, which fall into 6 orbits. Figure 13 shows both the adjacency graph of the maximal cones of the GIT-fan and that of their orbits under the S_5-action. This GIT-fan has also been discussed in [2, 8, 26]. Note that by considering orbits of cones not only the computation of the fan is considerably simplified, but also the theoretical understanding of the geometry becomes easier.

To summarize, Algorithm 1 requires the following key computational techniques from commutative algebra, convex geometry, and group theory:

- Gröbner basis computations,
- convex hull computations, and
- orbit decomposition.

These techniques are provided by SINGULAR, POLYMAKE, and GAP. At the current stage, POLYMAKE can be used from SINGULAR in a convenient way through `polymake.so`. An interface to use GAP functionality directly from SINGULAR is subject to future development.

References

1. Arnold, E.A.: Modular algorithms for computing Gröbner bases. J. Symbolic Comput. **35**(4), 403–419 (2003)
2. Arzhantsev, I.V., Hausen, J.: Geometric invariant theory via Cox rings. J. Pure Appl. Algebra **213**(1), 154–172 (2009)
3. Barakat, M.: Computations of unitary groups in characteristic 2 (2014). http://www.mathematik.uni-kl.de/~barakat/forJPSerre/UnitaryGroup.pdf
4. Behrends, R.: Shared memory concurrency for GAP. Comput. Algebra Rundbrief **55**, 27–29 (2014)
5. Behrends, R., Hammond, K., Janjic, V., Konovalov, A., Linton, S., Loidl, H.-W., Maier, P., Trinder, P.: HPC-GAP: engineering a 21st-century high-performance computer algebra system. Concurrency Comput. Pract. Experience (2016). cpe.3746
6. Behrends, R., Konovalov, A., Linton, S., Lübeck, F., Neunhöffer, M.: Parallelising the computational algebra system GAP. In: Proceedings of the 4th International Workshop on Parallel and Symbolic Computation, PASCO 2010, pp. 177–178. ACM, New York (2010)
7. Berchtold, F., Hausen, J.: GIT equivalence beyond the ample cone. Michigan Math. J. **54**(3), 483–515 (2006)
8. Bernal Guillén, M.M.: Relations in the Cox Ring of $\overline{M}_{0,6}$. Ph.D. thesis, University of Warwick (2012)
9. Bierstone, E., Milman, P.D.: Canonical desingularization in characteristic zero by blowing up the maximum strata of a local invariant. Invent. Math. **128**(2), 207–302 (1997)
10. Böhm, J., Decker, W., Laplagne, S., Pfister, G.: Computing integral bases via localization and Hensel lifting (2015). http://arxiv.org/abs/1505.05054
11. Böhm, J., Decker, W., Laplagne, S., Pfister, G.: Local to global algorithms for the Gorenstein adjoint ideal of a curve (2015). http://arxiv.org/abs/1505.05040

12. Böhm, J., Decker, W., Laplagne, S., Pfister, G., Steenpaß, A., Steidel, S.: locnormal.lib - A Singular library for a local-to-global approach to normalization (2013). Available in the Singular distribution, http://www.singular.uni-kl.de
13. Böhm, J., Decker, W., Laplagne, S., Pfister, G., Steenpaß, A., Steidel, S.: modnormal.lib - A Singular library for a modular approach to normalization (2013). Available in the Singular distribution, http://www.singular.uni-kl.de
14. Böhm, J., Decker, W., Laplagne, S., Pfister, G., Steenpaß, A., Steidel, S.: Parallel algorithms for normalization. J. Symbolic Comput. **51**, 99–114 (2013)
15. Böhm, J., Decker, W., Laplagne, S., Seelisch, F.: paraplanecurves.lib - A Singular library for the parametrization of rational curves (2013). Available in the Singular distribution, http://www.singular.uni-kl.de
16. Böhm, J., Keicher, S., Ren, Y.: gitfan.lib - A Singular library for computing the GIT fan (2015). Available in the Singular distribution, http://www.mathematik. uni-kl.de/~boehm/gitfan
17. Böhm, J., Keicher, S., Ren, Y.: Computing GIT-fans with symmetry and the Mori chamber decomposition of $\overline{M}_{0,6}$ (2016)
18. Boku, D.K., Decker, W., Fieker, C., Steenpass, A.: Gröbner bases over algebraic number fields. In: Proceedings of the International Workshop on Parallel Symbolic Computation, PASCO 2015, pp. 16–24. ACM, New York (2015)
19. Bosma, W., Cannon, J., Playout, C.: The Magma algebra system I: The user language. J. Symbolic Comput. **24**(3–4), 235–265 (1997). Computational algebra and number theory (London, 1993)
20. Bravo, A.M., Encinas, S., Villamayor U., O.: A simplified proof of desingularization and applications. Rev. Mat. Iberoamericana **21**(2), 349–458 (2005)
21. Bruns, W., Ichim, B.: Normaliz: algorithms for affine monoids and rational cones. J. Algebra **324**(5), 1098–1113 (2010)
22. Buchberger, B.: Ein Algorithmus zum Auffinden der Basiselemente des Restklassenring nach einem nulldimensionalen Polynomideal. Dissertation, Universität Innsbruck (1965)
23. de Jong, T.: An algorithm for computing the integral closure. J. Symbolic Comput. **26**(3), 273–277 (1998)
24. Decker, W., de Jong, T., Greuel, G.-M., Pfister, G.: The normalization: a new algorithm, implementation and comparisons. In: Dräxler, P., Ringel, C.M., Michler, G.O. (eds.) Computational Methods for Representations of Groups and Algebras. Progress in Mathematics, vol. 173, pp. 177–185. Birkhäuser, Basel (1999)
25. Decker, W., Greuel, G.-M., Pfister, G., Schönemann, H.: SINGULAR 4-0-2 — A computer algebra system for polynomial computations (2015). http://www.singular. uni-kl.de
26. Dolgachev, I.V., Hu, Y.: Variation of geometric invariant theory quotients. (With an appendix: "An example of a thick wall" by Nicolas Ressayre). Publ. Math. Inst. Hautes Étud. Sci. **87**, 5–56 (1998)
27. Eisenbud, D.: Commutative Algebra: With a View Toward Algebraic Geometry. Graduate Texts in Mathematics, vol. 150. Springer, New York (1995)
28. Eisenbud, D.: The Geometry of Syzygies: A Second Course in Commutative Algebra and Algebraic Geometry. Graduate Texts in Mathematics, vol. 229. Springer, New York (2005)
29. Eisenbud, D., Fløystad, G., Schreyer, F.-O.: Sheaf cohomology and free resolutions over exterior algebras. Trans. Am. Math. Soc. **355**(11), 4397–4426 (2003)
30. Encinas, S., Hauser, H.: Strong resolution of singularities in characteristic zero. Comment. Math. Helv. **77**(4), 821–845 (2002)

31. Erocal, B., Motsak, O., Schreyer, F.-O., Steenpass, A.: Refined algorithms to compute syzygies. J. Symb. Comput **74**, 308–327 (2016)
32. Frühbis-Krüger, A.: Computational aspects of singularities. In: Singularities in Geometry and Topology, pp. 253–327. World Sci. Publ., Hackensack (2007)
33. Frühbis-Krüger, A.: resolve.lib - A Singular library for the resolution of singularities (2015). Available in the Singular distribution, http://www.singular.uni-kl.de
34. The GAP Group. GAP - Groups, Algorithms, and Programming, Version 4.7.9 (2015)
35. Gawrilow, E., Joswig, M.: Polymake: a framework for analyzing convex polytopes. In: Kalai, G., Ziegler, G.M. (eds.) Polytopes – Combinatorics and Computation, pp. 43–74. Birkhäuser, Basel (2000)
36. Grauert, H., Remmert, R., Stellenalgebren, A.: Analytische Stellenalgebren. Springer, New York (1971). Unter Mitarbeit von O. Riemenschneider, Die Grundlehren der
37. Grayson, D.R., Stillman, M.E.: Macaulay2, a software system for research in algebraic geometry. http://www.math.uiuc.edu/Macaulay2/
38. Greuel, G.-M., Laplagne, S., Seelisch, F.: Normalization of rings. J. Symbolic Comput. **45**(9), 887–901 (2010)
39. Greuel, G.-M., Laplagne, S., Seelisch, F.: normal.lib - A Singular library for normalization (2010). Available in the Singular distribution, http://www.singular.uni-kl.de
40. Hampe, S.: a-tint: a polymake extension for algorithmic tropical intersection theory. European J. Combin. **36**, 579–607 (2014)
41. Hart, B.: ANTIC: Algebraic number theory in C. Comput. Algebra Rundbrief **56**, 10–12 (2015)
42. Hart, W., Johansson, F., Pancratz, S.: FLINT: Fast Library for Number Theory (2013). Version 2.4.0, http://flintlib.org
43. Huneke, C., Swanson, I.: Integral Closure of Ideals, Rings, and Modules. London Mathematical Society Lecture Note Series, vol. 336. Cambridge University Press, Cambridge (2006)
44. Keicher, S.: Computing the GIT-fan. Internat. J. Algebra Comput. **22**(7), 11 (2012). Article ID 1250064
45. Ren, Y.: polymake.so - A Singular module for interfacing with polymake (2015). Available in the Singular distribution, http://www.singular.uni-kl.de
46. Schreyer, F.-O.: Die Berechnung von Syzygien mit dem verallgemeinerten Weierstraßschen Divisionssatz und eine Anwendung auf analytische Cohen-Macaulay-Stellenalgebren minimaler Multiplizität. Diploma thesis, Universität Hamburg (1980)
47. Serre, J.-P.: Bases normales autoduales et groupes unitaires en caractéristique 2. Transform. Groups **19**(2), 643–698 (2014)
48. Steenpaß, A.: parallel.lib - A Singular library for parallel computations (2015). Available in the Singular distribution, https://www.singular.uni-kl.de
49. The homalg project authors. The homalg project - Algorithmic Homological Algebra (2003–2014). http://homalg.math.rwth-aachen.de/

Exploiting Structure
in Floating-Point Arithmetic

Claude-Pierre Jeannerod[✉]

Inria, Laboratoire LIP (U. Lyon, CNRS, ENSL, Inria, UCBL),
ENS de Lyon, 46 allée d'Italie, 69364 Lyon Cedex 07, France
claude-pierre.jeannerod@inria.fr

Abstract. The analysis of algorithms in IEEE floating-point arithmetic is most often carried out via repeated applications of the so-called standard model, which bounds the relative error of each basic operation by a common epsilon depending only on the format. While this approach has been eminently useful for establishing many accuracy and stability results, it fails to capture most of the low-level features that make floating-point arithmetic so highly *structured*. In this paper, we survey some of those properties and how to exploit them in rounding error analysis. In particular, we review some recent improvements of several classical, Wilkinson-style error bounds from linear algebra and complex arithmetic that all rely on such structure properties.

Keywords: Floating-point arithmetic · IEEE standard 754-2008 · Rounding error analysis · High relative accuracy

1 Introduction

When analyzing a priori the behaviour of a numerical algorithm in IEEE floating-point arithmetic, one most often relies exclusively on the so-called *standard model*: for base β, precision p, and rounding to nearest, this model says that the result $\hat{r}$ of each basic operation op $\in \{+, -, \times, /\}$ on two floating-point numbers x and y satisfies

$$\hat{r} = (x \operatorname{op} y)(1 + \delta), \qquad |\delta| \leqslant u \tag{1}$$

with $u = \frac{1}{2}\beta^{1-p}$ the *unit roundoff*. (Similar relations are also assumed for the square root and the fused multiply-add (FMA) operations.)

This model has been used long before the appearance of the first version of IEEE standard 754 [17,18], and the fact that it gives backward error results is already emphasized by Wilkinson [43]: considering for example floating-point addition, it is easily deduced from (1) that $\hat{r}$ is the exact sum of the slightly perturbed data $x(1 + \delta)$ and $y(1 + \delta)$, and, applying this repeatedly, that the computed approximation to the sum of n floating-point numbers x_i has the form $\sum_{i=1}^{n} \widetilde{x}_i$ with $|\widetilde{x}_i - x_i|/|x_i| \leqslant (1 + u)^{n-1} - 1 = (n - 1)u + O(u^2)$ for all i.

© Springer International Publishing Switzerland 2016
I.S. Kotsireas et al. (Eds.): MACIS 2015, LNCS 9582, pp. 25–34, 2016.
DOI: 10.1007/978-3-319-32859-1_2

Backward error analysis based on the standard model (1) has developed far beyond this basic example and turned out to be eminently useful for establishing many accuracy and stability results, as Higham's treatise [14] shows.

Although the standard model holds for IEEE 754 arithmetic as long as underflow and overflow do not occur, it fails, however, to capture most of the low-level features that make this arithmetic so highly *structured*. For example, by ensuring a relative error less than one, (1) implies that $\hat{r}$ has the same sign as the exact value $x \operatorname{op} y$, but it does not say that δ should be zero when $x \operatorname{op} y$ is a floating-point number.

Such low-level features are direct consequences of the two main ingredients of IEEE standard arithmetic. The first ingredient is the set $\mathbb{F}$ of floating-point numbers, which (up to ignoring underflow and overflow) can be viewed as

$$\mathbb{F} = \{0\} \cup \{M\beta^e \ : \ M, e \in \mathbb{Z}, \ \beta^{p-1} \leqslant |M| < \beta^p\}. \tag{2}$$

The second ingredient is a rounding function $\mathrm{RN} : \mathbb{R} \to \mathbb{F}$, which maps any real number to a nearest element in $\mathbb{F}$:

$$|\mathrm{RN}(t) - t| = \min_{f \in \mathbb{F}} |f - t| \qquad \text{for all } t \in \mathbb{R}, \tag{3}$$

with ties broken according to a given rule (say, round *to nearest even*). This rounding function is then used by IEEE standard arithmetic to operate on floating-point data as follows: in the absence of underflow and overflow, $x \operatorname{op} y$ must be computed as

$$\hat{r} = \mathrm{RN}(x \operatorname{op} y).$$

This way of combining the structured data in (2) and the minimization property (3) implies that $\hat{r}$ enjoys many more mathematical properties than just (1).

The goal of this paper is to show the benefits of exploiting such lower level features in the context of rounding error analysis. We begin by recalling some of these features in Sect. 2. Although the list given there is by no means exhaustive (cf. Rump, Ogita, and Oishi [37, Sect. 2]), it should already give a good idea of what can be deduced from (2) and (3). We then review some recent improvements of several classical, Wilkinson-style error bounds from linear algebra and complex arithmetic that all rely on such structure properties. Specifically, we will see in Sect. 3 that various general algorithms (for summation, inner products, matrix factorization, polynomial evaluation, ...) now have a priori error bounds which are both simpler and sharper than the classical ones. In Sect. 4 we will focus on more specific algorithms for core computations like 2×2 determinants or complex products, and show that in such cases exploiting the low-level features of IEEE standard arithmetic leads to proofs of high relative accuracy and tight error bounds.

Throughout this paper we assume for simplicity that β is even, that RN rounds to nearest even, and that underflow and overflow do not occur. (For summation, however, the results presented here still hold in the presence of underflow, since then floating-point addition is known to be exact; see Hauser [13].)

For more on floating-point arithmetic, we refer to the complementary texts by Brent and Zimmermann [3, Sect. 3], Corless and Fillion [6, Appendix A], Demmel [9, Sect. 1.5], Goldberg [10], Golub and Van Loan [11, Sect. 2.7], Higham [14, Sect. 2], [15], Knuth [27, Sect. 4.2], Monniaux [29], Muller et al. [31], Overton [33], Priest [34], Trefethen [41], and Trefethen and Bau [42, Sect. 13].

2   Low-Level Properties

Structure of the Floating-Point Number Set. By construction, the set $\mathbb{F}$ contains zero, has the symmetry property $\mathbb{F} = -\mathbb{F}$, and is invariant under *scaling* (that is, multiplication by an integer power of the base): $x\beta^k \in \mathbb{F}$ for all $x \in \mathbb{F}$ and $k \in \mathbb{Z}$. More precisely, every element of $\mathbb{F}$ is a multiple (by some $\pm\beta^k$) of an element of the subset $\mathbb{F} \cap [1, \beta)$. The elements of this subset have the form $1 + j\beta^{1-p}$, where j is an integer such that $0 \leqslant j < (\beta - 1)\beta^{p-1}$ and, since $u = \frac{1}{2}\beta^{1-p}$, this can be expressed concisely as follows:

$$\mathbb{F} \cap [1, \beta) = \{1, 1 + 2u, 1 + 4u, 1 + 6u, \ldots\}.$$

The numbers lying exactly halfway between two consecutive elements of $\mathbb{F}$, such as for example $1 + u$ and $1 + 3u$, are called *midpoints* for $\mathbb{F}$.

Some First Consequences of Rounding to Nearest. Since by definition $|\mathrm{RN}(t) - t| \leqslant |f - t|$ for all f in $\mathbb{F}$, choosing $t = x + \epsilon$ with $x \in \mathbb{F}$ and $\epsilon \in \mathbb{R}$ gives $|\mathrm{RN}(x + \epsilon) - (x + \epsilon)| \leqslant |\epsilon|$. With $\epsilon = 0$ we recover the obvious property that rounding a floating-point number leaves it unchanged:

$$x \in \mathbb{F} \quad \Rightarrow \quad \mathrm{RN}(x) = x. \tag{4}$$

Setting $\epsilon = y$ with y in $\mathbb{F}$, we deduce further that for floating-point addition the error bound implied by the standard model (1) can be refined slightly:

$$x, y \in \mathbb{F} \quad \Rightarrow \quad |\mathrm{RN}(x + y) - (x + y)| \leqslant \min\{u|x + y|, |x|, |y|\}. \tag{5}$$

(Similarly, a sharper bound can be deduced for the FMA operation by taking $\epsilon = yz$.) We will see in Sect. 3 how to exploit such a refinement in the context of floating-point summation.

Besides (4), other basic features include the following ones:

$$t \in \mathbb{R} \quad \Rightarrow \quad |\mathrm{RN}(t)| = \mathrm{RN}(|t|), \tag{6}$$

$$t \in \mathbb{R}, \quad k \in \mathbb{Z} \quad \Rightarrow \quad \mathrm{RN}(t\beta^k) = \mathrm{RN}(t)\beta^k, \tag{7}$$

$$t, t' \in \mathbb{R}, \quad t \leqslant t' \quad \Rightarrow \quad \mathrm{RN}(t) \leqslant \mathrm{RN}(t'). \tag{8}$$

Combining (4) with the monotonicity property (8), we see for example that if $x \in \mathbb{F}$ satisfies $x \leqslant t$ for some real t, then $x \leqslant \mathrm{RN}(t)$.

As another example, we note that (4), (7), and (8) already suffice to prove that the classical approximation to the mean of two floating-point numbers behaves as expected in base 2 (but not in base 10): using (7) and then (4) gives $\hat{r} := \mathrm{RN}(\mathrm{RN}(x+y)/2) = \mathrm{RN}((x+y)/2)$; then, using $f := \min\{x, y\} \leqslant (x+y)/2 \leqslant \max\{x, y\} =: g$ together with (8), we deduce that $\mathrm{RN}(f) \leqslant \hat{r} \leqslant \mathrm{RN}(g)$ and, applying (4) again, we conclude that $f \leqslant \hat{r} \leqslant g$.

The Functions ufp and ulp. A very convenient tool to go beyond the standard model is provided by the notion of *unit in the first place* (ufp), defined in [37] as

$$\mathrm{ufp}(t) = \begin{cases} 0 & \text{if } t = 0, \\ \beta^{\lfloor \log_\beta |t| \rfloor} & \text{if } t \in \mathbb{R} \backslash \{0\}. \end{cases}$$

Its relationship with the classical notion of *unit in the last place* (ulp) is via the equality $\mathrm{ulp}(t) = 2u\,\mathrm{ufp}(t)$, and its definition implies immediately that

$$t \in \mathbb{R} \backslash \{0\} \quad \Rightarrow \quad \mathrm{ufp}(t) \leqslant |t| < \beta \mathrm{ufp}(t). \tag{9}$$

From (4), (6), (8), it then follows that

$$t \in \mathbb{R} \quad \Rightarrow \quad \mathrm{ufp}(t) \leqslant |\mathrm{RN}(t)| \leqslant \beta \mathrm{ufp}(t).$$

Thus, $\mathrm{RN}(t)$ belongs to a range for which the distance between two consecutive floating-point numbers is exactly $2u\,\mathrm{ufp}(t)$, and being nearest to t implies

$$|\mathrm{RN}(t) - t| \leqslant u\,\mathrm{ufp}(t).$$

In terms of ulp's, this is just the usual half-an-ulp absolute error bound (attained at every midpoint) and, dividing further by $|t| > 0$, we arrive at

$$t \in \mathbb{R} \backslash \{0\} \quad \Rightarrow \quad \frac{|\mathrm{RN}(t) - t|}{|t|} \leqslant u\frac{\mathrm{ufp}(t)}{|t|}. \tag{10}$$

This inequality is interesting for at least three reasons. First, recalling (9), it allows us to recover the uniform bound u claimed by the standard model (1). Second, it shows that the relative error can be bounded by about u/β instead of u when $|t|$ approaches its upper bound $\beta \mathrm{ufp}(t)$; this is related to a phenomenon called *wobbling precision* [14, p. 39] and indicates that when deriving sharp error bounds the most difficult cases are likely to occur when $|t|$ lies in the leftmost part of its range $[\mathrm{ufp}(t), \beta \mathrm{ufp}(t))$. Third, it makes it easy to check that the bound u is in fact never attained, as noted in [14, p. 38], since either $|t| = \mathrm{ufp}(t) \in \mathbb{F}$ or $\mathrm{ufp}(t)/|t| < 1$. Indeed, the following slightly stronger statement holds:

$$t \in \mathbb{R} \backslash \{0\} \quad \Rightarrow \quad \frac{|\mathrm{RN}(t) - t|}{|t|} \leqslant \frac{u}{1 + u}. \tag{11}$$

If $|t| \geqslant (1+u)\mathrm{ufp}(t)$, the above inequality follows directly from the one in (10). Else, rounding to nearest implies that $|\mathrm{RN}(t)| = \mathrm{ufp}(t) \leqslant |t| < (1+u)\mathrm{ufp}(t)$ and, recalling that t has the same sign as its rounded value, we conclude that

$$\frac{|\mathrm{RN}(t) - t|}{|t|} = 1 - \frac{\mathrm{ufp}(t)}{|t|} < 1 - \frac{1}{1 + u} = \frac{u}{1 + u}.$$

The bound in (11) is given by Knuth in [27, p. 232] and, in the special case where $t = x + y$ or $t = xy$ with $x, y \in \mathbb{F}$, it was already noted by Dekker [8] (in base 2) and then by Holm [16] (in any base). Furthermore, it turns out to be attained if and only if t is the midpoint $\pm(1 + u)\mathrm{ufp}(t)$; see [25]. This best possible bound refines the standard model (1) only slightly, but we shall see in the rest of this paper that it can be worth exploiting in various situations.

Exact Floating-Point Subtraction and EFTs. We now briefly review what can be obtained *exactly* using floating-point and rounding to nearest. A first classical result is Sterbenz' theorem [40, p. 138], which ensures that floating-point subtraction is exact when the two operands are close enough to each other:

$$x, y \in \mathbb{F}, \quad y/2 \leqslant x \leqslant 2y \quad \Rightarrow \quad x - y \in \mathbb{F}.$$

Another exactness property is that the absolute error due to floating-point addition or multiplication is itself a floating-point number:

$$x, y \in \mathbb{F}, \quad \mathrm{op} \in \{+, \times\} \quad \Rightarrow \quad x \, \mathrm{op} \, y - \mathrm{RN}(x \, \mathrm{op} \, y) \in \mathbb{F}.$$

Furthermore, various floating-point algorithms are available for computing simultaneously the rounded value $\hat{r} = \mathrm{RN}(x \, \mathrm{op} \, y)$ and the exact value of the associated rounding error $e = x \, \mathrm{op} \, y - \hat{r}$. For addition, these are the Fast2Sum algorithm of Kahan [26] and Dekker [8], and the more general 2Sum algorithm of Knuth [27] and Møller [28]. For multiplication, it suffices to use the FMA operation as follows:

$$\hat{r} \leftarrow \mathrm{RN}(xy), \qquad e \leftarrow \mathrm{RN}(xy - \hat{r}). \tag{12}$$

(If no FMA is available, the pair $(\hat{r}, e)$ can be obtained using 7 multiplications and 10 additions, as shown by Dekker in [8].) These algorithms define in each case a so-called *error-free transformation* (EFT) [32], which maps $(x, y) \in \mathbb{F}^2$ to $(\hat{r}, e) \in \mathbb{F}^2$ such that $x \, \mathrm{op} \, y = \hat{r} + e$. In Sect. 4 we will see in particular how to exploit the transformation given by (12), possibly in combination with Sterbenz's theorem. For more examples of EFT-based, provably accurate algorithms—especially in the context of summation and elementary function evaluation—we refer to [35] and [31] and the references therein.

3 Revisiting Some Classical Wilkinson-Style Error Bounds

3.1 Summation

Given $x_1, \ldots, x_n \in \mathbb{F}$, we consider first the evaluation of the sum $\sum_{i=1}^{n} x_i$ by means of $n - 1$ floating-point additions, in any order. Following Wilkinson [43], we may apply the standard model (1) repeatedly in order to obtain the backward error result shown in Sect. 1, from which a forward error bound for the computed value $\hat{r}$ then follows directly:

$$\left| \hat{r} - \sum_{i=1}^{n} x_i \right| \leqslant \alpha \sum_{i=1}^{n} |x_i|, \qquad \alpha = (1 + u)^{n-1} - 1. \tag{13}$$

Such a bound is easy to derive, valid for any order, and a priori essentially best possible since there exist special values of the x_i for which the ratio error/(error bound) tends to 1 as $u \to 0$. The expression giving α, however, is somehow unwieldy and it is now common practice to have it replaced by the concise yet rigorous upper bound γ_{n-1}, using Higham's γ_k notation "$\gamma_k = ku/(1 - ku)$ if $ku < 1$" [14, p. 63]. Both bounds have the form $(n-1)u + O(u^2)$ and the second one further assumes implicitly that the dimension n satisfies $(n-1)u < 1$.

Recently, it was shown by Rump [36] that for recursive summation one can in fact always replace α in (13) by the simpler and sharper expression

$$\alpha = (n-1)u.$$

In other words, the terms of order $O(u^2)$ can be removed, and this without any restriction on n. The proof given in [36, p. 206] aims to bound the forward error $|\hat{r} - \sum_{i=1}^{n} x_i|$ directly, focusing on the last addition and proceeding by induction on n; in particular, one key ingredient is the refined model (5) of floating-point addition, which is used here to handle the case $|x_n| \leqslant u \sum_{i=1}^{n-1} |x_i|$. As noted in [24, Sect. 3], this proof technique is in fact not restricted to recursive summation, so the constant $(n-1)u$ eventually holds for any summation order.

3.2   Other Examples of $O(u^2)$-Free Error Bounds

Similar improvements have been obtained for the error bounds of several other computational problems, which we summarize in Table 1. The algorithms for which these new bounds hold are the classical ones (described for example in [14]) and the role played by α depends on the problem as follows: for dot products, α should be such that $|\hat{r} - x^T y| \leqslant \alpha |x|^T |y|$ with $x, y \in \mathbb{F}^n$ and $\hat{r}$ denoting the computed value; for matrix multiplication, $|\hat{C} - AB| \leqslant \alpha |A||B|$ with $A \in \mathbb{F}^{*\times n}$ and $B \in \mathbb{F}^{n\times *}$; for Euclidean norms (in dimension n), powers, and products, $|\hat{r} - r| \leqslant \alpha |r|$; for triangular system solving and LU and Cholesky matrix factorizations, we consider the usual backward error bounds $|\Delta T| \leqslant \alpha |T|$ for $(T + \Delta T)\hat{x} = b$, $|\Delta A| \leqslant \alpha |\hat{L}||\hat{U}|$ for $\hat{L}\hat{U} = A + \Delta A$, and $|\Delta A| \leqslant \alpha |\hat{R}^T||\hat{R}|$ for $\hat{R}^T \hat{R} = A + \Delta A$. (Here the matrices T, $\hat{U}$, $\hat{R}$ have dimensions $n \times n$, and $\hat{L}$ has dimensions $m \times n$ with $m \geqslant n$.) Finally, for the evaluation of $a(x) = \sum_{i=0}^{n} a_i x^i$ with Horner's rule, α is such that $|\hat{r} - a(x)| \leqslant \alpha \sum_{i=0}^{n} |a_i x^i|$.

The new values of α shown in Table 1 are free of any $O(u^2)$ term and thus *simpler and sharper* than the classical ones. In the last three cases, the price to be paid for those refined constants is some mild restriction on n; we refer to [38] for a precise condition and an example showing that it is indeed necessary.

4   Provably Accurate Numerical Kernels

4.1   Computation of $ab + cd$

As a first example of such kernels, let us consider the evaluation of $ab + cd$ for $a, b, c, d \in \mathbb{F}$. This operation occurs frequently in practice and is especially useful for complex arithmetic, discriminants, and robust orientation predicates. Since it is not part of the set of core IEEE 754-2008 functions for which correct rounding is required or recommended (and despite the existence of hardware designs as the one by Brunie [4, Sect. 3.3.2]), this operation will in general be implemented

Table 1. Some classical Wilkinson-style constants made simpler and sharper. Unless otherwise stated these results hold for any ordering, and $(\star)$ means "if $n \lesssim u^{-1/2}$".

Problem	Classical α	New α		Reference(s)
summation	$(n-1)u + O(u^2)$	$(n-1)u$		[24,36]
dot prod., mat. mul.	$nu + O(u^2)$	nu		[24]
Euclidean norm	$(\frac{n}{2}+1)u + O(u^2)$	$(\frac{n}{2}+1)u$		[25]
$Tx = b$, $A = LU$	$nu + O(u^2)$	nu		[39]
$A = R^T R$	$(n+1)u + O(u^2)$	$(n+1)u$		[39]
x^n (recursive, $\beta = 2$)	$(n-1)u + O(u^2)$	$(n-1)u$	$(\star)$	[12]
product $x_1 x_2 \cdots x_n$	$(n-1)u + O(u^2)$	$(n-1)u$	$(\star)$	[38]
poly. eval. (Horner)	$2nu + O(u^2)$	$2nu$	$(\star)$	[38]

in software using basic floating-point arithmetic. When doing so, however, some care is needed and a classical scheme like $\mathrm{RN}(\mathrm{RN}(ab) + \mathrm{RN}(cd))$ or, if an FMA is available, $\mathrm{RN}(ab + \mathrm{RN}(cd))$ can produce a highly inaccurate answer.

To avoid this, the following sequence of four operations was suggested by Kahan (see [14, p. 60]):

$$\hat{w} := \mathrm{RN}(cd); \quad \hat{f} := \mathrm{RN}(ab + \hat{w}); \quad e := \mathrm{RN}(cd - \hat{w}); \quad \hat{r} := \mathrm{RN}(\hat{f} + e).$$

Here the FMA operation is used to produce $\hat{f}$ and also to implement an EFT for the product cd, as in (12), thus giving $e = cd - \hat{w}$ exactly. By applying to $\hat{w}$, $\hat{f}$, and $\hat{r}$ the refined standard model given by (11) it is then easy to prove that

$$\frac{|\hat{r} - r|}{|r|} \leqslant 2u(1 + \psi), \qquad r = ab + cd, \qquad \psi = \frac{u|cd|}{2|r|}. \tag{14}$$

This kind of analysis (already done by Higham in the 1996 edition of [14]) shows that Kahan's algorithm computes $ab + cd$ with high relative accuracy as long as $\psi \not\gg 1$. The latter condition, however, does not always hold, as there exist inputs for which ψ is of the order of u^{-1} and the relative error bound $2u(1+\psi)$ is larger than 1.

This classical analysis was refined in [21], where we show that Kahan's algorithm above is in fact *always* highly accurate: first, a careful analysis of the absolute errors $\epsilon_1 = \hat{f} - (ab + \hat{w})$ and $\epsilon_2 = \hat{r} - (\hat{f} + e)$ using the ufp (or ulp) function gives $|\epsilon_1|, |\epsilon_2| \leqslant \beta u \, \mathrm{ufp}(r)$, so that $|\hat{r} - r| = |\epsilon_1 + \epsilon_2| \leqslant 2\beta u|r|$; then, by studying ϵ_1 and ϵ_2 simultaneously via a case analysis comparing $|\epsilon_2|$ to $u \, \mathrm{ufp}(r)$, we deduce that the constant $2\beta u$ can be replaced by $2u$ (that is, the term ψ can in fact be removed from the bound in (14)); third, we show that this bound is asymptotically optimal (as $u \to 0$) by defining

$$a = b = \beta^{p-1} + 1, \qquad c = \beta^{p-1} + \frac{\beta}{2}\beta^{p-2}, \qquad d = 2\beta^{p-1} + \frac{\beta}{2}\beta^{p-2},$$

and checking (by hand or, since recently, using a dedicated Maple library [22]) that the error committed for such inputs has the form $2u - 4u^2 + O(u^3)$.

A similar scheme was proposed by Cornea, Harrison, and Tang [7, p. 273], which ensures further that the value returned for $ab+cd$ is the same as for $cd+ab$. (Such a feature may be desirable when, say, implementing complex arithmetic.) We refer to [19,30] for sharp error analyzes combining ufp-based arguments, the refined bound $u/(1+u)$, and Sterbenz' theorem.

4.2 Complex Multiplication

Another important numerical kernel is the evaluation of the real and imaginary parts $R = ac - bd$ and $I = ad + bc$ of the complex product $z = (a+ib)(c+id)$. Consider first the conventional way, which produces $\hat{R} = \text{RN}(\text{RN}(ac) - \text{RN}(bd))$ and $\hat{I} = \text{RN}(\text{RN}(ad) + \text{RN}(bc))$. Although $\hat{R}$ or $\hat{I}$ can be completely inaccurate, it is known that high relative accuracy holds in the *normwise* sense: Brent, Percival, and Zimmermann [2] showed that $\hat{z} = \hat{R} + i\hat{I}$ satisfies

$$\frac{|\hat{z} - z|}{|z|} \leqslant \sqrt{5}u$$

and that this bound is asymptotically optimal (at least in base 2); in particular, the constant $\sqrt{5} = 2.23\ldots$ improves upon classical and earlier ones like $\sqrt{8} = 2.82\ldots$ by Wilkinson [44, p. 447] and $1 + \sqrt{2} = 2.41\ldots$ by Champagne [5].

Assume now that an FMA is available. In this case, $\hat{R}$ can be obtained as $\text{RN}(ac - \text{RN}(bd))$ or $\text{RN}(\text{RN}(ac) - bd)$, and similarly for $\hat{I}$, so that z can be evaluated using four different schemes. We showed in [20] that for each of these schemes the bound $\sqrt{5}u$ mentioned above can be reduced further to $2u$ and that this new bound is asymptotically optimal. We also proved that this normwise bound $2u$ remains sharp even if both $\hat{R}$ and $\hat{I}$ are computed with high relative accuracy as in Sect. 4.1.

The bound $\sqrt{5}u$ was obtained in [2] via a careful ulp-based case analysis. For the bound $2u$ we have proceeded similarly in [20, Sect. 3] but, as we observe in [25], in this case a much shorter proof follows from using just the refined standard model given by (11).

A direct application of these error bounds is to complex division: as noted by Baudin in [1], if αu bounds the normwise relative error of multiplication, then the bound $(\alpha+3)u+O(u^2)$ holds for division—assuming the classical formula $x/y = (x\overline{y})/(y\overline{y})$—and thus we can take $\alpha + 3 = 5$ or $5.23\ldots$ depending on whether the FMA operation is available or not. However, despite this and some recent progress made in the case of complex inversion [23], the best possible constants for complex division (with or without an FMA) remain to be determined.

Acknowledgements. I am grateful to Ilias Kotsireas, Siegfried M. Rump, and Chee Yap for giving me the opportunity to write this survey. This work was supported in part by the French National Research Agency, under grant ANR-13-INSE-0007 (MetaLibm).

References

1. Baudin, M.: Error bounds of complex arithmetic, June 2011. http://forge.scilab. org/upload/compdiv/files/complexerrorbounds_v0.2.pdf
2. Brent, R.P., Percival, C., Zimmermann, P.: Error bounds on complex floating-point multiplication. Math. Comput. **76**, 1469–1481 (2007)
3. Brent, R.P., Zimmerman, P.: Modern Computer Arithmetic. Cambridge University Press, Cambridge (2010)
4. Brunie, N.: Contributions to Computer Arithmetic and Applications to Embedded Systems. Ph.D. thesis, École Normale Supérieure de Lyon, Lyon, France, May 2014. https://tel.archives-ouvertes.fr/tel-01078204
5. Champagne, W.P.: On finding roots of polynomials by hook or by crook. Master's thesis, University of Texas, Austin, Texas (1964)
6. Corless, R.M., Fillion, N.: A Graduate Introduction to Numerical Methods, From the Viewpoint of Backward Error Analysis. Springer, New York (2013)
7. Cornea, M., Harrison, J., Tang, P.T.P.: Scientific Computing on Itanium®-based Systems. Intel Press, Hillsboro (2002)
8. Dekker, T.J.: A floating-point technique for extending the available precision. Numer. Math. **18**, 224–242 (1971)
9. Demmel, J.W.: Applied Numerical Linear Algebra. SIAM, Philadelphia (1997)
10. Goldberg, D.: What every computer scientist should know about floating-point arithmetic. ACM Comput. Surv. **23**(1), 5–48 (1991)
11. Golub, G.H., Van Loan, C.F.: Matrix Computations, 4th edn. Johns Hopkins University Press, Baltimore (2013)
12. Graillat, S., Lefèvre, V., Muller, J.M.: On the maximum relative error when computing integer powers by iterated multiplications in floating-point arithmetic. Numer. Algorithms **70**, 653–667 (2015). http://link.springer.com/article/10.1007/ s11075-015-9967-8
13. Hauser, J.R.: Handling floating-point exceptions in numeric programs. ACM Trans. Program. Lang. Syst. **18**(2), 139–174 (1996)
14. Higham, N.J.: Accuracy and Stability of Numerical Algorithms, 2nd edn. SIAM, Philadelphia (2002)
15. Higham, N.J.: Floating-point arithmetic. In: Higham, N.J., Dennis, M.R., Glendinning, P., Martin, P.A., Santosa, F., Tanner, J. (eds.) The Princeton Companion to Applied Mathematics, pp. 96–97. Princeton University Press, Princeton (2015)
16. Holm, J.E.: Floating-Point Arithmetic and Program Correctness Proofs. Ph.D. thesis, Cornell University, Ithaca, NY, USA, August 1980
17. IEEE Computer Society: IEEE Standard for Binary Floating-Point Arithmetic, ANSI/IEEE Standard 754–1985. IEEE Computer Society, New York (1985)
18. IEEE Computer Society: IEEE Standard for Floating-Point Arithmetic, IEEE Standard 754–2008. IEEE Computer Society, New York (2008)
19. Jeannerod, C.P.: A radix-independent error analysis of the Cornea-Harrison-Tang method, to appear in ACM Trans. Math. Softw. https://hal.inria.fr/hal-01050021
20. Jeannerod, C.P., Kornerup, P., Louvet, N., Muller, J.M.: Error bounds on complex floating-point multiplication with an FMA, to appear in Math. Comput. https:// hal.inria.fr/hal-00867040v4
21. Jeannerod, C.P., Louvet, N., Muller, J.M.: Further analysis of Kahan's algorithm for the accurate computation of 2×2 determinants. Math. Comput. **82**(284), 2245–2264 (2013)

22. Jeannerod, C.P., Louvet, N., Muller, J.M., Plet, A.: A library for symbolic floating-point arithmetic (2015). https://hal.inria.fr/hal-01232159
23. Jeannerod, C.-P., Louvet, N., Muller, J.-M., Plet, A.: Sharp error bounds for complex floating-point inversion. Numer. Algorithms 1–26 (2016). https://hal-ens-lyon.archives-ouvertes.fr/ensl-01195625
24. Jeannerod, C.P., Rump, S.M.: Improved error bounds for inner products in floating-point arithmetic. SIAM J. Matrix Anal. Appl. **34**(2), 338–344 (2013)
25. Jeannerod, C.P., Rump, S.M.: On relative errors of floating-point operations: optimal bounds and applications (2014). https://hal.inria.fr/hal-00934443
26. Kahan, W.: Further remarks on reducing truncation errors. Commun. ACM **8**(1), 40 (1965)
27. Knuth, D.E.: The Art of Computer Programming. Seminumerical Algorithms, vol. 2, 3rd edn. Addison-Wesley, Reading (1998)
28. Møller, O.: Quasi double-precision in floating point addition. BIT **5**, 37–50 (1965)
29. Monniaux, D.: The pitfalls of verifying floating-point computations. ACM Trans. Program. Lang. Syst. **30**(3), 12:1–12:41 (2008)
30. Muller, J.M.: On the error of computing $ab + cd$ using Cornea, Harrison and Tang's method. ACM Trans. Math. Softw. **41**(2), 7:1–7:8 (2015)
31. Muller, J.M., Brisebarre, N., de Dinechin, F., Jeannerod, C.P., Lefèvre, V., Melquiond, G., Revol, N., Stehlé, D., Torres, S.: Handbook of Floating-Point Arithmetic. Birkhäuser, Boston (2010)
32. Ogita, T., Rump, S.M., Oishi, S.: Accurate sum and dot product. SIAM J. Sci. Comput. **26**(6), 1955–1988 (2005)
33. Overton, M.L.: Numerical Computing with IEEE Floating Point Arithmetic: Including One Theorem, One Rule of Thumb, and One Hundred and One Exercises. Society for Industrial and Applied Mathematics, Philadelphia (2001)
34. Priest, D.M.: On Properties of Floating Point Arithmetics: Numerical Stability and the Cost of Accurate Computations. Ph.D. thesis, Mathematics Department, University of California, Berkeley, CA, USA, November 1992
35. Rump, S.M.: Ultimately fast accurate summation. SIAM J. Sci. Comput. **31**(5), 3466–3502 (2009)
36. Rump, S.M.: Error estimation of floating-point summation and dot product. BIT **52**(1), 201–220 (2012)
37. Rump, S.M., Ogita, T., Oishi, S.: Accurate floating-point summation part I: faithful rounding. SIAM J. Sci. Comput. **31**(1), 189–224 (2008)
38. Rump, S.M., Bünger, F., Jeannerod, C.P.: Improved error bounds for floating-point products and Horner's scheme. BIT (2015). http://link.springer.com/article/10.1007/s10543-015-0555-z
39. Rump, S.M., Jeannerod, C.P.: Improved backward error bounds for LU and Cholesky factorizations. SIAM J. Matrix Anal. Appl. **35**(2), 684–698 (2014)
40. Sterbenz, P.H.: Floating-Point Computation. Prentice-Hall, Englewood Cliffs (1974)
41. Trefethen, L.N.: Computing numerically with functions instead of numbers. Math. Comput. Sci. **1**(1), 9–19 (2007)
42. Trefethen, L.N., Bau III, D.: Numerical Linear Algebra. SIAM, Philadelphia (1997)
43. Wilkinson, J.H.: Error analysis of floating-point computation. Numer. Math. **2**, 319–340 (1960)
44. Wilkinson, J.H.: The Algebraic Eigenvalue Problem. Oxford University Press, Oxford (1965)

Symbolic Geometric Reasoning with Advanced Invariant Algebras

Hongbo Li[✉]

Key Laboratory of Mathematics Mechanization, Academy of Mathematics
and Systems Science, Chinese Academy of Sciences, Beijing 100190, China
hli@mmrc.iss.ac.cn

Abstract. In symbolic geometric reasoning, the output of an algebraic
method is expected to be geometrically interpretable, and the size of
the middle steps is expected to be sufficiently small for computational
efficiency. Invariant algebras often perform well in meeting the two expectations for relatively simple geometric problems. For example in classical geometry, symbolic manipulations based on basic invariants such as
squared distances, areas and volumes often have great performance in
generating readable proofs. For more complicated geometric problems,
the basic invariants are still insufficient and may not generate geometrically meaningful results.

An advanced invariant is a monomial in an "advanced algebra", and
can be expanded into a polynomial of basic invariants that are also
included in the algebra. In projective incidence geometry, Grassmann-
Cayley algebra and Cayley bracket algebra are an advanced algebra in
which the basic invariants are determinants of homogeneous coordinates
of points, and the advanced invariants are Cayley brackets. In Euclidean conformal geometry, Conformal Geometric Algebra and null bracket
algebra are an advanced algebra where the basic invariants are squared
distances between points and and signed volumes of simplexes, and the
advanced invariants are Clifford brackets.

This paper introduces the above advanced invariant algebras together
with their applications in automated geometric theorem proving. These
algebras are capable of generating extremely short and readable proofs.
For projective incidence theorems, the proofs generated are usually two-
termed in that the conclusion expression maintains two-termed during
symbolic manipulations. For Euclidean geometry, the proofs generated
are mostly one-termed or two-termed.

Keywords: Grassmann-Cayley algebra · Cayley bracket algebra ·
Conformal Geometric Algebra · Null bracket algebra · Automated
geometric theorem proving

1 Algebraic Approach to Geometric Reasoning

In classical geometry, besides the Euclidean approach to geometric reasoning,
the algebraic approach can be described by the following diagram:

© Springer International Publishing Switzerland 2016
I.S. Kotsireas et al. (Eds.): MACIS 2015, LNCS 9582, pp. 35–49, 2016.
DOI: 10.1007/978-3-319-32859-1_3

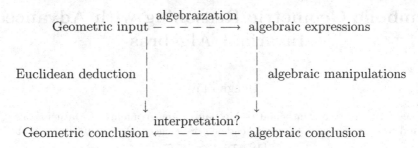

Geometric input $\overset{\text{algebraization}}{-\;-\;-\;-\;-\;\longrightarrow}$ algebraic expressions

Euclidean deduction $|$ $|$ algebraic manipulations

Geometric conclusion $\overset{\text{interpretation?}}{\longleftarrow\;-\;-\;-\;-\;-}$ algebraic conclusion

The following are key aspects in evaluating the algebraic approach:

- Symbolic algebraic manipulations: how-to and *efficiency*.
- Geometric *interpretation* of an algebraic conclusion: how-to.
- *Completeness*: Is the diagram commutative for arbitrary input of a given class?

While efficiency is very important for algorithms, geometric interpretability is vital for the algebraic approach to be geometrically successful, and completeness measures the scope of applicability. We use the following example to illustrate the problem of geometric interpretability by several algebraic methods for geometric theorem proving.

Example 1 (33rd M. Putnam Math Competition, 1972). A quadrilateral in space with equal opposite angles has equal opposite edges (Fig. 1).

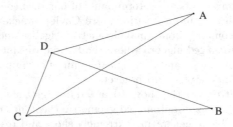

Fig. 1. Example 1.

In the planar case, the conclusion is obviously wrong. This example can can be used to test algebraic methods for the ability of generating geometrically meaningful non-degeneracy conditions, preferably the non-coplanarity of the quadrilateral.

The canonical algebraization is by coordinatization:

$$\mathbf{A} = (0,0,0), \qquad\qquad \mathbf{B} = (x_1, 0, 0),$$
$$\mathbf{C} = (x_1 + x_2, x_3, 0), \qquad \mathbf{D} = (x_1 + x_2 + x_4, x_3 + x_5, x_6),$$
$$d_1 = |\mathbf{AB}|, \;\; d_2 = |\mathbf{BC}|, \;\; d_3 = |\mathbf{CD}|, \;\; d_4 = |\mathbf{DA}|.$$

Hypothesis:

- 6 equalities: 2 of angles, 4 of squared distances by coordinates. The equality of two angles is represented by the equality of their cosines.
- 4 inequalities of distances being nonzero.

Conclusion:
$$g_1 := d_1 - d_3 = 0,$$
$$g_2 := d_2 - d_4 = 0.$$

Method 1. Characteristic Set [21]. When computing the characteristic set of the 6 equalities under the order of variables $d_i \prec x_j$ by the Maple package *wsolve* of Dingkang Wang, Maple returns "Error, (in expand/bigprod) object too large".

When computing the characteristic set of the 6 equalities together with the following 4 equalities obtained from the 4 inequalities by introducing variables $y_k, k = 1..4$ such that
$$h_k := y_k^2 d_k - 1 = 0,$$

then under $y_k \prec d_i \prec x_j$, 301 branches are generated in a flash with a laptop, among which 243 (resp. 249) branches do not pseudo-reduce g_1 (resp. g_2) to zero. Then non-degeneracy conditions are obtained by investigating the initials of the remaining branches. They are complicated polynomials without clear geometric meaning.

Method 2. Gröbner Basis [3]. Computing the Gröbner basis of the ideal $\mathcal{I}$ of the 6 equalities under the same order of variables $d_i \prec x_j$ in deglex ordering of monomials is easy for Maple; the result does not reduce any of g_1, g_2 to zero.

To find a non-degeneracy condition, by the method of F. Winkler [20], we compute a Gröbner basis of the saturation ideal $(\mathcal{I} : g_1^\infty)$, and obtain 20 elements, 5 of which are not in $\sqrt{\mathcal{I}}$. The simplest one is

$$p = d_2\{4(d_1 + d_3)x_2^2 + 4(d_1 + d_3)x_2 x_4 + 2(2d_1^2 + 2d_1 d_3 + d_2^2 - d_4^2)x_2$$
$$+ 2(d_1^2 + d_2^2 + d_1 d_3 + d_2 d_4)x_4 + d_1^3 + d_1^2 d_3 + d_1 d_2^2 + d_1 d_3^2 - d_1 d_4^2$$
$$- d_2^2 d_3 - 2d_2 d_3 d_4 + d_3^3 - d_3 d_4^2\}.$$

So a non-degeneracy condition of $g_1 = 0$ is $p \neq 0$. As $d_2 \neq 0$ by the hypothesis, the second factor in the expression of p, denoted by p/d_2, is a simpler non-degeneracy condition.

Similarly, a Gröbner basis of the second saturation ideal $(\mathcal{I} : g_2^\infty)$ is computed, which contains 19 elements, 5 of which are not in $\sqrt{\mathcal{I}}$. The second factor of p is one of the 5 elements, so it is a non-degeneracy condition for $g_2 = 0$.

By the Gröbner basis method, we get the result that the conclusion is true if $p/d_2 \neq 0$. The geometric meaning of polynomial p/d_2 is not clear.

Method 3. Vector Algebra [7]. Introduce 4 unit vectors $\mathbf{e}_1, \mathbf{e}_2, \mathbf{e}_3, \mathbf{e}_4$ to represent the directions of the 4 sides:

$$\mathbf{B} - \mathbf{A} = d_1\mathbf{e}_1, \quad \mathbf{C} - \mathbf{B} = d_2\mathbf{e}_2, \quad \mathbf{D} - \mathbf{C} = d_3\mathbf{e}_3, \quad \mathbf{A} - \mathbf{D} = d_4\mathbf{e}_4.$$

Hypothesis:

$$d_1\mathbf{e}_1 + d_2\mathbf{e}_2 + d_3\mathbf{e}_3 + d_4\mathbf{e}_4 = 0,$$
$$\mathbf{e}_1 \cdot \mathbf{e}_2 = \mathbf{e}_3 \cdot \mathbf{e}_4,$$
$$\mathbf{e}_1 \cdot \mathbf{e}_4 = \mathbf{e}_2 \cdot \mathbf{e}_3, \tag{1.1}$$
$$\mathbf{e}_1^2 = \mathbf{e}_2^2 = \mathbf{e}_3^2 = \mathbf{e}_4^2 = 1,$$
$$d_1, d_2, d_3, d_4 \neq 0.$$

Conclusion:

$$d_1 = d_3, \quad d_2 = d_4.$$

Triangulating (1.1) by vectorial equation solving [8]: under the order of variables $d_i \prec \mathbf{e}_j$, the characteristic set in vectorial equation form contains only 1 branch of maximal dimension – an algebraic variety defined by 8 scalar-valued polynomials, as following:

$$\begin{aligned} &d_4\mathbf{e}_4 + d_3\mathbf{e}_3 + d_2\mathbf{e}_2 + d_1\mathbf{e}_1, \\ &\mathbf{e}_3^2 - 1, \\ &\mathbf{e}_2^2 - 1, \\ &\mathbf{e}_1^2 - 1, \\ &d_4 - d_2, \\ &d_3 - d_1. \end{aligned} \tag{1.2}$$

Both conclusions are already in it. All other branches each have scalar-valued components of at least 9 polynomials, and the conclusions are false on them. Furthermore, $\mathbf{A}, \mathbf{B}, \mathbf{C}, \mathbf{D}$ are coplanar on all other branches.

So the theorem is generically true with non-degeneracy condition: the non-coplanarity of the quadrilateral. The conclusion is automatically discovered during triangulation.

What happened during vectorial equation solving? To uncover the myth, we show the procedure of triangulating the equations led by vector variable $\mathbf{e}_3$ after the elimination of $\mathbf{e}_4$, the latter being trivial by the first equation of (1.1). The input equations led by $\mathbf{e}_3$ are

$$\mathbf{e}_3^2 = 1,$$
$$2(d_2d_3 - d_1d_4)\mathbf{e}_2 \cdot \mathbf{e}_3 = d_1^2 + d_4^2 - d_2^2 - d_3^2,$$
$$2(d_2d_3 - d_1d_4)(d_1d_2 - d_3d_4)\mathbf{e}_1 \cdot \mathbf{e}_3 = d_2^4 - 2d_2^2d_4^2 + d_4^4 - d_1^2d_2^2 - d_1^2d_4^2 \\ - d_2^2d_3^2 - d_3^2d_4^2 + 4d_1d_2d_3d_4.$$

Solving for $\mathbf{e}_3$ by the following identity in vector algebra:

$$(\mathbf{e}_1 \times \mathbf{e}_2)^2\mathbf{e}_3 = -(\mathbf{e}_2 \cdot \mathbf{e}_3)\mathbf{e}_1 \times (\mathbf{e}_1 \times \mathbf{e}_2) + (\mathbf{e}_3 \cdot \mathbf{e}_1)\mathbf{e}_2 \times (\mathbf{e}_1 \times \mathbf{e}_2) + \lambda\mathbf{e}_1 \times \mathbf{e}_2, \tag{1.3}$$

where local parameter/coordinate λ satisfies

$$\lambda^2 = (\mathbf{e}_1 \times \mathbf{e}_2)^2\mathbf{e}_3^2 - (\mathbf{e}_2 \cdot \mathbf{e}_3)^2\mathbf{e}_1^2 - (\mathbf{e}_3 \cdot \mathbf{e}_1)^2\mathbf{e}_2^2 + 2(\mathbf{e}_1 \cdot \mathbf{e}_2)(\mathbf{e}_2 \cdot \mathbf{e}_3)(\mathbf{e}_3 \cdot \mathbf{e}_1),$$

we get

$$(d_1 + d_2 + d_3 + d_4)\overbrace{(d_1 + d_2 - d_3 - d_4)(d_1 + d_3 - d_2 - d_4)}\overbrace{(d_1 + d_4 - d_2 - d_3)}$$
$$\{(d_1d_4 - d_2d_3)\mathbf{e}_3 - (d_1d_2 - d_3d_4)\mathbf{e}_1 - (d_2^2 - d_4^2)\mathbf{e}_2\} = 0.$$

The two overbraced factors each lead to the maximal branch (1.2), while the other factors lead to lower dimensional configurations.

The vector algebra method leads to the beautiful triangulation result (1.2). When recalling the coordinate approach, one gets the mixed feeling that on one hand, by decomposing high dimensional geometry into a sequence of one dimensional geometries, Descartes' introduction of coordinates greatly facilitates the representation and manipulation of geometric objects; on the other hand, however, this factitious decomposition induces two big problems:

1. Results from algebraic computations are either difficult to interpret geometrically, or geometrically meaningless. Their dependencies upon the specific coordinate systems are either difficult or impossible to separate from the geometric properties they represent.
2. Middle expression swell: Both the input expression and the output expression are small in size, but the middle expressions are huge. Some computations are possible only theoretically.

Vectors and invariants, or more accurately, the coordinate-free version of *covariants*, have obvious representational advantage over coordinates, but do not necessarily lead to any manipulational advantage. The reason is that invariant indeterminates are not algebraically independent, and a generic algebraic relation among invariants is called a *syzygy*. In invariant-theoretic method, people do not get rid of algebraic dependencies, otherwise it becomes a traditional coordinate method. Although a monomial of basic invariants is geometrically meaningful, it is not so for a polynomial of basic invariants. With the presence of syzygies, the classical approach to normalizing an invariant, Young's *straightening algorithm* [22], has no control of middle expression swell.

It remains a challenge how geometric reasoning with covariants can be done more efficiently while preserving geometric meaning and controlling middle expression size. To meet the challenge, advanced invariant algebras are called for.

In this paper, two advanced invariant algebras are introduced: Grassmann-Cayley algebra and Cayley bracket algebra for projective incidence geometry, and Conformal Geometric Algebra and null bracket algebra for Euclidean conformal geometry. They are capable of generating extremely short and readable proofs in automated theorem proving.

2 Geometric Reasoning by Basic and Advanced Invariants

We start with a typical example in projective incidence geometry.

Example 2 (2D Desargues' Theorem, valid for nD). If lines $\mathbf{11'}, \mathbf{22'}, \mathbf{33'}$ concur, then points $\mathbf{a} = \mathbf{12} \cap \mathbf{1'2'}$, $\mathbf{b} = \mathbf{13} \cap \mathbf{1'3'}$, $\mathbf{c} = \mathbf{23} \cap \mathbf{2'3'}$ are collinear (Fig. 2).

Method 4. Area Method [4]. A 3D vector represents the homogeneous coordinates of a point in the projective plane. In the affine model of projective plane,

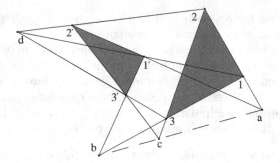

Fig. 2. Desargues' Theorem.

the determinant of the homogeneous coordinates of three affine points $\mathbf{a}, \mathbf{b}, \mathbf{c}$, denoted by $[\mathbf{abc}]$, is related to the signed area $S_{\mathbf{abc}}$ of triangle $\mathbf{abc}$ by

$$[\mathbf{abc}] := \det(\mathbf{a}, \mathbf{b}, \mathbf{c}) = 2S_{\mathbf{abc}}.$$

Chou, Gao and Zhang developed a readable proof generating method based on basic geometric invariants such as areas, ratios, squared distances, etc., and a set of elimination rules from geometric constructions. For Desargues' Theorem, the area method can generate an elegant *rational monomial proof*, *i.e.*, in each step of manipulating the conclusion expression, the expression remains a rational function whose numerator and denominator are both monomials of basic invariants. Details can be found in [4].

Method 5. Biquadratic Final Polynomials [2,5]. The method of Bokowski, Sturmfels and Richter-Gebert is based on the theory of *biquadratic final polynomials*. It searches for all kinds of geometric constraints that can be expressed by biquadratic bracket equalities, and then finds a subset of such equalities whose multiplication produces a biquadratic binomial representation of the conclusion after canceling common bracket factors.

For Desargues' Theorem, let $\mathbf{d} = \mathbf{11'} \cap \mathbf{22'}$, then a binomial proof is given by this method as following:

$$
\begin{array}{llll}
\mathbf{3'c, 1'a, 2d} & \text{concur} & \Longrightarrow & [\mathbf{23'd}][\mathbf{1'ac}] = -[\mathbf{2cd}][\mathbf{1'3'a}] \\
\mathbf{1'd, 2a, 3b} & \text{concur} & \Longrightarrow & [\mathbf{2ab}][\mathbf{31'd}] = [\mathbf{23a}][\mathbf{1'bd}] \\
\mathbf{3, 3', d} & \text{collinear} & \Longrightarrow & [\mathbf{23d}][\mathbf{1'3'd}] = -[\mathbf{23'd}][\mathbf{31'd}] \\
\mathbf{1', 3', b} & \text{collinear} & \Longrightarrow & [\mathbf{1'bd}][\mathbf{1'3'a}] = -[\mathbf{1'ab}][\mathbf{1'3'd}] \\
\mathbf{2, 3, c} & \text{collinear} & \Longrightarrow & [\mathbf{23a}][\mathbf{2cd}] = -[\mathbf{23d}][\mathbf{2ac}] \\
& & & \qquad\quad \times \qquad\qquad\qquad \times \\
& & & \qquad\quad \Downarrow \qquad\qquad\qquad \Downarrow \\
\mathbf{a, b, c} & \text{collinear} & \Longleftarrow & [\mathbf{2ab}][\mathbf{1'ac}] = [\mathbf{2ac}][\mathbf{1'ab}].
\end{array}
$$

Method 6. Cayley Bracket Algebra [10]. The method of Li and Wu is based on Grassmann-Cayley algebra, Cayley expansion and factorization. A Grassmann algebra is obtained by extending a base vector space with the outer (exterior) product. While a projective point **a** is represented by a vector of homogeneous coordinates and denoted by the same symbol **a**, line **ab** is represented by $\mathbf{B} := \mathbf{a} \wedge \mathbf{b}$, so that point **c** is on the line if and only if $\mathbf{c} \wedge \mathbf{B} = 0$. Similarly, plane **abc** is represented by $\mathbf{C} := \mathbf{a} \wedge \mathbf{b} \wedge \mathbf{c}$, and point **d** is on the plane if and only if $\mathbf{d} \wedge \mathbf{C} = 0$.

The *meet product* "$\vee$" is the dual of the outer product "$\wedge$". Let "$\sim$" be the Hodge dual operator [12] of the Grassmann algebra, then

$$(\mathbf{B} \vee \mathbf{C})^{\sim} := \mathbf{B}^{\sim} \wedge \mathbf{C}^{\sim}.$$

Grassmann-Cayley algebra is Grassmann algebra equipped also with the meet product and the dual operator. It is an algebra of span and intersection of linear subspaces. For example, the intersection of lines $\mathbf{12}, \mathbf{1'2'}$ is represented by $(\mathbf{1} \wedge \mathbf{2}) \vee (\mathbf{1'} \wedge \mathbf{2'})$. As a point on both line $\mathbf{1'2'}$ and line $\mathbf{12}$, the linear combination representations of the intersection on the two lines respectively are given by the following *shuffle formula*:

$$(\mathbf{1} \wedge \mathbf{2}) \vee (\mathbf{1'} \wedge \mathbf{2'}) = [\mathbf{122'}]\mathbf{1'} - [\mathbf{121'}]\mathbf{2'} = [\mathbf{11'2'}]\mathbf{2} - [\mathbf{21'2'}]\mathbf{1}. \tag{2.1}$$

A *Cayley bracket* is a scalar-valued monomial in Grassmann-Cayley algebra. *Cayley bracket algebra* is a commutative algebra generated by Cayley brackets. It is an algebra of advanced projective invariants, and includes bracket algebra as a subalgebra.

Cayley expansion [10] refers changing an expression of Cayley bracket algebra into bracket algebra. The purpose is to make simplification by eliminating all meet products. As Grassmann-Cayley algebra is neither associative nor commutative, converting an expression from Cayley bracket algebra into bracket algebra is a simplification from the algebraic viewpoint. The shuffle formula from left to right is a typical example.

Cayley factorization [19] is the inverse of Cayley expansion. It converts a bracket polynomial into a (rational) Cayley bracket, so that an incidence construction interpretation of the bracket polynomial can be read from the resulting Cayley bracket expression.

For Desargues' Theorem, the hypothesis is the concurrence of lines $\mathbf{11'}, \mathbf{22'}$, $\mathbf{33'}$, whose representation in Cayley bracket algebra is

$$(\mathbf{1} \wedge \mathbf{1'}) \vee (\mathbf{2} \wedge \mathbf{2'}) \vee (\mathbf{3} \wedge \mathbf{3'}) = 0.$$

The expression has 48 terms if expanded into homogeneous coordinate variables. The conclusion that intersections $\mathbf{12} \cap \mathbf{1'2'}$, $\mathbf{13} \cap \mathbf{1'3'}$, $\mathbf{23} \cap \mathbf{2'3'}$ are collinear is represented by

$$[\{(\mathbf{1} \wedge \mathbf{2}) \vee (\mathbf{1'} \wedge \mathbf{2'})\}\{(\mathbf{1} \wedge \mathbf{3}) \vee (\mathbf{1'} \wedge \mathbf{3'})\}\{(\mathbf{2} \wedge \mathbf{3}) \vee (\mathbf{2'} \wedge \mathbf{3'})\}] = 0.$$

It has 1290 terms if expanded into homogeneous coordinate variables.

The following binomial proof is valid also for the nD case. It is by simplifying the conclusion expression with Cayley expansion and factorization:

$$[\{(\underline{1} \wedge \underline{2}) \vee (1' \wedge 2')\}\{(1 \wedge 3) \vee (1' \wedge 3')\}\{(2 \wedge 3) \vee (2' \wedge 3')\}]$$

$$\overset{binomial}{=} \{(\underline{1} \wedge 2) \vee (\underline{1} \wedge 3) \vee (1' \wedge 3')\}\{(1' \wedge \underline{2'}) \vee (2 \wedge 3) \vee (\underline{2'} \wedge 3')\}$$
$$-\{(1 \wedge \underline{2}) \vee (\underline{2} \wedge 3) \vee (2' \wedge 3')\}\{(\underline{1'} \wedge 2') \vee (1 \wedge 3) \vee (\underline{1'} \wedge 3')\} \quad (2.2)$$

$$\overset{monomial}{=} [123][1'2'3'](-[11'3'][232'] + [131'][22'3'])$$

$$\overset{factor}{=} -[123][1'2'3'](1 \wedge 1') \vee (2 \wedge 2') \vee (3 \wedge 3').$$

Remark on monomial expansion: $(\underline{1} \wedge 2) \vee (\underline{1} \wedge 3) = [123]1$, as according to (2.1), the other term vanishes as a result of $[121] = 0$.

It turns out that with Cayley expansion and Cayley factorization, all projective incidence theorems tested so far have *robust binomial proofs*, in that when there is more than one monomial/binomial Cayley expansion available, then any such expansion leads to a binomial proof ultimately. The features of the Cayley bracket algebra method includes:

– easy and robust steps, no peculiar choice necessary in manipulations;
– short terms;
– input and output geometrically meaningful;
– hypothesis and conclusion expressions interrelated quantitatively;
– geometric theorems expressed as algebraic identities: easy to apply.

For example, in the proof (2.2), the hypothesis is in fact not used, and we get the following identity in Cayley bracket algebra:

$$[\{(1 \wedge 2) \vee (1' \wedge 2')\}\{(1 \wedge 3) \vee (1' \wedge 3')\}\{(2 \wedge 3) \vee (2' \wedge 3')\}]$$
$$= -[123][1'2'3'](1 \wedge 1') \vee (2 \wedge 2') \vee (3 \wedge 3').$$

It provides a quantitative description of the relation between the hypothesis expression and the conclusion expression, and is a much more general result than the original theorem. The identity from right to left is the converse of Desargues' Theorem, with the non-degeneracy conditions $[123][1'2'3'] \neq 0$ (the non-degeneracy of triangles 123 and $1'2'3'$) occurring naturally. Desargues' Theorem and its converse represented in algebraic identity form can be applied directly as term rewriting rules in symbolic manipulations.

A highlight of advanced invariant computing method is that advanced invariants are manipulated by their own mechanism, without resorting to low-level invariants or coordinates. The number of terms of the hypothesis expression and the conclusion expression when expanded into coordinates clearly indicates huge middle expression swell if manipulations are done in coordinates.

In history, Descartes' introduction of coordinates is a key step from qualitative description to quantitative analysis of geometric configurations. However, coordinates are sequences of numbers, they have no geometric meaning by themselves. Leibniz once dreamed of a geometric calculus dealing directly with geometric objects rather than with sequences of numbers. He needed an algebra that is so close to geometry that every expression has clear geometric meaning, and every algebraic manipulation corresponds to geometric transformation. Such an algebra, if exists, is rightly called geometric algebra, and its elements called geometric numbers.

Leibniz's dream in projective incidence geometry is realized by Grassmann-Cayley algebra and Cayley bracket algebra. Despite the efficiency, there are still limitations with this geometric algebra, for example the following:

- Inefficient representation of Euclidean metric structure.
  In Grassmann-Cayley algebra, for a vector $\mathbf{a}$ representing a point, both $\mathbf{a} \wedge \mathbf{a} = 0$ and $\mathbf{a} \vee \mathbf{a} = 0$. So this algebra cannot describe metric structure without extending the base numbers field.
- Lack of associativity between the outer product and the meet product.
- Inefficiency in handling nonlinear geometric objects.

In the past few years, at least for 3D projective geometry, with the introduction of an associative algebra for modeling projective line geometry, the second limitation is significantly alleviated [16]. The first limitation calls for advanced invariant algebras of Euclidean geometry.

3   Euclidean Invariants: From Basic to Advanced Ones

All Euclidean invariants are functions of distances. For algebraic invariants, there are two basic ones: the squared distances between two points (or equivalently, the inner products of two difference vectors), and the signed volumes of simplexes.

In history, there has been an advanced Euclidean invariant algebra: Cayley-Menger determinants (or bi-determinants) [1,18]. Besides invariant algebras, there also have been advanced algebras of covariants. For 3D geometry, vector algebra and quaternionic-variable (non-commutative) polynomial algebra are two advanced algebras of covariants. Their nD generalization is the Clifford algebra over $\mathbb{R}^n$.

In Euclidean geometry, a line segment $\mathbf{ab}$ has length $d_{\mathbf{ab}}$. In vector form, for points $\mathbf{a}, \mathbf{b}$ represented by vectors,

$$d_{\mathbf{ab}}^2 = (\mathbf{a} - \mathbf{b})^2 = (\mathbf{a} - \mathbf{b}) \cdot (\mathbf{a} - \mathbf{b}) = \mathbf{a}^2 + \mathbf{b}^2 - 2\mathbf{a} \cdot \mathbf{b}, \qquad (3.1)$$

where the dot symbol denotes the inner product. What are these: $\mathbf{a}^2$, $\mathbf{b}^2$ and $\mathbf{a} \cdot \mathbf{b}$? They always depend on the reference point (the origin of all vectors), and are geometrically meaningless. None of the above mentioned algebras of invariants or covariants makes $\mathbf{a}^2$ geometrically meaningful for "point" $\mathbf{a}$.

What do we expect from $\mathbf{a} \cdot \mathbf{b}$? It should reflect some relation between the two points. The only candidate is – the distance! Then $\mathbf{a} \cdot \mathbf{a}$ has to be zero, *i.e.*, $\mathbf{a}$ must be a *null vector*. Then from (3.1) we get

$$\mathbf{a} \cdot \mathbf{b} = -\frac{(\mathbf{a} - \mathbf{b})^2}{2} = -\frac{d_{\mathbf{ab}}^2}{2}. \qquad (3.2)$$

In history, Wachter (a student of Gauss) proposed embedding $\mathbb{R}^3$ isometrically into the 5D Minkowski space $\mathbb{R}^{4,1}$, so that all vectors of $\mathbb{R}^3$ are represented by null vectors of $\mathbb{R}^{4,1}$, and (3.2) is satisfied. Let $\mathbf{e}_1, \mathbf{e}_2, \mathbf{e}_3, \mathbf{e}, \mathbf{e}_0$ be a basis of $\mathbb{R}^{4,1}$, with metric $\begin{pmatrix} 1 & & & & \\ & 1 & & & \\ & & 1 & & \\ & & & 0 & -1 \\ & & & -1 & 0 \end{pmatrix}$. Then Wachter's isometric embedding is

$$(x, y, z) \in \mathbb{R}^3 \mapsto \left(x, y, z, 1, -\frac{x^2 + y^2 + z^2}{2} \right) \in \mathbb{R}^{4,1}.$$

This model changes conformal transformations of $\mathbb{R}^3$ into orthogonal transformations of $\mathbb{R}^{4,1}$, hence inducing a pin group representation of 3D Euclidean conformal transformations.

In [9], Li, Hestenes and Rockwood further studied the above model and proposed a sequence of Grassmann-Cayley algebraic representations of Euclidean conformal constructions. For example, the following list is on Minkowski representations of planes/lines/circles/spheres in $\mathbb{R}^n$:

- Conformal point at infinity: $\mathbf{e}$, the unique extra point for one-point compactification of $\mathbb{R}^n$.
- Points: null vectors $\mathbf{x}$, where $\mathbf{x} \cdot \mathbf{e} = -1$ for "point" $\mathbf{x}$.
- Line $\mathbf{ab}$: $\mathbf{e} \wedge \mathbf{a} \wedge \mathbf{b}$.   Point $\mathbf{d}$ on line $\mathbf{C}$: $\mathbf{d} \wedge \mathbf{C} = 0$.
- Plane $\mathbf{abc}$: $\mathbf{e} \wedge \mathbf{a} \wedge \mathbf{b} \wedge \mathbf{c}$.
- Circle $\mathbf{abc}$ (circum-circle of triangle $\mathbf{abc}$): $\mathbf{a} \wedge \mathbf{b} \wedge \mathbf{c}$.
- Sphere $\mathbf{abcd}$ (circum-sphere): $\mathbf{a} \wedge \mathbf{b} \wedge \mathbf{c} \wedge \mathbf{d}$.

Besides the Minkowski representations, there are also the dual representations of conformal objects, affine representations and dual affine representations of affine objects [12].

In [11], the Grassmann-Cayley algebraic representations of Euclidean conformal constructions are further extended to include the *reduced meet product* for representing the *second point of intersection of two circles/lines*. For two circles/lines $\mathbf{ab}_1\mathbf{c}_1$ and $\mathbf{ab}_2\mathbf{c}_2$, "point" $\mathbf{a}$ is obviously a point of intersection (if both are lines, then $\mathbf{a} = \mathbf{e}$ is the conformal point at infinity). Besides this trivial point of intersection, there is another point of intersection, denoted by $\mathbf{b}_1\mathbf{c}_1 \cap_\mathbf{a} \mathbf{b}_2\mathbf{c}_2$. In particular when $\mathbf{b}_1\mathbf{c}_1 \cap_\mathbf{a} \mathbf{b}_2\mathbf{c}_2 = \mathbf{a}$, the two circles/lines are tangent to each other at "point" $\mathbf{a}$ (in the case of two lines, that they are "tangent" to each other at the conformal point at infinity means they are parallel to each other).

The meet product of two circles/lines has the outer product factorization

$$(\mathbf{a} \wedge \mathbf{b}_1 \wedge \mathbf{c}_1) \vee (\mathbf{a} \wedge \mathbf{b}_2 \wedge \mathbf{c}_2) = \mathbf{a} \wedge \{(\mathbf{b}_1 \wedge \mathbf{c}_1) \vee_\mathbf{a} (\mathbf{b}_2 \wedge \mathbf{c}_2)\}. \qquad (3.3)$$

The second factor of the outer product is called the *reduced meet product*:

$$(\mathbf{b}_1 \wedge \mathbf{c}_1) \vee_\mathbf{a} (\mathbf{b}_2 \wedge \mathbf{c}_2) := [\mathbf{ab}_1\mathbf{c}_1\mathbf{c}_2]\mathbf{b}_2 - [\mathbf{ab}_1\mathbf{c}_1\mathbf{b}_2]\mathbf{c}_2 = [\mathbf{ab}_1\mathbf{b}_2\mathbf{c}_2]\mathbf{c}_1 - [\mathbf{ac}_1\mathbf{b}_2\mathbf{c}_2]\mathbf{b}_1. \qquad (3.4)$$

Vector (3.4) is not null, yet it must relate to the second point of intersection, as the meet product (3.3) equals the outer product of $\mathbf{a}$ with the second point of intersection. It turns out that the null vector representation of $\mathbf{b}_1\mathbf{c}_1 \cap_\mathbf{a} \mathbf{b}_2\mathbf{c}_2$ is the reflection of null vector $\mathbf{a}$ with respect to invertible vector $(\mathbf{b}_1 \wedge \mathbf{c}_1) \vee_\mathbf{a} (\mathbf{b}_2 \wedge \mathbf{c}_2)$.

To manipulate reflection multiplicatively (or "monomially" to control expression size), Clifford algebra is resorted to. This is an algebra obtained by extending a base inner-product space with the Clifford multiplication (or *geometric product*), which is associative, multilinear, and satisfies $\mathbf{aa} = \mathbf{a}^2 := \mathbf{a} \cdot \mathbf{a}$. In fact, $\mathbf{ab} = \mathbf{a} \cdot \mathbf{b} + \mathbf{a} \wedge \mathbf{b}$, where the juxtaposition denotes the Clifford multiplication.

In Hestenes' viewpoint, Clifford algebra is best constructed from Grassmann-Cayley algebra by also equipping it with the Clifford multiplication, just like constructing Grassmann-Cayley algebra from Grassmann algebra. This version of Clifford algebra is very suitable for describing and manipulating geometric constructions, and is nowadays called *Geometric Algebra* [6].

In Clifford algebra, the reflection of vector $\mathbf{b}$ with respect to vector $\mathbf{a}$ is represented by

$$\mathbf{b} \mapsto Ad_\mathbf{a}(\mathbf{b}) := -\mathbf{aba}^{-1}.$$

In our setting of representing the second point of intersection, the following *homogeneous reflection* is more convenient:

$$\mathbf{b} \mapsto N_\mathbf{b}(\mathbf{a}) := \frac{1}{2}\mathbf{aba}.$$

So $\mathbf{23} \cap_1 \mathbf{2'3'}$ is represented by $N_\mathbf{1}((\mathbf{2} \wedge \mathbf{3}) \vee_1 (\mathbf{2'} \wedge \mathbf{3'}))$.

Conformal Geometric Algebra (CGA) refers to the Clifford algebra over the Minkowski space $\mathbb{R}^{n+1,1}$ for representing conformal transformations of $\mathbb{R}^n$ by acting upon the Minkowski representation of conformal objects, together with other alternatives of the Minkowski representation such as the dual representation, affine representation, dual affine representation, reduced meet product representation, etc. Conformal Geometric Algebra and the null bracket algebra to be introduced below realize Leibniz's dream in Euclidean conformal geometry.

An nD *Clifford bracket algebra* is the commutative ring generated by the "hyper-determinants" and "hyper-inner products", which are obtained respectively by prolonging nD brackets and inner products of vector pairs with the Clifford multiplication, as following:

$$\mathbf{a}_1 \cdot \mathbf{a}_2 = \langle \mathbf{a}_1\mathbf{a}_2 \rangle \text{ prolonged to } \langle \mathbf{a}_1\mathbf{a}_2 \ldots \mathbf{a}_{2k} \rangle := \langle \mathbf{a}_1\mathbf{a}_2 \ldots \mathbf{a}_{2k} \rangle_0,$$
$$[\mathbf{a}_1 \ldots \mathbf{a}_n] = (\mathbf{a}_1 \wedge \cdots \wedge \mathbf{a}_n)^\sim \text{ prolonged to } [\mathbf{a}_1\mathbf{a}_2 \ldots \mathbf{a}_{n+2l}] := \langle \mathbf{a}_1\mathbf{a}_2 \ldots \mathbf{a}_{n+2l} \rangle_n^\sim,$$

where "$\langle \ \rangle_i$" denotes the i-grading operator: extracting the i-graded part (in Grassmann algebra) of the argument.

Instead of anti-commutativity and commutativity, the two long brackets have the following symmetries:

– Reversion:

$$\langle \mathbf{a}_1\mathbf{a}_2 \cdots \mathbf{a}_{2k} \rangle = \langle \mathbf{a}_{2k}\mathbf{a}_{2k-1} \cdots \mathbf{a}_1 \rangle,$$
$$[\mathbf{a}_1\mathbf{a}_2 \cdots \mathbf{a}_{n+2l}] = (-1)^{\frac{n(n-1)}{2}}[\mathbf{a}_{n+2l}\mathbf{a}_{n+2l-1} \cdots \mathbf{a}_1].$$

– Shift:
$$\langle a_1 a_2 \cdots a_{2k} \rangle = \langle a_{2k} a_1 a_2 \cdots a_{2k-1} \rangle,$$
$$[a_1 a_2 \cdots a_{n+2l}] = (-1)^{n-1}[a_{n+2l} a_1 \cdots a_{n+2l-1}].$$

Null bracket algebra is a Clifford bracket algebra generated by null vector variables. The property $aa = 0$ for null vector a provides great benefits in expression size control, in addition to adding more symmetries.

The following are geometric interpretations of the long brackets in terms of 2D trigonometry:

$$\langle a_1 a_2 \cdots a_{2l+2} \rangle = -\frac{d_{a_1 a_2} d_{a_2 a_3} \cdots d_{a_{2l+1} a_{2l+2}} d_{a_{2l+2} a_1}}{2} \cos(\angle(a_1 a_2 a_3, a_1 a_3 a_4)$$
$$+ \angle(a_1 a_4 a_5, a_1 a_5 a_6) + \cdots + \angle(a_1 a_{2l} a_{2l+1}, a_1 a_{2l+1} a_{2l+2}));$$
$$[a_1 a_2 \cdots a_{2l+2}] = -\frac{d_{a_1 a_2} d_{a_2 a_3} \cdots d_{a_{2l+1} a_{2l+2}} d_{a_{2l+2} a_1}}{2} \sin(\angle(a_1 a_2 a_3, a_1 a_3 a_4)$$
$$+ \angle(a_1 a_4 a_5, a_1 a_5 a_6) + \cdots + \angle(a_1 a_{2l} a_{2l+1}, a_1 a_{2l+1} a_{2l+2})),$$
$$\tag{3.5}$$

where $\angle(123, 134)$ denotes the angle of rotation from the tangent direction of oriented circle 123 at point 1 to the tangent direction of oriented circle 134 at the same point (Fig. 3(a)).

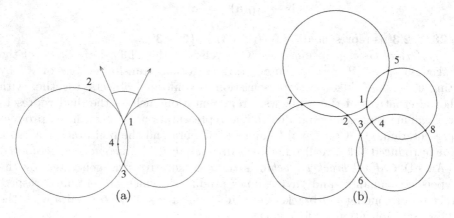

Fig. 3. $\angle(123, 134)$ (left); Miquel's 4-Circle Theorem (right).

For example when $l = 1$ in (3.5), then $2[a_1 a_2 a_3 a_4] = -d_{a_1 a_2} d_{a_2 a_3} d_{a_3 a_4} d_{a_4 a_1}$ $\sin \angle(a_1 a_2 a_3, a_1 a_3 a_4)$. So $[a_1 a_2 a_3 a_4] = 0$ if and only if the four points are cocircular/collinear.

Example 3 (Miquel's 4-Circle Theorem). Four circles intersect at eight points cyclically. If $1, 2, 3, 4$ are cocircular, so are $5, 6, 7, 8$ (Fig. 3(b)).

Method 7. Conformal Geometric Algebra + Null Bracket Algebra.
Similar to Example 2, we remove the cocircularity of points $1, 2, 3, 4$ from the

hypothesis, and see how the conclusion expression varies, with the hope that the removed constraint comes up automatically as a factor in the result.

New hypothesis:

– Free points: $1, 2, 3, 4, 5, 7$;
– Intersections: $6 = 15 \cap_2 37$, and $8 = 15 \cap_4 37$.

In Conformal Geometric Algebra, the two intersections are represented by

$$6 = 2^{-1}\{(1 \wedge 5) \vee_2 (3 \wedge 7)\}2\{(1 \wedge 5) \vee_2 (3 \wedge 7)\};$$
$$8 = 2^{-1}\{(1 \wedge 5) \vee_4 (3 \wedge 7)\}4\{(1 \wedge 5) \vee_4 (3 \wedge 7)\}.$$

Conclusion expression: $[\mathbf{5678}]$. It equals zero if and only if $5, 6, 7, 8$ are cocircular/collinear.

The following is an elegant monomial proof:

$$[\mathbf{5678}] \quad \overset{6,8}{=} \quad 2^{-2}[5\{(1 \wedge 5) \vee_2 (3 \wedge 7)\}2\{(1 \wedge 5) \vee_2 (3 \wedge 7)\}$$
$$7\{(1 \wedge 5) \vee_4 (3 \wedge 7)\}4\{(1 \wedge 5) \vee_4 (3 \wedge 7)\}]$$
$$\overset{expand}{=} -2^{-2}[\mathbf{1257}][\mathbf{1457}][\mathbf{2357}][\mathbf{3457}][\mathbf{51237341}]$$
$$\overset{monomial}{=} (1 \cdot 5)(3 \cdot 7)[\mathbf{1234}][\mathbf{1257}][\mathbf{1457}][\mathbf{2357}][\mathbf{3457}],$$

where the monomial factorization is: $\mathbf{aba} = 2(\mathbf{a} \cdot \mathbf{b})\mathbf{a}$.

As anticipated, bracket $[\mathbf{1234}]$ representing the missing constraint occurs automatically in the result. Thus we get a quantitative (hence stronger) version of Miquel's 4-Circle Theorem: If $6 = 15 \cap_2 37$ and $8 = 15 \cap_4 37$, then

$$\frac{[\mathbf{5678}]}{(5 \cdot 6)(7 \cdot 8)} = \frac{[\mathbf{1234}]}{(1 \cdot 2)(3 \cdot 4)} \frac{[\mathbf{1257}][\mathbf{3457}]}{[\mathbf{1457}][\mathbf{2357}]}.$$

By now over one hundred theorems in Euclidean geometry have been tested, and

– about 4/5 are given robust monomial or binomial proofs;
– more than 1/3 are given monomial proofs;
– by removing one or several equality constraints from the hypothesis, usually the missing constraints can be recovered from the conclusion expression;
– the computing steps are short and easy;
– input and output are geometrically meaningful;
– quantitative description of the relationship between the conclusion and some (in some cases, even all) equality constraints of the hypothesis, can be obtained; the experiment is more or less like playing a game, exciting and fun.

4   Conclusion

Advanced invariants of projective incidence geometry and Euclidean conformal geometry help to achieve tremendous simplifications in automated theorem proving, discovering, and extending.

Why can the computing be so short? A partial explanation is that many syzygies among basic invariants are integrated into symmetries within advanced invariants, and handling the latter is much easier.

Then how to compute? The answer includes expansion, factorization, normalization, and division of invariant polynomials and non-commutative covariant polynomials. In this paper we have used examples to illustrate expansion and factorization sufficiently. We do not have space left to talk about normalization and division, but refer to [14,15] for some new advances.

Finally, to what extent is the proving method complete? This is a fundamental problem of coordinate-free geometric reasoning. The completeness issue can be stated as follows: can every geometrically meaningful conclusion deduced from algebraic manipulations of the coordinate polynomials representing given geometric constraints be also deducible from symbolic manipulations of the (non-commutative) polynomials of (advanced) invariants and covariants?

For example for 3D Euclidean geometry and vector algebra, the question is raised as following: If in 3D geometric reasoning by coordinate variables x_i, y_j, z_k and basis vectors $\mathbf{e}_1, \mathbf{e}_2, \mathbf{e}_3$, both the input and the output are polynomial functions of the vector variables $\mathbf{v}_i = x_i\mathbf{e}_1 + y_i\mathbf{e}_2 + z_i\mathbf{e}_3$, and the algebraic manipulations include only polynomial addition, subtraction, multiplication and division, can the output also be deducible in the vector algebra of the variables $\mathbf{v}_i$, without further introducing any coordinate or parameter?

For invariant algebras, the answer to the completeness question is generally affirmative, while for covariant algebras, although a lot of efforts have been taken in recent years, even for Grassmann-Cayley algebra and vector algebra, the answer is not clear. By now, we have only reached the following conclusion for 3D covariant algebras [13,17]: *When compared with coordinate polynomials with arithmetic operations, the following covariant associative algebras are complete:*

1. *non-commutative polynomial ring in quaternionic variables;*
2. *non-commutative polynomial ring in quaternionic vector variables;*
3. *non-degenerate Clifford polynomial ring in 3D vector variables.*

References

1. Blumenthal, L.M.: Theory and Applications of Distance Geometry. Cambridge University Press, Cambridge (1953)
2. Bokowski, J., Sturmfels, B.: Computational Synthetic Geometry. LNM, vol. 1355. Springer, Heidelberg (1989)
3. Buchberger, B.: Application of Gröbner basis in non-linear computational geometry. In: Rice, J. (ed.) Scientific Software. Springer, New York (1988)
4. Chou, S.C., Gao, X.S., Zhang, J.Z.: Machine Proofs in Geometry. World Scientific, Singapore (1994)
5. Crapo, H., Richter-Gebert, J.: Automatic proving of geometric theorems. In: White, N. (ed.) Invariant Methods in Discrete and Computational Geometry, pp. 107–139. Kluwer, Dordrecht (1994)
6. Hestenes, D., Sobczyk, G.: Clifford Algebra to Geometric Calculus. Kluwer, Dordrecht (1984)

7. Li, H.: New Explorations of Automated Theorem Proving in Geometries. Ph.D. Thesis, Peking University, Beijing (1994)
8. Li, H.: Vectorial equation-solving for mechanical geometry theorem proving. J. Autom. Reasoning **25**, 83–121 (2000)
9. Li, H., Hestenes, D., Rockwood, A.: Generalized homogeneous coordinates for computational geometry. In: Sommer, G. (ed.) Geometric Computing with Clifford Algebras, pp. 27–60. Springer, Heidelberg (2001)
10. Li, H., Wu, Y.: Automated short proof generation in projective geometry with cayley and bracket algebras I. Incidence geometry. J. Symbolic Comput. **36**(5), 717–762 (2003)
11. Li, H.: A recipe for symbolic geometric computing: long geometric product, BREEFS and clifford factorization. In: Brown, C.W. (ed.) Proceedings of the ISSAC 2007, pp. 261–268. ACM, New York (2007)
12. Li, H.: Invariant Algebras and Geometric Reasoning. World Scientific, Singapore (2008)
13. Li, H., Huang, L., Liu, Y.: Normalization of quaternionic polynomials. arxiv: 1301.5338v1 [math.RA] (2013)
14. Li, H.: Normalization of Polynomials in Algebraic Invariants of Three-Dimensional Orthogonal Geometry. arxiv: 1302.7194v1 [cs.SC] (2013)
15. Li, H., Shao, C., Huang, L., Liu, Y.: Reduction among bracket polynomials. In: Proceedings of the ISSAC 2014, pp. 304–311. ACM Press (2014)
16. Li, H., Huang, L., Shao, C., Dong, L.: Three-Dimensional Projective Geometry with Geometric Algebra. arxiv: 1507.06634v1 [math.MG] (2015)
17. Liu, Y.: Normalization of Quaternionic-Variable Polynomials, Ph.D. Dissertation, AMSS, Chinese Academy of Sciences, May 2015
18. Mourrain, B., Stolfi, N.: Computational symbolic geometry. In: White, N.L. (ed.) Invariant Methods in Discrete and Computational Geometry, pp. 107–139. D. Reidel, Dordrecht (1995)
19. White, N.: Multilinear Cayley factorization. J. Symbolic Comput. **11**, 421–438 (1991)
20. Winkler, F.: Gröbner bases in geometry theorem proving and simplest degeneracy conditions. Math. Pannonica **1**(1), 15–32 (1990)
21. Wu, W.T.: Basics Principles of Mechanical Theorem Proving in Geometries I: Part of Elementary Geometries. Science Press, Beijing (1984). Springer, Wien (1994)
22. Young, A.: The Collected Papers of Alfred Young, 1873–1940. University of Toronto Press, Toronto (1977)

Congruence Testing of Point Sets in Three and Four Dimensions
Results and Techniques

Günter Rote[✉]

Institut für Informatik, Freie Universität Berlin, Takustraße 9, 14195 Berlin, Germany
rote@inf.fu-berlin.de

Abstract. I will survey algorithms for testing whether two point sets are congruent, that is, equal up to an Euclidean isometry. I will introduce the important techniques for congruence testing, namely dimension reduction and pruning, or more generally, condensation. I will illustrate these techniques on the three-dimensional version of the problem, and indicate how they lead for the first time to an algorithm for four dimensions with near-linear running time (joint work with Heuna Kim). On the way, we will encounter some beautiful and symmetric mathematical structures, like the regular polytopes, and Hopf-fibrations of the three-dimensional sphere in four dimensions.

1 Problem Statement

Given two n-point sets $A, B \subset \mathbb{R}^d$, we want to decide whether there is a translation vector t and an orthogonal matrix R such that $RA + t := \{ Ra + t \mid a \in A \}$ equals B, that is, A and B are *congruent*. Congruence asks whether two objects are the same up to Euclidean transformations, or in other words, whether they are considered equal from a geometric viewpoint. Congruence is therefore one of the fundamental basic notions.

The translation vector t can be easily eliminated from the problem by initially translating the two sets A and B such that their centers of gravity lie at the origin O.

If we do not restrict the dimension d, congruence becomes equivalent to graph isomorphism: a given graph $G = (V, E)$ with n vertices $v_1, \ldots, v_n$ can be represented by $n + |E|$ points in n dimensions. We simply take the n standard unit vectors $e_1, \ldots, e_n$ and add a point $(e_i + e_j)/2$ for each edge $v_i v_j \in E$. Then two graphs are isomorphic if and only if their corresponding point sets are congruent.

We thus restrict our attention to small dimensions. In two and three dimensions, algorithms with a running time of $O(n \log n)$ have been known. We review some of these algorithms, because their techniques are also important for higher dimensions.

The Computational Model: Exact Real Arithmetic. We use the Real Random-Access Machine (Real-RAM) model, as is common in Computational Geometry.

© Springer International Publishing Switzerland 2016
I.S. Kotsireas et al. (Eds.): MACIS 2015, LNCS 9582, pp. 50–59, 2016.
DOI: 10.1007/978-3-319-32859-1_4

We assume that we can compute arithmetic operations and square roots of real numbers exactly in constant time. The reason for this choice is not so much convenience, but the range of possible input instances. With rational inputs, for example, one cannot even realize a regular pentagon. Thus, the difficult problem instances, which are the symmetric ones, as we will see, would disappear.

It makes sense to ask for approximate congruence within some given tolerance ε. This problem is, however, NP-hard already in two dimensions (Iwanowski 1991). It becomes polynomial when the input points are sufficiently separated in relation to ε, and thus there is hope to solve the approximate congruence problem in higher dimensions, under suitable assumptions and at least in an approximate sense. This is left for future work.

2 Two Dimensions

In the plane, congruence can be tested by string-matching techniques (Manacher 1976). We sort the points clockwise around the origin, in $O(n \log n)$ time, and represent the point set as a cyclic string alternating between n distances from the origin and n angular distances between successive points. Two n-point sets A and B are then congruent if and only if their string representations α and β are cyclic shifts of each other. This is equivalent to asking whether α is a substring of $\beta\beta$, and it can be tested in linear time.

This idea can be extended to symmetry detection for a single set A: We find the lexicographically smallest cyclic reordering of the string. The starting point of this string, together with the cyclic shifts which yield the same string, gives rise to a set of p equidistant rays starting from the origin, which we call the *canonical axes*. Then the set A has a rotational symmetry group of order p, consisting of all rotations that leave the set of canonical axes invariant.

3 Three Dimensions

For testing congruence in space, there are several algorithms, which use different tools (Sugihara 1984; Atkinson 1987; Alt et al. 1988). We describe a variation which is very simple and illustrates the principal techniques that are used in this area: *dimension reduction*, *pruning*, and *condensation*.

Pruning and condensation tries to successively reduce A to a smaller and smaller point set A' while not losing any symmetries that A might have. Initially, we set $A' := A$. We compute the convex hull $H(A')$ of A' in $O(|A'| \log |A'|)$ time. Let $\bar{A}'$ denote the set of vertices of the polytope $H(A')$. We classify the points of $\bar{A}'$ by degree in the graph of $H(A')$. In case there are at least two different degrees in the graph, we replace A' by the smallest degree class in $\bar{A}'$ and repeat the convex-hull computation. In each iteration, the size of A' is reduced to half or less. We simultaneously carry out all steps for the set B. If at any stage, we notice an obvious difference between A' and B', for example, if $|A'| \neq |B'|$, we conclude that A and B are not congruent, and we terminate.

This pruning loop ends when all vertices in $H(A')$, and also in $H(B')$, have the same degree. At first glance, this procedure looks dangerous because we have thrown away points (including all points interior to the hulls of A and B) and have thereby *thrown away information*: The sets A' and B' might be congruent, whereas the original sets A and B are not. However, the prime goal of successive pruning steps is to eventually reduce the points sets to some sets A' and B' which are so small that we can afford to try all possibilities of mapping a fixed chosen point $u_0 \in A'$ to some point $v \in B'$. This is done as follows:

Once we have picked the point v, we can *reduce the dimension* of the problem by one: we choose some rotation R that brings u_0 to v. We denote by P the plane perpendicular to the axis through $Ru_0 = v$, and we project the sets RA and B onto P. (Here we must take the *original* sets A and B again.) To each projected point, we attach the signed distance from P as a label. We then look for two-dimensional congruences in P, but for *labeled* point sets. The labeling information can be easily incorporated into the algorithm of Sect. 2.

Thus, when $|A'| = |B'|$ is small, we can finish the problem by $|A'|$ instances of two-dimensional congruence in $O(|A'|n \log n)$ time.

Let us now see how we continue when our pruning process gets stuck. We will describe the steps only for the set A', but the reader has to keep in mind that they are carried out for the set B' in parallel. If the convex hull $H(A')$ is one-dimensional or two-dimensional, then we have found an axis or a plane with a corresponding axis or plane in $H(B')$. This allows us to reduce the question to one or two-dimensional problems, as described above.

We are left with the case that $H(A')$ is a three-dimensional polytope. By pruning, we can assume that all vertices of the graph of $H(A')$ have the same degree d. By Euler's formula, d can be 3, 4, or 5. Euler's formula also yields the number of faces F in terms of the number n' of vertices of $H(A')$: $|F| = (d-2)/2 \cdot n' + 2 \leq \frac{3}{2}n' + 2$. We now try to prune the *faces* by face degrees. If there are at least two different face degrees, the smallest degree class F' of faces has at most $\frac{3}{4}n' + 1$ elements. This number is smaller than n' unless $n' = 4$ and $H(A')$ is a tetrahedron. We compute the centers of gravity of the faces in F', and replace A' by the set of these centers. We call this procedure a *condensation*. Like pruning, it reduces A' to a smaller set, but in contrast to pruning, the smaller set is not necessarily a subset of A'.

With the new condensed set A' we restart the whole procedure from scratch, beginning with the convex hull computation. The only case where neither condensation, nor pruning, nor dimension reduction is possible is a convex polytope $H(A')$ in which all vertices and all faces have the same degree. Such a polytope must have the combinatorics of one of the five regular polytopes (Platonic solids): the tetrahedron, the octahedron, the icosahedron, the cube, or the dodecahedron. We know therefore that $|A'| \leq 20$, and we can resort to dimension reduction, which leads to at most 20 two-dimensional instances.

In all the above-mentioned pruning and reduction steps, we must avoid that the reduced set A' contains only the origin. When such a case would arise, we artificially select a different class of vertices or faces.

4 Pruning and Condensation

Pruning is very versatile: we can use any criterion of points that we can think of, as long as it is not too expensive to compute. For example, in our algorithm for four dimensions, we will build the closest-pair graph G, which connects all pairs of points of A' whose distance equals the smallest inter-point distance in the set, and try to prune by degree in this graph. If, however, all vertices happen to have degree 1 in G, thus forming a perfect matching of A', we condense A' to the set of midpoints of the matching edges.

The power of the pruning technique is that we can concentrate on those cases where pruning fails. These instances are highly symmetric and regular, and we will capitalize on this regularity to extract structures from the point set that allow us to proceed.

Formally, a condensation procedure is a mapping F that maps a set A to a set $A' = F(A)$. This mapping must be *equivariant* under rotations:

$$R \cdot F(A) = F(R \cdot A), \text{ for all rotations } R$$

A pruning procedure is the special case where $F(A) \subseteq A$. We say that condensation is *successful* if $F(A)$ is smaller than A and $F(A)$ is not the empty set or just the origin. We will be able to ensure a reduction by a constant factor for successful condensation steps, and thus we need not worry about the time for iterating the condensation, because the size of A' decreases at least geometrically.

5 The Three-Dimensional Point Groups

We have seen that congruence testing is closely connected to symmetry: "Random" point sets have no symmetries and are easy to check for congruence. The hard cases are the symmetric ones. It is therefore no surprise that congruence testing algorithms can tell us something about the symmetry groups of point sets.

In Sect. 3, we have stopped condensation as soon as we reached the combinatorial structure of a Platonic solid. By further condensation, based the edge lengths, we can achieve that the only remaining cases must also have the *geometry* of a Platonic solid, see Algorithm K in Kim and Rote (2016) for details. From this we can conclude the following theorem.

Theorem 1. *The symmetry group of a finite three-dimensional set of points is either*

1. *the symmetry group of one of the five Platonic solids,*
2. *the symmetry group of a prism over a regular polygon,*
3. *or a subgroup of one of the above groups.* ☐

These groups are the discrete subgroups of the orthogonal group $O(3)$ of 3×3 orthogonal matrices, and they are called the *three-dimensional point groups*. Case 2 covers the *reducible groups* (and their subgroups), those groups that are direct products of lower-dimensional point groups. They come from the case when our algorithm used dimension reduction. Theorem 1 is not very explicit, and quite redundant: The octahedron and the cube are dual to each other and have the same symmetries, and so do the dodecahedron and the icosahedron. The tetrahedral group is contained both in the group of the cube and of the dodeca-hedron. With some work, the explicit list of groups can be worked out from this theorem. However, the resulting classification of three-dimensional point groups was already known in the 19th century (Hessel's Theorem). We will mention potential extensions to four dimensions in Sect. 8.

6   General Dimensions

The best algorithms for general dimension d are a deterministic algorithm of Brass and Knauer (2002) and a randomized algorithm of Akutsu (1998). They reduce the dimensionality d of the problem by three, respectively four dimensions at a time, and achieve running times of $O(n^{\lceil d/3 \rceil} \log n)$ and $O(n^{\lfloor d/2 \rfloor / 2} \log n)$, respectively, for high enough dimensions.

7   Four Dimensions

We have recently managed to solve congruence testing in four dimensions in optimal $O(n \log n)$ time.

Theorem 2. *Given two sets A and B of n points in four dimensions, it can be decided in $O(n \log n)$ time and $O(n)$ space whether A and B are congruent.*

The algorithm is based on condensation and dimension reduction, but the details are quite involved, see Kim and Rote (2016). We can therefore give only rough overview, referring to the following flowchart, and glossing over many details.

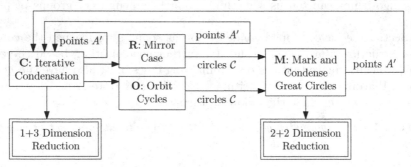

7.1   Iterative Pruning and Condensation Using the Closest-Pair Graph (Algorithm C)

After pruning by distance from the origin, we can assume that A lies on the three-dimensional sphere $\mathbb{S}^3 \subset \mathbb{R}^4$. As in Atkinson (1987), we compute the closest distance $\delta = \min\{\, \|a - a'\| : a, a' \in A, a \neq a' \,\}$ and the *closest-pair graph* G on the vertex set A, which connects all pairs of points whose distance is δ. The vertex degrees in H are bounded by the *kissing number* $K_3 = 12$, the maximum number of equal balls with disjoint interiors that can simultaneously touch a ball of the same size on $\mathbb{S}^3$. The closest-pair graph can be computed by divide and conquer in $O(n \log n)$ time in any fixed dimension (Bentley and Shamos 1976).

Now we start an iterative pruning and condensation process on G, first based on vertex degrees, and working its way up to higher and higher orders of regularity. In the end, we will have pruned G to such a degree that all *directed edge figures*, consisting of some edge uv and all adjacent edges, are congruent. This allows us to conclude that copies of a certain pattern can be found "everywhere" in G. This pattern is a path $t_0 u_0 v_0 w_0$ with the property that its three edges $t_0 u_0$, $u_0 v_0$, and $v_0 w_0$ have the same length δ and the two angles $t_0 u_0 v_0$ and $u_0 v_0 w_0$ are equal. In G, we can then define a nonempty set S of paths $a a' a''$ with the following property.

For every path $a_1 a_2 a_3 \in S$, there is a (unique) edge $a_3 a_4 \in G$ such that $a_2 a_3 a_4 \in S$ and $a_1 a_2 a_3 a_4$ is congruent to $t_0 u_0 v_0 w_0$.

7.2 Generating Orbit Cycles (Algorithm O)

By repeatedly applying this property, we can conclude:

For every triple $a_1 a_2 a_3 \in S$, there is a unique cyclic sequence $a_1 a_2 \dots a_\ell$ such that $a_i a_{i+1} a_{i+2} a_{i+3}$ is congruent to $t_0 u_0 v_0 w_0$ for all i. (Indices are taken modulo l.)
Moreover, there is a rotation matrix R such that $a_{i+1} = R a_i$. In other words, $a_1 a_2 \dots a_\ell$ is the orbit of a_1 under the rotation R.

We call such a cyclic sequence an *orbit cycle*. If the points A would live in $\mathbb{R}^3$, the geometric situation is easy to imagine: If the points $t_0 u_0 v_0 w_0$ lie in a plane, then the orbit cycle lies on a circle. Otherwise, they form an infinite helix that winds around an axis. This intuition is not misleading: on $\mathbb{S}^3$, the situation is the same, except that the axis of the helix is a great circle instead of a line.

The last case it the most interesting case for us: If the points $t_0 u_0 v_0 w_0$ do not lie in a plane, we can extract the axis circle from each orbit cycle. We will then work with the set $\mathcal{C}$ of these circles.

7.3 Marking and Condensation of Great Circles (Algorithm M)

We are given a set $\mathcal{C}$ of great circles in $\mathbb{S}^3$. We will treat these circles as objects in their own right, independent of the point set A from which they came.

The Distance Between Circles. We start by computing the closest-pair graph on $\mathcal{C}$. To to this, we have to define a distance between great circles. We do this by embedding them in the 5-sphere $\mathbb{S}^5 \subset \mathbb{R}^6$. Great circles in the 3-sphere can be equivalently regarded as 2-dimensional planes through the origin in 4-space, and we can use Plücker coordinates to represent them. (Planes in 4-space can be equivalently regarded as lines in (projective) 3-space, and this is the most familiar type of Plücker coordinates.) The Plücker coordinates are a 6-tuple of numbers in projective 6-space. We normalize them and represent each circle as a pair of antipodal points on $\mathbb{S}^5$, and define the *Plücker distance* between two circles as the smallest distance between the four representative points. This distance is a geometric invariant: In a different coordinate system, a plane will have different Plücker coordinates, but Plücker distances are unchanged.

Other distances have been considered in the literature. Conway et al. (1996) have tried to pack lines, planes, etc. in Grassmannian spaces, using the *chordal distance* (which comes from representing a plane as a symmetric 4×4 projection matrix) and the *geodesic distance* on the Plücker surface. For our case, the Plücker distance gives the embedding of lowest dimension and is therefore preferable.

The closest-pair graph $G(\mathcal{C})$ is thus computed in 6 dimensions. The number of neighbors is bounded by the kissing number K_5 in 5 dimensions, which is known to be bounded by 44.

We now look at each pair C, D of adjacent circles in $G(\mathcal{C})$. When projecting D on the plane of C, the image will generically be an ellipse D'. We use the major axis of D' to *mark* two points on C. Similarly, we project C to the plane of D and generate two markers on D. Repeating this for all edges of $G(\mathcal{C})$ produces at most $2K_5 \leq 88$ markers on each circle of $\mathcal{C}$. These markers form a new set of points A', and we start the whole algorithm from scratch with this set of points.

We argue that the new set A' is smaller than the original set A from which the orbit cycles are generated. We know that every point of A can belong only to a bounded number of orbit cycles, by the degree constraint in $G(A)$. If all orbit cycles are *long* enough, meaning that they contain sufficiently many points, we can therefore guarantee that the number of orbit cycles is small, say $|\mathcal{C}| \leq |A|/200$, and then $|A'| \leq 88 \cdot |\mathcal{C}|$ will be a successful condensation of A. If the orbit cycles are short, it means that the closest distance δ must be longer than some threshold δ_0. Then, by a straightforward packing argument on $\mathbb{S}^3$, the size of A is bounded by a constant, and we can "trivially" solve the problem by dimension reduction.

Isoclinic Circles and Hopf Bundles. The above procedure fails to generate markers if all projected ellipses turn out to be circles. Such planes C, D are called *isoclinic*. They come in two variations, *left-isoclinic* and *right-isoclinic*. It turns out that being isoclinic imposes a strong structure on the involved circles. We formulate their properties for right-isoclinic pairs; analogous statements hold for left-isoclinic pairs.

Proposition 3. *1. The relation of being right-isoclinic is transitive (as well as reflexive and symmetric). An equivalence class is called a* right Hopf bundle.
2. For each right Hopf bundle, there is a right Hopf map h that maps the circles of this bundle to points on $\mathbb{S}^2$.
3. By this map, two isoclinic circles with Plücker distance $\sqrt{2} \sin \alpha$ are mapped to points at angular distance 2α on the "Hopf sphere" $\mathbb{S}^2$.

4. *A circle can have at most $K_2 = 5$ closest neighbors on the Plücker sphere $\mathbb{S}^5$ that are right-isoclinic.*

The right Hopf map in Property 2 is obtained as follows (Hopf 1931, Sect. 5): Choose a positively oriented coordinate system x_1, y_1, x_2, y_2 for which some circle C_0 of the bundle lies in the x_1y_1-plane. Then the map $h \colon \mathbb{S}^3 \to \mathbb{S}^2$ defined by

$$h(x_1, y_1, x_2, y_2) = \big(2(x_1y_2 - y_1x_2),\ 2(x_1x_2 + y_1y_2),\ 1 - 2(x_2^2 + y_2^2)\big)$$

maps all points on a circle of the bundle to the same point on $\mathbb{S}^2$. A different choice of C_0 would lead to a different map, but by Property 3, the images are related by an isometry of $\mathbb{S}^2$. The constant $K_2 = 5$ in Property 4 is the kissing number on the 2-sphere. Property 4 is a direct consequence of Properties 2 and 3.

We use Proposition 3 in the following way: If all pairs of circles in a component of the closest-pair graph $G(\mathcal{C})$ are right-isoclinic, we know that they must belong to a common Hopf bundle. We then use a condensation procedure on the Hopf sphere, similar to the one described in Sect. 3, to condense the set of circles, and repeat the construction of the closest-pair graph.

If a circle C has both a left-isoclinic neighbor D and a right-isoclinic neighbor D', we conclude by Property 1 that D and D' cannot be isoclinic. We can therefore mark points on D and D'.

To summarize, we repeatedly condense the set $\mathcal{C}$ of circles until we can mark some points A' on them, or until the number of circles in $\mathcal{C}$ gets smaller than some threshold. In the latter case, we apply 2+2 Dimension Reduction, as described below in Sect. 7.5

7.4   The Mirror-Symmetric Case (Algorithm R)

The generation of orbit cycles requires that the points $t_0u_0v_0w_0$ don't lie in a plane. We can guarantee that such 4-tuples exist, unless the edge figures are perfectly mirror-symmetric: The perpendicular bisector of every edge uv in G acts as a mirror, reflecting the neighbors t of u to the neighbors w of v. Since each edge tu and each edge vw has the same mirror-symmetry, the mirror images of the mirrors are also mirrors. It follows that the component of G that contains u is the orbit of u under the group generated by the mirror reflections for the edges incident to u.

Such groups, groups that are generated by reflections, are called Coxeter groups, and they have been classified in all dimensions, cf. Coxeter (1973), Table 4 on p. 297. In four dimensions, there are eight such groups, which are related to the regular polytopes of 4-space, plus an infinite class of *reducible* groups, which are direct products two-dimensional Coxeter groups.

We deal with the Coxeter groups as follows. For each group Γ in the finite list, we determine the smallest distance δ such that the neighbors of a point $u \in \mathbb{S}^3$ at distance δ can generate the group Γ. The smallest value $\delta_{\min}$ of these bound implies, by a packing argument, that $|A|$ is bounded by a constant, and thus we can resort to dimension reduction.

For the infinite family of reducible groups, we are able to identify the two complementary 2-dimensional planes corresponding to the two factor groups, and thus we can replace each component of $G(A)$ by two circles. We process these circles like the the circles that result from orbit cycles (Algorithm M, Sect. 7.3).

7.5   2+2 Dimension Reduction

The classical dimension reduction procedure applies when the image of a *point* in A (or a line through the origin) is known. The image of the complementary 3-dimensional hyperplane is then also known, and we call this *1+3 dimension reduction*. By contrast, in *2+2 dimension reduction*, we have identified a two-dimensional plane P for the point set A and another two-dimensional plane Q for B, and we are looking for congruences that map P to Q, besides mapping A to B.

We first choose a joint coordinate system x_1, y_1, x_2, y_2 in which P and Q coincide with the $x_1 y_1$-plane. The allowable rotations are therefore restricted to independent rotations in the $x_1 y_1$-plane (by some angle φ) and in the complementary $x_2 y_2$-plane (by some angle ψ). After introducing polar coordinates in the two planes, the problem reduces to *translational* congruence between two point sets $\hat{A}$ and $\hat{B}$ on the two-dimensional torus $[0, 2\pi)^2$. The distance components of the polar coordinates are attached as a *label* to each point on the torus, and only points with equal label can be mapped to each other.

We now apply a sequence of condensation and relabeling steps, using Voronoi diagrams on the torus, which eventually lead to *canonical sets* $\hat{A}_0$ and $\hat{B}_0$. These sets play the same role as the canonical axes of Sect. 2 for the problem of a single rotation (or "translation on the one-dimensional torus"): If A and B are congruent (under the constraint of mapping P to Q), then we can choose arbitrary points $a \in \hat{A}_0$ and $b \in \hat{B}_0$, and the unique rotation that maps a to b will map A to B. We therefore have to test only a single candidate rotation.

8   The Four-Dimensional Point Groups

It is tempting to extend the high-level "characterization" of three-dimensional point groups of Theorem 1 to four dimensions:

Conjecture 4. A four-dimensional point group is either

1. the symmetry group of one of the five four-dimensional regular solids,
2. a direct product of lower-dimensional point groups,
3. or a subgroup of one of the above groups.

The four-dimensional point groups have been enumerated, first by Threlfall and Seifert (1931) for the case of direct congruences only (determinant $+1$), and most lately by Conway and Smith (2003). The book of Conway and Smith gives an explicit list of these groups (Tables 4.1–4.3, pp. 44–47). Thus, in principle, it should be a trivial matter to settle Conjecture 4. However, these groups are specified algebraically, and it is not easy to see geometrically what they are.

When we started our work, we hoped that our techniques would shed light on Conjecture 4, as was the case for three dimensions (Theorem 1), but so far, the implications of our algorithm are not so strong. (On the other hand, the analysis of our algorithm *uses* the classification of four-dimensional finite Coxeter groups, i.e., those point groups that are *generated by reflections*.)

It would also be interesting to see to what extent Conjecture 4 generalizes to higher dimensions. The regular polytopes are known in all dimensions. However, in eight dimensions, the root lattice E_8 has symmetries that don't come from regular polytopes, thus providing counterexamples to a straightforward generalization of Conjecture 4 for eight dimensions, and most likely also for six and seven dimensions.

References

Akutsu, T.: On determining the congruence of point sets in d dimensions. Comput. Geom.: Theory Appl. **4**(9), 247–256 (1998)

Alt, H., Mehlhorn, K., Wagener, H., Welzl, E.: Congruence, similarity, and symmetries of geometric objects. Discrete Comput. Geom. **3**(1), 237–256 (1988). http://dx.doi.org/10.1007/BF02187910

Atkinson, M.D.: An optimal algorithm for geometrical congruence. J. Algorithms **8**(2), 159–172 (1987). http://dx.doi.org/10.1016/0196-6774(87)90036-8

Brass, P., Knauer, C.: Testing the congruence of d-dimensional point sets. Int. J. Comput. Geom. Appl. **12**(1–2), 115–124 (2002). http://dx.doi.org/10.1142/S0218195902000761

Bentley, J.L., Shamos, M.I.: Divide-and-conquer in multidimensional space. In: Proceedings of the Eighth Annual ACM Symposium on Theory of Computing, STOC 1976, pp. 220–230. ACM, New York (1976). http://doi.acm.org/10.1145/800113.803652

Conway, J.H., Hardin, R.H., Sloane, N.J.A.: Packing lines, planes, etc.: packings in grassmannian spaces. Exp. Math. **5**, 139–159 (1996). https://projecteuclid.org/euclid.em/1047565645

Conway, J.H., Smith, D.A.: On Quaternions and Octonions. A K Peters, Natick (2003)

Coxeter, H.S.M.: Regular Polytopes, 3rd edn. Dover Publications, New York (1973)

Hopf, H.: Über die Abbildungen der dreidimensionalen Sphäre auf die Kugelfläche. Math. Ann. **104**, 637–665 (1931). http://www.digizeitschriften.de/dms/img/?PID=GDZPPN002274760

Iwanowski, S.: Testing approximate symmetry in the plane is NP-hard. Theor. Comput. Sci. **80**(2), 227–262 (1991). http://dx.doi.org/10.1016/0304-3975(91)90389-J

Kim, H., Rote, G.: Congruence testing of point sets in 4-space. In: Proceedings of the 32st International Symposium on Computational Geometry (SoCG 2016), LIPIcs (2016, to appear)

Manacher, G.: An application of pattern matching to a problem in geometrical complexity. Inf. Process. Lett. **5**(1), 6–7 (1976). http://dx.doi.org/10.1016/0020-0190(76)90092-2

Sugihara, K.: An $n \log n$ algorithm for determining the congruity of polyhedra. J. Comput. Syst. Sci. **29**(1), 36–47 (1984). http://dx.doi.org/10.1016/0022-0000(84)90011-4

Threlfall, W., Seifert, H.: Topologische Untersuchung der Diskontinuitätsbereiche endlicher Bewegungsgruppen des dreidimensionalen sphärischen Raumes. Math. Ann. **104**(1), 1–70 (1931). http://dx.doi.org/10.1007/BF01457920

Curves and Surfaces

Mesh Reduction to Exterior Surface Parts via Random Convex-Edge Affine Features

Andreas Beyer[1(✉)], Yu Liu[2,3], Hubert Mara[1], and Susanne Krömker[1]

[1] Interdisciplinary Center for Scientific Computing, Heidelberg University,
Heidelberg, Germany
andreas.beyer@iwr.uni-heidelberg.de
[2] Empa, Swiss Federal Laboratories for Materials Science and Technology,
Dübendorf, Switzerland
[3] Swiss Federal Institute of Technology, ETH Zurich, Zurich, Switzerland

Abstract. Data fusion of inputs from fundamentally different imaging techniques requires the identification of a common subset to allow for registration and alignment. In this paper, we describe how to reduce the isosurface of a volumetric object representation to its exterior surface, as this is the equivalent amount of data an optical surface scan of the very same specimen provides. Based on this, the alignment accuracy is improved, since only the overlap of both inputs has to be considered. Our approach allows for a rigorous reduction below 1 % of the original surface while preserving salient features and landmarks needed for further processing. The presented algorithm utilizes neighborhood queries from random points on an ellipsoid enclosing the specimen to identify data points in the mesh. Results for a real world object show a significant increase in alignment accuracy after reduction, compared to the alignment of the original representations via standard approaches.

1 Motivation

In the field of non-destructive testing, Computed Tomography (CT), as well as optical scans, is widely used for quality inspection of industrial parts.

Figure 1 shows a typical example of a real-world industrial object, as acquired via optical 3D imaging (a) and Computed Tomography, i.e., the extracted isosurface (b). Both imaging techniques have their own strengths and weaknesses. Data fusion now requires to align those representations, shown in (c), which is in principle feasible through standard approaches. Unfortunately, due to the characteristics of the acquired data sets, alignment algorithms are prone to introduce errors, which we address in the following. We introduce the notation $\mathcal{M}_{CT}$ for any isosurface mesh generated from CT data and the notation $\mathcal{M}_{opt}$ for a surface acquired with an optical 3D scanner.

Optical Acquisition systems typically apply fringe pattern projection and stereoscopic scanning. The field of view in which data points are acquired, is restricted to the focal area of the camera system. Depth information per data point is computed by triangulation via disparity in camera views and displacement of

© Springer International Publishing Switzerland 2016
I.S. Kotsireas et al. (Eds.): MACIS 2015, LNCS 9582, pp. 63–77, 2016.
DOI: 10.1007/978-3-319-32859-1_5

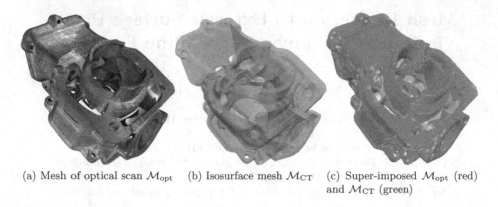

(a) Mesh of optical scan $\mathcal{M}_{\mathrm{opt}}$ (b) Isosurface mesh $\mathcal{M}_{\mathrm{CT}}$ (c) Super-imposed $\mathcal{M}_{\mathrm{opt}}$ (red) and $\mathcal{M}_{\mathrm{CT}}$ (green)

Fig. 1. Industrial example of a cylinder cast in different representations. (Color figure online)

the projected pattern. Thus, any data point acquired by optical systems must be visible either from both cameras or the projector and a camera.

Limitations of optical scanning arise since optical surface scanners are unable to acquire data points in narrow cavities or deep trenches. Also, $\mathcal{M}_{\mathrm{opt}}$ cannot reveal any interior structure or under cuttings. Therefore, $\mathcal{M}_{\mathrm{opt}}$ may have defects on the captured surface. They manifest as holes in the mesh. Other holes are due to reflective, translucent or matt black surfaces which are very difficult to acquire.

Any vertex v in $\mathcal{M}_{\mathrm{opt}}$ satisfies

$$\mathcal{M}_{\mathrm{opt}} = \{v \mid \exists \triangle(v, d_1, d_2) \text{ with } \angle(v) > \phi \wedge \text{object} \cap \triangle(v, d_1, d_2) = \emptyset\}. \quad (1)$$

This implies the condition of an unblocked view from cameras d_1 and d_2 to any point on the object. The opening angle ϕ of the triangle $\triangle(v, d_1, d_2)$ depends on the specific setup of the optical scanning system and describes the disparity angle of one camera and the projector or both cameras. Therefore, the minimal opening angle of any cavity of the object defines which data points can be acquired.

Cone Beam X-ray Computed Tomography (CBCT) is a cross sectional imaging technique derived from conventional X-ray imaging. X-rays emit from a point source, forming a cone shape, and interact with the object under investigation. The interaction follows the Beer-Lambert law according to which the transmission of the X-ray is related to the line integral of the attenuation coefficients of the object along a ray. A planar detector placed behind the specimen, perpendicular to the central ray, measures the intensity of each ray. The resulting 2D image corresponds to a conventional X-ray image and is referred to as a projection. In CBCT, a series of such projections are acquired while the source and detector pair is moving along a predefined trajectory with respect to the object. In a legacy CT, the trajectory is a full circle around the object while the center of the circle lies in the object.

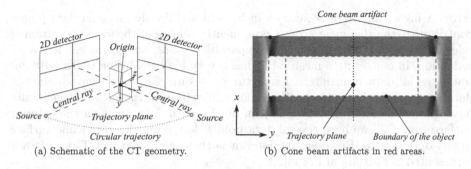

(a) Schematic of the CT geometry.   (b) Cone beam artifacts in red areas.

Fig. 2. CT system arrangement and visualization of common defect. (Color figure online)

With the projection images and a full circular trajectory, the attenuation coefficients in the illuminated area which contains the whole object can be computed using reconstruction algorithms such as the Feldkamp-Davis-Kress (FDK) method [FDK84]. Since the attenuation coefficients in the area are not homogeneous, the result is often represented as a 3D grid of voxels, which leads to a problem when trying to fuse data from optical scanning represented as polygons. Either a Marching Cubes algorithm [LC87] or in our case Volume Enclosing Surface Extraction Algorithm (VESTA) [Sch12] is applied to generate a watertight surface mesh $\mathcal{M}_{CT}$ from the scalar data on the dense voxel grid as reconstructed from the CT scans.

Limitations and artifacts of CT are related to Tuy's sufficiency condition [Tuy81], which suggests that only the attenuation coefficients in the circular trajectory plane can be exactly reconstructed. In the rest of the volume, cone beam artifacts arise due to the uncertainty of the attenuation coefficients.

The surfaces that are parallel to the trajectory plane are blurred by this effect. This leads to a reduced spatial resolution in y-direction which further causes segmentation problems. As shown in Fig. 2b, the boundary of the object (blue) is not properly reconstructed within red areas. A limited-angle CT scan uses a trajectory that is less than a full circle which violates Tuy's condition. The reconstruction from limited-angle scans is an underdetermined problem which has non-unique solutions [Ram91]. To mitigate the artifacts and to narrow down the solution set, a regularization term is used during reconstruction [LSFS14]. As prior information, we include the optical scan result to improve the output of the reconstruction algorithm.

The Key Problem in aligning object representations of fundamentally disparate imaging techniques is caused by the imbalance of represented information and difference in spatial resolution. For full-angle CT data and accompanying optical scan data of the very same object, inaccuracies in alignment do occur [BMK14]. Due to the corresponding artifacts, limited-angle data presents an even greater challenge. The data fusion task requires a very accurate alignment, which in turn is not feasible as long as internal structures contribute to the alignment error. A higher degree of accuracy can be achieved if only essential data points contribute to the alignment error. Consequently, aiming at the for preservation of relevant parts and the omission of incomparable regions, we need an efficient data reduction. In our setup, a mesh $\mathcal{M}_{opt}$ has a very high resolution up to $\sim 10\,\mu m$, but lacks all data from internal structures. Isosurfaces $\mathcal{M}_{CT}$ from volumetric data sets contain, in contrast, all interior and exterior structures, but generally have a lower accuracy of only $\sim 75\,\mu m$. So, for each data point on the exterior surface from CT, we have several data points describing the very same surface in the optical scan. However, the interior surface contained in CT data is not represented in optical data at all.

The exterior surface of an object, in our context, includes all surface parts visible from the outside. According to our definition in Eq. (1), $\mathcal{M}_{opt}$ is only a fraction of the complete exterior surface, which in turn is a fraction of all the data included in $\mathcal{M}_{CT}$. $\mathcal{M}_{opt}$ and $\mathcal{M}_{CT}$ provide different representations of the identical object, and to extract suitable subsets for alignment, the very same reduction can be applied to both. The key contributions of our approach are:

– identification of vertices, guaranteed to be on the exterior surface,
– reduction of $\mathcal{M}_{CT}$ and $\mathcal{M}_{opt}$ to corresponding subsets, and
– improvement of alignment, by omitting vertices which only contribute to error.

2   Alignment Algorithms

The alignment of mesh-based object representations usually follows one of two principles, either continuously evaluating randomly generated transformations or iteratively converging to a solution. Whereas the former is implemented in our project, the latter is applied via Meshlab[1].

RANdom Sample And Consensus (RANSAC) is an alignment scheme generating various hypotheses and verifying or falsifying those hypotheses based on random sample surveys. Our implementation follows Winckelbach et al. [WMW06] and selects a vertex pair v_1 and v_2 in each iteration. A 4D-vector c characterizing those vertices is computed from the vector $\overrightarrow{v_1v_2}$, the normal vector n_1 of vertex v_1, and n_2 of vertex v_2. The four components of c are:

1. the length of $\overrightarrow{v_1v_2}$,
2. the rotation angle between n_1 and n_2 around $\overrightarrow{v_1v_2}$,
3. the inclination angle between $\overrightarrow{v_1v_2}$ and n_1 around $\overrightarrow{v_1v_2} \times n_1$,
4. the inclination angle between $\overrightarrow{v_1v_2}$ and n_2 around $\overrightarrow{v_1v_2} \times n_2$.

For each iteration the vertex pair is selected alternately from $\mathcal{M}_{opt}$ and $\mathcal{M}_{CT}$. The computed c vector is stored along with the selected point pair in a database for this mesh, e.g., a hash table. In addition the database of the other mesh is searched for a similar c vector. If a similar vector is already stored in the

---

[1] Software provided by: Visual Computing Lab, CNR-ISTI, Pisa, Italy: http:// meshlab.sourceforge.net/

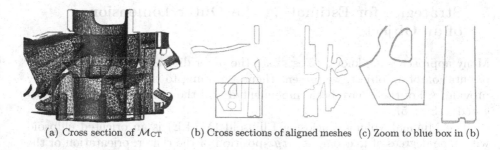

(a) Cross section of $\mathcal{M}_{CT}$ (b) Cross sections of aligned meshes   (c) Zoom to blue box in (b)

Fig. 3. ICP offset of cross sections from $\mathcal{M}_{CT}$ (green) and from $\mathcal{M}_{opt}$ (red). (Color figure online)

database of the other mesh, the characteristics of the corresponding vertex pairs are alike. In this case a hypothesis is formulated, i.e., a transformation matrix mapping one vertex pair on the other is computed. The hypothesis is tested by applying the transformation matrix to a random sample of vertices of one mesh and measuring the distance of the transformed vertices to the surface of the other mesh. If the root mean square error (RMSE) of the transformed vertices is below a given threshold the hypothesis is accepted as a global solution for the alignment task, if not the hypothesis is rejected and a new iteration starts.

This approach is very reliable and converges quickly to a suitable solution if both meshes are from the same imaging technique or at least have similar spatial resolution. For $\mathcal{M}_{opt}$ and $\mathcal{M}_{CT}$ this is generally not the case as shown by Beyer et al., which either causes the absence of hypotheses at all due to the lack of sufficiently similar c vectors and therefore no convergence. In case the similarity condition and the verification threshold are relaxed, the approach converges to alignments which are not accurate enough for our scenario. Thus, the presented implementation of RANSAC is preferred to, e.g., align partial mesh representations, as an optical scanner provides them, to construct the complete scan result, but is not suited for aligning $\mathcal{M}_{opt}$ and $\mathcal{M}_{CT}$.

Iterative Closest Point (ICP) algorithms successively minimize the error in rotation and translation between two meshes, i.e. between the two sets of vertices, to find an alignment [BM92]. The approach works fine for data points generated from the same imaging technique and, contrarily to RANSAC, does not seem to suffer from the difference in spatial resolution for $\mathcal{M}_{opt}$ and $\mathcal{M}_{CT}$. On the downside, due to the imbalance of information as described above, ICP tends to introduce a drift in the resulting transformation. This is caused by the internal structures only represented in $\mathcal{M}_{CT}$ and the attempt to minimize the distance per vertex between the meshes. Since those vertices do not have a suitable counterpart in $\mathcal{M}_{opt}$ the introduced drift can be seen as over-compensation. Figure 3 presents the offset as cross sections of both meshes.

3   Strategies for Estimating the Outer Dimensions of an Object

Many approaches are known to estimate the outer dimensions of the mesh representation of an object; we present them according to the level of detail they provide. All of them have been implemented and the relevant ones are investigated in Sect. 5.

The Minimal Volume Enclosing Ellipsoid (MVEE) is an oriented ellipsoid with nine degrees of freedom, i.e., xyz-position of the center, orientation of the three perpendicular axes and the three radii along these axes. The implementation based on Todd et al. [TY07] computes a parametric form of an enclosing primitive around the object. If the object is not already known to be roughly cuboid, this presents a better estimation of the dimensions in the general case (Fig. 4a).

The Convex Hull (CH) is the smallest convex set of vertices of an object which contains the object itself. It is an even better estimation of the object's dimensions than MVEE and usually is the basis of calculating MVEE, since it reduces the problem size drastically. However, it lacks a parametric form (Fig. 4b).

Alpha Shapes (AS) define a shape around the object, but this shape does not need to be convex. So far, it is the best approximation of the object's dimensions and commonly compared to shrink-wrapping or gift-wrapping an object. The Delaunay triangulation of all object vertices [Joe91] provides a basis to compute the α-complex [EM94] and in turn the α-shape as shown in Fig. 4c. Depending on the chosen α-value, the surface varies, i.e., the value defines how tight $\mathcal{M}_{AS}$ approximates the input mesh. We choose α such that the tightest hull is computed which still produces one connected component. Any deviation results in $\mathcal{M}_{AS}$ either loosely fitting the input mesh, or containing several unconnected surface parts.

The evaluation is shown in Sect. 5 after the presentation of our method in the following section.

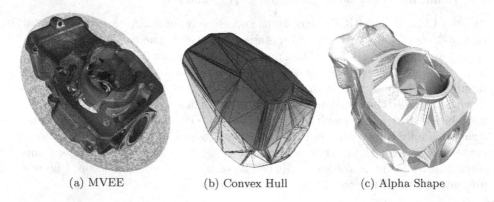

(a) MVEE   (b) Convex Hull   (c) Alpha Shape

Fig. 4. Different hulls around $\mathcal{M}_{opt}$.

4   Random Convex-Edge Affine Feature (RanCEAF) Selection

In our approach, the selection of surface points is performed via nearest neighbor search. The seeds of those queries are randomly distributed on an enclosing ellipsoid around the mesh $\mathcal{M}$. Thus, the seeds are guaranteed to be above the mesh itself and unrelated to the resolution of the underlying mesh.

Nearest neighbor searches (NNS) can be efficiently carried out by a suitable data structure, e.g., a k-d tree storing all vertices of the mesh under investigation. The seed vertex s of our query is above the exterior surface and the nearest neighbor v is chosen as:

$$v \in \mathcal{M} \text{ s.t. } \|v - s\|_2 = \min_{p \in \mathcal{M}} \left(\|p - s\|_2 \right). \tag{2}$$

Thus, v is the one vertex from the mesh, which is closest to the seed vertex s, and it is also ensured that v is not below the exterior surface.

A randomized distribution for seeds s is generated via spherical coordinates θ and ϕ. A Mersenne Twister pseudo-random generator of 32-bit numbers with a state size of 19937 bits is employed to provide a uniform distribution of $u, v \in [0, 1]$, with

$$\theta = 2\pi u \text{ and } \phi = \cos^{-1} (2v - 1). \tag{3}$$

In combination with a given radius r, the relation of spherical coordinates and Cartesian coordinates is established. In case of $r = 1$, the distribution contains points on a unit sphere such that any small area on the sphere is expected to hold the same number of points.[2] Let MVEE be described by its center c_{MVEE}, perpendicular axes a_1, a_2, a_3, and the respective radii r_1, r_2, r_3, which are derived from an eigenvalue decomposition to get a parametric form [TY07]. The xyz-coordinates of a point $q' = (q'_x, q'_y, q'_z)$ are based on θ and ϕ as follows:

$$q'_x = r_1 \sin(\theta) \cos(\phi), \quad q'_y = r_2 \sin(\theta) \sin(\phi), \quad q'_z = r_3 \cos(\theta). \tag{4}$$

This formulation respects the radii of the ellipsoid but not its orientation and location, all points q' in (4) are located on an axis-aligned ellipsoid centered at the origin of the Cartesian coordinate system. A transformation t given by a 4×4 matrix A_t is computed from a rotation to axes a_1, a_2, a_3 and the translation to the center c_{MVEE} of the MVEE. Thus, after applying (4), any point q' is transformed by A_t to its final position q on the surface of the MVEE around $\mathcal{M}$.

Locally convex regions in the underlying mesh serve as attractors for NNS if they represent a protruding structure on the exterior surface. To expand their scope in answering NNS queries, each generated point q on the MVEE is shifted for simplicity by a factor $e = 2$, such that its distance to c_{MVEE} is doubled. This finally represents the seed location s as shown in Fig. 5b. The last step is necessary to prevent local maxima of the mesh, contributing to the CH and

---

[2] Eric W. Weisstein, Sphere Point Picking: http://mathworld.wolfram.com/SpherePointPicking.html.

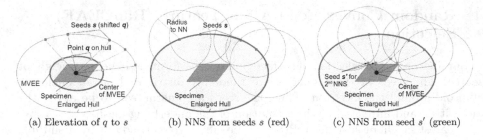

(a) Elevation of q to s (b) NNS from seeds s (red)   (c) NNS from seed s' (green)

Fig. 5. NNS without and with shifted seed vertices (2D example). (Color figure online)

defining the size of the MVEE, from only being selected by an NNS query in case the randomized seed vertex s is identical to this extreme point of $\mathcal{M}$. Any factor $e > 1$ is sufficient, since the chosen value only effects the initial query, and is already compensated after the first seed-shift operation.

Seed-shift operations allow for the extraction of larger surface parts, gradually relaxing the constraints on proximity to the MVEE and therefore the original seed vertex s. As shown in Fig. 5c, subsequent NNS with shifted seed vertices s' allow for bypassing the most prominent and most protruding structures and expanding the selected exterior surface parts. In this case any seed vertex s is shifted towards the center c_{MVEE} by the distance $\|v - s\|_2$, which equals the distance to its nearest neighbor as it was returned from the initial query. It is still not possible to penetrate the exterior surface since the only vertex $p \in \mathcal{M}$ which can be reached from s by shifting it to position s' is v itself—and therefore a vertex on the exterior surface. The benefit of this operation is, that less prominent but still salient, locally convex regions on the exterior surface can be added to the extracted subset.

The attributes of the extracted data are that both reduced meshes

- only contain those parts visible from the outside, i.e., the exterior surface,
- exclude narrow cavities and covered regions behind obstacles,
- include samples distributed over the whole object, preferably from salient regions,
- only contain measurement results, and no kind of smoothing, collapsing or averaging.

With this Random Convex-Edge Affine Feature (RanCEAF) selection, we present an approach to extract almost the same meaningful subset from each of the meshes $\mathcal{M}_{opt}$ and $\mathcal{M}_{CT}$ as a pre-processing step to allow for efficient and robust alignment.

5   Evaluation

Approaches like AABB and MVBB identify six vertices each which is not sufficient for providing an alignment. Likewise, the MVEE is calculated from the

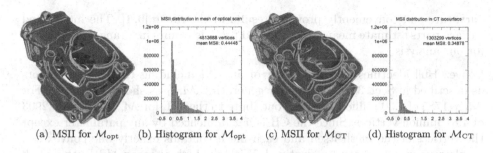

(a) MSII for $\mathcal{M}_{opt}$ (b) Histogram for $\mathcal{M}_{opt}$ (c) MSII for $\mathcal{M}_{CT}$ (d) Histogram for $\mathcal{M}_{CT}$

Fig. 6. Curvature visualized via Euclidean distance of MSII feature vectors.

set of vertices in the CH, but defined by eight points. Thus, they only allow for an estimation of object dimensions, but there is no identification of the exterior surface is performed.

Exterior surface identification is expected to include salient regions visible from the outside, i.e., from the perspective of an optical scanner. Internal structures and parts of the mesh covered by obstacles shall not be included. We measure curvature as salience via Multi-scale Integral Invariants [MKJB10], determine the fraction of the total surface included in the extracted surface and the salience of all vertices within this subset.

Multi-scale Integral Invariants (MSII) are computed from the intersection of the surface $\mathcal{M}$ and a set of n isocentric spheres with different radii, i.e., scales. The analysis is performed for each vertex v of the mesh, i.e., while each v defines the center of the nested spheres. The largest sphere S_0 has the radius r_0 depending on the size of the desired features. For the nested spheres $S_1...S_{n-1}$ the radii are equidistantly chosen such that radius r_x of each sphere equals $r_x = r_0 - x\frac{r_0}{n}$. In our case, $n = 16$ spheres are computed, which is heuristically a good trade-off between accuracy and performance. Two variants are implemented for the analysis, computing either (a) the fraction of the volume of S_x and the enclosed volume as intersection of S_x and the volume below the intersected surface area of the mesh, or (b) the fraction of the surface of a disc with the radius r_x and the surface area of the intersection of mesh and sphere.

A 16D feature vector holding the results per sphere is computed per vertex $v \in \mathcal{M}$. The output of these computations are in the range $]0, \frac{4}{3}\pi r^3[$ for analysis based on enclosed volume and $]0, +\infty[$ for analysis based on enclosed surface. After normalization for each of the radii, the feature vector contains entries in the range $]0, +\infty[$ for enclosed surface and $]0, 1[$ for enclosed volume. MSII provides invariant curvature information on various scales, i.e., it is prone to translation and rotation of the mesh and provides robust results for different resolution levels of the mesh. Therefore, it is highly suitable for analyzing the very same object represented as $\mathcal{M}_{CT}$ in lower resolution and $\mathcal{M}_{opt}$ in higher resolution while computing comparable feature vectors as shown in Fig. 6a and c.

The following evaluation of surface extraction methods considers the MSII analysis based on intersected volume since it is closer related to Gaussian curvature and conveniently provides results in the range $]0, 1[$. The intersected surface parts estimate mean curvature and provide results in a range not suitable for our analysis.

Convex Hull identifies the convex set of any $\mathcal{M}$ including the extreme points as described in Sect. 3, and allows for generating $\mathcal{M}_{CH}$ as shown in Fig. 4b. For $\mathcal{M}_{CT}$ 1785 of 1.30 million vertices contribute to the CH, for $\mathcal{M}_{opt}$ these are 2903 of 4.81 million vertices. Since the CH is not influenced by any parameter except the vertices of the mesh itself and each mesh contains exactly one convex set, no alternative subset can be identified. With the highest mean MSII value of all subsets and the absence of vertices with a MSII value close to zero, the result as shown in Fig. 7c is a sufficient feature extraction. The CH never contains internal structures but contains only the most prominent protruding structures. It is therefore not suitable to provide the basis of an accurate alignment in general. Intuitively, it seems sufficient in the presented case, but the applied ICP algorithm cannot compute a valid transformation.

Alpha Shape generates a surface $\mathcal{M}_{AS}$ for $\mathcal{M}_{CT}$. As shown in Fig. 8, AS does not identify the exterior surface, since interior structures are covered by $\mathcal{M}_{AS}$ and therefore included in the resulting subset. The same holds for $\mathcal{M}_{AS}$ of $\mathcal{M}_{opt}$ shown in Fig. 4c. The mean MSII values in the subsets are in the region of the corresponding original meshes (see Table 1) and the histogram in Fig. 8c is dominated by MSII values close to zero which makes AS unsuitable for feature extraction. Experiments with lower α-values did not improve the result.

RanCEAF subset of $\mathcal{M}_{opt}$ (Fig. 9a) for 50 k seeds contains 7474 vertices of all 4.81 million vertices. The RanCEAF subset of $\mathcal{M}_{CT}$ (Fig. 9b) contains 5023 vertices of all 1.30 million vertices. As the extracted surface parts in both cases represent less than 0.05 % of the vertices $p \in \mathcal{M}$, only the most prominent structures have been selected. The number of (removed) duplicates within the selection indicates that a small fraction of the exterior surface dominates the result by answering multiple NNS queries each. Thus, for sample sizes larger that 50 k seeds no drastic change in the extracted subset is expected since we already over-sampled this

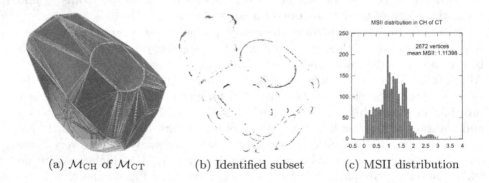

(a) $\mathcal{M}_{CH}$ of $\mathcal{M}_{CT}$ (b) Identified subset   (c) MSII distribution

Fig. 7. Convex Hull algorithm applied to $\mathcal{M}_{CT}$.

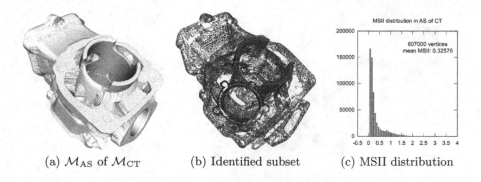

(a) $\mathcal{M}_{AS}$ of $\mathcal{M}_{CT}$ (b) Identified subset   (c) MSII distribution

Fig. 8. Alpha Shape algorithm applied to $\mathcal{M}_{CT}$.

subset by one order of magnitude. To allow for scalable mesh reduction, shift-seed operations (Fig. 5) provide sufficient data for alignment. The mean MSII values provided in both subsets are second highest after CH, which makes RanCEAF a suitable method for feature extraction. None of the presented RanCEAF results include interior structures and only after the third seed-shift operation MSII values close to zero dominate the histogram (Figs. 10f and 11f). For illustration purposes, Figs. 9, 10, and 11 show the extracted set of vertices and their connected faces. Via region growth in each vertex of the subset, more faces can be included to extract a larger portion of the exterior surface.

Seed-shift operations, as applied to $\mathcal{M}_{opt}$ in Fig. 10, and to $\mathcal{M}_{CT}$ in Fig. 11 expand the regions from which exterior surface points are selected and still provide a higher mean MSII value than the original meshes in Fig. 6. As the percentage of vertices with an MSII value ≥ 1.0 in Table 1 indicates, expanding the subset does not over-represent regions with low MSII values. In our experiments, the best increase in alignment accuracy was based on the output of the second seed-shift operation for $\mathcal{M}_{opt}$ and $\mathcal{M}_{CT}$.

6   Results

We have shown that our proposed method RanCEAF efficiently identifies the exterior surface of a given mesh. Furthermore, it allows to over-represent convex areas since they serve as attractors for regional queries from seeds on the enclosing ellipsoid. The protruding areas include the local maxima of the object under investigation and the resulting subset of all data points is suitable for alignment. The presented approach does not—in contrast to AS—introduce additional faces or require any further post-processing. The proposed method provides a reliable surface reduction, which can be iteratively expanded by applying multiple seed-shift operations. In general, the RanCEAF algorithm only relies on the vertices of the mesh $\mathcal{M}$ and therefore can be applied to point clouds. Only for the analysis based on MSII, faces are required in a pre-processing step and only for the sake

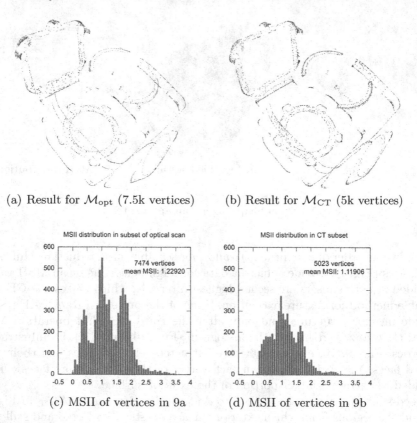

(a) Result for $\mathcal{M}_{\mathrm{opt}}$ (7.5k vertices)   (b) Result for $\mathcal{M}_{\mathrm{CT}}$ (5k vertices)

(c) MSII of vertices in 9a   (d) MSII of vertices in 9b

Fig. 9. RanCEAF result for 50 k seeds.

of evaluating our approach. The comparison of size and salience of extracted subsets, as shown in Table 1, indicates that the subsets extracted by our method are sufficiently large to serve as input for computing an alignment, and yet salient enough to grasp the essential structures of the presented geometry. The inherent parallelism of our approach is easily exploited (in our evaluation on an *Intel Xeon E7-4870*) and therefore not corrected for comparison to single-threaded algorithms in Table 1. For the presented object, the alignment of the complete meshes $\mathcal{M}_{\mathrm{opt}}$ and $\mathcal{M}_{\mathrm{CT}}$ via the ICP algorithm in Meshlab resulted in an RMSE of 2.736 mm. Computing the transformation matrix based on the extracted surfaces of both meshes and applying the obtained transformation to $\mathcal{M}_{\mathrm{opt}}$ and $\mathcal{M}_{\mathrm{CT}}$, provided a RMSE of 2.722 mm. The increase in accuracy reads as 0.5 % or an RMSE reduction of 14 µm, which potentially affects the selection of cells on the dense voxel grid as reconstructed from CT scans. Notice that there is no perfect alignment for both meshes. Therefore, the RMSE cannot be zero and the real increase in accuracy is higher than 0.5 %.

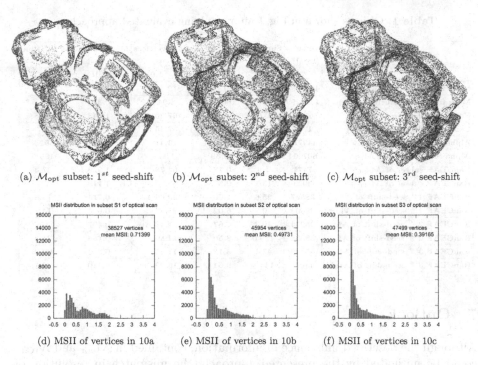

(a) $\mathcal{M}_{opt}$ subset: 1st seed-shift   (b) $\mathcal{M}_{opt}$ subset: 2nd seed-shift   (c) $\mathcal{M}_{opt}$ subset: 3rd seed-shift

(d) MSII of vertices in 10a   (e) MSII of vertices in 10b   (f) MSII of vertices in 10c

Fig. 10. RanCEAF with seed-shifts applied to $\mathcal{M}_{opt}$.

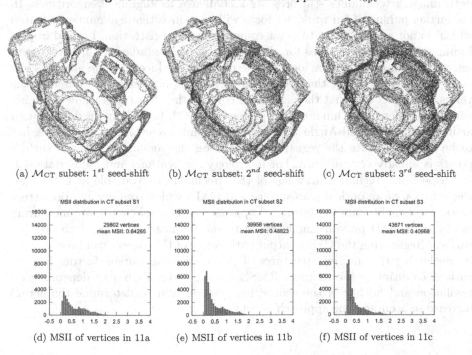

(a) $\mathcal{M}_{CT}$ subset: 1st seed-shift   (b) $\mathcal{M}_{CT}$ subset: 2nd seed-shift   (c) $\mathcal{M}_{CT}$ subset: 3rd seed-shift

(d) MSII of vertices in 11a   (e) MSII of vertices in 11b   (f) MSII of vertices in 11c

Fig. 11. RanCEAF with seed-shifts applied to $\mathcal{M}_{CT}$.

Table 1. Objects shown in Fig. 1 after applying evaluated approaches.

	Vertices (total)	Surface area (in cm^2)	Surface coverage (in %)	Salient vertices (in % with MSII $\geq$ 1.0)	Mean salience (in subset via MSII)	CPU time (in sec)
Mesh from optical scan $\mathcal{M}_{opt}$	4813688	1042.9	69.19	12.47	0.444	–
Mesh from CT isosurface $\mathcal{M}_{CT}$	1303299	1507.3	100.00	4.94	0.349	–
Convex Hull $\mathcal{M}_{CH}$ of $\mathcal{M}_{opt}$	2903	0.5	0.04	67.30	1.343	35.8
Convex Hull $\mathcal{M}_{CH}$ of $\mathcal{M}_{CT}$	1785	4.4	0.29	57.45	1.114	8.7
Alpha Shape $\mathcal{M}_{AS}$ of $\mathcal{M}_{opt}$	449773	16.6	1.10	13.41	0.464	551.9
Alpha Shape $\mathcal{M}_{AS}$ of $\mathcal{M}_{CT}$	607004	593.7	39.39	1.17	0.326	135.5
RanCEAF subset of $\mathcal{M}_{opt}$	7474	1.9	0.13	64.78	1.229	37.4
RanCEAF subset of $\mathcal{M}_{CT}$	5023	13.1	0.87	55.15	1.119	7.1
RanCEAF 1st seed-shift of $\mathcal{M}_{opt}$	38527	25.6	1.70	29.66	0.714	76.8
RanCEAF 1st seed-shift of $\mathcal{M}_{CT}$	29802	90.7	6.02	23.87	0.643	14.1
RanCEAF 2nd seed-shift of $\mathcal{M}_{opt}$	45954	40.2	2.67	17.42	0.497	116.3
RanCEAF 2nd seed-shift of $\mathcal{M}_{CT}$	39958	129.0	8.56	17.80	0.488	21.3
RanCEAF 3rd seed-shift of $\mathcal{M}_{opt}$	47499	47.1	3.12	11.45	0.392	156.6
RanCEAF 3rd seed-shift of $\mathcal{M}_{CT}$	43671	144.9	9.61	16.29	0.410	28.1

7   Outlook

Although the described imbalance in information, contained in $\mathcal{M}_{opt}$ and $\mathcal{M}_{CT}$, could be mitigated by the presented approach, the mismatch in resolution of both imaging techniques still presents a challenge to alignment algorithms. In the further pursuit of our work, our focus will be to investigate alignment schemes which do not rely on point to point comparison for registration. Instead of performing seed-shift operations for all seed vertices alike, adaptive application to selected seeds, based on the local geometry, would reduce runtime and preserve more features. The fact that both representations are known to describe the very same object and that they also both contain the object as a whole, matches with the challenges within our joint project ILATO[3]. Especially for dealing with artifacts from Limited-Angle CT, any data point irrelevant for alignment has to be neglected since the registration based on the remaining exterior surface points is already very difficult. For metrology applications and industrial quality inspections, technical drawings of the specimen are available as CAD files. Registration of an optical scan surface and CAD, which contains interior structures, can benefit from the presented approach. Likewise, coordinate-measuring machines (CMM) provide highly accurate tactile measurements of an object's surface. Registering the CMM output with $\mathcal{M}_{CT}$ of this object can be enhanced by neglecting the interior structures of $\mathcal{M}_{CT}$. We will pursue further investigations to estimate the minimal RMSE for the given alignment depending on resolution and fidelity of the acquisition systems and to determine the actual increase in accuracy our approach provides.

---

[3] http://www.iwr.uni-heidelberg.de/groups/ngg/ILATO/.

Acknowledgements. This joint project is funded by the Deutsche Forschungsge-meinschaft (DFG), grant number *BO 864/17-1*, and by the Swiss National Science Foundation (SNF), grant number *200021L 141311*. The Heidelberg Graduate School of Mathematical and Computational Methods for the Sciences (HGS MathComp) provides the optical scanning system as well as assistants to operate it. We thank our colleague Filip Sadlo for great help in improving the presentation of our work and implementing the reviewers comments. We also want to thank our project partners at the Swiss Federal Laboratories for Materials Science and Technology (Empa) for providing their expertise in metrology, the acquisition of numerous CT scans, and for having many fruitful discussions in frequent virtual or physical meetings. Above all, we thank Philipp Schütz, Urs Sennhauser, Jürgen Hofmann and Alexander Flisch.

References

[BM92] Besl, P.J., McKay, N.D.: A method for registration of 3D shapes. IEEE Trans. Pattern Anal. Mach. Intell. **14**(2), 239–256 (1992)

[BMK14] Beyer, A., Mara, H., Krömker, S.: ILATO project: fusion of optical surface models and volumetric CT data (2014). CoRR abs/1404.6583

[EM94] Edelsbrunner, H., Mücke, E.P.: Three-dimensional alpha shapes. ACM Trans. Graphics (TOG) **13**(1), 43–72 (1994)

[FDK84] Feldkamp, L.A., Davis, L.C., Kress, J.W.: Practical Cone-Beam algorithm. J. Opt. Soc. America A **1**(6), 612–619 (1984)

[Joe91] Joe, B.: Construction of three-dimensional delaunay triangulations using local transformations. Comput. Aided Geom. Des. **8**(2), 123–142 (1991)

[LC87] Lorensen, W.E., Cline, H.E.: Marching cubes: a high resolution 3D surface construction algorithm. SIGGRAPH Comput. Graph. **21**(4), 163–169 (1987)

[LSFS14] Liu, Y., Schuetz, P., Flisch, A., Sennhauser, U.: Exploring the limits of limited-angle computed tomography complemented with surface data. In: Proc. of the 11th Eur. Conf. on Non-Destructive Testing (ECNDT) (2014)

[MKJB10] Mara, H., Krömker, S., Jakob, S., Breuckmann, B.: GigaMesh and Gilgamesh - 3D multiscale integral invariant cuneiform character extraction. In: Proc. of the 11th Intl. Conf. on Virtual Reality, Archaeology and Cultural Heritage, pp. 131–138. Eurographics Association (2010)

[Ram91] Ramm, A.G.: Inversion of limited-angle tomographic data. Comput. Math. Appl. **22**(4–5), 101–111 (1991)

[Sch12] Schlei, B.: Extraction, volume-enclosing surface. Comput. Graph. **36**(2), 111–130 (2012)

[Tuy81] Tuy, H.: Reconstruction of a three-dimensional object from a limited range of views. J. Math. Anal. Appl. **80**(2), 598–616 (1981)

[TY07] Todd, M.J., Yıldırım, E.A.: On Khachiyan's algorithm for the computation of minimum-volume enclosing ellipsoids. Discrete Appl. Math. **155**(13), 1731–1744 (2007)

[WMW06] Winkelbach, S., Molkenstruck, S., Wahl, F.M.: Low-cost laser range scanner and fast surface registration approach. In: Franke, K., Müller, K.-R., Nickolay, B., Schäfer, R. (eds.) DAGM 2006. LNCS, vol. 4174, pp. 718–728. Springer, Heidelberg (2006)

Numeric and Certified Isolation
of the Singularities of the Projection
of a Smooth Space Curve

Rémi Imbach[(✉)], Guillaume Moroz, and Marc Pouget

LORIA Laboratory, INRIA Nancy Grand Est, Nancy, France
{remi.imbach,guillaume.moroz,marc.pouget}@inria.fr

Abstract. Let $\mathcal{C}_{P \cap Q}$ be a smooth real analytic curve embedded in $\mathbb{R}^3$, defined as the solutions of real analytic equations of the form $P(x,y,z) = Q(x,y,z) = 0$ or $P(x,y,z) = \frac{\partial P}{\partial z} = 0$. Our main objective is to describe its projection $\mathcal{C}$ onto the (x,y)-plane. In general, the curve $\mathcal{C}$ is not a regular submanifold of $\mathbb{R}^2$ and describing it requires to isolate the points of its singularity locus Σ. After describing the types of singularities that can arise under some assumptions on P and Q, we present a new method to isolate the points of Σ. We experimented our method on pairs of independent random polynomials (P, Q) and on pairs of random polynomials of the form $(P, \frac{\partial P}{\partial z})$ and got promising results.

Keywords: Topology of analytic real curve · Apparent contour · Singularities isolation · Numeric certified methods

1 Introduction

Consider a smooth analytic curve $\mathcal{C}_{P \cap Q} \subset \mathbb{R}^3$ defined by $P(x,y,z) = Q(x,y,z) = 0$ with P, Q analytic functions, and its projection $\mathcal{C} \subset \mathbb{R}^2$ on the (x,y)-plane. Computing the topology of $\mathcal{C}$, or computing a graph topologically equivalent to $\mathcal{C}$, requires computing the set Σ of its singularities (see Sect. 1.2 for a rigorous definition). In a second step, the study of the complement of Σ allows one to recover the topology of the curve. This fundamental problem arises in fields such as mechanical design, robotics and biology. A specific case of interest is when $Q = P_z$ (where P_z is the partial derivative $\frac{\partial P}{\partial z}$). In this case, the curve $\mathcal{C}$ is the apparent contour of the surface $P(x,y,z) = 0$. This case has been intensively studied and extended in the framework of the catastrophe theory (see [10] and references therein). Moreover, determining the topology of a projection of a space curve is an important step to compute its topology [7,11]. Similarly determining the topology of the apparent contour of a surface is an important step to compute its topology [1,5].

The goal of this paper is to take advantage of the specific structure of the singularities Σ and to propose a characterization allowing to isolate them efficiently. Since we do not restrict our work to the case $P = P_z = 0$, we also give

© Springer International Publishing Switzerland 2016
I.S. Kotsireas et al. (Eds.): MACIS 2015, LNCS 9582, pp. 78–92, 2016.
DOI: 10.1007/978-3-319-32859-1_6

a mathematical description of the types of singularities arising in the projection of curves defined by $P = Q = 0$ under some generic assumptions.

Our approach to isolating the singularities Σ is to construct a new system so-called *ball system*, the roots of which are in a one-to-one correspondence with the points of Σ. As shown with experimental results, this system suits numerical certified solvers such as subdivision methods or homotopy solvers in the polynomial case.

The rest of the paper is organized as follows. Section 2 classifies the singularities of $\mathcal{C}$ and relates them to the points where the projection Π_{xy} is not a diffeomorphism. The construction of the ball system and a proof of regularity of its solutions are exhibited in Sect. 3. Section 4 is dedicated to experiments. The rest of this section presents previous and related works, and gives explicitly the assumptions on P and Q for our method.

1.1 Previous Works

State-of-the-art symbolic methods that compute topology of real plane curves defined by polynomials are closely related to bivariate system solving.

Symbolic methods mainly rely on resultant and sub-resultant theory to isolate critical points, see for instance the book chapter [23] and references within. There are some alternatives, using for instance Gröbner bases and rational univariate representations [6,27].

Numerical methods can be used together with interval arithmetic to compute and certify the topology of a non-singular curve when the interest area is a compact subset of the plane [15,19,26]. However they fail near any singular point of the curve. Isolating singularities of a plane curve $f(x, y) = 0$ with a numerical method is a challenge since it is described by the non-square system $f = f_x = f_y = 0$, and singularities are not necessarily regular solutions of this system.

Non-regular solutions can be handled through deflation systems (see for instance [3,12,13,17,18,25]), but the resulting systems are usually still overdetermined or contain spurious solutions. Overdetermined systems can be translated into square systems using combinations of their equations with first derivatives [8]. Another deflation adapted to the singularities of the projection of a generic algebraic space curve using sub-resultant theory was proposed in [14]. In this paper we present a new deflation square system that can handle analytic curves.

Square systems with regular solutions can be solved by numerical approaches. Classical homotopy solvers [21] find all complex solutions of latter systems when their equations are polynomials. Subdivision methods [20,22,24,28] are numeric certified approaches to find all real solutions lying in an initial bounded domain of a system of analytic equations. When the latter are polynomial, these approaches can be extended to unbounded initial domains [24,28].

Starting with the work of Whitney [29], the catastrophe theory was developed to classify the singularities arising while deforming generic mappings (see [2,10] for example). From an algorithmic point of view, the authors of [9] use elements of the catastrophe theory to derive an algorithm isolating the singularities arising in mappings from $\mathbb{R}^2$ to $\mathbb{R}^2$.

1.2 Notations and Assumptions

In the following, $\mathcal{C}_{P \cap Q}$ denotes the curve defined as the zero set of the real analytic functions $P(x, y, z)$ and $Q(x, y, z)$ and B_0 is an open subset of $\mathbb{R}^2$. We will denote by Π_{xy} the projection from $\mathcal{C}_{P \cap Q}$ to the (x, y)-plane, and by $\mathcal{C}$ the projection $\Pi_{xy}(\mathcal{C}_{P \cap Q})$.

Regular Points and $A_k^{\pm}$ Singularities. A point p of the curve $\mathcal{C}$ is *regular* if there is a small neighborhood U of p in $\mathbb{R}^2$ such that $\mathcal{C} \cap U$ is a *regular submanifold* of $\mathbb{R}^2$. Otherwise it is *singular*. A singular point p of a curve $\mathcal{C}$ is of type $A_k^{\pm}$ if and only if $\mathcal{C}$ is equal to the solutions of the equation $x^2 \pm y^{k+1} = 0$ on a neighborhood U of p, up to a diffeomorphism from $U \subset \mathbb{R}^2$ to $V \subset \mathbb{R}^2$ [2, Sect. 9.8]. Remark that those are not the only type of singularities that can appear on a plane curve. Notice that the types A_{2k}^+ and A_{2k}^- are equivalent and simply denoted by A_{2k}. We will call *node* a singularity of type A_1^- or equivalently a transverse intersection of two real curve branches. We also call *cusp* a singularity of type A_{2k} and *ordinary cusp* the singularity A_2. With this notation, a point p of $\mathcal{C}$ is regular if and only if it is of type A_0.

In Sect. 2, we will describe the types of singularities of $\mathcal{C}$ assuming that :

(A_1). The curve $\mathcal{C}_{P \cap Q}$ is smooth above B_0.
(A_2). For any (α, β) in B_0, the system $P(\alpha, \beta, z) = Q(\alpha, \beta, z) = 0$ has at most 2 real roots counted with multiplicities.
(A_3). There is at most a discrete set of points (α, β) in B_0 such that $P(\alpha, \beta, z) = Q(\alpha, \beta, z) = 0$ has 2 real roots counted with multiplicities.
(A_4). Π_{xy} is a proper map from $\mathcal{C}_{P \cap Q} \cap (B_0 \times \mathbb{R})$ to its image, that is the inverse image of a compact subset is compact.

Then in Sect. 3, we will introduce the system of analytic equations that we will use to compute the singularities of $\mathcal{C}$. The solutions of this system will be regular under the following additional assumption:

(A_5). The singularities of the curve $\mathcal{C}$ are either nodes or ordinary cusps.

Notice that Thom Transversality Theorem implies that $(A_1), (A_2), (A_3)$ and (A_5) hold for generic analytic maps P, Q defining $\mathcal{C}_{P \cap Q}$ (see [10, Theorem 3.9.7 and Sect. 4.7]), and (A_4) holds at least for generic polynomial maps. If we assume only that the curve is smooth (assumption (A_1)), it would be interesting to prove that all the other assumptions hold after a generic linear change of coordinates.

If P, Q are polynomials, a semi-algorithm checking these conditions is given in [14, Semi-Algorithm 1]. Otherwise when P, Q are analytic maps, the latter semi-algorithm can be adapted only when B_0 is bounded.

2 Description of the Singularity Locus Σ

The different types of singularities of a plane curve have been classified in [2] for example. We describe in this section the types of singularities that can arise on the curve $\mathcal{C}$ under the Assumptions $(A_1) - (A_4)$, and we relate those singularities with the projection mapping Π_{xy}. More precisely, using Arnold's notation recalled below, we show that under the Assumptions $(A_1) - (A_4)$, the singularities of $\mathcal{C}$ are of type $A_k^{\pm}$ (Lemma 2 and Corollary 1). Moreover, we show that a singular point of $\mathcal{C}$ is either a critical value of Π_{xy}, or the image of two distinct points of $\mathcal{C}_{P \cap Q}$ by Π_{xy}.

Singularities of $\mathcal{C}$ and Critical Points of Π_{xy}. The critical points of Π_{xy} are the points of $\mathcal{C}_{P \cap Q}$ where the tangent to the curve is vertical, i.e. aligned with the z-axis. Assuming that the conditions $(A_1) - (A_4)$ are satisfied by the curve $\mathcal{C}_{P \cap Q}$, we show that for p a point on the curve $\Pi_{xy}(\mathcal{C}_{P \cap Q})$:

1. if p is a critical point of Π_{xy}, then it is a cusp point of $\mathcal{C}$ (singularity of type $A_{2(k+1)}$);
2. if p is the image of two distinct points of $\mathcal{C}_{P \cap Q}$, then it is a singularity of type A_{2k+1}^-;
3. otherwise, it is a regular point.

In particular, this implies that a point p is singular if and only if it is a critical value of Π_{xy} or it has two antecedents by Π_{xy}.

Lemma 1. *Let p be a point of $\mathcal{C}$. If p is not a critical value of Π_{xy} and $\Pi_{xy}^{-1}(p)$ has only one antecedent, then p is a regular point of $\mathcal{C}$.*

Proof. For U an open set of $\mathbb{R}^2$, we will denote by Π_{xy}^U the restriction of Π_{xy} to $\mathcal{C}_{P \cap Q} \cap \Pi_{xy}^{-1}(U)$. Since p is not a critical value of Π_{xy}, there exists a neighborhood U of p such that U does not contain any critical value of Π_{xy}, such that Π_{xy}^U is an immersion. Then, since p has a unique antecedent, (A_3) ensures that there is a neighborhood V of p such that Π_{xy}^V is a homeomorphism. Thus $\Pi_{xy}^{U \cap V}$ is an embedding and p is a regular point. $\qquad\square$

Lemma 2. *Let p be a point of $\mathcal{C}$. If p has two antecedents by Π_{xy}, then p is a singularity of $\mathcal{C}$ of type A_{2k+1}^- with $k \geq 0$.*

Proof. If $\Pi_{xy}^{-1}(p)$ contains more than one antecedent of p, then (A_2) implies that p has exactly two antecedents q_u and q_v. Since Π_{xy} is proper by Assumption (A_4) and $\mathcal{C}_{P \cap Q}$ is smooth by Assumption (A_1), for a small enough neighborhood U of p, $\Pi_{xy}^{-1}(U)$ is bounded and is the union of two smooth connected branches of $\mathcal{C}_{P \cap Q}$. And (A_3) implies that in a small enough neighborhood of p, p is the only point with two antecedents. Let $u = (u_x, u_y, u_z)$ and $v = (v_x, v_y, v_z)$ be the two vectors tangent to $\mathcal{C}_{P \cap Q}$ at the antecedents q_u and q_v of p. Assumption (A_2) implies that neither u nor v are vertical, hence $\tilde{u} = (u_x, u_y)$ and $\tilde{v} = (v_x, v_y)$ are non-zero vectors of $\mathbb{R}^2$. We now distinguish two cases.

First, $\tilde{u}$ and $\tilde{v}$ are independent vectors. In this case, the mapping $\left(\begin{smallmatrix} X \\ Y \end{smallmatrix}\right) = \left(\begin{smallmatrix} u_x & u_y \\ v_x & v_y \end{smallmatrix}\right)^{-1} \cdot \left(\begin{smallmatrix} x \\ y \end{smallmatrix}\right)$ is a diffeomorphic change of coordinates. Moreover $\left(\begin{smallmatrix} P_X(q_u) \\ Q_X(q_u) \end{smallmatrix}\right) = \left(\begin{smallmatrix} 0 \\ 0 \end{smallmatrix}\right)$ and $\left(\begin{smallmatrix} P_Y(q_u) \\ Q_Y(q_u) \end{smallmatrix}\right) \neq \left(\begin{smallmatrix} 0 \\ 0 \end{smallmatrix}\right)$. Thus by the analytic implicit function theorem, there exists an analytic function $f : \mathbb{R} \mapsto \mathbb{R}$ such that $Y = f(X)$ and $f(0) = f'(0) = 0$ such that the projection of the branch at q_u has an equation of the form $Y = X^2 \tilde{f}(X)$. Symmetrically, the projection of the branch at q_v has an equation of the form $X = Y^2 \tilde{g}(Y)$. Thus, up to a diffeomorphism of $\mathbb{R}^2$, the curve $\mathcal{C}$ around p has an equation of the form $(Y - X^2 \tilde{f}(X))(X - Y^2 \tilde{g}(Y)) = 0$, or equivalently $(X + Y - X^2 \tilde{f}(X) - Y^2 \tilde{g}(Y))^2 - (X - Y - X^2 \tilde{f}(X) + Y^2 \tilde{G}(Y))^2 = 0$. That is, p is a singularity of type A_1^-, also called a node.

In the case where $\tilde{u}$ and $\tilde{v}$ are co-linear, we follow the same approach, using this time the diffeomorphic change of coordinate $\left(\begin{smallmatrix} X \\ Y \end{smallmatrix}\right) = \left(\begin{smallmatrix} u_x & u_y \\ -u_y & u_x \end{smallmatrix}\right)^{-1} \cdot \left(\begin{smallmatrix} x \\ y \end{smallmatrix}\right)$. Moreover $\left(\begin{smallmatrix} P_X(q_u) \\ Q_X(q_u) \end{smallmatrix}\right) = \left(\begin{smallmatrix} 0 \\ 0 \end{smallmatrix}\right)$. As in the previous case, we use the analytic implicit function theorem at q_u and q_v, and we conclude that there exist two analytic functions f and g such that on a neighborhood of p, the curve $\mathcal{C}$ is given by the equation $(Y - X^2 f(X))(Y - X^2 g(X)) = 0$. That can be rewritten as $(2Y - X^2(f(X) + g(X)))^2 - X^4(g(X) - f(X))^2 = 0$. Assumption (A_3) ensures that the projections of the 2 branches have only one common point, such that $g(X) - f(X)$ does not vanish identically. Then, denoting by k the valuation of $f(X) - g(X)$, p is a singularity of type A_{2k+3}^-.   □

Finally, if p is a critical value of Π_{xy} we use Arnold's classification of singularities and prove that p is a singular point of type $A_{2(k+1)}$ with $k \geq 0$.

Lemma 3. *Assume that the curve $\mathcal{C}_{P \cap Q}$ satisfies $(A_1) - (A_3)$. Let q be a critical point of Π_{xy}. Then, there exists a neighborhood U of q and an invertible 2×2 matrix M of real analytic functions such that:*

$$\left(\begin{smallmatrix} P \\ Q \end{smallmatrix}\right) = M \cdot \left(\begin{smallmatrix} X - Z^{3+2k} \\ Y - Z^2 \end{smallmatrix}\right) \circ \Phi(x, y, z) \tag{1}$$

where $\Phi : (x, y, z) \mapsto (\phi(x, y), \psi(z))$ is a diffeomorphism and k is a natural integer.

Corollary 1. *Let p be a point of $\mathcal{C}$. If p is a critical value of Π_{xy}, then p is a cusp of $\mathcal{C}$ of type $A_{2(k+1)}$ with $k \geq 0$.*

Proof (of the corollary). Let q be the critical point associated to p and denote π_{xy} the projection from $\mathbb{R}^3$ to $\mathbb{R}^2$. First we show that it is sufficient to study the behavior of $\mathcal{C}_{P \cap Q}$ in a neighborhood of q to describe the curve $\mathcal{C}$ in a neighborhood of p. Indeed, Assumptions (A_2) and (A_4) imply that above a small enough neighborhood of p, the curve $\mathcal{C}_{P \cap Q}$ has a unique connected branch. In particular for any neighborhood U of the critical point q there exists a neighborhood $V \subset U$ such that $\pi_{xy}(V) \cap \mathcal{C} \subset \Pi_{xy}(U \cap \mathcal{C}_{P \cap Q})$.

Then, Lemma 3 shows that there exists a neighborhood U of q and a diffeomorphism ϕ from $\pi_{xy}(U) \subset \mathbb{R}^2$ to $V \subset \mathbb{R}^2$ a neighborhood of $(0,0)$ such that

$\phi(\Pi_{xy}(\mathcal{C}_{P \cap Q} \cap U)) = \{(X, Y) \in V \mid X^2 - Y^{3+2k}\}$. In particular, p is a singularity of type $A_{2(k+1)}$ with $k \geq 0$, that is a cusp. □

Proof (of Lemma 3). This lemma is essentially a consequence of the analytic implicit function theorem, combined with our assumptions. First, q is a critical point thus $\mathcal{C}_{P \cap Q}$ has a vertical tangent at q, up to a translation, we assume $q = (0, 0, 0)$. Since $\mathcal{C}_{P \cap Q}$ is non-singular (Assumption (A_1)), the matrix $\begin{pmatrix} P_x(q) & P_y(q) \\ Q_x(q) & Q_y(q) \end{pmatrix}$ is invertible. Using the analytic implicit function theorem ([16] or [10, Corollary 2.7.3]), there exist two real analytic functions f, g from $\mathbb{R}$ to $\mathbb{R}$ such that $P(f(z), g(z), z) = Q(f(z), g(z), z) = 0$ on a small enough neighborhood of 0. In particular, letting $\tilde{x} := x - f(z)$ and $\tilde{y} := y - g(z)$ we have $P = P(\tilde{x} + f(z), \tilde{y} + g(z), z)$ and $Q = Q(\tilde{x} + f(z), \tilde{y} + g(z), z)$. Using Hadamard's lemma ([10, Proposition 4.2.3]), there exist real analytic functions a, b, c, d such that $P = a \cdot \tilde{x} + b \cdot \tilde{y}$ and $Q = c \cdot \tilde{x} + d \cdot \tilde{y}$. Moreover, since $\begin{pmatrix} P_x(q) & P_y(q) \\ Q_x(q) & Q_y(q) \end{pmatrix}$ is invertible, the matrix $\begin{pmatrix} a(q) & b(q) \\ c(q) & d(q) \end{pmatrix}$ is also invertible. Let M_1 be the inverse of $\begin{pmatrix} a & b \\ c & d \end{pmatrix}$ on a small enough neighborhood of q. Then we have:

$$\begin{pmatrix} \tilde{P} \\ \tilde{Q} \end{pmatrix} := \begin{pmatrix} x - f(z) \\ y - g(z) \end{pmatrix} = M_1 \cdot \begin{pmatrix} P \\ Q \end{pmatrix}. \tag{2}$$

Moreover, since the curve has a vertical tangent at q, we have $f_z(0) = g_z(0) = 0$. And according to Assumption (A_2), either $f_{zz}(0)$ or $g_{zz}(0)$ is not zero. Without restriction of generality, assume $\mu := g_{zz}(0) \neq 0$. Up to a scale of the variable z, we can assume that $\mu = 2$. Thus, there exist analytic functions u, v such that f and g are of the form $f(z) = z^2 u(z)$ and $g(z) = z^2(1 + zv(z))$. Letting $\psi : z \mapsto Z := z\sqrt{1 + zv(z)}$, we have $\tilde{Q}(x, y, \psi^{-1}(Z)) = y - Z^2 = 0$. In particular, the function $\tilde{P} = x - z^2 u(z)$ can be rewritten as $\tilde{P}(x, y, \psi^{-1}(Z)) = x - Z^2(s(Z^2) + Zt(Z^2))$ with s and t two real analytic functions. Note that t cannot have all its derivatives vanishing at 0 since otherwise there would be a strictly positive dimensional set of points with two or more antecedents, contradicting Assumption (A_3). Let $k \in \mathbb{N}$ be the valuation of t, i.e. its first non vanishing derivative at 0. Then, there exists t' an analytic function such that $t(Z^2)$ is of the form $Z^{2k}(\eta + Z^2 t'(Z^2))$. The function $\tilde{P}(x, y, \psi^{-1}(Z))$ is of the form $x - Z^2(s(Z^2) + Z^{1+2k}(\eta + Z^2 t'(Z^2)))$. Using $\tilde{Q}$ to substitute $\psi(z)^2$ by y in $\tilde{P}$, there exists a matrix $M_2 := \begin{pmatrix} 1 & e \\ 0 & 1 \end{pmatrix}$ where e is an analytic function, such that:

$$\begin{pmatrix} x - \frac{s(y)}{y} - \psi(z)^{3+2k}(\eta + yt(y)) \\ y - \psi(z)^2 \end{pmatrix} = M_2 \cdot M_1 \cdot \begin{pmatrix} P \\ Q \end{pmatrix}. \tag{3}$$

Finally we recover (1) with:

$$\phi(x, y) = \begin{pmatrix} x - \frac{s(y)}{y} \\ \eta + yt(y) \end{pmatrix}, \quad \psi(z) = z\sqrt{1 + zv(z)}, \quad M = M_1^{-1} \cdot M_2^{-1} \cdot \begin{pmatrix} \frac{1}{\eta + yt(y)} & 0 \\ 0 & 1 \end{pmatrix}. \quad □$$

3   Modeling System

Following the result of Sect. 2, a naive approach to represent the singularities Σ of $\mathcal{C}$ is to use the two following systems.

1. For $(x, y, z_1, z_2) \in B_0 \times \mathbb{R}^2$:

$$P(x, y, z_1) = P(x, y, z_2) = Q(x, y, z_1) = Q(x, y, z_2) = 0 \text{ and } z_1 \neq z_2.$$

2. For $(x, y, z) \in B_0 \times \mathbb{R}$:

$$P(x, y, z) = Q(x, y, z) = P_z(x, y, z) = Q_z(x, y, z) = 0.$$

However, the first system is numerically unstable near the set $z_1 = z_2$ and the second one is over-determined. Instead, we will introduce an unified system. First we define the operators that will be used to construct our system.

3.1 Ball System

Definition 1. *Let $A(x, y, z)$ be a real analytic function. We denote by $S.A$ and $D.A$ the functions:*

$$S.A(x, y, c, r_2) = \begin{cases} \dfrac{1}{2}(A(x, y, c + \sqrt{r_2}) + A(x, y, c - \sqrt{r_2})) & \text{if } r_2 > 0 \\ A(x, y, c) & \text{if } r_2 = 0 \quad (4) \\ \dfrac{1}{2}(A(x, y, c + i\sqrt{-r_2}) + A(x, y, c - i\sqrt{-r_2})) & \text{if } r_2 < 0 \end{cases}$$

$$D.A(x, y, c, r_2) = \begin{cases} \dfrac{1}{2\sqrt{r_2}}(A(x, y, c + \sqrt{r_2}) - A(x, y, c - \sqrt{r_2})) & \text{if } r_2 > 0 \\ A_z(x, y, c) & \text{if } r_2 = 0. \quad (5) \\ \dfrac{1}{2\sqrt{-r_2}}(A(x, y, c + i\sqrt{-r_2}) - A(x, y, c - i\sqrt{-r_2})) & \text{if } r_2 < 0 \end{cases}$$

By abuse of notation, if M is a matrix of real analytic functions, $S.M$ and $D.M$ denote the matrices with the operator applied on each entry.

If A is a real analytic function, then $S.A$ and $D.A$ are also real analytic functions (see Lemma 6). This allows us to introduce the so-called *ball system* that we will use to compute Σ. In this system we map two solutions (x, y, z_1) and (x, y, z_2) of $P = Q = 0$ (or $P = P_z = 0$) to their center (x, y, c) and the square of their radius $r_2 = r^2$, with $r = |z_1 - c| = |z_2 - c|$. Figure 1 illustrates this mapping for singularities of the apparent contour of a torus. Its left part shows the surface $P = 0$, its set of z-critical points $\mathcal{C}_{P \cap P_z}$ and the apparent contour $\mathcal{C} = \Pi_{xy}(\mathcal{C}_{P \cap P_z})$. Its right part shows, for nodes and ordinary cusp singularities, their respective antecedents by Π_{xy}, centers c and radii r.

Lemma 4. *Let S be the set of solutions of the so-called* ball system:

$$\begin{cases} S.P(x, y, c, r_2) = 0 \\ S.Q(x, y, c, r_2) = 0 \\ D.P(x, y, c, r_2) = 0 \\ D.Q(x, y, c, r_2) = 0 \end{cases} \quad (6)$$

in $B_0 \times \mathbb{R} \times \mathbb{R}^+$. Then $\Pi'_{xy}(S) = \Sigma$, where Π'_{xy} is the projection from $\mathbb{R}^4$ to the (x, y)-plane.

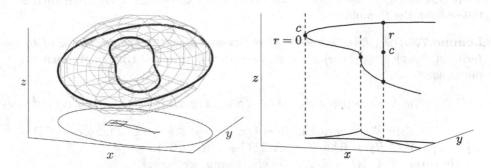

Fig. 1. Left: a torus, in bold line its set of z-critical points, its apparent contour, and the zoom zone corresponding to the right figure. Right: a detail, with antecedents, centers and radius corresponding to singularities.

Proof. According to Sect. 2, the singularity locus of $\mathcal{C}$ is exactly the union of the critical values of Π_{xy} and of the points that have several antecedents. They correspond respectively to the solutions of S such that $r = 0$ and such that $r > 0$.   □

One of the main advantage of this system is that its solutions are regular when the condition (A_5) is satisfied, and thus can be solved using certified numerical algorithms such as homotopy or subdivision methods (see Sect. 4).

Lemma 5. *Under the Assumptions $(A_1) - (A_4)$, all the solutions of the system $S.P = S.Q = D.P = D.Q = 0$ in $B_0 \times \mathbb{R} \times \mathbb{R}^+$ are regular if and only if (A_5) is satisfied.*

The next subsection is dedicated to the proof of this lemma.

3.2 Regularity Condition

Lemma 6. *If A is a real analytic function, then $S.A$ and $D.A$ are real analytic functions. Moreover, the derivatives of $S.A$ with respect to x, y, c, r_2 are respectively $S.A_x, S.A_y, S.A_z, \frac{1}{2} D.A_z$. The derivative of $D.A$ with respect to x, y, c, r_2 are respectively $D.A_x, D.A_y, D.A_z$ and $\frac{S.A_z - D.A}{2r_2}$ if $r_2 > 0$ and $\frac{1}{6} A_{zzz}$ if $r_2 = 0$.*

Proof. First, on a neighborhood of $r_2 > 0$, $S.A$ and $D.A$ are compositions of analytic functions, and thus are analytic. Likewise, for $r_2 < 0$, $S.A$ and $D.A$ are analytic functions, and all the coefficients of their series expansions are real, thus they are real valued analytic functions. Finally, on a neighborhood of $(x, y, c, 0)$, if $A(x, y, c + r) = \sum_{n=0}^{\infty} a_n(x, y, c) r^n$, the series expansions of $S.A$ and $D.A$ for $r_2 < 0$, $r_2 = 0$ and $r_2 > 0$ coincide as:

$$S.A(x, y, c, r_2) = \sum_{n=0}^{\infty} a_{2n}(x, y, c) r_2^n, \quad D.A(x, y, c, r_2) = \sum_{n=0}^{\infty} a_{2n+1}(x, y, c) r_2^n.$$

Thus $S.A$ and $D.A$ are analytic functions. The expressions of their derivatives follow from the formulas.   □

Lemma 7. *If* $\psi : U \subset \mathbb{R}^3 \mapsto V \subset \mathbb{R}^3$ *is an analytic diffeomorphism of the form* $\psi(x,y,z) = (\psi_1(x,y), \psi_2(x,y), \psi_3(x,y,z))$, *so-called* triangular, *then the mapping:*

$$SD.\psi : (x,y,c,r_2) \mapsto (\psi_1(x,y), \psi_2(x,y), S.\psi_3(x,y,c,r_2), r_2(D.\psi_3(x,y,c,r_2))^2)$$

is a real analytic diffeomorphism from $\{(x,y,c,r_2) \in \mathbb{R}^3 \times \mathbb{R}^+ \mid (x,y,c+\sqrt{r_2}) \in U\}$ *to* $\{(X,Y,C,R_2) \in \mathbb{R}^3 \times \mathbb{R}^+ \mid (X,Y,C+\sqrt{R_2}) \in V\}$.
Moreover, if $A : \mathbb{R}^3 \to \mathbb{R}$ *is an analytic map, we have:*

$$S.(A \circ \psi) = (S.A) \circ (SD.\psi)$$
$$D.(A \circ \psi) = (D.A) \circ (SD.\psi) \times D.\psi_3.$$

Proof. According to the previous lemma, $SD.\psi$ is analytic. Moreover, since ψ^{-1} is analytic, $SD.(\psi^{-1})$ is also analytic. Assuming that the inequalities at the end of the lemma are correct, we can use them to check that $SD.(\psi^{-1}) \circ SD.\psi$ is the identity by developing the formula. Such that $SD.\psi$ is a diffeomorphism.

To prove the final identities of the lemma, let $(X,Y,C,R_2) = SD.\psi(x,y,c,r_2)$. We can observe that $\psi_3(x,y,c+\sqrt{r_2}) = C + \sqrt{R_2}$ and $\psi_3(x,y,c-\sqrt{r_2}) = C - \sqrt{R_2}$ by expanding $S.\psi_3 + \sqrt{r_2(D.\psi_3)^2}$ and $S.\psi_3 - \sqrt{r_2(D.\psi_3)^2}$. Using these formula, we can deduce the identities by expanding the right and left hand side of the equalities.   □

Lemma 8. *Let* P, Q *be two analytic functions from* $U \subset \mathbb{R}^3$ *to* $\mathbb{R}$ *and assume that there exist two analytic functions* $\tilde{P}, \tilde{Q}$, *a* 2×2 *invertible matrix of analytic functions and a triangular diffeomorphism* $\phi : U \to V \subset \mathbb{R}^3$ *such that* $\begin{pmatrix} P \\ Q \end{pmatrix} = M \cdot \begin{pmatrix} \tilde{P} \\ \tilde{Q} \end{pmatrix} \circ \phi$. *Then we have:*

$$\begin{pmatrix} S.P \\ S.Q \\ D.P \\ D.Q \end{pmatrix} = \underbrace{\begin{pmatrix} S.M & r_2 D.\phi_3 D.M \\ \\ D.M & D.\phi_3 S.M \end{pmatrix}}_{T} \begin{pmatrix} S.\tilde{P} \\ S.\tilde{Q} \\ D.\tilde{P} \\ D.\tilde{Q} \end{pmatrix} \circ SD.\phi$$

where the matrix T *is invertible, of inverse* $\tilde{T} := \begin{pmatrix} S.M^{-1} & r_2 D.M^{-1} \\ D.M^{-1}/D.\phi_3 & S.M^{-1}/D.\phi_3 \end{pmatrix}$.

Proof. First, using the identity $ab + cd = \frac{1}{2}(a+c)(b+d) + \frac{1}{2}(a-c)(b-d)$, we can deduce:

$$\begin{pmatrix} S.P \\ S.Q \\ D.P \\ D.Q \end{pmatrix} = \begin{pmatrix} S.M & r_2 D.M \\ D.M & S.M \end{pmatrix} \begin{pmatrix} S.(\tilde{P} \circ \phi) \\ S.(\tilde{Q} \circ \phi) \\ D.(\tilde{P} \circ \phi) \\ D.(\tilde{Q} \circ \phi) \end{pmatrix}$$

Finally, expanding the operators in the right hand side vector using the formula in Lemma 7, we prove the desired identity. Finally, since ϕ is a triangular diffeomorphism, we can use the formula of Lemma 7 with $A = (\phi^{-1})_3$ to get

$1 = D((\phi^{-1})_3 \circ \phi) = D.(\phi^{-1})_3 \circ SD.\phi \times D.\phi_3$. In particular, $D.\phi_3$ is never 0 and $\tilde{T}$ is well defined. Expanding $\tilde{T} \cdot T$, we get the identity, such that $\tilde{T}$ is the inverse of T. $\qquad\square$

Corollary 2. *A point p solution of the system $S.P = S.Q = D.P = D.Q = 0$ is regular if and only if the point $SD.\phi(p)$ is regular in the system $S.\tilde{P} = S.\tilde{Q} = D.\tilde{P} = D.\tilde{Q} = 0$.*

Proof. The claim of the lemma can be verified by developing the product vector. For the corollary, it is sufficient to observe that on a point p solution of the system, the Jacobian matrices satisfy the relation:

$$\mathrm{Jac}_p \begin{pmatrix} S.P \\ S.Q \\ D.P \\ D.Q \end{pmatrix}(p) = T \cdot \mathrm{Jac}_{SD.\phi(p)} \begin{pmatrix} S.\tilde{P} \\ S.\tilde{Q} \\ D.\tilde{P} \\ D.\tilde{Q} \end{pmatrix} \cdot \mathrm{Jac}_p(SD.\phi). \qquad\square$$

We have now all the tools necessary to prove Lemma 5.

Proof (of Lemma 5). First, let q be a solution of our system with $r_2 = 0$. Then, according to Lemma 3, there exists an invertible matrix M and a triangular diffeomorphism ϕ such that on a neighborhood of q we have:

$$\begin{pmatrix} P \\ Q \end{pmatrix} = M \cdot \begin{pmatrix} X - Z^{3+2k} \\ Y - Z^2 \end{pmatrix} \circ \Phi(x, y, z).$$

Thus, the point q is regular in the ball system if and only if $(0, 0, 0)$ is regular in the ball system generated by $X - Z^{3+2k}$ and $Y - Z^2$ (Corollary 2). Computing the associated Jacobian matrix, we can check that q is regular if and only if $k = 0$, that is, if and only if its projection p is an ordinary cusp.

Now, let $q = (x, y, c, r_2)$ be a solution of the ball system with $r_2 > 0$. In this case q represents two points $q_1 = (x, y, c + \sqrt{r_2})$ and $q_2 = (x, y, c - \sqrt{r_2})$ of $C_{P \cap Q}$ with the same projection.

According to Lemma 6 the equation $\det \mathrm{Jac}_{(x,y,c,r_2)}(S.P, S.Q, D.P, D.Q) = 0$ can be written

$$\begin{vmatrix} S.P_x & S.P_y & S.P_z & \frac{D.P_z}{2} \\ S.Q_x & S.Q_y & S.Q_z & \frac{D.Q_z}{2} \\ D.P_x & D.P_y & D.P_z & \frac{S.P_z - D.P}{2r_2} \\ D.Q_x & D.Q_y & D.Q_z & \frac{S.Q_z - D.Q}{2r_2} \end{vmatrix} = 0$$

This determinant simplifies using the facts that (a) $D.P = D.Q = 0$ at the solutions, (b) one can multiply lines 3 and 4 by $\sqrt{r_2}$ and column 4 by $2\sqrt{r_2}$, c) one can replace lines ℓ_1, ℓ_3 by $\ell_1 + \ell_3, \ell_1 - \ell_3$ and ℓ_2, ℓ_4 by $\ell_2 + \ell_4, \ell_2 - \ell_4$. The equation is then equivalent to

$$\begin{vmatrix} P_x(q_1) & P_y(q_1) & P_z(q_1) & P_z(q_1) \\ Q_x(q_1) & Q_y(q_1) & Q_z(q_1) & Q_z(q_1) \\ P_x(q_2) & P_y(q_2) & P_z(q_2) & -P_z(q_2) \\ Q_x(q_2) & Q_y(q_2) & Q_z(q_2) & -Q_z(q_2) \end{vmatrix} = 0$$

Expending this expression, one can check that it is equivalent to

$$\begin{vmatrix} P_y(q_1)Q_z(q_1) - P_z(q_1)Q_y(q_1) & P_y(q_2)Q_z(q_2) - P_z(q_2)Q_y(q_2) \\ P_z(q_1)Q_x(q_1) - P_x(q_1)Q_z(q_1) & P_z(q_2)Q_x(q_2) - P_x(q_2)Q_z(q_2) \end{vmatrix} = 0$$

The later expression is equivalent to the condition that projection on the (x, y) plane of the tangent vectors of the 3D curve $\mathcal{C}_{P \cap Q}$ at the points q_1 and q_2 are collinear. Thus in the case where $r_2 > 0$, a solution of the ball system is regular iff it projects to a node.   □

4 Experiments

We propose some quantitative results on the isolation of the singularities of the projection $\mathcal{C}$ of a space real curve $\mathcal{C}_{P \cap Q}$ (or $\mathcal{C}_{P \cap P_z}$ in the case of an apparent contour) by solving the ball system proposed in this paper. We consider here that P and Q are polynomials, hence the equations of the ball system are polynomials and $\mathcal{C}$ admits at most finitely many singularities in $\mathbb{R}^2$. Under our assumptions, the curve $\mathcal{C}'$ defined as the resultant of P and Q with respect to z ($Q = P_z$ in the case of an apparent contour) is the union of $\mathcal{C}$ and a finite set of isolated points. Its singularities can be characterized as real solutions of a bivariate system based on the sub-resultant chain of P and Q (or P_z) (see [14]). We compare the resolution with three state-of-the-art methods of the sub-resultant system, denoted by $\mathcal{S}_2$ in what follows, and the ball system $S.P = S.Q = D.P = D.Q = 0$ defined in Subsect. 3.1, denoted by $\mathcal{S}_4$.

Experimental Data are random dense polynomials P, Q generated with degree d and integer coefficients chosen uniformly in $[\![-2^8, 2^8]\!]$.

Unless explicitly stated, the given running times are averages over five instances for a given degree d.

Testing Environment is a Intel(R) Xeon(R) CPU L5640 @ 2.27GHz machine with Linux.

4.1 Resolution Methods

Gröbner Basis and Rational Univariate Representations allow one to find all real roots of a system of polynomials. The routine `Isolate` of the mathematical software `Maple` implements this approach.

Homotopy Continuation provides all the complex solutions of a system of polynomials and relies on a numerical path-tracking step. Among available open-source software implementing homotopy, we chose `Bertini`[1] notably because it handles both double precision (DP) and an Adaptive Multi-Precision (AMP) arithmetics [4]. This is necessary to prevent the loss of solutions in system $\mathcal{S}_2$ which coefficients are quotients of big integers (see Table 2).

Subdivision uses interval arithmetic (see [20,24,28] for an introduction) to compute for a given system all its regular solutions lying in an initial open

---

[1] https://bertini.nd.edu/.

box $B_0 \subset \mathbb{R}^n$. Here $n = 2$ for system $\mathcal{S}_2$ and $n = 4$ for system $\mathcal{S}_4$. When P, Q are polynomials, the initial box can be $\mathbb{R}^n$ (see [24, p. 210] or [28, p. 233]). Otherwise, B_0 is bounded, and the number of singularities is finite. Since we focus on singularities induced by projection of real parts of the curve $\mathcal{C}_{P \cap Q}$ or $\mathcal{C}_{P \cap P_z}$, we did only research solutions of the ball system having $r_2 \geq 0$. We implemented a subdivision solver in c++, using the boost or mpfi interval arithmetic library. The implementation is described with more details in [14].

4.2   Singularities Isolation: Comments on Tables 1, 2 and 3

Tables 1, 2 and 3 report the sequential running times (columns t) in seconds to compute the singularities of projection and apparent contour curves, using system $\mathcal{S}_2$ or system $\mathcal{S}_4$ to represent their singularities.

Table 1 shows that for Isolate running times are better when solving system $\mathcal{S}_2$, due to its lower number of variables.

Table 2 refers to resolution with Bertini, using DP and AMP arithmetics. In addition to running times, it reports the number of missed solutions (columns Mis. Sols.) when using DP arithmetic. The resolution by homotopy in DP of system $\mathcal{S}_2$ is not satisfactory due to the high number of missed solutions. The use of AMP arithmetic resolves this problem: for all systems we tested, all solutions were found. But it induces an important additional cost. System $\mathcal{S}_4$ seems better suited to homotopy resolution. In DP arithmetic, fewer solutions are missed and the cost of AMP arithmetic is more acceptable. Notice however that for three examples, a solution was missed both with DP and AMP arithmetic due to the truncation of a path considered as converging to a solution at infinity.

Table 3 reports results obtained with our implementation of subdivision. For a given degree, resolution times are subject to an important variance. For low degrees it is more efficient to solve system $\mathcal{S}_2$ than system $\mathcal{S}_4$ due to the higher dimension (i.e. 4 instead of 2) of the research space in the latter case. The difference of running times decreases when d increases, due to the size (in terms of degree, number of monomials and bit-size of coefficients) of the resultant and sub-resultant polynomials that have to be evaluated to solve system $\mathcal{S}_2$.

Table 1. Isolating singularities of projection and apparent contour curves with the routine Isolate of Maple. Input polynomials have degree d. The running times are in seconds. (a) Fails with error.

	Projection		Apparent contour	
	system $\mathcal{S}_2$	system $\mathcal{S}_4$	system $\mathcal{S}_2$	system $\mathcal{S}_4$
d	t	t	t	t
4	1.321	4.293	0.206	0.1874
5	26.92	100.4	5.439	6.501
6	(a)	(a)	98.59	155.8
7	(a)	(a)	(a)	(a)

Table 2. Isolating singularities of projection and apparent contour curves with `Bertini` using DP and AMP arithmetic. Input polynomials have degree d. The running times are in seconds. (b) Has been run on a unique example. (c) Solution(s) is (are) missing due to infinite path(s) truncation.

	Bertini with DP arithmetic											
	Projection						Apparent contour					
	system S_2		system S_4				system S_2		system S_4			
d	t	Mis. Sols.	t	Mis. Sols.			t	Mis. Sols.	t	Mis. Sols.		
4	0.864	0	1.376	1		(c)	0.174	0	0.46	1		
5	16.03	3	8.326	0			3.638	0	3.818	2		(c)
6	177.6	2	40.21	0			54.49	1	20.80	1		
7	1458	193	152.1	1		(c)	617.9	6	88.50	0		
8	≥ 3000	599	(b)	508.5	3		2799	885	319.3	0		
9	≥ 3000	1389	(b)	1429	7		≥ 3000	1178	(b)	935.6	2	

	Bertini with AMP arithmetic			
	Projection		Apparent contour	
	system S_2	system S_4	system S_2	system S_4
d	t	t	t	t
4	2.332	1.804 (c)	2.332	1.434
5	147.8	13.888	147.852	15.01 (c)
6	≥ 3000	123.41	1005	165.7
7	≥ 3000	1089 (c)	≥ 3000	1147
8	≥ 3000	≥ 3000	≥ 3000	≥ 3000

Table 3. Isolating singularities of projection and apparent contour curves with subdivision. Input polynomials have degree d. The average running times t are given in seconds together with the standard deviation σ.

	Projection		Apparent contour	
	system S_2	system S_4	system S_2	system S_4
d	$t \pm \sigma$	$t \pm \sigma$	$t \pm \sigma$	$t \pm \sigma$
4	0.078 ± 0.03	0.759 ± 0.02	0.040 ± 0.02	1.509 ± 1.97
5	0.351 ± 0.13	1.973 ± 0.72	0.251 ± 0.23	25.34 ± 47.5
6	1.918 ± 0.55	6.442 ± 3.07	1.353 ± 0.57	11.38 ± 6.98
7	9.528 ± 3.92	22.43 ± 8.36	124.1 ± 142	54.21 ± 50.3
8	42.69 ± 16.8	57.00 ± 16.4	57.72 ± 63.7	99.22 ± 89.3
9	163.3 ± 111	137.5 ± 93	54.74 ± 33.3	95.11 ± 44.5

5 Conclusion

Given an analytic curve $\mathcal{C}_{P \cap Q}$ satisfying some specific generic assumptions, we have described the different possible types of singularities Σ of its projection $\mathcal{C} = \Pi_{xy}(\mathcal{C}_{P \cap Q})$. Moreover we have shown that these singularities can be computed as the regular solutions of a new so-called *ball system*.

Even if our characterization increases the number of variables of the system to solve in order to compute Σ, we have shown with experimental results that the ball system can be solved with numerical methods. With homotopy it is more often complete and faster to solve the latter system than the sub-resultant system. A certified resolution is provided by a subdivision solver. In term of computational cost, such solvers are known to suffer from the increase of the dimension of the research space. However for high degrees of input polynomials, the price to pay for solving the sub-resultant system seems higher than the one induced by the increasing of number of variables.

References

1. Alberti, L., Mourrain, B., Técourt, J.P.: Isotopic triangulation of a real algebraic surface. J. Symb. Comput. **44**(9), 1291–1310 (2009)
2. Arnold, V.I., Varchenko, A., Gusein-Zade, S.: Singularities of Differentiable Maps: Volume I: The Classification of Critical Points Caustics and Wave Fronts. Springer, Heidelberg (1988)
3. Bank, B., Giusti, M., Heintz, J., Lecerf, G., Matera, G., Solernó, P.: Degeneracy loci and polynomial equation solving. Found. Comput. Math. **15**(1), 159–184 (2015). http://dx.doi.org/0.1007/s10208-014-9214-z
4. Bates, D.J., Hauenstein, J.D., Sommese, A.J., Wampler, C.W.: Adaptive multi-precision path tracking. SIAM J. Numer. Anal. **46**(2), 722–746 (2008)
5. Berberich, E., Kerber, M., Sagraloff, M.: An efficient algorithm for the stratification and triangulation of an algebraic surface. Comput. Geom. Theory Appl. **43**(3), 257–278 (2010)
6. Cheng, J., Lazard, S., Peñaranda, L., Pouget, M., Rouillier, F., Tsigaridas, E.: On the topology of real algebraic plane curves. Math. Comput. Sci. **4**, 113–137 (2010)
7. Daouda, D.N., Mourrain, B., Ruatta, O.: On the computation of the topology of a non-reduced implicit space curve. In: Proceedings of the Twenty-First International Symposium on Symbolic and Algebraic Computation, ISSAC 2008, pp. 47–54. ACM, New York (2008)
8. Dedieu, J.: Points fixes, zéros et la méthode de Newton. Mathématiques et Applications. Springer, Heidelberg (2006)
9. Delanoue, N., Lagrange, S.: A numerical approach to compute the topology of the apparent contour of a smooth mapping from R^2 to R^2. J. Comput. Appl. Math. **271**, 267–284 (2014)
10. Demazure, M.: Bifurcations and Catastrophes: Geometry of Solutions to Nonlinear Problems. Universitext. Springer, Heidelberg (2000). École Polytechnique
11. El Kahoui, M.: Topology of real algebraic space curves. J. Symb. Comput. **43**(4), 235–258 (2008)
12. Giusti, M., Lecerf, G., Salvy, B., Yakoubsohn, J.C.: On location and approximation of clusters of zeros: Case of embedding dimension one. Found. Comput. Math. **7**(1), 1–58 (2007). http://dx.doi.org/10.1007/s10208-004-0159-5

13. Hauenstein, J.D., Mourrain, B., Szanto, A.: Certifying isolated singular points and their multiplicity structure. In: Proceedings of the 2015 ACM on International Symposium on Symbolic and Algebraic Computation, ISSAC 2015, pp. 213–220. ACM, New York (2015).http://doi.acm.org/10.1145/2755996.2756645

14. Imbach, R., Moroz, G., Pouget, M.: Numeric certified algorithm for the topology of resultant and discriminant curves. Research Report RR-8653. Inria, April 2015

15. Kearfott, R., Xing, Z.: An interval step control for continuation methods. SIAM J. Numer. Anal. **31**(3), 892–914 (1994)

16. Krantz, S.G., Parks, H.R.: A Primer of Real Analytic Functions. Basler Lehrbücher, vol. 4. Birkhäuser Basel, Boston (1992)

17. Leykin, A., Verschelde, J., Zhao, A.: Newton's method with deflation for isolated singularities of polynomial systems. Theoret. Comput. Sci. **359**(13), 111–122 (2006)

18. Mantzaflaris, A., Mourrain, B.: Deflation and certified isolation of singular zeros of polynomial systems. In: Proceedings of the 36th International Symposium on Symbolic and Algebraic Computation, ISSAC 2011, pp. 249–256. ACM, New York (2011). http://doi.acm.org/10.1145/1993886.1993925

19. Martin, B., Goldsztejn, A., Granvilliers, L., Jermann, C.: Certified parallelotope continuation for one-manifolds. SIAM J. Numer. Anal. **51**(6), 3373–3401 (2013)

20. Moore, R.E., Kearfott, R.B., Cloud, M.J.: Introduction to Interval Analysis. Society for Industrial and Applied Mathematics, Philadelphia (2009)

21. Morgan, A.: Solving Polynominal Systems Using Continuation for Engineering and Scientific Problems. Society for Industrial and Applied Mathematics, Philadelphia (2009)

22. Mourrain, B., Pavone, J.: Subdivision methods for solving polynomial equations. J. Symbolic Comput. **44**(3), 292–306 (2009). http://www.sciencedirect.com/science/article/pii/S0747717108001168

23. Mourrain, B., Pion, S., Schmitt, S., Técourt, J.P., Tsigaridas, E.P., Wolpert, N.: Algebraic issues in computational geometry. In: Boissonnat, J.D., Teillaud, M. (eds.) Effective Computational Geometry for Curves and Surfaces, chap. 3. Mathematics and Visualization, pp. 117–155. Springer, Heidelberg (2006)

24. Neumaier, A.: Interval Methods for Systems of Equations. Cambridge University Press, Cambridge (1990)

25. Ojika, T., Watanabe, S., Mitsui, T.: Deflation algorithm for the multiple roots of a system of nonlinear equations. J. Math. Anal. Appl. **96**(2), 463–479 (1983)

26. Plantinga, S., Vegter, G.: Isotopic approximation of implicit curves and surfaces. In: Eurographics/ACM SIGGRAPH Symposium on Geometry Processing, SGP 2004, pp. 245–254 (2004)

27. Rouillier, F.: Solving zero-dimensional systems through the rational univariate representation. J. Appl. Algebra Eng. Commun. Comput. **9**(5), 433–461 (1999)

28. Stahl, V.: Interval Methods for Bounding the Range of Polynomials and Solving Systems of Nonlinear Equations. Ph.D. thesis, Johannes Kepler University, Linz (1995)

29. Whitney, H.: On singularities of mappings of euclidean spaces. I. mappings of the plane into the plane. Ann. Math. **62**(3), 374–410 (1955)

Linear k-Monotonicity Preserving Algorithms and Their Approximation Properties

S.P. Sidorov[✉]

Department of Mechanics and Mathematics, Saratov State University, Saratov,
Russian Federation
SidorovSP@info.sgu.ru

Abstract. This paper examines the problem of finding the linear algorithm (operator) of finite rank n (i.e. with a n-dimensional range) which gives the minimal error of approximation of identity operator on some set over all finite rank n linear operators preserving the cone of k-monotonicity functions. We introduce the notion of linear relative (shape-preserving) n-width and find asymptotic estimates of linear relative n-widths for linear operators preserving k-monotonicity in the space $C^k[0,1]$. The estimates show that if linear operator with finite rank n preserves k-monotonicity, the degree of simultaneous approximation of derivative of order $0 \leq i \leq k$ of continuous functions by derivatives of this operator cannot be better than n^{-2} even on the set of algebraic polynomials of degree $k + 2$ (as well as on bounded subsets of Sobolev space $W_\infty^{(k+2)}[0,1]$).

1 Introduction

Different applications of computer-aided geometric design require to approximate functions with preservation of such properties as monotonicity, convexity, concavity and the like. The part of approximation theory that deals with this type of problem is known as the theory of shape preserving approximation. Over the past 30 years extensive study in the theory of shape-preserving approximation has brought about new results, the most substantial of which were outlined in [8,14,16].

The interest to the theory of shape-preserving approximation is caused primarily by the fact that its results have a number of applications, most of which relate to the use in computer-aided graphical design (CAGD) for which the preservation of shape of graphics is essential. CAGD often considers the task of creating a complex shape of the body surface (e.g., the fuselage of the aircraft, engine parts, architectural structures) as a discrete set of points. To represent the body, it is necessary to arrange these points on a curve or surface. A change in the derivative sign or any discontinuities of the first and even second derivative are visible to the human eye. For this reason, not without interest is smooth

The results were obtained within the framework of the state task of Russian Ministry of Education and Science (project 1.1520.2014K).

I.S. Kotsireas et al. (Eds.): MACIS 2015, LNCS 9582, pp. 93–106, 2016.
DOI: 10.1007/978-3-319-32859-1_7

approximation which retains the shape of the data. Spline methods of approximation inheriting such geometric properties were considered, in particular, in the works [17,24,31–33]. The review can be found in the book [23].

If function f from X has a shape property, it usually means that element f belongs to a certain cone in X. For example, if we would like to approximate a curve f from $C[0,1]$ with the preservation of monotonicity, then we should find an approximating function from the cone of all non-decreasing functions defined on [0,1], i.e. from the cone $\Delta^1 = \{f \in C[0,1] : f(x_1) \leq f(x_2) \text{ for any } 0 \leq x_1 < x_2 \leq 1\}$.

Let X be a normed linear space, V be a cone in X (a convex set, closed under nonnegative scalar multiplication). We will say that $f \in X$ has a shape in the sense of V whenever $f \in V$. Let X_n be a n-dimensional subset of X and $A \subset X$ be a set, $A \cap V \neq \varnothing$. Classical problems of approximation theory are of interest in the theory of shape-preserving approximation; they include estimation of (nonlinear) relative n-widths of $A \cap V$ with the constraint V in X

$$d_n(A \cap V, V)_X = \inf_{X_n} \sup_{f \in A \cap V} \inf_{g \in X_n \cap V} \|f - g\|_X, \tag{1}$$

the left-most infimum is taken over all affine subsets X_n of dimension $\leq n$, such that $X_n \cap V \neq \varnothing$.

The notion of relative n-width (1) was introduced by Konovalov in 1984 [13] and some estimates of relative shape-preserving n-widths have been obtained in papers [9,11,12]. A good introduction of n-widths can be found in [20].

In CAGD it is often necessary to find shape-preserving representations of curves (functions from a linear normed space X) with small errors of approximation in X, which can be easily treated on a computer. These representations could be taken from a parametric family of simple functions with a set of parameters which can be varied by computers to make changes in the curve. It can be achieved by using linear combinations of functions from a given n-dimensional subspace X_n of X with coefficients as the set of parameters. In applications X_n is often chosen to be either the set of all algebraic polynomials of degree $n-1$ or the n-dimensional subspace of algebraic splines. The main task is to construct algorithms that calculate these parameters automatically. Moreover, one is often interested in *linear* approximation algorithms which may be more practical and easier to calculate than best approximations which are typically non-linear.

The input of such linear approximation algorithms is usually the vector of the values of the approximated function at a certain finite number of points, while the resulting functions are from a given finite dimensional linear subspace of X. One of the most well-known algorithms of such type is the Bernstein operator $B_n f(x) := \sum_{i=0}^{n} \binom{n}{i} x^i (1-x)^{n-i} f(\frac{i}{n})$ for which the values of the function f at $n+1$ equidistant points in [0,1] determine the result $B_n f$ of operating on that function, which is from the $n+1$-dimensional subspace of all algebraic polynomials of degree $\leq n$.

Recall that a linear operator mapping X into a linear space of finite dimension n is called an *operator of finite rank n*. Bernstein operator B_n is the operator of rank $n + 1$.

Let A be a subset of X and $L : X \to X$ be a linear operator. The value

$$e(A, L) := \sup_{f \in A} \|f - Lf\|_X = \sup_{f \in A} \|(I - L)f\|_X$$

is the error of approximation of the identity operator I by the operator L on the set A.

Let $L : X \to X$ be a linear operator and V be a cone in X, $V \neq \varnothing$. We will say that the operator L preserves the shape in the sense of V, if $L(V) \subset V$. One might consider the problem of finding (if it exists) a linear operator of finite rank n which gives the minimal error of approximation of the identity operator on some set over all finite rank n linear operators preserving the shape in the sense of V. This leads us naturally to the notion of *linear* relative n-width. We introduce the definition of linear relative width based on Konovalov's ideas. A different definition of linear relative width based on the ideas of Korovkin was given in [27].

Let $V \subset X$ be a cone and $A \subset X$ be a set, $A \cap V \neq \varnothing$. Let us define *Konovalov linear relative n–width* of the set $A \cap V$ in X with the constraint V by

$$\delta_n(A \cap V, V)_X := \inf_{L_n(V) \subset V} \sup_{f \in A \cap V} \|f - L_n f\|_X,$$

where infimum is taken over all linear continuous operators $L_n : X \to X$ of finite rank n preserving the shape in the sense V, i.e. satisfying $L_n(V) \subset V$.

Estimation of linear relative n-widths is of interest in the theory of shape-preserving approximation as, knowing the value of relative linear n-width, we can judge the quality of approximation (in terms of optimality) this or that finite-dimensional method with shape-preserving property $L_n(V) \subset V$ is. The importance of relative linear widths is connected with the following property: if $L_n : X \to X$ is a linear continuous operator of finite rank at most n, such that $L_n(V) \subset V$, then

$$\sup_{f \in A \cap V} \|f - L_n f\|_X \geq \delta_n(A \cap V, V)_X.$$

In the other words, linear relative n-width $\delta_n(A, V)_X$ provides a lower bound on the degree of approximation of any linear operator L of finite rank n preserving the shape in the sense of cone V. Thus, the value of $\delta_n(A \cap V, V)_X$ tells us how well a given linear operator L approximates functions from $A \cap V$ relative to the theoretical lower bound.

A continuous function $f : [0, 1] \to \mathbb{R}$ is said to be k-monotone, $k \geq 1$, on $[0, 1]$ if and only if for all choices of $k + 1$ distinct $t_0, \dots, t_k$ in $[0, 1]$ the inequality $[t_0, \dots, t_k]f \geq 0$ holds, where $[t_0, \dots, t_k]f$ denotes the k-th divided difference of f at $0 \leq t_0 < t_1 < \dots < t_k \leq 1$. Let Δ^k denote the set of all k-monotone functions defined on [0,1]. Note that 2-monotone functions are just convex functions and 1-monotone functions are non-decreasing functions.

Despite the successful and active development of the shape-preserving approximation theory as well as a large amount of publications on the topic, the subject of the *linear* shape-preserving approximation is still insufficiently studied. In particular, the problems of the quantitative estimation of the convergence rate for linear approximation methods with shape preservation are not fully covered in the literature. In this paper we will try to fill this gap.

It is well-known [21] that Bernstein operator B_n preserves k-monotonicity for any $k \geq 0$ and has the convergence rate n^{-1} on the set of all twice differentiable functions [7]. In this paper we will find asymptotic estimates of linear relative n-widths of subsets of Sobolev-type spaces in the space $C^k[0,1]$ with the constraint Δ^k. It enables us to investigate how "bad" Bernstein operator is (in terms of the rate of convergence), compared with the best possible rate of convergence determined by the value of the Konovalov linear relative n–width with the constraint Δ^k.

2   The Example of Preservation k-Monotonocity

Denote by $C^k[0,1]$, $k \geq 0$, the space of all real-valued and k-times continuously differentiable functions defined on $[0,1]$, equipped with the norm

$$\|f\|_{C^k[0,1]} = \sum_{0 \leq i \leq k} \frac{1}{i!} \sup_{x \in [0,1]} |D^i f(x)|, \tag{2}$$

where D^i denotes the i-th differential operator, $D^i f(x) = d^i f(x)/dx^i$, and $D^0 = I$ is the identity operator, and the derivatives are taken from the right at 0 and from the left at 1. If $f \in C^k[0,1]$, then $f \in \Delta^k$ iff $f^{(k)}(t) \geq 0$, $t \in [0,1]$.

It is said that a linear operator L of $C[0,1]$ into $C[0,1]$ preserves k-monotonicity, if $L(\Delta^k) \subset \Delta^k$, i.e. a linear operator (algorithm) L preserves k-monotonicity if for each k-monotone function f the resulting function Lf is also k-monotone.

Denote by $B^k[0,1]$, $k \geq 0$, the space of all real-valued functions, whose k-th derivative is bounded on $[0,1]$ endowed with the sup-norm (2).

Let $W_\infty^{(k+2)}[0,1]$ be the Sobolev space of all real-valued, $(k + 1)$-times differentiable functions whose derivative of order $(k + 1)$ is absolutely continuous and whose derivative of order $k + 2$ is in $L^\infty[0,1]$, $\|f\|_\infty := \text{ess sup}_{x \in [0,1]} |f(x)|$. Denote $B_\infty^{(k+2)} := \{f \in W_\infty^{(k+2)}[0,1] : \|D^{k+2}f\|_\infty \leq 1\}$.

One of the main example of shape-preserving operator is Bernstein operator. T. Popoviciu [21] proved that if f is k-monotone on [0,1], then Bernstein polynomial $B_n f(x) := \sum_{i=0}^{n} \binom{n}{i} x^i (1-x)^{n-i} f(\frac{i}{n})$ also is monotone of order k on [0,1]. The papers [1,4–6,22] investigate the shape preserving and convergence properties of sequences of linear Bernstein-type operators. On the other hand, it is well-known that one of the shortcomings for Bernstein-type approximation is the low order of approximation [7]. In this section we present an operator with the higher order of approximation n^{-2}.

Let $k, n \in \mathbb{N}$, $n \geq k+2$, $z_j = j/n$, $j = 0, 1, \ldots, n$, and let $\Lambda_{k,n} : C^k[0,1] \to C^k[0,1]$ be the linear operator defined in steps from left to right by (see also [25])

$$\Lambda_{k,n} f(x) = \sum_{l=0}^{k} \frac{D^l f(0)}{l!} x^l + \frac{nx^{k+1}}{(k+1)!} \left[D^k f(z_1) - D^k f(0) \right], \quad x \in [0, z_1], \qquad (3)$$

$$\Lambda_{k,n} f(x) = \sum_{l=0}^{k} \frac{D^l \Lambda_{k,n} f(z_j)}{l!} (x - z_j)^l$$
$$+ \frac{n(x - z_j)^{k+1}}{(k+1)!} [D^k f(z_{j+1}) - D^k f(z_j)],$$
$$x \in (z_j, z_{j+1}], \ j = 1, \ldots, n-1. \qquad (4)$$

In the simplest case $k = 0$, the cone Δ^0 is the cone of all non-negative functions defined on [0,1] and $\Lambda_{0,n}$ is defined by

$$\Lambda_{0,n} f(x) = f(0) + nx \left(f\left(\frac{1}{n} \right) - f(0) \right), \quad x \in [0, 1/n],$$

$$\Lambda_{0,n} f(x) = \Lambda_{0,n} f\left(\frac{j}{n} \right) \left(x - \frac{j}{n} \right)$$
$$+ n \left(x - \frac{j}{n} \right) \left(f\left(\frac{j+1}{n} \right) - f\left(\frac{j}{n} \right) \right), \quad x \in \left(\frac{j}{n}, \frac{j+1}{n} \right], \ j = 1, \ldots, n-1.$$

The resulting function $\Lambda_{0,n} f$ is a piecewise linear function on [0,1] with breakpoints $(j/n, f(j/n))$, $j = 0, \ldots, n$, and linear operator $\Lambda_{0,n}$ preserves positivity of approximated functions (see Fig. 1), i.e. $\Lambda_{0,n}$ is a linear positive operator.

Lemma 1. $\Lambda_{k,n} : C^k[0,1] \to C^k[0,1]$ *is a continuous linear operator of finite rank $n + k + 1$, such that*

1. $\Lambda_{k,n}(\Delta^k) \subset \Delta^k$;
2. *there exists a constant $0 < c \leq 2^{k-3}/k!$ not depending on n such that*

$$\sup_{f \in B_\infty^{(k+2)} \cap \Delta^k} \| \Lambda_{k,n} f - f \|_{C^k[0,1]} \leq cn^{-2}. \qquad (5)$$

Proof. Since $D^k(\Lambda_{k,n} f)$ is a piecewise linear function on [0,1] with the set of breakpoints $\{(z_j, D^k f(z_j))\}_{j=0,\ldots,n}$, then for every $f \in C^k[0,1]$ such that $D^k f \geq 0$ the inequality $D^k(\Lambda_{k,n} f) \geq 0$ holds, i.e. $\Lambda_{k,n}(\Delta^k) \subset \Delta^k$.

Denote $e_i(x) = x^i$, $i = 0, 1, \ldots$. It can be easily verified that $\Lambda_{k,n} e_p = e_p$ for all $p = 0, 1, \ldots, k+1$ and if $x \in [z_j, z_{j+1})$ for some $0 \leq j \leq n-1$, then

$$(D^k(\Lambda_{k,n} e_{k+2}) - D^k e_{k+2})(x) = (k+2)! (z_{j+1} - x)(x - z_j)/2.$$

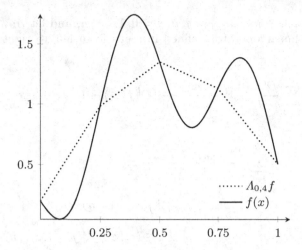

Fig. 1. Function $f(x) = 0.2 + sin(4\pi x) + sin(\pi x) + 0.3x$ (solid line) and the resulting function $\Lambda_{k,n}f$, $n = 4$, $k = 0$, (dotted line) defined on [0,1]

Let f be a function from $B_\infty^{(k+2)} \cap \Delta^k$. Let $x \in [z_j, z_{j+1}]$. Then $D^k f \in W_\infty^{(2)}[0,1]$ can be represented as

$$D^k f(x) = D^k f(z_j) + \frac{D^{k+1} f(z_j)}{1!}(x - z_j) + \int_{z_j}^1 (x - t)_+ D^{k+2} f(t)\, dt. \qquad (6)$$

where $y_+ := \max\{y, 0\}$. Similarly, if $x \in [z_j, z_{j+1}]$ then

$$D^k(\Lambda_{k,n}f)(x) = D^k(\Lambda_{k,n}f)(z_j) + \frac{D_+^{k+1}\Lambda_{k,n}f(z_j)}{1!}(x - z_j)$$
$$+ \int_{z_j}^1 (x - t)_+ D^{k+2}\Lambda_{k,n}f(t)\, dt, \qquad (7)$$

where $D_+^{k+1}\Lambda_{k,n}f(z_j)$ is the right-hand side derivative of $D^k\Lambda_{k,n}f$ at point z_j. It follows from (6) and (7) that if $x \in [z_j, z_{j+1}]$ then

$$\left(D^k(\Lambda_{k,n}f) - D^k f\right)(x) = (x - z_j)\left[n\left(D^k f(z_{j+1}) - D^k f(z_j)\right) - D^{k+1}f(z_j)\right]$$
$$- \int_{z_j}^1 (x - t)_+ D^{k+2}f(t)\, dt = \int_{z_j}^1 \left(n(x - z_j)(z_{j+1} - t)_+ - (x - t)_+\right) D^{k+2}f(t)\, dt.$$

Since $\|D^{k+2}f\|_\infty \le 1$, we have

$$\sup_{x \in [z_j, z_j]} \left|D^k(\Lambda_{k,n}f)(x) - D^k f(x)\right|$$

$$\le \sup_{x \in [0, \frac{1}{n}]} \int_0^{\frac{1}{n}} \left|nx\left(\frac{1}{n} - t\right)_+ - (x - t)_+\right| dt \le \sup_{x \in [0, \frac{1}{n}]} \frac{1}{2}x\left(\frac{1}{n} - x\right) = \frac{1}{8n^2}. \qquad (8)$$

It follows from (8) that

$$\left|D^k(\Lambda_{k,n}f)(x) - D^k f(x)\right| \le \frac{1}{8n^2} \text{ for every } x \in [0,1].$$

Since $D^i(\Lambda_{k,n}f - f)(0) = 0$ for all $i = 0,\dots,k$, we have by induction for $i = k-1,\dots,0$ and $x \in [0,1]$

$$|D^i(\Lambda_{k,n}f - f)(x)| = \left| D^i(\Lambda_{k,n}f - f)(0) + \int_0^x D^{i+1}(\Lambda_{k,n}f - f)(t)\,dt \right|$$

$$\le \frac{1}{8n^2} \frac{x^{k-i}}{(k-i)!}. \tag{9}$$

We have used the fact that if $g \in C[0,1]$ and there exists a constant $a \in \mathbb{R}$ such that $|g| \le a$ on [0,1], then

$$0 \le \int_0^x \int_0^{t_{p-1}} \dots \int_0^{t_1} |g(t_1)|\,dt_1 \dots dt_p \le a\frac{x^p}{p!}$$

for every $p \in \mathbb{N}$.

Then (9) implies $\|D^i(\Lambda_{k,n}f) - D^i f\|_{C[0,1]} \le \frac{1}{8n^2} \frac{1}{(k-i)!}$ and Lemma is proved with $c \le \frac{1}{8} \sum_{i=0}^k \frac{1}{i!(k-i)!} = 2^{k-3}/k!$.   □

Note that linear operator $\Lambda_{k,n}$ defined in (3) and (4) is the minimal shape-preserving projection [18] on the first interval $[0, \frac{1}{n}]$, and then it is smoothly extended to the next intervals. The paper [2] presents the example of linear finite-dimensional approximation method that preserves k-monotonicity of approximated functions and uses the values of function at equidistant points on [0,1] (rather than values of derivatives as it is in the definition of $\Lambda_{k,n}$).

Figure 2 plots the comparison for errors of approximation of exponential function $f(x) = e^x$ on interval [0,1] by Bernstein operator B_n and operator $\Lambda_{k,n}$ for different n and $k = 1$. Line (1) of the plot is $B_{10}f - f$, line (2) of Fig. 2 is the error $B_{20}f - f$, lines (3) and (4) plot the differences $\Lambda_{1,10}f - f$ and $\Lambda_{1,20}f - f$ respectively.

3   The Main Result

We need the preliminary lemma (see [26]).

Lemma 2. *Let $\Phi : C^k[0,1] \to R$ be a linear functional that has the following property: $\Phi(f) \ge 0$ for every $f \in C^k[0,1]$ such that $f \in \Delta^k$. Let $\langle \cdot, \cdot \rangle : C^k[0,1] \times C^k[0,1] \to \mathbb{R}$ be the bi-functional generated by a functional Φ in the following way: for every $f,g \in C^k[0,1]$ we suppose $\langle f,g \rangle = \Phi(h)$ with $h \in C^k[0,1]$ so that $D^k h = D^k f D^k g$ and $D^i h(0) = 0$, $i = 0,1,\dots,k-1$. Then*

$$|\langle f,g \rangle| \le [\langle f,f \rangle]^{\frac{1}{2}} [\langle g,g \rangle]^{\frac{1}{2}}, \quad f,g \in C^k[0,1]. \tag{10}$$

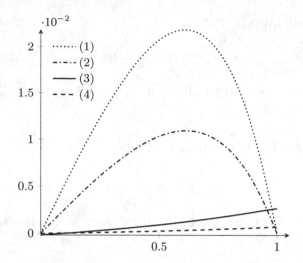

Fig. 2. Errors of approximation of function $f(x) = \exp(x)$ on $[0,1]$ by (1) Bernstein operator B_n, $n = 10$; (2) Bernstein operator B_n, $n = 20$; (3) operator $\Lambda_{k,n}$, $n = 10$, $k = 1$; (4) operator $\Lambda_{k,n}$, $n = 20$, $k = 1$

Using ideas of [29] let us prove the analogue of the main theorem in [25] with the omitted requirement $D^k L_n e_k = D^k e_k$, $e_k(x) := x^k$.

Lemma 3. *Let* $L_n : C^k[0,1] \to B^k[0,1]$ *be a linear operator of finite rank* n, $n > k+2$, *such that*

$$L_n(\Delta^k) \subset \Delta^k. \tag{11}$$

Then

$$\sup_{x\in[0,1]} \left(\frac{2}{(k+2)!} \left| D^k L_n e_{k+2}(x) - D^k e_{k+2}(x) \right| \right.$$

$$+ \frac{2}{(k+1)!} \left| D^k L_n e_{k+1}(x) - D^k e_{k+1}(x) \right|$$

$$\left. + \frac{1}{k!} \left| D^k L_n e_k(x) - D^k e_k(x) \right| \right) \geq \frac{1}{4n^2} \left(1 - \frac{1}{n^2} \right). \tag{12}$$

Proof. Let $\{v_1, \ldots, v_n\}$ be the system of functions generating the linear space $\{D^k L_n f : f \in C^k[0,1]\}$, i.e. $\mathrm{span}\{v_1, \ldots, v_n\} = \{D^k L_n f : f \in C^k[0,1]\}$. Consider the matrix $A = (v_j(z_i))_{j=1,\ldots,n}^{i=0,\ldots,n}$, where $z_i = i/n$, $i = 0, \ldots, n$.. We will assume that the rank of matrix A does not equal to 0, $\mathrm{rank}\, A \neq 0$. Indeed, if $\mathrm{rank}\, A = 0$, then $D^k L_n f(z_i) = \sum_{j=1}^{n} a_j(f) v_j(z_i) = 0$, $i = 0, \ldots, n$, for every $f \in C^k[0,1]$, and consequently $\|D^k L_n e_k - D^k e_k\|_{B[0,1]} \geq 1$. Therefore we will exclude the case $\mathrm{rank}\, A = 0$ from our analysis.

Take a non-trivial vector $\delta = (\delta_0, \ldots, \delta_n) \in R^{n+1}$, such that

$$\sum_{i=0}^{n} |\delta_i| = 1, \quad \sum_{i=0}^{n} \delta_i v_j(z_i) = 0, \quad j = 1, \ldots, n.$$

Let $h \in C^k[0, 1]$ be such that

1. $D^k h(z_i) = \operatorname{sgn} \delta_i, \; i = 0, \ldots, n;$
2. $D^k h$ is linear on each $[z_0, z_1], \ldots, [z_{n-1}, z_n];$
3. $D^i h(0) = 0, \; i = 0, 1, \ldots, k - 1.$

It follows from $D^k L_n h \in \operatorname{span} \{v_1, \ldots, v_n\}$ that

$$\sum_{i=0}^{n} \delta_i D^k L_n h(z_i) = 0.$$

Then

$$1 = \sum_{i=0}^{n} |\delta_i| = \sum_{i=0}^{n} \delta_i D^k h(z_i) = \sum_{i=0}^{n} \delta_i (D^k h(z_i) - D^k L_n h(z_i))$$

$$\leq \sum_{i=0}^{n} |\delta_i| |D^k L_n h(z_i) - D^k h(z_i)| \leq \|D^k L_n h - D^k h\|_{B[0,1]}. \qquad (13)$$

For $x \in [0, 1]$ we have

$$|D^k L_n h(x) - D^k h(x)| = \left| D^k L_n h(x) - D^k h(x) \frac{1}{k!} D^k L_n e_k(x) \right.$$

$$\left. + D^k h(x) \frac{1}{k!} D^k L_n e_k(x) - D^k h(x) \frac{1}{k!} D^k e_k(x) \right|$$

$$\leq \left| D^k L_n \left(h - D^k h(x) \frac{1}{k!} e_k \right)(x) \right|$$

$$+ \frac{1}{k!} |D^k h(x)| |D^k L_n e_k(x) - D^k e_k(x)|. \qquad (14)$$

Let $p_x \in C^k[0, 1]$ be such that

$$D^k p_x = \left| D^k \left(h - D^k h(x) \frac{1}{k!} e_k \right) \right|, \quad D^i p_x(0) = 0, \; i = 0, 1, \ldots, k - 1.$$

We get $D^k (h - D^k h(x) \frac{1}{k!} e_k) \leq D^k p_x$ and $D^k (-(h - D^k h(x) \frac{1}{k!} e_k)) \leq D^k p_x$. It follows from $L_n(\Delta^k) \subset \Delta^k$ that

$$D^k L_n (h - D^k h(x) \frac{1}{k!} e_k)(x) \leq D^k L_n p_x(x) \qquad (15)$$

and

$$- D^k L_n (h - D^k h(x) \frac{1}{k!} e_k)(x) \leq D^k L_n p_x(x) \qquad (16)$$

It follows from (15) and (16) that

$$\left| D^k L_n \left(h - D^k h(x) \frac{1}{k!} e_k \right)(x) \right| \le D^k L_n p_x(x). \tag{17}$$

Let $q_x \in C^k[0,1]$ be such that

$$D^k q_x(t) = |t - x| \text{ and } D^i q_x(0) = 0, \ i = 0, 1, \ldots, k-1.$$

We have

$$D^k p_x(t) = \left| D^k \left(h(t) - D^k h(x) \frac{1}{k!} t^k \right) \right| = |D^k h(t) - D^k h(x)|$$

$$\le 2n|t - x| = 2n D^k q_x(t).$$

We get $D^k(2n q_x - p_x) \ge 0$, then it follows from the shape-preserving property $L_n(\Delta^k) \subset \Delta^k$ that $D^k L_n(2n q_x - p_x)(x) \ge 0$, i.e.

$$D^k L_n p_x(x) \le 2n D^k L_n q_x(x). \tag{18}$$

It follows from Lemma 2 that

$$D^k L_n q_x(x) \le [D^k L_n g_x(x)]^{\frac{1}{2}} \left[\frac{1}{k!} D^k L_n e_k(x) \right]^{\frac{1}{2}}$$

$$\le [D^k L_n g_x(x)]^{\frac{1}{2}} \left[1 + \frac{1}{k!} |D^k L_n e_k(x) - D^k e_k(x)| \right]^{\frac{1}{2}}, \tag{19}$$

where

$$g_x = \frac{2}{(k+2)!} e_{k+2} - \frac{2}{(k+1)!} x e_{k+1} + \frac{1}{k!} x^2 e_k.$$

We have

$$D^k L_n g_x(x) = \frac{2}{(k+2)!} (D^k L_n e_{k+2} - D^k e_{k+2})(x)$$

$$- \frac{2}{(k+1)!} x(D^k L_n e_{k+1} - D^k e_{k+1})(x) + \frac{1}{k!} x^2 (D^k L_n e_k - D^k e_k)(x)$$

$$+ \frac{2}{(k+2)!} D^k e_{k+2}(x) - \frac{2}{(k+1)!} x D^k e_{k+1}(x) + \frac{1}{k!} x^2 D^k e_k(x)$$

$$\le \frac{2}{(k+2)!} \left| D^k L_n e_{k+2}(x) - D^k e_{k+2}(x) \right|$$

$$+ \frac{2}{(k+1)!} \left| D^k L_n e_{k+1}(x) - D^k e_{k+1}(x) \right|$$

$$+ \frac{1}{k!} \left| D^k L_n e_k(x) - D^k e_k(x) \right|. \tag{20}$$

If

$$\frac{1}{k!}\|D^k L_n e_k - D^k e_k\|_{B[0,1]} \leq \frac{1}{4n^2}. \tag{21}$$

then it follows from (13), (14), (17), (18), (19) that

$$1 - \frac{1}{k!}\|D^k L_n e_k - D^k e_k\|_{C[0,1]} \leq 2n \left(\sup_{x \in [0,1]} D^k L_n g_x(x) \right)^{\frac{1}{2}} \left(1 + \frac{1}{4n^2} \right)^{\frac{1}{2}}, \tag{22}$$

Both sides of the inequality are positive, therefore we get

$$4n^2 \sup_{x \in [0,1]} D^k L_n g_x(x) \geq \frac{(1 - \frac{1}{4n^2})^2}{1 + \frac{1}{4n^2}} \geq 1 - \frac{1}{n^2}, \; n \geq k+2,$$

and then (12) follows from (20). If (21) does not hold, then (12) holds a fortiori. □

The positive preserving approximation ($k = 0$) has the same order of approximation n^{-2} [30]. It is should be noted that k-monotonicity preserving results can not be obtained from positive preserving results since $\|D^k(L_n f) - D^k f\| = \|L_n(D^k f) - D^k f\|$ is not hold in general.

Denote $\Pi_m := \text{span}\{e_0, e_1, \ldots, e_m\}$, i.e. Π_m is the set of all algebraic polynomials of degree $\leq m$. Denote $P_m := \{f \in \Pi_m : \|f\|_{C^m[0,1]} \leq 1\}$, i.e. P_m is the set of all algebraic polynomials of degree $\leq m$ whose norm in $C^m[0,1]$ is bounded by 1.

Theorem 1. *Let $n \geq k+2$. Then there exists $0 < c_1 \leq 2^{k-3}/k!$ not depending on n such that*

$$\delta_n(B_\infty^{(k+2)} \cap \Delta^k, \Delta^k)_{C^k[0,1]} < c_1 n^{-2}. \tag{23}$$

The estimate (23) can not be improved even on the set of algebraic polynomials P_{k+2} bounded in $C^{k+2}[0,1]$, i.e. there exists $c_2 > 0$ not depending on n such that

$$c_2 n^{-2} < \delta_n(P_{k+2} \cap \Delta^k, \Delta^k)_{C^k[0,1]}. \tag{24}$$

Proof. The inequality (23) follows from Lemma 1. The estimate (24) follows from Lemma 3. □

The proof of Lemma 1 shows that the finite rank n linear operator $\Lambda_{k,n}$ satisfies $\Lambda_{k,n}(\Delta^k) \subset \Delta^k$ and

$$\sup_{f \in P_m} \|D^i(\Lambda_{k,n}f) - D^i f\|_{B[0,1]} = 0, \; 0 \leq m \leq k+1,$$

for all $0 \leq i \leq k$. Therefore

$$\delta_n(P_m \cap \Delta^k, \Delta^k)_{C^k[0,1]} = 0$$

for all $0 \leq m \leq k+1$.

Konovalov linear relative n-width $\delta_n(B_\infty^{(k+2)} \cap \Delta^k, \Delta^k)_{C^k[0,1]}$ provides a lower bound on the degree of approximation of any linear operator of finite rank n preserving the shape in the sense of cone Δ^k on the subset of $(k+2)$-times differentiable functions. The theorem states that the theoretical lower bound is n^{-2}. Thus, the degree of approximation by Bernstein polynomials B_n (which is equal to n^{-1}) is not optimal relative to the best possible for linear operators of finite rank n preserving k-monotonicity. We can remark that algorithm $\Lambda_{k,n}$ defined by (3)–(4) has the optimal order of approximation n^{-2}.

4   Conclusion

Software developers and designers often need mathematical and computational methods for the representation of geometric objects with preserving the shape of the data as they arise in areas ranging from industrial design, scientific visualization, CAD/CAM to robotics. Another application of shape-preserving algorithms is in the optimization theory and the theory of dynamic optimization. In particular, the paper [3] presents algorithms for solving the dynamic programming problems based on shape-preserving methods of approximation and shows the applicability of the cone-preserving algorithms for the optimal growth problem.

The paper shows that if linear operator with finite rank n preserves k-monotonicity, the degree of simultaneous approximation of derivative of order $0 \le i \le k$ of continuous functions by derivatives of this operator cannot be better than n^{-2} even on the set P_{k+2} (as well as on the ball $B_\infty^{(k+2)}$). Results show that the shape-preserving property of operators is negative in the sense that the error of approximation of such operators does not decrease with the increase of smoothness of approximated functions. In other words, there is saturation effect for linear finite-rank operators preserving k-monotonicity (see also [28]). It is worth noting that non-linear approximation preserving k-monotonicity does not have this shortcoming [15]. On the other hand, for sequences of linear operators preserving k-monotonicity (as well as intersections of cones) there are [10,19] simple convergence conditions (Korovkin type results).

References

1. Barnabas, B., Coroianu, L., Gal, S.G.: Approximation and shape preserving properties of the Bernstein operator of max-product kind. Int. J. Math. Math. Sci. **2009**, 26 (2009). Article ID 590589
2. Boytsov, D.I., Sidorov, S.P.: Linear approximation method preserving k-monotonicity. Siberian Electron. Math. Rep. **12**, 21–27 (2015)
3. Cai, Y., Judd, K.L.: Shape-preserving dynamic programming. Math. Meth. Oper. Res. **77**, 407–421 (2013)
4. Cárdenas-Morales, D., Garrancho, P., Raşa, I.: Bernstein-type operators which preserve polynomials. Comput. Math. Appl. **62**, 158–163 (2011)
5. Cárdenas-Morales, D., Muñoz-Delgado, F.J.: Improving certain Bernstein-type approximation processes. Math. Comput. Simul. **77**, 170–178 (2008)

6. Cárdenas-Morales, D., Muñoz-Delgado, F.J., Garrancho, P.: Shape preserving approximation by Bernstein-type operators which fix polynomials. Appl. Math. Comput. **182**, 1615–1622 (2006)
7. Floater, M.S.: On the convergence of derivatives of Bernstein approximation. J. Approximation Theor. **134**(1), 130–135 (2005)
8. Gal, S.G.: Shape-Preserving Approximation by Real and Complex Polynomials. Springer, Boston (2008)
9. Gilewicz, J., Konovalov, V.N., Leviatan, D.: Widths and shape-preserving widths of sobolev-type classes of s-monotone functions. J. Approx. Theor. **140**(2), 101–126 (2006)
10. Gonska, H.H.: Quantitative Korovkin type theorems on simultaneous approximation. Math. Z. **186**(3), 419–433 (1984)
11. Konovalov, V., Leviatan, D.: Shape preserving widths of Sobolev-type classes of k-monotone functions on a finite interval. Isr. J. Math. **133**, 239–268 (2003)
12. Konovalov, V., Leviatan, D.: Shape-preserving widths of weighted Sobolev-type classes of positive, monotone, and convex functions on a finite interval. Constructive Approximation **19**, 23–58 (2008)
13. Konovalov, V.N.: Estimates of diameters of Kolmogorov type for classes of differentiable periodic functions. Mat. Zametki **35**(3), 369–380 (1984)
14. Kopotun, K.A., Leviatan, D., Prymak, A., Shevchuk, I.A.: Uniform and pointwise shape preserving approximation by algebraic polynomials. Surv. Approximation Theor. **6**, 24–74 (2011)
15. Kopotun, K., Shadrin, A.: On k-monotone approximation by free knot splines. SIAM J. Math. Anal. **34**, 901–924 (2003)
16. Kvasov, B.I.: Methods of Shape Preserving Spline Approximation. World Scientific Publ. Co., Pte. Ltd., Singapore (2000)
17. Kvasov, B.: Monotone and convex interpolation by weighted cubic splines. Comput. Math. Math. Phys. **53**(10), 1428–1439 (2013)
18. Lewicki, G., Prophet, M.P.: Minimal shape-preserving projections onto π_n: generalizations and extensions. Numer. Funct. Anal. Optim. **27**(7–8), 847–873 (2006)
19. Muñoz-Delgado, F.J., Ramírez-González, V., Cárdenas-Morales, D.: Qualitative Korovkin-type results on conservative approximation. J. Approx. Theor. **94**, 144–159 (1998)
20. Pinkus, A.: nWidths in Approximation Theory. Springer, Heidelberg (1985)
21. Popoviciu, T.: About the Best Polynomial Approximation of Continuous Functions. Mathematical Monography. Sect. Mat. Univ. Cluj, (In Romanian), fasc. III (1937)
22. Păltănea, R.: A generalization of Kantorovich operators and a shape-preserving property of Bernstein operators. Bull. Transilvania Univ. of Braşov, Ser. III: Math. Inf. Phys. **5**(54), 65–68 (2012)
23. Shevaldin, V.T.: Local Approximation by Splines. UrO RAN, Ekaterinburg (2014)
24. Shevaldin, V., Strelkova, E., Volkov, Y.: Local approximation by splines with displacement of nodes. Siberian Adv. Math. **23**(1), 69–75 (2013)
25. Sidorov, S.P.: On the order of approximation by linear shape-preserving operators of finite rank. East J. Approximations **7**(1), 1–8 (2001)
26. Sidorov, S.P.: Basic properties of linear shape-preserving operators. Int. J. Math. Anal. **5**(37–40), 1841–1849 (2011)
27. Sidorov, S.P.: Linear relative n-widths for linear operators preserving an intersection of cones. Int. J. Math. Math. Sci. **2014**, 7 (2014). Article ID 409219
28. Sidorov, S.: On the saturation effect for linear shape-preserving approximation in Sobolev spaces. Miskolc Mathematical Notes 16 (2015)

29. Vasiliev, R.K., Guendouz, F.: On the order of approximation of continuous functions by positive linear operators of finite rank. J. Approx. Theor. **69**(2), 133–140 (1992)
30. Vidensky, V.S.: On the exact inequality for linear positive operators of finite rank. Dokl. Akad. Nauk Tadzhik. SSR **24**, 715–717 (1981)
31. Volkov, Y.S., Shevaldin, V.T.: Shape preserving conditions for quadratic spline interpolation in the sense of subbotin and marsden. Trudy Inst. Mat. i Mekh. UrO RAN **18**(4), 145–152 (2012)
32. Volkov, Y., Bogdanov, V., Miroshnichenko, V., Shevaldin, V.: Shape-preserving interpolation by cubic splines. Math. Notes **88**(5–6), 798–805 (2010)
33. Volkov, Y., Galkin, V.: On the choice of approximations in direct problems of nozzle design. Comput. Math. Math. Phys. **47**(5), 882–894 (2007)

Applied Algebraic Geometry

Workspace Multiplicity and Fault Tolerance of Cooperating Robots

Daniel A. Brake[1], Daniel J. Bates[2]([✉]), Vakhtang Putkaradze[3],
and Anthony A. Maciejewski[4]

[1] Department of Applied and Computational Mathematics and Statistics,
University of Notre Dame, Notre Dame, USA
[2] Department of Mathematics, Colorado State University, Fort Collins, USA
bates@math.colostate.edu
[3] Department of Mathematical and Statistical Sciences, University of Alberta,
Edmonton, Canada
[4] Department of Electrical and Computer Engineering, Colorado State University,
Fort Collins, USA

Abstract. Cooperating robotic systems, especially in the context of
fault-tolerance of complex robotic mechanisms, is an important question for theoretical and applied studies. In this paper, we focus on one
measure of fault tolerance in robots, namely, the *multiplicity* of the configurations for reaching a particular point in the workspace, which is
difficult to measure using traditional methods. As a particular example,
we consider the case of a free-swinging failure of a robotic arm that is
handled by having a cooperating functional robot grasp the link adjacent to the failed joint. We present an efficient method to compute the
multiplicity measure of the workspace, based on the tools from numerical
algebraic geometry, applied to the inverse kinematics problem re-cast in
the form of a polynomial system. To emphasize the difference between
our methods and more traditional approaches, we compute the measure
of workspace based on the multiplicity of configurations, and optimize
placement of synergistic robot arms and the optimal grasp point on each
link of the broken robot based on this measure.

Keywords: Workspace mapping · Joint failure · Homotopy
continuation · Monte Carlo methods

1 Introduction

Fault-tolerance and robustness play important roles in the design of autonomous
systems, including robotic arms. Often times, this has been achieved with either

D.A. Brake—Partially supported by grants NSF-DMS-09087551 and NSF-IIS-
0812437.
D.J. Bates—Partially supported by grants NSF DMS–0914674 and NSF DMS–
1115668.
V. Putkaradze—Partially supported by grant NSF-DMS-09087551.
A.A. Maciejewski—Partially supported by grant NSF-IIS-0812437.

© Springer International Publishing Switzerland 2016
I.S. Kotsireas et al. (Eds.): MACIS 2015, LNCS 9582, pp. 109–123, 2016.
DOI: 10.1007/978-3-319-32859-1_8

redundancy in drive systems, or cooperating robotic arms. Consideration of fault-tolerance is essential in dangerous locations, and continues to be studied extensively. Determining the reliability of a robot via fault-tree analysis appeared in [1], which permits quantification of weaknesses in design. The detection and isolation of faults, in various components of a robot, such as sensors, controllers, and actuators, as well as collision with environmental elements, was treated in [2–5]. Failure-tolerant regions of a workspace were defined in [6], which also gave a method for computing a fail-tolerance measure which allows a robot to operate in real-time within this fail-tolerant region. Other measures of fail-tolerance include analysis of the singular values of the Jacobian matrix for a robot [7] and a 'manipulability index' as defined in [8].

Anticipation of failures has also been explored considerably; decomposing a task into primary and secondary goals is one method of dealing with these problems. Robots can be designed and operated with fail-tolerance in mind, such as dual actuators at joints [9], kinematic redundancy [9–13], and reconfigurability [14]. Prioritization of tasks subject to other constraints (e.g., environmental) was treated in [15], restriction of operation to a fault-tolerant region in [16], and anticipation of free-swinging joint failures in [17]. The overarching theme in this field is graceful degradation of performance.

In this paper, we consider autonomously operated robotic systems that are deployed into a hazardous or remote location, such that the repair of a failed joint is impractical or impossible. If the system has to operate for an extended period of time, we want the system to operate to the best of its remaining physical capability after a joint failure. Certainly, there are many ways for a robotic system to fail. Sensors give false or no readings; electronic components break; controllers fault; actuators seize or give way. We aim to supplement present methods regarding graceful joint failure, by informing designers and operators of possibilities for assistance to a broken robot by a functional one, should one be available. In the event of a free-swinging failure, our method provides information about optimal placement of a predetermined socket or other suitable apparatus on an articulated arm to allow assistance from a functional robot. Even with an actuator with non-free-swinging failure, such as unexpected resistance or friction, our method could contribute to mission success. The failure-proof of systems has been particularly important for applications such as space-faring vehicles [11].

To make our consideration more realistic, suppose a system operating in a remote environment possesses several robotic arms for various tasks. Should one joint in one arm fail, the presence of a second arm may open the way to preserve some of the workspace of the failed arm. In particular, if the second arm can grasp the first, perhaps some of the lost workspace can be restored. In this work, we shall assume that the relative position of the bases for robotic arms cannot change, and only one joint fails at a time.

As far as fault-tolerance consideration of a *single* robot goes, the multiplicity of configurations may not be a relevant quantity. For the purpose of restoring workspace using *cooperating* robotic arms the multiplicity of configurations is of crucial importance. As we shall see below, the workspaces of cooperating robots tend to consist of several isolated sets.

The main goal of this paper is to outline a method that is alternative to the traditional ways of workspace computation. The main difference is that we use sampling *directly in the workspace*, and therefore are not subject to failures of a Jacobian method due to possible singularities of the forward kinematics mapping. Our method, which is based on the applications of ideas in algebraic geometry (homotopy continuation) to the solution of polynomial problems, can provide the exact number of solutions for each particular point of the workspace. In particular, the inverse kinematics problem can be cast as a polynomial system, and homotopy continuation provides a means for efficiently producing numerical approximations of *all* isolated complex and real-valued solutions of the polynomial system. Methods of algebraic geometry have been applied in the kinematic description of robots, see for example [18–23]. However, analysis of the cooperation of multiple arms via the tools of numerical algebraic geometry has never been explored.

We shall outline a particular application of the method: computation of workspace and optimization of the grasping point for two cooperating robots, in case of joint failure of the first robot. This example was chosen as a realistic demonstration, leading to interesting shapes of workspaces with several sets of multiple solutions. For the optimization, we ask two questions: *(a)* What is the best placement of the two robots in relation to each other? *(b)* Is there a best place for the second arm to grasp the first? Clearly, the answer to both questions depends on the choice of measure for the post-failure workspace, which is problem dependent. In this paper, we introduce a multiplicity-weighted workspace measure, depending on both the number of robot configurations and the geometric size of the workspace. Combining this on the pre- and post-failure workspaces gives a single measure, parameterized by a user-determined weighting factor, and the distance between robot bases and grasp point.

The key feature of this manuscript is the application of methods of algebraic geometry to the description of fault tolerance of two cooperating robots. These techniques are guaranteed to give all isolated solutions to algebraic equations describing the positions of robotic arms, and thus robustly find solutions in arbitrarily complex settings, such as multiple joint values in isolated regions of joint-space.

In Sect. 2, we provide a formal statement of the general problem we are considering. Details about our method are described in Sect. 3, and background regarding the numerical solution of polynomial systems may be found in Sect. 4. Sections 5.1 and 5.2 respectively contain two and three dimensional examples.

2   Formal Problem Statement

Given two articulated arms, we optimize the placement of grasping sockets to maximize post-failure workspace (we assume a socket attachment mechanism of the two arms as considered in [14]). That is, if one robot has a free swinging joint failure as in [24,25], we would like to ensure that when the functional robot joins at the socket to assist in completion of tasks, the resulting cooperative workspace is as large as possible. A simplified example of this problem is in Fig. 1.

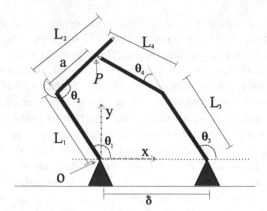

Fig. 1. Two robots in grasping configuration, with functional robot contacting the disabled unit at point P. This is a simplified 2D model of the general 6D position and orientation scenario. Parameter δ is the distance between the robot bases.

Each robot has its own pre-failure workspace, $W_1, W_2 \subset \mathbb{R}^n \times SO(N)$, where n and N depend on the particular mechanism type. When the two robots are placed near enough, there is an intersection workspace $W_\cap = W_1 \cap W_2$. In a grasping configuration, there is a post-failure workspace W_f, which contains all remaining workspace locations Robot 1 can reach, if Robot 2 attaches to a socket location on Robot 1. The measures $|W_\cap|$ and $|W_f|$ developed below in Eq. (2) depend on the separation between the robots, as well as their orientation. Note that the intersection workspace does not depend on the grasping point, as it reflects the workspace prior to failure for each robot. In both pre- and post-failure workspaces, we take into account joint limits, which are an important consideration for practical applications. Indeed, robots typically have limited range of movement for each joint, which reduces the size of all workspaces. These limits introduce a dependence on the relative rotations of the robots to each other, and to the measures of the various workspaces.

The parameters describing an optimal configuration of two cooperating robots include: separation between cooperating arms, relative orientation of the bases, and location of grasping point. In this paper, we do not optimize with respect to base position of each robot, as we consider that the position of the bases must be given from design limitations, *e.g.* the placement of power cords and motors on the apparatus. Nevertheless, our method is capable of optimizing socket placement for different base separations, and this is demonstrated in the examples.

In order to quantify post-failure workspace size and find an optimal configuration, we introduce a maximizing objective function Ω. There are many possible objective functions one can imagine, and the right choice depends on the application. In this paper, we define a multiplicity-weighted measure of workspace W and objective function Ω_λ through the multiplicity of configurations (i.e., the number of inverse kinematic solutions placing the robot in the specified configuration) at a given point x, denoted $m(x)$:

$$|W| = \int_W m(x)\mathrm{d}x, \tag{1}$$

$$\Omega_\lambda = (1 - \lambda)|W_f| + \lambda(|W_1| - |W_\cap|), \tag{2}$$

where λ is a weighting factor to be chosen by the user. We have chosen Eq. (2) for an example objective function in order to balance between the benefit of having a second robot, and the maximization of the post-failure workspace. A configuration which imparts entirely distinct workspaces would maximize $\lambda(|W_1| - |W_\cap|)$, but W_f would be an empty set. Contrarily, we might find a configuration which results in full restoration of W_f upon entering grasping stance, but which makes the pre-failure workspaces overlap greatly, so $\lambda(|W_1| - |W_\cap|)$ would be small. We prefer to balance between pre- and post-failure benefits of having two robots, and Eq. (2) is one way of doing this.

It seems that one obvious way to maximize $|W_f|$ could be to set $\delta = 0$ so the two robots have exactly the same base point, and make the two robots identical. However, this may be an impractical situation. First of all, there may be much greater benefit by compromising and having smaller intersection workspaces by placing the robots further apart; second, depending on robot geometry, it could be awkward or impossible for two identical robots to operate from the same base point.

In fact, the number of configurations could be infinite at some points in the workspace. However, this is an algebraic condition so this set of all points in the workspace for which this is true has measure zero. Such points are kinematic singularities and should be excluded in Eq. (2) above. More practically, the Monte Carlo methods used later in the paper miss such points with probability one.

3   Workspace Computation

In this section, we describe our solution to the problem outlined in Sect. 2. We start with notation. Let the separation between bases be δ and without loss of generality let this translation between coplanar bases be entirely along the x-axis; we sort out-of-limit solutions in a post-processing procedure. We consider $0 < \delta < \delta_{\max}$, the largest value of δ corresponding to the sum of the lengths of each robot fully extended. The normalized distance to the grabbing point or socket, measured from the failed joint, is denoted by a, with $0 < a < 1$; the point at which a socket is attached to the left robot, hereafter Robot 1, is P; a test point in space is Q; and the origin at the base of Robot 1. Finally, let the fully functional machine be known as Robot 2. A simplified version of this notation is found in Fig. 1.

There is some probability that each joint could fail, so it is important to take into consideration the placement of a socket on each link (of non-zero length). We use the Denavit-Hartenberg (DH) convention to describe the configurations of the two robots, and describe the contact point P in terms of the DH parameters for Robot 1, as P is some distance a from the origin of the previous frame. A spatial sample Q can be said to be in the post-failure workspace of Robot

1 for a parameter pair (δ, a), if there exists a set of joint angles such that end effector of Robot 1 is at Q, and the end effector of Robot 2 is at P.

The equations for the inverse kinematics problem for each step of the method are first solved via a standard homotopy run at $\bar{p}^*$, a random point in complex parameter space. Then all subsequent runs at points in the workspace are treated as parameter homotopies, beginning at $\bar{p}^*$. For W_i, the parameters are the x, y, z coordinates of Q; for $W_\cap$, we add parameter δ; and for W_f, we add a. The software package *Bertini* [28] can be used to compute numerical approximations of the solutions of polynomial systems. *Bertini* works over $\mathbb{C}$, so we find solutions for each point regardless of whether it lies inside the workspace. However, the points that have solutions for which all variables are numerically real-valued are those lying within the workspace, while those with nonzero imaginary component lie outside.

We note that the inverse kinematics equations for a robot with rotary joints are trigonometric in nature. In order to use polynomial homotopy continuation, we write the equations in polynomial form by treating each sine-cosine pair as a separate variable, mapping $\cos(\theta_j) = c_j$ and $\sin(\theta_j) = s_j$ and coupling with the algebraic condition $c_j^2 + s_j^2 - 1 = 0$.

We compute the initial pre-failure workspaces for each robot via a random sampling method on an ambient set S of the workspace guaranteed to contain the workspace, and estimate the measure of the workspaces as in Eq. (2) using the number of samples in the workspace pointwise multiplied by multiplicity factor for each point,

$$|W| = \lim_{r \to \infty} |S| \frac{1}{r} \sum_{j=1}^{r} m(x_j), \tag{3}$$

where $|S|$ is the typical Euclidean measure of the sampling space.

The method for computing optimal $\delta_{\max}$ and $a_{\max}$ is described in Algorithm 1. Starting from S, we compute each of the necessary workspaces. After computing W_1 we no longer need S, because $W_\cap, W_f \subset W_1$. Instead, all further computations are over W_1. Once we have the data for the multiplicities of the workspaces, we simply compute Ω_λ, and return its maximizing parameter values.

For the case of $N \geq 3$ joints working in 3 dimensions without orientation, there are 3 algebraic inverse kinematic equations in $2N$ variables, coupled with N Pythagorean identities, when computing W_i. As long as the number of joints equals the number of degrees of freedom in the ambient workspace, *Bertini* will find all solutions, and we will be able to measure $|W_i|$. For kinematically redundant robots, the joint space consists of a set of higher $N - 3$ dimensional manifolds, which could be described by defining a mesh of the same dimensionality as coordinates on the joints. Our method readily applies to this higher dimensional problem. However, the issues we are facing are related to the curse of dimensionality, and hence in the computation of the data as a whole, not in computing the solutions at a particular point in space, which is fast. For example, with $N = 5$ joints, and a 3 dimension workspace, each point in the workspace corresponds to a two dimensional manifold in joint space. To cut down the manifold to be 0-dimensional for solving via *Bertini*, we could discretely sample each pair of joints possible, and solve for the remaining three. However, combinatorial growth issues arise, *e.g.* if we wanted to solve each spatial sample for each possible combination of joints.

4 Homotopy Continuation

In general, given an arbitrary (not necessarily robotic) set of polynomials $\bar{f} = \{f_1, \ldots, f_N\}$ in N variables for which we seek the solutions, homotopy methods begin by choosing and solving some other related polynomial system $\bar{g} = \{g_1, \ldots, g_N\}$ for which the solutions are easily found. By varying the coefficients in $\bar{g}$ to those of $\bar{f}$, statements in algebraic geometry guarantee that, with probability one, the solutions will vary continuously, thus forming *solution curves* or *paths* from the solutions of $\bar{g}$ to those of $\bar{f}$. These solution curves may then be tracked numerically with standard predictor/corrector methods, such as a combination of Runge-Kutta and Newton's method. Further details may be found in [19, 26, 27].

There is a setting in which homotopy methods are particularly effective, and of which we make use in this paper. If instead of solving one polynomial system $\bar{f}$ we aim to solve a large number of polynomial systems that differ only in coefficients (i.e., we aim to solve $\bar{f}(\bar{p})$, where $\bar{p}$ is some set of parameters), there is an especially efficient homotopy method known as a *parameter homotopy*. In this setting, we first solve $\bar{f}(\bar{p})$ at a single instance of $\bar{p}^*$ (typically chosen as random complex numbers, for theoretical reasons, as described in Sect. 7.1 of [19]). This stage may require the tracking of a number of superfluous, divergent paths. However, all other instances of $\bar{f}(\bar{p})$ may then be solved by simply following the handful of finite solutions at $\bar{p}^*$ to any other choice of $\bar{p}$. For example, the system used to solve W_f, grasping on the third link, for an initial random complex parameter choice, requires following 20,736 paths to find the *sixteen* solutions of interest. Then, for all other points in the parameter space, it suffices to follow just sixteen paths.

Again, there is much theory and detail underlying these methods, most of which may be found in [19, 27]. During the process of homotopy continuation, a certain number of paths will fail as they near a singularity in parameter space. In the context of robotic workspaces, these failed paths indicate proximity to workspace boundary or kinematic singularity. Right now, we are not

---

Input : DH parameters for each robot, λ, samplings of δ, a, S
Output: Optimal $\delta_{\max}, a_{\max}, \Omega_\lambda(\delta_{\max}, a_{\max})$

Compute $W_1 = \{x \in S \mid m(x) > 0\}$
Compute $W_\cap(\delta)$ using W_1
Compute $W_f(\delta, a)$ using W_1
Evaluate $\Omega_\lambda(\delta, a)$ using Equations (2) and (3)
Find $\delta_{\max}, a_{\max}$ maximizing Ω_λ
return Maximizers $\delta_{\max}, a_{\max}$ and value $\Omega_\lambda(\delta_{\max}, a_{\max})$;

Algorithm 1. Optimization of Ω_λ

using this information, and simply ignore failed paths; however, this property could ultimately be used for more accurate and efficient prediction of, for example, workspace boundaries.

In this paper, it is adequate to accept homotopy continuation as a numerical method that will quickly provide accurate approximations to all isolated solutions of a polynomial system. Several software packages are available for these sorts of computations. We use a software package named *Bertini* [28], which has been under development for the past decade by D. Bates, J. Hauenstein, A. Sommese, and C. Wampler. The repeated calls to *Bertini* were parallelized for efficiency using the method described in [29].

5   Two Examples

5.1   2D Case: Two Link Planar Robots

We start with the illustration of the method for the two dimensional example that is shown in Fig. 1. This was first considered this [30]; the previous work has been expanded to include more general method and examples, as in Sect. 5.2, where the more challenging three-dimensional problem is considered.

The robots are identical, both having two joints, all link lengths having length one; the DH parameters are summarized in Table 1.

Table 1. Denavit-Hartenberg Parameters for two-link planar robot.

θ_j	α_j	a_j	d_j	$\theta_{j,min}$	$\theta_{j,max}$
θ_1	0	1	0	$-120°$	$120°$
θ_2	0	1	0	$-120°$	$120°$

Here, we define c_j and s_j to be the cosine and sine of the joint θ_j values, respectively, with $j = 1, 2$ corresponding to the failed robot and $j = 3, 4$ corresponding to the assisting robot. Let joint index $j \in \{1, 2, 3, 4\}$, so that we have a sine-cosine pair for each of the two joints in the two robots. In this notation, our equations for this step are,

$$0 = \begin{cases} c_1 c_2 - s_1 s_2 + c_1 - x = 0 \\ s_1 c_2 + c_1 s_2 + s_1 - y = 0 \\ a c_1 c_2 - a s_1 s_2 + c_1 - (c_3 c_4 - s_3 s_4 + c_3 + \delta) = 0 \\ a s_1 c_2 + a c_1 s_2 + s_1 - (s_3 c_4 + c_3 s_4 + s_3) = 0 \\ s_j^2 + c_j^2 - 1 = 0 \quad \forall j \end{cases} \tag{4}$$

To demonstrate the method, we show results for an initial sampling of $r = 10^4$ points, taken from the two dimensional rectangle $Q \in [-2, 2] \times [-2, 2]$. Of these points, 7836 had at least one real solution; the estimated *Euclidean* size of the

Table 2. Estimates of 2D workspaces and $\Omega_{1/3}$, grasping on the second link.

δ	a	$|W_1|$	$|W_\cap|$	$|W_f|$	$\Omega_{1/3}$
0.276	0.076	12.79	7.5810	4.424	4.687
0.276	0.406	12.79	7.5810	7.946	7.035
0.276	0.736	12.79	7.5810	9.927	8.356
1.606	0.076	12.79	1.8904	6.241	7.795
1.606	0.406	12.79	1.8904	5.622	7.383
1.606	0.736	12.79	1.8904	4.718	6.780
2.936	0.076	12.79	0.2031	2.119	5.610
2.936	0.406	12.79	0.2031	2.829	6.083
2.936	0.736	12.79	0.2031	3.635	6.621

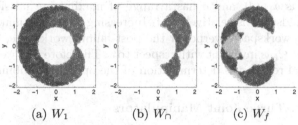

(a) W_1 (b) $W_\cap$ (c) W_f

Fig. 2. Examples of joint-limited workspaces, for $\delta = 1.87$, $a = 1$, and grasping on the *first* link, rotated relative to one another. Red indicates a point having one solution, blue indicates two, and cyan indicates four (Color figure online).

workspace would be $4^2 \times 7836/10000 \approx 12.5 \approx 4\pi$. The multiplicity measure of the joint-limited robot, sieved during post-processing, is ≈ 12.79.

The first step in the computation is to estimate the pre-failure workspaces for each robot, as described in Algorithm 1. Secondly, we intersect and obtain $W_\cap = W_1 \cap W_2$, for a specified set of δ values. Finally, for each pair of values (δ, a) for which we estimate the post-failure workspace, we solve Eq. (4).

As the separation increases so that the workspaces barely overlap, Robot 2 may only grasp Robot 1 near the end effector, and the resulting W_f is small. See Fig. 2, and column four of Table 2. In these plots, cyan area is fully accessible from 4 configurations, blue corresponds to 2, and red is reachable from only one configuration. The inaccessible area generally increases when δ is increasing, and larger δ lead to smaller post-failure workspaces for fixed a. The converse is not true: $|W_f|$ is not a monotonic function of a for a fixed δ. It is interesting to note that our method computes explicitly the number of configurations reaching the desired point in a post-failure workspace, even in the case when the configurations belong to isolated domains in the workspace. Such problems are usually challenging for traditional methods of workspace computation.

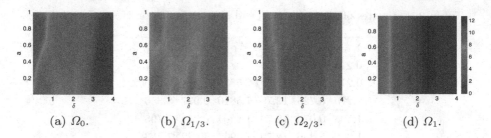

(a) Ω_0.   (b) $\Omega_{1/3}$.   (c) $\Omega_{2/3}$.   (d) Ω_1.

Fig. 3. Objective function contours for a grid of (δ, a) pairs for 2D cooperating robots, with the broken robot being grasped on the second link.

Finally, in Fig. 3 we show Ω_λ from Eq. (2) to determine the optimal ball joint placement. For weighting factor $\lambda = 0$, Ω_0 is simply the value of $|W_f|$, and the maximizer is $\delta \approx 0$, and a has a range of maximizing possibilities. With $\lambda = 1/3$, the landscape is relatively flat. Increasing λ further makes the size of the intersection workspace overpower the post-failure workspace, and by $\lambda = 1$ the maximum of Ω_1 is invariant with respect to a. Therefore, the weighting factor λ plays a critical role in the determination of the optimal grabbing point.

5.2   3D Case: Three Joint Manipulators

Equations determining the kinematics of spatial robots are more complicated than those of planar manipulators. This concerns both the higher number of relevant equations, due to the higher number of links in a typical 3D robots, and the structure of equations describing the motions. Yet, the method for optimizing cooperative workspaces remains fundamentally the same, as those equations can be brought to an algebraic form and then solved using the methods outlined in Sect. 3. Thus, in this section we only write a sketch of the method in 3D.

Consider two cooperating PUMA 6-degree-of-freedom robots, ignoring orientation of the end effector. In this paper, we use DH parameters defined as in Table 3. Treating the first three links of the PUMA as our robot, we ignore the wrist; we do this to reduce the dimension and to reduce the number of points to analyze. Instead, we use a tool frame to translate from the arm to the end effector. The frame we used retains the orientation of the arm, and translates along the final z-axis.

Each of the workspaces W_1, W_2, $W_\cap$ will have six equations in six variables, which are solved via *Bertini* after bringing the trigonometric part of these equations into algebraic form. Correspondingly, there will be 12 equations defining the post-failure workspaces with 12 variables. In order to find $W_{1,2}$ we sample randomly an oversized rectangular box surrounding the robot. In order to determine good bounds for the workspaces, we first do forward kinematics by randomly sampling the three joint variables $\theta_j \in [0, 2\pi]$. The result of this estimate is that the cubic box $[-0.9, 0.9] \times [-0.9, 0.9] \times [-0.9, 0.9]$ contains W_i, and is thus a good starting point for finding the cooperative workspaces in which we are interested.

Table 3. Denavit-Hartenberg Parameters for PUMA robot.

θ_i	α_i	a_i (m)	d_i (m)	$\theta_{i,min}$	$\theta_{i,max}$
θ_1	0	0	0	$-160°$	$160°$
θ_2	$-\pi/2$	0.4318	0.2435	$-225°$	$45°$
θ_3	0	-0.0203	-0.0934	$-45°$	$225°$
θ_4	$\pi/2$	0	0.4331	$-110°$	$170°$
θ_5	$-\pi/2$	0	0	$-100°$	$100°$
θ_6	$\pi/2$	0	0.5625	$-266°$	$266°$

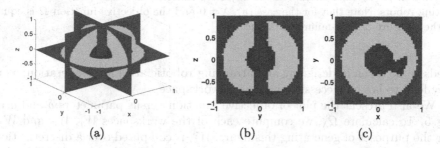

(a)   (b)   (c)

Fig. 4. Top: Slices of the pre-failure PUMA workspace W_1 by the planes $x = 0$ (b) and $z = 0$ (c). (a): Combined picture of workspace with three slices $x = 0$, $y = 0$, $z = 0$. Yellow color: areas accessible the angles satisfying joint limits given in Table 3. Red color: real solutions violating joint limits. Dark blue region: inaccessible (Color figure online).

The PUMA has joint limits that reduce its workspaces significantly. The limits we use for this example appear in Table 3. Because joint limits are written as inequalities $\theta_{j,min} \leq \theta_j \leq \theta_{j,max}$, they are not algebraic equations. We simply use these joint limits in post-processing, only selecting the suitable joint angles among all real solutions.

Results for various workspace computations are shown in Fig. 4. These data are based on 10^4 points in (x, y, z) space. Of particular interest is the presence of voids inside of workspaces, plotted by gridding the Monte Carlo data. Unreachable area is represented by the dark blue color. The red and yellow colors show the accessible work space, with the yellow areas being accessible only if the joint limits are satisfied. The resulting workspace is essentially a torus, although the realization we have chosen makes it hard to visualize as a volume in the 3D space, because of the relative narrowness of the "hole". Instead, we have chosen to represent the workspace through the slices. Two slices by the planes $x = 0$ and $z = 0$ are shown in the top of the figure, and the combined figure presenting the slices is shown in the bottom. The voids in W_i come from the offsets of the arm, and they expand for cooperating robots. Our results show that great care

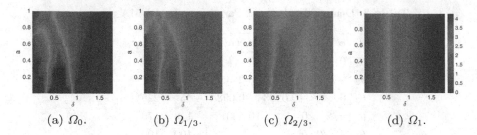

(a) Ω_0.　　(b) $\Omega_{1/3}$.　　(c) $\Omega_{2/3}$.　　(d) Ω_1.

Fig. 5. Considering grasping on the second link to restore workspace after failure of joint 2 of a PUMA robot. (a) Measurement of the intersection workspace $|W_\cap|$, as a function of δ. (b)–(c) Contours of the objective function Ω versus (δ, a) for 3D cooperating robots. Note that for the case (a), $\lambda = 0$ and the objective function Ω is equal to the measure of post-failure workspace $|W_f|$.

needs to be taken in designing and arranging robotic arms for cooperation, as it may lead to large inaccessible regions of workspace.

We also present the plot of objective function versus parameters δ and a in Fig. 5. To calculate Ω_λ, we compute each of the workspaces W_1, $W_\cap$ and W_f. For the purposes of generating this figure, $W_\cap$ is computed over a discretization of $0 \leq \delta \leq 1.8$ into 16 values. Finally, we compute W_f, for a particular value of δ and a. The results are repeated over a 16×16 regular grid of $0 \leq \delta \leq 1.8$, and $0 \leq a \leq 1$, and considering grasping the failed robot on each of the three possible links. Thus, the total number of parameter combinations we considered for W_f is $7.68 \cdot 10^6$.

The Ω_λ landscapes presented in Fig. 5 for the PUMA robot are similar to those for the 2D planar robot above in Fig. 3. Because each joint could fail independently of the others, we consider a grasping location for each link of the robot; hence, we have an objective function landscape for each of the three links of the PUMA. However, the landscapes for each joint are similar, so only those for the second link are presented here.

As with the 2D example above, the weighting factor λ plays a crucial role in optimizing. The limiting cases of $\lambda = 1$ and $\lambda = 0$ return simply $(W_1 - |W_\cap|)$ or $|W_f|$ respectively. The maximum values of Ω occur with $\delta \approx 0.9\,\mathrm{m}$ between the robots. Around $\lambda = 1/3$, we put more emphasis on increasing the remainder of workspace accessible in grasping configuration; Ω clearly indicates to grasp near the end of the second link. In any case, the optimal distance and grasping location depends on the link and on λ, making user preference crucial in the optimization procedure.

6   Conclusion

We used homotopy continuation, as implemented in *Bertini*, to estimate the size of workspaces, the intersection of workspaces, and post-failure grasping workspaces in the case of having two serial robots placed near one another,

in two and three dimensions. We also solve the problem of the optimal configuration for these robots for one example of a user-defined objective function.
A general algorithm for solving the problem of finding optimal placement and
configuration of two such robots was also presented.

By using algebraic geometric methods, we avoid issues such as isolated
domains and multiple solutions to inverse kinematics equations. Knowing multiple solutions may be important, especially for obstacle avoidance, where one
or more solutions may not be collision free. The homotopy continuation algorithms will not encounter any difficulty in that case, whereas the Jacobian control
method will need to be augmented with the specific knowledge from the problem
and yet may fail to find all isolated solutions. Our algorithm can be extended to
more general cases. For example, it will be relatively straightforward to account
for different designs for the two robots (including unequal link lengths) in the
objective function. Also, methodological choices in the algorithm, such as the
use of homotopy continuation, have been made to make the generalization to
higher-dimensional workspaces possible.

The method we present could be complemented by the software in [31];
while their focus is not on failure tolerance, their interactive CAD workspace
mapper uses a Jacobian method to find singularities and determine workspace
boundaries of parallel manipulators, which could be supplemented by homotopy
continuation.

It should be noted that the methods of numerical algebraic geometry may
be used to compute *complex* positive-dimensional components of the solutions
sets of polynomial systems. Until recently, it was nearly impossible to detect
positive-dimensional *real* solutions. However, recent advances such as the software Bertini_real [32,33] which implements numerical real algebraic curve and
surface decompositions, have made it possible to compute algebraic objects
including singularities. Above dimension two, the curse of dimensionality continues to constrain techniques.

References

1. Carreras, C., Walker, I.D.: Interval methods for fault-tree analysis in robotics.
   IEEE Trans. Robot. Autom. **50**(1), 3–11 (2001)
2. Anand, M., Selvaraj, T., Kumanan, S., Janarthanan, J.: A hybrid fuzzy logic artificial neural network algorithm-based fault detection and isolation for industrial
   robot manipulators. Int. J. Manuf. Res. **2**(3), 279–302 (2007)
3. Ji, M., Sarkar, N.: Supervisory fault adaptive control of a mobile robot and its
   application in sensor-fault accommodation. IEEE Trans. Robot. **23**(1), 174–178
   (2007)
4. De Luca, A., Ferrajoli, L.: A modified Newton-Euler method for dynamic computations in robot fault detection and control. In: IEEE International Conference on
   Robotics and Automation, pp. 3359–3364, May 2009
5. Brambilla, D., Capisani, L., Ferrara, A., Pisu, P.: Fault detection for robot manipulators via second-order sliding modes. IEEE Trans. Ind. Electron. **55**(11), 3954–
   3963 (2008)

6. Groom, K.N., Maciejewski, A.A., Balakrishnan, V.: Real-time failure-tolerant control of kinematically redundant manipulators. IEEE Trans. Robot. Autom. **15**(6), 1109–1116 (1999)
7. Maciejewski, A.A.: Fault tolerant properties of kinematically redundant manipulators. In Proceedings IEEE International Conference on Robotics and Automation, Cincinatti, OH, USA, pp. 638–642 (1990)
8. Roberts, R.G., Maciejewski, A.A.: A local measure of fault tolerance for kinematically redundant manipulators. IEEE Trans. Robot. Autom. **12**(4), 543–552 (1996)
9. Hassan, M., Notash, L.: Optimizing fault tolerance to joint jam in the design of parallel robot manipulators. Mech. Mach. Theory **42**, 1401–1407 (2007)
10. McInroy, J.E., O'Brien, J.F., Neat, G.W.: Precise, fault-tolerant pointing using a Stewart platform. IEEE/ASME Trans. Mechatronics **4**(1), 91–95 (1999)
11. Wu, E.C., Hwang, J.C., Chladek, J.T.: Fault-tolerant joint development for the space shuttle remote manipulator system: analysis and experiment. IEEE Trans. Robot. Autom. **9**(5), 675–684 (1993)
12. Yi, Y., McInroy, J.E., Chen, Y.: Fault tolerance of parallel manipulators using task space and kinematic redundancy. IEEE Trans. Robot. **22**(5), 1017–1021 (2006)
13. Paredis, C.J.J., Khosla, P.K.: Designing fault-tolerant manipulators: how many degrees of freedom? Int. J. Robot. Res. **15**(6), 611–628 (1996)
14. Aghili, F., Parsa, K.: A reconfigurable robot with lockable cylindrical joints. IEEE Trans. Robot. **25**(4), 785–797 (2009)
15. Chen, Y., McInroy, J.E., Yi, Y.: Optimal, fault-tolerant mappings to achieve secondary goals without compromising primary performance. IEEE Trans. Robot. **19**(4), 680–691 (2003)
16. Lewis, C.L., Maciejewski, A.A.: Fault tolerant operation of kinematically redundant manipulators for locked joint failures. IEEE Trans. Robot. Autom. **13**(4), 622–629 (1997)
17. English, J.D., Maciejewski, A.A.: Fault tolerance for kinematically redundant manipulators: anticipating free-swinging joint failures. IEEE Trans. Robot. Autom. **14**(4), 566–575 (1998)
18. Sommese, A.J., Wampler, C.W.: Numerical algebraic geometry and algebraic kinematics. Acta Numerica **20**, 469–567 (2011)
19. Sommese, A.J., Wampler, C.W.: The Numerical Solution to Systems of Polynomials Arising in Engineering and Science. World Scientific, Singapore (2005)
20. Sommese, A., Verschelde, J., Wampler, C.W.: Advances in polynomial continuation for solving problems in kinematics. J. Mech. Des. **126**(2), 262–268 (2004)
21. Wampler, C.W., Morgan, A.P.: Solving the kinematics of general 6R manipulators using polynomial continuation. In: Warwick, K. (ed.) Robotics: Applied Mathematics and Computational Aspects, pp. 57–69. Clarendon Press, Oxford (1993)
22. Wampler, C.W., Morgan, A.P., Sommese, A.J.: Complete solution of the nine-point path synthesis problem for four-bar linkages. J. Mech. Des. **114**, 153–159 (1992)
23. Wampler, C.W., Hauenstein, J.D., Sommese, A.J.: Mechanism mobility and a local dimension test. Mech. Mach. Theory **46**(9), 1193–1206 (2011)
24. English, J.D., Maciejewski, A.A.: Measuring and reducing the Euclidean-space measures of robotic joint failures. IEEE Trans. Robot. Autom. **16**(1), 20–28 (2000)
25. English, J.D., Maciejewski, A.A.: Failure tolerance through active braking: a kinematic approach. Int. J. Rob. Res. **20**(4), 287–299 (2001)
26. Allgower, E.L., Georg, K.: Numerical Continuation Methods: An Introduction. Springer, Berlin (1990)
27. Bates, D.J., Hauenstein, J.D., Sommese, A.J., Wampler, C.W.: Numerically Solving Polynomial Systems with Bertini. SIAM, Philadelphia (2013)

28. Bates, D.J., Hauenstein, J.D., Sommese, A.J., Wampler, C.W.: Bertini: Software for Numerical Algebaic Geometry (2015). http://bertini.nd.edu
29. Bates, D.J., Brake, D.A., Niemerg, M.: Paramatopy: Parallel parameter homotopy via Bertini (2015). http://www.paramotopy.com
30. Brake, D.A., Bates, D.J., Putkaradze, V., Maciejewski, A.A.: Illustration of numerical algebraic methods for workspace estimation of cooperating robots after joint failure. In: IASTED Technology Conferences, Pittsburg, PN, USA, November 2010
31. Macho, E., Pinto, C., Amezua, E., Hernndez, A.: Software tool to compute, analyze and visualize workspaces of parallel kinematics robots. Adv. Robot. **25**, 675–698 (2011)
32. Brake, D.A., Bates, D.J., Hao, W., Hauenstein, J.D., Sommese, A., Wampler, C.W.: Bertini_real: Software for real algebraic sets (2015). http://www.bertinireal.com
33. Brake, D.A., Bates, D.J., Hao, W., Hauenstein, J.D., Sommese, A.J., Wampler, C.W.: Bertini_real: software for one- and two-dimensional real algebraic sets. In: Hong, H., Yap, C. (eds.) ICMS 2014. LNCS, vol. 8592, pp. 175–182. Springer, Heidelberg (2014)

Numerical Local Irreducible Decomposition

Daniel A. Brake$^{(\boxtimes)}$, Jonathan D. Hauenstein, and Andrew J. Sommese

Department of Applied and Computational Mathematics and Statistics,
University of Notre Dame, Notre Dame, IN 46556, USA
{dbrake,hauenstein,sommese}@nd.edu
http://www.nd.edu/~jhauenst
http://www.nd.edu/~sommese

Abstract. Globally, the solution set of a system of polynomial equations with complex coefficients can be decomposed into irreducible components. Using numerical algebraic geometry, each irreducible component is represented using a witness set thereby yielding a numerical irreducible decomposition of the solution set. Locally, the irreducible decomposition can be refined to produce a local irreducible decomposition. We define local witness sets and describe a numerical algebraic geometric approach for computing a numerical local irreducible decomposition for polynomial systems. Several examples are presented.

Keywords: Numerical algebraic geometry · Numerical irreducible decomposition · Local irreducible decomposition · Numerical local irreducible decomposition

1 Introduction

For a polynomial system $f : \mathbb{C}^N \to \mathbb{C}^n$, the *algebraic set* defined by f is the set $\mathcal{V}(f) = \{x \in \mathbb{C}^N \mid f(x) = 0\}$. An algebraic set V is *reducible* if there exist nonempty algebraic sets $V_1, V_2 \subsetneq V$ such that $V = V_1 \cup V_2$ and for $i \neq j$, $V_i \not\subset V_j$. If V is not reducible, it is *irreducible*. For $\mathcal{V}(f)$, there exist irreducible algebraic sets $V_1, \ldots, V_k$, called *irreducible components*, such that $\mathcal{V}(f) = \bigcup_{i=1}^{k} V_i$ and $V_j \not\subset \bigcup_{i \neq j} V_i$. The irreducible components $V_1, \ldots, V_k$ are said to form the *irreducible decomposition* of $\mathcal{V}(f)$.

A fundamental computation in numerical algebraic geometry is the *numerical irreducible decomposition* (NID), that is, computing a *witness set* for each of the irreducible components; e.g., see [2, Chap. 10]. For an irreducible component $V \subset \mathcal{V}(f) \subset \mathbb{C}^N$ of dimension d and degree r, a witness set for V is the triple $\{f, \mathcal{L}, W\}$ where $\mathcal{L} \subset \mathbb{C}^N$, called a *witness slice*, is a general linear space of codimension d and $W = V \cap \mathcal{L}$, called a *witness point set*, is a set of r points.

D.A. Brake—Supported in part by supported in part by NSF ACI-1460032.

J.D. Hauenstein—Supported in part by Army Young Investigator Program (YIP), a Sloan Research Fellowship, and NSF ACI-1460032.

A.J. Sommese—Supported in part by the Vincent J. and Annamarie Micus Duncan Chair of Mathematics and NSF ACI-1440607.

© Springer International Publishing Switzerland 2016
I.S. Kotsireas et al. (Eds.): MACIS 2015, LNCS 9582, pp. 124–129, 2016.
DOI: 10.1007/978-3-319-32859-1_9

One can naturally extend the global notions of reducibility, irreducible components, and irreducible decomposition to the local case (e.g., see [5, Chap. B]). Moreover, one can locally extend to the case that f is holomorphic in an open neighborhood. Our main contribution is to extend the numerical algebraic geometric notions to the local case via *local witness sets* and a *numerical local irreducible decomposition*, defined in Sect. 2, the computation of which is described in Sect. 3, and demonstrated on several examples in Sect. 4 using Bertini [1].

2 Local Witness Sets

Let $f : \mathbb{C}^N \to \mathbb{C}^n$ be a polynomial system, $V_1, \ldots, V_k$ be the irreducible components of $\mathcal{V}(f)$, and $x^* \in \mathcal{V}(f)$. If $x^* \in V_i$, then V_i localized at x^* can be decomposed uniquely, up to reordering, into a finite union of locally irreducible components $T_{i,1}, \ldots, T_{i,m_i}$, e.g., see Theorem 7 of [5, Chap. B]. If $x^* \notin V_i$, then V_i localized at x^* is empty, i.e., $m_i = 0$. Hence, the local irreducible decomposition of $\mathcal{V}(f)$ at x^* is $\bigcup_{i=1}^k \bigcup_{j=1}^{m_i} T_{i,j}$.

Example 1. Consider the irreducible polynomial $f(x) = x_1^2 - x_2^2 + x_2^4$. Hence, for a general $x^* \in \mathcal{V}(f) \subset \mathbb{C}^2$, the irreducible curve $\mathcal{V}(f)$ is locally irreducible at x^*. The origin arises as a self-crossing of the curve $\mathcal{V}(f)$ and hence decomposes into two locally irreducible components at the origin, say

$$T_{1,1}, T_{1,2} = \left\{ \left(x_1, \pm\sqrt{1 - \sqrt{1 - 4x_1^2}}\big/\sqrt{2} \right) \;\middle|\; x_1 \text{ near } 0 \right\}.$$

As with the global case, where witness sets form the key data structure in formulating a NID, *local witness sets* will be used to formulate a *numerical local irreducible decomposition* (NLID). The two key differences between a witness set and a local witness set, which we formally define below, are:

1. a local witness set is only well-defined on a neighborhood of x^*; and
2. all points in the local witness point set converge to x^* as the witness slice deforms to slice through x^*.

The key to understanding the local structure of an analytic set is the local parameterization theorem (see [5, Chap. C,D,E] and [6]). For a pure d-dimensional reduced analytic set $V \subset \mathbb{C}^N$ containing x^*, the local parameterization theorem implies (among other things) that there is an open ball $\mathcal{U} \subset \mathbb{C}^N$ centered at x^* such that given a general linear projection $\pi : \mathbb{C}^N \to \mathbb{C}^d$ and any open ball $B_\epsilon(\pi(x^*))$ with $\epsilon > 0$ small enough, the map $\pi_{\widehat{V}}$ is a proper branched covering from $\widehat{V} := V \cap \pi^{-1}(B_\epsilon(\pi(x^*))) \cap \mathcal{U}$ onto $B_\epsilon(\pi(x^*))$. Moreover, the sheet number is the multiplicity of the point x^* on V, denoted μ_{x^*}.

Remark 1. Since $\pi_{\widehat{V}}$ is proper, the Remmert proper mapping theorem implies that there is an analytic set $R \subset B_\epsilon(\pi(x^*))$ with $\dim R < d$ such that $\pi_{\widehat{V} \backslash \pi^{-1}(R)}$ is an unbranched μ_{x^*}-sheeted cover from $\widehat{V} \backslash \pi^{-1}(R)$ onto $B_\epsilon(\pi(x^*)) \backslash R$. Hence, if V is locally irreducible at x^*, then $\widehat{V} \backslash \pi^{-1}(R)$ is connected and the monodromy action on any fiber of $\pi_{\widehat{V} \backslash \pi^{-1}(R)}$ is transitive.

The local parameterization theorem is a local version of the Noether Normalization Theorem. For a pure d-dimensional algebraic set $V \subset \mathbb{C}^N$, the Noether Normalization Theorem states that the restriction π_V to V of a general linear projection $\pi : \mathbb{C}^N \to \mathbb{C}^d$ is a proper $\deg V$-to-one map of V onto $\mathbb{C}^d$. Given a general codimension d linear space $\mathcal{L}$ containing x^*, it follows that $\mathcal{L} \cap V$ consists of x^* and $\deg V - \mu_{x^*}$ smooth points. Given a preassigned open set $\mathcal{O}$ around $\mathcal{L} \cap V$, the intersection of any d codimensional linear space $\mathcal{L}'$ sufficiently near $\mathcal{L}$ will have $\mathcal{L}' \cap V \subset \mathcal{O}$. By choosing $\mathcal{O}$ as the intersection of V with $\deg V - \mu_{x^*} + 1$ disjoint small open balls, we see that the $\mathcal{L}' \cap V$ has precisely μ_{x^*} points near x^*.

Definition 1. *Let $f : \mathbb{C}^N \to \mathbb{C}^n$ be a system of functions which are holomorphic in a neighborhood of $x^* \in \mathbb{C}^N$ with $f(x^*) = 0$. Let $V \subset \mathbb{C}^N$ be a locally irreducible component of $\mathcal{V}(f)$ at x^* of dimension d and $\ell_1, \dots, \ell_d : \mathbb{C}^N \to \mathbb{C}$ be general linear polynomials such that $\ell_i(x^*) = 0$. For $u \in \mathbb{C}^d$, let $\mathcal{L}_u \subset \mathbb{C}^N$ be the linear space defined by $\ell_i(x) = u_i$ for $i = 1, \dots, d$. A local witness set for V is the triple $\{f, \mathcal{L}_{u^*}, W\}$ defined in a neighborhood $U \subset \mathbb{C}^d$ of the origin for general $u^* \in U$ and W is the finite subset of points in $V \cap \mathcal{L}_{u^*}$ which are the start points of the paths defined by $V \cap \mathcal{L}_{u(t)}$ where $u : [0,1] \to U$ is any path with $u(0) = 0$ and $u(1) = u^*$ which converge to x^* as $t \to 0$.*

Remark 2. The choice of points W inside of $V \cap \mathcal{L}_{u^*}$ is well-defined and equal to the multiplicity μ_{x^*} of V at x^*. We call μ_{x^*} the *local degree* of V at x^*.

Remark 3. When V is a curve, the neighborhood U is often referred to as the *endgame operating zone*, e.g., see [2, Sect. 3.3.1]. For all cases, we will call U the *generalized endgame operating zone*.

As Remark 1 suggests, one can perform monodromy loops using local witness sets similarly to classical witness sets. Local witness sets can also be used to sample components and to perform local membership testing.

In particular, a *numerical local irreducible decomposition* consists of a formal union of local witness sets, one for each local irreducible component.

Example 2. Reconsider f from Example 1 with $x^* = (0,0)$. For simplicity, we take $\ell_1(x) = x_1$ which then defines the neighborhood $U = \{u \in \mathbb{C} \mid |u| < 1/2\}$. We arbitrarily select $u^* = 1/6$ which implies that

$$\mathcal{V}(f) \cap \mathcal{L}_{u^*} = \left\{ \left(\tfrac{1}{6}, \pm\sqrt{\tfrac{1}{2} - \tfrac{\sqrt{2}}{3}} \right), \left(\tfrac{1}{6}, \pm\sqrt{\tfrac{1}{2} + \tfrac{\sqrt{2}}{3}} \right) \right\}.$$

As u (and hence x_1) deforms to 0, the first two points in $\mathcal{V}(f) \cap \mathcal{L}_{u^*}$ converge to x^* while the last two converge to $(0, \pm 1)$, respectively. For local irreducible components $T_{1,1}$ and $T_{1,2}$ of $\mathcal{V}(f)$ at x^*, local witness sets are

$$\mathcal{W}_1 = \left\{ f, \mathcal{L}_{u^*}, \left\{ \left(\tfrac{1}{6}, \sqrt{\tfrac{1}{2} - \tfrac{\sqrt{2}}{3}} \right) \right\} \right\} \text{ and } \mathcal{W}_2 = \left\{ f, \mathcal{L}_{u^*}, \left\{ \left(\tfrac{1}{6}, -\sqrt{\tfrac{1}{2} - \tfrac{\sqrt{2}}{3}} \right) \right\} \right\},$$

with each $T_{1,i}$ having local degree 1. Since $T_{1,1} \cup T_{1,2}$ form a local irreducible decomposition of $\mathcal{V}(f)$ at x^*, the formal union $\mathcal{W}_1 \cup \mathcal{W}_2$ is a NLID.

3   Computing Numerical Local Irreducible Decompositions

When decomposing a pure-dimensional set into its irreducible components, one simplification is to reduce down to the curve case. That is, if $V \subset \mathbb{C}^N$ is pure d-dimensional and $\mathcal{M} \subset \mathbb{C}^N$ is a general linear space of codimension $d - 1$, then the irreducible components of V correspond with the irreducible components of $V \cap \mathcal{M}$. Unfortunately, this need not hold for the local case.

Example 3. Consider $V = \mathcal{V}(x_1^2 + x_2^2 + x_3^2) \subset \mathbb{C}^3$ which is irreducible at the origin. For a general complex plane $\mathcal{L} = \mathcal{V}(a_1 x_1 + a_2 x_2 - x_3)$ through the origin, it is easy to check that $V \cap \mathcal{L}$ consists of two lines through the origin.

The following outlines a procedure for computing a NID that follows from Sect. 2. We assume that we are given a polynomial system $f : \mathbb{C}^N \to \mathbb{C}^n$ and a point $x^* \in \mathcal{V}(f)$. Since we can loop over the irreducible components of $\mathcal{V}(f)$, the key computation is to compute the NLID for an irreducible component $V \subset \mathcal{V}(f)$ given a witness set $\{f, \mathcal{L}, W\}$ for V with $d = \dim V$.

1. Select random linear polynomials $\ell_i : \mathbb{C}^N \to \mathbb{C}$ with $\ell_i(x^*) = 0$.
2. Pick random $u^* \in \mathbb{C}^d$ in the generalized endgame operating zone. Construct the linear spaces $\mathcal{L}_{u^*}$ and $\mathcal{L}_0$ defined by $\ell_i = u_i^*$ and $\ell_i = 0$, respectively. Compute $W' = V \cap \mathcal{L}_{u^*}$ via the homotopy defined by $V \cap (t \cdot \mathcal{L} + (1-t) \cdot \mathcal{L}_{u^*})$.
3. Compute W_{x^*} consisting of points $w \in W'$ such that the path defined by the homotopy $V \cap \mathcal{L}_{t \cdot u^*}$ starting at w at $t = 1$ limit to x^* as $t \to 0$.
4. Use monodromy loops inside the generalized endgame operating zone to compute the local monodromy group which partitions $W_{x^*} = W_1 \sqcup \cdots \sqcup W_s$. The NLID for V at x^* is defined by the formal union $\bigcup_{i=1}^{s} \{f, \mathcal{L}_{u^*}, W_i\}$.

Remark 4. The key to performing the same computation in the holomorphic case is to compute the finite set W_{x^*} in Item 3. The number of such points in W_{x^*} can be computed via a local multiplicity computation using Macaulay dual spaces [3,9] in certain cases. For example, if $x^* \in \mathbb{C}^N$ and $f : \mathbb{C}^N \to \mathbb{C}^{N-d}$ is a system of holomorphic functions at x^* such that the local dimension of $\mathcal{V}(f)$ at x^* is d, it follows from [4, pg. 158] that the multiplicity of $\{f, \ell_1, \ldots, \ell_d\}$ at x^* is equal to the number of points in W_{x^*}.

4   Examples

4.1   Illustrative Example

Consider the irreducible curve $V = \mathcal{V}(x_1^5 + 2x_2^5 - 3x_1 x_2(x_1 - x_2)(x_2 - x_1^2)) \subset \mathbb{C}^2$ with Fig. 1(a) plotting the real points of V and $x^* = (0,0)$. For simplicity, we take $\ell_1(x) = 2x_1 + 3x_2$, $u^* = 1/8$, and $\mathcal{L}_u$ defined by $\ell_1(x) = u$. Hence, $V \cap \mathcal{L}_{u^*}$ consists of five points, with four of the paths defined by the homotopy $V \cap \mathcal{L}_{t \cdot u^*}$ limiting to x^* as $t \to 0$. Therefore, W_{x^*} in Item 3 consists of 4 points.

We now perform monodromy loops which, in the curve case, means looping around 0. We observe that this loop breaks into 3 distinct cycles, two remain on their own branch and two interchange. Therefore, there are 3 local irreducible components as shown in Fig. 1(b), two of local degree 1 and one of local degree 2.

(a)   (b)

Fig. 1. Plot of (a) the real points of an irreducible quintic curve and (b) the real points near the origin, which locally decomposes into three components.

4.2   Local Irreducibility and Real Solutions

If the polynomial system f has real coefficients, the complex conjugate, $\text{conj}(V)$, of an irreducible component $V \subset \mathcal{V}(f)$ is also an irreducible component. If $V \neq \text{conj}(V)$, then all real points on V must be contained in $V \cap \text{conj}(V)$ where $\dim V > \dim(V \cap \text{conj}(V))$. For example, the "home" position of a cubic-center 12-bar mechanism [11], as presented in [10, Fig. 3], can be shown to be rigid, i.e., isolated over the real numbers, by observing that the only two irreducible components containing the "home" position are two sextic curves which are conjugates of each other [7].

   The NID is not always sufficient to reveal structure at singularities. Consider the Whitney umbrella $V = \mathcal{V}(x_1^2 - x_2^2 x_3) \subset \mathbb{C}^3$, which is an irreducible surface. For a random point on the "handle," i.e., $x^* = (0,0,\alpha)$ for random $\alpha \in \mathbb{C}$, the NLID reveals that V at x^* has two local irreducible components, each of local degree 1. At the origin, the NLID reveals that it is irreducible of local degree 2. When $\alpha < 0$, say $x^* = (0,0,-1)$, global information is not enough to observe that the real local dimension is smaller than the complex local dimension. However, the local viewpoint does indeed reveal that the two local irreducible components are complex conjugates of each other showing a smaller real local dimension.

4.3   Foldable Griffis-Duffy Platform

In our last example, we consider the "folded" pose, as shown in [8, Fig. 3], of a foldable Griffis-Duffy platform with the polynomial system available at [1] (see also [2, Chap. 8]). Our local approach verifies that the local irreducible decomposition of the "folded" pose consists of three double lines and a self-crossing of a quartic curve as mentioned in [8,10].

References

1. Bates, D.J., Hauenstein, J.D., Sommese, A.J., Wampler, C.W.: Bertini: Software for numerical algebraic geometry. http://bertini.nd.edu
2. Bates, D.J., Hauenstein, J.D., Sommese, A.J., Wampler, C.W.: Numerically Solving Polynomial Systems with Bertini. Software, Environments, and Tools, vol. 25. Society for Industrial and Applied Mathematics, Philadelphia (2013)

3. Dayton, B., Zeng, Z.: Computing the multiplicity structure in solving polynomial systems. In: Proceedings of ISSAC, pp. 166–123. ACM, New York (2005)
4. Fischer, G.: Complex Analytic Geometry. Lecture Notes in Mathematics, vol. 538. Springer, Berlin-New York (1976)
5. Gunning, R.C.: Introduction to Holomorphic Functions of Several Variables. Vol. II: Local Theory. Wadsworth & Brooks/Cole Advanced Books & Software, Monterey, CA (1990)
6. Gunning, R.C.: Lectures on Complex Analytic Varieties: The Local Parametrization Theorem. Mathematical Notes Princeton University Press, Princeton; University of Tokyo Press, Tokyo (1970)
7. Hauenstein, J.D.: Numerically computing real points on algebraic sets. Acta Appl. Math. **125**(1), 105–119 (2013)
8. Lu, Y., Bates, D.J., Sommese, A.J., Wampler, C.W.: Finding all real points of a complex curve. Contemp. Math. **448**, 183–205 (2007)
9. Macaulay, F.S.: The Algebraic Theory of Modular Systems. Cambridge University Press, Cambridge (1916)
10. Wampler, C.W., Hauenstein, J.D., Sommese, A.J.: Mechanism mobility and a local dimension test. Mech. Mach. Theory **46**(9), 1193–1206 (2011)
11. Wampler, C., Larson, B., Edrman, A.: A new mobility formula for spatial mechanisms. In: Proceedings of DETC/Mechanisms and Robotics Conference, Las Vegas, NV (CDROM), 4–7 September 2007

Computing the Chow Variety of Quadratic Space Curves

Peter Bürgisser[1], Kathlén Kohn[1]([✉]), Pierre Lairez[1], and Bernd Sturmfels[1,2]

[1] Institute of Mathematics, Technische Universität Berlin, Berlin, Germany
kohn@math.tu-berlin.de
[2] Department of Mathematics, University of California, Berkeley, USA

Abstract. Quadrics in the Grassmannian of lines in 3-space form a 19-dimensional projective space. We study the subvariety of coisotropic hypersurfaces. Following Gel'fand, Kapranov and Zelevinsky, it decomposes into Chow forms of plane conics, Chow forms of pairs of lines, and Hurwitz forms of quadric surfaces. We compute the ideals of these loci.

Keywords: Chow variety · Coisotropic hypersurface · Grassmannian · Space curve · Computation

1 Introduction

The Chow variety, introduced in 1937 by Chow and van der Waerden [4], parameterizes algebraic cycles of any fixed dimension and degree in a projective space, each given by its Chow form. The case of curves in $\mathbb{P}^3$ goes back to an 1848 paper by Cayley [3]. A fundamental problem, addressed by Green and Morrison [8] as well as Gel'fand, Kapranov and Zelevinsky [6, Sect. 4.3], is to describe the equations defining Chow varieties. We present a definitive computational solution for the smallest non-trivial case, namely for cycles of dimension 1 and degree 2 in $\mathbb{P}^3$.

The *Chow form* of a cycle of degree 2 is a quadratic form in the Plücker coordinates of the Grassmannian $G(2,4)$ of lines in $\mathbb{P}^3$. Such a quadric in $G(2,4)$ represents the set of all lines that intersect the given cycle. Quadratic forms in Plücker coordinates form a projective space $\mathbb{P}^{19}$. The Chow variety we are interested in, denoted $G(2,2,4)$, is the set of all Chow forms in that $\mathbb{P}^{19}$. The aim of this note is to make the concepts in [3,4,8] and [6, Sect. 4.3] completely explicit.

We start with the 9-dimensional subvariety of $\mathbb{P}^{19}$ whose points are the *coisotropic quadrics* in $G(2,4)$. By [6, Sect. 4.3, Theorem 3.14], this decomposes as the Chow variety and the variety of Hurwitz forms [9], representing lines

P. Bürgisser and P. Lairez were partially supported by DFG grant BU 1371/2-2.
K. Kohn was supported by a Fellowship from the Einstein Foundation Berlin.
B. Sturmfels was supported by the US National Science Foundation and the Einstein Foundation Berlin.

© Springer International Publishing Switzerland 2016
I.S. Kotsireas et al. (Eds.): MACIS 2015, LNCS 9582, pp. 130–136, 2016.
DOI: 10.1007/978-3-319-32859-1_10

that are tangent to a quadric surface in $\mathbb{P}^3$. Section 2 studies the ideal generated by the coisotropy conditions. We work in a polynomial ring in 20 variables, one for each quadratic Plücker monomial on $G(2,4)$ minus one for the Plücker relation. We derive the coisotropic ideal from the differential characterization of coisotropy. Proposition 1 exhibits the decomposition of this ideal into three minimal primes. In particular, this shows that the coisotropic ideal is radical, and it hence resolves the degree 2 case of a problem posed in 1986 by Green and Morrison [8]. They wrote: 'We do not know whether [the differential characterization of coisotropy] generates the full ideal of these Chow variables.'

Section 3 derives the radical ideal of the Chow variety $G(2,2,4)$ in $\mathbb{P}^{19}$. Its two minimal primes represent Chow forms of plane conics and Chow forms of pairs of lines. We also study the characterization of Chow forms among all coisotropic quadrics by the vanishing of certain differential forms. These represent the integrability of the α-distribution in [6, Sect. 4.3, Theorem 3.22]. After saturation by the irrelevant ideal, the integrability ideal is found to be radical.

2   Coisotropic Quadrics

The Grassmannian $G(2,4)$ is a quadric in $\mathbb{P}^5$. Its points are lines in $\mathbb{P}^3$. We represent these lines using dual Plücker coordinates $\boldsymbol{p} = (p_{01}, p_{02}, p_{03}, p_{12}, p_{13}, p_{23})$ subject to the Plücker relation $p_{01}p_{23} - p_{02}p_{13} + p_{03}p_{12}$. Following [9, Sect. 2], by *dual* coordinates we mean that p_{ij} is the ij-minor of a 2×4-matrix whose rows span the line. The generic quadric in $G(2,4)$ is written as a generic quadratic form

$$Q(\boldsymbol{p}) = \boldsymbol{p} \cdot \begin{pmatrix} c_0 & c_1 & c_2 & c_3 & c_4 & c_5 \\ c_1 & c_6 & c_7 & c_8 & c_9 & c_{10} \\ c_2 & c_7 & c_{11} & c_{12} & c_{13} & c_{14} \\ c_3 & c_8 & c_{12} & c_{15} & c_{16} & c_{17} \\ c_4 & c_9 & c_{13} & c_{16} & c_{18} & c_{19} \\ c_5 & c_{10} & c_{14} & c_{17} & c_{19} & c_{20} \end{pmatrix} \cdot \boldsymbol{p}^T . \tag{1}$$

The quadric $Q(\boldsymbol{p})$ is an element in $V := \mathbb{C}[\boldsymbol{p}]_2/\mathbb{C}\{p_{01}p_{23} - p_{02}p_{13} + p_{03}p_{12}\} \cong \mathbb{C}^{21}/\mathbb{C}$. Hence, $\boldsymbol{c} = (c_0, c_1, \ldots, c_{20})$ serves as homogeneous coordinates on $\mathbb{P}^{19} = \mathbb{P}(V)$, which – due to the Plücker relation – need to be understood modulo

$$c_5 \mapsto c_5 + \lambda, \quad c_9 \mapsto c_9 - \lambda, \quad c_{12} \mapsto c_{12} + \lambda . \tag{2}$$

The coordinate ring $\mathbb{Q}[V]$ is a subring of $\mathbb{Q}[c_0, c_1, \ldots, c_{20}]$, namely it is the invariant ring of the additive group action (2). Hence $\mathbb{Q}[V]$ is the polynomial ring in 20 variables $c_0, c_1, c_2, c_3, c_4, c_5 - c_{12}, c_6, c_7, c_8, c_9 + c_{12}, c_{10}, c_{11}, c_{13}, \ldots, c_{20}$.

We are interested in the $\boldsymbol{c}$'s that lead to *coisotropic* hypersurfaces of $G(2,4)$. For these, the tangent space at any point ℓ, considered as a subspace of $T_\ell G(2,4) = \mathrm{Hom}(\ell, \mathbb{C}^4/\ell)$, has the form $\{\varphi \mid \varphi(a) = 0\} + \{\varphi \mid \mathrm{im}(\varphi) \subset M\}$, for some $a \in \ell \backslash \{0\}$ and some plane M in $\mathbb{C}^4/\ell$. By [6, Sect. 4.3, (3.24)], the quadric hypersurface $\{Q(\boldsymbol{p}) = 0\}$ in $G(2,4)$ is coisotropic if and only if there exist $s, t \in \mathbb{C}$ such that

$$\frac{\partial Q}{\partial p_{01}} \cdot \frac{\partial Q}{\partial p_{23}} - \frac{\partial Q}{\partial p_{02}} \cdot \frac{\partial Q}{\partial p_{13}} + \frac{\partial Q}{\partial p_{03}} \cdot \frac{\partial Q}{\partial p_{12}} = s \cdot Q + t \cdot (p_{01}p_{23} - p_{02}p_{13} + p_{03}p_{12}) . \quad (3)$$

Equivalently, the vector $(t, s, -1)^T$ is in the kernel of the 21×3 matrix in Fig. 1. The 3×3 minors of this matrix are all in the subring $\mathbb{Q}[V]$. The *coisotropic ideal I* is the ideal of $\mathbb{Q}[V]$ generated by these minors. The subscheme $V(I)$ of $\mathbb{P}^{19} = \mathbb{P}(V)$ represents all coisotropic hypersurfaces $\{Q = 0\}$ of degree two in $G(2,4)$. Using computations with Maple and Macaulay2 [7], we found that I has codimension 10, degree 92 and is minimally generated by 175 cubics. Besides, $V(I)$ is the reduced union of three components, of dimensions nine, eight and five.

Proposition 1. *The coisotropic ideal is the intersection of three prime ideals:*

$$I = P_{\text{Hurwitz}} \cap P_{\text{ChowLines}} \cap P_{\text{Squares}}. \quad (4)$$

So, I is radical. The prime P_{Hurwitz} has codimension 10 and degree 92, it is minimally generated by 20 quadrics, and its variety $V(P_{\text{Hurwitz}})$ consists of Hurwitz forms of quadric surfaces in $\mathbb{P}^3$. The prime $P_{\text{ChowLines}}$ has codimension 11 and degree 140, it is minimally generated by 265 cubics, and $V(P_{\text{ChowLines}})$ consists of Chow forms of pairs of lines in $\mathbb{P}^3$. The prime P_{Squares} has codimension 14 and degree 32, it is minimally generated by 84 quadrics, and $V(P_{\text{Squares}})$ consists of all quadrics $Q(\boldsymbol{p})$ that are squares modulo the Plücker relation.

$$
\begin{pmatrix}
0 & c_0 & 2c_0c_5 - 2c_1c_4 + 2c_2c_3 \\
0 & c_1 & c_0c_{10} - c_1c_9 + c_2c_8 + c_3c_7 - c_4c_6 + c_1c_5 \\
0 & c_2 & c_0c_{14} - c_1c_{13} + c_2c_{12} + c_3c_{11} - c_4c_7 + c_2c_5 \\
0 & c_3 & c_0c_{17} - c_1c_{16} + c_2c_{15} + c_3c_{12} - c_4c_8 + c_3c_5 \\
0 & c_4 & c_0c_{19} - c_1c_{18} + c_2c_{16} + c_3c_{13} - c_4c_9 + c_4c_5 \\
1 & c_5 & c_0c_{20} - c_1c_{19} + c_2c_{17} + c_3c_{14} - c_4c_{10} + c_5^2 \\
0 & c_6 & 2c_1c_{10} - 2c_6c_9 + 2c_7c_8 \\
0 & c_7 & c_1c_{14} - c_6c_{13} + c_7c_{12} + c_8c_{11} + c_2c_{10} - c_7c_9 \\
0 & c_8 & c_1c_{17} - c_6c_{16} + c_7c_{15} + c_8c_{12} + c_3c_{10} - c_8c_9 \\
-1 & c_9 & c_1c_{19} - c_6c_{18} + c_7c_{16} + c_8c_{13} + c_4c_{10} - c_9^2 \\
0 & c_{10} & c_1c_{20} - c_6c_{19} + c_7c_{17} + c_8c_{14} - c_9c_{10} + c_5c_{10} \\
0 & c_{11} & 2c_2c_{14} - 2c_7c_{13} + 2c_{11}c_{12} \\
1 & c_{12} & c_2c_{17} - c_7c_{16} + c_{11}c_{15} + c_3c_{14} - c_8c_{13} + c_{12}^2 \\
0 & c_{13} & c_2c_{19} - c_7c_{18} + c_{11}c_{16} + c_4c_{14} + c_{12}c_{13} - c_9c_{13} \\
0 & c_{14} & c_2c_{20} - c_7c_{19} + c_{11}c_{17} + c_{12}c_{14} + c_5c_{14} - c_{10}c_{13} \\
0 & c_{15} & 2c_3c_{17} - 2c_8c_{16} + 2c_{12}c_{15} \\
0 & c_{16} & c_3c_{19} - c_8c_{18} + c_4c_{17} + c_{12}c_{16} - c_9c_{16} + c_{13}c_{15} \\
0 & c_{17} & c_3c_{20} - c_8c_{19} + c_{12}c_{17} + c_5c_{17} - c_{10}c_{16} + c_{14}c_{15} \\
0 & c_{18} & 2c_4c_{19} - 2c_9c_{18} + 2c_{13}c_{16} \\
0 & c_{19} & c_4c_{20} - c_9c_{19} + c_5c_{19} - c_{10}c_{18} + c_{13}c_{17} + c_{14}c_{16} \\
0 & c_{20} & 2c_5c_{20} - 2c_{10}c_{19} + 2c_{14}c_{17}
\end{pmatrix}
$$

Fig. 1. This matrix has rank ≤ 2 if and only if the quadric given by $\boldsymbol{c}$ is coisotropic.

This proposition answers a question due to Green and Morrison, who had asked in [8] whether I is radical. To derive the prime decomposition (4), we computed the three prime ideals as kernels of homomorphisms of polynomial rings, each expressing the relevant geometric condition. This construction ensures that the ideals are prime. We then verified that their intersection equals I. For details, check our computations, using the link given at the end of this article.

From the geometric perspective of [6], the third prime P_{Squares} is extraneous, because nonreduced hypersurfaces in $G(2,4)$ are excluded by Gel'fand, Kapranov and Zelevinsky. Theorem 3.14 in [6, Sect. 4.3] concerns irreducible hypersurfaces, and the identification of Chow forms within the coisotropic hypersurfaces [6, Sect. 4.3, Theorem 3.22] assumes the corresponding polynomial to be squarefree. With this, the following would be the correct ideal for the coisotropic variety in $\mathbb{P}^{19}$:

$$P_{\text{Hurwitz}} \cap P_{\text{ChowLines}} = (I : P_{\text{Squares}}). \tag{5}$$

This means that the reduced coisotropic quadrics in $G(2,4)$ are either Chow forms of curves or Hurwitz forms of surfaces. The ideal in (5) has codimension 10, degree 92, and is minimally generated by 175 cubics and 20 quartics in $\mathbb{Q}[V]$.

A slightly different point of view on the coisotropic ideal is presented in a recent paper of Catanese [2]. He derives a variety in $\mathbb{P}^{20} = \mathbb{P}(\mathbb{C}[\boldsymbol{p}]_2)$ which projects isomorphically onto our variety $V(I) \subset \mathbb{P}^{19}$. The center of projection is the Plücker quadric. To be precise, Proposition 4.1 in [2] states the following: For every $Q \in \mathbb{C}[\boldsymbol{p}]_2 \backslash \mathbb{C} (p_{01}p_{23} - p_{02}p_{13} + p_{03}p_{12})$ satisfying (3) there is a unique $\lambda \in \mathbb{C}$ such that the quadric $Q_\lambda := Q + \lambda \cdot (p_{01}p_{23} - p_{02}p_{13} + p_{03}p_{12})$ satisfies

$$\frac{\partial Q_\lambda}{\partial p_{01}} \cdot \frac{\partial Q_\lambda}{\partial p_{23}} - \frac{\partial Q_\lambda}{\partial p_{02}} \cdot \frac{\partial Q_\lambda}{\partial p_{13}} + \frac{\partial Q_\lambda}{\partial p_{03}} \cdot \frac{\partial Q_\lambda}{\partial p_{12}} = t \cdot (p_{01}p_{23} - p_{02}p_{13} + p_{03}p_{12}) \tag{6}$$

for some $t \in \mathbb{C}$. This implies that $V(I)$ is isomorphic to the variety of all $Q \in \mathbb{P}(\mathbb{C}[\boldsymbol{p}]_2) \backslash \{p_{01}p_{23} - p_{02}p_{13} + p_{03}p_{12}\}$ satisfying (6). Let I_2 be generated by the 2×2 minors of the 21×2 matrix that is obtained by deleting the middle column of the matrix in Fig. 1. Then $V(I_2)$ contains exactly those $Q \in \mathbb{P}(\mathbb{C}[\boldsymbol{p}]_2)$ satisfying (6), and $V(I)$ is the projection of $V(I_2)$ from the center $(p_{01}p_{23} - p_{02}p_{13} + p_{03}p_{12})$. The ideal I_2 has codimension 11, degree 92, and is minimally generated by 20 quadrics. Interestingly, Catanese shows furthermore in [2, Theorem 3.3] that a hypersurface in $G(2,4)$ is coisotropic if and only if it is selfdual in $\mathbb{P}^5$ with respect to the inner product given by the Plücker quadric.

3   The Chow Variety

In this section we study the Chow variety $G(2,2,4)$ of one-dimensional algebraic cycles of degree two in $\mathbb{P}^3$. By [6, Sect. 4.1, Example 1.3], the Chow variety $G(2,2,4)$ is the union of two irreducible components of dimension eight in $\mathbb{P}^{19}$, one corresponding to planar quadrics and the other to pairs of lines. Formally, this means that $G(2,2,4) = V(P_{\text{ChowConic}}) \cup V(P_{\text{ChowLines}})$, where $P_{\text{ChowConic}}$ is the homogeneous prime ideal in $\mathbb{Q}[V]$ whose variety comprises the Chow forms of irreducible curves of degree two in $\mathbb{P}^3$. The ideal $P_{\text{ChowConic}}$ has codimension 11 and degree 92, and it is minimally generated by 21 quadrics and 35 cubics. The radical ideal $P_{\text{ChowConic}} \cap P_{\text{ChowLines}}$ has codimension 11, degree $232 = 92 + 140$, and it is minimally generated by 230 cubics.

Since $G(2, 2, 4)$ should be contained in the coisotropic variety $V(I)$, it seems that $P_{\text{ChowConic}}$ is missing from the decomposition (4). Here is the explanation:

Proposition 2. *Every Chow form of a plane conic in $\mathbb{P}^3$ is also a Hurwitz form. In symbols, $P_{\text{Hurwitz}} \subset P_{\text{ChowConic}}$ and thus $V(P_{\text{ChowConic}}) \subset V(P_{\text{Hurwitz}})$.*

Our first proof is by computer: just check the inclusion of ideals in Macaulay2. For a conceptual proof, we consider a 4×4-symmetric matrix $M = M_0 + \epsilon M_1$, where $\text{rank}(M_0) = 1$. By [9, Equation (1)], the Hurwitz form of the corresponding quadric surface in $\mathbb{P}^3$ is $Q(\boldsymbol{p}) = \boldsymbol{p}(\wedge_2 M)\boldsymbol{p}^T$. Divide by ϵ and let $\epsilon \to 0$. The limit is the Chow form of the plane conic defined by restricting M_1 to $\ker(M_0) \simeq \mathbb{P}^2$. This type of degeneration is familiar from the study of complete quadrics [5]. Proposition 2 explains why the locus of irreducible curves is not visible in (4).

Gel'fand, Kapranov and Zelevinsky [6, Sect. 4.3] introduce a class of differential forms in order to discriminate Chow forms among all coisotropic hypersurfaces. In their setup, these forms represent the integrability of the α-distribution $\mathcal{E}_{\alpha, Z}$. We shall apply the tools of computational commutative algebra to shed some light on the characterization of Chow forms via integrability of α-distributions.

For this, we use local affine coordinates instead of Plücker coordinates. A point in the Grassmannian $G(2, 4)$ is represented as the row space of the matrix

$$\begin{pmatrix} 1 & 0 & a_2 & a_3 \\ 0 & 1 & b_2 & b_3 \end{pmatrix}. \tag{7}$$

We express the quadrics Q in (1) in terms of the local coordinates a_2, a_3, b_2, b_3, by substituting the Plücker coordinates with the minors of the matrix (7), i.e.,

$$p_{01} = 1, \quad p_{02} = b_2, \quad p_{03} = b_3, \quad p_{12} = -a_2, \quad p_{13} = -a_3, \quad p_{23} = a_2 b_3 - b_2 a_3. \tag{8}$$

We consider the following differential 1-forms on affine 4-space:

$$\alpha_1^1 := \frac{\partial Q}{\partial a_2} da_2 + \frac{\partial Q}{\partial a_3} da_3, \quad \alpha_2^1 := \frac{\partial Q}{\partial a_2} db_2 + \frac{\partial Q}{\partial a_3} db_3,$$

$$\alpha_1^2 := \frac{\partial Q}{\partial b_2} da_2 + \frac{\partial Q}{\partial b_3} da_3, \quad \alpha_2^2 := \frac{\partial Q}{\partial b_2} db_2 + \frac{\partial Q}{\partial b_3} db_3.$$

By taking wedge products, we derive the 16 differential 4-forms

$$dQ \wedge d\alpha_j^i \wedge \alpha_l^k = q_{ijkl} \cdot da_2 \wedge da_3 \wedge db_2 \wedge db_3 \quad \text{for } i, j, k, l \in \{1, 2\}. \tag{9}$$

Here the expressions q_{ijkl} are certain polynomials in $\mathbb{Q}[V][a_2, a_3, b_2, b_3]$.

Theorems 3.19 and 3.22 in [6, Sect. 4.3] state that a squarefree coisotropic quadric Q is a Chow form if and only if all 16 coefficients q_{ijkl} are multiples of Q. By taking normal forms of the polynomials q_{ijkl} modulo the principal ideal $\langle Q \rangle$, we obtain a collection of 720 homogeneous polynomials in c. Among these, 58 have degree three, 340 have degree four, and 322 have degree five. The aforementioned result implies that these 720 polynomials cut out $G(2,2,4)$ as a subset of $\mathbb{P}^{19}$.

The *integrability ideal* $J \subset \mathbb{Q}[V]$ is generated by these 720 polynomials and their analogues from other affine charts of the Grassmannian, obtained by permuting columns in (7). We know that $V(J)$ equals the union of $G(2,2,4)$ with all double hyperplanes in $G(2,4)$ (corresponding to P_{Squares}) set-theoretically. Maple, Macaulay2 and Magma verified for us that it holds scheme-theoretically:

Proposition 3. *The integrability ideal J is minimally generated by 210 cubics. Writing $\mathfrak{m}$ for the irrelevant ideal $\langle c_0, c_1, \ldots, c_{20} \rangle$ of $\mathbb{Q}[V]$, we have*

$$\sqrt{J} = (J : \mathfrak{m}) = P_{\text{ChowConic}} \cap P_{\text{ChowLines}} \cap P_{\text{Squares}}. \tag{10}$$

4 Conclusion

We reported on computational experiments with hypersurfaces in the Grassmannian $G(2,4)$ that are associated to curves and surfaces in $\mathbb{P}^3$. For degree 2, all relevant parameter spaces were described by explicit polynomials in 20 variables. All ideals and computations discussed in this note can be obtained at

www3.math.tu-berlin.de/algebra/static/pluecker/

Many possibilities exist for future work. Obvious next milestones are the ideals for the Chow varieties of degree 3 cycles in $\mathbb{P}^3$, and degree 2 cycles in $\mathbb{P}^4$. Methods from representation theory promise a compact encoding of their generators, in terms of irreducible GL(4)-modules. Another question we aim to pursue is motivated by the geometry of condition numbers [1]: express the volume of a tubular neighborhood of a coisotropic quadric in $G(2,4)$ as a function of c.

References

1. Bürgisser, P., Cucker, F.: Condition: The Geometry of Numerical Algorithms. Springer, Heidelberg (2013)
2. Catanese, F.: Cayley forms and self-dual varieties. Proc. Edinb. Math. Soc. **57**, 89–109 (2014)
3. Cayley, A.: On the theory of elimination. Camb. Dublin Math. J. **3**, 116–120 (1848)
4. Chow, W.-L., van der Waerden, B.L.: Zur algebraischen Geometrie. IX. Über zugeordnete Formen und algebraische Systeme von algebraischen Mannigfaltigkeiten. Math. Ann. **113**, 696–708 (1937)
5. DeConcini, C., Goresky, M., MacPherson, R., Procesi, C.: On the geometry of quadrics and their degenerations. Comment. Math. Helvetici **63**, 337–413 (1988)
6. Gel'fand, I.M., Kapranov, M.M., Zelevinsky, A.V.: Discriminants, Resultants and Multidimensional Determinants. Birkhäuser, Boston (1994)

7. Grayson, D., Stillman, M.: Macaulay2, a software system for research in algebraic geometry. www.math.uiuc.edu/Macaulay2/
8. Green, M., Morrison, I.: The equations defining Chow varieties. Duke Math. J. **53**, 733–747 (1986)
9. Sturmfels, B.: The Hurwitz form of a projective variety. arXiv:1410.6703

Numerically Testing Generically Reduced Projective Schemes for the Arithmetic Gorenstein Property

Noah S. Daleo[1](✉) and Jonathan D. Hauenstein[2]

[1] Department of Mathematics, Worcester State University,
Worcester, MA 01602, USA
ndaleo@worcester.edu

[2] Department of Applied and Computational Mathematics and Statistics,
University of Notre Dame, Notre Dame, IN 46556, USA
hauenstein@nd.edu
http://www.worcester.edu/noah-daleo
http://www.nd.edu/~jhauenst

Abstract. Let $X \subset \mathbb{P}^n$ be a generically reduced projective scheme. A fundamental goal in computational algebraic geometry is to compute information about X even when defining equations for X are not known. We use numerical algebraic geometry to develop a test for deciding if X is arithmetically Gorenstein and apply it to three secant varieties.

1 Introduction

When the defining ideal of a generically reduced projective scheme $X \subset \mathbb{P}^n$ is unknown, numerical methods based on sample points may be used to determine properties of X. In [4], numerical algebraic geometry was used to decide if X is arithmetically Cohen-Macaulay based on the Hilbert functions of subschemes of X. In our present work, we expand this to decide if X is arithmetically Gorenstein. Our method relies on numerically interpolating points approximately lying on a general curve section of X as well as a *witness point set* for X, which is defined in Sect. 2.4. This test does not assume that one has access to polynomials vanishing on X, e.g., X may be the image of an algebraic set under a polynomial map. In such cases, our method is an example of *numerical elimination theory* (see [2, Chap. 16] and [3]).

Much of the literature regarding arithmetically Gorenstein schemes focuses on the case in which the codimension is at most 3 (see, e.g., [6,8,10]), but less is known for larger codimensions. Our test is applicable to schemes of any codimension. For example, Sects. 4.2 and 4.3 consider schemes of codimension 6.

The rest of this article is organized as follows. In Sect. 2, we provide prerequisite background material. In Sect. 3, we describe a numerical test for whether or not a scheme is arithmetically Gorenstein. In Sect. 4, we demonstrate this test on three examples.

© Springer International Publishing Switzerland 2016
I.S. Kotsireas et al. (Eds.): MACIS 2015, LNCS 9582, pp. 137–142, 2016.
DOI: 10.1007/978-3-319-32859-1_11

2   Background

2.1   Arithmetically Cohen-Macaulay and Arithmetically Gorenstein

If $X \subset \mathbb{P}^n$ is a projective scheme with ideal sheaf $\mathcal{I}_X$, then X is said to be *arithmetically Cohen-Macaulay (aCM)* if

$$H_*^i(\mathcal{I}_X) = 0 \text{ for } 1 \leq i \leq \dim X$$

where $H_*^i(\mathcal{I}_X)$ is the ith cohomology module of $\mathcal{I}_X$. In particular, all zero-dimensional schemes are aCM and every aCM scheme is pure-dimensional. If X is aCM, then its *Cohen-Macaulay type* is the rank of the last free module in a minimal free resolution of $\mathcal{I}_X$. An aCM scheme X is said to be *arithmetically Gorenstein (aG)* if X has Cohen-Macaulay type 1.

We will make use of the following fact about Cohen-Macaulay type [11, Corollary 1.3.8].

Theorem 1. *Let $X \subset \mathbb{P}^n$ be an aCM scheme with $\dim X \geq 1$ and $H \subset \mathbb{P}^n$ be a general hypersurface of degree $d \geq 1$. Then $X \cap H$ is aCM and has the same Cohen-Macaulay type as X.*

2.2   Hilbert Functions

Suppose that $X \subset \mathbb{P}^n$ is a nonempty scheme and consider the corresponding homogeneous ideal $I \subset \mathbb{C}[x_0, \ldots, x_n]$. Let $\mathbb{C}[x_0, \ldots, x_n]_t$ denote the vector space of homogeneous polynomials of degree t, which has dimension $\binom{n+t}{t}$, and let $I_t = I \cap \mathbb{C}[x_0, \ldots, x_n]_t$. Then, the *Hilbert function of X* is the function $HF_X : \mathbb{Z} \to \mathbb{Z}$ defined by

$$HF_X(t) = \begin{cases} 0 & \text{if } t < 0 \\ \binom{n+t}{t} - \dim I_t & \text{otherwise.} \end{cases}$$

The *Hilbert series of X*, denoted HS_X, is the generating function of HF_X, namely,

$$HS_X(t) = \sum_{j=0}^{\infty} HF_X(j) \cdot t^j.$$

There is a polynomial $P(t) = c_0 + c_1 t + c_2 t^2 + \cdots + c_r t^r$ with $\deg X = P(1)$ such that

$$HS_X(t) = \frac{P(t)}{(1-t)^{\dim X+1}}.$$

The vector of coefficients $[c_0 \ c_1 \ c_2 \ \cdots \ c_r]$ is called the *h-vector* of X. If X is aG, i.e., aCM of Cohen-Macaulay type 1, then the *h*-vector of X is symmetric: $c_i = c_{r-i}$ [13, Theorem 4.1]. Therefore, two necessary conditions on X to be aG are pure-dimensionality and a symmetric *h*-vector. These conditions can be used to identify schemes which are not aG, e.g., see Sect. 4.2.

2.3   Cayley-Bacharach Property

Let $Z \subset \mathbb{P}^n$ be a nonempty reduced zero-dimensional scheme with h-vector $[c_0 \ c_1 \ c_2 \ \cdots \ c_r]$. The scheme Z is said to have the *Cayley-Bacharach (C-B) property* if, for every subset $Y \subset Z$ with $|Y| = |Z| - 1$, $HF_Y(r-1) = HF_Z(r-1)$. The following, which is [5, Theorem 5], relates the C-B property to aG schemes.

Theorem 2. *If $Z \subset \mathbb{P}^n$ is a nonempty reduced zero-dimensional scheme, Z is arithmetically Gorenstein if and only if Z has the Cayley-Bacharach property and its h-vector is symmetric.*

2.4   Witness Point Sets

For a pure-dimensional generically reduced scheme $X \subset \mathbb{P}^n$, let $\mathcal{L} \subset \mathbb{P}^n$ be a general linear space with $\dim \mathcal{L} = \operatorname{codim} X$. The set $W = X \cap \mathcal{L}$ is called a *witness point set* for X.

3   Method

For a pure-dimensional generically reduced scheme $X \subset \mathbb{P}^n$, one can determine that X is arithmetically Gorenstein by combining Theorems 1 and 2. We describe the zero-dimensional and positive-dimensional cases below. A generalization of this approach, using Macaulay dual spaces, for pure-dimensional schemes that are not generically reduced is currently being written by the authors and will be presented elsewhere.

3.1   Reduced Zero-Dimensional Schemes

If $\dim X = 0$, we can simply apply Theorem 2 to determine if X is aG. That is, given a numerical approximation of each point in X, we use the numerical interpolation approach described in [7] to compute the Hilbert function of X. In particular, there is an integer $\rho_X \geq 0$, which is called the *index of regularity* of X, such that

$$0 = HF_X(-1) < 1 = HF_X(0) < \cdots < HF_X(\rho_X - 1) < HF_X(\rho_X) = HF_X(\rho_X + 1) = \cdots = |X|.$$

The h-vector for X is $[c_0 \ c_1 \ \cdots \ c_{\rho_X}]$ where $c_t = HF_X(t) - HF_X(t-1)$. Thus, we can now test for symmetry of the h-vector, i.e., $c_i = c_{\rho_X - i}$.

If the h-vector is symmetric, we then test for the Cayley-Bacharach property. That is, for each $Y \subset X$ with $|Y| = |X| - 1$, we use [7] to compute $HF_Y(\rho_X - 1)$. If $HF_Y(\rho_X - 1) = HF_X(\rho_X - 1)$ for every such subset Y, then X has the C-B property. Hence, if the h-vector is symmetric and X has the C-B property, then X is aG.

Example 1. Consider $X = \{[0, 1, 1], [0, 1, 2], [0, 1, 3], [1, 1, -1]\} \subset \mathbb{P}^2$. It is easy to verify that $\rho_X = 2$ and the h-vector for X is $[1 \ 2 \ 1]$, which is symmetric. However, X does not have the Cayley-Bacharach property and thus is not aG, since $HF_Y(1) = 2 \neq 3 = HF_X(1)$ for $Y = \{[0, 1, 1], [0, 1, 2], [0, 1, 3]\}$.

3.2   Generically Reduced Positive-Dimensional Schemes

If $\dim X \geq 1$, Theorems 1 and 2 show that X is aG if and only if X is aCM and a witness point set for X is aG, i.e., has a symmetric h-vector and has the C-B property. We start with the witness point set condition and then summarize the aCM test presented in [4].

Let $W = X \cap \mathcal{L}$ be a witness point set for X defined by the general linear slice $\mathcal{L}$. We apply the strategy of Sect. 3.1 to W with one simplification for deciding that W has the C-B property. This simplification arises from the fact that witness point sets for an irreducible scheme have the so-called *uniform position property*. That is, if X is irreducible, then W has the C-B property if and only if $HF_Y(\rho_W - 1) = HF_W(\rho_W - 1)$ for *any* $Y \subset W$ with $|Y| = |W| - 1$. In general, if X has k irreducible components, say $X_1, \ldots, X_k$ with $W_i = X_i \cap \mathcal{L}$, then the witness point set W has the C-B property if and only if, for $i = 1, \ldots, k$, $HF_{Z_i}(\rho_W - 1) = HF_W(\rho_W - 1)$ where $Z_i = \bigcup_{j \neq i} W_j \cup Y_i$ for *any* $Y_i \subset W_i$ with $|Y_i| = |W_i| - 1$.

If W is aG, then X is aG if and only if X is aCM. The arithmetically Cohen-Macaulayness of X is decided using the approach of [4] by comparing the Hilbert function of W and the Hilbert function of a general curve section of X as follows. Let $\mathcal{M} \subset \mathbb{P}^n$ be a general linear space with $\dim \mathcal{M} = \operatorname{codim} X + 1$ and $C = X \cap \mathcal{M}$, i.e., $\dim C = 1$. By numerically sampling points approximately lying on C, we compute $HF_C(t)$ via [7] for $t = 1, \ldots, \rho_W + 1$. The following is a version of [4, Corollary 3.3] that decides the arithmetically Cohen-Macaulayness of X via HF_W and HF_C.

Theorem 3. *With the setup given above, X is arithmetically Cohen-Macaulay if and only if $HF_W(t) = HF_C(t) - HF_C(t - 1)$ for $t = 1, \ldots, \rho_W + 1$.*

4   Examples

It has been speculated that the homogeneous coordinate ring of any secant variety of any Segre product of projective spaces is Cohen-Macaulay [12], but some examples of such secant varieties are known to not be arithmetically Gorenstein [9]. We demonstrate our test on two such secant varieties in Sects. 4.1 and 4.2. Section 4.3 considers a secant variety of a Veronese variety.

4.1 $\sigma_3(\mathbb{P}^1 \times \mathbb{P}^1 \times \mathbb{P}^1 \times \mathbb{P}^1)$

Let $X = \sigma_3(\mathbb{P}^1 \times \mathbb{P}^1 \times \mathbb{P}^1 \times \mathbb{P}^1) \subset \mathbb{P}^{15}$, which is the third secant variety to the Segre product of $\mathbb{P}^1 \times \mathbb{P}^1 \times \mathbb{P}^1 \times \mathbb{P}^1$ with $\dim X = 13$. We computed a witness point set W for X using Bertini [1] and found that $\deg X = 16$. Using [7], we compute

$$\rho_W = 6, \quad HF_W = 1, 3, 6, 10, 13, 15, 16, 16, \quad \text{and} \quad h = [1\ 2\ 3\ 4\ 3\ 2\ 1].$$

Clearly, the h-vector for W is symmetric. Since X is irreducible, we selected one subset $Y \subset W$ consisting of 15 points. The witness point set W has the Cayley-Bacharach property since $HF_Y(5) = 15 = HF_W(5)$ and thus we conclude W is arithmetically Gorenstein by Theorem 2.

Next, we consider the arithmetically Cohen-Macaulayness of X. Let $\mathcal{M} \subset \mathbb{P}^{15}$ be a general linear space with $\dim \mathcal{M} = 3$ and $C = X \cap \mathcal{M}$. Via sampling C, we find that

$$HF_C = 1, 4, 10, 20, 33, 48, 64, 80.$$

Therefore, by Theorem 3, X is arithmetically Cohen-Macaulay and, hence, we can conclude it is arithmetically Gorenstein by Theorem 1. In fact, since X is aCM, we can observe from HF_W that two polynomials of degree 4 must vanish on X. We found that these two polynomials generate the ideal of X meaning that X is actually a complete intersection.

4.2 $\sigma_3(\mathbb{P}^1 \times \mathbb{P}^1 \times \mathbb{P}^1 \times \mathbb{P}^2)$

Let $X = \sigma_3(\mathbb{P}^1 \times \mathbb{P}^1 \times \mathbb{P}^1 \times \mathbb{P}^2) \subset \mathbb{P}^{23}$, where $\dim X = 17$. We computed a witness point set W for X using **Bertini** and found that $\deg X = 316$. Using [7], we compute

$$\rho_W = 6, \quad HF_W = 1, 7, 28, 84, 171, 261, 316, 316, \quad \text{and} \quad h = [1\ 6\ 21\ 56\ 87\ 90\ 55].$$

Since h is not symmetric, we conclude that W and, hence, X are not arithmetically Gorenstein.

Remark 1. Although the lack of symmetry in h is sufficient to show that W is not aG, we note that W satisfies the Cayley-Bacharach property and X is aCM. Since X is aCM, we can observe from HF_W that 39 polynomials of degree 4 must vanish on X which generate the ideal of X.

4.3 $\sigma_3(\nu_4(\mathbb{P}^2))$

Let ν_4 be the degree 4 Veronese embedding of $\mathbb{P}^2$ into $\mathbb{P}^{14}$ and consider the scheme $X = \sigma_3(\nu_4(\mathbb{P}^2)) \subset \mathbb{P}^{14}$, where $\dim X = 8$. We computed a witness point set W for X using **Bertini** and found that $\deg X = 112$. Using [7], we compute

$$\rho_W = 6, \quad HF_W = 1, 7, 28, 84, 105, 111, 112, 112, \quad \text{and} \quad h = [1\ 6\ 21\ 56\ 21\ 6\ 1].$$

Clearly, the h-vector for W is symmetric. Since X is irreducible, we selected one subset $Y \subset W$ consisting of 111 points. The witness point set W has the Cayley-Bacharach property since $HF_Y(5) = 111 = HF_W(5)$ and thus we conclude W is arithmetically Gorenstein by Theorem 2.

Next, we consider the arithmetically Cohen-Macaulayness of X. Let $\mathcal{M} \subset \mathbb{P}^{14}$ be a general linear space with $\dim \mathcal{M} = 7$ and $C = X \cap \mathcal{M}$. Via sampling C, we find that

$$HF_C = 1, 8, 36, 120, 225, 336, 448, 560.$$

Therefore, by Theorem 3, X is arithmetically Cohen-Macaulay and, hence, we can conclude it is arithmetically Gorenstein by Theorem 1. In fact, since X is aCM, we can observe from HF_W that 105 polynomials of degree 4 must vanish on X and they generate the ideal of X.

Acknowledgments. The authors would like to thank Luke Oeding for helpful discussions. Both authors were supported in part by DARPA Young Faculty Award (YFA) and NSF grant DMS-1262428. JDH was also supported by Sloan Research Fellowship and NSF ACI-1460032.

References

1. Bates, D.J., Hauenstein, J.D., Sommese, A.J., Wampler, C.W.: Bertini: Software for numerical algebraic geometry. http://bertini.nd.edu
2. Bates, D.J., Hauenstein, J.D., Sommese, A.J., Wampler, C.W.: Numerically solving polynomial systems with Bertini. Software, Environments, and Tools, vol. 25. Society for Industrial and Applied Mathematics (SIAM), Philadelphia (2013)
3. Daleo, N.S.: Algorithms and applications in numerical elimination theory. Ph.D. Dissertation. North Carolina State University (2015)
4. Daleo, N.S., Hauenstein, J.D.: Numerically deciding the arithmetically Cohen-Macaulayness of a projective scheme. J. Symb. Comp. **72**, 128–146 (2016)
5. Davis, E.D., Geramita, A.V., Orecchia, F.: Gorenstein algebras and the Cayley-Bacharach theorem. Proc. Am. Math. Soc. **93**(4), 593–597 (1985)
6. Geramita, A., Migliore, J.: Reduced Gorenstein codimension three subschemes of projective space. Proc. Am. Math. Soc. **125**(4), 943–950 (1997)
7. Griffin, Z.A., Hauenstein, J.D., Peterson, C., Sommese, A.J.: Numerical computation of the Hilbert function and regularity of a zero dimensional scheme. In: Springer Proceedings in Mathematics and Statistics, vol. 76, pp. 235–250. Springer, New York (2014)
8. Hartshorne, R., Sabadini, I., Schlesinger, E.: Codimension 3 arithmetically Gorenstein subschemes of projective n-space. Annales de l'institut Fourier **58**, 2037–2073 (2008)
9. Michalek, M., Oeding, L., Zwiernik, P.: Secant cumulants and toric geometry. Int. Math. Res. Notices **2015**(12), 4019–4063 (2015)
10. Migliore, J., Peterson, C.: A construction of codimension three arithmetically Gorenstein subschemes of projective space. Trans. Am. Math. Soc. **349**(9), 3803–3821 (1997)
11. Migliore, J.: Introduction to Liaison Theory and Deficiency Modules. Birkhäuser, Boston (1998)
12. Oeding, L., Sam, S.V.: Equations for the fifth secant variety of Segre products of projective spaces (2015). arxiv:1502.00203
13. Stanley, R.P.: Hilbert functions of graded algebras. Adv. Math. **28**(1), 57–83 (1978)

Some Results Concerning
the Explicit Isomorphism Problem
over Number Fields

Péter Kutas[✉]

Department of Mathematics and its Applications,
Central European University, Budapest, Hungary
kutas_peter@phd.ceu.edu

Abstract. We consider two problems. First let u be an element of a quaternion algebra B over $\mathbb{Q}(\sqrt{d})$ such that u is non-central and $u^2 \in \mathbb{Q}$. We relate the complexity of finding an element v' such that $uv' = -v'u$ and $v'^2 \in \mathbb{Q}$ to a fundamental problem studied earlier. For the second problem assume that $A \cong M_2(\mathbb{Q}(\sqrt{d}))$. We propose a polynomial (randomized) algorithm which finds a non-central element $l \in A$ such that $l^2 \in \mathbb{Q}$. Our results rely on the connection between solving quadratic forms over $\mathbb{Q}$ and splitting quaternion algebras over $\mathbb{Q}$ [4], and Castel's algorithm [1] which finds a rational solution to a non-degenerate quadratic form over $\mathbb{Q}$ in 6 dimensions in randomized polynomial time. We use these two results to construct a four dimensional subalgebra over $\mathbb{Q}$ of A which is a quaternion algebra. We also apply our results to analyze the complexity of constructing involutions.

1 Introduction

We consider the following algorithmic problem, which we call explicit isomorphism problem: let K be a field, A an associative algebra over K, given by structure constants over K. Suppose that A is isomorphic to the full matrix algebra $M_n(K)$. Construct explicitly an isomorphism $A \to M_n(K)$. Or, equivalently, give an irreducible A-module.

Recall, that for an algebra A over a field K and for a K-basis $a_1, \ldots, a_m$ of A over K the products $a_i a_j$ can be expressed as linear combinations of the a_i:

$$a_i a_j = \gamma_{ij1} a_1 + \gamma_{ij2} a_2 + \cdots + \gamma_{ijm} a_m.$$

The elements $\gamma_{ijk} \in K$ are called structure constants. In this paper an algebra is considered to be given by a collection of structure constants.

Let K be an algebraic number field. In [4] Rónyai proved that the task of factoring square-free integers can be reduced in randomized polynomial time to the explicit isomorphism problem for quaternion algebras over K. Let us recall the notion of an ff-algorithm. This is an algorithm which is allowed to call oracles for factoring integers and polynomials over finite fields. The cost of the call is the size of the input. In [2] Ivanyos, Rónyai and Schicho proposed an

© Springer International Publishing Switzerland 2016
I.S. Kotsireas et al. (Eds.): MACIS 2015, LNCS 9582, pp. 143–148, 2016.
DOI: 10.1007/978-3-319-32859-1_12

ff-algorithm which solves the explicit isomorphism problem in polynomial time if the dimension of the matrix algebra, the degree of K over $\mathbb{Q}$ and the discriminant of K are bounded. The running time of the algorithm depends exponentially on the first two parameters and polynomially on the third (note that in order for the algorithm to be polynomial, it has to be polynomial in the logarithm of the discriminant). An important research problem would be to create an algorithm which would also run in polynomial time when the degree of the number field is not assumed to be bounded or at least its running time depends polynomially on the logarithm of the discriminant. We assume that $A \cong M_n(K)$. An interesting approach to the explicit isomorphism problem would be to find a polynomial algorithm which finds a subalgebra B of A which is isomorphic to $M_n(\mathbb{Q})$. This would immediately result in an algorithm which depends only polynomially on the degree of the number field.

Let $A \cong M_2(\mathbb{Q}(\sqrt{d}))$. In this short paper we give some results towards showing that finding a subalgebra B of A which is isomorphic to $M_2(\mathbb{Q})$ is at least as hard as factoring integers. On the other hand we can construct a four dimensional subalgebra of A over $\mathbb{Q}$ which is a quaternion algebra. Our algorithm is randomized and runs in polynomial time if one is allowed to call oracles for factoring integers. Note that this does not follow from the algorithm of [2] since the algorithm there is not polynomial in the logarithm of the discriminant.

We also give an application of our results. We construct a unitary involution in a quaternion algebra over $\mathbb{Q}(\sqrt{d})$. Our algorithm is randomized and runs in polynomial time assuming we can call oracles for integer factorisation. Note that finding an involution of the first kind can be achieved in polynomial time [4].

2   Quadratic Forms

We denote by $H_K(\alpha, \beta)$ the quaternion algebra over the field K ($char(K) \neq 2$) with parameters α, β (i.e. it has a quaternion basis $1, u, v, uv$ such that $u^2 = \alpha, v^2 = \beta$ and $uv = -vu$). In this section we consider two problems. The first question is the following. Let us assume that B is $H_{\mathbb{Q}(\sqrt{d})}(a, b + c\sqrt{d})$ where $a, b, c \in \mathbb{Q}$. The quaternion basis is $1, u, v, uv$. This means that u^2 is not just in $\mathbb{Q}(\sqrt{d})$ but also in $\mathbb{Q}$. What is the complexity of finding an element v' such that $v'^2 \in \mathbb{Q}$ and $v'u + uv' = 0$? One can assume that $c \neq 0$ otherwise v would suffice.

Theorem 1. *Let $B = H_{\mathbb{Q}(\sqrt{d})}(a, b + c\sqrt{d})$ given by: $u^2 = a, v^2 = b + c\sqrt{d}$, where $a, b, c \in \mathbb{Q}, c \neq 0$. Then finding an element v' such that $uv' + v'u = 0$ and v'^2 is a rational multiple of the identity is equivalent to the explicit isomorphism problem for the quaternion algebra $H_{\mathbb{Q}}(d - (\frac{b}{c})^2, a)$.*

Proof. Since v' anticommutes with u (i.e. $uv' + v'u = 0$) it must be a $\mathbb{Q}(\sqrt{d})$-linear combination of v and uv. This means we have to search for $s_1, s_2, s_3, s_4 \in \mathbb{Q}$ such that:

$$((s_1 + s_2\sqrt{d})v + (s_3 + s_4\sqrt{d})uv)^2 \in \mathbb{Q}$$

Expanding this expression we obtain the following:

$$((s_1 + s_2\sqrt{d})v + (s_3 + s_4\sqrt{d})uv)^2 = (s_1^2 + s_2^2 d + 2s_1 s_2\sqrt{d})(b + c\sqrt{d})$$
$$- (s_3^2 + s_4^2 d + 2s_3 s_4\sqrt{d})a(b + c\sqrt{d})$$

In order for this to be rational, the coefficient of $\sqrt{d}$ has to be zero. So we obtain the following equation:

$$c(s_1^2 + s_2^2 d) + 2bs_1 s_2 - ac(s_3^2 + s_4^2 d) - 2abs_3 s_4 = 0$$

First we divide by c. Note that c is nonzero. Let $f = b/c$.

$$s_1^2 + s_2^2 d + 2f s_1 s_2 - a(s_3^2 + s_4^2 d) - 2af s_3 s_4 = 0$$

Now consider the following change of variables: $x := s_1 + f s_2$, $y := s_2, z := s_3 + s_4 f$, $w := s_4$. Note that the transition matrix of this change is an upper triangular matrix with 1-s in the diagonal so it has determinant 1 (this means that these two equations are "equivalent"). In terms of these new variables the equation takes the following form:

$$x^2 + (d - f^2)y^2 - az^2 - a(d - f^2)w^2 = 0.$$

Finding a solution of this is equivalent to finding a zero divisor in the quaternion algebra $H(d - f^2, a)$ (see [4] or [1, Chap. 1]).   □

Now we turn to the following problem. Let us assume that $A \cong M_2(\mathbb{Q}(\sqrt{d}))$ is given by structure constants. Can one find a non-central element in A whose square is in $\mathbb{Q}$ in (randomized) polynomial time?

Proposition 1. *Let $A \cong M_2(\mathbb{Q}\sqrt{d})$ be given by structure constants. Then there exists a randomized polynomial algorithm which finds a non-central element l, such that $l^2 \in \mathbb{Q}$.*

Proof. First we construct a quaternion basis w and w' of A. We have the following:

$$w^2 = r_1 + t_1\sqrt{d}, \quad w'^2 = r_2 + t_2\sqrt{d}$$

If t_1 or t_2 is 0 then w or w' will be a suitable element. If $r_1 t_2 + r_2 t_1 = 0$ then ww' satisfies the conditions above. From now on we assume that all three quantities are non-zero. First observe that if t_1 or t_2 is zero then we are done (w and w' are not in the center since $ww' = -w'w$). So from now on we assume that neither of them is zero. In order to ensure that the square of l is in $\mathbb{Q}(\sqrt{d})$ it has to be in the $\mathbb{Q}(\sqrt{d})$-subspace generated by w, w' and ww'. The condition $l^2 \in \mathbb{Q}$ gives the following equation ($s_1, \ldots, s_6 \in \mathbb{Q}$):

$$((s_1 + s_2\sqrt{d})w + (s_3 + s_4\sqrt{d})w' + (s_5 + s_6\sqrt{d})ww')^2 \in \mathbb{Q}$$

If we expand this we obtain:

$$((s_1 + s_2\sqrt{d})w + (s_3 + s_4\sqrt{d})w' + (s_5 + s_6\sqrt{d})ww')^2 = (s_1^2 + ds_2^2 + 2s_1s_2\sqrt{d})$$
$$(r_1 + t_1\sqrt{d}) + (s_3^2 + ds_4^2 + 2s_3s_4\sqrt{d})(r_2 + t_2\sqrt{d}) - (s_5^2 + ds_6^2 + 2s_5s_6\sqrt{d})$$
$$(r_1 + t_1\sqrt{d})(r_2 + t_2\sqrt{d})$$

In order for this to be in $\mathbb{Q}$ the coefficient of $\sqrt{d}$ has to be zero:

$$t_1s_1^2 + t_1ds_2^2 + 2r_1s_1s_2 + t_2s_3^2 + t_2ds_4^2 + 2r_2s_3s_4 - (r_1t_2 + t_1r_2)s_5^2$$
$$-(r_1t_2 + t_1r_2)ds_6^2 - 2(r_1r_2 + t_1t_2d)s_5s_6 = 0$$

The left hand side of this equation is a quadratic form in 6 variables. First we calculate its determinant. Its matrix is block diagonal with three 2×2 blocks. So the determinant is the product of these determinants. The first determinant is $t_1^2 d - r_1^2$ which is nonzero since d is not a square (note that t_1 is nonzero). The second is $t_2^2 d - r_2^2$ which is nonzero also (note that t_2 is nonzero). The third is $(r_1t_2+t_1r_2)^2 d - (r_1r_2+t_1t_2d)^2$ which is nonzero due to the same reason (note that the coefficient of d is nonzero due to discussion at the beginning of the proof). This is a non-degenerate quadratic form in dimension 6, so it can be solved by Castel's algorithm [1]. This algorithm runs in randomized polynomial time. Note that it must have a solution since A is a full matrix algebra over $\mathbb{Q}(\sqrt{d})$.   □

There is a nice consequence of this result.

Corollary 1. *Let $A \cong M_2(\mathbb{Q}(\sqrt{d}))$ be given by structure constants. Then one can find a four dimensional subalgebra over $\mathbb{Q}$ which is a quaternion algebra by a randomized algorithm which runs in polynomial time if we are allowed to call oracles for factoring integers.*

Proof. First we find an element l such that $l^2 \in \mathbb{Q}$. Then one finds an element l' such that $ll' + l'l = 0$ and $l'^2 \in \mathbb{Q}$. These can be done using the method of Theorem 1 and Proposition 1 combined with the algorithm from [2]. The only thing we need to show is that for any l such that $l^2 \in \mathbb{Q}$ there exists a four dimensional subalgebra over $\mathbb{Q}$ which is a quaternion algebra and contains l. Indeed, since splitting a quaternion algebra over $\mathbb{Q}$ can be achieved by an ff-algorithm which runs is polynomial time [2].

There exists a subalgebra A_0 in A which is isomorphic to $M_2(\mathbb{Q})$. In this subalgebra there is an element l' for which l and l' have the same minimal polynomial over $\mathbb{Q}(\sqrt{d})$. This means there exists an $m \in A$ such that $l = m^{-1}l'm$. Hence $m^{-1}A_0m$ will contain l.   □

3   Constructing Involutions

In this section we consider the complexity of constructing involutions in a central simple algebra. For definitions and basic facts from the theory of involutions the reader is referred to [3].

Let A be a central simple algebra over K. Recall that an involution σ is a map $\sigma : A \to A$ with the following properties:

1. $\sigma(x + y) = \sigma(x) + \sigma(y)$ for all $x, y \in A$.
2. $\sigma(xy) = \sigma(y)\sigma(x)$ for all $x, y \in A$.
3. $\sigma(\sigma(x)) = x$ for all $x \in A$.

The restriction of an involution to the center of the algebra is an automorphism of K of order at most two. If it is the trivial automorphism then the involution is called an **involution of the first kind** otherwise it is called **unitary** or an **involution of the second kind**.

Let H be a quaternion algebra over a field of characteristic different from 2 given by structure constants. Then one can construct an involution of the first kind easily. We compute a quaternion representation of the algebra. An algorithm for this task is described in [4]. Let $1, i, j, k$ be a quaternion basis. Then the following map is an involution of the first kind:

$$a + bi + cj + dk \mapsto a - bi - cj - dk$$

Let $A \cong M_2(\mathbb{Q}(\sqrt{d}))$ be given by structure constants. How hard is it to construct a unitary involution on A? Theorem 1 would suggest that it is hard to do in polynomial time. The reason for this is that if we compose a unitary involution with an involution of the first kind, and look at the fixed elements, we obtain a four dimensional subalgebra over $\mathbb{Q}$ which is a quaternion algebra [3, Proposition 2.22]Unfortunately it is not known whether finding such quaternion subalgebra over $\mathbb{Q}$ is hard or not.

However using Corollary 1 one can construct a unitary involution in randomized polynomial time if one is allowed to call oracles for factoring integers.

Corollary 2. *Let $A \cong M_2(\mathbb{Q}(\sqrt{d}))$ be given by structure constants. One can construct a unitary involution in randomized polynomial time with oracle calls for factoring integers.*

Proof. Let A' be a quaternion subalgebra of A over $\mathbb{Q}$ given by a quaternion basis $1, u, v, uv$ ($u^2 = a, v^2 = b, a, b \in \mathbb{Q}$). Note that every element in A is a $\mathbb{Q}(\sqrt{d})$-linear combination of $1, u, v, uv$. Then consider the following map:

$$\sigma : \alpha + \beta \cdot u + \gamma \cdot v + \delta \cdot uv \mapsto \overline{\alpha} - \overline{\beta} \cdot u - \overline{\gamma} \cdot v - \overline{\delta} \cdot uv$$

One can easily check that this is a unitary involution. □

One can also show that if $A \cong M_3(\mathbb{Q})$ is given by structure constants then constructing an involution of the first kind in A is as hard as finding an explicit version of this isomorphism. One may use that fact that every skew-symmetric element (i.e. an element x such that $\sigma(x) = -x$) is a zero divisor.

References

1. Castel, P.: Un algorithme de résolution des équations quadratiques en dimension 5 sans factorisation, Ph.D. thesis, October 2011
2. Ivanyos, G., Rónyai, L., Schicho, J.: Splitting full matrix algebras over algebraic number fields. J. Algebra **354**, 211–223 (2012)
3. Knus, M.-A., Merkurjev, A., Rost, M., Tignol, J.-P.: The book of involutions. AMS Colloquium Publications, vol. 44, p. 593 (1998)
4. Rónyai, L.: Simple algebras are difficult. In: Proceedings of the 19th Annual ACM Symposium on the Theory of Computing, New York, pp. 398–408 (1987)

Cryptography

Implementing Cryptographic Pairings on Accumulator Based Smart Card Architectures

Peter Günther[1](✉) and Volker Krummel[2]

[1] University of Paderborn, Paderborn, Germany
peter.guenther@uni-paderborn.de
[2] Wincor Nixdorf International GmbH, Paderborn, Germany
volker.krummel@wincor-nixdorf.de

Abstract. In this paper, we show how bilinear pairings can be implemented on modern smart card architectures. We do this by providing a memory-efficient implementation of the eta pairing on accumulator based cryptographic coprocessors. We provide timing results for different key-sizes on a state of the art smart card, the Infineon SLE 78. On one hand, our results show that pairings can efficiently be computed on smart cards. On the other hand, our results identify bottlenecks that have to be considered for future smart card designs.

1 Introduction

Since the invention of the first fully functional identity based encryption (IBE) scheme [5], that was based on bilinear pairings, pairings have become an important tool in cryptography. Today numerous schemes such as hierarchical identity-based encryption, attribute based encryption (ABE), and identity based signatures use pairings as their main building blocks. Many pairing based schemes are very well suited to embedded applications. For example with IBE, the expensive public key infrastructure of large scale systems like the internet of things can be significantly simplified [9]. Hence, efficient implementations of pairings on embedded and resource constrained devices will become important in the future. In many pairing based schemes the secret key is one argument of the pairing. To protect this secret in an adversarial environment, implementations on smart cards are the standard solution. This raises the question if such constrained platforms are able to compute pairings with acceptable performance. Furthermore, bottlenecks of current architectures have to be identified.

Our Contribution

Towards answering this question, we provide an implementation of the eta pairing for fields of characteristic 2. As hardware platform, we use the Infineon SLE 78 controller [8] that has a dedicated coprocessor for finite field arithmetic.

This work was partially supported by the German Ministry of Education and Research, grant 16KIS0062.

© Springer International Publishing Switzerland 2016
I.S. Kotsireas et al. (Eds.): MACIS 2015, LNCS 9582, pp. 151–165, 2016.
DOI: 10.1007/978-3-319-32859-1_13

Because memory is a bottleneck for efficient pairing implementations on state of the art coprocessors we base our implementation on fields of characteristic 2 that provide especially memory efficient arithmetic. This allows us to remove the memory bottleneck and evaluate the performance of the coprocessor for fields of size up to 2000 bits, although recent results question the applicability of small characteristic fields for cryptography [1]. At 33 MHz, we are able to compute the eta pairing in 60 ms for fields of size 1000 bits, in 100 ms for fields of size 1500 bits, and in 160 ms for fields of size 2000 bits.

Because of the insecurity of characteristic 2 fields, we regard our results as proof of concept to show that pairings can be computed efficiently on smart cards, but only if the pairing is carefully selected according to the available resources. Furthermore, efficient implementations for fields of large prime characteristic will become very important. Our analysis also indicates how size and organization of memory have to be adapted for future cryptographic coprocessors in order to support those implementations.

Previous Work

Previous results already show that it is indeed possible to compute pairings on existing smart card controllers [3,11,12]. We need to distinguish between standard controllers and controllers with dedicated hardware support of finite field arithmetic. Regarding the former, [11] shows that the eta pairing for fields of size approximately 1000 bits can be computed on an Atmel AVR controller in less than 2 s at a CPU clock of less than 8 MHz. In [3] it is shown that the Tate pairing over fields of size 1000 bits can be computed on an STMicroelectronics ST22 controller at 33 MHz in 750 ms.

Because arithmetic in finite fields is one of the major ingredients of a pairing computation, better results on controllers with hardware support of finite field arithmetic can be expected. In [12], different pairings are implemented on the Philips HiPerSmart that offers special instruction set enhancements for cryptographic applications. And indeed, it is shown that the Tate pairing for fields of size 1000 bits can be computed in less than 500 ms at 20.57 MHz. Furthermore, the eta pairing for fields characteristic 2 and of size 1500 bits can be computed in 220 ms at 20.57 MHz. Partially, this efficient implementation is achieved by assuming that the secret argument of the pairing is constant. Hence, intermediate values solely depending on the secret are precomputed and stored in memory. We remark that our implementation is not subject to this severe and often impractical restriction.

Organization of this Work

The paper is organized as follows. We start with some background on pairings in Sect. 2. In Sect. 3 we define a generic accumulator based architecture of a cryptographic coprocessor to abstract from the concrete hardware. Then, in Sect. 4 we outline our implementation of the eta pairing on this architecture. We analyze the implementation in terms of memory requirements and required base field multiplications. In Sect. 5 we give timing results of our implementation for a concrete instantiation and for various key-sizes on the SLE 78 smart card controller. Then we will point to bottlenecks of current cryptographic coprocessors for the computation of pairings. Finally, we will conclude in Sect. 6.

2 Background

In this section we first give a short introduction into elliptic curves and pairings. Then, we motivate the eta pairing for our implementation. Finally, we will give the necessary background on the eta pairing.

2.1 Definition of Pairings and the Embedding Degree

This section is only a very brief introduction. For more details on elliptic curves and pairings we refer, for example, to [9].

Let E denote an elliptic curve that is defined over a finite field $\mathbb{F}_q$, where $q = p^m$ for some prime p and $m \geq 1$. Based on the chord and tangent law, we define an additive group $(E, +)$. For a point $U \in E$ we write $U = (x_U, y_U)$ to reference its x and y coordinate. With aU we denote scalar multiplication of U with $a \in \mathbb{Z}$. Let $\mathbb{F}_{q^k}$ be an extension field of $\mathbb{F}_q$. With $E(\mathbb{F}_{q^k})[r]$ we denote the $\mathbb{F}_{q^k}$-rational r-torsion points of E, i.e., the points defined over $\mathbb{F}_{q^k}$ and of order dividing r.

For $U, V \in E$, let $l_{U,V}$ denote the equation of the line through U and V. With g_U we denote the equation of the tangent line through U at E. Hence, $l_{U,V}$ and g_U are the lines that occur while computing $U + V$ and $2U$, respectively. For $n \in \mathbb{N}$ and $P \in E$, we recursively define the function $f_{n,P}$ as follows:

$$f_{1,P} = 1 \qquad\qquad f_{n+1,P} = f_{n,P} \frac{l_{P,nP}}{l_{-(n+1)P,(n+1)P}}. \qquad (1)$$

Miller presented an algorithm to evaluate $f_{n,P}$ efficiently at points on E [10].

If k is the smallest integer such that r divides $q^k - 1$, then we call k the *embedding degree* of q with respect to r. Let $\mathbb{G}_1, \mathbb{G}_2$, and $\mathbb{G}_T$ be groups of order r with $\mathbb{G}_1, \mathbb{G}_2 \subseteq E(\mathbb{F}_{q^k})[r]$ and $\mathbb{G}_T \subseteq \mathbb{F}_{q^k}^*$. Then a pairing is an efficiently computable, non-degenerate bilinear map $e : \mathbb{G}_1 \times \mathbb{G}_2 \to \mathbb{G}_T$. For cryptographic applications the embedding degree has to be chosen such that the complexity of computing discrete logarithms in $\mathbb{G}_1$ and $\mathbb{G}_2$ and the complexity of computing discrete logarithms in $\mathbb{G}_T$ are balanced. Note that the complexity in the former groups is conjectured to be exponential while the complexity in the latter group is sub-exponential. Table 1 shows some current state of the art choices of k for large prime characteristic q.

2.2 Motivation for the Eta Pairing

Memory is a central concern when implementing cryptographic pairings on smart cards. The basic building block of pairing calculations is the Miller algorithm [10].

Table 1. Relation of subgroup size r, base field size q, extension field size q^k, and embedding degree k to match the complexity of computing discrete logarithms in $E(\mathbb{F}_q)$ and in $\mathbb{F}_{q^k}^*$ for a large prime q. Source: Table 1.1 of [6].

Security level	$\log r$	$\log q$	$k \log q$	k
80 bits	160	160	960–1280	4–6
80 bits	160	320	960–1280	2–4
112 bits	224	224	2200–3600	10–16
112 bits	224	448	2200–3600	5–8
128 bits	256	256	3000–5000	12–20
128 bits	256	512	3000–5000	6–10

We can think of it as an interleaved computation of a double-and-add algorithm for scalar multiplication on $E(\mathbb{F}_q)$ and a square-and-multiply algorithm for exponentiation in $\mathbb{F}_{q^k}$. Hence, to compute the pairing, we need to store elements in $\mathbb{F}_q$ *and* in $\mathbb{F}_{q^k}$ simultaneously. This results in an overall higher memory consumption compared to standard elliptic curve cryptography (ECC) or RSA. Hence, in order to implement pairing based cryptography (PBC) on a smart card that has been designed for standard ECC or RSA, the memory consumption has to be reduced as far as possible.

Here, we achieve this by two means. First, we choose the eta pairing over fields of characteristic $p = 2$ that allows very memory efficient implementations. Secondly, we optimize our implementation for memory efficiency, if necessary at the expense of extra field additions. By reducing the memory consumption, we are able to remove the memory bottleneck of current smart card controllers.

2.3   The Eta Pairing for Fields of Characteristic 2

We now define the eta pairing according to [2, Sect. 6.1]. We stick to the notation of [2] as far as possible. Let $\mathbb{F}_{2^m}$ be a finite field of characteristic 2 and size 2^m. Define the extension field $\mathbb{F}_{q^k} = \mathbb{F}_{2^{4m}} = \mathbb{F}_{2^m}(s, t)$ with $s^2 = s + 1$ and $t^2 = t + s$. Furthermore, define the elliptic curve $E : y^2 + y = x^3 + x + b$, where $b \in \mathbb{F}_2$ and $\#E(\mathbb{F}_q) = 2^m + 1 + \epsilon a$. Here, $a = 2^{(m+1)/2}$ and $\epsilon = (-1)^b$ when $m = 1, 7 \mod 8$, and $\epsilon = -(-1)^b$ when $m = 3, 5 \mod 8$.

Define the distortion map by

$$\psi : E(\mathbb{F}_{2^m}) \to E(\mathbb{F}_{2^{4m}}) \tag{2}$$

$$(x, y) \mapsto (x + s^2, y + sx + t). \tag{3}$$

Now, we are able to define the eta pairing:

Definition 1. *Define* $T = -\epsilon 2^{(m+1)/2} - 1$ *and*

$$M = (2^{4m} - 1)/(2^m + 1 + \epsilon a) = (2^{2m} - 1)(2^m - \epsilon 2^{(m+1)/2} + 1). \tag{4}$$

Then the eta pairing with parameter T is defined as

$$\eta_T : E(\mathbb{F}_{2^m}) \times E(\mathbb{F}_{2^m}) \to \mathbb{F}_{2^{4m}}^* \qquad (5)$$

$$(P, Q) \mapsto f_{T,P}(\psi(Q))^M. \qquad (6)$$

Here, the exponentiation with M is also called *final exponentiation*.

The following theorem will allow us to simplify the computation of $f_{T,P}(\psi(Q))$.

Theorem 1. *Let $P' = -\epsilon P$. The map η_T is a non-degenerate, bilinear map that can be computed as*

$$\eta_T(P, Q) = (f_{a,P'}(\psi(Q)) l_{aP',\epsilon P'}(\psi(Q)))^M.$$

Proof. See [2, Sect. 6.1]

With the Miller algorithm (cf. [10]) the function $f_{a,P'}$ can be efficiently evaluated at $\psi(Q)$. Because a is a power of 2, for the eta pairing, this computation reduces mainly to point doubling and the evaluation of g_R at $\psi(Q)$ where R is of the form $2^i P'$. The simplification is shown in Algorithm 1.

Algorithm 1. Miller algorithm for computing $f_{T,P}(\psi(Q))$.

Require: Elliptic curve $E : y^2 + y = x^3 + x + b$ with $b \in \mathbb{F}_2$, $P, Q \in E(\mathbb{F}_{2^m})$ and
$\quad T = -\epsilon 2^{(m+1)/2} - 1$.
Ensure: $f_{T,P}(\psi(Q)) \in \mathbb{F}_{2^{4m}}$
1: $(f, P', R) \leftarrow (1, -\epsilon P, -\epsilon P)$
2: **for** $i \leftarrow (m-1)/2 \ldots 0$ **do**
3: $\quad f \leftarrow f^2$
4: $\quad (f, R) \leftarrow (f \cdot g_R(\psi(Q)), 2R)$
5: **end for**
6: $f \leftarrow f \cdot l_{aP',\epsilon P'}(\psi(Q))$
7: **return** f

---

3   The Architecture

In this section, we define a model for the underlying computer architecture of our implementation. It acts as an abstraction from the concrete architecture used for execution of the implementation.

Our implementation is based on a platform that consists of a CPU, the main memory (also denoted as RAM), program memory, and a big integer unit (BIU). We assume that the CPU and the BIU are connected to the RAM via a data bus of width w_b. The BIU is an accumulator based cryptographic coprocessor that supports operations in $\mathbb{F}_p$ and $\mathbb{F}_{2^m}$ and consists of the following components:

1. An accumulator register ACC of width w_r
2. The BIU internal memory organized as operand registers $R_0, \ldots, R_{n-1}$, each of width w_r.

Table 2. Instruction set of the BIU

Instruction	Operand Op	Result
`AddRed`	$R_i, RMOD, RMUL$	$ACC \leftarrow ACC + Op \mod RMOD$
`SubRed`	$R_i, RMOD, RMUL$	$ACC \leftarrow ACC - Op \mod RMOD$
`MultRed`	R_i	$ACC \leftarrow Op \cdot RMUL \mod RMOD$
`Load`	$R_i, RMOD, RMUL$	$ACC \leftarrow Op$
`Store`	$R_i, RMOD, RMUL$	$Op \leftarrow ACC$
`Exch`	$R_i, RMOD, RMUL$	$(ACC, Op) \leftarrow (Op, ACC)$

3. A multiplier register `RMUL` of width w_r.
4. A modulus register `RMOD` of width w_r.
5. An algorithmic arithmetic unit (ALU) for performing arithmetic in $\mathbb{F}_p$ or in $\mathbb{F}_{2^m}$ with instructions defined in Table 2.

The instruction set of the BIU is given in Table 2. Arithmetic in finite fields is supported. In the case of $\mathbb{F}_p$, the register `RMOD` stores the modulus p. In the case of $\mathbb{F}_{2^m}$, the register `RMOD` stores the irreducible polynomial $f(X)$ of degree m defining $\mathbb{F}_{2^m} = \mathbb{F}_2/(f(X))$.

Because w_b is assumed to be much smaller than w_r, loading a register from RAM or saving a register to RAM is much slower than the corresponding instructions `Load`, `Store`, and `Exch` for transferring data between internal registers. We also assume that the internal memory of the BIU has better hardware-protection against side channel attacks when compared to the RAM. Altogether, RAM access is an expensive operation. This motivates our requirement to allocate all intermediate variables within the internal memory of the BIU during the execution of the pairing.

4   Implementation on the Generic Architecture

In this section, we provide the details of our implementation of the eta pairing from Definition 1 on the architecture described in Sect. 3. The following theorem describes our implementation of Algorithm 1 with respect to the required resources:

Theorem 2. *Algorithm 1 can be computed on the BIU defined in Sect. 3 with $n = 12$ general purpose operand registers of width $w_r = m$ and requires $15 \cdot (m - 1)/2 + 20$ multiplications in $\mathbb{F}_{2^m}$. Furthermore, no access of the BIU to RAM is required.*

The second theorem is about the implementation of the final exponentiation from (6).

Theorem 3. *Let $\alpha \in \mathbb{F}_{2^{4m}}$ and M as defined in (4). Then on the BIU from Sect. 3, exponentiation α^M can be computed with $n = 12$ general purpose operand registers of width $w_r = m$ and requires $4m + 40$ multiplications in $\mathbb{F}_{2^m}$. Furthermore, no access of the BIU to RAM is required.*

The remainder of this section is structured as follows. In Sect. 4.1 we outline how squaring of the Miller variable f in Line 3 of Algorithm 1 is implemented. Then in Sect. 4.2 we give details on the computation of $g_R(\psi(Q))$ and $l_{aP',\epsilon P'}(\psi(Q))$ in Line 4 and Line 6, respectively. In Sect. 4.3 we show how the multiplication of $g_R(\psi(Q))$ and $l_{aP',\epsilon P'}(\psi(Q))$ with f is implemented. Section 4.4 combines the previous results to obtain an analysis of the complete Miller algorithm. Finally, in Sect. 4.5 we explain how we compute the final exponentiation with M.

4.1   Squaring the Miller Variable

In Line 3 of Algorithm 1 we need to square the Miller variable f. Squaring is a linear operation in fields of characteristic 2. This results in the following lemma that summarizes the costs for the squaring in Line 3 of Algorithm 1:

Lemma 1. *On the BIU from Sect. 3, the squaring of the Miller variable f in Algorithm 1, Line 3 requires $k = 4$ multiplications in $\mathbb{F}_{2^m}$. Furthermore, not more than $k = 4$ operand registers are required to store the arguments, the output, and all intermediate results.*

Remark 1. Note the major difference to fields of characteristic p, where p is a large prime. For squaring elements in $\mathbb{F}_{p^k}$, we cannot compute the coefficients in-place. Instead k registers for storing the input and k registers for accumulating the result are required. This results in a doubling of the required number of registers to $2k$.

We further remark that we propose another implementation than in [4]. There, an optimization is applied that requires the computation of square roots of x_R and y_R [2]. In theory, the computation of square roots in characteristic 2 is efficient. In practice, square root algorithms in $\mathbb{F}_{2^m}$ require bit manipulations that are inefficient without dedicated hardware [4]. This bit-fiddling would not support our strategy of high throughput at the BIU. In [12] this problem does not arise since P is assumed to be constant. Hence, the value of $\sqrt{x_R}$ and $\sqrt{y_R}$ can be precomputed for each iteration of Algorithm 1. To support variable arguments P we avoid square root computations and do not apply this optimization.

4.2   Point Doubling and Line Functions

In this section we show how doubling of R and the value of $g_R(\psi(Q))$ from Line 4 of Algorithm 1 can be computed in a combined way. It will become clear from Lemmas 2 and 4 that the computation of the tangent $g_R(\psi(Q))$ and the computation of the line $l_{aP',\epsilon P'}(\psi(Q))$ from Line 6 have a lot of code in common. Consequently, with Algorithm 2 we provide an implementation that is able to handle both functions. This helps to decrease code size that is also critical for smart cards.

Computing the Tangent $g_R(\psi(Q))$: The following lemma basically adapts [2, Lemma 8] to our notation.

Lemma 2. *Let* $P', Q \in E(\mathbb{F}_{2^m})$. *Define* $x_i = x_{P'}^{2^{2i}} \in \mathbb{F}_{2^m}$, $y_i = y_{P'}^{2^{2i}} \in \mathbb{F}_{2^m}$, $\delta_0, \delta_1 \in \{0, 1\}$ *such that* $\delta_0 = 1$ *if and only if* $i = 0, 3 \mod 4$, *and* $\delta_1 = 1 + i \mod 2$. *Define* $a_0, a_1 \in \mathbb{F}_{2^m}$ *as* $a_1 = x_i^2 + x_Q$ *and* $a_0 := x_i^2(x_i + x_Q) + x_i + y_i + y_Q + \delta_1 a_1$. *Then the tangent line through* $R = 2^i P'$ *at* $\psi(Q)$ *is given as*

$$g_R(\psi(x_Q, y_Q)) = a_0 + \delta_0 + (a_1 + \delta_1) s + t. \tag{7}$$

We call elements in $\mathbb{F}_{2^{4m}}$ of the form (7) sparse as only 2 instead of 4 elements in $\mathbb{F}_{2^m}$ are required for their representation. We show in Sect. 4.3 that multiplication with these elements is also more memory efficient than with arbitrary elements of $\mathbb{F}_{2^{4m}}$.

Lemma 3. *Let* $x_i, y_i, x_Q, y_Q, a_0,$ *and* a_1 *be defined as in Lemma 2. Then the simultaneous computation of* $a_0, a_1, x_{i+1},$ *and* y_{i+1} *requires 5 multiplications in* $\mathbb{F}_{2^m}$ *on the BIU defined in Sect. 3. Furthermore, not more than 4 operand registers are required to store all arguments, the result, and all intermediate values.*

Proof. By setting $\delta_2 = 1$, Algorithm 2 shows how $a_0, a_1, x_i^2,$ and y_i^2 are computed with 3 multiplications in $\mathbb{F}_{2^m}$ and the 4 registers $R_0, \ldots, R_3$ on the BIU. Furthermore, the computation of $x_{i+1} = x_i^4$ and $y_{i+1} = y_i^4$ requires two additional multiplications.

Computing the Line $l_{aP', \epsilon P'}(\psi(Q))$: We now show that Algorithm 2 can also be used to compute $l_{aP', \epsilon P'}(\psi(Q))$ (cf. [2, Lemma 7]):

Lemma 4. *Let* $P', Q \in E(\mathbb{F}_{2^m})$. *Define* $\delta_0, \delta_1 \in \mathbb{F}_2$ *such that* $\delta_0 = (m - \epsilon)/2 \mod 2$ *and* $\delta_1 = (m - 1)/2 \mod 2$. *Define* $a_0, a_1 \in \mathbb{F}_{2^m}$ *as* $a_1 := x_{P'} + x_Q$ *and* $a_0 := x_{P'}(x_{P'} + x_Q) + x_{P'} + y_{P'} + y_Q + \delta_1 a_1$. *Then the tangent line through* $aP' = 2^{(m+1)/2} P'$ *and* $\epsilon P'$ *at* $\psi(Q)$ *is given as*

$$l_{aP', \epsilon P'}(\psi(Q)) = a_0 + \delta_0 + (a_1 + \delta_1) s + t. \tag{8}$$

With respect to the required resources we obtain:

Lemma 5. *Let* $x_{P'}, y_{P'}, x_Q, y_Q, a_0,$ *and* a_1 *be defined as in Lemma 4. Then the simultaneous computation of* a_0 *and* a_1 *requires 1 multiplication in* $\mathbb{F}_{2^m}$ *on the BIU defined in Sect. 3. Furthermore, not more than 4 operand registers are required to store all arguments, the result, and intermediate values.*

Proof. The lemma directly follows by setting $x_i = x_{P'}$, $y_i = y_{P'}$, and $\delta_2 = 0$ in Algorithm 2.

Remark 2. Because of the special form of a in the case of the eta pairing the input P' does not need to be saved beyond Line 1 of Algorithm 1. This is because $x_{P'} = x_0 = (x_{(m-1)/2})^2$ and $y_{P'} = y_0 = (y_{(m-1)/2})^2$. Hence, $x_{P'}$ and $y_{P'}$ equal the output of Algorithm 2 in the last round of Algorithm 1. We cannot use this trick for general pairings and hence two additional registers for storing $x_{P'}$ and $y_{P'}$ are required in the general case.

Algorithm 2. LineFunction: simultaneous computation of $a_1 = x_i^{1+\delta_2} + x_Q$,
$a_0 = x_i^{1+\delta_2}(x_i + x_Q) + x_i + y_i + y_Q + \delta_1 a_1$, $x_i^{1+\delta_2}$, and $y_i^{1+\delta_2}$

Require: $\delta_1, \delta_2 \in \{0,1\}$, $R_0 \leftarrow x_i$, $R_1 \leftarrow y_i$, $R_2 \leftarrow y_Q$, $R_3 \leftarrow x_Q$
Ensure: $R_0 \leftarrow x_i^{1+\delta_2}$, $R_1 \leftarrow y_i^{1+\delta_2}$, $R_2 \leftarrow a_0 = x_i^{1+\delta_2}(x_i + x_Q) + x_i + y_i + y_Q + \delta_1 a_1$,
   $R_3 \leftarrow a_1 = x_i^{1+\delta_2} + x_Q$

1: **procedure** LineFunction(δ_1, δ_2)
2:     Load R_0 ▷ ACC $\leftarrow x_i$
3:     Store RMUL   ▷ RMUL $\leftarrow x_i$
4:     **if** $\delta_2 = 1$ **then**
5:         MultRed R_0 ▷ ACC $\leftarrow x_i^2$
6:     **end if**
7:     Store RMUL   ▷ RMUL $\leftarrow x_i^{1+\delta_2}$
8:     Load R_0 ▷ ACC $\leftarrow x_i$
9:     AddRed R_1 ▷ ACC $\leftarrow x_i + y_i$
10:    AddRed R_2 ▷ ACC $\leftarrow x_i + y_i + y_Q$
11:    Store R_2 ▷ $R_2 \leftarrow x_i + y_i + y_Q$
12:    Load R_0 ▷ ACC $\leftarrow x_i$
13:    AddRed R_3 ▷ ACC $\leftarrow x_i + x_Q$
14:    Store R_0 ▷ $R_0 \leftarrow x_i + x_Q$
15:    MultRed R_0 ▷ ACC $\leftarrow x_i^{1+\delta_2}(x_i + x_Q)$
16:    AddRed R_2 ▷ ACC $\leftarrow x_i^{1+\delta_2}(x_i + x_Q) + x_i + y_i + y_Q$
17:    Store R_2 ▷ $R_2 \leftarrow x_i^{1+\delta_2}(x_i + x_Q) + x_i + y_i + y_Q$
18:    Load RMUL   ▷ ACC $\leftarrow x_i^{1+\delta_2}$
19:    Store R_0 ▷ $R_0 \leftarrow x_i^{1+\delta_2}$
20:    AddRed R_3 ▷ ACC $\leftarrow x_i^{1+\delta_2} + x_Q = a_1$
21:    Store R_3 ▷ $R_3 \leftarrow a_1$
22:    **if** $\delta_1 = 1$ **then**
23:        AddRed R_2 ▷ ACC $\leftarrow x_i^{1+\delta_2}(x_i + x_Q) + x_i + y_i + y_Q + \delta_1 a_1 = a_0$
24:        Store R_2 ▷ $R_2 \leftarrow x_i^{1+\delta_2}(x_i + x_Q) + x_i + y_i + y_Q + \delta_1 a_1 = a_0$
25:    **end if**
26:    **if** $\delta_2 = 1$ **then**
27:        Load R_1 ▷ ACC $\leftarrow y_i$
28:        Store RMUL   ▷ RMUL $\leftarrow y_i$
29:        MultRed R_1 ▷ ACC $\leftarrow y_i^2$
30:        Store R_1 ▷ $R_1 \leftarrow y_i^2$
31:    **end if**
32: **end procedure**

4.3   Sparse Multiplication with the Line Function

In this section, we show how we compute the products in Line 4 and Line 6 of Algorithm 1. In Sect. 4.2, we showed in Lemmas 2 and 5 that one operand occurring in these multiplications has the sparse form $a_0 + a_1 s + (\delta_0 + \delta_1 s) + t \in \mathbb{F}_{2^m}(s, t)$. This motivates the following lemma:

Lemma 6. *Define $A_0 = (a_0 + a_1 s)$ with $a_i, b_i \in \mathbb{F}_{2^m}$ and $\delta_0, \delta_1 \in \mathbb{F}_2$. Let $B \in \mathbb{F}_{2^m}(s, t)$. On the BIU from Sect. 3, the computation of*

$$(A_0 + \delta_0 + \delta_1 s + t) \cdot B \tag{9}$$

requires 6 multiplications in $\mathbb{F}_{2^m}$. *Furthermore, not more than 8 operand registers are required to store the arguments, the output, and all intermediate results.*

One may wonder why we do not absorb $\delta_0 + \delta_1 s$ into A_0. The reason is that this requires two additions with constants in $\mathbb{F}_2$. This is very in-efficient on the BIU because the BIU has to be stopped to load δ_0 and δ_1 into the operand registers. We now show that we can do better by spending a few additional additions in $\mathbb{F}_{2^m}$.

With $B = B_0 + B_1 t$ where $B_i = b_{2i} + b_{2i+1} s$ we obtain for the sparse multiplication in (9):

$$A_0 B_0 + \delta_0 B_0 + \delta_1 B_0 s + B_1 s + (A_0 B_1 + B_0 + (\delta_0 + 1) B_1 + \delta_1 B_1 s) t. \qquad (10)$$

To perform multiplications in $\mathbb{F}_{2^m}(s)$ we use Karatsuba's trick for extensions of degree 2. This allows us to perform a multiplication in $\mathbb{F}_{2^m}(s)$ at the cost of only three multiplications in $\mathbb{F}_{2^m}$:

$$A_0 B_i = a_0 b_{2i} + ((a_0 + a_1)(b_{2i} + b_{2i+1}) - a_0 b_{2i} - a_1 b_{2i+1}) s + a_1 b_{2i+1} s^2 \qquad (11)$$

$$= a_0 b_{2i} + a_1 b_{2i+1} + ((a_0 + a_1)(b_{2i} + b_{2i+1}) - a_0 b_{2i}) s. \qquad (12)$$

Our implementation of (12) on the BIU requires only four operand registers $\mathsf{R}_{2i}, \mathsf{R}_{2i+1}, \mathsf{R}_{2j}$, and R_{2j+1} and the special registers RMUL and ACC. Since $\delta_0, \delta_1 \in \mathbb{F}_2$ multiplication with δ_i in (10) is just a conditional execution of the corresponding addition. In total, our implementation requires 8 registers to compute (10). Furthermore, (12) has to be evaluated for $i \in \{0, 1\}$ to compute $A_0 B_0$ and $A_0 B_1$. This sums up to 6 multiplications in $\mathbb{F}_{2^m}$ for the computation of (10) and completes the proof of Lemma 6.

Remark 3. For the eta pairing, the sparseness of $l_{aP', \epsilon P'}(\psi(Q))$ and $g_R(\psi(Q))$ resulted in a multiplication that requires only $2k = 8$ registers. In the general case the image of $g_R(x, y)$ is a full $\mathbb{F}_{q^k}$ element. Hence, in general, the multiplication in Line 4 of Algorithm 1 requires at least $3k$ registers for storing the two arguments, all intermediate values, and the result.

4.4   The Complete Miller Algorithm

In the previous sections we defined the individual components that are required to compute Algorithm 1 on the BIU architecture. Table 3 gives an overview of the required multiplications and the required number of registers of the individual steps. If we combine all components into an implementation of Algorithm 1 this sums up to $(m - 1)/2 \cdot 15 + 20$ multiplications in $\mathbb{F}_{2^m}$. Furthermore, we see that not more than the $n = 12$ general purpose registers $\mathsf{R}_0, \ldots, \mathsf{R}_{11}$ are required for the complete computation. This proves Theorem 2.

Remark 4. Remarks 1 and 2 show that implementations of a pairing in large prime characteristic requires at least $2 + k$ additional registers compared to the implementation of the eta pairing. Here, 2 additional registers are required to store the input P (see Remark 2) and k additional registers are required to perform a squaring in fields of large prime characteristic (see Remark 1).

Table 3. Overview of the computational costs of Algorithm 1.

Line	Computation	Multiplications in $\mathbb{F}_{2^m}$	Registers	Reference
Line 3	f^2	4	4	Lemma 1
Line 4	$g_R(\psi(Q))$	3	4	Lemma 3
Line 4	$2R$	2	1	Lemma 3
Line 4	$f \cdot g_R(\psi(Q))$	6	8	Lemma 6
Line 6	$l_{aP',\epsilon P'}(\psi(Q))$	1	4	Lemma 5
Line 6	$f \cdot l_{aP',\epsilon P'}(\psi(Q)$	6	8	Lemma 6

4.5   The Final Exponentiation

In this section, we give an intuition how we obtain the result of Theorem 3 for computing the final exponentiation with M from (4). As, for example, in [4], we perform the exponentiation in two steps. In the first step, we compute the exponentiation with $2^{2m} - 1$ and in the second step, we compute exponentiation with $2^m - \epsilon 2^{(m+1)/2} + 1$. The exponentiation with $2^{2m} - 1$ requires an application of the 2^{2m}-th power Frobenius automorphism and an inversion. With the norm map from $\mathbb{F}_{2^{4m}}$ to $\mathbb{F}_{2^m}$ we can reduce the inversion in $\mathbb{F}_{2^{4m}}$ to an inversion in $\mathbb{F}_{2^m}$ plus a constant number of multiplication ins $\mathbb{F}_{2^{4m}}$. Then we perform inversion in $\mathbb{F}_{2^m}$ with Fermat's little theorem that requires $2(m - 2) + 1$ multiplications in $\mathbb{F}_{2^m}$.

The exponentiation with $2^m - \epsilon 2^{(m+1)/2} + 1$ reduces to an application of the 2^m-th power Frobenius automorphism, an exponentiation with $2^{(m+1)/2}$, and a constant number of multiplications. Because squaring is linear, we can perform exponentiation with $2^{(m+1)/2}$ in $\mathbb{F}_{2^{4m}}$ with $2(m - 1)$ multiplications in $\mathbb{F}_{2^m}$. A detailed analysis shows that $4m + 40$ multiplications in $\mathbb{F}_{2^m}$ are required for exponentiation with M.

The required number of $n = 12$ registers results from the multiplications in $\mathbb{F}_{2^{4m}}$ because for one multiplication, we need to store 3 elements in $\mathbb{F}_{2^{4m}}$: the two factors and the product.

5   Performance on Real Hardware

To evaluate the performance of our implementation from the previous section, we instantiated the implementation on the Infineon SLE 78 smart card controller. In this section, we present timing results of the complete pairing computation for different key sizes of practical relevance. Furthermore, we outline limitations of current hardware for supporting fields with large characteristic.

5.1   The SLE 78 Smart Card

The CPU of the SLE 78 controller [8] is an improvement of the well-known 80251 controller and supports frequencies up to 33 MHz. It implements a 16 bit reduced instruction set architecture. The RAM is connected via a 32 bit memory bus to the system. Peripherals like the cryptographic coprocessor are controlled via a $w_b = 16$ bit wide peripheral bus to the system that is used for configuration and data transfer.

The SLE 78 is equipped with a coprocessor for big integer arithmetic called Crypto@2304T that supports 4096 bit RSA and 521 bit ECC. It is possible to instantiate our implementation on the coprocessor as long as the size of the base field does not exceed these 521 bit significantly.

5.2   Measurement Setup and Results

Our measurements are based on ISO7816-4 command/response pairs exchanged between a card reader and the SLE 78. The reader sends a command to the SLE 78 that initiates the pairing computation. The SLE 78 computes the pairing and responds with the result. We measure the time between command and response. Hence, our measurements include the timing of the pairing and a small offset that is introduced by the communication.

We performed our experiments for base fields of size 271, 379, and 523 bits, i.e. extension fields of size 1084, 1516, and 2092 bits, respectively. Our curves are defined as $E : y^2 + y = x^3 + x + b$ with $b = 0$ for $m = 271$ and $b = 1$ for $m \in \{379, 523\}$. From the timing results in Table 4 we see that the SLE 78 is able to compute the eta pairing in 61 ms for fields of size 1084 bit at 33 MHz. In [11] the same field and the same pairing is analyzed. There, a pairing computation takes 1.9 s at 7.3 MHz CPU frequency. This corresponds to a computation time of more than 420 ms at 33 MHz. Hence, we conclude that the hardware support of the Crypto@2304T enables a significant improvement.

Table 4. Measurement and simulation results of the pairing computation on the SLE 78 for different base fields $\mathbb{F}_{2^m}$ at a CPU clock of 33 MHz.

Measurement			Simulation		
m	$k.m$	Complete pairing	Miller Alg.	Final exp.	Complete pairing
271	1084	61 ms	36 ms (72 %)	14 ms (28 %)	50 ms (100 %)
379	1516	98 ms	58 ms (70 %)	25 ms (30 %)	83 ms (100 %)
523	2092	163 ms	99 ms (70 %)	43 ms (30 %)	142 ms (100 %)

For fields of size 1500 bits, we are able to compute the eta pairing in 100 ms. In [12], the same pairing for the same field but with one fixed argument is computed in 220 ms at 20.57 MHz. This corresponds to a computation time of more than 137 ms at 33 MHz. Note, that our implementation can handle two variable arguments without any precomputations in less time.

For fields of size 2000 bits, we are able to compute the eta pairing in 163 ms. We see that if we double the field size from 1084 to 2092 bits, the execution time increases only by a factor of 2.7. This is remarkable since the asymptotic running time of the pairing computation is cubic in m. To explain this effect, note that according to Theorems 2 and 3 the number of $\mathbb{F}_{2^m}$ multiplications is linear in m. We conclude that the asymptotic quadratic complexity of the $\mathbb{F}_{2^m}$ multiplications is partially compensated by the cryptographic coprocessor.

We are also interested in the individual execution times of the Miller loop and the final exponentiation. Our implementation does not support measurements of the two separate steps. Therefore, we use a simulator of the SLE 78 that is provided by the manufacturer for debugging. During a simulation it is possible to set break-points at arbitrary instructions and to obtain timing simulations for the code execution. The results also listed in Table 4. We see that in the simulation, the ratio of the computation time of the Miller algorithm is fixed at roughly 70 %. This is supported by Theorems 2 and 3 that show that the ratio of required multiplications in $\mathbb{F}_{2^m}$ of the Miller algorithm is at approximately $7.5/(7.5+4) = 65\,\%$. We see that even though our theoretical analysis in Sect. 4 neglects $\mathbb{F}_{2^m}$ additions and any control overhead it predicts the simulations results correctly.

5.3   Limitations of Today's Cryptographic Coprocessors

From the timing results of the previous section we conclude that currently, the processing power of available cryptographic coprocessors is not the major bottleneck. But our case study also shows that the internal memory of coprocessors that were designed for RSA and standard ECC is a limitation for PBC. Furthermore, in large prime characteristic fields, implementations of Algorithm 1 are less memory efficient. To give a rough idea how much memory is necessary to compute pairings for those fields, we estimate the minimum memory requirements for computing pairings over fields of large prime characteristic q with $m = \lceil \log(q) \rceil$. For efficiency reasons, we assume that $P \in E(\mathbb{F}_q)$ and $Q \in E(\mathbb{F}_{q^k})$ [9, Remark II.19]. We argue that roughly $(4+5k)m$ bits of memory are required to compute the pairing:

- $4m$ bits to store the coordinates of P and R
- $2km$ bits to store the coordinates of Q
- km bits to store f
- km bits to store $l_{R,P}(Q)$ or $g_R(Q)$
- km bits to store intermediate results during the computation of $f \leftarrow f^2$, $f \leftarrow f \cdot l_{R,P}(Q)$, or $f \leftarrow f \cdot g_R(Q)$

For the case of so-called type 3 pairings [7], we can reduce the size of Q, $l_{R,P}(Q)$, and $g_R(Q)$ by a factor of $\gcd(k,6)$ by using sextic twists [9]. This results in a lower estimation of $(4 + 3k/6 + 2k)m$ required bits of memory.

Next we consider some examples from Table 1. As explained in Sect. 2.1, the embedding degree k is used to match the difficulty of the discrete logarithm problem in $E(\mathbb{F}_q)$ and in $\mathbb{F}_{q^k}^*$. Hence, for a fixed security level, we will be given m and k such that computing discrete logarithms has approximately the same complexity in both groups. To balance the hardness at the 80 bit security level with $m = 160$ we need extension fields of degree $4 \leq k \leq 6$. Our estimation from above shows that for $k = 4$ at least $(4 + 5 \cdot 4) \cdot 160 = 3840$ bits are required to compute the pairing. If we can use type 3 pairings, this reduces to at least $(4 + 3 + 2 \cdot 4) \cdot 160 = 2400$ bits of memory. This amount of memory is already available on current cryptographic coprocessors. To balance the hardness at the 112 bit security level with $m = 224$, we require extension fields of degree $10 \leq k \leq 16$. Hence, for $k = 10$ at least $(4 + 50) \cdot 224 = 12096$ bits are required in the general case and with $k = 12$ we require $(4 + 6 + 24) \cdot 224 = 7616$ bits for type 3 pairings. At the 128 bit security level with $m = 256$ and $k = 12$, 16384 bits in the general case or 8704 bits in the case of type 3 pairings are required. We remark that these estimates are very optimistic. They assume a perfect memory organization and no specific optimizations. For example, the use of projective coordinates or Karatsuba multiplication in $\mathbb{F}_{q^k}$ would require additional memory.

6   Conclusion

In this work, we analyzed an implementation of the eta pairing that we optimized for memory constrained devices. We demonstrated its efficiency by giving timing results for the execution on the Infineon SLE 78 smart card controller. Our results show that this controller allows the implementation of pairings in less than 100 ms.

But based on our analysis, we also argue that the memory of the cryptographic coprocessors is critical for performing PBC beyond the 80 bit security level. Designed for ECC and RSA, there is a gap between the performance of the coprocessor and the size of its memory when used for the computation of pairings. Especially in the smart card setting, memory efficient implementations are an important topic of further research.

References

1. Barbulescu, R., Gaudry, P., Joux, A., Thomé, E.: A heuristic quasi-polynomial algorithm for discrete logarithm in finite fields of small characteristic. In: Nguyen, P.Q., Oswald, E. (eds.) EUROCRYPT 2014. LNCS, vol. 8441, pp. 1–16. Springer, Heidelberg (2014)
2. Barreto, P.S.L.M., Galbraith, S.D., O'Eigeartaigh, C., Scott, M.: Efficient pairing computation on supersingular Abelian varieties. Des. Codes Crypt. **42**(3), 239–271 (2007)
3. Bertoni, G., Breveglieri, L., Chen, L., Fragneto, P., Harrison, K.A., Pelosi, G.: A pairing SW implementation for Smart-Cards. J. Syst. Softw. **81**(7), 1240–1247 (2008)
4. Beuchat, J.L., Brisebarre, N., Detrey, J., Okamoto, E., Rodríguez-Henríquez, F.: A Comparison between hardware accelerators for the modified tate pairing over F2m and F3m. IACR Cryptology ePrint Archive 2008, 115 (2008). http://eprint.iacr.org/
5. Boneh, D., Franklin, M.: Identity-based encryption from the Weil pairing. In: Kilian, J. (ed.) CRYPTO 2001. LNCS, vol. 2139, pp. 213–229. Springer, Heidelberg (2001)

6. Freeman, D., Scott, M., Teske, E.: A taxonomy of pairing-friendly elliptic curves. J. Cryptology **23**, 224–280 (2010)
7. Galbraith, S.D., Paterson, K.G., Smart, N.P.: Pairings for cryptographers. Discrete Appl. Math. **156**(16), 3113–3121 (2008)
8. Infineon Technologies AG: Product Brief SLE 78 (PB_SLE78CXxxxP.pdf), January 2014
9. Joye, M., Neven, G. (eds.): Identity-Based Cryptography, Cryptology and Information Security, vol. 2. IOS Press, Amsterdam (2009)
10. Miller, V.S.: The Weil pairing, and its efficient calculation. J. Cryptology **17**(4), 235–261 (2004)
11. Oliveira, L.B., Aranha, D.F., Gouvêa, C.P.L., Scott, M., Câmara, D.F., López, J., Dahab, R.: TinyPBC: pairings for authenticated identity-based non-interactive key distribution in sensor networks. Comput. Commun. **34**(3), 485–493 (2011)
12. Scott, M., Costigan, N., Abdulwahab, W.: Implementing cryptographic pairings on smartcards. In: Goubin, L., Matsui, M. (eds.) CHES 2006. LNCS, vol. 4249, pp. 134–147. Springer, Heidelberg (2006)

Short Group Signatures
with Distributed Traceability

Johannes Blömer$^{(\boxtimes)}$, Jakob Juhnke, and Nils Löken

Department of Computer Science, Paderborn University, Paderborn, Germany
{bloemer,juhnke,nilo}@mail.uni-paderborn.de

1 Introduction

Group signatures, introduced by Chaum and van Heyst [15], are an important primitive in cryptography. In group signature schemes every group member can anonymously sign messages on behalf of the group. In case of disputes a dedicated opening manager is able to trace signatures - he can extract the identity of the producer of a given signature. A formal model for static group signatures schemes and their security is defined by Bellare, Micciancio, and Warinschi [4], the case of dynamic groups is considered by Bellare, Shi, and Zhang [5]. Both models define group signature schemes with a single opening manager. The main difference between these models is that the number of group members in static schemes is fixed, while in dynamic schemes group members can join the group over time.

Important techniques to design group signature schemes were first described by Ateniese et al. [1]. In [4,5] generic constructions of group signature schemes are presented. The main building blocks of those constructions are generic digital signature schemes, encryption schemes, and non-interactive zero-knowledge proof systems. Concrete realizations, for example [10] as a static and [17] as a dynamic scheme, use efficient instantiations of these techniques to obtain efficient and short group signature schemes. Beside efficient constructions different extensions of group signatures have been considered. Schemes supporting verifier-local revocation [11,28] or linkability [21,23,27] demonstrate the flexibility of group signatures and inspired cryptographers to use group signatures as a tool for more complex primitives, for example e-cash systems [2,13], credential systems [12,29] or reputation systems [7].

Related Work. Having a single opening manager that can identify signers requires a lot of trust in this manager. Several techniques have been considered to deal with this problem. Manulis [26] defines a variant of dynamic group signatures, called *democratic group signatures*, which completely get rid of the opening manager and where every user can trace signatures on his own. In this

J. Blömer and N. Löken—Partially supported by the German Research Foundation (DFG) within the Collaborative Research Centre On-The-Fly Computing (SFB 901). J. Juhnke—Supported by the Ministry of Education and Research, grant 16SV7055, project "KogniHome".

© Springer International Publishing Switzerland 2016
I.S. Kotsireas et al. (Eds.): MACIS 2015, LNCS 9582, pp. 166–180, 2016.
DOI: 10.1007/978-3-319-32859-1_14

model anonymity is only guaranteed against outsiders, not against other group members. Zheng et al. [32] extend the model of Manulis such that not every group member can trace signatures, but a set of cooperating group members is able to reveal a signers identity. This is achieved using threshold public key encryption (TPKE).

Another variant of distributed tracing, but also using TPKE, is considered by Benjumea, Choi, Lopez, and Yung [6]. They define dynamic multi-group signatures with fair tracing. In such systems every user is a member of different groups and only cooperating opening managers can reveal a signers identity. These systems are strongly related to credential systems.

A model for dynamic group signatures supporting distributed tracing is defined by Ghadafi [22]. Ghadafi also gives a generic construction for schemes with distributed tracing.

Our Contribution. In this paper we construct a simple variant of static group signatures with distributed traceability. Our construction is more efficient than the generic construction by Ghadafi [22]. We use threshold public key encryption to distribute the opener's secret key and prove the security of our scheme, including anonymity, in the random oracle model. Our scheme is an extension of the scheme by Boneh, Boyen, and Shacham [10]. However, our technique can be applied to other group signature schemes to obtain distributed traceability. Our basic construction of a group signature scheme with distributed traceability guarantees CPA-full-anonymity, as defined in [10] and suitably extended for distributed traceability. In the last section of this paper we briefly present three extensions of our basic result. First, we describe how to combine our technique with a construction due to Fischlin [20] to obtain a group signature scheme with distributed traceability that achieves the stronger and most desirable security notion of CCA-full-anonymity. Second, in our scheme, as well as in Ghadafi's scheme, there is a so called threshold t such that in order to identify a signer at least t opening managers have to cooperate. We show how to generalize our construction such that it supports monotone access structures for traceability, i.e. any authorized set of opening managers in the access structure can identify signers. Last, we show how to generalize our scheme to dynamic group signatures with distributed traceability.

2   Preliminaries

In this section we introduce the building blocks for our group signature scheme. Similar to [8], we define *bilinear groups* as follows.

Definition 1 (Bilinear Group Pair). *Let* $\mathbb{G}_1, \mathbb{G}_2, \mathbb{G}_T$ *be groups of prime order* p *with efficiently computable group operations, let* $\psi\colon \mathbb{G}_2 \to \mathbb{G}_1$ *be an efficiently computable isomorphism, and let* $\mathrm{e}\colon \mathbb{G}_1 \times \mathbb{G}_2 \to \mathbb{G}_T$ *be an efficiently computable mapping with the following properties:*

– *Bilinearity: for all $u \in \mathbb{G}_1$, $v \in \mathbb{G}_2$ and $a, b \in \mathbb{Z}_p$: $e(u^a, v^b) = e(u, v)^{ab}$*
– *Non-degeneracy: $e(g_1, g_2) \neq 1_{G_T}$.*

Then we call $(\mathbb{G}_1, \mathbb{G}_2)$ a bilinear group pair.

Definition 2 (Bilinear Group Generator). *A bilinear group generator $\mathcal{G}$ is a probabilistic polynomial time algorithm that, on input 1^λ, outputs a description of a bilinear group pair $(\mathbb{G}_1, \mathbb{G}_2)$. We denote the output of $\mathcal{G}$ by $\mathbb{GD} = (\mathbb{G}_1, \mathbb{G}_2, \mathbb{G}_T, g_1, g_2, p, \psi, e)$.*

Throughout this paper we will assume that $g_1 = \psi(g_2)$.

Since we will use bilinear group pairs in our construction of group signature schemes, we define the used computational assumptions with respect to bilinear group generators $\mathcal{G}$.

Definition 3 (Decision Linear Problem – D-Linear2). *Let $(\mathbb{G}_1, \mathbb{G}_2)$ be a bilinear group pair. Given $g_2, g_2^\alpha, g_2^\beta, g_1^{\alpha\gamma}, g_1^{\beta\delta}, g_1^\varepsilon \in \mathbb{G}_2^3 \times \mathbb{G}_1^3$, where $\alpha, \beta, \gamma, \delta \overset{\$}{\leftarrow} \mathbb{Z}_p$, the Decision Linear Problem is to decide whether $\varepsilon = \gamma + \delta$.*

Definition 4. *We say the Decision Linear assumption holds for bilinear group generator $\mathcal{G}$ if for all probabilistic polynomial time algorithms $\mathcal{A}$ there exists a negligible function* negl *such that*

$$\left| \Pr\left[\mathcal{A}(\mathbb{GD}, g_2, g_2^\alpha, g_2^\beta, g_1^{\alpha\gamma}, g_1^{\beta\delta}, g_1^\varepsilon) = 1 \right] \right.$$
$$\left. - \Pr\left[\mathcal{A}(\mathbb{GD}, g_2, g_2^\alpha, g_2^\beta, g_1^{\alpha\gamma}, g_1^{\beta\delta}, g_1^{\gamma+\delta}) = 1 \right] \right| \leq \mathrm{negl}(\lambda),$$

where the probabilities are taken over random bits used by $\mathcal{G}$, $\mathcal{A}$, and the random choices of $\alpha, \beta, \gamma, \delta, \varepsilon \overset{\$}{\leftarrow} \mathbb{Z}_p$.

Definition 5 (q-Strong Diffie-Hellman Problem – q-SDH). *Let $(\mathbb{G}_1, \mathbb{G}_2)$ be a bilinear group pair. Given a tuple $\left(g_2^\gamma, g_2^{(\gamma^2)}, \ldots, g_2^{(\gamma^q)} \right)$, the q-Strong Diffie-Hellman Problem is to output a pair $\left(g_1^{\frac{1}{x+\gamma}}, x \right)$, where $x \in \mathbb{Z}_p$.*

Definition 6. *We say the SDH assumption holds for bilinear group generator $\mathcal{G}$ if for all probabilistic polynomial time algorithms $\mathcal{A}$ and for every polynomial bounded function $q : \mathbb{Z} \to \mathbb{Z}$ there exists a negligible function* negl *such that*

$$\Pr\left[\mathcal{A}\left(\mathbb{GD}, g_2^\gamma, g_2^{(\gamma^2)}, \ldots, g_2^{(\gamma^{q(\lambda)})} \right) = \left(g_1^{\frac{1}{x+\gamma}}, x \right) \right] \leq \mathrm{negl}(\lambda),$$

where the probability is taken over the random bits used by $\mathcal{G}$, $\mathcal{A}$, and the random choice of $\gamma \overset{\$}{\leftarrow} \mathbb{Z}_p$.

For our construction we will use a variant of q-SDH called *extended q-SDH*: given $\left(h, g_2^\gamma, \ldots, g_2^{(\gamma^{q(\lambda)})} \right)$, for $h \overset{\$}{\leftarrow} \mathbb{G}_1$, output $\left((g_1 \cdot h^y)^{\frac{1}{x+\gamma}}, x, y \right) \in \mathbb{G}_1 \times \mathbb{Z}_p^2$.

It is not hard to see that the following lemma holds.

Lemma 1. *Let $\mathcal{A}$ be an algorithm that solves extended q-SDH in polynomial time with non-negligible probability ε. Then there exists an algorithm $\mathcal{B}$ that solves q-SDH in polynomial time with non-negligible probability ε.*

3   Group Signature Schemes and Distributed Traceability

Group signature schemes [4] are signature schemes that provide signer anonymity by forming a group of signers that share a common public key. In case of misbehaving users, signer anonymity can be revoked by an opening manager. With *distributed traceability*, the task of anonymity revocation is distributed among several opening servers who need to cooperate to identify signers.

Definition 7 ($\mathsf{DOMS}(t, m, n)$). *A group signature scheme with t-out-of-m-distributed traceability (with $t \leq m$) for n users consists of six probabilistic polynomial time algorithms* Setup, Sign, SignatureVerify, ShareOpen, ShareVerify *and* ShareCombine, *and a protocol* Join.

- Setup$(1^\lambda, m, t) \to (\mathrm{PK}, \mathrm{VK}, \mathrm{SK}, \mathrm{IK})$*: on input 1^λ (where λ is the security parameter), a number m of opening management servers, and a threshold t, it outputs a public key PK, a verification key VK, a vector SK of opening management server private keys, and a key issuer private key IK.*
- Join$(\mathrm{PK}, \mathrm{IK}; \mathrm{PK}) \to \mathrm{uk_{id}}$*: This protocol is executed by the key issuer and a user id $\in \{1, \ldots, n\}$. The key issuer takes as input PK and IK, the user takes as input PK. It outputs a membership certificate of user id to the key issuer and a user private key $\mathrm{uk_{id}}$ to the user. The key issuer adds the membership certificate to a registration list $\mathrm{RegList}$.*
- Sign$(\mathrm{PK}, \mathrm{uk_{id}}, M) \to \sigma$*: on input PK, $\mathrm{uk_{id}}$ and a message M, it outputs a signature σ.*
- SignatureVerify$(\mathrm{PK}, M, \sigma) \to v_s \in \{0, 1\}$*: on input PK, M and σ it outputs a bit v_s.*
- ShareOpen$(\mathrm{PK}, \mathrm{sk}_i, M, \sigma) \to \theta_i / \perp$*: on input PK, an opening management server private key sk_i, M and σ, it outputs an open share θ_i or an error symbol $\perp$.*
- ShareVerify$(\mathrm{PK}, \mathrm{VK}, M, \sigma, \theta_i) \to v_o \in \{0, 1\}$*: on input PK, VK, M, σ and θ_i, it outputs a bit v_o.*
- ShareCombine$(\mathrm{PK}, \mathrm{VK}, M, \sigma, \Theta, \mathrm{RegList}) \to \mathrm{id} / \perp$*: on input PK, VK, M, σ, a set Θ of t open shares, and the registration list $\mathrm{RegList}$, it outputs a user identifier id or an error symbol $\perp$.*

For consistency we require that, for keys $\mathrm{PK}, \mathrm{VK}, \mathrm{SK} = (\mathrm{sk}_1, \ldots, \mathrm{sk}_m), \mathrm{IK}$ generated during Setup *and for every user private key $\mathrm{uk_{id}}$ generated by executing* Join *for user identifier id with respect to IK, the following properties hold for every message M:*

- SignatureVerify$(\mathrm{PK}, M, \mathsf{Sign}(\mathrm{PK}, \mathrm{uk_{id}}, M)) = 1$.
- *For every valid signature σ on M and every open share θ_i that is output by* ShareOpen$(\mathrm{PK}, \mathrm{sk}_i, M, \sigma)$ *it holds:* ShareVerify$(\mathrm{PK}, \mathrm{VK}, M, \sigma, \theta_i) = 1$.
- *For every σ that is output by* Sign$(\mathrm{PK}, \mathrm{uk_{id}}, M)$ *and every set Θ of t valid open shares it holds:* ShareCombine$(\mathrm{PK}, \mathrm{VK}, M, \sigma, \Theta, \mathrm{RegList}) = \mathrm{id}$.

The security notions for group signature schemes without distributed traceability, as defined in [4], include full-anonymity and full-traceability. Full-anonymity means that nobody except the opening manager can tell who generated a given signature. Full-traceability means that nobody is able to generate signatures (1) on behalf of honest users and (2) which can not be traced back to an existing user. As a drawback, the key issuer must be honest. In DOMS we introduce a Join protocol to achieve strong-exculpability. This splits the definition of full-traceability into two different security properties: traceability and strong-exculpability. Here, traceability means that, even when the key issuer is corrupted, it is not possible to generate signatures that can not be traced back to an existing user, while strong-exculpability means that nobody can generate signatures on behalf of honest users. These properties have already been considered in the context of group signatures without distributed traceability [1,24]. For DOMS the definitions of traceability and strong-exculpability can be left unchanged. However, due to the distributed opening of signatures we have to give an adapted definition for anonymity. In this section we give a definition of anonymity that is an extension of the CPA-full-anonymity defined in [10]. We discuss the stronger notion of CCA-full-anonymity in Sect. 5. By incorporating our technique into Fischlin's CCA-fully-anonymous variant of the Boneh, Boyen, and Shacham group signature [20], in Sect. 5 we also show how to strengthen our basic construction to achieve CCA-full-anonymity.

Definition 8 (Anonymity - $\mathrm{Exp}_{\mathcal{A},DOMS}^{\mathrm{anon-b}}(\lambda, t, m, n)$). *Given a threshold group signature scheme with t-out-of-m-distributed traceability for n users, consider the following t-out-of-m-threshold chosen-plaintext anonymity game:*

1. *The adversary $\mathcal{A}$ chooses $t - 1$ different indices $s_1, \ldots, s_{t-1} \subset \{1, \ldots, m\}$.*
2. *$\mathcal{C}$ executes Algorithm Setup to compute the key material. The public key PK, the verification key VK, the corrupted management servers' private keys $\mathrm{sk}_{s_1}, \ldots, \mathrm{sk}_{s_{t-1}}$, and the key issuer's private key IK are given to $\mathcal{A}$. Then, $\mathcal{C}$ and $\mathcal{A}$ engage in n executions of protocol Join with $\mathcal{A}$ playing the user's role, $\mathcal{C}$ playing the key issuers' role. After this step, $\mathcal{A}$ holds user private keys $\mathrm{uk}_1, \ldots, \mathrm{uk}_n$.*
3. *Eventually, $\mathcal{A}$ outputs two user indices $\mathrm{id}_0, \mathrm{id}_1$ and a message M upon which it wants to be challenged.*
4. *The challenger computes $\sigma \leftarrow \mathsf{Sign}(\mathrm{PK}, \mathrm{uk}_{\mathrm{id}_b}, M)$ and returns σ to $\mathcal{A}$.*
5. *When $\mathcal{A}$ outputs a bit b', the output of the experiment is also b'.*

Definition 9. *A group signature scheme with t-out-of-m-distributed traceability DOMS(t, m, n) is anonymous, if for all probabilistic polynomial time algorithms $\mathcal{A}$ there exists a negligible function negl such that*

$$\left| \Pr\left[\mathrm{Exp}_{\mathcal{A},DOMS}^{\mathrm{anon-1}}(\lambda, t, m, n) = 1 \right] - \Pr\left[\mathrm{Exp}_{\mathcal{A},DOMS}^{\mathrm{anon-0}}(\lambda, t, m, n) = 1 \right] \right| \leq \mathrm{negl}(\lambda).$$

The probability is over the random bits of $\mathcal{A}$, as well as the random bits used in the experiment.

Our definition of anonymity only ensures selective security: the adversary has to decide which management servers he wants to corrupt *before* he has access to the public keys. This simplifies our construction and the proof of security.

On Achieving Distributed Traceability. Many group signature schemes achieve signer anonymity and signer identification by including a cipher on the signer's identity in the signature. So, to identify the signer the cipher needs to be decrypted. Our approach to distribute traceability works on all group signature schemes of this kind for which the used encryption scheme has a *threshold* variant. Threshold encryption schemes differ from other public key encryption schemes as they require multiple servers to cooperate during decryption. We apply our technique to a specific group signature scheme by Boneh, Boyen, and Shacham [10], but it is straightforward to adapt the technique to other group signature schemes.

4 A Group Signature Scheme with Distributed Traceability

Our technique to achieve distributed traceability requires a group signature scheme that includes ciphers on signer identities in signatures and uses an encryption scheme to which a threshold variant can be constructed. We illustrate our technique using the group signature scheme given by Boneh, Boyen and Shacham [10], but it can be used for other schemes as well. First, we present the encryption scheme and its threshold variant, then we present protocols to add users to the group and to prove group membership. Finally, we present DOMS, a group signature scheme with t-out-of-m-distributed traceability.

Threshold Public Key Encryption. We will use Threshold Public Key Encryption (TPKE) [9,14,16,18,25] to achieve distributed traceability of our group signature scheme. This idea was already proposed by [6,22].

Definition 10 (Threshold Public Key Encryption). *A Threshold Public Key Encryption Scheme* TPKE *consists of five probabilistic polynomial time algorithms* (KeyGen, Encrypt, ShareDec, ShareVerify, Combine), *where*

- KeyGen$(1^\lambda, m, t) \rightarrow$ (PK, VK, SK): *on input* 1^λ *(where* λ *is the security parameter), the number of decryption servers* m *and a threshold parameter* t *($t \le m$), it outputs a tuple* (PK, VK, SK), *where* PK *is a public key,* VK *is a verification key and* SK $= (sk_1, \ldots, sk_m)$ *is a vector of* m *private key shares.*
- Encrypt(PK, M) $\rightarrow c$: *on input* PK *and message* M, *it outputs ciphertext* c.
- ShareDec(PK, c, sk_i) $\rightarrow \theta_i /\ \bot$: *on input* PK, c *and the* i'*th private key share* sk_i, *it outputs a decryption share* θ_i *or a special error symbol* $\bot$.
- ShareVerify(PK, VK, c, θ_i) $\rightarrow v \in \{0, 1\}$: *on input* PK, VK, c, *and* θ_i, *it outputs a bit* $v \in \{0, 1\}$.
- Combine(PK, VK, c, Θ) $\rightarrow M /\ \bot$: *on input* PK, VK, c *and a set* Θ *of* t *decryption shares* θ_i, *it outputs a message* M *or a special error symbol* $\bot$.

As consistency requirements, for all (PK, VK, SK) *output by* KeyGen($1^\lambda, t, m$) *the following two properties must hold:*

1. *For every c as output of* Encrypt(PK, M) *and all* $i \in \{1, \ldots, m\}$ *it holds: if* θ_i *is the output of* ShareDec(PK, c, sk$_i$)*, then* ShareVerify(PK, VK, c, θ_i) = 1.
2. *For every c as output of* Encrypt(PK, M) *and every set* Θ *of t valid decryption shares it holds:* Combine(PK, VK, c, Θ) = M.

Definition 11 (Threshold CPA - $\mathrm{Exp}_{\mathcal{A},TPKE}^{\mathrm{tcpa}-b}(\lambda, t, m)$). *The Threshold Chosen Plaintext Attack is defined using the following game between a challenger and an adversary* $\mathcal{A}$ *both with input* (λ, t, m):

1. $\mathcal{A}$ *outputs a set* $S \subset \{1, \ldots, m\}$ *of size* $|S| = t - 1$.
2. (PK, VK, SK) $\leftarrow$ KeyGen($1^\lambda, t, m$) *is run by the challenger. Then* PK, VK *and all* sk$_i$ *for* $i \in S$ *are given to* $\mathcal{A}$.
3. $\mathcal{A}$ *outputs two messages* M_0, M_1 *of equal length, and receives a ciphertext* $c \leftarrow$ Encrypt(PK, M_b) *from its challenger.*
4. *When* $\mathcal{A}$ *outputs a bit* $b' \in \{0, 1\}$*, return* b'.

Definition 12. *A Threshold Public Key Encryption Scheme is semantically secure against chosen plaintext attacks, if for all probabilistic polynomial time algorithms* $\mathcal{A}$ *there exists a negligible function* negl *such that*

$$\left| \Pr[\mathrm{Exp}_{\mathcal{A},TPKE}^{\mathrm{tcpa}-1}(\lambda, t, m) = 1] - \Pr[\mathrm{Exp}_{\mathcal{A},TPKE}^{\mathrm{tcpa}-0}(\lambda, t, m) = 1] \right| \leq \mathrm{negl}(\lambda),$$

where the probabilities are taken over the random bits used by $\mathcal{A}$ *and in the experiments.*

Analogously to the definition of anonymity we define selective security for the threshold encryption. Using the techniques of [25] adaptively secure threshold encryption schemes can be constructed.

Linear Encryption and its Threshold Variant. Linear Encryption is a public key encryption scheme that was introduced by Boneh, Boyen, and Shacham in [10]. It is defined as follows:

- (PK, SK) $\leftarrow$ KeyGen(1^λ), where PK:=($\mathbb{GD}, \hat{u}, \hat{v}, \hat{h}$) is the public key consisting of a bilinear group pair ($\mathbb{G}_1, \mathbb{G}_2$) and generators $(\hat{u}, \hat{v}, \hat{h}) \in \mathbb{G}_2^3$, and SK:=$(\xi_1, \xi_2) \in \mathbb{Z}_p^2$ is the secret key such that $\hat{u}^{\xi_1} = \hat{v}^{\xi_2} = \hat{h}$.
- Encrypt(PK, M): choose $\alpha, \beta \xleftarrow{\$} \mathbb{Z}_p$ and set $c:=(\psi(\hat{u})^\alpha, \psi(\hat{v})^\beta, M \cdot \psi(\hat{h})^{\alpha+\beta})$.
- Decrypt(PK, SK, c): parse c as (T_1, T_2, T_3) and compute $M:=T_3/(T_1^{\xi_1} \cdot T_2^{\xi_2})$.

It is not hard to see that Linear Encryption is correct and secure under the Decision Linear assumption. We could also define the encryption scheme in $\mathbb{G}_1$, without $\mathbb{G}_2$, the isomorphism ψ, and the pairing e. However, we need ψ and e for the reduction of the Threshold Linear Encryption defined below. Furthermore, we need the following lemma to prove security of our system.

Lemma 2. *Let* $\mathbb{G} = \langle g \rangle$ *be a group of prime order p and $f(x)$ an arbitrary polynomial over $\mathbb{Z}_p$ of degree $t - 1$. Define $F \colon \mathbb{Z}_p \to \mathbb{G}$ as $F(x) := g^{f(x)}$. Suppose $X \subset \mathbb{Z}_p$ such that $|X| = t$. Then, given $\{(x_i, F(x_i)\}_{x_i \in X}$, one can evaluate F at any $x \in \mathbb{Z}_p$ using Lagrange interpolation (LI).*

Proof.

$$F(x) = g^{f(x)} \overset{LI}{=} g^{\sum_{x_i \in X} \Delta_{x_i, X}(x) f(x_i)} = \prod_{x_i \in X} \left(g^{f(x_i)} \right)^{\Delta_{x_i, X}(x)} = \prod_{x_i \in X} F(x_i)^{\Delta_{x_i, X}(x)}$$

Hence, using the last expression one can evaluate $F(x)$. □

Using Shamir's secret sharing technique [31] we obtain a t-out-of-m Threshold Public Key Encryption scheme, called *Threshold Linear Encryption* (TLE):

- $(\mathrm{PK}, \mathrm{SK}, \mathrm{VK}) \leftarrow \mathsf{KeyGen}(1^\lambda, t, m)$, where $\mathrm{PK} := (\mathbb{GD}, \hat{u}, \hat{v}, \hat{h})$ is the public key consisting of a bilinear group pair $(\mathbb{G}_1, \mathbb{G}_2)$ and generators $(\hat{u}, \hat{v}, \hat{h}) \in \mathbb{G}_2^3$, $\mathrm{SK} := ((\xi_1(1), \xi_2(1)), \ldots, (\xi_1(m), \xi_2(m)))$ for uniformly at random chosen polynomials ξ_1, ξ_2 of degree $t-1$ over $\mathbb{Z}_p$ such that $\hat{u}^{\xi_1(0)} = \hat{v}^{\xi_2(0)} = \hat{h}$ are the secret key shares, and $\mathrm{VK} := ((\hat{u}^{\xi_1(1)}, \hat{v}^{\xi_2(1)}), \ldots, (\hat{u}^{\xi_1(m)}, \hat{v}^{\xi_2(m)}))$ is the verification key.
- $\mathsf{Encrypt}(\mathrm{PK}, M)$: choose $\alpha, \beta \overset{\$}{\leftarrow} \mathbb{Z}_p$ and set $c := (\psi(\hat{u})^\alpha, \psi(\hat{v})^\beta, M \cdot \psi(\hat{h})^{\alpha+\beta})$.
- $\mathsf{ShareDec}(\mathrm{PK}, \mathrm{sk}_i, c)$: set $\theta_{i,1} := T_1^{\xi_1(i)}$, $\theta_{i,2} := T_2^{\xi_2(i)}$, and $\theta_i := (\theta_{i,1}, \theta_{i,2})$.
- $\mathsf{ShareVerify}(\mathrm{PK}, \mathrm{VK}, c, \theta_i)$: output 1, if and only if $e(\theta_{i,1}, \hat{u}) = e(T_1, \mathrm{vk}_{i,1})$ and $e(\theta_{i,2}, \hat{v}) = e(T_2, \mathrm{vk}_{i,2})$.
- $\mathsf{Combine}(\mathrm{PK}, \mathrm{VK}, c, \Theta)$: If all decryption shares $\theta_i \in \Theta$ are valid, for $i \in S \subseteq \{1, \ldots, m\}, |S| \geq t$, then use the Lagrange polynomial interpolation to decrypt and output the message M. This can be done by computing $M := T_3 / \prod_{i \in S} (\theta_{i,1} \cdot \theta_{i,2})^{\Delta_{i,S}(0)}$. Otherwise, output $\bot$.

Correctness of Threshold Linear Encryption:

1. For every ciphertext $c = (T_1, T_2, T_3) = (\psi(\hat{u})^\alpha, \psi(\hat{v})^\beta, M \cdot \psi(\hat{h})^{\alpha+\beta})$ as output of $\mathsf{Encrypt}(\mathrm{PK}, M)$ and all $i \in \{1, \ldots, m\}$ it holds: let θ_i be the output of $\mathsf{ShareDec}(\mathrm{PK}, \mathrm{sk}_i, c)$, then $e(\theta_{i,1}, \hat{u}) = e(T_1^{\xi_1(i)}, \hat{u}) = e(T_1, \hat{u}^{\xi_1(i)}) = e(T_1, \mathrm{vk}_{i,1})$ and $e(\theta_{i,2}, \hat{v}) = e(T_2^{\xi_2(i)}, \hat{v}) = e(T_2, \hat{v}^{\xi_2(i)}) = e(T_2, \mathrm{vk}_{i,2})$.
2. For every $c = (T_1, T_2, T_3)$ as output of $\mathsf{Encrypt}(\mathrm{PK}, M)$ and every set Θ of t valid decryption shares it holds: $\mathsf{Combine}(\mathrm{PK}, \mathrm{VK}, c, \Theta) = M$, which follows from Lemma 2.

Lemma 3. *If the Linear Encryption scheme is semantically secure against chosen plaintext attacks, then the Threshold Linear Encryption scheme $\mathsf{TLE}(t, m)$ is secure against threshold chosen-plaintext attacks.*

The proof will be given in the full version of this paper.

4.1   Construction of Our Group Signature Scheme

Our Group Signature Scheme is based on the construction given in [10]. We define a zero-knowledge protocol which will be transformed into a group signature scheme using the Fiat-Shamir heuristic [19].

The Basic Zero-Knowledge Protocol. Given $(\mathbb{GD}, h_1, w = g_2^\gamma)$, where $\mathbb{GD} \leftarrow \mathcal{G}(1^\lambda)$ is the output of a bilinear group generator (such that the SDH assumption holds), $h_1 \in \mathbb{G}_1$ and $\gamma \xleftarrow{\$} \mathbb{Z}_p$ is some unknown value, the secret of a prover is the tuple $(A, x, y) = \left((g_1 \cdot h_1^y)^{\frac{1}{x+\gamma}}, x, y\right)$, where $x, y \in \mathbb{Z}_p$. To prove possession of such a tuple, the prover can use the bilinear map e which is contained in $\mathbb{GD}$: $e(A, g_2)^x \cdot e(A, g_2^\gamma) \cdot e(h_1, g_2)^{-y} = e(g_1, g_2)$.

Protocol 1. Compute a Threshold Linear Encryption (T_1, T_2, T_3) of A and helper values:

$$\alpha, \beta \xleftarrow{\$} \mathbb{Z}_p, \quad T_1 := u^\alpha, \quad T_2 := v^\beta, \quad T_3 := A \cdot h^{\alpha+\beta}, \quad \delta_1 := x\alpha, \quad \delta_2 := x\beta.$$

Choose blinding values $r_\alpha, r_\beta, r_x, r_y, r_{\delta_1}, r_{\delta_2} \xleftarrow{\$} \mathbb{Z}_p$ and compute

$$R_1 := u^{r_\alpha}, \qquad\qquad R_2 := v^{r_\beta},$$

$$R_3 := e(T_3, g_2)^{r_x} \cdot e(h, w)^{-r_\alpha - r_\beta} \cdot e(h, g_2)^{-r_{\delta_1} - r_{\delta_2}} \cdot e(h_1, g_2)^{-r_y},$$

$$R_4 := T_1^{r_x} \cdot u^{r_{\delta_1}}, \qquad\qquad R_5 := T_2^{r_x} \cdot v^{r_{\delta_2}}.$$

Given $T_1, T_2, T_3, R_1, R_2, R_3, R_4, R_5$, the verifier responds with a challenge $c \xleftarrow{\$} \mathbb{Z}_p$. The prover computes

$$s_\alpha := r_\alpha + c\alpha, \qquad s_\beta := r_\beta + c\beta, \qquad s_x := r_x + cx,$$
$$s_y := r_y + cy, \qquad s_{\delta_1} := r_{\delta_1} + c\delta_1, \qquad s_{\delta_2} := r_{\delta_2} + c\delta_2$$

and sends them to the verifier who then checks the following five equations

$$u^{s_\alpha} \stackrel{?}{=} T_1^c \cdot R_1 \qquad\qquad v^{s_\beta} \stackrel{?}{=} T_2^c \cdot R_2$$

$$e(T_3, g_2)^{s_x} \cdot e(h, w)^{-s_\alpha - s_\beta} \cdot e(h, g_2)^{-s_{\delta_1} - s_{\delta_2}} \cdot e(h_1, g_2)^{-s_y} \stackrel{?}{=} (e(g_1, g_2)/e(T_3, w))^c \cdot R_3$$

$$T_1^{s_x} \cdot u^{-s_{\delta_1}} \stackrel{?}{=} R_4 \qquad\qquad T_2^{s_x} \cdot v^{-s_{\delta_2}} \stackrel{?}{=} R_5$$

and accepts, if all five equations hold.

Lemma 4. *The above protocol is complete (a verifier accepts all interactions with an honest prover), zero-knowledge (there is simulator for transcripts of protocol executions) and a proof of knowledge (there is an extractor) under the Decision Linear assumption.*

   The proof is similar to that in [10] and is given in the full version of this paper.

Construction 2. The $\mathsf{DOMS}(t, m, n)$ group signature scheme with distributed traceability works as follows:

- $\mathsf{Setup}(1^\lambda, t, m, n)$: run $\mathcal{G}(1^\lambda)$ to obtain $\mathbb{GD}$, compute TLE keys $\mathsf{PK}_{\mathsf{TLE}} = (\hat{u}, \hat{v}, \hat{h}) \in \mathbb{G}_1^3$, $\mathsf{VK}_{\mathsf{TLE}}$ and $\mathsf{SK}_{\mathsf{TLE}} = ((\xi_1(1), \xi_2(1)), \ldots, (\xi_1(m), \xi_2(m)))$, and choose element $\gamma \xleftarrow{\$} \mathbb{Z}_p$ and generator $h_1 \xleftarrow{\$} \mathbb{G}_1$. Furthermore, fix some hash function $H \colon \{0, 1\}^* \to \mathbb{Z}_p$. Set $\mathsf{PK}_{\mathsf{DOMS}} := (\mathbb{GD}, \hat{u}, \hat{v}, \hat{h}, g_1, h_1, g_2, w = g_2^\gamma, H)$, $\mathsf{VK}_{\mathsf{DOMS}} := \mathsf{VK}_{\mathsf{TLE}}$, $\mathsf{SK}_{\mathsf{DOMS}} := \mathsf{SK}_{\mathsf{TLE}}$ and $\mathsf{IK}_{\mathsf{DOMS}} := \gamma$. Publish $\mathsf{PK}_{\mathsf{DOMS}}$ and $\mathsf{VK}_{\mathsf{DOMS}}$, give $\mathsf{IK}_{\mathsf{DOMS}}$ to the key issuer and distribute $\mathsf{SK}_{\mathsf{DOMS}}$ amongst the opening management servers.
- $\mathsf{Join}(\mathsf{PK}_{\mathsf{DOMS}}, \mathsf{IK}_{\mathsf{DOMS}})$: the user id picks an element $y_{\mathsf{id}} \xleftarrow{\$} \mathbb{Z}_p$ and sends $h_1^{y_{\mathsf{id}}}$ to the key issuer. The key issuer chooses $x_{\mathsf{id}} \xleftarrow{\$} \mathbb{Z}_p$ and computes $A_{\mathsf{id}} := (g_1 \cdot h_1^{y_{\mathsf{id}}})^{1/(\gamma + x_{\mathsf{id}})}$. The membership certificate $(\mathsf{id}, A_{\mathsf{id}}, x_{\mathsf{id}})$ is permanently linked to the user id and sent to her. Additionally, the key issuer stores $(\mathsf{id}, A_{\mathsf{id}}, x_{\mathsf{id}})$ in RegList. The user sets the private key to $\mathsf{uk}_{\mathsf{id}} := (A_{\mathsf{id}}, x_{\mathsf{id}}, y_{\mathsf{id}})$.
- $\mathsf{Sign}(\mathsf{PK}_{\mathsf{DOMS}}, \mathsf{uk}_i, M)$: apply the Fiat-Shamir heuristic to Protocol 1 and compute T_1, T_2, T_3, R_1, R_2, R_3, R_4, R_5 as defined in the protocol, use the hash function H to compute the challenge value $c := H(M, T_1, T_2, T_3, R_1, R_2, R_3, R_4, R_5)$ and compute response-values as in the protocol to obtain signature $\sigma := (T_1, T_2, T_3, c, s_\alpha, s_\beta, s_x, s_y, s_{\delta_1}, s_{\delta_2})$.
- $\mathsf{SignatureVerify}(\mathsf{PK}_{\mathsf{DOMS}}, M, \sigma)$: compute the R-values using the verification equations from Protocol 1 and output 1, if $c \overset{?}{=} H(M, T_1, T_2, T_3, R_1, R_2, R_3, R_4, R_5)$. Otherwise, output 0.
- $\mathsf{ShareOpen}(\mathsf{PK}_{\mathsf{DOMS}}, \mathsf{sk}_i \in \mathsf{SK}_{\mathsf{DOMS}}, M, \sigma)$: verify that σ is a valid signature on M. If so, output a TLE decryption share for (T_1, T_2, T_3). Otherwise, output $\bot$.
- $\mathsf{ShareVerify}(\mathsf{PK}_{\mathsf{DOMS}}, \mathsf{VK}_{\mathsf{DOMS}}, \theta, M, \sigma)$: if σ is a valid signature on M and θ is a valid TLE decryption share, output valid. Otherwise, output invalid.
- $\mathsf{ShareCombine}(\mathsf{PK}_{\mathsf{DOMS}}, \mathsf{VK}_{\mathsf{DOMS}}, \theta_1, \ldots, \theta_t, M, \sigma, \mathsf{RegList})$: if σ is not a valid signature on M and any θ_i is not a valid open share, output $\bot$. Otherwise, combine the open shares/TLE decryption shares to obtain value $\tilde{A}$. Use RegList to identify the user linked to $\tilde{A}$.

It is not hard to see that DOMS satisfies the requirements imposed on group signature schemes with t-out-of-m-distributed traceability.

4.2 Proof of Anonymity

Lemma 5. *If the Threshold Linear Encryption* $\mathsf{TLE}(t, m)$ *is semantically secure against threshold chosen-plaintext attacks, then* $\mathsf{DOMS}(t, m, n)$ *is anonymous in the random oracle model.*

Proof. Assume $\mathcal{A}$ is an algorithm that breaks the anonymity of $\mathsf{DOMS}(t, m, n)$. Then we can construct an algorithm $\mathcal{B}$ that breaks the threshold chosen-plaintext security of $\mathsf{TLE}(t, m)$.

In the first step algorithm $\mathcal{A}$ outputs a set S of management server indices, where $|S| = t - 1$. This set is forwarded by $\mathcal{B}$ to its threshold chosen-plaintext challenger. Then $\mathcal{B}$ is given $(\mathrm{PK}, \mathrm{SK}, \mathrm{VK})$, where $\mathrm{PK} = (\mathbb{GD}, \hat{u}, \hat{v}, \hat{h})$, $\mathrm{SK} = \{(\xi_1(i), \xi_2(i))\}_{i \in S}$, and $\mathrm{VK} = ((\hat{u}^{\xi_1(1)}, \hat{v}^{\xi_2(1)}), \ldots, (\hat{u}^{\xi_1(m)}, \hat{v}^{\xi_2(m)}))$. The key issuer private key IK and elements h_1, w for the public key are generated by setting $\mathrm{IK}{:=}\gamma$ for $\gamma \xleftarrow{\$} \mathbb{Z}_p$ and $h_1 \xleftarrow{\$} \mathbb{G}_1$. Furthermore, $\mathcal{B}$ sets $w{:=}g_2^\gamma$ and gives $\mathrm{PK}, \mathrm{VK}, \mathrm{SK}, \mathrm{IK}$ to $\mathcal{A}$. Then, n simulations of the Join protocol are executed between $\mathcal{A}$ and $\mathcal{B}$, so $\mathcal{A}$ gets user private keys $\mathrm{uk}_1, \ldots, \mathrm{uk}_n$ with $\mathrm{uk}_i{:=}(A_i, x_i, y_i)$.

During the interaction $\mathcal{A}$ is allowed to query the random oracle H. $\mathcal{B}$ responses to those queries by returning some $r \xleftarrow{\$} \mathbb{Z}_p$, ensuring to respond to identical queries with the same value.

When $\mathcal{A}$ outputs user indices id_0 and id_1 and a message M upon which it wants to be challenged, $\mathcal{B}$ requests a challenge from the TLE chosen-plaintext challenger on messages A_{id_0} and A_{id_1}. Based on the challenge ciphertext (T_1, T_2, T_3) $\mathcal{B}$ generates a transcript $(T_1, T_2, T_3, R_1, R_2, R_3, R_4, R_5, c, s_\alpha, s_\beta, s_x, s_y, s_{\delta_1}, s_{\delta_2})$ of Protocol 1 using the zero-knowledge simulator (Lemma 4). Then $\mathcal{B}$ patches the random oracle $H(M, T_1, T_2, T_3, R_1, R_2, R_3, R_4, R_5){:=}c$. If the patch fails, $\mathcal{B}$ outputs $\bot$ and exits, but this only happens with negligible probability. Otherwise, $\mathcal{B}$ generates the challenge signature $\sigma{:=}(T_1, T_2, T_3, c, s_\alpha, s_\beta, s_x, s_y, s_{\delta_1}, s_{\delta_2})$ for $\mathcal{A}$ based on the transcript.

When $\mathcal{A}$ outputs a bit b' as its guess on the identity used to generate the challenge signature $\mathcal{B}$ outputs b' as its guess on the message encrypted in its challenge ciphertext.

Since $\mathcal{B}$ generates a valid challenge signature for user id_b with the same distribution as in the real group signature scheme, $\mathcal{B}$'s guess on b is correct, whenever $\mathcal{A}$'s guess is correct. Hence, $\mathcal{B}$ breaks the threshold chosen-plaintext security of TLE with the same probability as $\mathcal{A}$ breaks the anonymity of DOMS. □

4.3   Further Properties

Our DOMS scheme also provides *traceability* and *strong-exculpability*. Traceability can be shown assuming the SDH assumption holds. The proof informally works as follows: given an instance of q-SDH we use the technique from [8] to generate up to q different group membership certificates in the Join protocol. If an adversary against traceability outputs a signature that can not be opened to an existing user, we use the Forking Lemma [30] to extract a complete membership certificate (A, x, y). This certificate can then be transformed into a solution to the original q-SDH instance using the technique of [8].

To prove strong-exculpability we have to assume that computing discrete logarithms (DLog) is hard. This assumption is implied by D-Linear2 and SDH. The given DLog instance is used within the Join protocol when an honest user id sends $h_1^{y_{\mathrm{id}}}$ to the key issuer. If an adversary against strong-exculpability outputs a signature that can be traced back to user id, we use the Forking Lemma [30] to extract the complete membership certificate $(A_{\mathrm{id}}, x_{\mathrm{id}}, y_{\mathrm{id}})$, which includes the discrete logarithm y_{id} as the solution to the original DLog instance.

5   Extensions and Modifications

In this section we briefly discuss further adaptions of our group signature scheme. More detailed descriptions are given in the full version of this paper.

Achieving CCA-Full-Anonymity. So far, we have restricted ourselves to CPA-full-anonymous group signatures. Here we show how to modify our scheme to achieve CCA-full-anonymity. CCA-full-anonymity is defined analogously to CPA-full-anonymity, except that in the anonymity experiment (Definition 8) the adversary is also given oracle access to Open (used to open the identity of signers in case of centralized traceability) or ShareOpen(·), ShareVerify(·), and ShareCombine(·) (in case of distributed traceability). Among other things, the construction of Boneh, Boyen, and Shacham, as well as ours, crucially depends on the structure of Linear Encryption. Hence, turning our CPA-fully-anonymous group signature into CCA-fully-anonymous group signature cannot be achieved by simply using a CCA-secure variant of Linear Encryption. Instead, Linear Encryption must be replaced by a CCA-secure variant while at the same time preserving its basic structure.

Fischlin [20] shows how to transform any Σ-protocol into a non-interactive zero-knowledge proof of knowledge (NIZK) with an online extractor. As an application of this technique, Fischlin obtains a CCA-fully-anonymous group signature scheme based on the Boneh-Boyen-Shacham scheme. The only modification to the original scheme is that signatures σ are extended to include a NIZK proof of knowledge π for the values α, β used to compute the ciphertext (T_1, T_2, T_3), i.e. the first three elements of a signature. Intuitively, this leads to CCA-full-anonymity, since an adversary that submits a valid signature to the Open oracle must already know the values α, β used to hide the identity of the group member. Hence, the Open oracle is useless to an adversary. More precisely, in a simulation of an adversary the simulator can use the online extractor for the NIZK proof of knowledge π to compute the values α, β, use these to recover the identity of the group member, and answer Open queries correctly. By the same reasoning, combined with Lemma 2, one sees that by incorporating Fischlin's extension in Construction 2 we get a CCA-fully-anonymous group signature scheme with distributed traceability.

The proofs in Fischlin's NIZK proofs of knowledge are not of constant size (in term of the number of group elements). Therefore, unlike the original scheme the CCA-full-anonymous variant of the Boneh, Boyen, Shacham group signature scheme no longer has signatures of constant size. However, as Fischlin points out, the signatures are still moderately large, and for reasonable parameters shorter than signatures in RSA-based schemes. The same remarks apply to our CPA-fully-anonymous and CCA-fully-anonymous schemes with distributed traceability.

Considering Dynamic Groups. Bellare, Shi, and Zhang define in [5] a model for dynamic group signatures. In this model the number of group members is not fixed in advance - group members join the group over time. Hence, the Join protocol must ensure that no adversary obtains information about the membership certificates, even under concurrent executions of the protocol. To achieve these properties, Join can be implemented as a concurrent zero-knowledge protocol. In [17] such a protocol is defined, which can also be used in our system.

Additionally to the flexible group joining mechanism in dynamic groups, the Join protocol has to ensure another security property - the *non-frameability*. During the Join protocol a user commits to a personal public key upk_{id}. Using this user public key the opener of signatures has to prove that the claimed user really generated the signature in question. Non-frameability then guarantees that the opener is not able to forge such proofs.

To achieve non-frameability in our construction, we add a new algorithm Judge and modify the algorithm ShareCombine. As user public key upk_{id} we use $h_1^{y_{id}}$, and we let ShareCombine include the proof string $\tau := x_{id}$. Recall that this value is known to the share combiner since it is part of the membership certificate list RegList created by the key issuer during Join. Algorithm Judge then checks the outcome of the combining procedure by verifying the equation $e(g_1, g_2) \overset{?}{=} e(A_{id}, wg_2^{x_{id}}) \cdot e(upk_{id}^{-1}, g_2)$. If the equation does not hold, the algorithm *rejects*, which means that the combiner tried to blame the user id.

Other Variants of Distributed Traceability. In the group signature with distributed traceability, any set of management servers large enough can identify signers. This implies that any management server has the same rights and powers. However, it may be useful to have management servers with different rights. This can be modeled with monotone access structures and realized similarly to the construction given above by replacing Shamir's secret sharing schemes with secret sharing schemes based on monotone span programs (MSP) [3]. We briefly discuss this generalization. Informally, a monotone access structure $\mathfrak{A}$ over a universe $U = \{1, \ldots, m\}$ is a subset of the power set of U. Elements in $\mathfrak{A}$ are called authorized sets. An access structure is called monotone if and only if every superset of an authorized set is authorized. In group signatures with monotone group management we are given a monotone access structure $\mathfrak{A}$ over the set of management servers and we require that every authorized set of servers in $\mathfrak{A}$ must be able to identify signers. This allows us to express different levels of power of the opening managers. Group signatures with monotone group management can be realized with secret sharing schemes for monotone access structures (see again [3]), i.e. the secret opening key is shared such that every authorized set of opening managers can reveal the signers identity. Secret sharing schemes for monotone access structures can be realized with monotone span programs. To construct group signatures with monotone group management, given an access structure $\mathfrak{A}$ we construct an MSP-based variant of Linear Encryption such that decryption of ciphertexts is possible if and only if decryption shares from an authorized set in $\mathfrak{A}$ are available. The security of the resulting scheme can be proved using a reduction to the Linear Encryption. Replacing in our construction Threshold Linear Encryption with a MSP-based Linear Encryption for access structure $\mathfrak{A}$ yields a group signature scheme with monotone group management for access structure $\mathfrak{A}$. In particular, as long as an adversary does not corrupt an authorized set of management servers he does not obtain any information about encrypted identities, i.e. signers. Replacing Threshold Linear Encryption with MSP-based Linear Encryption for the MSP-variant in our group signature scheme increases flexibility of our constructions. This generalization only influences the decryption and combine algorithms. Protocol 1 must not be modified and signatures remain unchanged.

Acknowledgements. We thank the anonymous reviewers for their helpful comments which greatly improved the paper.

References

1. Ateniese, G., Camenisch, J.L., Joye, M., Tsudik, G.: A practical and provably secure coalition-resistant group signature scheme. In: Bellare, M. (ed.) CRYPTO 2000. LNCS, vol. 1880, pp. 255–270. Springer, Heidelberg (2000)
2. Baldimtsi, F., Chase, M., Fuchsbauer, G., Kohlweiss, M.: Anonymous transferable e-cash. In: Katz, J. (ed.) PKC 2015. LNCS, vol. 9020, pp. 101–124. Springer, Heidelberg (2015)
3. Beimel, A.: Secret-sharing schemes: a survey. In: Chee, Y.M., Guo, Z., Ling, S., Shao, F., Tang, Y., Wang, H., Xing, C. (eds.) IWCC 2011. LNCS, vol. 6639, pp. 11–46. Springer, Heidelberg (2011)
4. Bellare, M., Micciancio, D., Warinschi, B.: Foundations of group signatures: formal definitions, simplified requirements, and a construction based on general assumptions. In: Biham, E. (ed.) EUROCRYPT 2003. LNCS, vol. 2656, pp. 614–629. Springer, Heidelberg (2003)
5. Bellare, M., Shi, H., Zhang, C.: Foundations of group signatures: the case of dynamic groups. In: Menezes, A. (ed.) CT-RSA 2005. LNCS, vol. 3376, pp. 136–153. Springer, Heidelberg (2005)
6. Benjumea, V., Choi, S.G., Lopez, J., Yung, M.: Fair traceable multi-group signatures. In: Tsudik, G. (ed.) FC 2008. LNCS, vol. 5143, pp. 231–246. Springer, Heidelberg (2008)
7. Blömer, J., Juhnke, J., Kolb, C.: Anonymous and publicly linkable reputation systems. In: Böhme, R., Okamoto, T. (eds.) FC 2015. LNCS, vol. 8975, pp. 478–488. Springer, Heidelberg (2015)
8. Boneh, D., Boyen, X.: Short signatures without random oracles and the SDH assumption in bilinear groups. J. Cryptology **21**(2), 149–177 (2008)
9. Boneh, D., Boyen, X., Halevi, S.: Chosen ciphertext secure public key threshold encryption without random oracles. In: Pointcheval, D. (ed.) CT-RSA 2006. LNCS, vol. 3860, pp. 226–243. Springer, Heidelberg (2006)
10. Boneh, D., Boyen, X., Shacham, H.: Short group signatures. In: Franklin, M. (ed.) CRYPTO 2004. LNCS, vol. 3152, pp. 41–55. Springer, Heidelberg (2004)
11. Boneh, D., Shacham, H.: Group signatures with verifier-local revocation. In: CCS 2004, pp. 168–177. ACM (2004)
12. Camenisch, J.L., Lysyanskaya, A.: Signature schemes and anonymous credentials from bilinear maps. In: Franklin, M. (ed.) CRYPTO 2004. LNCS, vol. 3152, pp. 56–72. Springer, Heidelberg (2004)
13. Canard, S., Pointcheval, D., Sanders, O., Traoré, J.: Divisible e-cash made practical. In: Katz, J. (ed.) PKC 2015. LNCS, vol. 9020, pp. 77–100. Springer, Heidelberg (2015)
14. Canetti, R., Goldwasser, S.: An efficient *threshold* public key cryptosystem secure against adaptive chosen ciphertext attack. In: Stern, J. (ed.) EUROCRYPT 1999. LNCS, vol. 1592, pp. 90–106. Springer, Heidelberg (1999)

15. Chaum, D., van Heyst, E.: Group signatures. In: Davies, D.W. (ed.) EUROCRYPT 1991. LNCS, vol. 547, pp. 257–265. Springer, Heidelberg (1991)
16. De Santis, A., Desmedt, Y., Frankel, Y., Yung, M.: How to share a function securely. In: Proceedings of the Twenty-Sixth Annual ACM Symposium on Theory of Computing, STOC 1994, pp. 522–533. ACM (1994)
17. Delerablée, C., Pointcheval, D.: Dynamic fully anonymous short group signatures. In: Nguyên, P.Q. (ed.) VIETCRYPT 2006. LNCS, vol. 4341, pp. 193–210. Springer, Heidelberg (2006)
18. Desmedt, Y.G., Frankel, Y.: Threshold cryptosystems. In: Brassard, G. (ed.) CRYPTO 1989. LNCS, vol. 435, pp. 307–315. Springer, Heidelberg (1990)
19. Fiat, A., Shamir, A.: How to prove yourself: practical solutions to identification and signature problems. In: Odlyzko, A.M. (ed.) CRYPTO 1986. LNCS, vol. 263, pp. 186–194. Springer, Heidelberg (1987)
20. Fischlin, M.: Communication-efficient non-interactive proofs of knowledge with online extractors. In: Shoup, V. (ed.) CRYPTO 2005. LNCS, vol. 3621, pp. 152–168. Springer, Heidelberg (2005)
21. Franklin, M., Zhang, H.: Unique group signatures. In: Foresti, S., Yung, M., Martinelli, F. (eds.) ESORICS 2012. LNCS, vol. 7459, pp. 643–660. Springer, Heidelberg (2012)
22. Ghadafi, E.: Efficient distributed tag-based encryption and its application to group signatures with efficient distributed traceability. In: Aranha, D.F., Menezes, A. (eds.) LATINCRYPT 2014. LNCS, vol. 8895, pp. 327–347. Springer, Heidelberg (2015)
23. Hwang, J.Y., Lee, S., Chung, B.H., Cho, H.S., Nyang, D.: Group signatures with controllable linkability for dynamic membership. Inf. Sci. **222**, 761–778 (2013)
24. Kiayias, A., Yung, M.: Group signatures: provable security, efficient constructions and anonymity from trapdoor-holders. IACR Cryptology ePrint Archive 2004, 76 (2004). http://eprint.iacr.org/2004/076
25. Libert, B., Yung, M.: Non-interactive CCA-secure threshold cryptosystems with adaptive security: new framework and constructions. In: Cramer, R. (ed.) TCC 2012. LNCS, vol. 7194, pp. 75–93. Springer, Heidelberg (2012)
26. Manulis, M.: Democratic group signatures: on an example of joint ventures. In: ASIACCS 2006, p. 365. ACM (2006)
27. Manulis, M., Sadeghi, A.-R., Schwenk, J.: Linkable democratic group signatures. In: Chen, K., Deng, R., Lai, X., Zhou, J. (eds.) ISPEC 2006. LNCS, vol. 3903, pp. 187–201. Springer, Heidelberg (2006)
28. Nakanishi, T., Funabiki, N.: A short verifier-local revocation group signature scheme with backward unlinkability. In: Yoshiura, H., Sakurai, K., Rannenberg, K., Murayama, Y., Kawamura, S. (eds.) IWSEC 2006. LNCS, vol. 4266, pp. 17–32. Springer, Heidelberg (2006)
29. Persiano, G., Visconti, I.: An Efficient and usable multi-show non-transferable anonymous credential system. In: Juels, A. (ed.) FC 2004. LNCS, vol. 3110, pp. 196–211. Springer, Heidelberg (2004)
30. Pointcheval, D., Stern, J.: Security arguments for digital signatures and blind signatures. J. Cryptology **13**, 361–396 (2000)
31. Shamir, A.: How to share a secret. Commun. ACM **22**, 612–613 (1979)
32. Zheng, D., Li, X., Ma, C., Chen, K., Li, J.: Democratic group signatures with threshold traceability. IACR Cryptology ePrint Archive 2008, 112 (2008). http://eprint.iacr.org/2008/112

On the Optimality of Differential
Fault Analyses on CLEFIA

Ágnes Kiss[1]([⊠]), Juliane Krämer[1,2], and Anke Stüber[2]

[1] TU Darmstadt, Darmstadt, Germany
agnes.kiss@ec-spride.de, jkraemer@cdc.informatik.tu-darmstadt.de
[2] TU Berlin, Berlin, Germany
anke@sec.t-labs.tu-berlin.de

Abstract. In 2012, several Differential Fault Analyses on the AES cipher were analyzed from an information-theoretic perspective. This analysis exposed whether or not the leaked information was fully exploited. We apply the same approach to all existing Differential Fault Analyses on the CLEFIA cipher. We show that only some of these attacks are already optimal. We improve those analyses which did not exploit all information. With one exception, all attacks against CLEFIA-128 reach the theoretical limit after our improvement. Our improvement of an attack against CLEFIA-192 and CLEFIA-256 reduces the number of fault injections to the lowest possible number reached so far.

Keywords: CLEFIA · Differential fault analysis · Fault attack

1 Introduction

An attack which actively alters the computation of a cryptographic algorithm by inducing software or hardware faults is called fault attack. A Differential Fault Analysis (DFA) is a specific form of a fault attack. After inducing a fault into one or several computations of a cryptographic algorithm, the secret key of this algorithm is revealed by analyzing the difference between correct and faulty results of the computation. For symmetric algorithms, Differential Fault Analyses were first described in 1997 [3]. Since then, they were successfully applied to various symmetric ciphers and their key schedule, e.g., DES [3], AES [9], and CLEFIA. They were also applied to other cryptographic algorithms, such as stream ciphers [7] and hash functions [6].

In 2012, Sakiyama et al. analyzed the information-theoretic optimality of seven Differential Fault Analyses on the Advanced Encryption Standard

This work has been co-funded by the DFG as part of projects P1 and E3 within the CRC 1119 CROSSING and by the European Union's Seventh Framework Program (FP7/2007-2013) under grant agreement n. 609611 (PRACTICE). The authors would like to thank TU Berlin, especially the Chair for Security in Telecommunications and Jean-Pierre Seifert, for valuable support.

© Springer International Publishing Switzerland 2016
I.S. Kotsireas et al. (Eds.): MACIS 2015, LNCS 9582, pp. 181–196, 2016.
DOI: 10.1007/978-3-319-32859-1_15

(AES) [11]. They developed a model which quantifies the amount of information a certain fault can deliver. Information-theoretic optimality does not imply that an attack is also optimal from other points of view, e.g., a non-optimal method might be easier to conduct in practice. However, an attack which is not optimal can still be improved in the given framework, i.e., the key space can be further reduced or a key space with the same size can be determined with less fault injections.

We apply the approach from Sakiyama et al. to analyze the information-theoretic optimality of Differential Fault Analyses on CLEFIA. The CLEFIA cipher is a 128-bit block cipher proposed by Sony Corporation in 2007 [13]. Since then, several attacks against it have been published, including side channel attacks which exploit cache accesses [10], Impossible Differential Attacks [17], and novel methods such as the Improbable Differential Attack [16]. In this paper, we analyze six published Differential Fault Analyses on CLEFIA [1,2,4,14,15, 18]. To the best of our knowledge, these are all DFAs on CLEFIA until today.

Contribution: We analyze all published Differential Fault Analyses on CLEFIA from an information-theoretic perspective with the techniques introduced in [11]. These DFAs are described in Sect. 3. The methodology and the results of our analysis are described in Sect. 4. Our results show that some of the attacks are optimal, while others do not exploit all available information. With one exception, we optimized all attacks against CLEFIA-128 which proved not to be optimal. The optimized attacks reach the theoretical limits and thus exploit all available information. For longer keys, all DFAs were shown not to be optimal in our analysis. We considerably improved one of them. The improved attack is concretely the best known attack against CLEFIA-192 and CLEFIA-256. In Sect. 5, we explain how we optimized the non-optimal attacks against CLEFIA-128 and describe our improved attack against CLEFIA-192/256.

2   Background

We first explain Differential Fault Analysis. Then, we present the CLEFIA cipher and provide background knowledge on information theory.

2.1   Differential Fault Analysis

For a Differential Fault Analysis (DFA), an attacker needs at least one correct ciphertext and one faulty ciphertext. Thus, she has to have the ability to induce faults on the cryptographic primitive level. These faults can be described in detailed fault models, which include the location and the timing of the fault, and the number of bits and bytes which are affected by the fault. A fault can, for example, affect one byte in the register storing the first four bytes of the state (location) in the penultimate round (timing). The assumed fault model gives the attacker partial information about the difference between certain states of the correct and the faulty computations, although she will not know the concrete value of the fault in most scenarios. Since the attacker also knows the correct

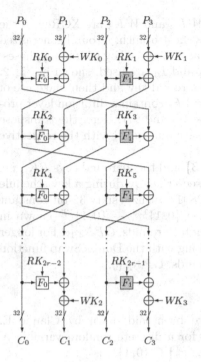

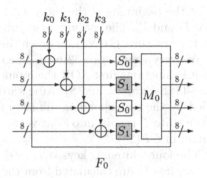

Fig. 2. F-function F_0 (F_1 with reversed S-boxes and M_1 diffusion matrix)

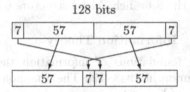

Fig. 1. CLEFIA encryption algorithm

Fig. 3. DoubleSwap function Σ

and faulty ciphertext, and thereby their difference, she can deduce information about the secret key. Small differences in the fault models might crucially affect the capabilities and the complexity of the attacks [3]. For the attacks analyzed in this work, the attacker is assumed to have full control on the timing and the location of the fault, and is able to induce not permanent, but transient faults.

2.2 CLEFIA

The 128-bit block cipher CLEFIA was developed by Sony Corporation and presented in 2007 [13][1]. To be compatible with AES, CLEFIA supports key lengths of 128, 192, and 256 bits. CLEFIA is a Feistel cipher with four 32-bit data lines which are used during r rounds throughout the encryption and decryption processes. Corresponding to the increasing key lengths, the number of rounds are 18, 22, and 26. According to the four data lines, $P_i \in \{0,1\}^{32}$, $i \in \{0,\ldots,3\}$ denote the four 32-bit parts of the plaintext P, so that $P = P_0|P_1|P_2|P_3$. Similarly, the state is denoted by $T = T_0|T_1|T_2|T_3$ and the ciphertext by $C = C_0|C_1|C_2|C_3$.

As shown in Fig. 1, CLEFIA uses $2r$ round keys during encryption. In the kth round, RK_{2k} and RK_{2k+1} are used for $k \in \{0,\ldots,r-1\}$. Moreover, four whitening keys are used, from which WK_0 and WK_1 are XORed with P_1 and

---

[1] All figures in this section are taken from [13].

P_3 at the beginning of the encryption, while WK_2 and WK_3 are XORed to the final T_1 and T_3. The encryption algorithm uses the 4-branch, r-round generalized Feistel structure, $GFN_{4,r}$, between the initial and the final key whitening phases.

In every round, two 32-bit F-functions F_0 and F_1 are used, shown in Fig. 2. Both F-functions first XOR the input with the round key and then make use of the 8-bit S-boxes S_0 and S_1. Afterwards, F_0 and F_1 contain a diffusion layer provided by the corresponding diffusion matrices M_0 and M_1. Here, the transpose of the state T is split into 8-bit vectors which are multiplied with the respective matrix in $GF(2^8)$.

The four whitening keys WK_i, $i \in \{0, \ldots, 3\}$ and the $2r$ round keys RK_i, $i \in \{0, \ldots, 2r-1\}$ are calculated from the initial secret key K during a key schedule procedure. The key scheduling for CLEFIA-128 [13] utilizes sixty 32-bit constant values, the so-called DoubleSwap function $\Sigma : \{0,1\}^{128} \rightarrow \{0,1\}^{128}$ shown in Fig. 3, and the 4-branch Feistel structure through 12 rounds, $GFN_{4,12}$. For longer keys, altogether 84 constant values are used along with the DoubleSwap function and the 8-branch Feistel structure with 10 rounds, $GFN_{8,10}$.

2.3   Information Theory

The foundations of information theory have been laid down by Claude E. Shannon in 1948 [12]. The Shannon entropy for a discrete random variable X with $p(X = x_i) = p_i$, $i = 1, \ldots, n$ and $\{x_1, \ldots, x_n\} \subsetneq \{0,1\}^*$, is

$$H(X) := -\sum_{i=1}^{n} p_i \log_2(p_i). \tag{1}$$

It quantifies the uncertainty when the value of X is to be predicted. For two discrete random variables X and Y, their joint entropy can be defined as well as the respective conditional entropies. The joint entropy of X and Y, $H(XY)$ or $H(X,Y)$, is

$$H(XY) := -\sum_{i=1}^{n}\sum_{j=1}^{m} p(X = x_i, Y = y_j) \log_2(p(X = x_i, Y = y_j)). \tag{2}$$

We have $H(XY) \leq H(X)+H(Y)$, while for stochastically independent X and Y, equality holds. Following the definition by Shannon, the conditional entropy of X is the average of the entropy of X for each value of Y, weighted according to the probability of getting that particular Y [12]. Thus, the entropy of X conditioned on Y, i.e., the conditional entropy $H(X \mid Y)$, is defined as

$$H(X \mid Y) := \sum_{j=1}^{m} p(Y = y_j) H(X \mid Y = y_j), \text{with} \tag{3}$$

$$H(X \mid Y = y_j) := -\sum_{i=1}^{n} p(X = x_i | Y = y_j) \log_2(p(X = x_i | Y = y_j)). \tag{4}$$

Using the definition of the conditional probability [5], Eq. 3 becomes

$$H(X \mid Y) = H(XY) - H(Y). \tag{5}$$

3 Differential Fault Analyses on CLEFIA

We analyzed six DFAs on CLEFIA which are, to the best of our knowledge, all published DFAs on CLEFIA. They were published between 2007 and 2013.

The first DFA was presented already in 2007 [4]. The attack against CLEFIA-128 uses independent random byte faults at six different positions of the algorithm. These are induced in T_0 and T_2 in rounds 15, 16 and 17 and help to reveal RK_{30}, RK_{31}, $RK_{32} \oplus WK_3$, $RK_{33} \oplus WK_2$, RK_{34} and RK_{35}. Thus, the attack needs at least 6 faulty encryptions. However, the authors state that the fault inductions have to be repeated until all bytes are recovered. They had to induce at least 18 faults in their simulations. Based on the recovered round keys, the original secret key can be revealed by analyzing the key scheduling algorithm. If the key size is 192 or 256, the same procedure has to be applied in rounds $r - 9$ to $r - 1$. Here, the simulated attacks need 54 faults to be successful.

One year later, these results were improved [14]. Takahashi and Fukunaga encrypt a random plaintext, which does not have to be known, with the same secret key three times with CLEFIA-128. They insert four-byte faults in the 16th round in two of these encryptions, one into F_0 and one into F_1. The authors use the fact that "a fault corrupts the intermediate values of the fault-injection round and the subsequent rounds". Thus, they obtain more information out of a single fault, since they also analyze how the differences propagate through the next two rounds. After analyzing rounds 16 to 18, 2^{19} candidates for the round keys are left. By applying the inverse of the DoubleSwap function and $GFN_{4,12}^{-1}$ to all round key candidates, the 128-bit secret key can be uniquely identified. In 2010, the authors adapted the same attack to keys with 192 and 256 bits, where 10.78 faults are needed on average [15].

Still in 2010, multiple-byte Differential Fault Analyses were described for CLEFIA-128 [18]. The authors propose three attack models, including the first attack which exploits fault injections in the final round. This attack exploits faults in the inputs of F_0 and F_1 in rounds 18, 17, and 16. The authors consider multiple-byte faults, so that each single fault can affect up to four bytes. The second attack builds on [4] and induces faults into the inputs of F_0 and F_1 in both the 15th and 17th round, but extends their fault model to multiple bytes. For the third attack, which targets the 16th round, however, the strict four-byte fault model of [14] has been loosened to a one-to-four-byte model, so that the attack presented in [14] can be seen as a special case of this more general attack description. According to the authors, the minimum amount of faulty ciphertexts for these three attacks to be successful is 5 to 6, 6 to 8, and 2. For the first two attacks, these faults help to reveal the secret key completely. In the third attack, 2^{19} candidates for the secret key remain. Unfortunately, the description of these three attacks is not always easy to follow, and several statements in [18] are inconsistent with one another. The description of these attacks in the work at hand is to the best of our knowledge.

In 2012 it was analyzed if protecting the last four rounds of CLEFIA-128 would counter DFAs [1]. The authors show that it is sufficient to induce two random byte-faults in the computation of F_0 and F_1 in round 14. It is assumed that the faults are induced before the diffusion operation of the F-functions, cf. Fig. 3 in [1]. These two faults are enough to uniquely reveal the secret key by exploiting the propagation of the faults in the final four rounds. Thus, it is not sufficient to protect the last four rounds of CLEFIA against such attacks. One year later, the same authors extended their attack to CLEFIA-192 and CLEFIA-256 [2]. Here, they also start to scrutinize how much information a certain fault can provide. They induce two random faults each in the computation of F_0 and F_1, in rounds $r - 4$ and $r - 8$. Thus, they induce eight faults altogether. They can reveal the whole secret key of both CLEFIA-192 and CLEFIA-256.

4   Information-Theoretic Analysis of DFAs on CLEFIA

We adapt the information-theoretic methodology described in [11] (and recapitulated in the full version of this paper [8, Sect. 4.1]) to the CLEFIA block cipher. Then, we evaluate the optimality of all published DFAs against CLEFIA.

4.1   General Methodology Adapted to CLEFIA

We refine the general model of [11] for the 8-bit model of CLEFIA with S_0 and S_1 S-boxes. The refined model is shown in Fig. 4. It shows the XORing of the round key and the application of the S-boxes S_0 and S_1 within the F-functions F_0 and F_1 of the cipher CLEFIA. x and y denote the 8-bit input and the 8-bit output of the S-box. z is the 8-bit value for which $x = z \oplus k$ with the key k. Given a correct and a faulty execution of the encryption algorithm, the 8-bit differences are $\Delta x = x_1 \oplus x_2$, $\Delta y = y_1 \oplus y_2$, and $\Delta z = z_1 \oplus z_2$. The attacker can gain the values $x_1, x_2, \Delta x, y_1, y_2, \Delta y$, and Δz. $K, X_1, X_2, Y_1, Y_2, \Delta X$, and ΔY are discrete random variables with possible 8-bit values of $k, x_1, x_2, y_1, y_2, \Delta x$, and Δy.

To analyze the values of $H(X_1 \mid \Delta X \Delta Y)$ and $H(\Delta X \mid \Delta Y)$, we take a look at the number of possible solutions for x_1 that satisfy

$$\Delta y = S_i(x_1) \oplus S_i(x_1 \oplus \Delta x) \tag{6}$$

for $S_i \in \{S_0, S_1\}$. The number of possible solutions to Eq. 6 for x_1 when Δx and Δy are given can be derived from the values of Table 1 in [14]: we divide the values by 256 since we do not regard the values for the 8-bit part k of the round key but solve the equation for x_1 instead. Furthermore, we discard the values for $\Delta x, \Delta y = 0$ since we assume them to be nonzero. Hence, we discard

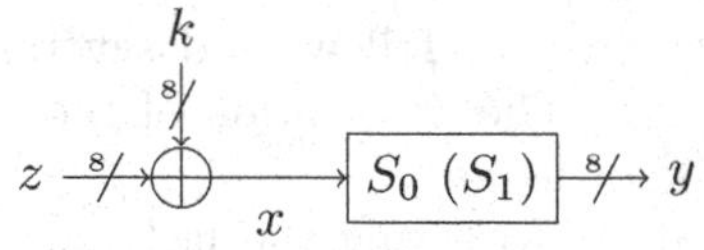

Fig. 4. Simple cipher model using an 8-bit CLEFIA S-box $S_0(S_1)$.

Table 1. Solutions to Eq. 6 for S_0 and x_1 with fixed Δx and Δy.

Solutions for x_1	0	2	4	6	8	10	Total
Occurrences	39511	19501	5037	848	119	9	65025 ($= 255^2$)

510 occurrences of 0 and one occurrence of 256 possible values for x_1. For the S-box S_0, the amount of occurrences of 0 to 10 possible solutions can be found in Table 1. For the S-box S_1, for every Δy we obtain 0 for 128 values of Δx, 2 for 126 values of Δx, and 4 for one value of Δx.

Now we calculate $H\left(K \mid Y_1Y_2\right)$ for the case that we have no information on the fault model, i.e., X_1, ΔX, and K are independent and identically distributed. Then, we calculate $H\left(K \mid Y_1Y_2\right)$ with additional information on the fault model.

Applying the information from Table 1, we yield for S_0

$$H_{S_0}\left(\Delta X \mid \Delta Y\right) = \sum_{i=1}^{255} P\left(\Delta Y = \Delta y_i\right) H_{S_0}\left(\Delta X \mid \Delta Y = \Delta y_i\right) \approx 6.535, \quad (7)$$

$$H_{S_0}\left(X_1 \mid \Delta X \Delta Y\right) = \sum_{i=1}^{255} P\left(\Delta Y = \Delta y_i\right) H_{S_0}\left(X_1 \mid \Delta X \Delta Y = \Delta y_i\right) \approx 1.465, \quad (8)$$

$$H_{S_0}\left(K \mid Y_1Y_2\right) \overset{([8])}{=} 6.535 + 1.465 = 8 = H\left(K\right). \quad (9)$$

For S_1, since its differential property is equal to the one of the AES S-box, the calculations are analogous to calculations 5 and 6 from [11] and are thus omitted in the present work. We have $H_{S_1}\left(\Delta X \mid \Delta Y\right) = \frac{447}{64}$, $H_{S_1}\left(X_1 \mid \Delta X \Delta Y\right) = \frac{65}{64}$, and $H_{S_1}\left(K \mid Y_1Y_2\right) = \frac{447}{64} + \frac{65}{64} = 8 = H\left(K\right)$.

Since $H_{S_0}\left(K \mid Y_1Y_2\right) = H_{S_1}\left(K \mid Y_1Y_2\right) = H\left(K\right)$, no information on K can be obtained without information on the fault model. We repeat the calculation using some assumptions on the fault model. Let $\mathcal{X} \subseteq \{0,1\}^n$ be the set of values that ΔX can take in the employed fault model.

As a coarse estimate for $H_{S_0}\left(\Delta X \Delta Y\right)$, we consider only the values for Δx and Δy that allow at least one possible solution for x_1, i.e., $19501 + 5037 + 848 + 119 + 9 = 25514$ of $255^2 = 65025$ values. Let X_1 and ΔX be independent and identically distributed over $\{0,1\}^n$ and $\mathcal{X}$, and ΔY identically distributed over $\{0,1\}^n$. We have $H_{S_0}\left(\Delta Y\right) \approx n$ and

$$H_{S_0}\left(\Delta X \Delta Y\right) \approx \log_2\left(\frac{25514 \cdot |\mathcal{X}| \cdot 2^n}{65025}\right) \text{ and therefore} \quad (10)$$

$$H_{S_0}\left(\Delta X \mid \Delta Y\right) \approx \log_2\left(|\mathcal{X}|\right) - 1.349. \quad (11)$$

We have $H_{S_0}\left(X_1 \mid \Delta X \Delta Y\right) \approx 1.465$ and $H_{S_0}\left(K \mid Y_1Y_2\right) \overset{([8])}{\approx} \log_2\left(|\mathcal{X}|\right) + 0.115$.

Without information on the fault model we have $\mathcal{X} = \{0,1\}^n$, so maximally $H_{S_0}\left(\Delta X \mid \Delta Y\right) \approx n - 1.349$. Hence, we define $m_{S_0} := (n - 1.349) - H_{S_0}\left(\Delta X \mid \Delta Y\right) \overset{([11])}{\approx} n - \log_2\left(|\mathcal{X}|\right)$ as the amount of information leaked from

Table 2. Information-theoretic optimality of DFAs against CLEFIA-128. The location and the timing of the faults describe the fault model. The number of key bits that can be learned from a single fault is denoted with m, and t denotes the number of faults the authors use to reduce the key space to $|\mathcal{K}|$ candidates.

Differential fault attack	Location	Timing	m	t	$|\mathcal{K}|$	Optimality
Chen et al. [4]	1 random byte	15, 16, 17	118.006	18	1	2 faults suffice
Takahashi, Fukunaga [14,15]	4 known bytes	16	96.023	2	1	optimal
Zhao et al. [18]	4 known bytes	16, 17, 18	96.023	12	1	2 faults suffice
	4 known bytes	15, 17	96.023	8	1	2 faults suffice
	4 known bytes	16	96.023	2	2^{19}	$|\mathcal{K}|$ can be 1
Ali, Mukhopadhyay [1,2]	1 known byte	14	120.006	2	1	optimal
Improvement on [4]	1 random byte	**15**	118.006	**2**	1	**optimal**
Improvement on [18]	4 known bytes	**16**	96.023	**2**	**1**	**optimal**

a fault injected before the application of the S-box S_0. Thus, the amount of information a certain fault can yield from an information-theoretic perspective depends on the amount of values the fault can attain. For S_1, as before, the calculation is analogous to the one from [11] and we use the definition $m_{S_1} := (n-1) - H_{S_1}(\Delta X \mid \Delta Y) \approx n - \log_2(|\mathcal{X}|)$ for the amount of information leaked from a fault injected before the application of the S-box S_1. Since the estimations for both S_0 and S_1 lead to the same definition for the amount of leaked information, we define

$$m \approx n - \log_2(|\mathcal{X}|). \tag{12}$$

4.2   Results of the Information-Theoretic Analysis

We will now present the results of our information-theoretic analysis regarding the optimality of all existing Differential Fault Attacks against CLEFIA. We calculate the amount of leaked information m by means of Eq. 12. The results of our analysis are summarized in Table 2 for CLEFIA-128 and in Table 3 for CLEFIA-192 and CLEFIA-256.

Attacks Against CLEFIA-128. The first Differential Fault Attack against CLEFIA-128 [4] uses 18 faults that are injected in one random byte of a four-byte register, so we have $2^8 - 1$ possible faults in four possible locations and the size of the set of possible values for ΔX is $|\mathcal{X}| = (2^8 - 1) \cdot 4$. We get $m \approx 128 - \log_2((2^8 - 1) \cdot 4) \approx 118.006$. Thus, in theory one fault is sufficient to reduce the key space to 2^{10} and two faults leak enough information to uniquely identify the key. Since the attack in [4] needs 18 faults, it is not information-theoretically optimal.

Table 3. Information-theoretic optimality of DFAs against CLEFIA-192/256. The location and the timing of the faults describe the fault model. The number of key bits that can be learned from a single fault is denoted with m, and t denotes the number of faults the authors use to reduce the key space to $|\mathcal{K}|$ candidates. According to our analysis, 2 or 3 faults suffice for achieving optimality.

Differential fault attack	Location	Timing	m	t	$|\mathcal{K}|$
Chen et al. [4]	1 random byte	$r-9,\ldots,r-1$	118.006	54	1
Takahashi, Fukunaga [15]	4 known bytes	$r-8, r-5, r-2$	96.023	10.78	1
Ali, Mukhopadhyay [2]	1 known byte	$r-8, r-4$	120.006	8	1
Improvement on [4]	1 random byte	$\mathbf{r-7, r-4}$	118.006	**8**	1

The second attack against CLEFIA-128 uses two faults which are injected in four bytes with known position [14]. In the first step, they reduce the key space to $2^{19.02}$. Then, they recover the key through an exhaustive search utilizing the key schedule, but no plaintexts. For this fault model $|\mathcal{X}| = \left(2^8 - 1\right)^4$, so $m \approx 128 - \log_2\left(\left(2^8 - 1\right)^4\right) \approx 96.023$ for a single fault. Therefore, two faults are needed to uniquely identify the key. As the attack uses only two faults we consider it optimal.

The next three DFAs by Zhao et al. [18] from 2010 are described for multiple-byte faults that affect one to four known bytes in the calculation. For these fault models, we have $|\mathcal{X}| = \left(2^8 - 1\right)^i$, $i \in \{1, \ldots, 4\}$, and we get

$$m \approx 128 - \log_2\left(\left(2^8 - 1\right)^i\right), \tag{13}$$

which implies for $i = 1, 2, 3, 4$, the amount of leaked information $m \approx 120.006$, $112.011, 104.017, 96.023$, respectively.

The authors give results only for the case of four-byte faults. Hence, we included only this case in Table 2. The authors state that their first attack uses six faults that affect eight bytes, but since the faults are injected in four-byte registers and their description of the attack also mentions one-to-four-byte faults, we assume four-byte faults. In case they really simulated eight-byte faults, these would have been two independent four-byte faults during one computation. Thus, their six faults count as twelve faults in our model, cf. Table 2. However, from Eq. 13 we find that only two faults are needed to recover the secret key. As the attack needs more faults, it is not optimal from an information-theoretic perspective. The second attack from [18] uses eight faults. From Eq. 13 we know that two faults leak enough information to uniquely identify the key, so this attack is not optimal either. The third proposed attack equates to the attack from Takahashi and Fukunaga in the case of four-byte faults. Zhao et al. state that with two faults the attack reduces the key space to 2^{19}. Since in theory two faults are enough to uniquely identify the key, this attack is not optimal.

Another DFA against CLEFIA-128 was described by Ali and Mukhopadhyay [1,2]. It uses faults that affect one known byte, so we have $|\mathcal{X}| = 2^8 - 1$.

Since we have $m \approx 128 - \log_2 \left(2^8 - 1\right) \approx 120.006$, two faults are needed to leak enough information to uniquely identify the key. As their attack succeeds with only two faults, it optimally exploits the information leaked from these faults.

Attacks Against CLEFIA-192 and CLEFIA-256. We observed that all previously presented DFAs against CLEFIA-192 and CLEFIA-256 are not optimal from an information-theoretic perspective.

The first Differential Fault Attack against CLEFIA-192 and CLEFIA-256 was published by Chen et al. in 2007 [4]. As in their attack against CLEFIA-128, the faults are injected in one random byte in a four-byte register. For $2^8 - 1$ possible faults in four possible locations and thereby $|\mathcal{X}| = \left(2^8 - 1\right) \cdot 4$, we get $m \approx 128 - \log_2 \left(\left(2^8 - 1\right) \cdot 4\right) \approx 118.006$. Thus, in theory two and three faults are sufficient to uniquely identify the 192-bit and 256-bit key, respectively. As the attack needs 54 faults, it is not optimal.

In 2010 Takahashi and Fukunaga adapted their Differential Fault Attack against CLEFIA-128 to longer keys [15]. Their faults affect four known bytes, so the fault model gives $|\mathcal{X}| = \left(2^8 - 1\right)^4$ and we have $m \approx 128 - \log_2 \left(\left(2^8 - 1\right)^4\right) \approx 96.023$, so in theory two and three faults are enough to recover the 192-bit and 256-bit key, respectively. However, since $2 \cdot 96 = 192$, an attack with only two faults will most probably not succeed in revealing a 192-bit key. Nevertheless, since the attack needs 10.78 faults on average, it is not optimal.

The most recent Differential Fault Attack against CLEFIA-192 and CLEFIA-256 was published in 2013 by Ali and Mukhopadhyay [2]. Analogously to their attack against CLEFIA-128, it works with faults that affect one known byte, so we have $|\mathcal{X}| = 2^8 - 1$. The attack needs eight faults to recover the key. We have $m \approx 128 - \log_2 \left(2^8 - 1\right) \approx 120.006$. Again, two and three faults leak enough information to uniquely identify the 192- and 256-bit key, respectively. Thus, this attack is not optimal, but information-theoretically the best known DFA against CLEFIA-192 and CLEFIA-256.

5   Improvement of the Non-Optimal DFAs

For the non-optimal attacks, we seek for an improved DFA in the same fault model, i.e., we use a subset of the faults injected in the original attack in order to reveal the secret key. We show that with one exception, all previously non-optimal attacks against CLEFIA-128 can be improved to be optimal from an information-theoretic perspective. For CLEFIA-192 and CLEFIA-256, we achieve a considerable improvement in one of the algorithms. The improved version requires significantly less fault injections than before and achieves the best success rate with this low fault number. Our experimental results are presented in Tables 4 and 5 and for our methodology used for the validation, the reader is referred to the full version of the paper [8, Sect. 5.3].

We first describe the basic idea of DFA methods on CLEFIA that is used to reveal the round keys and thereafter the secret key K. Analyzing the jth round, the attacker calculates the input Z_k^j of the F-function F_k, where $k \in \{0, 1\}$. It can be calculated by means of the correct ciphertext and some of the round keys of later rounds. The difference ΔZ_k^j of the inputs of the F-function is calculated with a correct and a faulty ciphertext. $\Delta Z_{k,i}^j$ with $i \in \{0, \ldots, 3\}$ denotes the input difference of one of the four S-boxes used in F_k. In order to obtain the output differences $\Delta Y_{k,i}^j$ of the S-boxes, the inverse of the corresponding diffusion matrix M_k is applied to the 32-bit output difference of F_k. The input-output differences for the S-boxes are retrieved, using which differential equations for all 8-bit states are deduced:

$$\Delta Y_{k,i}^j = \mathrm{S}(Z_{k,i}^j \oplus RK_{2j-2+k,i} \oplus \Delta Z_{k,i}^j) \oplus \mathrm{S}(Z_{k,i}^j \oplus RK_{2j-2+k,i}). \qquad (14)$$

Here, we have $i \in \{0, \ldots, 3\}$ and S denotes the S-box used in the state, as shown in Fig. 2. The difference distribution table of an S-box stores all the values of $Z_{k,i}^j \oplus RK_{2j-2+k,i}$ corresponding to a choice $(\Delta Z_{k,i}^j, \Delta Y_{k,i}^j)$. Table 1 shows the possible numbers of solutions for $Z_{k,i}^j \oplus RK_{2j-2+k,i}$ in case of S_0, and we described the case of S_1 in Sect. 4. If j is odd, a whitening key is also XORed to this value. Therefore, after using the difference distribution tables, a limited number of candidates remains for each of the four 8-bit parts of the round key in case the input-output differences are nonzero values, because the fault affected the round. After recovering the necessary round keys, the original secret key K can be deduced by analyzing the key scheduling of CLEFIA.

5.1 Improvements on CLEFIA-128

For the deduction of the 128-bit secret key K, the most efficient algorithm uses the values of RK_{30}, RK_{31}, $RK_{32} \oplus WK_3$, $RK_{33} \oplus WK_2$, RK_{34} and RK_{35}. Thus, we need to recover these by examining the input-output differences in the last three rounds. Two of the analyzed six attacks against CLEFIA-128 are already optimal, and we improved three of the remaining four. In Table 2 we see the attacks already existing along with our proposed improvements.

Optimization of the Attack by Chen et al.: The fault model used by Chen et al. in [4] is the byte-oriented model of random faults. A one-byte fault is induced into the register composed of four bytes in an intermediate step. The attacker knows the register into which the fault is injected, but does not have any knowledge of the concrete location or the value of the fault. Each fault is injected before a diffusion matrix in a certain round, so that a single random byte fault causes four-byte faults in the next round. In their original attack, they inject three faults into each of six locations in the 15th, 16th, and 17th round.

To make this attack optimal, we show that only two of the six faults induced in the 15th round are enough to uniquely reveal the secret key. In this fault model, the analysis presented by Takahashi and Fukunaga in Sect. 6 of their paper [14] can be borrowed. Takahashi and Fukunaga claim that in their attack,

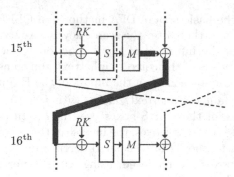

Fig. 5. Fault injection areas in the attack against CLEFIA-128 from [14,15]. (Figure taken from [14].)

the fault injection area can be chosen from two areas. One is the area in the 15th round within the dashed rectangle in Fig. 5, where any bit in any byte can be corrupted. The other area is a total of four bytes in the region after the diffusion matrix of round 15, denoted by a bold line in the same figure. Afterwards, on the bold line all four bytes are corrupted due to the fault propagation. Since the first injection area is the same as the one used by Chen et al. in [4], which means that if we use the two random byte faults injected into the 15th round, we can borrow the key retrieval technique of Takahashi and Fukunaga from [14]. Their method exploits the property of the CLEFIA key schedule procedure that two of the whitening keys (WK_2, WK_3) store the last two words of the original secret key (K_2, K_3), and thus, it uniquely verifies the original secret key. An attacker can recover a limited number of round key candidates and from each combination of these, a possible secret key can be calculated. Then, among $2^{19.02}$ candidates, the original secret key is verified uniquely.

Optimization of the Attacks by Zhao et al.: Zhao et al. use a different fault model in [18], exploiting one to four random byte-faults. The attacker does not have any knowledge of the concrete location or the value of the faults. With these looser conditions the authors claim their attacks to be more practical. Despite this, they analyze their attack only with four-byte faults and thus we also include these results in Table 2. When injecting faults according to their original fault model, much more faults are necessary, since at each step less bytes of the input difference are nonzero. The authors present three attacks: the first uses 12 faults in the last three rounds, the second uses 8 faults in the penultimate round and two rounds above, and the third attack only 2 faults in the round before the penultimate round.

Their third attack uses only two faults if four bytes are disturbed in the 16th round. After identifying candidates for the round keys, they deduce the secret key candidates and verify one of them as the original key. Since this verification process is not described, we assume that they do a brute-force search on a known plaintext-ciphertext pair. This type of exhaustive search is not necessary, since

Table 4. Experimental results on existing and proposed Differential Fault Analyses on CLEFIA-128. With t faults, we obtained the reduced key space $\mathcal{K}$ in 100 or 2000 simulation experiments with the given success rate.

Differential fault attack	Timing	t	$|\mathcal{K}|$	Experiments	Success
Chen et al. [4]	15, 16, 17	18	1	2000	99.1 %
Takahashi, Fukunaga [14,15]	16	2	1	100	97 %
Zhao et al. [18]	16, 17, 18	12	1	2000	81.3 %
	15, 17	8	1	2000	68.7 %
	16	2	2^{19}	100	97 %
Ali, Mukhopadhyay [1,2]	14	2	1	2000	91.45 %
Improvement on [4]	15	2	1	100	97 %
Improvement on [18]	16	2	1	100	97 %

the verification process from other attacks can be applied [1,4,14]. With this technique, the attack is information-theoretically optimal.

In case of their first attack, Zhao et al. inject 12 four-byte faults into the 18th, 17th and 16th rounds, and by means of these faults, they identify the secret key uniquely. If we use the analysis from the above described and improved third attack, we use only two of these faults, the ones injected into the 16th round. Therefore, we reduced the number of faults injected to two, which is claimed in Table 2 in order to achieve optimality for this attack.

Their second attack uses faults in two rounds. First, they induce four four-byte faults into two locations in the 17th round, by means of which they deduce the last four round keys. After this, they inject faults into the 15th round and compute the remaining two round keys necessary to reveal the secret key. By examining these injection points, no algorithm can reveal the secret key using only two of the faults. The faults injected in the 17th round can only recover the last two round keys, since they do not affect the 16th round input-output differences. An analysis with two four-byte faults injected in the 15th round is not possible with the existing techniques, lacking the knowledge on the value of the fault. If the value of the fault was known or a 32-bit brute-force search was allowed, the method by Ali and Mukhopadhyay [1] could be used with the fault value instead of the fault pattern. Here, we consider it impractical, since the complexity of the attack would be $2^{32} \cdot 2^{25.507} = 2^{57.507}$.

5.2 Improvements on CLEFIA-192 and CLEFIA-256

Table 3 shows that there is no existing attack against CLEFIA-192/256 which is information-theoretically optimal. In order to deduce the secret key, the most efficient algorithm needs to recover RK_{30}, RK_{31}, $RK_{32} \oplus WK_3$, $RK_{33} \oplus WK_2$, RK_{34}, RK_{35}, $RK_{36} \oplus WK_2$, $RK_{37} \oplus WK_3$, RK_{38}, RK_{39}, $RK_{40} \oplus WK_3$, $RK_{41} \oplus WK_2$, RK_{42}, and RK_{43}. Thus, a successful attack needs to calculate the input-output differences of at least the last seven rounds. All proposed attacks identify the

Table 5. Experimental results on existing and proposed Differential Fault Analyses on CLEFIA-192/256. With t faults, we obtained the reduced key space $\mathcal{K}$ in 100 or 2000 simulation experiments with the given success rate.

Differential fault attack	Timing	t	$|\mathcal{K}|$	Experiments	Success
Chen et al. [4]	$r-9,\ldots,r-1$	54	1	2000	98.3%
Takahashi, Fukunaga [15]	$r-8, r-5, r-2$	10	1	100	51%
Ali, Mukhopadhyay [2]	$r-8, r-4$	8	1	2000	43.4%
Improvement on [4]	$r-7, r-4$	8	1	2000	51.2%

secret key uniquely, yet the best attack from an information-theoretic perspective is the last proposed method by Ali and Mukhopadhyay [2]. Their technique uses the value of the faults they inject strictly into the first byte of a given register. This register is found before the diffusion matrix of round $r-4$ and $r-8$, so the fault propagates with a given fault pattern shown in [2, Fig. 5]. The attack uses this fault pattern during the calculations of eight round keys. By means of this method, we improve the analysis described by Chen et al. in [4].

Improvement of the Attack by Chen et al.: Chen et al. inject the faults in the same area of a round as Ali and Mukhopadhyay [2], though not strictly in the first, but randomly into one of the four bytes of the register. They induce altogether 54 faults into rounds $r-9$ to $r-1$, i.e., 6 faults per round. Half of the faults are induced in T_0, and half of the faults are induced in T_2. We, instead, mix the analyses of Ali and Mukhopadhyay [2] and Takahashi and Fukunaga [15], and apply this mixed technique to the fault model of Chen et al.

We first use four faults injected only into round $r-4$ (two into T_0, two into T_2). An injected fault f implies one of four fault patterns in case of both diffusion matrices M_0 and M_1, depending on which byte the fault was induced into. After calculating the fault patterns, the algorithm from [2] can be borrowed. When we use the fault pattern for the input-output differences, two times 16 checks are necessary, since there are four possible patterns for both the fault injections.

After determining the first eight necessary round keys, we use another four faults. We do not inject them four, but three rounds earlier, into round $r-7$. Here, we use the analysis technique from [15] to recover the rest of the necessary round keys. In Sect. 5.1, it is explained why this attack can be directly applied to the fault model of Chen et al.

Originally, Chen et al. injected 6 faults in 9 rounds each, altogether 54 faults. After using only 8 of these faults injected into rounds $r-4$ and $r-7$, we have all the necessary information to calculate the secret key. This way we reduced the number of fault injections to the lowest possible number reached to date for CLEFIA-192 and CLEFIA-256. Our attack cannot be prevented by protecting only the last four rounds of the algorithm. Moreover, as shown in Table 5, it shows a better success rate than the DFA from [2].

6   Conclusion

Our analysis of CLEFIA shows that an attacker needs at least two faults to fully reveal the secret 128-bit key. Based on these findings, we improved all but one attack against CLEFIA-128. From an information-theoretic perspective, the improved Differential Fault Analyses all reach the theoretical limit.

For longer keys, we considerably improved one of the existing attacks. Our proposed attack reaches the lowest number of faults reached so far.

References

1. Ali, S., Mukhopadhyay, D.: Protecting last four rounds of CLEFIA is not enough against differential fault analysis. IACR Cryptology ePrint Archive, p. 286 (2012)
2. Ali, S., Mukhopadhyay, D.: Improved differential fault analysis of CLEFIA. In: Workshop on Fault Diagnosis and Tolerance in Cryptography (FDTC 2013), pp. 60–70. IEEE (2013)
3. Biham, E., Shamir, A.: Differential fault analysis of secret key cryptosystems. In: Kaliski Jr., B.S. (ed.) CRYPTO 1997. LNCS, vol. 1294, pp. 513–525. Springer, Heidelberg (1997)
4. Chen, H., Wu, W., Feng, D.: Differential fault analysis on CLEFIA. In: Qing, S., Imai, H., Wang, G. (eds.) ICICS 2007. LNCS, vol. 4861, pp. 284–295. Springer, Heidelberg (2007)
5. Feller, W.: An Introduction to Probability Theory and Its Applications, vol. 1. Wiley, New York (1968)
6. Fischer, W., Reuter, C.A.: Differential fault analysis on Grøstl. In: Workshop on Fault Diagnosis and Tolerance in Cryptography (FDTC 2012), pp. 44–54. IEEE (2012)
7. Karmakar, S., Chowdhury, D.R.: Differential fault analysis of MICKEY-128 2.0. In: Workshop on Fault Diagnosis and Tolerance in Cryptography (FDTC 2013), pp. 52–59. IEEE (2013)
8. Krämer, J., Stüber, A., Kiss, Á.: On the optimality of differential fault analyses on CLEFIA. IACR Cryptology ePrint Archive 2014, p. 572 (2014)
9. Piret, G., Quisquater, J.-J.: A differential fault attack technique against SPN structures, with application to the AES and KHAZAD. In: Walter, C.D., Koç, Ç.K., Paar, C. (eds.) CHES 2003. LNCS, vol. 2779, pp. 77–88. Springer, Heidelberg (2003)
10. Rebeiro, C., Poddar, R., Datta, A., Mukhopadhyay, D.: An enhanced differential cache attack on CLEFIA for large cache Lines. In: Bernstein, D.J., Chatterjee, S. (eds.) INDOCRYPT 2011. LNCS, vol. 7107, pp. 58–75. Springer, Heidelberg (2011)
11. Sakiyama, K., Li, Y., Iwamoto, M., Ohta, K.: Information-theoretic approach to optimal differential fault analysis. IEEE Trans. Inf. Forensics Secur. 7, 109–120 (2012)
12. Shannon, C.: A mathematical theory of communication. Bell Syst. Tech. J. 27(379–423), 623–656 (1948)
13. Shirai, T., Shibutani, K., Akishita, T., Moriai, S., Iwata, T.: The 128-bit blockcipher CLEFIA (Extended Abstract). In: Biryukov, A. (ed.) FSE 2007. LNCS, vol. 4593, pp. 181–195. Springer, Heidelberg (2007)

14. Takahashi, J., Fukunaga, T.: Improved differential fault analysis on CLEFIA. In: Workshop on Fault Diagnosis and Tolerance in Cryptography (FDTC 2008), pp. 25–34. IEEE (2008)
15. Takahashi, J., Fukunaga, T.: Differential fault analysis on CLEFIA with 128, 192, and 256-bit keys. IEICE Trans. **93–A**, 136–143 (2010)
16. Tezcan, C.: The improbable differential attack: cryptanalysis of reduced round CLEFIA. In: Gong, G., Gupta, K.C. (eds.) INDOCRYPT 2010. LNCS, vol. 6498, pp. 197–209. Springer, Heidelberg (2010)
17. Tsunoo, Y., Tsujihara, E., Shigeri, M., Saito, T., Suzaki, T., Kubo, H.: Impossible differential cryptanalysis of CLEFIA. In: Nyberg, K. (ed.) FSE 2008. LNCS, vol. 5086, pp. 398–411. Springer, Heidelberg (2008)
18. Zhao, X., Wang, T., Gao, J.: Multiple bytes differential fault analysis on CLEFIA. IACR Cryptology ePrint Archive, p. 78 (2010)

Verified Numerical Computation

H^3 and H^4 Regularities of the Poisson Equation on Polygonal Domains

Takehiko Kinoshita[1,2], Yoshitaka Watanabe[3,4(✉)], and Mitsuhiro T. Nakao[5]

[1] Center for the Promotion of Interdisciplinary Education and Research,
Kyoto University, Kyoto 606-8501, Japan
[2] Research Institute for Mathematical Sciences, Kyoto University,
Kyoto 606-8502, Japan
[3] Research Institute for Information Technology, Kyushu University,
Fukuoka 812-8581, Japan
watanabe@cc.kyushu-u.ac.jp
[4] CREST, Japan Science and Technology Agency, Kawaguchi, Saitama, Japan
[5] National Institute of Technology, Sasebo College, Nagasaki 857-1193, Japan

Abstract. This paper presents two equalities of H^3 and H^4 semi-norms for the solutions of the Poisson equation in a two-dimensional polygonal domain. These equalities enable us to obtain higher order constructive a priori error estimates for finite element approximation of the Poisson equation with validated computing.

Keywords: Poisson equation · A priori estimates

1 Introduction

Consider the Poisson equation

$$\begin{cases} -\triangle u = f & \text{in } \Omega, \\ u = 0 & \text{on } \partial\Omega \end{cases}$$

(1a)

(1b)

with a multiply-connected polygonal domain $\Omega \subset \mathbb{R}^2$. The regularities of solutions of the equation (1a)–(1b) depend on the shape of Ω and f. For example, when Ω is convex and $f \in L^2(\Omega)$, it is well-known (e.g. Grisvard [1]) that there exists a unique solution $u \in H_0^1(\Omega) \cap H^2(\Omega)$ of (1a)–(1b).

Recently, Hell, Ostermann and Sandbichler [2, Lemma 2.4], and Hell and Ostermann [3, Proposition 3] showed the following results.

Lemma 1. *Let $\Omega = (0,1)^2$. Then all solutions to (1a)–(1b) lie in $H^3(\Omega)$ for $f \in H_0^1(\Omega)$. Moreover, for $f \in H_0^1(\Omega) \cap H^2(\Omega)$ the solution of (1a)–(1b) lies in $H^4(\Omega)$.*

Remark 1. The assumption $f \in H_0^1(\Omega)$ is essential at Lemma 1. For example, Hell and Ostermann [3] pointed out that, in the case of $f = 1$, the solution is not in $H^3(\Omega)$ even though $f \in C^\infty(\Omega)$.

© Springer International Publishing Switzerland 2016
I.S. Kotsireas et al. (Eds.): MACIS 2015, LNCS 9582, pp. 199–201, 2016.
DOI: 10.1007/978-3-319-32859-1_16

2 A Priori Error Estimations

Higher regularities of the solutions for the Poisson equation such as Lemma 1 will lead us to higher order error estimations for finite element approximate solutions of (1a)–(1b). For example, a result by Nakao, Yamamoto and Kimura [4] strongly suggests that when $f \in H_0^1(\Omega)$ and a solution u of (1a)–(1b) lies in $H^3(\Omega)$, for P2 (or Q2) finite element approximation u_h of u, there exists numerically determined $C_2 > 0$ satisfying

$$\|u - u_h\|_{H_0^1(\Omega)} \leq C_2 h^2 \, |u|_{H^3(\Omega)}. \tag{2}$$

Here, h shows the mesh size, $\|u\|_{H_0^1(\Omega)}$ and $|u|_{H^3(\Omega)}$ are H_0^1 norm and H^3 semi-norm of u defined by

$$\|u\|_{H_0^1(\Omega)} := |u|_{H^1(\Omega)} = \|\nabla u\|_{L^2(\Omega)^2} = \sqrt{\|u_{x_1}\|_{L^2(\Omega)}^2 + \|u_{x_2}\|_{L^2(\Omega)}^2},$$

$$|u|_{H^3(\Omega)} := \sqrt{\|u_{x_1 x_1 x_1}\|_{L^2}^2 + 3\|u_{x_1 x_1 x_2}\|_{L^2}^2 + 3\|u_{x_1 x_2 x_2}\|_{L^2}^2 + \|u_{x_2 x_2 x_2}\|_{L^2}^2},$$

respectively. Moreover, if u has sufficient regularities and u_h is a P3 (or Q3) finite element approximation, there also exists $C_3 > 0$ such that

$$\|u - u_h\|_{H_0^1(\Omega)} \leq C_3 h^3 \, |u|_{H^4(\Omega)}, \tag{3}$$

where $|u|_{H^4(\Omega)}$ is H^4 semi-norm of u defined by

$$|u|_{H^4(\Omega)} := \Big(\|u_{x_1 x_1 x_1 x_1}\|_{L^2(\Omega)}^2 + 4\|u_{x_1 x_1 x_1 x_2}\|_{L^2(\Omega)}^2$$

$$+ 6\|u_{x_1 x_1 x_2 x_2}\|_{L^2(\Omega)}^2 + 4\|u_{x_1 x_2 x_2 x_2}\|_{L^2(\Omega)}^2 + \|u_{x_2 x_2 x_2 x_2}\|_{L^2(\Omega)}^2 \Big)^{\frac{1}{2}}.$$

3 Main Theorem

We present a priori estimates replaced by f in the right-hand side of (2) and (3) instead of H^3 and H^4 semi-norms of u, respectively.

Let $D^1(-\triangle)$ and $D^2(-\triangle) \subset H_0^1(\Omega)$ be the Banach spaces defined by

$$D^1(-\triangle) := \big\{ u \in H_0^1(\Omega) \; ; \; -\triangle u \in H_0^1(\Omega) \big\},$$

$$D^2(-\triangle) := \big\{ u \in H_0^1(\Omega) \; ; \; -\triangle u \in H_0^1(\Omega) \cap H^2(\Omega) \big\},$$

respectively. Note that $D^n(-\triangle)$ ($n \in \{1, 2\}$) is the set of solutions of the Poisson equation (1a)–(1b). We assume that $D^k(-\triangle) \cap C^\infty(\overline{\Omega})$ is dense in $D^k(-\triangle) \cap H^{k+2}(\Omega)$ for $k = 1, 2$.

Theorem 1. *It is true that*

$$|u|_{H^3(\Omega)} = \|\nabla(\triangle u)\|_{L^2(\Omega)^2}, \quad \forall u \in D^1(-\triangle) \cap H^3(\Omega). \tag{4}$$

Remark 2. Using (2) and (4) we obtain an a priori error estimate with $O(h^2)$:

$$\|u - u_h\|_{H_0^1(\Omega)} \leq C_2 h^2 \|f\|_{H_0^1(\Omega)}.$$

Theorem 2. *It is true that*

$$|u|_{H^4(\Omega)} = \|\Delta^2 u\|_{L^2(\Omega)}, \quad \forall u \in D^2(-\Delta) \cap H^4(\Omega). \tag{5}$$

Remark 3. Using (3) and (5) we obtain an a priori error estimate with $O(h^3)$:

$$\|u - u_h\|_{H_0^1(\Omega)} \leq C_3 h^3 \|\Delta f\|_{L^2(\Omega)}.$$

Acknowledgments. This work was supported by the Grant-in-Aid from the Ministry of Education, Culture, Sports, Science and Technology of Japan (Nos. 15H03637, 15K05012) and supported by Program for Leading Graduate Schools "Training Program of Leaders for Integrated Medical System for Fruitful Healthy-Longevity Society."

References

1. Grisvard, P.: Elliptic Problems in Nonsmooth Domains. Pitman, Boston (1985)
2. Hell, T., Ostermann, A., Sandbichler, M.: Modification of dimension-splitting methods - overcoming the order reduction due to corner singularities. IMA J. Numer. Anal. **35**, 1078–1091 (2015)
3. Hell, T., Ostermann, A.: Compatibility conditions for dirichlet and neumann problems of poisson's equation on a rectangle. J. Math. Anal. Appl. **420**, 1005–1023 (2014)
4. Nakao, M.T., Yamamoto, N., Kimura, S.: On best constant in the optimal error stimates for the H_0^1-projection into piecewise polynomial spaces. J. Approx. Theor. **93**, 491–500 (1998)

Explicit Error Bound for Modified Numerical Iterated Integration by Means of Sinc Methods

Tomoaki Okayama$^{(\boxtimes)}$

Hiroshima City University, 3-4-1, Ozuka-higashi, Asaminami-ku, Hiroshima, Japan
okayama@hiroshima-cu.ac.jp

Abstract. This paper reinforces numerical iterated integration developed by Muhammad–Mori in the following two points: (1) the approximation formula is modified so that it can achieve a better convergence rate in more general cases, and (2) an explicit error bound is given in a computable form for the modified formula. The formula works quite efficiently, especially if the integrand is of a product type. Numerical examples that confirm it are also presented.

Keywords: Sinc quadrature · Sinc indefinite integration · Repeated integral · Verified numerical integration · Double-exponential transformation

1 Introduction

The concern of this paper is efficient approximation of a two-dimensional iterated integral

$$I = \int_a^b \left(\int_A^{q(x)} f(x,y)\,\mathrm{d}y \right) \mathrm{d}x, \tag{1}$$

with an a priori rigorous error bound. Here, $q(x)$ is a monotone function that may have derivative singularity at the endpoints of $[a, b]$, and the integrand $f(x,y)$ also may have singularity on the boundary of the square region $[a, b] \times [A, B]$ (see also Figs. 1 and 2). In this case, a Cartesian product rule of a well known one-dimensional quadrature formula (such as the Gaussian formula and the Clenshaw–Curtis formula) does not work properly, or at least its mathematically-rigorous error bound is quite difficult to obtain, because such formulas require analyticity of the integrand in a neighbourhood of the boundary [1].

Promising quadrature formulas that do not require analyticity at the endpoints may include the tanh formula [15], the IMT formula [3,4], and the double-exponential formula [20], which enjoy *exponential convergence* whether the integrand has such singularity or not. Actually, based on the IMT formula, an automatic integration algorithm for (1) was developed [12]. Further improved version was developed as d2lri [2] and r2d2lri [13], where the lattice rule is employed with the IMT transformation [3,4] or the Sidi transformation [16,17].

© Springer International Publishing Switzerland 2016
I.S. Kotsireas et al. (Eds.): MACIS 2015, LNCS 9582, pp. 202–217, 2016.
DOI: 10.1007/978-3-319-32859-1_17

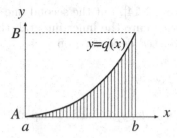

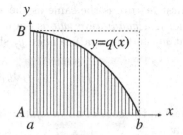

Fig. 1. The domain of integration (1) when $q'(x) \geq 0$.

Fig. 2. The domain of integration (1) when $q'(x) \leq 0$.

As a related study, based on the double-exponential formula, an automatic integration algorithm over a sphere was developed [14], which also intended to deal with such integrand singularities. The efficiency of those algorithms is also suggested by their numerical experiments.

From a mathematical viewpoint, however, those algorithms do not guarantee the accuracy of the approximation in reality. In order to estimate the error (for giving a stopping criterion), Robinson and de Doncker [12] considered the sequence of the number of function evaluation points $\{N_m\}_m$ and that of approximation values $\{I_{N_m}\}$, and made the important assumption:

$$D_{N_m} := |I_{N_m} - I_{N_{m-1}}| \simeq |I - I_{N_{m-1}}|, \tag{2}$$

which enables the error estimation $|I - I_{N_m}| \simeq D_{N_m}^2 / D_{N_{m-1}}$. A similar approach was taken in the studies described above [2,13,14]. The problem here is that it is quite difficult to guarantee the validity of (2), although it had been widely accepted as a realistic practical assumption for constructing automatic quadrature routines in that period. The recent trend is that the approximation error is bounded by a *strict* inequality (instead of estimation '$\simeq$') as

$$|I - I_N| \leq E_N,$$

where E_N is given in a *computable* form (see, for example, Petras [11]). Such an explicit error bound is desired for constructing a more reliable, *verified numerical integration* routine. In addition to mathematical rigor, such a bound gives us another advantage: the sufficient number N for the required precision, say N_0, can be known without generating the sequence $\{I_N\}$. This means low computational cost, since we do not have to compute for any N with $N < N_0$.

The objective of this study is to give such an explicit error bound for the numerical integration method developed by Muhammad–Mori [7]. Their method is based on Sinc methods [18,19] combined with the double-exponential transformation [5,20], and it has the following two features:

1. it has beautiful *exponential accuracy* even if $f(x, y)$ or $q(x)$ has boundary singularity, and
2. it employs an *indefinite integration formula* instead of a quadrature formula for the inner integral.

The first feature is the same as the studies above [12,14], but the second one is unique. If a *quadrature rule* is employed to approximate the inner integral, the weight w_j and quadrature node y_j should be adjusted depending on x as

$$\int_A^{q(x)} f(x,y)\,\mathrm{d}y \approx \sum_j w_j(x) f(x, y_j(x)),$$

whereas in the case of an *indefinite integration formula*, y_j is fixed (independent of x) as

$$\int_A^{q(x)} f(x,y)\,\mathrm{d}y \approx \sum_j w_j(x) f(x, y_j).$$

This independence of x is quite useful to check mathematical assumptions on the integrand $f(x, y)$ for the exponential accuracy. Furthermore, as a special case, when the integrand is of a product type: $f(x, y) = X(x)Y(y)$, the number of function evaluations to approximate (1) is drastically dropped from $\mathrm{O}(n \times n)$ to $\mathrm{O}(n + n)$, where n denotes the number of the terms of $\sum$ (it is also emphasized in the original paper [7]).

However, rigorous error analysis is not given for the formula, and there is room for improvement in the convergence rate. Moreover, it cannot handle the case $q'(x) \leq 0$ (only the case $q'(x) \geq 0$ is considered). In order to reinforce their formula, this study contributes in the following points:

3. their formula is modified so that it can achieve a better convergence rate in both cases (i.e., the case $q'(x) \geq 0$ and $q'(x) \leq 0$), and
4. a rigorous, explicit error bound is given for the modified formula.

The error bound shows that the convergence rate of the formula is generally $\mathrm{O}(\exp(-c\sqrt{n}/\log(\sqrt{n})))$, and if $f(x, y) = X(x)Y(y)$, it becomes $\mathrm{O}(\exp(-c'n/\log n))$.

The remainder of this paper is organized as follows. In Sect. 2, after the review of basic formulas of Sinc methods, Muhammad–Mori's original formula [7] is described. Then, the formula is modified in Sect. 3, and its explicit error bound is also presented. Its proof is given in Sect. 5. Numerical examples are shown in Sect. 4. Section 6 is devoted to conclusion.

2   Review of Muhammad–Mori's Approximation Formula

2.1   Sinc Quadrature and Sinc Indefinite Integration Combined with the DE Transformation

The Sinc quadrature and Sinc indefinite integration are approximation formulas for definite and indefinite integration on $\mathbb{R}$, respectively, expressed as

$$\int_{-\infty}^{\infty} G(\xi)\,\mathrm{d}\xi \approx \tilde{h} \sum_{i=-M_-}^{M_+} G(i\tilde{h}), \tag{3}$$

$$\int_{-\infty}^{\xi} G(\eta)\,\mathrm{d}\eta \approx \sum_{j=-N_-}^{N_+} G(jh) J(j, h)(\xi), \tag{4}$$

where $J(j, h)$ is defined by using the sine integral $\mathrm{Si}(x) = \int_0^x \{(\sin \sigma)/\sigma\}\, d\sigma$ as

$$J(j,h)(\xi) = h \left\{ \frac{1}{2} + \frac{1}{\pi} \mathrm{Si}[\pi(\xi/h - j)] \right\}.$$

Although the formulas (3) and (4) are approximations on $\mathbb{R}$, they can be used on the finite interval (a, b) by using the Double-Exponential (DE) transformation

$$x = \psi(\xi) = \frac{b-a}{2} \tanh \left(\frac{\pi}{2} \sinh \xi \right) + \frac{b+a}{2}.$$

Since $\psi : \mathbb{R} \to (a, b)$, we can apply the formulas (3) and (4) in the case of finite intervals combining the DE transformation as

$$\int_a^b g(x)\, dx = \int_{-\infty}^{\infty} g(\psi(\xi))\psi'(\xi)\, d\xi \approx \tilde{h} \sum_{i=-M_-}^{M_+} g(\psi(i\tilde{h}))\psi'(i\tilde{h}), \tag{5}$$

$$\int_a^x g(y)\, dy = \int_{-\infty}^{\psi^{-1}(x)} g(\psi(\eta))\psi'(\eta)\, d\eta \approx \sum_{j=-N_-}^{N_+} g(\psi(jh))\psi'(jh)J(j,h)(\psi^{-1}(x)), \tag{6}$$

which are called the "DE-Sinc quadrature" and the "DE-Sinc indefinite integration," proposed by Takahasi–Mori [20] and Muhammad–Mori [6], respectively.

2.2   Muhammad–Mori's Approximation Formula

Let the domain of integration (1) be as in Fig. 1, i.e., $q(a) = A$, $q(b) = B$, and $q'(x) \geq 0$. Using the monotonicity of $q(x)$, Muhammad–Mori [7] rewrote the given integral I by applying $y = q(s)$ as

$$I = \int_a^b \left(\int_A^{q(x)} f(x,y)\, dy \right) dx = \int_a^b \left(\int_a^x f(x, q(s))q'(s)\, ds \right) dx. \tag{7}$$

Note that $s \in (a, b)$ (i.e., not (A, B)). Then, they applied (5) and (6), taking $\tilde{h} = h$, $M_- = M_+ = m$, and $N_- = N_+ = n$ for simplicity, as follows:

$$I \approx h \sum_{i=-m}^{m} \psi'(ih) \left(\int_a^{\psi(ih)} f(\psi(ih), q(s))q'(s)\, ds \right)$$

$$\approx h \sum_{i=-m}^{m} \psi'(ih) \left\{ \sum_{j=-n}^{n} f(\psi(ih), q(\psi(jh)))q'(\psi(jh))\psi'(jh)J(j,h)(ih) \right\}.$$

If we introduce $x_i = \psi(ih)$, $w_j = \pi \cosh(jh) \mathrm{sech}^2(\pi \sinh(jh)/2)/4$, and $\sigma_k = \mathrm{Si}[\pi k]/\pi$, which can be calculated and stored prior to computation (see also a value table for σ_k [18, Table 1.10.1]), the formula is rewritten as

$$I \approx (b-a)^2 h^2 \sum_{i=-m}^{m} w_i \left\{ \sum_{j=-n}^{n} f(x_i, q(x_j))q'(x_j)w_j \left(\frac{1}{2} + \sigma_{i-j} \right) \right\}. \tag{8}$$

The total number of function evaluations, say N_{total}, of this formula is $N_{\text{total}} = (2m + 1) \times (2n + 1)$. As a special case, if the integrand is of a product type: $f(x, y) = X(x)Y(y)$, the formula is rewritten as

$$
I \approx (b - a)^2 h^2 \sum_{i=-m}^{m} U(i) \left\{ \sum_{j=-n}^{n} V(j) \left(\frac{1}{2} + \sigma_{i-j} \right) \right\},
\tag{9}
$$

where $U(i) = X(x_i)w_i$ and $V(j) = Y(q(x_j))q'(x_j)w_j$. In this case, $N_{\text{total}} = (2m + 1) + (2n + 1)$, which is significantly smaller than $(2m + 1) \times (2n + 1)$.

They [7] also roughly discussed the error rate of the formula (8) as follows. Let $\mathscr{D}_d$ be a strip domain defined by $\mathscr{D}_d = \{\zeta \in \mathbb{C} : |\operatorname{Im} \zeta| < d\}$ for $d > 0$. Assume that the integrand g in (5) and (6) is analytic on $\psi(\mathscr{D}_d)$ (which means $g(\psi(\cdot))$ is analytic on $\mathscr{D}_d$), and further assume that $g(x)$ behaves $\mathrm{O}(((x - a)(b - x))^{\nu-1})$ ($\nu > 0$) as $x \to a$ and $x \to b$. Under those assumptions with some additional mild conditions, it is known that the approximation (5) converges with $\mathrm{O}(e^{-2\pi d/h})$, and the approximation (6) converges with $\mathrm{O}(h\,e^{-\pi d/h})$, by taking $h = \tilde{h}$ and

$$
M_+ = M_- = m = \left\lceil \frac{1}{h} \log\left(\frac{4d}{(\nu - \epsilon)h} \right) \right\rceil, \quad N_+ = N_- = n = \left\lceil \frac{1}{h} \log\left(\frac{2d}{(\nu - \epsilon)h} \right) \right\rceil,
$$

where ϵ is an arbitrary small positive number. Therefore, if the same assumptions are satisfied for both approximations in (8), it enjoys exponential accuracy: $\mathrm{O}(h\,e^{-\pi d/h})$. Since $m \simeq n \simeq \sqrt{N_{\text{total}}/4}$ and $h \simeq \log(cn)/n$ (where $c = 2d/(\nu - \epsilon)$), this can be interpreted in terms of N_{total} as

$$
\mathrm{O}\left(\frac{\log(c\sqrt{N_{\text{total}}/4})}{\sqrt{N_{\text{total}}/4}} \exp\left[\frac{-\pi d\sqrt{N_{\text{total}}/4}}{\log(c\sqrt{N_{\text{total}}/4})} \right] \right).
\tag{10}
$$

If the integrand is of a product type, since $m \simeq n \simeq N_{\text{total}}/4$, it becomes

$$
\mathrm{O}\left(\frac{\log(cN_{\text{total}}/4)}{N_{\text{total}}/4} \exp\left[\frac{-\pi d(N_{\text{total}}/4)}{\log(cN_{\text{total}}/4)} \right] \right).
\tag{11}
$$

Although the convergence rate was roughly discussed as above, the quantity of the approximation error cannot be obtained because a rigorous error bound was not given. Moreover, the case $q'(x) \leq 0$ (cf. Fig. 2) is not considered. This situation will be improved in the next section.

3   Main Results: Modified Approximation Formula and Its Explicit Error Bound

This section is devoted to a description of a new approximation formula and its error bound. The proof of the error bound is given in Sect. 5.

3.1   Modified Approximation Formula

In the approximations (5) and (6), Muhammad–Mori [7] set the mesh size as $\tilde{h} = h$ for simplicity, but here, $\tilde{h}$ is selected as $\tilde{h} = 2h$. Furthermore, both $M_- = M_+$ and $N_- = N_+$ are *not* assumed. Then, after applying $y = q(s)$ as in (7), the modified formula is derived as

$$
I \approx 2h \sum_{i=-M_-}^{M_+} \psi'(2ih) \left(\int_a^{\psi(2ih)} f(\psi(2ih), q(s)) q'(s)\, ds \right)
$$

$$
\approx 2h \sum_{i=-M_-}^{M_+} \psi'(2ih) \left\{ \sum_{j=-N_-}^{N_+} f(\psi(2ih), q(\psi(jh))) q'(\psi(jh)) \psi'(jh) J(j,h)(2ih) \right\},
$$

which can be rewritten as

$$
I \approx I_{\mathrm{DE}}^{\mathrm{inc}}(h) := 2(b-a)^2 h^2 \sum_{i=-M_-}^{M_+} w_{2i} \left\{ \sum_{i=-N_-}^{N_+} f(x_{2i}, q(x_j)) q'(x_j) w_j \left(\frac{1}{2} + \sigma_{2i-j} \right) \right\}.
$$

$$\tag{12}$$

The positive integers $M_\pm$ and $N_\pm$ are also selected depending on h, which is explained in the subsequent theorem that states the error bound.

The formula (12) is derived in the case $q'(x) \geq 0$ (cf. Fig. 1), but in the case $q'(x) \leq 0$ (cf. Fig. 2) as well, we can derive the similar formula as follows. First, applying $y = q(s)$, we have

$$
I = \int_a^b \left(\int_A^{q(x)} f(x,y)\, dy \right) dx = \int_a^b \left(\int_x^b f(x, q(s))\{-q'(s)\}\, ds \right) dx
$$

$$
= \int_a^b \left(\int_a^b f(x, q(s))\{-q'(s)\}\, ds - \int_a^x f(x, q(s))\{-q'(s)\}\, ds \right) dx.
$$

Then, apply (5) and (6) to obtain

$$
I \approx 2h \sum_{i=-M_-}^{M_+} \psi'(2ih) \left\{ \sum_{j=-N_-}^{N_+} f(x_{2i}, q(x_j))\{-q'(x_j)\} \psi'(jh)\, (h - J(j,h)(2ih)) \right\}.
$$

Here, $\lim_{\xi \to \infty} J(j,h)(\xi) = h$ is used. The right-hand side can be rewritten as

$$
I_{\mathrm{DE}}^{\mathrm{dec}}(h) := 2(b-a)^2 h^2 \sum_{i=-M_-}^{M_+} w_{2i} \left\{ \sum_{i=-N_-}^{N_+} f(x_{2i}, q(x_j))\{-q'(x_j)\} w_j \left(\frac{1}{2} - \sigma_{2i-j} \right) \right\}.
$$

$$\tag{13}$$

The formulas (12) and (13) inherit the advantage of Muhammad–Mori's one in the sense that $N_{\text{total}} = (M_- + M_+ + 1) \times (N_- + N_+ + 1)$ in general, but if the integrand is of a product type: $f(x, y) = X(x)Y(y)$, it becomes $N_{\text{total}} = (M_- + M_+ + 1) + (N_- + N_+ + 1)$, which is easily confirmed by rewriting it in the same way as (9). Furthermore, it also inherits (or even enhances) the *exponential* accuracy, which is described next.

3.2 Explicit Error Bound of the Modified Formula

For positive constants κ, λ and d with $0 < d < \pi/2$, let us define $c_{\kappa,\lambda,d}$ as

$$c_{\kappa,\lambda,d} = \frac{1}{\cos^{\kappa+\lambda}(\frac{\pi}{2}\sin d)\cos d},$$

and define ρ_κ as

$$\rho_\kappa = \begin{cases} \arcsinh\left(\frac{\sqrt{1+\sqrt{1-(2\pi\kappa)^2}}}{2\pi\kappa}\right) & (0 < \kappa < 1/(2\pi)), \\ \arcsinh(1) & (1/(2\pi) \leq \kappa). \end{cases}$$

Then, the errors of $I_{\text{DE}}^{\text{inc}}(h)$ and $I_{\text{DE}}^{\text{dec}}(h)$ are estimated as stated below.

Theorem 1. *Let α, β, γ, δ, and K be positive constants, and d be a constant with $0 < d < \pi/2$. Assume the following conditions:*

1. q is analytic and bounded in $\psi(\mathscr{D}_d)$,
2. $f(\cdot, q(w))$ and $f(z, q(\cdot))$ are analytic in $\psi(\mathscr{D}_d)$ for all z, $w \in \psi(\mathscr{D}_d)$,
3. it holds for all z, $w \in \psi(\mathscr{D}_d)$ that

$$|f(z, q(w))q'(w)| \leq K|z - a|^{\alpha-1}|b - z|^{\beta-1}|w - a|^{\gamma-1}|b - w|^{\delta-1}. \tag{14}$$

Let $\mu = \min\{\alpha, \beta\}$, $\overline{\mu} = \max\{\alpha, \beta\}$, $\nu = \min\{\gamma, \delta\}$, $\overline{\nu} = \max\{\gamma, \delta\}$, let $\tilde{h} = 2h$, let n and m be positive integers defined by

$$n = \left\lceil \frac{1}{h}\log\left(\frac{2d}{\nu h}\right)\right\rceil, \quad m = \left\lceil\frac{1}{2}\left\{n + \frac{1}{h}\log\left(\frac{\mu}{\nu}\right)\right\}\right\rceil, \tag{15}$$

and let M_- and M_+ be positive integers defined by

$$\begin{cases} M_- = m, & M_+ = m - \lfloor\log(\beta/\alpha)/\tilde{h}\rfloor & (\text{if } \mu = \alpha), \\ M_+ = m, & M_- = m - \lfloor\log(\alpha/\beta)/\tilde{h}\rfloor & (\text{if } \mu = \beta), \end{cases} \tag{16}$$

and let N_- and N_+ be positive integers defined by

$$\begin{cases} N_- = n, & N_+ = n - \lfloor\log(\delta/\gamma)/h\rfloor & (\text{if } \nu = \gamma), \\ N_+ = n, & N_- = n - \lfloor\log(\gamma/\delta)/h\rfloor & (\text{if } \nu = \delta), \end{cases} \tag{17}$$

and let h (> 0) be taken sufficiently small so that

$$M_-\tilde{h} \geq \rho_\alpha, \quad M_+\tilde{h} \geq \rho_\beta, \quad N_-h \geq \rho_\gamma, \quad N_+h \geq \rho_\delta$$

are all satisfied. Then, if $q'(x) \geq 0$, it holds that

$$|I - I_{DE}^{inc}(h)|$$

$$\leq 2K(b-a)^{\alpha+\beta+\gamma+\delta-2} \left[\frac{B(\gamma,\delta)c_{\gamma,\delta,d}}{\mu} \left\{ e^{\frac{\pi}{2}\bar{\mu}} + \frac{2c_{\alpha,\beta,d}}{1 - e^{-\pi d/h}} \right\} \right.$$

$$\left. + \frac{1}{\nu} \left\{ B(\alpha,\beta) + \frac{4c_{\alpha,\beta,d}\, e^{-\pi d/h}}{\mu(1 - e^{-\pi d/h})} \right\} \left\{ 1.1 e^{\frac{\pi}{2}\bar{\nu}} + \frac{hc_{\gamma,\delta,d}}{d(1 - e^{-2\pi d/h})} \right\} \right] e^{-\pi d/h},$$

$$(18)$$

where $B(\kappa,\lambda)$ is the beta function. If $q'(x) \leq 0$, $|I - I_{DE}^{dec}(h)|$ is bounded by the same term on the right hand side of (18).

The convergence rate of (18) is $O(e^{-\pi d/h})$, which can be interpreted in terms of N_{total} as follows. Since $n \simeq N_- \simeq N_+$ and $m \simeq M_- \simeq M_+ \simeq (n/2)$, we can see $N_{total} \simeq ((n/2) + (n/2) + 1)(n+n+1) \simeq 2n^2$. From this and $h \simeq \log(c'n)/n$ (where $c' = 2d/\nu$), the convergence rate of the modified formula is

$$O\left(\exp\left[\frac{-\pi d\sqrt{N_{total}/2}}{\log(c'\sqrt{N_{total}/2})} \right] \right).$$

This rate is better than Muhammad–Mori's one (10). If the integrand is of a product type: $f(x,y) = X(x)Y(y)$, it becomes

$$O\left(\exp\left[\frac{-\pi d(N_{total}/3)}{\log(cN_{total}/3)} \right] \right),$$

since $N_{total} \simeq ((n/2) + (n/2) + 1) + (n+n+1) \simeq 3n$ in this case. This rate is also better than Muhammad–Mori's one (11).

Remark 1. The inequality (18) states the bound of the *absolute* error, say $E^{abs}(h)$. If necessary, the bound of the *relative* error $E^{rel}(h)$ is also obtained as follows:

$$E^{rel}(h) = \frac{|I - I_{DE}^{inc}(h)|}{|I|} \leq \frac{E^{abs}(h)}{|I|} \leq \frac{E^{abs}(h)}{||I_{DE}^{inc}(h)| - E^{abs}(h)|}.$$

4   Numerical Examples

In this section, numerical results of Muhammad–Mori's original formula [7] and modified formula are presented. The results of an existing library r2d2lri [13] are also shown. The computation was done on a Mac Pro with two 2.93 GHz 6-Core Xeon processors and 32 GB of DDR3 ECC SDRAM, running Mac OS X 10.6. The computation programs were implemented in C/C++ with double-precision floating-point arithmetic, and compiled by GCC 4.0.1 with no optimization. The following three examples were conducted.

Example 1. (The integrand and boundary function are smooth [7, Example 2]).

$$\int_0^{\sqrt{2}} \left(\int_0^{x^2/2} \frac{dy}{x + y + (1/2)} \right) dx = 2\log(1+\sqrt{2})^{1+\sqrt{2}} - \frac{\log(1+2\sqrt{2})^{1+2\sqrt{2}}}{2} - \sqrt{2}.$$

Example 2. (Derivative singularity exists in the integrand and boundary function [7, Example 1]).

$$\int_0^1 \left(\int_0^{\sqrt{1-(1-x)^2}} \sqrt{1-y^2}\, dy \right) dx = \frac{2}{3}.$$

Example 3. (The integrand is weakly singular at the origin [2, Example 27]).

$$\int_0^1 \left(\int_0^{1-x} \frac{dy}{\sqrt{xy}} \right) dx = \pi.$$

In the case of Example 1, the assumptions in Theorem 1 are satisfied with $\alpha = \beta = \delta = 1$, $\gamma = 2$, $d = \log(2)$, and $K = 16.6$. The results are shown in Figs. 3 and 4. In both figures, error bound (say $\tilde{E}^{\mathrm{rel}}(h)$) given by Theorem 1 surely includes the observed relative error $E^{\mathrm{rel}}(h)$ in the form $E^{\mathrm{rel}}(h) \leq \tilde{E}^{\mathrm{rel}}(h)$, which is also true in all the subsequent examples (note that such error bound is not given for Muhammad–Mori's original formula). In view of the performance, r2d2lri is better than the original/modified formulas, but its error estimate just claims $E^{\mathrm{rel}}(h) \approx \tilde{E}^{\mathrm{rel}}(h)$, and does not guarantee $E^{\mathrm{rel}}(h) \leq \tilde{E}^{\mathrm{rel}}(h)$.

In the case of Example 2, the assumptions in Theorem 1 are satisfied with $\alpha = \beta = 1$, $\gamma = 1/2$, $\delta = 3$, $d = 1$, and $K = 1.63$. The results are shown in Figs. 5 and 6. In this case, the convergence of the original/modified formulas is incredibly fast compared to r2d2lri. This is because the integrand is of a product type: $f(x, y) = X(x)Y(y)$.

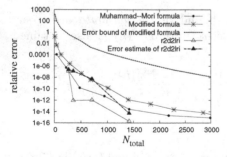

Fig. 3. Relative error with respect to N_{total} in Example 1.

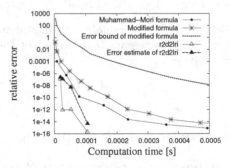

Fig. 4. Relative error with respect to computation time in Example 1 ("Error bound of modified formula" and "Error estimate of r2d2lri" show each computation time needed to obtain both the approximation value and its error bound/estimate).

The integrand of Example 3 is also of a product type. In this example, the assumptions in Theorem 1 are satisfied with $\alpha = \delta = 1/2$, $\beta = \gamma = 1$, $d = 4/3$, and $K = 1$. The results are shown in Figs. 7 and 8. In this case, the performance of r2d2lri is much worse than that in Example 2, which seems to be due to the singularity of the integrand. In contrast, the modified formula attains a similar convergence rate to that in Example 2. Muhammad–Mori's original formula cannot be used in this case since $q(x) = 1 - x$ does not satisfy $q'(x) \geq 0$.

5 Proofs

In this section, only the inequality (18) (for $|I - I_{DE}^{inc}(h)|$) is proved, since $|I - I_{DE}^{dec}(h)|$ is bounded in exactly the same way. Let us have a look at the sketch of the proof first.

5.1 Sketch of the Proof

The error $|I - I_{DE}^{inc}(h)|$ can be bounded by a sum of two terms as follows:

$$|I - I_{DE}^{inc}(h)| \leq \left| \int_a^b F(x)\,dx - \tilde{h} \sum_{i=-M_-}^{M_+} F(\psi(i\tilde{h}))\psi'(i\tilde{h}) \right|$$

$$+ \tilde{h} \sum_{i=-M_-}^{M_+} \psi'(i\tilde{h}) \left| \int_a^{\psi(i\tilde{h})} f_i(s)\,ds - \sum_{j=-N_-}^{N_+} f_i(\psi(jh))\psi'(jh)J(j,h)(i\tilde{h}) \right|,$$

where $F(x) = \int_a^x f(x, q(s))q'(s)\,ds$, $f_i(s) = f(\psi(i\tilde{h}), q(s))q'(s)$, and $\tilde{h} = 2h$. The first term (say E_1) and the second term (say E_2) are bounded as follows:

$$E_1 \leq \frac{B(\gamma,\delta)c_{\gamma,\delta,d}}{\mu} \left\{ e^{\frac{\pi}{2}\bar{\mu}} + \frac{2c_{\alpha,\beta,d}}{1 - e^{-2\pi d/\tilde{h}}} \right\} 2K(b-a)^{\alpha+\beta+\gamma+\delta-2} e^{-2\pi d/\tilde{h}}, \quad (19)$$

$$E_2 \leq \frac{1}{\nu} \left\{ B(\alpha,\beta) + \frac{4c_{\alpha,\beta,d}}{\mu} \frac{e^{-2\pi d/\tilde{h}}}{1 - e^{-2\pi d/\tilde{h}}} \right\} \left\{ 1.1\, e^{\frac{\pi}{2}\bar{\nu}} + \frac{hc_{\gamma,\delta,d}}{d(1 - e^{-2\pi d/h})} \right\}$$

$$\times 2K(b-a)^{\alpha+\beta+\gamma+\delta-2} e^{-\pi d/h}. \quad (20)$$

Then, taking $\tilde{h} = 2h$, we get the desired inequality (18). In what follows, the inequalities (19) and (20) are shown in Sects. 5.2 and 5.3, respectively.

5.2 Bound of E_1 (Error of the DE-Sinc Quadrature)

The following two lemmas are important results for this project.

Lemma 1 (Okayama et al. [10, Lemma 4.16]). *Let $\tilde{L}$, α, and β be positive constants, and let $\mu = \min\{\alpha, \beta\}$. Let F be analytic on $\psi(\mathscr{D}_d)$ for d with*

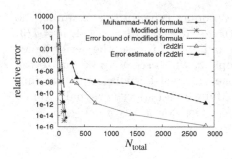

Fig. 5. Relative error with respect to N_{total} in Example 2.

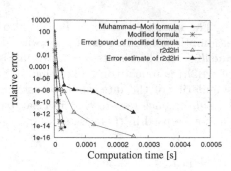

Fig. 6. Relative error with respect to computation time in Example 2.

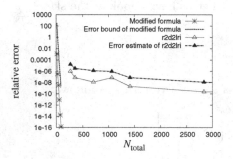

Fig. 7. Relative error with respect to N_{total} in Example 3.

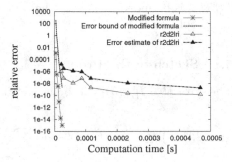

Fig. 8. Relative error with respect to computation time in Example 3.

$0 < d < \pi/2$, and satisfy $|F(z)| \le \tilde{L}|z-a|^{\alpha-1}|b-z|^{\beta-1}$ for all $z \in \psi(\mathscr{D}_d)$. Then it holds that

$$\left| \int_a^b F(x)\,\mathrm{d}x - \tilde{h} \sum_{i=-\infty}^{\infty} F(\psi(i\tilde{h}))\psi'(i\tilde{h}) \right| \le \tilde{C}_1 \tilde{C}_2 \frac{\mathrm{e}^{-2\pi d/\tilde{h}}}{1-\mathrm{e}^{-2\pi d/\tilde{h}}},$$

where the constants $\tilde{C}_1$ and $\tilde{C}_2$ are defined by

$$\tilde{C}_1 = \frac{2\tilde{L}(b-a)^{\alpha+\beta-1}}{\mu}, \quad \tilde{C}_2 = 2c_{\alpha,\beta,d}. \tag{21}$$

Lemma 2 (Okayama et al. [10, Lemma 4.18]). *Let the assumptions in Lemma 1 be fulfilled. Furthermore, let $\overline{\mu} = \max\{\alpha,\,\beta\}$, let m be a positive integer, let M_- and M_+ be positive integers defined by (16), and let m be taken sufficiently large so that $M_-\tilde{h} \ge \rho_\alpha$ and $M_+\tilde{h} \ge \rho_\beta$ hold. Then it holds that*

$$\left| \tilde{h} \sum_{i=-\infty}^{-(M_-+1)} F(\psi(i\tilde{h}))\psi'(i\tilde{h}) + \tilde{h} \sum_{i=M_++1}^{\infty} F(\psi(i\tilde{h}))\psi'(i\tilde{h}) \right| \le \mathrm{e}^{\frac{\pi}{2}\overline{\mu}} \tilde{C}_1 \,\mathrm{e}^{-\frac{\pi}{2}\mu\exp(m\tilde{h})},$$

where $\tilde{C}_1$ is a constant defined in (21).

What should be checked here is whether the conditions of those two lemmas are satisfied under the assumptions in Theorem 1. The next lemma answers to this question.

Lemma 3. *Let the assumptions in Theorem 1 be fulfilled, and let F be defined as $F(z) = \int_a^z f(z, q(w))q'(w)\,\mathrm{d}w$. Then, the assumptions of Lemmas 1 and 2 are satisfied with $\tilde{L} = K(b-a)^{\gamma+\delta-1}\,\mathrm{B}(\gamma, \delta)c_{\gamma,\delta,d}$.*

If this lemma is proved, combining Lemmas 1 and 2, and using the relations (15)–(17), we get the desired inequality (19). For the proof of Lemma 3, we need the following inequalities.

Lemma 4 (Okayama et al. [10, Lemma 4.22]). *Let x and y be real numbers with $|y| < \pi/2$. Then we have*

$$\left| \frac{1}{1 + e^{\pi \sinh(x+\mathrm{i}y)}} \right| \leq \frac{1}{(1 + e^{\pi \sinh(x)\cos y})\cos(\frac{\pi}{2}\sin y)},$$

$$\left| \frac{1}{1 + e^{-\pi \sinh(x+\mathrm{i}y)}} \right| \leq \frac{1}{(1 + e^{-\pi \sinh(x)\cos y})\cos(\frac{\pi}{2}\sin y)}.$$

Lemma 5. *Let x, ξ, $y \in \mathbb{R}$ with $|y| < \pi/2$, let γ and δ be positive constants, and let us define a function $\psi^{(0,1)}(x,y)$ as*

$$\psi^{(0,1)}(x,y) = \frac{1}{2}\tanh\left(\frac{\pi \cos y}{2}\sinh x\right) + \frac{1}{2}.$$

Then it holds that

$$\int_{-\infty}^{\xi} \frac{\pi\,|\cosh(x+\mathrm{i}y)|\,\mathrm{d}x}{|1 + e^{-\pi\sinh(x+\mathrm{i}y)}|^{\gamma}|1 + e^{\pi\sinh(x+\mathrm{i}y)}|^{\delta}} \leq \frac{\mathrm{B}(\psi^{(0,1)}(\xi,y); \gamma, \delta)}{\cos^{\gamma+\delta}(\frac{\pi}{2}\sin y)\cos y},$$

where $\mathrm{B}(t; \kappa, \lambda)$ is the incomplete beta function.

Proof. From Lemma 4 and $|\cosh(x+\mathrm{i}y)| \leq \cosh(x)$, we obtain

$$\int_{-\infty}^{\xi} \frac{\pi\,|\cosh(x+\mathrm{i}y)|\,\mathrm{d}x}{|1 + e^{-\pi\sinh(x+\mathrm{i}y)}|^{\gamma}|1 + e^{\pi\sinh(x+\mathrm{i}y)}|^{\delta}}$$

$$\leq \frac{1}{\cos^{\gamma+\delta}(\frac{\pi}{2}\sin y)\cos y}\int_{-\infty}^{\xi} \frac{\pi\cosh(x)\cos(y)\,\mathrm{d}x}{(1 + e^{-\pi\sinh(x)\cos y})^{\gamma}(1 + e^{\pi\sinh(x)\cos y})^{\delta}}$$

$$= \frac{\mathrm{B}(\psi^{(0,1)}(\xi,y); \gamma, \delta)}{\cos^{\gamma+\delta}(\frac{\pi}{2}\sin y)\cos y}. \qquad \square$$

By using the estimates, Lemma 3 is proved as follows.

Proof. The estimate of the constant $\tilde{L}$ is essential. Let $\xi = \mathrm{Re}[\psi^{-1}(z)]$ and $y = \mathrm{Im}[\psi^{-1}(z)]$, i.e., $z = \psi(\xi + \mathrm{i}y)$. By applying $w = \psi(x + \mathrm{i}y)$, we have

$$|F(z)|$$

$$= \left| \int_{-\infty}^{\xi} f(z, q(\psi(x+\mathrm{i}\,y)))q'(\psi(x+\mathrm{i}\,y))\psi'(x+\mathrm{i}\,y)\,\mathrm{d}x \right|$$

$$\leq K|z-a|^{\alpha-1}|b-z|^{\beta-1} \int_{-\infty}^{\xi} |\psi(x+\mathrm{i}\,y)-a|^{\gamma-1}|b-\psi(x+\mathrm{i}\,y)|^{\delta-1}|\psi'(x+\mathrm{i}\,y)|\,\mathrm{d}x$$

$$= K|z-a|^{\alpha-1}|b-z|^{\beta-1}(b-a)^{\gamma+\delta-1} \int_{-\infty}^{\xi} \frac{\pi\,|\cosh(x+\mathrm{i}\,y)|\,\mathrm{d}x}{|1+\mathrm{e}^{-\pi\sinh(x+\mathrm{i}\,y)}\,|^{\gamma}|1+\mathrm{e}^{\pi\sinh(x+\mathrm{i}\,y)}\,|^{\delta}}.$$

Then, the desired bound of $\tilde{L}$ is obtained by using Lemma 5 and $\mathrm{B}(\psi^{(0,1)}(\xi,y);\gamma,\delta) \leq \mathrm{B}(\gamma,\delta)$. $\qquad\square$

5.3 Bound of E_2 (Error of the DE-Sinc Indefinite Integration)

The following two lemmas are important results for this project.

Lemma 6 (Okayama et al. [10, Lemma 4.19]). *Let L, γ, and δ be positive constants, and let $\nu = \min\{\gamma, \delta\}$. Let f be analytic on $\psi(\mathscr{D}_d)$ for d with $0 < d < \pi/2$, and satisfy $|f(w)| \leq L|w-a|^{\gamma-1}|b-w|^{\delta-1}$ for all $w \in \psi(\mathscr{D}_d)$. Then it holds that*

$$\sup_{x\in(a,b)} \left| \int_a^x f(s)\,\mathrm{d}s - \sum_{j=-\infty}^{\infty} f(\psi(jh))\psi'(jh)J(j,h)(\psi^{-1}(x)) \right| \leq \frac{C_1 C_2}{2d}\frac{h\,\mathrm{e}^{-\pi d/h}}{1-\mathrm{e}^{-2\pi d/h}},$$

where the constants C_1 and C_2 are defined by

$$C_1 = \frac{2L(b-a)^{\gamma+\delta-1}}{\nu}, \quad C_2 = 2c_{\gamma,\delta,d}. \tag{22}$$

Lemma 7 (Okayama et al. [10, Lemma 4.20]). *Let the assumptions in Lemma 6 be fulfilled. Furthermore, let $\bar{\nu} = \max\{\gamma, \delta\}$, let n be a positive integer, let N_- and N_+ be positive integers defined by (17), and let n be taken sufficiently large so that $N_-h \geq \rho_\gamma$ and $N_+h \geq \rho_\beta$ hold. Then it holds that*

$$\sup_{x\in(a,b)} \left| \sum_{j=-\infty}^{-(N_-+1)} G(jh)J(j,h)(\psi^{-1}(x)) + \sum_{j=N_++1}^{\infty} G(jh)J(j,h)(\psi^{-1}(x)) \right|$$

$$\leq 1.1\,\mathrm{e}^{\frac{\pi}{2}\bar{\nu}}\,C_1\,\mathrm{e}^{-\frac{\pi}{2}\nu\exp(nh)},$$

where $G(x) = f(\psi(x))\psi'(x)$, and C_1 is a constant defined in (22).

What should be checked here is whether the conditions of those two lemmas are satisfied under the assumptions in Theorem 1. The next lemma answers this question.

Lemma 8. *Let the assumptions in Theorem 1 be fulfilled, and let $f_i(z)$ be defined as $f_i(z) = f(\psi(i\tilde{h}), q(z))q'(z)$. Then, the assumptions of Lemmas 6 and 7 are satisfied with $f = f_i$ and $L = K(\psi(i\tilde{h}) - a)^{\alpha-1}(b - \psi(i\tilde{h}))^{\beta-1}$.*

The proof is omitted since it is obvious from (14). Combining Lemmas 6 and 7, and using the relations (15)–(17), we have

$$E_2 \le C_3 \times \frac{2K(b-a)^{\gamma+\delta-1}}{\nu}\left\{1.1\,\mathrm{e}^{\frac{\pi}{2}\bar{\nu}} + \frac{hc_{\gamma,\delta,d}}{d(1 - \mathrm{e}^{-2\pi d/h})}\right\}\mathrm{e}^{-\pi d/h},$$

where

$$C_3 = \tilde{h}\sum_{i=-M_-}^{M_+} \psi'(i\tilde{h})(\psi(i\tilde{h}) - a)^{\alpha-1}(b - \psi(i\tilde{h}))^{\beta-1}.$$

What is left is to bound the term C_3, which is done by the next lemma.

Lemma 9. *Let α and β be positive constants, and let $\mu = \min\{\alpha, \beta\}$. Then C_3 is bounded as*

$$C_3 \le (b-a)^{\alpha+\beta-1}\left\{\mathrm{B}(\alpha, \beta) + \frac{4c_{\alpha,\beta,d}}{\mu}\frac{\mathrm{e}^{-2\pi d/\tilde{h}}}{1 - \mathrm{e}^{-2\pi d/\tilde{h}}}\right\}.$$

Proof. Let us define F as $F(x) = (x - a)^{\alpha-1}(b - x)^{\beta-1}$. We readily see

$$\tilde{h}\sum_{i=-M_-}^{M_+} F(\psi(i\tilde{h}))\psi'(i\tilde{h}) \le \tilde{h}\sum_{i=-\infty}^{\infty} F(\psi(i\tilde{h}))\psi'(i\tilde{h})$$

$$\le \int_a^b F(x)\,\mathrm{d}x + \left|\int_a^b F(x)\,\mathrm{d}x - \tilde{h}\sum_{i=-\infty}^{\infty} F(\psi(i\tilde{h}))\psi'(i\tilde{h})\right|,$$

and we further see $\int_a^b F(x)\,\mathrm{d}x = (b - a)^{\alpha+\beta-1}\mathrm{B}(\alpha, \beta)$. For the second term, use Lemma 1 to obtain

$$\left|\int_a^b F(x)\,\mathrm{d}x - \tilde{h}\sum_{i=-\infty}^{\infty} F(\psi(i\tilde{h}))\psi'(i\tilde{h})\right| \le \frac{4(b-a)^{\alpha+\beta-1}c_{\alpha,\beta,d}}{\mu}\frac{\mathrm{e}^{-2\pi d/\tilde{h}}}{1 - \mathrm{e}^{-2\pi d/\tilde{h}}},$$

which completes the proof.   □

6 Concluding Remarks

Muhammad–Mori [7] proposed an approximation formula for (1), which can converge *exponentially* with respect to N_{total} even if $f(x, y)$ or $q(x)$ has boundary singularity. It is particularly worth noting that their formula is quite efficient if f is of a product type: $f(x, y) = X(x)Y(y)$. However, its convergence was not proved in a precise sense, and it cannot be used in the case $q'(x) \leq 0$ (only the case $q'(x) \geq 0$ was considered). This paper improved the formula in the sense that both cases $(q'(x) \geq 0$ and $q'(x) \leq 0)$ are taken into account, and it can achieve a better convergence rate. Furthermore, a rigorous error bound that is *computable* is given, enabling us to mathematically guarantee the accuracy of the approximation. Numerical results in Sect. 4 confirm the error bound and the exponential rate of convergence, and also suggest that the modified formula is incredibly accurate if f is of a product type, similar to the original formula. This is because, instead of a *definite* integration formula (quadrature rule), an *indefinite* integration formula is employed for the approximation of the inner integral.

However, as said in the original paper [7], the use of the *indefinite* integration formula has a drawback: it cannot be used when $f(x, y)$ has a singularity along $y = q(x)$, e.g.,

$$\int_a^b \left(\int_A^{q(x)} \frac{dy}{\sqrt{q(x) - y}} \right), \quad \int_a^b \left(\int_A^{q(x)} \sqrt{(q(x) - y)(q(x) + y)}\, dy \right),$$

and so on (f can have singularity at the endpoints $y = A$ and $y = B$, though). This is because the assumption of Theorem 1 (more precisely, Lemmas 6 and 7) is not satisfied in this case. In such a case, a *definite* integration formula should be employed for the approximation of the inner integral. Actually, such an approach was already successfully taken in some one-dimensional cases [8,9]. It also may work for (1), which will be considered in a future report.

References

1. Eiermann, M.C.: Automatic, guaranteed integration of analytic functions. BIT **29**, 270–282 (1989)
2. Hill, M., Robinson, I.: d2lri: a nonadaptive algorithm for two-dimensional cubature. J. Comput. Appl. Math. **112**, 121–145 (1999)
3. Iri, M., Moriguti, S., Takasawa, Y.: On a certain quadrature formula. RIMS Kōkyūroku, Kyoto Univ. **91**, 82–118 (1970). (in Japanese)
4. Iri, M., Moriguti, S., Takasawa, Y.: On a certain quadrature formula. J. Comput. Appl. Math. **17**, 3–20 (1987)
5. Mori, M., Sugihara, M.: The double-exponential transformation in numerical analysis. J. Comput. Appl. Math. **127**, 287–296 (2001)
6. Muhammad, M., Mori, M.: Double exponential formulas for numerical indefinite integration. J. Comput. Appl. Math. **161**, 431–448 (2003)
7. Muhammad, M., Mori, M.: Numerical iterated integration based on the double exponential transformation. Jpn. J. Indust. Appl. Math. **22**, 77–86 (2005)
8. Okayama, T., Matsuo, T., Sugihara, M.: Approximate formulae for fractional derivatives by means of Sinc methods. J. Concr. Appl. Math. **8**, 470–488 (2010)
9. Okayama, T., Matsuo, T., Sugihara, M.: Sinc-collocation methods for weakly singular Fredholm integral equations of the second kind. J. Comput. Appl. Math. **234**, 1211–1227 (2010)

10. Okayama, T., Matsuo, T., Sugihara, M.: Error estimates with explicit constants for Sinc approximation, Sinc quadrature and Sinc indefinite integration. Numer. Math. **124**, 361–394 (2013)
11. Petras, K.: Principles of verified numerical integration. J. Comput. Appl. Math. **199**, 317–328 (2007)
12. Robinson, I., de Doncker, E.: Algorithm 45. Automatic computation of improper integrals over a bounded or unbounded planar region. Computing **27**, 253–284 (1981)
13. Robinson, I., Hill, M.: Algorithm 816: r2d2lri: an algorithm for automatic two-dimensional cubature. ACM Trans. Math. Softw. **28**, 75–100 (2002)
14. Roose, D., de Doncker, E.: Automatic integration over a sphere. J. Comput. Appl. Math. **7**, 203–224 (1981)
15. Schwartz, C.: Numerical integration of analytic functions. J. Comput. Phys. **4**, 19–29 (1969)
16. Sidi, A.: A new variable transformation for numerical integration. In: Brass, H., Hämmerlin, G. (eds.) Numerical Integration IV. International Series of Numerical Mathematics, vol. 112, pp. 359–373. Birkhäuser Verlag, Basel (1993)
17. Sidi, A.: Extension of a class of periodizing variable transformations for numerical integration. Math. Comput. **75**, 327–343 (2006)
18. Stenger, F.: Numerical Methods Based on Sinc and Analytic Functions. Springer, New York (1993)
19. Stenger, F.: Summary of Sinc numerical methods. J. Comput. Appl. Math. **121**, 379–420 (2000)
20. Takahasi, H., Mori, M.: Double exponential formulas for numerical integration. Publ. RIMS, Kyoto Univ. **9**, 721–741 (1974)

Verified Computations for Solutions to Semilinear Parabolic Equations Using the Evolution Operator

Akitoshi Takayasu[1(✉)], Makoto Mizuguchi[2],
Takayuki Kubo[3], and Shin'ichi Oishi[4]

[1] Research Institute for Science and Engineering, Waseda University, Tokyo, Japan
takitoshi@aoni.waseda.jp
[2] Graduate School of Fundamental Science and Engineering,
Waseda University, Tokyo, Japan
[3] Institute of Mathematics, University of Tsukuba, Ibaraki, Japan
[4] Department of Applied Mathematics,
Waseda University and CREST, JST, Tokyo, Japan

Abstract. This article presents a theorem for guaranteeing existence of a solution for an initial-boundary value problem of semilinear parabolic equations. The sufficient condition of our main theorem is derived by a fixed-point formulation using the evolution operator. We note that the sufficient condition can be checked by verified numerical computations.

1 Introduction

Let $J := (t_0, t_1]$ $(0 \leq t_0 < t_1 < \infty)$ be a time interval and Ω a convex polygonal domain in $\mathbb{R}^2$. In this article we consider the following initial-boundary value problems of semilinear parabolic equations:

$$
\begin{cases}
\partial_t u - \Delta u = f(u) & \text{in } J \times \Omega, \\
u(t, x) = 0 & \text{on } J \times \partial\Omega, \\
u(t_0, x) = u_0(x) & \text{in } \Omega.
\end{cases}
\tag{1}
$$

Here, $\partial_t u = \frac{\partial u}{\partial t}$, $\Delta = \frac{\partial^2}{\partial x_1^2} + \frac{\partial^2}{\partial x_2^2}$ denotes the Laplacian, the domain of the Laplacian is $D(\Delta) = H^2(\Omega) \cap H_0^1(\Omega)$, $f(u)$ is a real-valued function in $J \times \Omega$ such that $f : H_0^1(\Omega) \to L^2(\Omega)$ is a twice Fréchet differentiable nonlinear mapping for $\forall t \in J$, and $u_0 \in H_0^1(\Omega)$ is an initial function. Let $\tau := t_1 - t_0$.

The main aim of this article is to present Theorem 1 for proving existence and local uniqueness of a solution to (1) in a neighborhood of an approximate solution. This approximate solution consists of two numerical solutions. Let V_h be a finite dimensional subspace of $D(A)$. For two numerical solutions $\hat{u}_0, \hat{u}_1 \in V_h$, we define the approximate solution $\omega(t)$ as

$$
\omega(t) = \hat{u}_0 \phi_0(t) + \hat{u}_1 \phi_1(t), \ t \in J,
\tag{2}
$$

© Springer International Publishing Switzerland 2016
I.S. Kotsireas et al. (Eds.): MACIS 2015, LNCS 9582, pp. 218–223, 2016.
DOI: 10.1007/978-3-319-32859-1_18

where $\phi_i(t)$ $(i = 0, 1)$ is a linear Lagrange basis such that $\phi_i(t_j) = \delta_{ij}$ (δ_{ij} is a Kronecker's delta for $j = 0, 1$).

The evolution operator is introduced by Tanabe and Sobolevskii [1, 2]. Using the evolution operator, studies of parabolic equation have been developed in the field of mathematical analysis (cf. [3, 4]).

In this article, we present a fixed-point form by using the evolution operator. Existence of its fixed-point is equivalent to that of the *mild solution* to (1). We then derive a sufficient condition for verifying existence of the fixed-point. By numerically checking whether the sufficient condition holds, existence and local uniqueness of the *mild solution* to (1) are proved.

2   Fixed-Point Formulation

Let us start from the following fact: the *mild solution* u of (1) exists if and only if the function $z = u - \omega$ is the *mild solution* of

$$\begin{cases} \partial_t z - \Delta z = f(z + \omega) - \partial_t \omega + \Delta \omega & \text{in } J \times \Omega, \\ z(t, x) = 0 & \text{on } J \times \partial \Omega, \\ z(t_0, x) = u_0(x) - \hat{u}_0(x) & \text{in } \Omega. \end{cases}$$

Suppose that $z = e^{\sigma(t - t_0)} v$ holds for a certain $\sigma > 0$. Then v is a solution of the following equation:

$$\begin{cases} \partial_t v + A(t)v = g(v) & \text{in } J \times \Omega, \\ v(t, x) = 0 & \text{on } J \times \partial \Omega, \\ v(t_0, x) = u_0(x) - \hat{u}_0(x) & \text{in } \Omega, \end{cases} \tag{3}$$

where

$$A(t) = -\Delta + \left(\sigma - f'[\omega(t)] \right),$$

$$g(v) = e^{-\sigma(t - t_0)} \left\{ f\left(\omega + e^{\sigma(t - t_0)} v \right) - f(\omega) - f'[\omega(t)] e^{\sigma(t - t_0)} v + f(\omega) - \partial_t \omega - \Delta \omega \right\}$$

holds. The operator $f'[\omega(t)] : H_0^1(\Omega) \to L^2(\Omega)$ denotes a Fréchet derivative of f at $\omega(t)$ for $t \in J$. We furthermore assume that $f'[\omega(t)]$ is a symmetric operator for $t \in J$.

From the definition of ω in (2), the domain of $A(t)$ becomes $D(A(t)) = D(\Delta)$ for each $t \in J$ ($D(A(t))$ is independent of $t \in J$). Let us fix $\mu > 0$. We define a norm of $V = H_0^1(\Omega)$ as

$$\|\phi\|_V = \left(\|\nabla \phi\|_{L^2}^2 + \mu \|\phi\|_{L^2}^2 \right)^{1/2} \text{ for } \phi \in V.$$

We determine the $\sigma > 0$ such that $\sigma - f'[\omega(t)] \geq \mu$ *a.e.* Ω for $\forall t \in J$. It then follows

$$|(A(t)u, v)_{L^2}| = |(\nabla u, \nabla v)_{L^2} + ((\sigma - f'[\omega(t)])u, v)_{L^2}|$$
$$= |(\nabla u, \nabla v)_{L^2} + \mu(u, v)_{L^2} + ((\sigma - f'[\omega(t)] - \mu)u, v)_{L^2}|$$
$$\leq \left(1 + C_\sigma C_\mu^2 \right) \|u\|_V \|v\|_V$$

and

$$(A(t)u, u)_{L^2} = |(\nabla u, \nabla u)_{L^2} + ((\sigma - f'[\omega(t)])u, u)_{L^2}|$$
$$\geq \|u\|_V^2,$$

where $C_\sigma = \sup_{x \in \Omega} |\sigma - f'[\omega(t)] - \mu|$, $C_\mu > 0$ such that $\|\phi\|_{L^2} \leq C_\mu \|\phi\|_V$. This yields $D(A(t)^{1/2}) = V$, i.e., the following holds for $\phi \in V$:

$$\|\phi\|_V \leq \|A(t)^{1/2}\phi\|_{L^2} \leq M\|\phi\|_V, \ M = \left(1 + C_\sigma C_\mu^2\right)^{1/2}.$$

For each $t \in J$, $-A(t)$ thus becomes the sectorial operator and generates a holomorphic semigroup $\left\{e^{-sA(t)}\right\}_{s \geq 0}$ over $L^2(\Omega)$. The eigenvalue of $-A(t)$ for $t \in J$ is bounded below by $\lambda_A = \lambda_{\min} + \mu > 0$, where $\lambda_{\min}$ denotes the minimum eigenvalue of $-\Delta$. Therefore, the operator $A(t)$ becomes a symmetric positive operator. Additionally, for $t, s \in J$ there exists $C > 0$ and $\alpha > 0$ such that

$$\|A(t)A(s)^{-1} - I\|_{L^2, L^2} = \|(A(t) - A(s))A(s)^{-1}\|_{L^2, L^2} \leq C|t - s|^\alpha,$$

where $\|\cdot\|_{L^2, L^2}$ denotes the operator norm over $L^2(\Omega)$.

From the above facts it is well-known [1–4] that the operator $-A(t)$ generates an evolution operator $\{U(t, s)\}_{t_0 \leq s \leq t \leq t_1}$ on $L^2(\Omega)$. The evolution operator is described by

$$U(t, s) = e^{-(t-s)A(s)} + \int_s^t e^{-(t-r)A(r)} R(r, s) dr \ (t_0 \leq s \leq r \leq t \leq t_1),$$

where $R(t, s)$ is the solution of the following integral equation:

$$\begin{cases} R(t, s) = R_1(t, s) + \int_s^t R_1(t, r) R(r, s) dr, \\ R_1(t, s) = -(A(t) - A(s)) e^{-(t-s)A(s)}. \end{cases} \quad (4)$$

By using the evolution operator $\{U(t, s)\}_{t_0 \leq s \leq t \leq t_1}$, we define a nonlinear operator $T : C(J; V) \to C(J; V)$ as

$$T(v) := U(t, t_0)v(t_0) + \int_{t_0}^t U(t, s)g(v(s)) ds \ (t_0 \leq s \leq t \leq t_1). \quad (5)$$

If v satisfies the fixed-point form $v = T(v)$ in $C(J; V)$, then there exists a solution of (3) that is described by the evolution operator. In the following we derive a sufficient condition for verifying existence of the solution to (3). If this sufficient condition holds, existence of the *mild solution* to (1) is also proved.

3   Main Theorem

Let us define a function space

$$X_\sigma := \left\{ v \in C(J; V) : \sup_{t \in J} e^{\sigma(t-t_0)} \|v(t)\|_V < \infty \right\}$$

with the norm $\|v\|_{X_\sigma} := \sup_{t \in J} e^{\sigma(t-t_0)} \|v(t)\|_V$. The following theorem gives a sufficient condition for guaranteeing existence and local uniqueness of a *mild solution* to (1) in

$$B_J(\omega, \rho) := \left\{ u \in C(J;V) : \|u - \omega\|_{C(J;V)} \le \rho \right\}.$$

Theorem 1. *Assume that* $\hat{u}_0 \in V_h$ *satisfies* $\|u_0 - \hat{u}_0\|_{H_0^1} \le \varepsilon_0$ *and* $0 \le \sigma < \frac{\lambda_A}{2}$ *holds. Assume that* ω *satisfies the following estimate:*

$$\|\partial_t \omega - \Delta \omega - f(\omega)\|_{C(J;L^2(\Omega))} \le \delta.$$

Assume also that there exists a monotonically non-decreasing function $L_\omega :$ $[0,\infty) \to [0,\infty)$ *corresponding to the first Fréchet derivative of* $f : H_0^1(\Omega) \to L^2(\Omega)$ *such that*

$$\|(f'[\omega + h] - f'[\omega])\phi\|_{C(J;L^2(\Omega))} \le L_\omega(\rho) \|\phi\|_{C(J;V)}, \quad \forall \phi \in C(J;V),$$

where $h \in X_\sigma$ *satisfying* $\|h\|_{X_\sigma} \le \rho$ *for a certain* $\rho > 0$. *If*

$$(M + O_1(\tau)) \varepsilon_0 + \sqrt{\frac{2\pi}{e(\lambda_A - 2\sigma)}} \operatorname{erf}\left(\sqrt{\frac{(\lambda_A - 2\sigma)\tau}{2}} \right) (1 + O_2(\tau)) (L_\omega(\rho)\rho^2 + \delta) < \rho \quad (6)$$

holds, then the mild solution $u(t) := u(t, \cdot)$, $t \in J$, *of* (1) *uniquely exists in the ball* $B_J(\omega, \rho)$. *Here,* $O_1(\tau)$ *and* $O_2(\tau)$ *in* (6) *are given by*

$$O_1(\tau) = 2C_\mu \sqrt{\frac{C_\omega}{e}} \tau^{\frac{1}{2}} \sinh\left(\sqrt{C_\omega} \tau \right) \ and \ O_2(\tau) = 2\sqrt{C_\omega} \tau \sinh\left(\sqrt{C_\omega} \tau \right),$$

respectively, if $R_1(t,s)$ *in* (4) *satisfies* $\|R_1(t,s)\|_{L^2,L^2} \le C_\omega(t-s)e^{-(t-s)\lambda_A}$.

Before we sketch a proof of the main theorem, some lemmas are necessary.

Lemma 1. *If* $R_1(t,s)$ *in* (4) *satisfies* $\|R_1(t,s)\|_{L^2,L^2} \le C_\omega(t-s)e^{-(t-s)\lambda_A}$, *it follows*

$$\|R(t,s)\|_{L^2,L^2} \le \sqrt{C_\omega} \sinh\left(\sqrt{C_\omega}(t-s) \right) e^{-(t-s)\lambda_A}.$$

Lemma 2. *For the evolution operator* $\{U(t,s)\}_{t_0 \le s \le t \le t_1}$ *generated by* $-A(t)$ *and* $v(t_0) = u_0 - \hat{u}_0$, *the following estimate holds:*

$$\|U(t,t_0)v(t_0)\|_V \le \left(Me^{-(t-t_0)\lambda_A} + O_1(\tau)e^{-\frac{1}{2}(t-t_0)\lambda_A} \right) \varepsilon_0.$$

Lemma 3. *For the evolution operator* $\{U(t,s)\}_{t_0 \le s \le t \le t_1}$ *generated by* $-A(t)$ *and* $g(v)$ *in* (3), *the following estimate holds:*

$$\|U(t,s)g(v(s))\|_V \le e^{-\frac{1}{2}}(t-s)^{-\frac{1}{2}}e^{-\frac{1}{2}(t-s)\lambda_A} \|g(v(s))\|_{L^2} (1 + O_2(\tau)).$$

Proofs of these lemmas are omitted for lack of space.

Sketch of the Proof. For $\rho > 0$ let $Z = \{v \in X_\sigma : \|v\|_{X_\sigma} \leq \rho\}$. Let us consider the fixed-point form (5). On the basis of Banach's fixed-point theorem, we give a sufficient condition of T having a fixed-point in Z. First, we derive a condition guaranteeing $T(Z) \subset Z$. For $v \in Z$, Lemmas 1 and 2 gives

$$\|T(v(t))\|_V \leq \left(M e^{-(t-t_0)\lambda_A} + O_1(\tau) e^{-\frac{1}{2}(t-t_0)\lambda_A} \right) \varepsilon_0$$
$$+ \int_{t_0}^t e^{-\frac{1}{2}}(t-s)^{-\frac{1}{2}} e^{-\frac{1}{2}(t-s)\lambda_A} \|g(v(s))\|_{L^2} (1 + O_2(\tau)) \, ds.$$

It follows

$$e^{\sigma(t-t_0)} \|T(v(t))\|_V \leq \left(M e^{-(t-t_0)(\lambda_A - \sigma)} + O_1(\tau) e^{-\frac{1}{2}(t-t_0)(\lambda_A - 2\sigma)} \right) \varepsilon_0$$
$$+ \int_{t_0}^t e^{-\frac{1}{2}}(t-s)^{-\frac{1}{2}} e^{-\frac{1}{2}(t-s)(\lambda_A - 2\sigma)} e^{\sigma(s-t_0)} \|g(v(s))\|_{L^2} (1 + O_2(\tau)) \, ds. \quad (7)$$

From (3) and the assumptions of the theorem, we have

$$e^{\sigma(s-t_0)} \|g(v(s))\|_{L^2} \leq \left\| f\left(\omega(s) + e^{\sigma(s-t_0)} v(s) \right) - f(\omega(s)) - f'[\omega(s)] e^{\sigma(s-t_0)} v(s) \right\|_{L^2}$$
$$+ \|f(\omega(s)) - \partial_s \omega(s) - \Delta \omega(s)\|_{L^2}$$
$$\leq L_\omega(\rho)\rho^2 + \delta.$$

The upper bound of (7) with respect to $t \in J$ is given by

$$\|T(v)\|_{X_\sigma} \leq (M + O_1(\tau)) \varepsilon_0$$
$$+ \sqrt{\frac{2\pi}{e(\lambda_A - 2\sigma)}} \operatorname{erf}\left(\sqrt{\frac{(\lambda_A - 2\sigma)\tau}{2}} \right) (1 + O_2(\tau)) (L_\omega(\rho)\rho^2 + \delta).$$

From (6) $\|T(v)\|_{X_\sigma} < \rho$ holds. Namely, we obtain $T(v) \in Z$.

Next, under the assumptions of the theorem, we show that T is a contraction mapping on Z. For $v_1, v_2 \in Z$, we have

$$\|T(v_1) - T(v_2)\|_{X_\sigma} \leq \sqrt{\frac{2\pi}{e(\lambda_A - 2\sigma)}} \operatorname{erf}\left(\sqrt{\frac{(\lambda_A - 2\sigma)\tau}{2}} \right)$$
$$(1 + O_2(\tau)) L_\omega(\rho)\rho \|v_1 - v_2\|_{X_\sigma}.$$

The assumption (6) also implies

$$\sqrt{\frac{2\pi}{e(\lambda_A - 2\sigma)}} \operatorname{erf}\left(\sqrt{\frac{(\lambda_A - 2\sigma)\tau}{2}} \right) (1 + O_2(\tau)) L_\omega(\rho)\rho < 1.$$

Therefore, T becomes a contraction mapping on Z. Banach's fixed-point theorem asserts that there uniquely exists a fixed-point $v = T(v)$ in Z. It yields that the *mild solution* of (1) uniquely exists in the ball $B_J(\omega, \rho)$.   □

References

1. Tanabe, H.: On the equations of evolution in a Banach space. Osaka Math. J. **12**(2), 363–376 (1960)
2. Sobolevskii, P.E.: On equations of parabolic type in Banach space with unbounded variable operator having a constant domain. Akad. Nauk Azerbaidzan. SSR Doki, 17:6 (1961). (in Russian)
3. Pazy, A.: Semigroups of Linear Operators and Applications to Partial Differential Equations. Springer, New York (1983)
4. Fujita, H., Saito, N., Suzuki, T.: Operator Theory and Numerical Methods. North Holland, Amsterdam (2001)

Verified Error Bounds for the Real Gamma Function Using Double Exponential Formula over Semi-infinite Interval

Naoya Yamanaka[1,4(✉)], Tomoaki Okayama[2], and Shin'ichi Oishi[3,4]

[1] Faculty of Modern Life, Teikyo Heisei University,
4-21-2, Nakano, Nakano, Tokyo 164-8530, Japan
n.yamanaka@thu.ac.jp
[2] Graduate School of Information Sciences, Hiroshima City University,
3-4-1, Ozuka-higashi, Asaminami, Hiroshima 731-3194, Japan
okayama@hiroshima-cu.ac.jp
[3] Faculty of Science and Engineering, Waseda University,
3-4-1, Okubo, Shinjuku, Tokyo 169-8555, Japan
oishi@waseda.jp
[4] CREST, Japan Science and Technology Agency,
4-1-8, Honcho, Kawaguchi, Saitama 332-0012, Japan

Abstract. An algorithm is presented for computing verified error bounds for the value of the real gamma function. It has been shown that the double exponential formula is one of the most efficient methods for calculating integrals of the form. Recently, an useful evaluation based on the double exponential formula over the semi-infinite interval has been proposed. However, the evaluation would be overflow when applied to the real gamma function directly. In this paper, we present a theorem so as to overcome the problem in such a case. Numerical results are presented for illustrating effectiveness of the proposed theorem in terms of the accuracy of the calculation.

Keywords: Gamma function · Verified bound · Double exponential formula

1 Introduction

This paper is concerned with a verified numerical computation of the real gamma function. The real gamma function is defined by

$$\Gamma(x) := \int_0^\infty u^{x-1} \exp(-u) du, \tag{1}$$

for all real x.

Several verified numerical algorithms have been proposed for the real gamma function [1,2]. Basically, these algorithms use the following properties of the gamma function,

© Springer International Publishing Switzerland 2016
I.S. Kotsireas et al. (Eds.): MACIS 2015, LNCS 9582, pp. 224–228, 2016.
DOI: 10.1007/978-3-319-32859-1_19

$$\Gamma(x+1) = x\Gamma(x), \tag{2}$$

$$-x\Gamma(-x)\Gamma(x) = \frac{\pi}{\sin \pi x}. \tag{3}$$

Thanks to the properties, it suffices to consider the following range of x:

$$1 \leqq x \leqq 2. \tag{4}$$

To calculate the integral of the range, Rump used the polynomial approximation for the range of the real gamma function [1], and Kashiwagi used the power series arithmetic after dividing the integral Eq. (1) into a part of a finite integral and an infinite integral [2].

The purpose of this paper is to present an efficient theorem based on verified numerical integration algorithm over semi-infinite interval. It has been shown that the double exponential formula proposed by Takahasi and Mori [3] is one of the most efficient methods for calculating an approximate value of such integrals. The idea of the double exponential formula is to transform a given integral into an integral over $(-\infty, \infty)$ via a change of variable $u = \varphi(t)$ as

$$\int_0^\infty f(u)du = \int_{-\infty}^\infty f(\varphi(t))\,\varphi'(t)dt. \tag{5}$$

Then, the integral on the right hand side of Eq. (5) is evaluated by the trapezoidal formula. In the present case, as the function $\varphi(t)$,

$$\varphi(t) = \log(1 + \exp(\pi \sinh(t))) \tag{6}$$

is appropriate [4]. Thus, the double exponential formula is explicitly written as

$$\int_0^\infty f(u)du \approx h \sum_{k=-M}^N f(\varphi(kh))\,\varphi'(kh). \tag{7}$$

Recently, Okayama [4] has given an error formula in which all constants are explicitly given in the case where f satisfied on a complex domain that

$$|f(z)| \leqq K \left| \frac{z}{1+z} \right|^{\alpha-1} |\exp(-\beta z)|, \tag{8}$$

where $K > 0$, $0 < \alpha \leqq 1$ and $\beta > 0$. This result is very useful in such a case, but it cannot be applied directly to the type of the real gamma function. In this paper, we present a theorem so as to overcome the problem in the present case. By numerical experiments it is shown that the proposed approach is efficient in terms of the accuracy of the results.

2   Error Bounds of the Double Exponential Formula

2.1   Preliminary

In this paper, we are concerned with a problem of calculating an inclusion of the real gamma function defined by Eq. (1). Here, we briefly review Okayama's result.

Let d be a positive constant. We define a domain $\mathscr{D}_d$ by

$$\mathscr{D}_d = \{z \in \mathbb{C} \ : \ |\text{Im } z| < d\}. \tag{9}$$

Theorem 1 (Okayama [4, Theorem 2.9, $\alpha = 1$]). *Let d be a constant with $0 < d < \pi/2$. Assume that f is analytic in $\varphi\,(\mathscr{D}_d)$, and there exist positive constants K and β ($\beta \leq 1$) such that*

$$|f(z)| \leq K\,|\exp(-\beta z)| \tag{10}$$

for all $z \in \varphi\,(\mathscr{D}_d)$. Let h be defined as

$$h = \frac{\log\,(4dn/\beta)}{n}, \tag{11}$$

and let M and N be defined as

$$N = n, \quad M = n - \lfloor \log(1/\beta)/h \rfloor, \tag{12}$$

and x_γ be defined for $\gamma > 0$ by

$$x_\gamma = \begin{cases} \operatorname{arcsinh}\left(\frac{\sqrt{1+\sqrt{1-(2\pi\gamma)^2}}}{2\pi\gamma}\right) & \left(if \ \ 0 < \gamma < \frac{1}{2\pi}\right), \\[2ex] \operatorname{arcsinh}(1) & \left(if \ \ \frac{1}{2\pi} \leq \gamma\right). \end{cases} \tag{13}$$

Furthermore, let n be taken sufficiently large so that

$$n \geq \frac{e}{4d}, \quad Mh \geq x_1, \quad and \quad Nh \geq x_\beta, \tag{14}$$

hold. Then it holds that

$$\left| \int_0^\infty f\,(u)\,du - h \sum_{k=-M}^N f\,(\varphi(kh))\,\varphi'(kh) \right| \leq C \exp\left(\frac{-2\pi dn}{\log(4dn/\beta)}\right), \tag{15}$$

where the constant C is defined by

$$C = \frac{2K}{\beta}\left\{\frac{2}{(1 - \exp(-\pi\beta e/2))\left\{\cos(\frac{\pi}{2}\sin(d))\right\}^{1+\beta}\cos(d)} + \exp\left(\frac{\pi}{2}\right)\right\}. \tag{16}$$

This theorem is very useful if f is bounded as Eq. (10). However, the upper bound K in the theorem would be overflow when applied with $\beta = 1$ to the following type of f:

$$f(u) = u^{x-1}\exp(-u). \tag{17}$$

2.2 Main Theorem

To overcome the problem, we present the following lemma.

Lemma 1 *Let x be a real number with $1 \leqq x \leqq 2$, let d be a constant with $0 < d < \pi/2$, and let β, γ be defined by*

$$\beta = 1 - \left(\frac{x-1}{2\pi}\right), \qquad \gamma = -\log\left(\cos\left(\frac{\pi}{2}\sin d\right)\right). \qquad (18)$$

Then it follows that

$$\sup_{z\in\varphi(\mathscr{D}_d)} \left|z^{x-1}\exp(-(1-\beta)z)\right| \leqq \left[(\gamma^2 + \pi^2)\exp(\gamma/\pi)\right]^{\frac{x-1}{2}}. \quad (=: C') \qquad (19)$$

By using the lemma, we will obtain the following theorem.

Theorem 2 *Let x be a real number with $1 \leq x \leq 2$, let d be a constant with $0 < d < \pi/2$, and let β, γ be defined by Eq. (18). Assume that f is analytic in $\varphi(\mathscr{D}_d)$ and there exists a positive constant K such that*

$$|f(z)| \leqq K\,|z|^{x-1}\,|\exp(-z)| \qquad (20)$$

for all $z \in \varphi(\mathscr{D}_d)$. Let h be defined as Eq. (11), and let M and N be defined as Eq. (12). Furthermore, let n be taken sufficiently large so that the conditions in Eq. (14) are met. Then it holds that

$$\left|\int_0^\infty f(u)\,du - h\sum_{k=-M}^{N} f(\varphi(kh))\,\varphi'(kh)\right| \leqq CC'\exp\left(\frac{-2\pi dn}{\log(4dn/\beta)}\right), \qquad (21)$$

where the constants C and C' are defined by Eqs. (16) and (19), respectively.

The theorem can be obtained by using the following inequality:

$$\left|z^{x-1}\exp(-z)\right| = \left|z^{x-1}\exp(-(1-\beta)z)\exp(-\beta z)\right|$$
$$\leqq \sup_{z\in\varphi(\mathscr{D}_d)} \left\{\left|z^{x-1}\exp(-(1-\beta)z)\right|\right\}|\exp(-\beta z)| \leq C'|\exp(-\beta z)|. \qquad (22)$$

3 Numerical Result

In this section, we present the numerical experiments. These experiments were done under the following computer environment: Mac OS X Yosemite (10.10), Memory 32 GB, Intel Core i7 4.0 GHz, Apple LLVM version 6.0 (clang-600.0.54) with kv library version 0.4.22 for the interval arithmetic [5] and MPFR library for high precision arithmetic [6].

In Fig. 1, the maximum of the error bound (the right-hand side of Eq. (21)) on the following points

$$x = 1,\ 1 + \frac{1}{32},\ \ldots,\ 1 + \frac{31}{32},\ 2 \qquad (23)$$

is shown with respect to the number n, in range of $[1, 150]$. Furthermore, we show the maximum errors of the double exponential formula on x in Eq. (23), using the floating points system having 53 bits, 106 bits and 212 bits mantissa. We chose $d = 1.5$ in the experiments. From the graph, we can observe that the error bound of the proposed theorem sharply includes the actual errors in the range of given mantissa. We can also see that the results by the double exponential formula are very stable, and only a few digits are lost throughout the calculation of the integral.

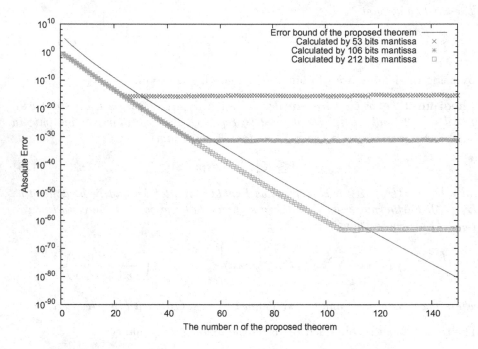

Fig. 1. The maximum error bound of $\Gamma(x)$ on the points in Eq. (23), and the maximum errors of the double exponential formula using the floating points system having 53 bits, 106 bits and 212 bits mantissa.

References

1. Rump, S.M.: Verified sharp bounds for the real gamma function over the entire floating-point range. Nonlinear Theor. Appl. IEICE **5**, 339–348 (2014)
2. Kashiwagi, M.: Verified algorithm for special functions (in Japanese). http://verifiedby.me/kv/special/
3. Takahasi, H., Mori, M.: Double exponential formulas for numerical integration. Publ. RIMS Kyoto Univ. **9**, 721–741 (1974)
4. Okayama, T.: Error estimates with explicit constants for Sinc quadrature and Sinc indefinite integration over infinite intervals. Reliable Comput. **19**, 45–65 (2013)
5. kv Library. http://verifiedby.me/kv/
6. The GNU MPFR Library: http://www.mpfr.org/

Polynomial System Solving

Improving a CGS-QE Algorithm

Ryoya Fukasaku[1], Hidenao Iwane[2], and Yosuke Sato[1(✉)]

[1] Tokyo University of Science, 1–3, Kagurazaka, Shinjuku-ku, Tokyo, Japan
`1414704@ed.tus.ac.jp, ysato@rs.kagu.tus.ac.jp`
[2] Fujitsu Laboratories Ltd, National Institute of Informatics,
2-1-2 Hitotsubashi, Chiyoda-ku, Tokyo, Japan
`iwane@jp.fujitsu.com`

Abstract. A real quantifier elimination algorithm based on computation of comprehensive Gröbner systems introduced by Weispfenning and recently improved by us has a weak point that it cannot handle a formula with many inequalities. In this paper, we further improve the algorithm so that we can handle more inequalities.

Keywords: QE · Comprehensive Gröbner system · Descartes' rule

1 Introduction

In the recent paper [2] we have introduced an improved version of Weispfenning's real quantifier elimination algorithm [5]. The algorithm, called a *CGS-QE* algorithm in this paper, is implemented in Maple and shown to be satisfactorily practical for a given quantified formula with many equations. When a given formula contains many inequalities, however, our algorithm produces huge polynomials on the way of its prosecution, which causes the computation not to terminate in feasible period of time. The size of such polynomials increases at a rate proportional to 2^l where l is the number of inequalities contained in the given formula. The main reason is that we directly apply Descartes' rule of signs for computing a signature of the characteristic polynomial $\chi_1^J(X)$. Its total degree is proportional to 2^l. In this short paper, we introduce a finer method to compute its signature using the factorization structure $\chi_1^J(2^l X) = c\Pi_{(e_1, e_2, \ldots, e_l) \in \{0,1\}^l} \chi_{h_1^{e_1} h_2^{e_2} \cdots h_l^{e_l}}^I(X)$. (Their definitions and detailed descriptions are given in Sect. 2.)

The paper is organized as follows. In Sect. 2, we give a minimal background concerning the CGS-QE algorithm to understand our work of this paper. Our new method is introduced in Sect. 3. Since our work is still on going, we have not completely implemented the new method yet. Nevertheless, we can see its effectiveness through the examples given in Sect. 4.

2 Preliminary

In the rest of the paper, $\mathbb{Q}$ and $\mathbb{R}$ denote the field of rational numbers and the field of real numbers respectively. $\bar{X}$ denotes some variables $X_1, \ldots, X_n$. $T(\bar{X})$

© Springer International Publishing Switzerland 2016
I.S. Kotsireas et al. (Eds.): MACIS 2015, LNCS 9582, pp. 231–235, 2016.
DOI: 10.1007/978-3-319-32859-1_20

denotes a set of terms in $\bar{X}$. For an ideal $I \subset \mathbb{Q}[\bar{X}]$, $V_{\mathbb{R}}(I)$ denotes the varieties of I in $\mathbb{R}$, i.e., $V_{\mathbb{R}}(I) = \{\bar{c} \in \mathbb{R}^n | \forall f \in I \ f(\bar{c}) = 0\}$. Let I be a zero dimensional ideal in a polynomial ring $\mathbb{Q}[\bar{X}]$. Considering the residue class ring $\mathbb{Q}[\bar{X}]/I$ as a vector space over $\mathbb{Q}$, let $t_1, \ldots, t_d$ be its basis. For an arbitrary $h \in \mathbb{Q}[\bar{X}]/I$ and each i, j $(1 \leq i, j \leq d)$ we define a linear map $\theta_{h,i,j}$ from $\mathbb{Q}[\bar{X}]/I$ to $\mathbb{Q}[\bar{X}]/I$ by $\theta_{h,i,j}(f) = ht_i t_j f$ for $f \in \mathbb{Q}[\bar{X}]/I$. Let $q_{h,i,j}$ be the trace of $\theta_{h,i,j}$ and M_h^I be a symmetric matrix such that the (i, j)-th component is given by $q_{h,i,j}$. The characteristic polynomial of M_h^I is denoted by $\chi_h^I(X)$. We abuse the notation $\sigma(f(X))$ to denote the number of positive real roots of $f(X) = 0$ minus the number of negative real roots of $f(X) = 0$ for a polynomial $f(X) \in \mathbb{Q}[X]$, and call it the signature of $f(X)$. The signature of M_h^I denoted $\sigma(M_h^I)$ is defined as the signature $\sigma(\chi_h^I(X))$ of its characteristic polynomial. The real root counting theorem found independently in [1,4] is the following assertion.

Theorem 1. $\sigma(M_h^I) = \#(\{\bar{c} \in V_{\mathbb{R}}(I) | h(\bar{c}) > 0\}) - \#(\{\bar{c} \in V_{\mathbb{R}}(I) | h(\bar{c}) < 0\}).$

We have the following corollary.

Corrollary 2. $\sigma(M_1^I) = \#(V_{\mathbb{R}}(I)).$

Using an obvious relation $h \geqslant 0 \Leftrightarrow \exists z \ z^2 = h$, we have the following fact.

Lemma 3. Let $h_1, \ldots, h_l$ be polynomials in $\mathbb{Q}[\bar{X}]$ and $\bar{Z} = Z_1, \ldots, Z_l$ be new variables. Using the same notations as above, let J be an ideal in $\mathbb{Q}[\bar{X}, \bar{Z}]$ defined by $J = I + \langle Z_1^2 - h_1, \ldots, Z_l^2 - h_l \rangle$. Then the following equation holds.

$$\#(V_{\mathbb{R}}(J)) = 2^l \#(\{\bar{c} \in V_{\mathbb{R}}(I) | h_1(\bar{c}) \geqslant 0, \ldots, h_l(\bar{c}) \geqslant 0\}).$$

The next theorem plays an important role in our CGS-QE algorithm of [2].

Theorem 4. Let I be a zero dimensional ideal of $\mathbb{Q}[\bar{X}]$ and $J = I + \langle Z_1^2 - h_1, \ldots, Z_l^2 - h_l \rangle$ be an ideal of $\mathbb{Q}[\bar{X}, \bar{Z}]$ with polynomials $h_1, \ldots, h_l \in \mathbb{Q}[\bar{X}]$. Let k be a dimension of $\mathbb{Q}[\bar{X}]/I$ and $\{t_1, \ldots, t_k\} \subset T(\bar{X})$ be a basis of the vector space $\mathbb{Q}[\bar{X}]/I$, then $\{t_1 Z_1^{e_1} Z_2^{e_2} \cdots Z_l^{e_l}, \ldots, t_k Z_1^{e_1} Z_2^{e_2} \cdots Z_l^{e_l} | (e_1, e_2, \ldots, e_l) \in \{0,1\}^l\}$ forms a basis of the vector space $\mathbb{Q}[\bar{X}, \bar{Z}]/J$. Let M_g^J denote a symmetric matrix and χ_g^J denote its characteristic polynomial for a polynomial $g \in \mathbb{Q}[\bar{X}]$ induced by the above basis of $\mathbb{Q}[\bar{X}, \bar{Z}]/J$. Let M_g^I denote a symmetric matrix and χ_g^I denote its characteristic polynomial for a polynomial $g \in \mathbb{Q}[\bar{X}]$ induced by the above basis of $\mathbb{Q}[\bar{X}]/I$. Then we have the following equation for some non-zero constant c.

$$\chi_g^J(2^l X) = c \Pi_{(e_1, e_2, \ldots, e_l) \in \{0,1\}^l} \chi_{gh_1^{e_1} h_2^{e_2} \cdots h_l^{e_l}}^I(X).$$

As an easy consequence of these results, we have the following fact.

Theorem 5. $\{\bar{c} \in V_{\mathbb{R}}(I) | h_1(\bar{c}) \geqslant 0, \ldots, h_l(\bar{c}) \geqslant 0\} \neq \varnothing \Leftrightarrow \sigma(\chi_1^J(X)) \neq 0 \Leftrightarrow \sigma(\Pi_{(e_1, e_2, \ldots, e_l) \in \{0,1\}^l} \chi_{h_1^{e_1} h_2^{e_2} \cdots h_l^{e_l}}^I(X)) \neq 0.$

3   Computation of Signatures

In order to apply Theorem 5 to CGS-QE, we need to express $\sigma(\chi_1^J(X)) \neq 0$ as a first order formula in terms of the coefficients of the parametric polynomial $\chi_1^J(X)$. In [2], we compute a simplified form of the formula $I_d(a_0, \ldots, a_{d-1})$ defined by $I_d(a_0, \ldots, a_{d-1}) \Leftrightarrow \sigma(f(X)) \neq 0$ for a polynomial $f(X) = X^d + a_{d-1}X^{d-1} + \cdots + a_0$ which has only real roots, up to $d = 12$ using Descartes' rule of signs. By Theorem 5, the degree of the characteristic polynomial $\chi_1^J(X)$ is $k2^l$. Even when $k = 2$, it exceeds 12 for $l = 3$. Furthermore, the size of its coefficients increases exponentially with l. Hence, we need some heuristic devise for handling more than 3 inequalities as long as we use the characteristic polynomial $\chi_1^J(X)$.

Using Theorems 4 and 5, we also have the following equation:

$$\sigma(\chi_1^J(X)) = \sum_{(e_1, e_2, \ldots, e_l) \in \{0,1\}^l} \sigma(\chi_{h_1^{e_1} h_2^{e_2} \cdots h_l^{e_l}}^I(X)).$$

By the equation we can re-describe $\sigma(\chi_1^J(X)) \neq 0$ in terms of $\sigma(\chi_{h_1^{e_1} h_2^{e_2} \cdots h_l^{e_l}}^I(X))$. When $l = 2$ and k(the dimension of I)$= 2$ for example, since $\sigma(\chi_1^I(X)) \geqslant 0$ by Corollary 2 and $2 \geqslant \sigma(\chi_1^I(X)), \sigma(\chi_{h_1}^I(X)), \sigma(\chi_{h_2}^I(X)), \sigma(\chi_{h_1 h_2}^I(X)) \geqslant -2$ because all of them are quadratic polynomials, the following relation holds:

$$\sigma(\chi_1^J(X)) \neq 0 \Leftrightarrow$$
$$\sigma(\chi_{h_1}^I(X)) + \sigma(\chi_{h_2}^I(X)) + \sigma(\chi_{h_1 h_2}^I(X)) > 0 \vee$$
$$\sigma(\chi_{h_1}^I(X)) + \sigma(\chi_{h_2}^I(X)) + \sigma(\chi_{h_1 h_2}^I(X)) = 0 \wedge \sigma(\chi_1^I(X)) \neq 0 \vee$$
$$\sigma(\chi_{h_1}^I(X)) + \sigma(\chi_{h_2}^I(X)) + \sigma(\chi_{h_1 h_2}^I(X)) = -1 \wedge \sigma(\chi_1^I(X)) = 2.$$

At a glance, the second formula looks more complicated. Remember that $\chi_1^J(X)$ is the product of 2^l polynomials $\chi_{h_1^{e_1} h_2^{e_2} \cdots h_l^{e_l}}^I(X)$ for $(e_1, e_2, \ldots, e_l) \in \{0,1\}^l$. Hence, each coefficient of $\chi_1^J(X)$ is a polynomial of coefficients of $\chi_{h_1^{e_1} h_2^{e_2} \cdots h_l^{e_l}}^I(X)$, some of which has total degree 2^l. On the other hand, if we use the second formula we get a formula which contains only linear polynomials of coefficients contained in some polynomial $\chi_{h_1^{e_1} h_2^{e_2} \cdots h_l^{e_l}}^I(X)$. In the next section, we explicitly give polynomial representations of the above two formulas.

4   Examples

Let $k = 2, l = 2$ and $\chi_1^I(X) = X^2 + a_0 X + b_0, \chi_{h_1}^I(X) = X^2 + a_1 X + b_1, \chi_{h_2}^I(X) = X^2 + a_2 X + b_2, \chi_{h_1 h_2}^I(X) = X^2 + a_3 X + b_3$. Example 6 is a polynomial representation of the formula $\sigma(\chi_1^J(X)) \neq 0$ obtained by $I_8(c_0, c_1, \ldots, c_7)$ where $c_i \; i = 1 \ldots, 7$ are the coefficient of X^i for the expanded polynomial of $(X^2 + a_0 X + b_0)(X^2 + a_1 X + b_1)(X^2 + a_1 X + b_1)(X^2 + a_2 X + b_2)$. Example 7 is a polynomial representation of the second formula in the previous section.

Example 6. $(a_2a_3b_0b_1 + a_1a_3b_0b_2 + a_0a_3b_1b_2 + b_0b_1b_2 + a_1a_2b_0b_3 + a_0a_2b_1b_3 + b_0b_1b_3 + a_0a_1b_2b_3 + b_0b_2b_3 + b_1b_2b_3 \leq 0 \wedge a_3b_0b_1b_2 + a_2b_0b_1b_3 + a_1b_0b_2b_3 + a_0b_1b_2b_3 \neq 0) \vee (0 \leq a_1a_2a_3b_0 + a_0a_2a_3b_1 + a_2b_0b_1 + a_3b_0b_1 + a_0a_1a_3b_2 + a_1b_0b_2 + a_3b_0b_2 + a_0b_1b_2 + a_3b_1b_2 + a_0a_1a_2b_3 + a_1b_0b_3 + a_2b_0b_3 + a_0b_1b_3 + a_2b_1b_3 + a_0b_2b_3 + a_1b_2b_3 \wedge 0 \leq a_0a_1a_2a_3 + a_1a_2b_0 + a_1a_3b_0 + a_2a_3b_0 + a_0a_2b_1 + a_0a_3b_1 + a_2a_3b_1 + b_0b_1 + a_0a_1b_2 + a_0a_3b_2 + a_1a_3b_2 + b_0b_2 + b_1b_2 + a_0a_1b_3 + a_0a_2b_3 + a_1a_2b_3 + b_0b_3 + b_1b_3 + b_2b_3 \wedge 0 \leq a_0a_1a_2 + a_0a_1a_3 + a_0a_2a_3 + a_1a_2a_3 + a_1b_0 + a_2b_0 + a_3b_0 + a_0b_1 + a_2b_1 + a_3b_1 + a_0b_2 + a_1b_2 + a_3b_2 + a_0b_3 + a_1b_3 + a_2b_3 \wedge 0 \leq a_0a_1a_2 + a_0a_1a_3 + a_0a_2a_3 + a_1a_2a_3 + a_1b_0 + a_2b_0 + a_3b_0 + a_0b_1 + a_2b_1 + a_3b_1 + a_0b_2 + a_1b_2 + a_3b_2 + a_0b_3 + a_1b_3 + a_2b_3 \wedge 0 < a_0 + a_1 + a_2 + a_3) \vee (a_1a_2a_3b_0 + a_0a_2a_3b_1 + a_2b_0b_1 + a_3b_0b_1 + a_0a_1a_3b_2 + a_1b_0b_2 + a_3b_0b_2 + a_0b_1b_2 + a_3b_1b_2 + a_0a_1a_2b_3 + a_1b_0b_3 + a_2b_0b_3 + a_0b_1b_3 + a_2b_1b_3 + a_0b_2b_3 + a_1b_2b_3 \leq 0 \wedge 0 \leq a_0a_1a_2a_3 + a_1a_2b_0 + a_1a_3b_0 + a_2a_3b_0 + a_0a_2b_1 + a_0a_3b_1 + a_2a_3b_1 + b_0b_1 + a_0a_1b_2 + a_0a_3b_2 + a_1a_3b_2 + b_0b_2 + b_1b_2 + a_0a_1b_3 + a_0a_2b_3 + a_1a_2b_3 + b_0b_3 + b_1b_3 + b_2b_3 \wedge a_0a_1a_2 + a_0a_1a_3 + a_0a_2a_3 + a_1a_2a_3 + a_1b_0 + a_2b_0 + a_3b_0 + a_0b_1 + a_2b_1 + a_3b_1 + a_0b_2 + a_1b_2 + a_3b_2 + a_0b_3 + a_1b_3 + a_2b_3 \leq 0 \wedge 0 \leq a_0a_1 + a_0a_2 + a_1a_2 + a_0a_3 + a_1a_3 + a_2a_3 + b_0 + b_1 + b_2 + b_3 \wedge a_0a_1a_2 + a_0a_1a_3 + a_0a_2a_3 + a_1a_2a_3 + a_1b_0 + a_2b_0 + a_3b_0 + a_0b_1 + a_2b_1 + a_3b_1 + a_0b_2 + a_1b_2 + a_3b_2 + a_0b_3 + a_1b_3 + a_2b_3 < 0) \vee (a_2a_3b_0b_1 + a_1a_3b_0b_2 + a_0a_3b_1b_2 + b_0b_1b_2 + a_1a_2b_0b_3 + a_0a_2b_1b_3 + b_0b_1b_3 + a_0a_1b_2b_3 + b_0b_2b_3 + b_1b_2b_3 \leq 0 \wedge 0 \leq a_1a_2a_3b_0 + a_0a_2a_3b_1 + a_2b_0b_1 + a_3b_0b_1 + a_0a_1a_3b_2 + a_1b_0b_2 + a_3b_0b_2 + a_0b_1b_2 + a_3b_1b_2 + a_0a_1a_2b_3 + a_1b_0b_3 + a_2b_0b_3 + a_0b_1b_3 + a_2b_1b_3 + a_0b_2b_3 + a_1b_2b_3 \leq 0 \wedge 0 \leq a_0a_1 + a_0a_2 + a_1a_2 + a_0a_3 + a_1a_3 + a_2a_3 + b_0 + b_1 + b_2 + b_3 \wedge a_0 + a_1 + a_2 + a_3 < 0) \vee (a_2a_3b_0b_1 +$

⋮

(66 lines)

⋮

$0) \vee (a_3b_0b_1b_2 + a_2b_0b_1b_3 + a_1b_0b_2b_3 + a_0b_1b_2b_3 \leq 0 \wedge 0 \leq a_2a_3b_0b_1 + a_1a_3b_0b_2 + a_0a_3b_1b_2 + b_0b_1b_2 + a_1a_2b_0b_3 + a_0a_2b_1b_3 + b_0b_1b_3 + a_0a_1b_2b_3 + b_0b_2b_3 + b_1b_2b_3 \wedge a_1a_2a_3b_0 + a_0a_2a_3b_1 + a_2b_0b_1 + a_3b_0b_1 + a_0a_1a_3b_2 + a_1b_0b_2 + a_3b_0b_2 + a_0b_1b_2 + a_3b_1b_2 + a_0a_1a_2b_3 + a_1b_0b_3 + a_2b_0b_3 + a_0b_1b_3 + a_2b_1b_3 + a_0b_2b_3 + a_1b_2b_3 < 0 \wedge 0 \leq a_0a_1a_2 + a_0a_1a_3 + a_0a_2a_3 + a_1a_2a_3 + a_1b_0 + a_2b_0 + a_3b_0 + a_0b_1 + a_2b_1 + a_3b_1 + a_0b_2 + a_1b_2 + a_3b_2 + a_0b_3 + a_1b_3 + a_2b_3 \wedge a_0 + a_1 + a_2 + a_3 \leq 0) \vee (a_3b_0b_1b_2 + a_2b_0b_1b_3 + a_1b_0b_2b_3 + a_0b_1b_2b_3 \leq 0 \wedge 0 \leq a_2a_3b_0b_1 + a_1a_3b_0b_2 + a_0a_3b_1b_2 + b_0b_1b_2 + a_1a_2b_0b_3 + a_0a_2b_1b_3 + b_0b_1b_3 + a_0a_1b_2b_3 + b_0b_2b_3 + b_1b_2b_3 \wedge a_1a_2a_3b_0 + a_0a_2a_3b_1 + a_2b_0b_1 + a_3b_0b_1 + a_0a_1a_3b_2 + a_1b_0b_2 + a_3b_0b_2 + a_0b_1b_2 + a_3b_1b_2 + a_0a_1a_2b_3 + a_1b_0b_3 + a_2b_0b_3 + a_0b_1b_3 + a_2b_1b_3 + a_0b_2b_3 + a_1b_2b_3 < 0 \wedge 0 \leq a_0a_1a_2 + a_0a_1a_3 + a_0a_2a_3 + a_1a_2a_3 + a_1b_0 + a_2b_0 + a_3b_0 + a_0b_1 + a_2b_1 + a_3b_1 + a_0b_2 + a_1b_2 + a_3b_2 + a_0b_3 + a_1b_3 + a_2b_3 \wedge a_0a_1 + a_0a_2 + a_1a_2 + a_0a_3 + a_1a_3 + a_2a_3 + b_0 + b_1 + b_2 + b_3 \leq 0) \vee (a_3b_0b_1b_2 + a_2b_0b_1b_3 + a_1b_0b_2b_3 + a_0b_1b_2b_3 \leq 0 \wedge 0 \leq a_2a_3b_0b_1 + a_1a_3b_0b_2 + a_0a_3b_1b_2 + b_0b_1b_2 + a_1a_2b_0b_3 + a_0a_2b_1b_3 + b_0b_1b_3 + a_0a_1b_2b_3 + b_0b_2b_3 + b_1b_2b_3 \wedge a_1a_2a_3b_0 + a_0a_2a_3b_1 + a_2b_0b_1 + a_3b_0b_1 + a_0a_1a_3b_2 + a_1b_0b_2 + a_3b_0b_2 + a_0b_1b_2 + a_3b_1b_2 + a_0a_1a_2b_3 + a_1b_0b_3 + a_2b_0b_3 + a_0b_1b_3 + a_2b_1b_3 + a_0b_2b_3 + a_1b_2b_3 < 0 \wedge 0 < a_0a_1a_2 + a_0a_1a_3 + a_0a_2a_3 + a_1a_2a_3 + a_1b_0 + a_2b_0 + a_3b_0 + a_0b_1 + a_2b_1 + a_3b_1 + a_0b_2 + a_1b_2 + a_3b_2 + a_0b_3 + a_1b_3 + a_2b_3 \wedge a_0a_1a_2a_3 + a_1a_2b_0 + a_1a_3b_0 + a_2a_3b_0 + a_0a_2b_1 + a_0a_3b_1 + a_2a_3b_1 + b_0b_1 + a_0a_1b_2 + a_0a_3b_2 + a_1a_3b_2 + b_0b_2 + b_1b_2 + a_0a_1b_3 + a_0a_2b_3 + a_1a_2b_3 + b_0b_3 + b_1b_3 + b_2b_3 \leq 0)$

Example 7. $(b1 > 0 \wedge a1 < 0 \wedge ((b2 \geqq 0 \wedge a2 < 0) \vee (b3 \geqq 0 \wedge a3 < 0))) \vee (b2 > 0 \wedge a2 < 0 \wedge ((b1 \geqq 0 \wedge a1 < 0) \vee (b3 \geqq 0 \wedge a3 < 0))) \vee (b3 > 0 \wedge a3 < 0 \wedge ((b2 \geqq 0 \wedge a2 < 0) \vee (b1 \geqq 0 \wedge a1 < 0))) \vee (b1 = 0 \wedge a1 < 0 \wedge b2 = 0 \wedge a2 < 0 \wedge \neg(b3 > 0 \wedge a3 > 0)) \vee (b3 = 0 \wedge a3 < 0 \wedge b2 = 0 \wedge a2 < 0 \wedge \neg(b1 > 0 \wedge a1 > 0)) \vee (b1 = 0 \wedge a1 < 0 \wedge b3 = 0 \wedge a3 < 0 \wedge \neg(b2 > 0 \wedge a2 > 0)) \vee (b1 = 0 \wedge a1 < 0 \wedge ((a2 = 0 \wedge b2 = 0) \vee b2 < 0) \wedge ((a3 = 0 \wedge b3 < 0)) \vee (b3 = 0 \wedge a3 < 0 \wedge (a2 = 0 \wedge b2 = 0) \vee b2 < 0) \wedge ((a1 = 0 \wedge b1 = 0) \vee b1 < 0)) \vee (b2 = 0 \wedge a2 < 0 \wedge ((a1 = 0 \wedge b1 = 0) \vee b1 < 0) \wedge ((a3 = 0 \wedge b3 = 0) \vee b3 < 0)) \vee (\neg((a0 = 0 \wedge b0 = 0) \vee b0 < 0) \wedge ((b1 > 0 \wedge a1 < 0 \wedge ((a2 = 0 \wedge b2 = 0) \vee b2 < 0) \wedge b3 > 0 \wedge a3 > 0) \vee (b1 > 0 \wedge a1 < 0 \wedge b2 > 0 \wedge a2 > 0 \wedge ((a3 = 0 \wedge b3 = 0) \vee b3 < 0)) \vee (b2 > 0 \wedge a2 < 0 \wedge ((a1 = 0 \wedge b1 = 0) \vee b1 < 0) \wedge b3 > 0 \wedge a3 > 0) \vee (b2 > 0 \wedge a2 < 0 \wedge b1 > 0 \wedge a1 > 0 \wedge ((a3 = 0 \wedge b3 = 0) \vee b3 < 0)) \vee (b3 > 0 \wedge a3 < 0 \wedge ((a2 = 0 \wedge b2 = 0) \vee b2 < 0) \wedge b1 > 0 \wedge a1 > 0) \vee (b3 > 0 \wedge a3 < 0 \wedge b2 > 0 \wedge a2 > 0 \wedge ((a1 = 0 \wedge b1 = 0) \vee b1 < 0)) \vee (b1 > 0 \wedge a1 < 0 \wedge b2 = 0 \wedge a2 > 0 \wedge b3 = 0 \wedge a3 > 0) \vee (b2 > 0 \wedge a2 < 0 \wedge b1 = 0 \wedge a1 > 0 \wedge b3 = 0 \wedge a3 > 0) \vee (b3 > 0 \wedge a3 < 0 \wedge b2 = 0 \wedge a2 = 0 \wedge b1 = 0 \wedge a1 > 0) \vee (b1 = 0 \wedge a1 < 0 \wedge b2 = 0 \wedge a2 > 0 \wedge b3 > 0 \wedge a3 > 0) \vee (b2 = 0 \wedge a2 < 0 \wedge b1 = 0 \wedge a1 > 0 \wedge b3 > 0 \wedge a3 > 0) \vee (b3 = 0 \wedge a3 < 0 \wedge b2 = 0 \wedge a2 > 0 \wedge b1 > 0 \wedge a1 > 0) \vee (b1 = 0 \wedge a1 < 0 \wedge b3 = 0 \wedge a3 > 0 \wedge ((a2 = 0 \wedge b2 = 0) \vee b2 < 0)) \vee (b2 = 0 \wedge a2 < 0 \wedge b1 = 0 \wedge a1 > 0 \wedge ((a3 = 0 \wedge b3 = 0) \vee b3 < 0)) \vee (b3 = 0 \wedge a3 < 0 \wedge b2 = 0 \wedge a2 > 0 \wedge ((a1 = 0 \wedge b1 = 0) \vee b1 < 0)) \vee (b3 = 0 \wedge a3 < 0 \wedge b1 = 0 \wedge a1 > 0 \wedge ((a2 = 0 \wedge b2 = 0) \vee b2 < 0)) \vee ((a1 = 0 \wedge b1 = 0) \vee b1 < 0) \wedge ((a2 = 0 \wedge b2 = 0) \vee b2 < 0) \wedge ((a3 = 0 \wedge b3 = 0) \vee b3 < 0)))) \vee (b0 > 0 \wedge a0 < 0 \wedge ((b1 > 0 \wedge a1 < 0 \wedge b2 > 0 \wedge a2 > 0 \wedge b3 = 0 \wedge a3 > 0) \vee (b1 > 0 \wedge a1 < 0 \wedge b3 > 0 \wedge a3 > 0 \wedge b2 = 0 \wedge a2 > 0) \vee (b2 > 0 \wedge a2 < 0 \wedge b3 > 0 \wedge a3 > 0 \wedge b1 = 0 \wedge a1 > 0) \vee (b2 > 0 \wedge a2 < 0 \wedge b1 = 0 \wedge a1 > 0 \wedge b3 > 0 \wedge a3 > 0) \vee (b3 > 0 \wedge a3 < 0 \wedge b1 = 0 \wedge a1 > 0 \wedge b2 = 0 \wedge a2 > 0) \vee (b1 = 0 \wedge a1 < 0 \wedge ((a2 = 0 \wedge b2 = 0) \vee b2 < 0) \wedge b3 > 0 \wedge a3 > 0) \vee (b1 = 0 \wedge a1 < 0 \wedge ((a3 = 0 \wedge b3 = 0) \vee b3 < 0) \wedge b2 > 0 \wedge a2 > 0) \vee (b2 = 0 \wedge a2 < 0 \wedge ((a1 = 0 \wedge b1 = 0) \vee b1 < 0) \wedge b3 > 0 \wedge a3 > 0) \vee (b3 = 0 \wedge a3 < 0 \wedge b1 > 0 \wedge a1 > 0 \wedge ((a1 = 0 \wedge b1 = 0) \vee b1 < 0) \wedge b2 > 0 \wedge a2 > 0) \vee (b3 = 0 \wedge a3 < 0 \wedge b2 = 0 \wedge a2 > 0 \wedge b1 > 0 \wedge a1 > 0) \vee (b3 = 0 \wedge a3 < 0 \wedge b2 = 0 \wedge a2 > 0 \wedge b1 > 0 \wedge a1 > 0) \vee (((a1 = 0 \wedge b1 = 0) \vee b1 < 0) \wedge b2 = 0 \wedge a2 > 0 \wedge ((a3 = 0 \wedge b3 = 0) \vee b3 < 0)) \vee (((a2 = 0 \wedge b2 = 0) \vee b2 < 0) \wedge b1 = 0 \wedge a1 > 0 \wedge ((a3 = 0 \wedge b3 = 0) \vee b3 < 0)) \vee (((a1 = 0 \wedge b1 = 0) \vee b1 < 0) \wedge b3 = 0 \wedge a3 > 0 \wedge ((a2 = 0 \wedge b2 = 0) \vee b2 < 0))))$

5 Conclusion and Remarks

When we apply our CGS-QE algorithm to a basic quantified formula:

$$\exists \bar{X}(f_1(\bar{Y}, \bar{X}) = 0 \wedge \cdots \wedge f_m(\bar{Y}, \bar{X}) = 0 \wedge h_1(\bar{Y}, \bar{X}) \geqslant 0 \wedge \cdots \wedge h_l(\bar{Y}, \bar{X}) \geqslant 0),$$

we first compute a comprehensive Gröbner system (CGS) of the parametric ideal $\langle f_1, \ldots, f_m \rangle$ of $\mathbb{Q}[\bar{X}]$ with parameters $\bar{Y}$. For a segment of $\bar{Y}$ such that the associated Gröbner basis is zero-dimensional we do not need any further CGS computations. We can compute an equivalent quantifier free formula in this segment even when l is not small. On the other hand, for a segment of $\bar{Y}$ such that the associated Gröbner basis is not zero-dimensional, we further proceed a recursive step of the CGS-QE algorithm. For such a computation the input basic quantified formula is constructed from a formula $I_d(a_0, \ldots, a_{d-1})$ for some d and $a_0, \ldots, a_{d-1}$. As long as we use the formula I_d, the size of input polynomials can be very big even for a small l. As we see in the previous section, the polynomial representation obtained from I_8 contains big polynomials of $a_0, b_0, a_1, b_1, a_2, b_2, a_3, b_3$ with total degree 4. For a polynomial ideal consisting of big and high degree polynomials it is usually impossible to compute its Gröbner basis. It is the main reason that our CGS-QE algorithm cannot deal with many inequalities. If we use the new polynomial representation introduced in the paper, we are not bothered by this phenomena anymore. We do not have explosion of the polynomial size.

For applying our method to a CGS-QE algorithm, we have to prepare a simplified algebraic representation of the formula

$$\sum_{(e_1, e_2, \ldots, e_l) \in \{0,1\}^l} \sigma(\chi^I_{h_1^{e_1} h_2^{e_2} \ldots h_l^{e_l}}(X)) \neq 0$$

$$(\chi^I_{h_1^{e_1} h_2^{e_2} \ldots h_l^{e_l}}(X) = X^k + c_{k-1}^{(e_1, e_2, \ldots, e_l)} X^{k-1} + \cdots + c_1^{(e_1, e_2, \ldots, e_l)} X + c_0^{(e_1, e_2, \ldots, e_l)})$$

in terms $c_{k-1}^{(e_1, e_2, \ldots, e_l)}, \ldots, c_1^{(e_1, e_2, \ldots, e_l)}, c_0^{(e_1, e_2, \ldots, e_l)}$ $(e_1, e_2, \ldots, e_l) \in \{0,1\}^l$ for each k and l. As we have obtained a simplified formula for I_d up to $d = 12$, we can also obtain them using a simplification method by Boolean function manipulation introduced in [3].

References

1. Becker, E., Wörmann, T.: On the trace formula for quadratic forms. Recent advances in real algebraic geometry and quadratic forms (Berkeley, CA, 1990/1991; San Francisco, CA, 1991), pp. 271–291, Contemp. Math., 155, Amer. Math. Soc., Providence, RI (1994)
2. Fukasaku, R., Iwane, H., Sato, Y.: Real quantifier elimination by computation of comprehensive gröbner systems. In: Proceedings of International Symposium on Symbolic and Algebraic Computation, pp. 173–180. ACM (2015)
3. Iwane, H., Higuchi, H., Anai, H.: An effective implementation of a special quantifier elimination for a sign definite condition by logical formula simplification. In: Gerdt, V.P., Koepf, W., Mayr, E.W., Vorozhtsov, E.V. (eds.) CASC 2013. LNCS, vol. 8136, pp. 194–208. Springer, Heidelberg (2013)
4. Pedersen, P., Roy, M.-F., Szpirglas, A.: Counting real zeroes in the multivariate case. In: Proceedings of the Effective Methods in Algebraic Geometry, pp. 203–224. Springer (1993)
5. Weispfenning, V.: A new approach to quantifier elimination for real algebra. In: Caviness, B.F., Johnson, J.R. (eds.) Quantifier Elimination and Cylindrical Algebraic Decomposition, pp. 376–392. Springer, Vienna (1998)

Efficient Subformula Orders for Real Quantifier Elimination of Non-prenex Formulas

Munehiro Kobayashi[1(✉)], Hidenao Iwane[2,3], Takuya Matsuzaki[3,4],
and Hirokazu Anai[2,3,5]

[1] University of Tsukuba, Tsukuba, Japan
munehiro-k@math.tsukuba.ac.jp
[2] Fujitsu Laboratories Ltd., Kawasaki, Japan
[3] National Institute of Informatics, Chiyoda, Japan
[4] Nagoya University, Nagoya, Japan
[5] Kyushu University, Fukuoka, Japan

Abstract. In this paper we study speeding up real quantifier elimination (QE) methods for non-prenex formulas. Our basic strategy is to solve non-prenex first-order formulas by performing QE for subformulas constituting the input non-prenex formula. We propose two types of methods (heuristic methods/machine learning based methods) to determine an appropriate ordering of QE computation for the subformulas. Then we empirically examine their effectiveness through experimental results over more than 2,000 non-trivial example problems. Our experiment results suggest machine learning can save much effort spent to design effective heuristics by trials and errors without losing efficiency of QE computation.

Keywords: Real quantifier elimination · Support vector machine · Non-prenex formulas

1 Introduction

In this paper we aim at speeding up quantifier elimination (QE) methods for *non-prenex formulas* over the reals by automatizing a heuristic procedure in QE computation using a machine learning method.

When we discuss real QE algorithms, we usually assume that input first-order formulas are in prenex form. We say a formula is in prenex form if it has the form of a string of quantifiers followed by a quantifier-free part. However, given formulas are not always prenex ones in practice. Transforming non-prenex formulas to prenex form causes to miss the algebraic independency between variables and then the QE computation tends to be hard. Hence, in view of practical computation, the following strategy is considered to be effective: dividing a given non-prenex formula into subformulas, applying QE to each subformulas, and then logically integrating the results. Furthermore, we can obtain further efficiency by utilizing intermediate QE results to simplify remaining formulas before applying QE to them.

© Springer International Publishing Switzerland 2016
I.S. Kotsireas et al. (Eds.): MACIS 2015, LNCS 9582, pp. 236–251, 2016.
DOI: 10.1007/978-3-319-32859-1_21

Here we face the problem to determine the order of performing QE for the subformulas. Total time required for achieving QE for the input non-prenex formula depends heavily on the order. Ordinarily, some heuristic methods to choose an appropriate order for subformulas are implemented (e.g., "easy to solve" subformula is first). For example, this strategy is successfully employed in [8].

In spite of its practical efficacy, studies on QE for non-prenex formulas are few [11]. Actually, the existing computer algebra systems that accept non-prenex formulas as inputs for QE execution are only Mathematica and SyNRAC [9] on Maple. We focus on QE for non-prenex formulas to improve efficiency of QE algorithm in practical use.

Our target problem is to determine an appropriate order for improving the efficiency of QE computation for non-prenex formulas. In this paper we design some heuristics methods and others based on machine learning, and examine their effectiveness through detailed analysis of the experimental results over 2,306 nontrivial QE problems.

The 2,306 nontrivial QE problems are provided from the activity in "Todai robot project – Can a Robot Get Into the University of Tokyo?" [1,10], which is a project to develop an artificial intelligence system to automatically solve natural language math problems of university entrance examinations. As 2,306 samples are sufficient to apply machine learning, the binary classification problem of if a formula is easy to perform QE for was learned with support vector machines (SVMs).

So far there are some attempts to propose heuristic methods for determining variable orders aiming at speeding up elimination methods in computer algebra. Dolzmann et al. gave heuristics that estimates the optimal projection order for cylindrical algebraic decomposition (CAD) [4]. Their study is based on a statistical analysis. Later, Huang et al. applied machine learning to choose heuristics for selecting projection order for CAD [6]. They showed SVMs, which are based on statistical learning theory, achieved a good switching of heuristics and the obtained mixture of heuristics performed better than any single heuristics.

While Huang et al. mainly measured the numbers of cells produced by CAD, we made our study more practical by dealing with time performance. We arranged this study in hope with a possible contribution to Todai robot project. Thus, among time performances, one of which is simple execution time, we most value the property that a QE computation finishes in a fixed time.

Our experimental results on more than 2,000 QE problems also indicate that machine learning based methods are promising for heuristic processes if we have sufficient amount of data (input problems and computational results). This fact is beneficial in efficiently designing an effective method for such heuristics parts since usually developing effective heuristic methods takes a lot of trials and errors.

The rest of this paper is organized as follows. Section 2 shows the problem we discuss. Section 3 provides some methods to determine an appropriate order of QE computation for subformulas from an input non-prenex formula. Section 4 is devoted to explain the computational results of the proposed methods and detailed analysis. The concluding remarks are made in Sect. 5.

2   Problem

In this paper, we study efficient QE for non-prenex formulas over the reals. The following formula φ gives an example of a non-prenex formula:

$$\varphi \equiv \varphi_1 \vee \varphi_2 \vee \varphi_3, \tag{1}$$

where

$\varphi_1 \equiv \exists x_0(-x_0 \leq -1),$

$\varphi_2 \equiv \exists x_1(x_1 \leq 0),$ and

$$\varphi_3 \equiv \exists x_2 \exists x_3 \exists x_4 \Big(\exists x_5 \big((x_2^3 x_4 x_5 - x_2^2 x_4 - x_2^2 x_5 + x_2 + 1 = 0 \vee$$
$$x_2^3 x_4 x_5 x_6 - x_2^2 x_4 x_5 - x_2^2 x_4 x_6 - x_2^2 x_5 x_6 + x_2 x_4 + x_2 x_5 + x_2 x_6 - x_6 = 0) \wedge$$
$$(x_2^3 x_4 x_6 - x_2^2 x_4 - x_2^2 x_6 + x_2 + 1 = 0 \vee$$
$$x_2^3 x_4 x_5 x_6 - x_2^2 x_4 x_5 - x_2^2 x_4 x_6 - x_2^2 x_5 x_6 + x_2 x_4 + x_2 x_5 + x_2 x_6 - x_5 = 0) \big) \wedge$$
$$\big(x_2^3 x_3 - x_2^2 x_3 - x_2^2 + x_2 + 1 = 0 \vee$$
$$x_2^3 x_3 x_4 - x_2^2 x_3 x_4 - x_2^2 x_3 - x_2^2 x_4 + x_2 x_3 + x_2 x_4 + x_2 - x_4 = 0 \big) \wedge$$
$$\big(x_2^3 x_4 - x_2^2 x_4 - x_2^2 + x_2 + 1 = 0 \vee$$
$$x_2^3 x_3 x_4 - x_2^2 x_3 x_4 - x_2^2 x_3 - x_2^2 x_4 + x_2 x_3 + x_2 x_4 + x_2 - x_3 = 0 \big) \Big).$$

Our approach is to modify the order of processing subformulas by varying ordering functions on formulas.

The paper [8] of Iwane et al. states that sort orderings on formulas affect the efficiency of QE computation. The paper treats the problem of constructing formulas equivalent to university entrance examination problems using the natural language processing, and the problem of solving them by performing QE for the constructed formulas. According to [8], construction of formulas by the current natural language processing results in large and complicated QE problems, and solving such problems simply by a general QE algorithm is not realistic. The following approaches are proposed by them to be effective in QE computation if the input QE problem is much redundant: use of special QE algorithms, simplification of intermediate formulas, and decomposition of the formula into separately solvable parts. Algorithm 1 (adapted from [8]) shows their procedure to perform QE for a non-prenex formula. QE_{prenex} in Algorithm 1 executes QE using traditional algorithms such as CAD, virtual term substitution, and QE by real root counting [7] in the form $\forall x(f(x) > 0)$ and $\forall x(x \geq 0 \rightarrow f(x) > 0)$. Their proposed QE algorithm rearranges an input formula into non-prenex form, makes sorting over the subformulas, and in that order performs QE recursively for each subformula. The conditions obtained from solved (i.e., quantifier-free) subformulas are immediately utilized to simplify the remaining subformulas. Hence subformulas should be sorted in the order from easiest to hardest. Our study investigated the effect of changing 'SORT' in Algorithm 1.

Algorithm 1. $\mathrm{QE}_{\mathrm{main}}(\varphi, \psi_{\mathrm{nec}}, \psi_{\mathrm{suf}})$

Input: a first-order formula φ, quantifier-free formulas ψ_{nec} and ψ_{suf}
Output: a quantifier-free formula φ' s.t. $\varphi \wedge \psi_{\mathrm{nec}} \vee \psi_{\mathrm{suf}}$ is equivalent to $\varphi' \wedge \psi_{\mathrm{nec}} \vee \psi_{\mathrm{suf}}$

1: **if** φ is quantifier-free **then**
2:     **return** $Simplify(\varphi, \psi_{\mathrm{nec}}, \psi_{\mathrm{suf}})$
3: **else if** φ is a prenex formula **then**
4:     **return** $\mathrm{QE}_{\mathrm{prenex}}(\varphi, \psi_{\mathrm{nec}}, \psi_{\mathrm{suf}})$
5: **else if** $\varphi \equiv Q_1 x_1 \cdots Q_m x_m \, \xi$ where $Q_j \in \{\exists, \forall\}$ **then**
6:     /* ξ is not quantifier-free */
7:     **return** $\mathrm{QE}_{\mathrm{prenex}}(Q_1 x_1 \cdots Q_m x_m \mathrm{QE}_{\mathrm{main}}(\xi, \psi_{\mathrm{nec}}, \psi_{\mathrm{suf}}), \psi_{\mathrm{nec}}, \psi_{\mathrm{suf}})$
8: **else if** $\varphi \equiv \vee \xi_i$ **then**
9:     **return** $\neg\mathrm{QE}_{\mathrm{main}}(\neg\varphi, \neg\psi_{\mathrm{suf}}, \neg\psi_{\mathrm{nec}})$
10: /* $\varphi \equiv \wedge \xi_i$ */
11: **if** $\xi_i \equiv (f_i = 0)$ and f_i is reducible ($f_i = \prod g_{i,j}$) **then**
12:     **return** $\mathrm{QE}_{\mathrm{main}}(\vee_j(g_{i,j} = 0 \wedge (\wedge_{k \neq i} \xi_k)), \psi_{\mathrm{nec}}, \psi_{\mathrm{suf}})$
13: $\{\varphi_i\}_{i=1}^n \leftarrow \mathrm{SORT}(\{\xi_i\}_{i=1}^n)$
14: **for all** i such that $1 \leq i \leq n$ **do**
15:     $\varphi_i \leftarrow \mathrm{QE}_{\mathrm{main}}(\varphi_i, \psi_{\mathrm{nec}}, \psi_{\mathrm{suf}})$
16:     $\psi_{\mathrm{nec}} \leftarrow \psi_{\mathrm{nec}} \wedge \varphi_i$
17: **return** $\wedge_i \varphi_i$

---

3 Methodology

In this section, we introduce the ordering functions. Our aim is to find an *ordering function* that makes QE computation efficient, and to develop a method to obtain such an ordering function systematically by using machine learning. Our ordering functions can be divided into three types – heuristics, random, and machine learning. The way SVMs are utilized to define ordering functions is also described.

3.1 Ordering Functions

We experimented with the sorting determined by the following ordering functions on formulas:

Heuristics. The measure described below is calculated from the formula, and the order of QE computation for subformulas was determined by comparing the measure. We identified 10 measures for heuristics:
    ADG [8]. The measure is a weighted sum of 1, 2, 3, 19, 27, 45 in Table 1.
    nvar, npoly, sotd, msotd, mtdeg, mdeg, and mterm. The measures are 1, 2, 3, 7, 8, 9, and 10 in Table 1 respectively.
    nvar_rev and npoly_rev. The measures are 1 and 2 in Table 1 respectively. These ordering functions give the reversed orders of 'nvar' and 'npoly' respectively, and were designed to investigate the worst cases.
Random. Each subformula is assigned random priority.

Machine Learning. We use an SVM to calculate the *decision value*, which is described in Subsect. 3.2, of a formula using the model obtained by training in advance. The formula which has the greater decision value is the smaller formula.

As the above configuration suggests, the ordering functions by heuristics and machine learning are similar in the way that both search for the best combination of features shown in Table 1 to obtain efficient sorting in QE computation.

3.2 Features and Labeling for Machine Learning

Features. In order to perform machine learning over formulas with SVM, we need to characterize a formula by a fixed length *feature vector* of numerical values. Each value of the vector is supposed to capture some feature of the formula. We chose 58 features shown in Table 1 to construct such a vector for a given formula. The features are taken from other studies [2,4,6] and derived from the conditions of special QE algorithms. While there are apparently overlapped features that are expressed with other features such as feature 4 expressed with feature 3 divided by feature 2, these dependent features are selected in the aim of emphasizing the nonlinear relation among the defining features. Although machine learning technique basically evaluates all features and their relations rather flatly, one way to make SVMs sensitive to a particular relation among features is adding a new dependent feature that perceive the target relation well. Another way to modify SVMs in focused features and their relations is designing an appropriate kernel function, but we decided to stick to the traditional radial basis function (RBF) kernel for the simplicity of implementation required for experiment. We also prepared diminished features of size 43 by omitting the rows 4, 14, 15, 16, and 18 in Table 1.

Labeling. We also need training samples labeled with either $+1$ or -1 to train an SVM. In our experiment, a feature vector in training samples was labeled $+1$ if the corresponding QE problem was solved in N seconds, and -1 otherwise. We experimented with N varying 1, 2, 3, 4, 5, 6, 7, 8, 9, 10, 20, 30 and 55. Among these thresholds, 55 s is the time that separates the least 90 % from the top 10 % of the execution times.

Use of SVMs for Ordering Function. We used SVMs to learn binary classification problems, which is the most elementary function of SVMs. There are two steps to make binary classification using SVMs. First, we train a model with a set of training samples which were vectors labeled either $+1$ (positive) or -1 (negative). Second, we request an SVM to predict if a subformula is likely to belong to the positive class or the negative class.

In the first step, an SVM calculates a hyperplane that divides positive and negative training samples in some vector space. In the second step, we can obtain a decision value, which is a kind of distance from a new sample to the dividing hyperplane. Decision values can be considered as a measure of the confidence

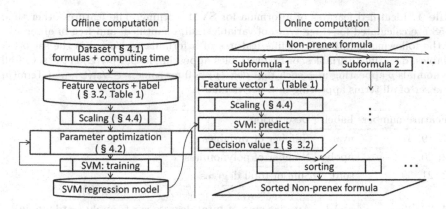

Fig. 1. Flow of sorting by using SVMs

in a correct prediction, and so the formula with the largest decision value was solved first. Figure 1 illustrates the procedure to make sorting with an ordering function determined using an SVM.

In order to configure ordering functions, we trained 17 models in total by changing labels on training data with varying threshold time, parameter selection of SVMs, and feature selection.

4 Computational Experiments

In this section, computational experiments performed are described. QE computations were made using the Maple-package SyNRAC [9]. For the machine learning experiment, we used LIBSVM [3], which is an implementation of an SVM written in C++. We ran all the computational experiments on a computer with an Intel(R) Xeon(R) CPU E7-4870 2.40 GHz and 1007 GB of memory.

4.1 Datasets

We have extracted 2,306 first-order formulas that were generated when the automated natural language math problem solving system (developed by the Todai robot project) solved the problems, which were also collected by the project. The problems were originally taken from three sources: "Chart-shiki", Japanese university entrance exams, and International Mathematical Olympiads. "Chart-shiki" is a popular problem book series for preparing university entrance examinations. The math problem solving system accepts problems in algebra, linear algebra, geometry, pre-calculus, calculus, combinatorics, Peano arithmetic, and higher order formula as its inputs. Among these problems, the problems which can be expressed in the language of real closed field were utilized to generate our dataset, where such kind of problems are mainly in algebra, linear algebra, and geometry.

Table 1. Identified features of a formula for SVM learning. The features other than 55–58 are calculated for three types of variables: all, quantified, and free in order. We use the following notations to define features of a formula φ: $\mathsf{Var}(\varphi)$ is the set of all variables, quantified variables, or free variables appearing in φ, $\mathsf{Poly}(\varphi)$ is the set of all polynomials p appearing in φ such that $\deg_v(p) > 0$ for some $v \in \mathsf{Var}(\varphi)$, and $\mathsf{Term}(p)$ is the set of all terms appearing in a polynomial p.

Feature number	Label	Description
1, 19, 37	nvar	Number of variables
2, 20, 38	npoly	Number of polynomials
3, 21, 39	sotd	Sum of total degrees $\left(\sum_{p \in \mathsf{Poly}(\varphi)} \sum_{t \in \mathsf{Term}(p)} \sum_{v \in \mathsf{Var}(\varphi)} \deg_v(t)\right)$
4, 22, 40	asotd	Average sum of total degrees w.r.t. npoly (sotd/npoly)
5, 23, 41	atdeg	Average total degree w.r.t. npoly $\left(\sum_{p \in \mathsf{Poly}(\varphi)} \max_{t \in \mathsf{Term}(p)}\left(\sum_{v \in \mathsf{Var}(\varphi)} \deg_v(t)\right)/\text{npoly}\right)$
6, 24, 42	aterm	Average number of terms w.r.t. npoly $\left(\sum_{p \in \mathsf{Poly}(\varphi)} \sum_{t \in \mathsf{Term}(p)} 1/\text{npoly}\right)$
7, 25, 43	msotd	Maximum sum of total degrees $\left(\max_{p \in \mathsf{Poly}(\varphi)} \sum_{t \in \mathsf{Term}(p)} \sum_{v \in \mathsf{Var}(\varphi)} \deg_v(t)\right)$
8, 26, 44	mtdeg	Maximum total degree $\left(\max_{p \in \mathsf{Poly}(\varphi)} \max_{t \in \mathsf{Term}(p)} \sum_{v \in \mathsf{Var}(\varphi)} \deg_v(t)\right)$
9, 27, 45	mdeg	Maximum degree $\left(\max_{p \in \mathsf{Poly}(\varphi)} \max_{v \in \mathsf{Var}(\varphi)} \deg_v(p)\right)$
10, 28, 46	mterm	Maximum number of terms $\left(\max_{p \in \mathsf{Poly}(\varphi)} \sum_{t \in \mathsf{Term}(p)} 1\right)$
11, 29, 47	ndeg1	Number of polynomials with degree 1 $\left(\sum_{p \in \mathsf{Poly}(\varphi), \max_{t \in \mathsf{Term}(p)} \sum_{v \in \mathsf{Var}(\varphi)} \deg_v(t)=1} 1\right)$
12, 30, 48	ndeg2	Number of polynomials with degree 2
13, 31, 49	ndeg3	Number of polynomials with degree 3
14, 32, 50	rdeg1	Ratio of ndeg1 and npoly (ndeg1/npoly)
15, 33, 51	rdeg2	Ratio of ndeg2 and npoly (ndeg2/npoly)
16, 34, 52	rdeg3	Ratio of ndeg3 and npoly (ndeg3/npoly)
17, 35, 53	mcoef	Maximum absolute value of coefficients $\left(\max_{p \in \mathsf{Poly}(\varphi)} \max_{c \in \mathsf{Coeff}(p)} |c|\right)$
18, 36, 54	acoef	Average absolute value of coefficients w.r.t. npoly (mcoef/npoly)
55		Ratio of the number of the symbol '$=$' and npoly
56		Ratio of the number of the symbol '$\neq$' and npoly
57		Ratio of the number of the symbol '$<$' and npoly
58		Ratio of the number of the symbol '$\leq$' and npoly

For each formula, we executed QE computation with SyNRAC and recorded the execution time, where we set the limit of timeout to 600 s. Table 2 shows the distribution of execution times for the 2,116 formulas for which QE computation

Table 2. Number of formulas for which QE computation took less than N seconds. These execution times were measured using the 'ADG' ordering function.

N	0.1	0.2	0.3	0.4	0.5	0.6	0.7	0.8	0.9
#	433	846	1055	1261	1379	1476	1627	1694	1725
N	1.0	2.0	3.0	4.0	5.0	6.0	7.0	8.0	9.0
#	1752	1884	1937	1958	1977	1990	1999	2005	2013
N	10	20	30	100	150	200	300	500	600
#	2015	2040	2057	2096	2100	2102	2109	2113	2116

stopped in less than 600 s. There are 187 formulas which took more than 600 s to finish QE computation. The other 4 problems caused error while computing QE with SyNRAC, which were omitted from the training set for machine learning. The size 2,302 of the training set is large enough for the application of machine learning with SVMs. For example, three cases of application of SVMs are introduced in [5], and the sizes of the training data are 3,089, 391, 1,243 while the numbers of features are 4, 20, 21 respectively.

4.2   Parameter Optimization of SVMs

We selected the RBF kernel for our machine learning, which is a popular choice when the number of features is smaller than the number of training set. In our case, the number of features was 58 and the number of training set was 2,302.

We used a python script "grid.py," which is distributed with LIBSVM, for grid search over parameters (C, γ) where C is the cost parameter and γ is the parameter of the RBF kernel. The script "grid.py" performs cross validation to estimate the *accuracy* of each parameter combination in the specified range, and outputs the parameter combination that recorded the highest accuracy. Here, accuracy is defined as the number of correct predictions divided by the total number of predictions. The script "grid.py" was executed twice to obtain the optimal parameters. We specified the options of "grid.py" -log2c, -log2g, -w1 and -w-1. The options -log2c and -log2g specify the range and the incremental width of C and γ respectively. The values of the options -log2c and -log2g were set to −5, 15, 2 and 3, −15, −2 respectively in the first execution, which are default, and in the second time the values were set to $C_0 - 1$, $C_0 + 1$, 0.2 and $\gamma_0 - 1$, $\gamma_0 + 1$, 0.2 respectively, where (C_0, γ_0) is the output of the first execution. The option -w1 and -w-1 are for handling unbalance in training data, and were set to the ratio of the number of training vector labeled +1 and −1.

4.3   An Illustrative Example

We again consider the non-prenex formula (1) in Sect. 2. What we do here is to perform QE for φ using an ordering function configured with an SVM. QE for φ can be divided into 3 QE subproblems for φ_1, for φ_2, and for φ_3. In this case, φ_1

Table 3. An example of QE execution times and feature vectors of formulas. Features with an asterisk are omitted in machine learning for '<9_dim'.

Feature no.	Time(sec)	1	2	3	4*	5	6	7	8	9	10	11	12	13
φ_1	0.078	1	1	1	1	1	2	1	1	1	2	1	0	0
φ_2	0.037	1	1	1	1	1	1	1	1	1	1	1	0	0
φ_3	1551.989	5	8	136	17	5	6.5	25	6	3	8	0	0	0

Feature no.	14*	15*	16*	17	18*	19	20	21	22*	23	24	25	26	27	28
φ_1	1	0	0	1	1	1	1	1	1	1	2	1	1	1	2
φ_2	1	0	0	1	1	1	1	1	1	1	1	1	1	1	1
φ_3	0	0	0	1	1	4	8	125	15.6	4.6	6.5	21	5	3	8

Feature no.	29	30	31	32*	33*	34*	35	36*	37	38	39	40*	41	42	43
φ_1	1	0	0	1	0	0	1	1	0	0	0	0	0	0	0
φ_2	1	0	0	1	0	0	1	1	0	0	0	0	0	0	0
φ_3	0	0	0	0	0	0	1	1	1	3	11	3.7	1	7	5

Feature no.	44	45	46	47	48	49	50*	51*	52*	53	54*	55	56	57	58
φ_1	0	0	0	0	0	0	0	0	0	0	0	0	0	0	1
φ_2	0	0	0	0	0	0	0	0	0	0	0	0	0	0	1
φ_3	1	1	8	3	0	0	1	0	0	1	1	1	0	0	0

and φ_2 are trivially true, and so φ is true. The order of processing subproblems are determined by decision values calculated with LIBSVM. Table 3 shows the QE computation time and the features of φ_1, φ_2, and φ_3. Table 4 shows the decision value and the order of QE subproblems. The QE computation for φ with ordering functions '<2', '<3', and '<7' solved the subproblem φ_3 first, and did not finish in 600 s. On the other hand, the QE computation with the other ordering functions started to solve the subproblems for φ_1 or φ_2, and finished computation in less than 1 s.

4.4 Experimental Results

We made two types of experiments. Experiment 1 investigated the efficiency of QE computation when the prepared ordering functions were used. Experiment 2 justifies Experiment 1 in the point that the apparently common training and test set did not cause to overestimate the result of ordering functions based on machine learning. The amount of computation is large as it took more than 40 h to make a series of QE computation with a single ordering function.

Experiment 1. We prepared 28 ordering functions in total. Among those, 10 functions were heuristics and 1 function was random, which are described in Subsect. 3.1. The rest 17 functions which were configured using machine learning

Table 4. Example of decision values

	φ_1	φ_2	φ_3	Order
<1	−4.143236	−4.126463	−7.026860	$\varphi_2 \to \varphi_1 \to \varphi_3$
<2	4.924164	4.659644	6.950127	$\varphi_3 \to \varphi_1 \to \varphi_2$
<3	2.092471	2.145601	4.441247	$\varphi_3 \to \varphi_2 \to \varphi_1$
<4	1.428065	1.430775	−0.009359	$\varphi_2 \to \varphi_1 \to \varphi_3$
<4_opt	1.891265	1.887521	−2.485134	$\varphi_1 \to \varphi_2 \to \varphi_3$
<5	1.981705	2.103514	0.978644	$\varphi_2 \to \varphi_1 \to \varphi_3$
<6	2.063262	2.127833	−2.001929	$\varphi_2 \to \varphi_1 \to \varphi_3$
<7	1.462456	1.469122	3.834284	$\varphi_3 \to \varphi_2 \to \varphi_1$
<8	4.010731	4.015434	0.297530	$\varphi_2 \to \varphi_1 \to \varphi_3$
<8_opt	23.705870	23.860115	−0.525661	$\varphi_2 \to \varphi_1 \to \varphi_3$
<9	25.655055	25.840633	−1.913830	$\varphi_2 \to \varphi_1 \to \varphi_3$
<9_dim	1.761298	1.933184	−0.091157	$\varphi_2 \to \varphi_1 \to \varphi_3$
<10	36.241697	36.340194	−3.770675	$\varphi_2 \to \varphi_1 \to \varphi_3$
<20	38.722068	38.077265	6.001291	$\varphi_1 \to \varphi_2 \to \varphi_3$
<30	6.523585	6.529640	0.220255	$\varphi_2 \to \varphi_1 \to \varphi_3$
<30_opt	21.553545	21.639367	1.056053	$\varphi_2 \to \varphi_1 \to \varphi_3$
<55	21.408608	21.551922	−1.358177	$\varphi_2 \to \varphi_1 \to \varphi_3$

consist of 13 functions named '<1', ..., '<55' that were designed by simply following the setting given in Subsects. 3.2 and 4.2, three functions named '<4_opt', '<8_opt', and '<30_opt' of which parameters were chosen by assuming the output of the first step (C_0, γ_0) in Subsect. 4.2 are $(10, -1)$, and 1 function named '<9_dim' where machine learning was performed on the diminished features described in Subsect. 3.2. After extracting feature vectors from 2,302 training samples, each feature was scaled to be between 0 and 1.

As baseline, we ran the experiment with randomized order of the subformulas. We repeated it with fourteen different seeds of the random number generator. We will show the average computation time over the fourteen runs as well as the best and the worst cases in the fourteen runs chosen for each problem.

We performed QE for the 2,306 formulas with the 28 prepared ordering functions. Table 5 shows the statistics on the results. The columns "solved", "error", and "timeout" give the number of formulas for which QE computation finished in less than 600 s, for which QE caused error, and for which computation took more than 600 s respectively. The errors occurred in the experiments were due to bugs in Maple and SyNRAC, or computer memory shortage. The columns "average", "average(log)", and "median" state respectively the average execution time of solved QE problems, the average logarithm of execution times of solved QE problems, and the median of execution times where timed out problems assumed to take infinite seconds to stop. The values of "average" and "median" are given in units of seconds. The row 'rand_aver' shows the results averaged over the four-teen runs with random subformula order. 'rand_best' (resp. 'rand_worst') shows the sum of the best (resp. worst) result in the fourteen runs for each problem.

An Overall Argument on Table 5. Here, we especially value the number of solved problems in less than 600 s, because the property if QE computation for a formula finishes in a fixed time or not is critical in practical use rather than whether a computation takes 0.1 or 1 s to finish. From this point of view, the results of 'nvar_rev', 'npoly_rev', 'random', '<1', and '<2' are considerably bad compared to the results of the other orderings. The reason '<1' and '<2' behaved poorly is presumed that the learning algorithm failed to set large weight on the hardest problems in the negative class since there were relatively easy problems mixed in the negative class.

An Effect of Parameter Selection. We also observe that the result '<4' and '<8' are relatively inferior to the orderings of close condition. Table 6 shows the para-meters used to train models. The rows "$\log_2 C$" and "$\log_2 \gamma$" list the logarithm to the base 2 of cost and gamma parameters used respectively, and the row "nSV" lists the number of support vectors. The parameters of '<4' and '<8' lie apart from those of '<3', '<5', '<6', '<7', '<9', '<10', and '<55'. Besides, the numbers of support vectors of '<4' and '<8' are larger than expected from other values. This note suggests that parameter selection affects the performance, and actu-ally '<4_opt' and '<8_opt', of which parameters were similar to '<9' etc., solves more problems in time. We used accuracy as the target of the hyperparameter optimization in this experiment, but it would be effective to replace accuracy with other known measures such as the Matthews correlation coefficient, which is reported to have performed well in the application of SVMs to the problem of the optimal projection order for CAD [6].

Minor Contribution of Omitted Features. Moreover, we see '<9' is just slightly better when compared with '<9_dim'. This means the omitted 15 features did not contribute much. The analysis what feature contributes much is left to further study.

Significance of the Achieved Result. Table 7 illustrates the contrast in QE compu-tation time among 4 heuristics 'ADG', 'nvar', 'npoly', and 'sotd', and 5 ordering functions using machine learning '<1', '<3', '<5', '<7', and '<9'. The column "id" shows the identification numbers of formulas, and each row shows the time consumed to make QE for the corresponding formula. The row "total" lists the number of the formulas for which QE computation finished in 600 s shown in Table 7. The example in Subsect. 4.3 corresponds to the problem with "id" 2,051 in Table 7. The problems which are not shown in Table 7 were either solved in 600 s or timed out regardless of the choice of ordering functions.

Notice it was only less than 40 problems out of 2,306 that the results of "solved" and "timeout" changed by ordering functions. If we consider the results of only a small number of problems changed, it is valuable to have achieved the ordering functions that solved more problems by machine learning such as '<9' compared with the ordering functions by heuristics such as 'ADG' despite the modest difference between them. From another view point of the computation time, '<9' can be evaluated to perform better than 'ADG' again. We can observe in Table 7 that it took relatively long time to perform QE for the problems of which results altered by ordering functions. Hence, solving more problems tends to cause larger average computation time since the averages are taken over solved problems. '<9' compared to 'ADG' recorded the average computation time of the same level, less average of logarithm of the computation time, and smaller median of the computation time. Thus '<9' can be stated to have achieved better performance in overall speed than 'ADG'. The fact that effective ordering functions were obtained by machine learning implies machine learning can save a lot of time because searching for the best application of features by trials and errors consumes long time, where it takes about 40 h to make QE computation for 2,306 problems in a single experiment. On the other hand, we also found that no single ordering function can be regarded as the best. This shows that there is still room for study in sort ordering for efficient QE for non-prenex formulas.

Experiment 2. In order to justify the result of Experiment 1, it is necessary to estimate the effect of the overlap of the training set and the test set. We achieved this by taking another experiment of cross validation. The 2,302 training samples were shuffled and divided into 6 groups, and 6 models were trained using 5 of the 6 groups as training data. For the machine learning, we set the threshold for labeling to be 9 s and parameters $(\log_2 C, \log_2 \gamma)$ to be $(10, -1)$ which were the same parameters for the case '<9'. QE computation with the ordering function using the model obtained was executed for each formula in the remaining group.

Table 8 shows the statistics on the results of the cross validation experiment. The row '<9' is already shown in Table 5, and is printed again for reference.

First, all the 4 problems which were excluded from cross validation because of the error also caused error or timed out with the ordering function '<9'. Therefore, the number of solved problems with 'cross_validation' was smaller by 3 than that of '<9' with the erroneous input excluded. From these results, the effect of ill generalization caused by the common training and test set is estimated to be quite minor in this study.

The purpose of machine learning is to obtain a model that estimates the general characteristics by analyzing given training data. In order to attain this goal, the characteristics that only appear in the training set need to be abstracted. We usually verify that the abstraction is properly made by examining if the model obtained from the training set is also valid on a validation set which contains different data from the training set.

The problem in the overlap of the training set and the test set in this study lies in the possibility that the results in Table 5 was obtained by making use of the characteristics special only to the 2,302 training data in solving the 2,306 QE problems, and the same method is not valid on new QE problems any more.

Table 5. Statistics on computation time for QE with varying ordering functions

	Solved	Error	Timeout	Average (sec)	Average (log)	Median (sec)
ADG	2116	3	187	4.770	−1.038	0.337
nvar	2115	4	187	4.741	−0.993	0.358
npoly	2114	4	188	4.594	−1.134	0.267
sotd	2114	4	188	4.269	−1.217	0.234
msotd	2114	4	188	4.240	−1.212	0.233
mtdeg	2114	4	188	4.172	−1.189	0.262
mdeg	2115	3	188	4.005	−1.230	0.222
mterm	2114	4	188	4.982	−1.150	0.256
nvar_rev	2097	2	207	4.857	−1.129	0.260
npoly_rev	2098	2	206	4.560	−1.119	0.268
rand_aver	2104.93	3.64	197.43	4.764	−1.182	0.241
rand_best	2121	2	183	4.569	−1.389	0.190
rand_worst	2087	2	217	4.531	−0.890	0.356
<1	2099	4	203	5.141	−0.934	0.300
<2	2102	3	201	5.320	−0.995	0.332
<3	2115	3	188	4.673	−0.928	0.360
<4	2112	4	190	4.732	−0.959	0.384
<4_opt	2118	2	186	5.020	−1.085	0.279
<5	2117	3	186	5.153	−0.928	0.392
<6	2115	2	189	4.540	−1.155	0.261
<7	2115	2	189	4.508	−1.178	0.244
<8	2113	3	190	4.806	−1.105	0.267
<8_opt	2117	2	187	4.714	−1.142	0.245
<9	2118	2	186	4.750	−1.165	0.239
<9_dim	2116	4	186	4.718	−0.984	0.363
<10	2117	2	187	4.578	−0.956	0.362
<20	2116	3	187	4.451	−1.200	0.235
<30	2114	2	190	4.886	−1.173	0.245
<30_opt	2115	3	188	4.628	−1.199	0.234
<55	2114	3	189	4.808	−0.993	0.351

If this case happened, the result of 'cross_validation' would be considerably bad, because characteristics special only to 5 groups of the training set should not be valid on the remaining group.

On the other hand, the difference between the number of solved problems with '<9' and 'cross_validation' were 3, and is possible to arise from parameter selection. This is because we can conclude that the performance of the ordering

Table 6. Employed parameters of machine learning and the number of support vectors

	<1	<2	<3	<4	<5	<6	<7	<8	<9	<10	<20	<30	<55
$\log_2 C$	12	15.4	9.0	5.4	9.6	11.6	10.8	3.4	10.0	10.8	15.2	7.2	11.8
$\log_2 \gamma$	−3.6	−4.8	−3.0	0.8	−1.8	−2.2	−3.0	1.0	−1.0	−0.8	−2.2	0.8	−1.4
nSV	538	427	428	462	374	354	390	492	324	297	276	350	274

	<4_opt	<8_opt	<30_opt	<9_dim
$\log_2 C$	10.4	9.2	10.6	13.4
$\log_2 \gamma$	−1.2	−0.8	−0.6	0
nSV	359	325	285	292

Table 7. Timing data (in seconds) for order functions. TO expresses the computation time was greater than 600 s, and ERR expresses the computation caused error.

id	ADG	nvar	npoly	sotd	<1	<3	<5	<7	<9
2051	0.12	0.13	0.12	0.12	0.19	TO	0.16	TO	0.15
1142	0.57	0.61	0.64	0.60	TO	0.63	0.64	0.65	0.63
1944	TO	TO	10.69	16.75	TO	6.92	8.24	17.40	4.61
889	5.01	5.74	5.52	5.56	TO	5.64	5.62	6.00	5.98
272	2.98	3.55	3.26	3.26	ERR	5.32	5.74	6.21	6.76
1024	8.15	11.87	10.20	10.22	TO	10.31	10.28	10.54	10.88
993	TO	9.74	TO	TO	TO	10.44	10.05	10.92	13.45
999	TO	TO	TO	TO	12.54	13.33	13.35	13.30	13.57
1528	22.28	25.46	24.57	23.90	TO	25.06	23.27	23.60	23.24
2167	28.49	114.20	21.46	21.31	TO	48.71	27.14	24.27	27.32
2294	37.91	TO	31.60	31.80	TO	48.95	28.88	27.41	31.93
1088	58.49	TO	TO	TO	TO	74.81	32.61	TO	32.43
1039	31.24	32.11	32.38	31.65	32.27	32.46	32.44	33.46	32.69
1112	40.31	39.73	39.03	40.14	120.55	40.05	39.67	TO	39.40
1234	45.87	49.48	41.81	42.72	14.92	39.83	39.78	15.17	44.14
32	42.21	43.80	292.48	44.66	TO	41.92	44.04	43.76	44.25
1239	39.87	48.73	43.35	44.19	13.44	44.93	40.2	13.79	44.94
2045	67.18	60.06	61.08	60.17	TO	71.15	TO	60.30	TO
1101	61.42	61.16	63.67	61.85	TO	62.39	61.29	60.74	62.39
2168	62.42	62.72	64.16	61.87	TO	60.44	153.39	62.74	74.04
33	69.83	75.44	76.99	76.54	TO	69.87	68.68	76.28	75.79
1040	80.14	87.35	107.34	87.68	TO	TO	89.18	89.93	87.61
28	199.02	208.30	TO	TO	TO	TO	202.83	210.17	209.45
151	239.64	218.08	244.91	246.53	TO	243.18	236.42	243.94	248.93
1125	TO	TO	TO	TO	TO	252.36	252.79	296.94	250.25
1092	266.09	153.79	TO	149.94	323.97	TO	TO	TO	275.69
5	279.42	285.78	295.58	289.15	TO	283.14	277.4	291.48	292.78
17	341.99	293.81	230.59	345.67	272.41	352.92	307.32	242.58	339.28
112	TO	TO	TO	TO	409.60	TO	TO	TO	TO
610	437.12	455.59	456.30	TO	393.91	TO	TO	TO	TO
1091	523.86	528.97	476.57	466.62	TO	475.28	521.76	363.77	472.13
2216	525.31	482.94	485.39	481.33	TO	573.73	566.82	482.50	487.02
150	513.11	592.43	589.52	596.24	586.34	571.58	548.01	586.74	586.56
total	28	27	26	26	11	27	29	27	30

function using machine learning is not expected to be local to the 2,306 QE problems used. The above argument justifies that in Experiment 1, the result of the ordering functions with machine learning are not overestimated and valid.

Table 8. Results of Experiment 2

	Solved	Error	Timeout	Average (sec)	Average (log)	Median (sec)
<9	2118	2	186	4.750	−1.165	0.239
cross_validation	2115	0	187	4.378	−1.109	0.273
group 1	356	0	28	5.509	−1.189	0.217
group 2	356	0	28	5.454	−1.067	0.281
group 3	346	0	38	5.487	−0.937	0.321
group 4	351	0	33	2.637	−1.364	0.214
group 5	350	0	33	4.868	−0.729	0.406
group 6	356	0	27	2.329	−1.318	0.188

5   Conclusion

We showed that ordering of subformulas affects the performance of QE computation for non-prenex formulas. The ordering determined by heuristics performed well compared to random baseline, but the ordering by machine learning performed still better when the training vectors were appropriately labeled. These results suggest that machine learning can save a lot of effort to design heuristics by trials and errors. Meanwhile, we also observed that improper labeling and parameter selection can cause poor performance.

Since our study focused on only one particular dataset of math problems, additional validation on other datasets is necessary for our method to be more convincing. Also, further study should be made to make full use of machine learning extracting valuable patterns in formulas. Our experiment is suggestive to proceed application of machine learning to heuristic portions in computer algebra algorithms.

Acknowledgments. We thank for the funding from Todai robot project.

References

1. Arai, N.H., Matsuzaki, T., Iwane, H., Anai, H.: Mathematics by machine. In: Proceedings of the 39th International Symposium on Symbolic and Algebraic Computation, pp. 1–8. ACM (2014)
2. Brown, C.W., Davenport, J.H.: The complexity of quantifier elimination and cylindrical algebraic decomposition. In: Proceedings of the 2007 International Symposium on Symbolic and Algebraic Computation, ISSAC 2007, pp. 54–60. ACM, New York (2007)
3. Chang, C.C., Lin, C.J.: LIBSVM: a library for support vector machines. ACM Trans. Intell. Syst. Technol. (TIST) **2**(3), 27 (2011)
4. Dolzmann, A., Seidl, A., Sturm, T.: Efficient projection orders for CAD. In: Proceedings of the 2004 International Symposium on Symbolic and Algebraic Computation, pp. 111–118. ACM (2004)

5. Hsu, C.W., Chang, C.C., Lin, C.J., et al.: A practical guide to support vector classification (2003)
6. Huang, Z., England, M., Wilson, D., Davenport, J.H., Paulson, L.C., Bridge, J.: Applying machine learning to the problem of choosing a heuristic to select the variable ordering for cylindrical algebraic decomposition. In: Watt, S.M., Davenport, J.H., Sexton, A.P., Sojka, P., Urban, J. (eds.) CICM 2014. LNCS, vol. 8543, pp. 92–107. Springer, Heidelberg (2014)
7. Iwane, H., Higuchi, H., Anai, H.: An effective implementation of a special quantifier elimination for a sign definite condition by logical formula simplification. In: Gerdt, V.P., Koepf, W., Mayr, E.W., Vorozhtsov, E.V. (eds.) CASC 2013. LNCS, vol. 8136, pp. 194–208. Springer, Heidelberg (2013)
8. Iwane, H., Matsuzaki, T., Arai, N., Anai, H.: Automated natural language geometry math problem solving by real quantifier elimination. In: Proceedings of the 10th International Workshop on Automated Deduction in Geometry, pp. 75–84 (2014)
9. Iwane, H., Yanami, H., Anai, H.: SyNRAC: a toolbox for solving real algebraic constraints. In: Hong, H., Yap, C. (eds.) ICMS 2014. LNCS, vol. 8592, pp. 518–522. Springer, Heidelberg (2014)
10. Matsuzaki, T., Iwane, H., Anai, H., Arai, N.H.: The most uncreative examinee: a first step toward wide coverage natural language math problem solving. In: Twenty-Eighth AAAI Conference on Artificial Intelligence, pp. 1098–1104 (2014)
11. Strzeboński, A.W.: Real quantifier elimination for non-prenex formulas (2011). unpublished manuscript

Solving Extended Ideal Membership Problems in Rings of Convergent Power Series via Gröbner Bases

Katsusuke Nabeshima[1(✉)] and Shinichi Tajima[2]

[1] Institute of Socio-Arts and Sciences, Tokushima University, 1-1 Minamijosanjima, Tokushima, Japan
nabeshima@tokushima-u.ac.jp
[2] Graduate School of Pure and Applied Sciences, University of Tsukuba, 1-1-1 Tennoudai, Tsukuba, Japan
tajima@math.tsukuba.ac.jp

Abstract. An extended ideal membership algorithm is considered in the ring of convergent power series. It is shown that the problem for zero-dimensional ideals in a local ring can be solved in a polynomial ring. The key of the proposed method is the use of ideal quotients in polynomial rings. A new algorithm is given to solve the extended ideal membership problems in local rings. A generalization of the resulting algorithm to ideals with parameters is also described.

Keywords: Gröbner bases · Extended ideal membership problems · Comprehensive Gröbner systems · Parametric syzygy systems

1 Introduction

Ideal membership problem is ubiquitous in many fields of mathematics. Several methods for solving the problems have been introduced, which lay a cornerstone of the foundation of computer algebra. The problem in local rings is also of considerable importance and algorithms for solving the problems that utilize Mora's tangent cone algorithm, standard bases or the normal form algorithm have been introduced [4,7,15,16]. In [19,20,27,29], the authors of the present paper proposed an alternative method to solve the ideal membership problems in local rings, which can be extended to handle parametric cases. However, the study of singularity, for instance, computation of Gauss-Manin connections [12, 22,23], b-functions [2,31], or computation of integral dependence relations [25], often requires solving extended ideal membership problems in local rings.

In this paper we consider, for zero-dimensional ideals, the extended ideal membership problem in the rings of convergent power series and we give a new method for solving them. The key of the proposed methods is the use of ideal quotients in polynomial rings. We show that the use of ideal quotients reduces the problem in local rings to computations in polynomial rings.

© Springer International Publishing Switzerland 2016
I.S. Kotsireas et al. (Eds.): MACIS 2015, LNCS 9582, pp. 252–267, 2016.
DOI: 10.1007/978-3-319-32859-1_22

In order to precisely state the problems and the results, let F be a set of s polynomials $f_1, f_2, \ldots, f_s$ in $\mathbb{C}[x_1, x_2, \ldots, x_n]$ and let $\langle F \rangle$ be the ideal in $\mathbb{C}[x_1, x_2, \ldots, x_n]$ generated by F. Let $\mathbb{V}(F)$ denote the variety in $\mathbb{C}^n$ defined by F:

$$\mathbb{V}(F) = \{x \in \mathbb{C}^n | f_1(x) = f_2(x) = \cdots = f_s(x) = 0\}.$$

Assume that the given set F has the origin $O \in \mathbb{C}^n$ as an isolated zero of $\mathbb{V}(F)$: there exists a neighborhood X of the origin $O \in \mathbb{C}^n$ s.t. $\mathbb{V}(F) \cap X = \{O\}$. Now let $\langle F \rangle_{\{O\}}$ denote the ideal in the ring of convergent power series generated by F and let h be a polynomial in $\mathbb{C}[x_1, x_2, \ldots, x_n]$. Assume, we know in advance, that the polynomial h as an convergent power series is in the ideal $\langle F \rangle_{\{O\}}$. Then, our extended ideal membership problem is to find $q_1, q_2, \ldots, q_s \in \mathbb{C}\{x_1, x_2, \ldots, x_n\}$ such that

$$h = q_1 f_1 + q_2 f_2 + \cdots + q_s f_s.$$

Let us consider an example for illustration. Let $f_1 = 3x^2 + 2y^8$, $f_2 = 16xy^7 + 10y^9$, and let $\langle f_1, f_2 \rangle_{\{O\}}$ be an ideal generated by f_1, f_2 in $\mathbb{C}\{x, y\}$. Then, it is easy to see, by using for instance a standard basis of $\langle f_1, f_2 \rangle_{\{O\}}$, or by using algebraic local cohomology classes w.r.t. $\langle f_1, f_2 \rangle_{\{O\}}$ that $h = xy^9$ belongs to the ideal $\langle f_1, f_2 \rangle_{\{O\}}$. (Incidentally, $xy^9 \notin \langle f_1, f_2 \rangle$ where $\langle f_1, f_2 \rangle$ is the ideal generated by f_1, f_2 in $\mathbb{C}[x, y]$). Indeed, xy^9 can be written as

$$xy^9 = \frac{-8y^7}{15} \left(1 - \frac{128}{75} y^4 + \frac{16384}{5625} y^8 - \cdots\right) f_1 + \frac{15x + 16y^6}{150} \left(1 - \frac{128}{75} y^4 + \frac{16384}{5625} y^8 - \cdots\right) f_2$$

$$= \frac{-8y^7}{15} \left(\sum_{k=0}^{\infty} \left(-\frac{128}{75} y^4\right)^k\right) f_1 + \frac{15x + 16y^6}{150} \left(\sum_{k=0}^{\infty} \left(-\frac{128}{75} y^4\right)^k\right) f_2.$$

As $\sum_{k=0}^{\infty} \left(-\frac{128}{75} y^4\right)^k = \frac{1}{1 + \frac{128}{75} y^4}$, $h = xy^9$ is rewritten as

$$xy^9 = q_1 f_1 + q_2 f_2, \quad \text{where } q_1 = \frac{-40y^7}{75 + 128y^4} \text{ and } q_2 = \frac{15x + 16y^6}{2(75 + 128y^4)}.$$

Algorithms which will be derived in the present paper compute the denominator $75 + 128y^4$ and the numerators.

The proposed method is simple and effective because the problems are easily solved in polynomial rings. This means in particular that, unlike tangent cone algorithms, the resulting algorithm does not invoke Mora's reduction [15,16]. Moreover, the resulting algorithm can be easily generalized, as will be shown in Sect. 3, to treat parametric cases according to its simplicity.

Throughout this paper, we use the notation x as the abbreviation of n variables $x_1, \ldots, x_n$. $\mathbb{C}[x]$ is a polynomial ring and $\mathbb{C}\{x\}$ is a convergent power series ring. The set of natural numbers $\mathbb{N}$ includes zero.

2   Solving Extended Ideal Membership Problems

Let F be a set of s polynomials $f_1, f_2, \ldots, f_s$ in $\mathbb{C}[x]$ s.t. $\{x \in X | f_1(x) = f_2(x) = \cdots = f_s(x) = 0\} = \{O\}$ where X is a neighborhood of the origin O of $\mathbb{C}^n$.

Let $\mathcal{I}_O = \langle f_1, \ldots, f_s \rangle_{\{O\}}$ be an ideal generated by $f_1, f_2, \ldots, f_s$ in $\mathbb{C}\{x\}$ and $I = \langle f_1, \ldots, f_s \rangle$ be an ideal generated by $f_1, f_2, \ldots, f_s$ in $\mathbb{C}[x]$. Note that the ideal membership problem of $\mathcal{I}_O$ in $\mathbb{C}\{x\}$ can be solved by a standard basis of $\mathcal{I}_O$ or a basis of algebraic local cohomology classes w.r.t. F. Assume that a polynomial $h \in \mathbb{C}[x]$ is a member of $\mathcal{I}_O$ in $\mathbb{C}\{x\}$. Then, there exist $q_1, q_2, \ldots, q_s \in \mathbb{C}\{x\}$ such that

$$h = q_1 f_1 + q_2 f_2 + \cdots + q_s f_s.$$

Here, we introduce a new algorithm for computing convergent power series $q_1, q_2, \ldots, q_s$. The idea of the new algorithm is based on the following lemma.

Lemma 1. *Let h be a polynomial in $\mathbb{C}[x]$, s.t. $h \in \mathcal{I}_O = \langle f_1, f_2, \ldots, f_s \rangle_{\{O\}} \subset \mathbb{C}\{x\}$. Then, there exists a polynomial $g \in I : \langle h \rangle$ s.t. $g \notin \mathfrak{m}$, where $I : \langle h \rangle = \{g \in \mathbb{C}[x] \mid gh \in I\}$ is the ideal quotient and $\mathfrak{m} = \langle x_1, x_2, \ldots, x_n \rangle$ is the maximal ideal in $\mathbb{C}\{x\}$.*

Proof. As I has a minimal primary decomposition and $\{x \in X | f_1(x) = \cdots = f_s(x) = 0\} = \{O\}$, I can be written as $I = I_0 \cap I_1 \cap I_2 \cap \cdots \cap I_r$ where $I_0, I_1, \ldots, I_r$ are primary ideals and $\mathbb{V}(I_0) = \{O\}$, $O \notin \mathbb{V}(I_i)$ for each $i \in \{1, \ldots, r\}$. Notice that $\mathcal{I}_O = \mathbb{C}\{x\} \otimes I_0$ where $\otimes$ is a tensor product. Recall that for any polynomial h, we have $\mathbb{V}(I : \langle h \rangle) = \bigcup_{i \in S} \mathbb{V}(I_i)$, where $S = \{i \mid h \notin I_i\}$. Since, $h \in I_0$, we have $\mathbb{V}(I : \langle h \rangle) \subseteq \bigcup_{1 \le i \le r} \mathbb{V}(I_i)$, which immediately implies that there exists a polynomial $g \in \mathbb{C}[x]$ s.t. $gh \in I$ and $g(O) \neq 0$. $\qquad\square$

Using the same notation as in the proof above, suppose that a polynomial h in $\mathbb{C}[x]$ s.t. $h \in \mathcal{I}_O$ is given. Then there exists $g \in I : \langle h \rangle$ with $g(O) \neq 0$. Since $gh \in I$, gh can be written as

$$gh = p_1 f_1 + p_2 f_2 + \cdots + p_s f_s$$

where $p_1, p_2, \ldots, p_r \in \mathbb{C}[x]$. The condition $g(O) \neq 0$ implies that $q_i = \frac{p_i}{g}$, is well-defined as an element of $\mathbb{C}\{x\}$ for each $i = 1, 2, \ldots, s$. That is, if we have polynomials $g, p_1, p_2, \ldots, p_s$, then we are able to solve the extended ideal membership problem of h as follows

$$h = \frac{p_1}{g} f_1 + \frac{p_2}{g} f_2 + \cdots + \frac{p_s}{g} f_s.$$

In fact, the denominator g can be obtained by using an algorithm for computing ideal quotients and polynomials $p_1, p_2, \ldots, p_s$ can also be obtained in $\mathbb{C}[x]$ by utilizing a Gröbner basis of I and the extended Gröbner basis algorithm (Chap. 5 [1]). The consideration above yields the following simple algorithm for solving the extended ideal membership problems in rings of convergent power series. In the algorithm, a monomial order is fix.

Algorithms for computing bases of ideal quotients (Chap. 4 of [3]) and extended Gröbner bases (Chap. 5 [1]) use Gröbner bases computation in $\mathbb{C}[x]$. Hence, the algorithm **ExtIMP** clearly terminates. The correctness of the algorithm follows from the discussion given after Lemma 1.

We illustrate the algorithm with the following example.

Algorithm 1. ExtIMP

Input: $f_1, f_2, \ldots, f_s$: polynomials in $\mathbb{C}[x]$ satisfying $\{x \in X | f_1(x) = f_2(x) = \cdots$
$= f_s(x) = 0\} = \{O\}$.
      h: a polynomial in $\mathbb{C}[x]$ satisfying $h \in \langle f_1, \ldots, f_s \rangle_{\{O\}} \subset \mathbb{C}\{x\}$.
Output: $q_1, \ldots, q_s$: convergent power series satisfying $h = q_1 f_1 + \cdots + q_s f_s$.
BEGIN

1. $Q \leftarrow$ Compute a basis of the ideal quotient $\langle f_1, \ldots, f_s \rangle : \langle h \rangle$ in $\mathbb{C}[x]$.
2. $g \leftarrow$ Take a polynomial g from Q satisfying $g(O) \neq 0$.
3. $p_1, \ldots, p_s \leftarrow$ Compute $p_1, \ldots, p_s \in \mathbb{C}[x]$ satisfying $gh = p_1 f_1 + p_2 f_2 + \cdots + p_s f_s$, by a Gröbner basis of $\langle f_1, f_2, \ldots, f_s \rangle$ and the extended Gröbner basis algorithm (Chap. 5 [1]) in $\mathbb{C}[x]$.
4. For each $i \in \{1, \ldots, s\}$, set $q_i = \dfrac{p_i}{g}$.

END

Example 1. Let us consider a polynomial $f = x_1^3 + 2x_1 x_2^8 + x_2^{10}$ that defines an isolated singularity at the origin O. Then, $x_1 x_2^9$ is a member of the ideal $\mathcal{I}_O = \langle \frac{\partial f}{\partial x_1}, \frac{\partial f}{\partial x_2} \rangle_{\{O\}}$ in $\mathbb{C}\{x_1, x_2\}$. Incidentally, $x_1 x_2^9 \notin I = \langle \frac{\partial f}{\partial x_1}, \frac{\partial f}{\partial x_2} \rangle$ in $\mathbb{C}[x_1, x_2]$.
    Let us execute the algorithm.

1. The algorithm for computing bases of ideal quotients gives $\{75 + 128x_2^4, 8x_1 + 5x_2^2\}$ as a basis of the ideal quotient $\langle \frac{\partial f}{\partial x_1}, \frac{\partial f}{\partial x_2} \rangle : \langle x_1 x_2^9 \rangle$.
2. Take $75 + 128x_2^4$ from the basis and set $g = 75 + 128x_2^4$ because $g(O) \neq 0$. Note that $(75 + 128x_2^4)x_1 x_2^9 \in I$ in $\mathbb{C}[x_1, x_2]$.
3. The extended Gröbner basis algorithm shows that the polynomial $(75 + 128x_2^4)(x_1 x_2^9)$ can be written as

$$(75 + 128x_2^4)(x_1 x_2^9) = 40x_2^7 \frac{\partial f}{\partial x_1} - \frac{1}{2}(15x_1 + 16x_2^6) \frac{\partial f}{\partial x_2}.$$

4. Therefore,

$$x_1 x_2^9 = \frac{-40x_2^7}{75 + 128x_2^4} \cdot \frac{\partial f}{\partial x_1} + \frac{(15x_1 + 16x_2^6)}{2(75 + 128x_2^4)} \cdot \frac{\partial f}{\partial x_2}$$

in $\mathbb{C}\{x_1, x_2\}$.   □

If one is familiar with computing syzygies, the third step of the algorithm can be carried out by computing syzygies.

Lemma 2. *The third step of the algorithm* **ExtIMP** *can be changed as follow.*

3-1. Syz ← Compute a Gröbner basis of a syzygy module of $\langle gh, f_1, \ldots, f_s \rangle$
     w.r.t a POT monomial order in $(\mathbb{C}[x])^{s+1}$.

3-2. $(c_0, c_1, \ldots, c_s)$ ← Take an element $(c_0, c_1, \ldots, c_s)$ from Syz whose first component is a nonzero constant where $c_0 gh + c_1 f_1 + \cdots + c_s f_s = 0$.

3-3. For each $i \in \{1, \ldots, s\}$, set $p_i = -\dfrac{c_i}{c_0}$.

The definition of POT *(position-over-term) monomial orders are from [4, 7].*

Proof. Since $gh \in \langle f_1, f_2, \ldots, f_s \rangle \subset \mathbb{C}[x]$, there exist $p_1, p_2, \ldots, p_s \in \mathbb{C}[x]$ s.t. $gh = p_1 f_1 + p_2 f_2 + \cdots + p_s f_s$. Let $(d_1, d_2, \ldots, d_s)$ be a syzygy of $f_1, f_2, \ldots, f_s$, i.e., $d_1 f_1 + d_2 f_2 + \cdots + d_s f_s = 0$. Then,

$$(gh - (p_1 f_1 + p_2 f_2 + \cdots + p_s f_s)) + (d_1 f_1 + d_2 f_2 + \cdots + d_s f_s) = 0,$$

i.e., $$gh + (d_1 - p_1)f_1 + (d_2 - p_2)f_2 + \cdots + (d_s - p_s)f_s = 0.$$

Hence, $(1, d_1 - p_1, d_2 - p_2, \ldots, d_s - p_s)$ is a syzygy of $gh, f_1, f_2, \ldots, f_s$. As Syz is a Gröbner basis of a syzygy module of $\langle gh, f_1, \ldots, f_s \rangle$ w.r.t a POT monomial order in $(\mathbb{C}[x])^{s+1}$ and $(1, d_1 - p_1, d_2 - p_2, \ldots, d_s - p_s) \in \langle \mathrm{Syz} \rangle$, there exists an element $(c_0, c_1, \ldots, c_s) \in \mathrm{Syz}$ such that the first component is a nonzero constant. $\square$

Example 2. Let us consider N_{25} singularity defined by $f = x_1^4 x_2 + x_2^8 + x_1^2 x_2^5$. Then, $h = x_1^7$ is a member of the ideal $\mathcal{I}_O = \langle \frac{\partial f}{\partial x_1}, \frac{\partial f}{\partial x_2} \rangle_{\{O\}}$ in $\mathbb{C}\{x_1, x_2\}$. Let $\succ_{lex}$ be the lexicographic monomial order such that $x_1 \succ_{lex} x_2$ in $\mathbb{C}[x_1, x_2]$ and $\succ_{POT}$ be a POT monomial order with $\succ_{lex}$ in $(\mathbb{C}[x_1, x_2])^3$.

By the algorithm for computing bases of ideal quotients, one has $\{524288 + 6561 x_1^2, 32 - 9x_2\}$ as a basis of the ideal quotient $\langle \frac{\partial f}{\partial x_1}, \frac{\partial f}{\partial x_2} \rangle : \langle x_1^7 \rangle$. Set $g = 32 - 9x_2$ and compute a Gröbner basis of the syzygy module of $\langle (32 - 9x_2) x_1^7, \frac{\partial f}{\partial x_1}, \frac{\partial f}{\partial x_2} \rangle$ w.r.t. $\succ_{POT}$ in $(\mathbb{C}[x])^3$. Then, the Gröbner basis is

$$\left\{ \left(2, 128 x_2^6 + 88 x_1^2 x_2^3 + 5 x_1^4, -32 x_1 x_2^4 - 64 x_1^3 - 2 x_1^3 x_2 \right), \right.$$
$$\left. \left(2, 128 x_2^6 - 40 x_2^7 + 88 x_1^2 x_2^3 - 25 x_1^2 x_2^4, -32 x_1 x_2^4 + 10 x_1 x_2^5 - 64 x_1^3 + 18 x_1^3 x_2 \right) \right\}.$$

If we select the first element of the Gröbner basis, then x_1^7 can be written as

$$x_1^7 = \frac{128 x_2^6 + 88 x_1^2 x_2^3 + 5 x_1^4}{-2(32 - 9x_2)} \cdot \frac{\partial f}{\partial x_1} + \frac{-32 x_1 x_2^4 - 64 x_1^3 - 2 x_1^3 x_2}{-2(32 - 9x_2)} \cdot \frac{\partial f}{\partial x_2}$$

in $\mathbb{C}\{x_1, x_2\}$. $\square$

The Fig. 1 represents an outline of the algorithm **ExtIMP**.

Assume that $h \in \mathcal{I}_O = \langle f_1, \ldots, f_s \rangle_{\{O\}}$ and $\{g_1, \ldots, g_k\}$ is a standard basis of $\mathcal{I}_O$. Then, there exist $g \in \mathbb{C}[x] \cap \mathbb{C}\{x\}$, $p_1, \ldots, p_k \in \mathbb{C}[x]$ such that

$$gh = p_1 g_1 + p_2 g_2 + \cdots + p_k g_k.$$

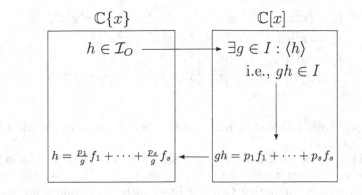

$$\mathbb{C}\{x\} \qquad\qquad \mathbb{C}[x]$$

Fig. 1. Outline of **ExtIMP**

In principle, the Mora's tangent cone algorithms [15, 16] provide g and $p_1, \dots, p_k$. Moreover, it is also possible to express g_i as a linear combination of given generators $f_1, \dots, f_s$ [7]. Thus, the extended ideal membership problems can be solved in the local ring by a standard basis and the Mora's tangent cone algorithms. However, the algorithm **ExtIMP** which is completely different from existing algorithms does not invoke the Mora's tangent cone algorithms. This means that the problems of rings of convergent power series can be solved in polynomial rings by the resulting algorithm. In this regard, the proposed method is easy to understand and implement. The algorithm **ExtIMP** is the main result of this paper.

The algorithm **ExtIMP** has been implemented in the computer algebra system SINGULAR [5]. Here we give the results of benchmark tests. Table 1 shows a comparison of the implementation of **ExtIMP**[1] with SINGULAR's command syz[2,3] in computation time (CPU time). x, y, z are variables. We used a PC [OS: Windows 7 (64bit), CPU: Intel(R) Core i-7-2600 CPU @ 3.40 GHz 3.40 GHz, RAM: 8 GB] and SINGULAR version 3-1-4. The time is given in second. In Table 1, $>20\,$m means it takes more than 20 min.

We use the following polynomials define an isolated singularity.

$$f_1 = (y^4 + xz^3 + x^3)^2 + y^8 + z^9 + 8xy^7,$$
$$f_2 = (y^{13} + x^3)^2 + 3y^{14} - 2x^3y^{20},$$
$$f_3 = (x^3 + xz^2 + xy^3 + zy^3)^3 + xz^8 + 4xy^{12},$$
$$f_4 = (x^5 + y^7)^2 + 3y^{14} + 3x^{10}y^5 - 2xy^{14},$$

---

[1] The degree reverse lex. monomial order with the coordinate (x, y) or (x, y, z), is used in the implementation of **ExtIMP**.

[2] $\mathrm{syz}(g_1, g_2, \dots, g_r)$ outputs a standard basis of the module of syzygies w.r.t. the generators $g_1, g_2, \dots, g_r$ where $g_1, g_2, \dots, g_r \in \mathbb{Q}[x]$. Thus, the command syz outputs the similar results. For each $i \in \{1, \dots, 8\}$, $\mathrm{syz}(h, \frac{\partial f_i}{\partial x}, \frac{\partial f_i}{\partial y}, \frac{\partial f_i}{\partial z})$ (or $\mathrm{syz}(h, \frac{\partial f_i}{\partial x}, \frac{\partial f_i}{\partial y})$) has been executed in Table 1.

[3] The negative degree reverse lex. monomial order with the coordinate (x, y) or (x, y, z), is used in SINGULAR's command syz.

$$f_5 = (x^2z + yz^2 + y^5 + y^3z)^2 + z^5 + x^6y + x^3y^2z^2,$$
$$f_6 = (x^2y + z^4 + y^5)^2 + x^5 + y^5z^4 + x^2y^3z^3,$$
$$f_7 = (y^4 + xz^3 + x^3)^2 + x^5 + y^5z^4,$$
$$f_8 = (y^4 + xz^3 + x^3)^2 + x^5 + y^5z^4 + x^2y^3z^3.$$

In each problem of Table 1, a polynomial h is a member of the ideal $\langle F \rangle$ in $\mathbb{C}\{x,y\}$ or $\mathbb{C}\{x,y,z\}$.

As is evident from the Table 1, the new algorithm **ExtIMP** results in better performance in contrast to SINGULAR's command **syz**. The essential point of the new algorithm is computing Gröbner bases instead of computing standard bases. In general, the computational complexity of Gröbner bases is smaller than the computational complexity of standard bases. That's why the new algorithm results in better performance.

3   Parametric Cases

Here, we generalize the algorithm **ExtIMP** to parametric cases, namely, we consider parametric polynomial ideals.

First, we briefly review the notion of algebraic local cohomology and the method to solve parametric ideal membership problems that exploits algebraic local cohomology classes. Second, by using comprehensive Gröbner systems, we present an algorithm for computing bases of parametric ideal quotients. Last, we generalize **ExtIMP** to handle parametric cases.

We use in this section the following notation for systems of parametric ideals. For $g_1, \ldots, g_r \in \mathbb{C}[t]$, $\mathbb{V}(g_1, \ldots, g_r) \subseteq \mathbb{C}^m$ denotes the affine variety of $g_1, \ldots, g_r$, i.e., $\mathbb{V}(g_1, \ldots, g_r) := \{t \in \mathbb{C}^m \mid g_1(t) = \cdots = g_r(t) = 0\}$ and $\mathbb{V}(0) := \mathbb{C}^m$ where $t = (t_1, t_2, \cdots, t_m)$ is the abbreviation of m variables which represents

Table 1. Comparison of **ExtIMP** with syz

	F	h	syz	**ExtIMP**
1	$\frac{\partial f_1}{\partial x}, \frac{\partial f_1}{\partial y}, \frac{\partial f_1}{\partial z}$	z^{20}	0.040	0.040
2	$\frac{\partial f_2}{\partial x}, \frac{\partial f_2}{\partial y}$	x^{16}	0.460	0.030
3	$\frac{\partial f_3}{\partial x}, \frac{\partial f_3}{\partial y}, \frac{\partial f_3}{\partial z}$	$x^{100} + y^{100}$	0.220	0.180
4	$\frac{\partial f_1}{\partial x}, \frac{\partial f_1}{\partial y}, \frac{\partial f_1}{\partial z}$	$5x^{10}y^{30} + x^3z^{20}$	88.510	1.090
5	$\frac{\partial f_2}{\partial x}, \frac{\partial f_2}{\partial y}$	$x^{15} + y^{30}$	>20 m	0.060
6	$\frac{\partial f_4}{\partial x}, \frac{\partial f_4}{\partial y}$	$2x^{15} - xy^{20}$	>20 m	1.420
7	$\frac{\partial f_5}{\partial x}, \frac{\partial f_5}{\partial y}, \frac{\partial f_5}{\partial z}$	$x^{15} + x^{16} + xy^{10}z^{14}$	>20 m	12.390
8	$\frac{\partial f_6}{\partial x}, \frac{\partial f_6}{\partial y}, \frac{\partial f_6}{\partial z}$	$2x^{12}y^{16}z^2 - 3z^{20}$	>20 m	19.090
9	$\frac{\partial f_7}{\partial x}, \frac{\partial f_7}{\partial y}, \frac{\partial f_7}{\partial z}$	$x^{30}y^{20}$	>20 m	124.175
10	$\frac{\partial f_8}{\partial x}, \frac{\partial f_8}{\partial y}, \frac{\partial f_8}{\partial z}$	$x^{20} + y^{20}$	>20 m	398.990

parameters. We call an algebraically constructible set of the form $\mathbb{V}(g_1, \ldots, g_r) \setminus \mathbb{V}(g'_1, \ldots, g'_{r'}) \subseteq \mathbb{C}^m$ with $g_1, \ldots, g_r, g'_1, \ldots, g'_{r'} \in \mathbb{C}[t]$, a *stratum*. For a stratum $\mathbb{A} \subseteq \mathbb{C}^m$, we define $\mathbb{C}[t]_\mathbb{A} = \{\frac{c}{b} \mid c, b \in \mathbb{C}[t], b(t) \neq 0 \text{ for } t \in \mathbb{A}\}$. Then for every $a \in \mathbb{A}$, the specialization homomorphism $\sigma_a : \mathbb{C}[t]_\mathbb{A}[x] \to \mathbb{C}[x]$ (or $\sigma_a : \mathbb{C}[t]_\mathbb{A}[\xi] \to \mathbb{C}[\xi]$) is defined as a map that substitutes a into m variables t [10]. When we say that $\sigma_a(h)$ makes sense for $h \in \mathbb{C}(t)[x]$, it has to be understood that $h \in \mathbb{C}[t]_\mathbb{A}[x]$ for some $\mathbb{A}$ with $a \in \mathbb{A}$ and for $F \subset \mathbb{C}[t]_\mathbb{A}[x]$, $\sigma_a(F) = \{\sigma_a(h) | h \in F\}$.

3.1   Algebraic Local Cohomology and Membership Problems

Here, we briefly review algebraic local cohomology classes and how to solve ideal membership problems by utilizing the algebraic local cohomology classes. The details are in [8, 9, 19, 26–29].

Let $H^n_{[O]}(\mathbb{C}[x])$ denote the set of algebraic local cohomology classes supported at the origin O with coefficients in $\mathbb{C}$, defined by

$$H^n_{[O]}(\mathbb{C}[x]) := \lim_{k \to \infty} \mathrm{Ext}^n_{\mathbb{C}[x]}(\mathbb{C}[x]/\langle x_1, x_2, \ldots, x_n \rangle^k, \mathbb{C}[x]).$$

We represent an algebraic local cohomology class, given by finite sum of the form $\sum c_\lambda \left[\frac{1}{x^{\lambda+1}}\right]$, as a polynomial in n variables $\sum c_\lambda \xi^\lambda$, which is called "polynomial representation", where $c_\lambda \in \mathbb{C}$, $\lambda \in \mathbb{N}^n$ and $\xi = (\xi_1, \ldots, \xi_n)$. The multiplication by x^α for polynomial representation is defined as

$$x^\alpha * \xi^\lambda := \begin{cases} \xi^{\lambda-\alpha}, & \lambda_i \geq \alpha_i, i = 1, \ldots, n, \\ \\ 0, & \text{otherwise}, \end{cases}$$

where $\alpha = (\alpha_1, \ldots, \alpha_n) \in \mathbb{N}^n$, $\lambda = (\lambda_1, \ldots, \lambda_n) \in \mathbb{N}^n$, and $\lambda - \alpha = (\lambda_1 - \alpha_1, \ldots, \lambda_n - \alpha_n)$. We use "$*$" as the multiplication.

Let fix a monomial order $\succ$ on $\mathbb{C}[\xi]$. For a given algebraic local cohomology class of the form

$$\psi = c_\lambda \xi^\lambda + \sum_{\xi^\lambda \succ \xi^{\lambda'}} c_{\lambda'} \xi^{\lambda'}, \quad c_\lambda \neq 0,$$

we call ξ^λ the *head term*, c_λ the *head coefficient*, $c_\lambda \xi^\lambda$ the *head monomial* and $\xi^{\lambda'}$ the *lower terms*. We denote the head term by $\mathrm{ht}(\psi)$, head coefficient by $\mathrm{hc}(\psi)$, head monomial by $\mathrm{hm}(\psi)$. Furthermore, we also denote the set of terms of ψ as $\mathrm{Term}(\psi) := \{\xi^\kappa | \psi = \sum_{\kappa \in \mathbb{N}^n} c_\kappa \xi^\kappa, c_\kappa \neq 0, c_\kappa \in \mathbb{C}\}$ and the set of lower terms of ψ as $\mathrm{LL}(\psi) := \{\xi^\kappa \in \mathrm{Term}(\psi) | \xi^\kappa \neq \mathrm{ht}(\psi)\}$. Let Ψ be a finite subset of $H^n_{[O]}(\mathbb{C}[x])$. We denote the set of head terms of Ψ as $\mathrm{ht}(\Psi) := \{\mathrm{ht}(\psi) | \psi \in \Psi\}$, the set of terms of Ψ as $\mathrm{Term}(\Psi) := \bigcup_{\psi \in \Psi} \mathrm{Term}(\psi)$ and the set of lower terms of Ψ as $\mathrm{LL}(\Psi) := \bigcup_{\psi \in \Psi} \mathrm{LL}(\psi)$.

Let F be a set of s polynomials $f_1, f_2, \ldots, f_s$ in $\mathbb{C}[x]$ s.t. $\{x \in X | f_1(x) = f_2(x) = \cdots = f_s(x) = 0\} = \{O\}$. We define a set H_F to be the set of algebraic local cohomology classes in $H^n_{[O]}(\mathbb{C}[x])$ that are annihilated by the ideal generated by F:

$$H_F = \{\psi \in H^n_{[O]}(\mathbb{C}[x]) \mid f_1 * \psi = f_2 * \psi = \cdots = f_s * \psi = 0\}.$$

In our works [20, 27–29], algorithms for computing a basis of the vector space H_F, have been introduced. Bases of the finite-dimensional vector spaces H_F can be computed by the algorithms implemented in a computer algebra system Risa/Asir [21].

For a given monomial in $\mathbb{C}[x]$, the next theorem tells us the normal form of the monomial modulo $\mathcal{I}_O$ w.r.t. a monomial order $\succ$ in $\mathbb{C}\{x\}$.

Theorem 1 ([27]). *Let $\succ$ be a global monomial order and Ψ be a basis of the vector space H_F. Suppose that an element of linear combination in Ψ is represented*

$$\xi^\tau + \sum_{\xi^\tau \succ \xi^\kappa} c_{(\tau,\kappa)}\xi^\kappa, \quad c_{(\tau,\kappa)} \in \mathbb{C}.$$

Then, the following properties hold.

1. *If $\xi^\lambda \in \mathrm{LL}(\Psi)x^\lambda \equiv \sum_{\xi^\kappa \in \mathrm{ht}(\Psi)} c_{(\kappa,\lambda)}x^\kappa \mod \mathcal{I}_O$ in $\mathbb{C}\{x\}$*
2. *If $\xi^\lambda \in \mathrm{ht}(\Psi)x^\lambda \equiv x^\lambda \mod \mathcal{I}_O$ in $\mathbb{C}\{x\}$*
3. *If $\xi^\lambda \notin \mathrm{LL}(\Psi)$ and $\xi^\lambda \notin \mathrm{ht}(\Psi)x^\lambda \equiv 0 \mod \mathcal{I}_O$*

As any polynomial h is a linear combination of finite number of monomials, the normal form of h modulo $\mathcal{I}_O$ w.r.t. a monomial order is computed by the theorem. If $h \equiv 0 \mod \mathcal{I}_O$, then h is a member of the ideal $\mathcal{I}_O$ in $\mathbb{C}\{x\}$.

We turn to the parametric cases. Let us assume that a set F of s polynomials $f_1, f_2, \ldots, f_s$ in $(\mathbb{C}[t])[x]$ satisfying **generically** $\{x \in X | f_1(x) = \cdots = f_s(x) = 0\} = \{O\}$ are given. Here, $t = (t_1, t_2, \ldots, t_m)$ are regarded as parameters and x, ξ are the main variables.

We define a set $H_F = \cup_{a \in \mathbb{C}^m} H_{\sigma_a(F)}$ to be the set of algebraic local cohomology classes in $H^n_{[O]}(\mathbb{C}[x])$ that are annihilated by $\langle F \rangle$, where

$$H_{\sigma_a(F)} = \{\psi \in H^n_{[O]}(\mathbb{C}[x]) \mid \sigma_a(f_1) * \psi = \sigma_a(f_2) * \psi = \cdots = \sigma_a(f_s) * \psi = 0\}.$$

The ideal $\langle F \rangle$ at $a \in \mathbb{C}^m$ is a zero-dimensional ideal if and only if $H_{\sigma_a(F)}$ is a finite-dimensional vector space.

Definition 1. *Let $\mathbb{A}_i, \mathbb{B}_j$ be strata in $\mathbb{C}^m$ and S_i a subset of $(\mathbb{C}[t]_{\mathbb{A}_i})[\xi]$ where $1 \le i \le \ell$ and $1 \le j \le k$. Set $\mathcal{S} = \{(\mathbb{A}_1, S_1), \ldots, (\mathbb{A}_\ell, S_\ell)\}$ and $\mathcal{D} = \{\mathbb{B}_1, \ldots, \mathbb{B}_k\}$. Then, a pair $(\mathcal{S}, \mathcal{D})$ is called a **parametric local cohomology system** of H_F on $\mathbb{A}_1 \cup \cdots \cup \mathbb{A}_\ell \cup \mathbb{B}_1 \cup \cdots \cup \mathbb{B}_k$, if for all $i \in \{1, \ldots, \ell\}$ and $a \in \mathbb{A}_i$, $\sigma_a(S_i)$ is a basis of the vector space $H_{\sigma_a(F)}$, and for all $j \in \{1, \ldots, k\}$ and $b \in \mathbb{B}_j$, $\{x \in X | \sigma_b(f_1)(x) = \sigma_b(f_2)(x) = \cdots = \sigma_b(f_s)(x) = 0\}$ is not zero-dimensional for any sufficiently small neighborhood X of O.*

In our works [19, 20], algorithms for computing a parametric local cohomology system of H_F, have been introduced and implemented in the computer algebra system Risa/Asir [21]. Hence, parametric ideal membership problems can be solved by a parametric local cohomology system of H_F and Theorem 1.

Example 3. Let $f = x_1^3 + t_1 x_1^2 x_2^2 + x_1 x_2$ and $F = \{\frac{\partial f}{\partial x_1}, \frac{\partial f}{\partial x_2}\}$ where x_1, x_2 are variables and t_1 is a parameter. Let $\succ$ be the total degree lexicographic monomial order s.t $x \succ y$. Then, our implementation outputs the following as a parametric local cohomology system of H_F. The variables ξ_1 and ξ_2 correspond to the variables x_1 and x_2, respectively.

- If a parameter t_1 belongs to $\mathbb{V}(t_1)$, then $S_1 = \{1, \xi_1, \xi_2, \xi_1\xi_2, \xi_2^2, \xi_2^3,$
  $\xi_1\xi_2^2, \xi_2^4, \xi_1\xi_2^3, \xi_2^5 - \frac{1}{3}\xi_1^2, \xi_2^6 - \frac{1}{3}\xi_1^2\xi_2, \xi^7 - \frac{1}{3}\xi_1^2\xi_2^2, \xi_2^8 - \frac{1}{3}\xi_1^2\xi_2^3\}$ is a basis of H_F.
- If a parameter t_1 belongs to $\mathbb{C}\backslash\mathbb{V}(t_1)$, then $S_2 = \{1, \xi_1, \xi_2, \xi_1\xi_2, \xi_2^2, \xi_2^3, \xi_2^4, \xi_1\xi_2^2 -$
  $\frac{2}{3}t_1\xi_1^2, \xi_2^5 - \frac{1}{3}\xi_1^2, \xi_2^6 - \frac{1}{2t_1}\xi_1\xi_2^3, \xi_2^7 - \frac{1}{2t_1}\xi_1\xi_2^4 - \frac{15}{8t_1^3}\xi_1\xi_2^3 + \frac{5}{4t_1^2}\xi_1^2\xi_2, \xi_2^8 - \frac{1}{2t_1}\xi_1\xi_2^5 -$
  $\frac{15}{8t_1^3}\xi_1\xi_2^3\frac{5}{4t_1^2}\xi_1^2\xi_2^2 - \frac{225}{32t_1^5}\xi_1\xi_2^3 - \frac{2}{3t_1}\xi_1^3 + \frac{75}{16t_1^4}\xi_1^2\xi_2\}$ is a basis of H_F.

Let us consider the ideal membership problems of $h = x_1^2 x_2^3$. If $t_1 = 0$, then it follows from $\xi_1^2\xi_2^3 \in \mathrm{LL}(S_1)$ and $x_1^2 x_2^3 \equiv -\frac{1}{3}\xi_1^8 \bmod \mathcal{I}_O$ that $x_1^2 x_2^3 \notin \mathcal{I}_O$ holds. In contrast, if $t_1 \neq 0$, then since $\xi_1^2\xi_2^3 \notin \mathrm{LL}(S_2)$ and $\xi_1^2\xi_2^3 \notin \mathrm{ht}(S_2)$, $x_1^2 x_2^3 \equiv 0 \bmod \mathcal{I}_O$, i.e. $x_1^2 x_2^3 \in \mathcal{I}_O$ holds.

Let us consider the ideal membership problem of $h = x_1^3 x_2$. Then, since $\xi_1^3\xi_2 \notin \mathrm{LL}(S_1)$, $\xi_1^3\xi_2 \notin \mathrm{ht}(S_1)$, $\xi_1^3\xi_2 \notin \mathrm{LL}(S_2)$ and $\xi_1^3\xi_2 \notin \mathrm{ht}(S_2)$, $x_1^3 x_2 \in \mathcal{I}_O$ on $\mathbb{C}$.

3.2 Comprehensive Gröbner Systems (CGS)

It is known that the notion of comprehensive Gröbner system is useful and indispensable for studying parametric ideals. Here, first, we quickly review a comprehensive Gröbner system of a parametric ideal. Second, we introduce an algorithm for computing bases of ideal quotients with parameters. Last, we give a notion of parametric syzygy systems.

Comprehensive Gröbner systems for parametric ideals were introduced, constructed, and studied by Weispfenning [30] in 1992. Since then, the algorithm for computing comprehensive Gröbner systems has been improved by several authors [11,13,14,18,24]. Now, there exist several implementations [6,11,14,18] for computing comprehensive Gröbner systems.

Definition 2 (CGS). *Let fix a monomial order. Let F be a subset of $(\mathbb{C}[t])[x]$, $\mathbb{A}_1, \ldots, \mathbb{A}_\ell$ strata in $\mathbb{C}^m$ and $G_1, \ldots, G_\ell$ subsets of $(\mathbb{C}[t])[x]$. A finite set $\mathcal{G} = \{(\mathbb{A}_1, G_1), \ldots, (\mathbb{A}_\ell, G_\ell)\}$ of pairs is called a **comprehensive Gröbner system** (**CGS**) on $\mathbb{A}_1 \cup \cdots \cup \mathbb{A}_\ell$ for F if $\sigma_a(G_i)$, $a \in \mathbb{A}_i$, is a Gröbner basis of the ideal $\langle\sigma_a(F)\rangle$ in $\mathbb{C}[x]$ for each $i = 1, \ldots, \ell$. Each $(\mathbb{A}_i, G_i)$ is called a **segment** of $\mathcal{G}$. We simply say $\mathcal{G}$ is a comprehensive Gröbner system for F if $\mathbb{A}_1 \cup \cdots \cup \mathbb{A}_\ell = \mathbb{C}^m$.*

The algorithm (Chap. 4 [3]) for computing bases of ideal quotients can be generalized to parametric cases. In the algorithm, a monomial order is fix.

Algorithm 2. ParaQuotient

Input: $F = \{f_1, \ldots, f_s\} \subset (\mathbb{C}[t])[x]$, $g \in (\mathbb{C}[t])[x]$. $\mathbb{A} \subset \mathbb{C}^m$.
Output: $\mathcal{G}$: a CGS of the ideal quotient $\langle F \rangle : \langle g \rangle$ on $\mathbb{A}$.
BEGIN
$\mathcal{G} \leftarrow \emptyset$;
$\mathcal{H} \leftarrow$ Compute a CGS of $\langle F \rangle \cap \langle g \rangle$ on $\mathbb{A}$;
while $\mathcal{H} \neq \emptyset$ **do**
  Select a segment $(\mathbb{A}', H')$ from $\mathcal{H}$; $\mathcal{H} \leftarrow \mathcal{H} \backslash \{(\mathbb{A}', H')\}$;
  $G' \leftarrow \{h/g \,|\, h \in H'\}$; $\mathcal{D} \leftarrow \mathcal{D} \cup \{(\mathbb{A}', G')\}$;
end-while
return $\mathcal{G}$;
END

Algorithms [11,13,14,18,24] for computing comprehensive Gröbner bases terminate. Thus, this algorithm terminates, too. As this is the natural extension to parametric cases by using CGSs, the correctness is clear.

Example 4. Let $f = x_1^3 + t_1 x_1^2 x_2^2 + x_1 x_2$ and $F = \{\frac{\partial f}{\partial x_1}, \frac{\partial f}{\partial x_2}\}$ where x_1, x_2 are variables and t_1 is a parameter. Let $\succ$ be the lexicographic monomial order s.t. $z \succ x_1 \succ x_2$ where z is an auxiliary variable. From Example 3, $x_1^3 x_2$ is a member of $\langle F \rangle_{\{0\}}$ on the parameter space $\mathbb{C}$.

In order to compute a CGS of $\langle F \rangle : \langle x_1^3 x_2 \rangle$, first we need to compute a basis of $\langle F \rangle \cap \langle g_i \rangle$ on $\mathbb{C}$. Since our implementation outputs the following

$$\{(\mathbb{V}(t_1), \{x_1^3 x_2, x_2^9 z, x_1 x_2^4 z, 3x_1^2 z + x_2^5 z\}), \ (\mathbb{C} \backslash \mathbb{V}(t_1), \{(75 x_2^2$$
$$- 16 t_1^2 x_2) x_1^3, -84375 x_1^4 x_2 - 2048 t_1^5 x_1^3 x_2, -3 t_1 x_1^3 x_2 - 10 x_2^9 z, 10125 x_1^3 x_2$$
$$- 256 t_1^6 x_1 x_2^3 z - 1800 t_1 x_2^8 z - 480 t_1^3 x_2^7 z - 128 t_1^5 x_2^6 z, 3x_1^2 z + 2 t_1 x_1 x_2^2 z + x_2^5 z\})\}$$

as the CGS of $\langle z \frac{\partial f}{\partial x_1}, z \frac{\partial}{\partial x_2}, (1-z) x_1^3 x_2 \rangle$ w.r.t. $\succ$, a CGS of $\langle F \rangle \cap \langle x_1^3 x_2 \rangle$ w.r.t. $\succ$ is constructed by discarding polynomials that contains the auxiliary variable z from the output, as follows

$$\{(\mathbb{V}(t_1), \{x_1^3 x_2\}), \ (\mathbb{C} \backslash \mathbb{V}(t_1), \{(75 x_2^2 - 16 t_1^2 x_2) x_1^3, -84375 x_1^4 x_2 - 2048 t_1^5 x_1^3 x_2\})\}.$$

Nextly, by dividing all elements of the sets of polynomials by $x_1^3 x_2$, the set

$$\{(\mathbb{V}(t_1), \{1\}), \ (\mathbb{C} \backslash \mathbb{V}(t_1), \{75 x_2 - 16 t_1^2, 84375 x_1 + 2048 t_1^5\})\}.$$

is obtained as the CGS of $\langle F \rangle : \langle x_1^3 x_2 \rangle$ w.r.t. $\succ$.   □

In order to construct an algorithm for solving extended ideal membership problems, we need an extended Gröbner bases algorithm or an algorithm for computing parametric syzygies. Both the algorithms have been published in [17]. Now we give the notion of parametric syzygy system.

Definition 3 (PSS). *Let fix a monomial order. Let $f_1, \ldots, f_s$ be polynomials in $\mathbb{C}[t][x]$, $\mathbb{A}_1, \ldots, \mathbb{A}_\ell$ strata in $\mathbb{C}^m$ and $G_1, \ldots, G_\ell$ subsets of $(\mathbb{C}[t][x])^s$. A finite set $\mathcal{G} = \{(\mathbb{A}_1, G_1), \ldots, (\mathbb{A}_\ell, G_\ell)\}$ of pairs is called a **parametric syzygy system***

(PSS) on $\mathbb{A}_1 \cup \cdots \cup \mathbb{A}_\ell$ for $f_1, \ldots, f_s$ if $\sigma_a(G_i)$, $a \in \mathbb{A}_i$, is a Gröbner basis of the syzygy module of $\langle \sigma_a(f_1), \ldots, \sigma_a(f_s) \rangle$ in $(\mathbb{C}[x])^s$ for each $i = 1, \ldots, \ell$. Each $(\mathbb{A}_i, G_i)$ is called a **segment** of $\mathcal{G}$. We simply say $\mathcal{G}$ is a parametric syzygy system for $f_1, \ldots, f_s$ if $\mathbb{A}_1 \cup \cdots \cup \mathbb{A}_\ell = \mathbb{C}^m$.

An algorithm for computing parametric syzygy systems [17], implemented in the computer algebra system Risa/Asir [21], is utilized in the Subsect. 3.3 to construct the algorithm for solving the extended ideal membership problems of parametric ideals.

3.3   Solving Extended Ideal Membership Problems

Here, we extend the algorithm **ExtIMP** to parametric cases by using the algorithms for computing parametric ideal quotients **ParaQuotient** and parametric syzygy systems. In the algorithm, a monomial order is fix.

Algorithm 3. ParaExtIMP

Input: $f_1, f_2, \ldots, f_s$: polynomials in $(\mathbb{C}[t])[x]$ satisfying $\{x \in X \mid f_1(x) = f_2(x) = \cdots = f_s(x) = 0\} = \{O\}$ on $\mathbb{A} \subset \mathbb{C}^m$.

$\quad$ h: a polynomial satisfying $h \in \langle f_1, \ldots, f_s \rangle_{\{O\}} \subset \mathbb{C}\{x\}$ on $\mathbb{A}$.

Output: $\{(\mathbb{A}_1, (q_{11}, \ldots, q_{1s})), \ldots, (\mathbb{A}_\ell, (q_{\ell 1}, \ldots, q_{\ell s}))\}$: $h = q_{j1}f_1 + \cdots + q_{js}f_s$ on $\mathbb{A}_j$

$\quad$ for each $j \in \{1, \ldots, \ell\}$ and $\mathbb{A} = \bigcup_{i=1}^{\ell} \mathbb{A}_i$.

BEGIN

$\mathcal{G} \leftarrow \emptyset$;

$\mathcal{Q} \leftarrow$ Compute a CGS of $\langle F \rangle : \langle h \rangle$ on $\mathbb{A}$ by **ParaQuotient**;

while $\mathcal{Q} \neq \emptyset$ **do**

$\quad$ Select a segment $(\mathbb{A}', G')$ from $\mathcal{Q}$; $\mathcal{Q} \leftarrow \mathcal{Q} \backslash \{(\mathbb{A}', G')\}$;

$\quad$ $g \leftarrow$ Take a polynomial g s.t. $g \neq 0$ and $g(O) \neq 0$ from G';

$\quad$ $\mathcal{S} \leftarrow$ Compute a PSS for $gh, f_1, \ldots, f_s$ on $\mathbb{A}'$;

$\quad$ **while** $\mathcal{S} \neq \emptyset$ **do**

$\quad\quad$ Select a segment $(\mathbb{A}'', G'')$ from $\mathcal{S}$; $\mathcal{S} \leftarrow \mathcal{S} \backslash \{(\mathbb{A}'', G'')\}$;

$\quad\quad$ $(c_0, c_1, \ldots, c_s) \leftarrow$ Take $(c_0, c_1, \ldots, c_s)$ s.t. c_0 is a nonzero constant from G'';

$\quad\quad$ $\mathcal{G} \leftarrow \mathcal{G} \cup \{(\mathbb{A}'', (-\frac{c_1}{c_0 g}, \ldots, -\frac{c_s}{c_0 g}))\}$;

$\quad$ **end-while**

end-while

return $\mathcal{G}$;

END

The termination and the correctness can be readily verified because this algorithm consists of an appropriate combination of **ExtIMP** and **ParaQuotient**.

Example 5. Let us consider Example 4, again. A set $\{(\mathbb{V}(t_1), \{1\}), (\mathbb{C} \backslash \mathbb{V}(t_1), \{75x_2 - 16t_1^2, 84375x_1 + 2048t_1^5\})\}$ is a CGS of $\langle F \rangle : \langle x_1^3 x_2 \rangle$ w.r.t. $\succ$.

Let us compute on $\mathbb{V}(t_1)$ a PSS for $x_1^2 x_2$, $\frac{\partial f}{\partial x_1}$, $\frac{\partial f}{\partial x_2}$ w.r.t. $\succ$. Then, our implementation for computing PSSs outputs

$$\{(\mathbb{V}(t_1), \{(-1, 0, x_1^2 x_2),\ (0, -x_1, 3x_1^2 + x_2)\})\}.$$

Take the first vector $(-1, 0, x_1^2 x_2)$ because the first component is a nonzero constant. Thus, if the parameter t_1 belongs to $\mathbb{V}(t_1)$, then $x_1^3 x_2$ can be written as

$$x_1^3 x_2 = 0 \cdot \frac{\partial f}{\partial x_1} + x_1^4 x_2^2 \cdot \frac{\partial f}{\partial x_2}.$$

Let us compute on $\mathbb{C} \backslash \mathbb{V}(t_1)$ a PSS for $x_1^3 x_2 (75 x_2 - 16 t_1^2)$, $\frac{\partial f}{\partial x_1}$, $\frac{\partial f}{\partial x_2}$ w.r.t. $\succ$. Then, our implementation outputs

$$\{(\mathbb{C} \backslash \mathbb{V}(t_1), \{(-3, (75 x_2^2 - 16 t_1^2 x_2) x_1, -75 x_2^3 + 16 t_1^2 x_2^2),$$
$$(0, 2 t_1 x_1^2 x_1 + x_1, -3 x_1^2 - 2 t_1 x_1 x_2^2 - x_2)\})\}.$$

Take the first vector $(-3, (75 x_2^2 - 16 t_1^2 x_2) x_1, -75 x_2^3 + 16 t_1^2 x_2^2)$ because the first component is a nonzero constant. Thus, if the parameter t_1 belongs to $\mathbb{C} \backslash \mathbb{V}(t_1)$, then $x_1^3 x_2$ can be written as

$$x_1^3 x_2 = \frac{75 x_1 x_2^2 - 16 t_1^2 x_1 x_2}{3(-16 t_1^2 + 75 x_2)} \cdot \frac{\partial f}{\partial x_1} + \frac{-75 x_2^3 + 16 t_1^2 x_2^2}{3(-16 t_1^2 + 75 x_2)} \cdot \frac{\partial f}{\partial x_2}. \qquad \square$$

The algorithm **ParaExtIMP** has been implemented in the computer algebra system Risa/Asir [21].

Example 6. Let $f = x_1^3 + x_1 x_2^8 + t_1 x_1 x_2^7 + t_2 x_2^{10}$ and $F = \{\frac{\partial f}{\partial x_1}, \frac{\partial f}{\partial x_2}\}$ where x_1, x_2 are variables and t_1, t_2 are parameters. Then, $x_1^3 x_2^2 \in \langle F \rangle_{\{O\}} \subset \mathbb{C}\{x_1, x_2\}$ can be checked by computing a parametric local cohomology system of F. Our implementation give us a solution of the extended ideal membership problem w.r.t. $\succ$ as follows.

- If the parameters t_1 and t_2 belong to $\mathbb{C}^2 \backslash \mathbb{V}(t_1 t_2)$, then $x_1^3 x_2^2$ can be written as

$$x_1^3 x_2^2 = \frac{(64 x_2^6 + 176 t_1 x_2^5 + 161 t_1^2 x_2^4 + 49 t_1^3 x_2^3 + 300 t_2^2 x_2^2) x_1 + 80 t_2 x_2^8 + 150 t_1 t_2 x_2^7 + 70 t_1^2 t_2 x_2^6}{3(300 t_2^2 + 49 t_1^3 x_2 + 161 t_1^2 x_2^2 + 176 t_1 x_2^3 + 64 x_2^4)} \cdot \frac{\partial f}{\partial x_1}$$
$$+ \frac{(-30 t_2 x_2 - 30 t_1 t_2) x_1 - 8 x_2^7 - 23 t_1 x_2^6 - 22 t_1^2 x_2^5 - 7 t_1^3 x_2^4}{3(300 t_2^2 + 49 t_1^3 x_2 + 161 t_1^2 x_2^2 + 176 t_1 x_2^3 + 64 x_2^4)} \cdot \frac{\partial f}{\partial x_2}.$$

- If the parameters t_1 and t_2 belong to $\mathbb{V}(t_1) \backslash \mathbb{V}(t_1, t_2)$, then $x_1^3 x_2^2$ can be written as

$$x_1^3 x_2^2 = \frac{(32 x_2^6 + 150 t_2^2 x_2^2) x_1 + 40 t_2 x_2^8}{6(75 t_2^2 + 16 x_2^4)} \cdot \frac{\partial f}{\partial x_1} + \frac{-15 t_1 x_1 x_2 - 4 x_2^7}{6(75 t_2^2 + 16 x_2^4)} \cdot \frac{\partial f}{\partial x_2}.$$

- If the parameters t_1 and t_2 belong to $\mathbb{V}(t_2) \backslash \mathbb{V}(t_1, t_2)$, then $x_1^3 x_2^2$ can be written as

$$x_1^3 x_2^2 = \frac{-50331648 x_2^2 x_1^3 + (-1404928 t_1^4 x_2^6 + 823543 t_1^8 x_2^2) x_1}{3(823543 t_1^8 - 50331648 x_1^2)} \cdot \frac{\partial f}{\partial x_1}$$
$$+ \frac{(\lozenge)}{3(823543 t_1^8 - 50331648 x_1^2)} \cdot \frac{\partial f}{\partial x_2},$$

where $(\lozenge) = (6291456 x_2^3 + 786432 t_1 x_2^2 - 688128 t_1^2 x_2 + 602112 t_1^3) x_1^2 + 175616 t_1^4 x_2^7 + 21952 t_1^5 x_2^6 - 19208 t_1^6 x_2^5 + 16807 t_1^7 x_2^4 - 117649 t_1^8 x_2^3$.

– If the parameters t_1 and t_2 belong to $\mathbb{V}(t_1, t_2)$, then $x_1^3 x_2^2$ can be written as

$$x_1^3 x_2^2 = \frac{x_1 x_2^2}{3} \cdot \frac{\partial f}{\partial x_1} + \frac{-x_2^3}{24} \cdot \frac{\partial f}{\partial x_2}.$$

It takes $0.047\,$s to get this result by our implementation. ([OS: Windows 7 (64bit), CPU: Intel(R) Core i-7-2600 CPU @ 3.40 GHz 3.40 GHz]) $\qquad\square$

4 Concluding Remarks

We have introduced an algorithm for solving extended ideal membership problems for zero-dimensional ideals in the rings of convergent power series. The key idea is to use ideal quotients in polynomial rings, which reduces the problem from local rings to polynomial rings. The proposed algorithm does not invoke the Mora's tangent cone algorithm, and it can be easily implemented. Furthermore, the new algorithm works well under the additional assumption that the origin is an isolated zero.

The anonymous referees have given a useful question about the generalization of the algorithm **ExtIMP**. If a set of polynomials has an isolated common root at a point of $\mathbb{C}^n$, then Lemma 1 can be extended to the following. Let $f_1, f_2, \ldots, f_s$ be polynomial in $\mathbb{C}[x]$ s.t. $\{x \in X \mid f_1(x) = f_2(x) = \cdots = f_s(x) = 0\} = \{A\}$ where X is a neighborhood of a point $A = (a_1, \ldots, a_n)$ of $\mathbb{C}^n$. Let $\mathcal{I}_A$ be an ideal generated by $f_1, f_2, \ldots, f_s$ in $\mathbb{C}\{x-a\} := \mathbb{C}\{x_1 - a_1, \ldots, x_n - a_n\}$ and $I = \langle f_1, \ldots, f_s \rangle$ be an ideal generated by $f_1, f_2, \ldots, f_s$ in $\mathbb{C}[x]$. Then, the following lemma holds.

Lemma 3. *Let h be a polynomial in $\mathbb{C}[x]$, s.t. $h \in \mathcal{I}_A \subset \mathbb{C}\{x - a\}$. Then, there exists a polynomial $g \in I : \langle h \rangle$ s.t. $g \notin \mathfrak{m}_A$, where $\mathfrak{m}_A = \langle x_1 - a_1, x_2 - a_2, \ldots, x_n - a_n \rangle$ in $\mathbb{C}\{x - a\}$.*

The proof of Lemma 3 is completely same as that of Lemma 1. By this extension, the algorithm **ExtIMP** and **ParaExtIMP** can be generalized to the case of $\mathcal{I}_A$. However, Theorem 1 cannot be applied directly to test an ideal membership to such a case.

Note that the extension of **ExtIMP** has not been implemented in a computer algebra system and compared with other existing algorithm, yet. We will report on the performance elsewhere.

Acknowledgments. We thank referees for careful reading our manuscript and for giving useful comments. This work has been partly supported by JSPS Grant-in-Aid for Young Scientists (B) (No.15K17513) and Grant-in-Aid for Scientific Research (C) (No.15K04891).

References

1. Becker, T., Weispfenning, V.: Gröbner Bases. Springer, New York (1992)
2. Briançon, J., Granger, M., Maisonobe, P., Miniconi, M.: Algorithme de calcul du polunôme du Bernstein : cas non dégénéré. Ann. Inst. Fourier **39**, 553–610 (1989)
3. Cox, D., Little, J., O'Shea, D.: Ideals, Varieties and Algorithms, 3rd edn. Springer, New York (2007)
4. Cox, D., Little, J., O'Shea, D.: Using Algebraic Geometry. Springer, New York (1998)
5. Decker, W., Greuel, G.-M., Pfister, G., Schönemann, H.: Singular 3-1-6 - A computer algebra system for polynomial computations (2012). http://www.singular.uni-kl.de
6. Dolzmann, A., Sturm, T.: Redlog: computer algebra meets computer logic. ACM SIGSAM Bull. **31**, 2–9 (1997)
7. Greuel, G.-M., Pfister, G.: A Singular Introduction to Commutative Algebra, 2nd edn. Springer, Heidelberg (2008)
8. Grothendieck, A.: Théorèmes de dualité pour les faisceaux algébriques cohérents. Séminaire Bourbaki 149 (1957)
9. Hartshorne, R., Grothendieck, A.: Local Cohomology; a Seminar. Lecture Notes in Mathematics, 41. Springer, New York (1967)
10. Kalkbrener, M.: On the stability of Gröbner bases under specializations. J. Symbolic Comput. **24**, 51–58 (1997)
11. Kapur, D., Sun, D., Wang, D.: A new algorithm for computing comprehensive Gröbner systems. In: Proceedings of the ISSAC 2010, pp. 29–36. ACM (2010)
12. Kulikov, V.S.: Mixed Hodge Structures and Singularities. Cambridge University Press, New York (1998)
13. Manubens, M., Montes, A.: Improving DISPGB algorithm using the discriminant ideal. J. Symbolic Comput. **41**, 1245–1263 (2006)
14. Montes, A., Wibmer, M.: Gröbner bases for polynomial systems with parameters. J. Symbolic Comput. **45**, 1391–1425 (2010)
15. Mora, T.: An algorithm to compute the equations of tangent cones. In: Calmet, Jacques (ed.) ISSAC 1982 and EUROCAM 1982. LNCS, vol. 144, pp. 158–165. Springer, Heidelberg (1982)
16. Mora, T., Pfister, G., Traverso, T.: An introduction to the tangent cone algorithm. Adv. Comput. Res. **6**, 199–270 (1992). Issued in robotics and nonlinear geometry
17. Nabeshima, K.: On the computation of parametric Gröbner bases for modules and syzygies. Jpn. J. Ind. Appl. Math. **27**, 217–238 (2010)
18. Nabeshima, K.: Stability conditions of monomial bases and comprehensive Gröbner systems. In: Gerdt, V.P., Koepf, W., Mayr, E.W., Vorozhtsov, E.V. (eds.) CASC 2012. LNCS, vol. 7442, pp. 248–259. Springer, Heidelberg (2012)
19. Nabeshima, K., Tajima, S.: On efficient algorithms for computing parametric local cohomology classes associated with semi-quasihomogeneous singularities and standard bases. In: Proceedings of the ISSAC 2014, pp. 351–358. ACM (2014)
20. Nabeshima, K., Tajima, S.: Algebraic local cohomology with parameters and parametric standard bases for zero-dimensional ideals (2015). arXiv:1508.06724
21. Noro, M., Takeshima, T.: Risa/Asir - a computer algebra system. In: Proceedings of the ISSAC 1992, pp. 387–396. ACM (1992). http://www.math.kobe-u.ac.jp/Asir/asir.html
22. Schulze, M.: Algorithms for the gauss-manin connections. J. Symbolic Comput. **32**, 549–564 (2001)

23. Schulze, M.: Algorithmic gauss-manin connection - algorithms to compute hodge-theoretic invariants of isolated hypersurface singularities. vom Fachbereich Mathematik der Universität Kaiserslautern zum Verleihyng des akademischen Grades Doktor der Naturwissenschaften (2002)

24. Suzuki, A., Sato, Y.: A simple algorithm to compute comprehensive Gröbner bases using Gröbner bases. In: Proceedings of the ISSAC 2006, pp. 326–331. ACM (2006)

25. Swanson, I., Huneke, C.: Integral Closure of Ideals, Rings and Modules. Cambridge University Press, Cambridge (2006)

26. Tajima, S., Nakamura, Y.: Algebraic local cohomology class attached to quasi-homogeneous isolated hypersurface singularities. Publ. Res. Inst. Math. Sci. **41**, 1–10 (2005)

27. Tajima, S., Nakamura, Y.: Annihilating ideals for an algebraic local cohomology class. J. Symbolic Comput. **44**, 435–448 (2009)

28. Tajima, S., Nakamura, Y.: Algebraic local cohomology classes attached to unimodal singularities. Publ. Res. Inst. Math. Sci. **48**, 21–43 (2012)

29. Tajima, S., Nakamura, Y., Nabeshima, K.: Standard bases and algebraic local cohomology for zero dimensional ideals. Adv. Stud. Pure Math. **56**, 341–361 (2009)

30. Weispfenning, V.: Comprehensive Gröbner bases. J. Symbolic Comput. **36**, 669–683 (1992)

31. Yano, T.: On the theory of b-functions. Publ. Res. Inst. Math. Sci. **14**, 111–202 (1978)

Advanced Algebraic Attack on Trivium

Frank-M. Quedenfeld[1]([⊠]) and Christopher Wolf[2]

[1] University of Technology Braunschweig, Braunschweig, Germany
frank.quedenfeld@googlemail.com
[2] Research Center Jülich, Jülich, Germany

Abstract. This paper presents an algebraic attack against Trivium that breaks 625 rounds using only 4096 bits of output in an overall time complexity of $2^{42.2}$ Trivium computations. While other attacks can do better in terms of rounds (799), this is a practical attack with a very low data usage (down from 2^{40} output bits) and low computation time (down from 2^{62}).

From another angle, our attack can be seen as a proof of concept: how far can algebraic attacks can be pushed when several known techniques are combined into one implementation? All attacks have been fully implemented and tested; our figures are therefore not the result of any potentially error-prone extrapolation, but results of practical experiments.

Keywords: Trivium · Algebraic modelling · Similar variables · ElimLin · Sparse multivariate algebra · Equation solving over $\mathbb{F}_2$

1 Introduction

Algebraic attacks against symmetric ciphers are more than a decade old. In fact, they can be traced back to Claude Shannon [26].

Recently, the *Elimination of linear variables* or ElimLin algorithm was used to attack several ciphers, in particular CTC2, LBlock and MIBS. According to [12] from FSE 2012, only 6 rounds can be broken for CTC2. This attack requires up to 180 h on a standard PC and requires 210 guessed bits and 64 chosen cipher texts (CC). Guessing 220 bits and 16 CP brings the attack down to 3 h. For LBlock, the paper reports 8 rounds (out of 32) for 32 guessed bits (out of 80) for 6 known plain texts (KP). For MIBS (32 rounds), the paper reports a break for 3 to 5 rounds with 0/16/20 guessed bits, respectively. An initial implementation of the ElimLin algorithm was employed on DES in [8]. Here, plain ElimLin could break 5 rounds of DES with 3 KP and 23 guessed bits; using a SAT solver, this number can be increased to 6 rounds for an unspecified number of KP (most likely 1) and 20 key bits fixed. In Sect. 1.2 we discuss this with more details and references.

In this article, we show that ElimLin can be greatly improved when employed together with other techniques from solving systems of equations like a proper

---

A full version of this paper can be found at [22].

© Springer International Publishing Switzerland 2016
I.S. Kotsireas et al. (Eds.): MACIS 2015, LNCS 9582, pp. 268–282, 2016.
DOI: 10.1007/978-3-319-32859-1_23

monomial ordering or a new variant of eXtended Linearization. In particular, we use the Trivium stream cipher as a testbed for algebraic attacks, mainly due to its simple algebraic structure and its good scalability: full Trivium has 1152 rounds, so we can see the effect of adding some component to our equation solver well, cf. Sect. 3 for all building blocks. In addition, we restricted ourselves to attacks that can be fully implemented on a current computer. Our implementation was able to break round-reduced Trivium with 625 rounds. In particular, our data complexity is far better than for non-algebraic attacks. Non-algebraic attacks need at least 2^{40} to 2^{45} output bits of Trivium with $767 - 799$ rounds as we present in Sect. 1.2. We are able to bring this down to 2^{11} or 2^{12}. Further non-algebraic attacks can use up to 2^{60} to 2^{72} Trivium computations, which is not feasible on modern computers. This is because they are guessing a huge number of variables.

In particular, we show that algebraic attacks become specifically efficient against Trivium if we do not use a lot of output for one instance, but few output bits for many instances. However, this new type of attack only works if we have access to a sparse equation solver over $\mathbb{F}_2$ that can deal with many variables and also many equations ($\approx 10^6$ in both cases). This sparse polynomial system solver is the second major contribution of this paper. To the best of our knowledge, such a solver does not yet exist in the open literature.

1.1   Organization

We start with a review of existing work in the area of algebraic cryptanalysis and specifically cryptanalysis of Trivium in Sect. 1.2. In addition, we discuss several ways to solve systems over $\mathbb{F}_2$. This is followed by a discussion of Trivium and the idea of *"similar variables"* in Sect. 2. The overall solver and the tweaks we need to deal with a full representation of Trivium are given in Sect. 3. This is followed by practical experiments on round reduced Trivium in Sect. 4. The paper concludes with some remarks, open questions and directions for further research in Sect. 5.

1.2   Related Work

Before going into details about our attack, we review related work in algebraic cryptanalysis, cryptanalysis of Trivium and solving systems of equations over $\mathbb{F}_2$.

Algebraic Cryptanalysis. Algebraic cryptanalysis works on a simple assumption: We are able to express any cryptographic primitive in simple non-linear equations over a finite field (*e.g.* $\mathbb{F}_2$ or $\mathbb{F}_{256}$), cf. [5] for an overview on some ciphers. This part of the attack is called *"modelling"*. If we now use this description and solve the overall system for a given output (stream ciphers) or a given plaintext/ciphertext pair (block cipher) we obtain the secret key.

For stream ciphers, algebraic attacks [3,10] seem to work fine, as for some public key systems [18,19] and other primitives [28]. We want to note that even round reduced variants of Trivium has escaped all efforts to be broken by purely algebraic methods.

Attacks on Trivium. We briefly sketch some of the most important attacks against Trivium. We want to stress that Trivium is still secure—despite its simple and elegant design; and the combined effort of the cryptanalytic community.

The attacks from [14,20] are both cube attacks. Cube attacks use the chosen IV scenario. In this scenario we can generate multiple output bits of a stream cipher using the same key but different initialization vectors (IVs). In a nutshell, cube attacks simplify the encryption function by generating the sum over this function for all 0/1-combinations of some IV variables as described in [14,23]. They recover the full key of a 799 round-reduced variant of Trivium in 2^{62} computations guessing 62 variables.

Other key recovery attacks in the chosen IV scenario are not that successful. For instance, [21] describes a linear attack on Trivium breaking a 288 round-reduced variant with a likelihood of 2^{-72}.

Previous algebraic attacks [24,25,27,29] use output bits generated by *one* unknown IV and one key. They are all based on the model described in [24] and fail even for round-reduced variants of Trivium because they attack the internal state of the cipher rather than the key. In these attacks the adversary is not able to use output bits generated by different IVs.

Solving Systems Over $\mathbb{F}_2$. As pointed out above, any algebraic cryptanalysis has two steps: The first is the algebraic modelling, the second is solving the corresponding system of equations. For simplicity, we assume that all equations are at most quadratic over $\mathbb{F}_2$. When we have an equation of total degree greater or equal 3 we can introduce intermediate variables to reduce the degree of the equation. Furthermore all our algorithms work on systems of equations with higher degree.

Basically, the most promising algorithms come from the F-family of Gröbner basis algorithms [15,16], see [2] for an overview. They have been successfully applied in the case of public key schemes [18,19], but also stream ciphers [17]. The main disadvantage is the high memory consumption. Although there are some counter examples for special cases, the running time of Gröbner basis algorithms is inherently exponential in the number of variables. Even worse, the memory consumption increases with $\mathcal{O}(n^r)$ for n the number of variables and r the *degree of regularity*. In particular the latter makes Gröbner bases unusable for our purpose as we have $r \geq 2$ and $n \approx 2^{20}$.

Another line of algorithms comes from the so called "XL—eXtended Linearization" [9]. Here, the main operation is multiplying all known equations with all monomials up to a certain degree. While these new equations are trivially true, some of them are linearly independent and can hence be used in a so-called *Macaulay matrix* to reduce the overall problem to linear algebra over $\mathbb{F}_2$. In a Macaulay matrix, the rows represent polynomials while the columns represent monomials, cf. [2] for an overview of the idea. Although it has been shown that techniques from the XL-family are strictly less efficient than from the F-family [4,13,30,31], XL does have its merits as it is easier to adapt to different settings. Hence we have used a specialized version of XL in our solver to improve its efficiency, cf. Sect. 3.

Last but not least, there is the `ElimLin` algorithm [8,12] where linear equations are used to eliminate variables. After that, the system is simplified with linear algebra techniques, cf. Sect. 3.1. It can handle large sparse polynomial systems with 3056 variables and 4331 monomials in 2900 equations and it is so simple that it can easily be tweaked for specific purposes. However, we want to stress that *plain* ElimLin without further modifications is not efficient enough to deal with systems that arise from the modelling of Trivium.

1.3 Our Contributions

Our first contribution are techniques to model many instances of Trivium as a quadratic equation system. We also introduce strategies to handle the large number of variables within this model. The modelling techniques and strategies can be applied to any symmetric cipher since the upcoming system of equation is structured according to the update function of the cipher.

The second contribution of this paper is a solver which is able to solve structured, sparse, quadratic equation systems. Based on ElimLin and eXtended Linearization we introduce a monomial order to have more control in the ElimLin-Step and change XL so that it preserves the monomial structure of the system.

With the above mentioned techniques we settle an attack on a round reduced variant of Trivium with $R = 625$ rounds in $2^{42.2}$ time and 2^{12} data complexity on an average computer.

2 Trivium

The main point of our attack is to get a suited algebraic system of equations over $\mathbb{F}_2$ for a considered cipher. As soon as the modelling part is done, we will solve the system of equations with a special purpose solver, cf. Sect. 3. We present our modelling techniques with the stream cipher Trivium from [7] as a testbed.

Trivium generates up to 2^{64} keystream bits from an 80 bit IV and an 80 bit key. The cipher consists of an initialization or "clocking" phase of R rounds and a keystream generation phase. For $R = 1152$ we obtain the *full* version as stated in [7], otherwise a *round-reduced* variant of Trivium, denoted by Trivium-R. There are several ways to describe Trivium—below we use the most compact one with three quadratic, recursive equations for the state bits and one linear equation to generate the output in the keystream generation phase.

2.1 Definition and Direct Considerations

Consider three shift registers $A := (a_i, \ldots, a_{i-92}), B := (b_i, \ldots, b_{i-83})$ and $C := (c_i, \ldots, c_{i-110})$ of length $93, 84$ and 111 respectively. They are called the *state* of Trivium.

First the state of Trivium is initialized with

$$A = (k_0, \ldots, k_{79}, 0, \ldots, 0)$$
$$B = (v_0, \ldots, v_{79}, 0, \ldots, 0) \text{ and}$$
$$C = (0, \ldots, 0, 1, 1, 1).$$

Here $(k_0, \ldots, k_{79})$ is the secret *key* and $(v_0, \ldots, v_{79})$ is the public *initialization vector (IV)* of Trivium.

Before output is generated Trivium is updated R rounds according to the following three *state update functions*:

$$b_i := a_{i-65} + a_{i-92} + a_{i-90}a_{i-91} + b_{i-77},$$
$$c_i := b_{i-68} + b_{i-83} + b_{i-81}b_{i-82} + c_{i-86},$$
$$a_i := c_{i-65} + c_{i-110} + c_{i-108}c_{i-109} + a_{i-68}.$$

For $R = 1152$ we obtain the original variant from [7]. Since this variant have not been broken, round-reduced variants are used to evaluate the security margin of Trivium.

After this initialization phase Trivium generates one output bit z_i per round by the function

$$z_i := c_{i-65} + c_{i-110} + a_{i-65} + a_{i-92} + b_{i-68} + b_{i-83}.$$

We produce n_o bits of output z_i for $i = (R+1) \ldots (R+n_o)$.

Recovering the initial vector A is the prime aim of a key recovery attack. Note that the vectors B, C are actually known to an attacker at the beginning of the initialization phase since the IV is public. Therefore we make the additional assumption that an attacker has control over the IV used within the cipher and obtain a stream of output bits for a fixed key and any choice of IV. This is called *chosen IV scenario* and is in line, *e.g.* with cube attacks.

To launch our attack, we use several output streams (*Trivium instances*) that share the same key but different values for the IV. We see in Sect. 4 that this will lead to successful attacks at Trivium-R.

Obviously, we need approx. $3RT + n_oT$ intermediate variables, respectively, if we want to represent T instances of Trivium-R with n_o output bits each. Before discussing strategies for solving such rather large systems, we start with an observation on Trivium.

2.2 Similar Variables

Previous algebraic attacks such as [25,27,29] are based on the algebraic representation of Trivium given in Sect. 2.1. Therefore, we would expect similar results. However, we do not consider only *one* instance of Trivium but several (thousand). Consequently, the relation between these instances becomes important for the overall success of our attack.

Let $I \subset V$ be a subset of all IV variables V. We call I the *master cube* of the attack. In addition, we consider the first n_o output bits of Trivium initialized with the same key and all 0/1-combinations for variables in the master cube I. All other IV variables are set to zero.

We set up all Trivium instances with symbolic key variables $k_0, \ldots, k_{79}$. Denote the current Trivium instance by $t \in \mathbb{N}$. We initialize these instances for a given number of rounds R and introduce three new variables for very round i for the entries $a_{t,i}, b_{t,i}$ and $c_{t,i}$ in the three registers A_t, B_t and C_t. This produces a quadratic system with a large number of variables and monomials.

Now we take a more general point of view and introduce similar variables for generalized systems of equations. In particular, we denote all intermediate variables by $y_0, y_1, \ldots$

Definition 1. *Let* $P = \mathbb{F}_2[k_0, \ldots, k_{79}, y_0, y_1, \ldots] =: \mathbb{F}_2[K, Y]$ *be the Boolean polynomial ring in the key variables* K *and all intermediate variables* Y.

We call the two intermediate variables y_i *and* y_j *similar iff* $y_i + y_j = p(K, Y \backslash \{y_i, y_j\})$, *where* $p(K, Y \backslash \{y_i, y_j\})$ *is a polynomial of degree* $\deg(p) \leq 1$.

Following the definition, we check for similar variables whenever we should introduce a new intermediate variable. If the new variable is similar to one or a linear combination of many existing variables we continue computations without a new variable. Instead of using a new intermediate variable for quadratic monomials already introduced, we use a linear combination of existing variables.

In generic systems, similar variables could be not of much use. However, if all equations stem from one algebraic model for one given cipher, we are likely to find many similarities. The following example illustrates how we work with similar variables in the case of Trivium.

Example 1. Consider the polynomials

$$f_0 = y_0 + k_{78}k_{79} + k_{53}$$
$$f_1 = y_1 + k_{77}k_{78} + k_{79} + k_{52}$$

in P. These polynomials define intermediate variables y_0, y_1 and form the system of equation F.

Assume we want to introduce the intermediate variable y_2 with

$$f_2 = y_2 + k_{79}k_{78} + k_{78}k_{77} + k_5 + k_{61}.$$

It holds that $y_2 = y_1 + y_0 + k_{53} + k_{79} + k_{52} + k_5 + k_{61}$. Therefore, we do not need to introduce y_2 and can continue computation with y_0 and y_1. This does not only save us one variable, but we also have replaced a quadratic equation by a (potentially more useful) linear one.   □

Note that there are different ways of considering similar variables. In any case, we need a solver that can first identify them and second make use of them by replacing all linear relations within a given system.

While the above definition captures any behavior for any system of equations, we see that it applies very well to Trivium, see Fig. 1 for some experimental results on Trivium-R. Here, we have generated $T = 32$ instances of Trivium with $n_o = 66$ output bits. On the x-axis we see the number of initialization rounds; on the y-axis we have the total number of variables in use. As we can see the number of intermediate variables has greatly decreased; even for only 32 instances of Trivium. For $R = 600$ rounds we also produce $66 \cdot 32 = 2112$

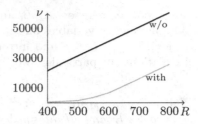

Fig. 1. Number of variables for $T = 32$ instances of Trivium and $n_o = 66$ output bits; Number of Rounds R against number of variables ν with and without similar variables

output equations. The model in [24] can just handle one instance of Trivium. So it would need $288 + 3 \cdot 2112 = 6624$ variables to produce that amount of output equations. When using more instances of Trivium we get even more efficient.

To produce the smallest number of monomials possible in our system we change the algorithm to generate the system. Instead of going forward and generating the Trivium instance we start with the output and go backwards and just generate the variables we need. This way, we just generate the variables and monomials that are needed. Note that this is contrary to earlier algebraic modellings of Trivium such as [24,25].

To evaluate our representation of Trivium, we made several experiments with the fast Gröbner basis PolyBoRi [6]. PolyBoRi is specialized on Gröbner basis for Boolean polynomial rings and uses a variant of Faugère's F4 algorithm (see [15]).

Overall, we could solve a system of Trivium-420 with a suited monomial ordering. The system of equations has 1300 variables and 3900 monomials in 1500 equations. For more details see [22]. Thus solving polynomial systems arising for large R is still out of reach for current Gröbner basis implementations like PolyBoRi. The memory consumption is simply too high.

We have hence designed a solver which can handle such large numbers of equations and variables and will describe it in the following section.

3   Solving the System

Before going into details for the experiments, we describe our strategy to solve rather sparse systems over $\mathbb{F}_2$ arising from the above representation of Trivium. We have based our solver on multivariate *quadratic* polynomials over $\mathbb{F}_2$ as this is generally enough to capture full Trivium. Specifically, our goal was to develop a *working implementation* than can handle around 10^6 variables and 10^6 equations, respectively, over $\mathbb{F}_2$. To the best of our knowledge, software with such special properties is not available at the moment. Our solver is organized around a specialized C++-core that natively handles quadratic polynomials over $\mathbb{F}_2$ and the `ElimLin` and `SL` algorithm. In addition, we have used several other building blocks which we describe in the remainder of the section. We report experimental results in Sect. 4.

3.1 Main Core

ElimLin or *Elimination of Linear variables* has been investigated in [8,12]. We generalize it with a monomial ordering, so the algorithm becomes

1. First we generate the Macaulay matrix for the system according to some monomial ordering τ.
2. Echelonize the matrix according to τ. This naturally splits up the system into linear equations L and quadratic equations Q.
3. For each element $p \in L$, use the leading term $\mathrm{LT}(p)$. If there is at least one equation in Q that also contains the variable $\mathrm{LT}(p)$, eliminate $\mathrm{LT}(p)$ in Q.
4. If we substitute at least one variable in Q, go back to step 2.

We want to stress that ElimLin preserves the overall degree of our system Q. In addition, it automatically detects all similar variables (see Definition 1). Moreover ElimLin is able to deal with rather large but sparse systems of equations.

The original ElimLin algorithm did not have any ordering but used heuristics to determine which variable to eliminate in the non-linear part of the overall system. We found this approach fine for small systems but difficult to use for larger ones: The likelihood to fall into local optima was simply too high—even with advanced heuristics. Determining the correct order proved to be challenging and required careful experiments as can be seen in Sect. 2.2. Hence, we used the *degree reverse lexical (degrevlex)* ordering. Note that this also works well in case of Gröbner basis algorithms. In the case of Trivium, we take the key variables first and sort the intermediate variables ascending according to rounds and instances of Trivium. We want to stress that the ordering is crucial in our analysis. Like in Gröbner techniques the results differ significantly depending on the ordering.

Sparse Linearization. ElimLin cannot conclude that $a \cdot b + a = 1 \Rightarrow a = 1, b = 0$ for some variables $a, b \in \mathbb{F}_2$ and therefore cannot solve arbitrary systems of equations. Hence we use a new variant of Extended Linearization that we call *Sparse Linearization*.

In the XL-algorithm we multiply all quadratic polynomials with any monomial up to a degree $D - 2$ that can be generated by the used variables. While these new equations are algebraically dependent, they can produce linearly independent rows in the Macaulay matrix. There is a variant for sparse systems called XSL in which systems of equations get multiplied by already used monomials. It has been used in the past to attack different cryptographic systems in [11].

Even XSL produces new monomials and increases the overall total degree of polynomial systems. Hence we may not use it for huge, structured systems of equations. The *Sparse Linearization (SL)* preserves the total degree and structure of the system by doing as follows:

We multiply each polynomial f in the polynomial system F with any variable ν contained in quadratic monomials of f. If the total degree of νf increases, we do not insert the new polynomial into F. Further we check if all monomials of νf are already used in F. If so, we insert νf into the system.

After SL is done we need to check for linear dependence of the new polynomials. We do this by the echelonization step of ElimLin.

Sparse Polynomial Core. The core of our algorithm is substitution of variables from linear equations and echelonization. While the first requires polynomials, the second needs linear matrices. In particular Gröbner basis algorithms would construct a so-called *Macaulay matrix* and go back and forth between a matrix and a polynomial representation, see [2] for an overview. In our implementation, we used the polynomials over $\mathbb{F}_2$ directly but also implemented matrix-like operations (*e.g.* row addition) directly for polynomials. To this aim, each polynomial is stored as a (sorted) list of monomials rather than sparse vectors over $\mathbb{F}_2$. To make computations fast, we also keep a dictionary of lead terms, monomials in use by each polynomial and also a list of variables in use per polynomial. This way, addition of two polynomials with the same lead term and elimination of variables does not depend on the overall number of polynomials anymore. For speed, this part of the code is written in C++ (approx. 2500 lines).

M4RI. While the sparse strategy from above turned out to be efficient for sparse matrices, it fails if the matrices become increasingly dense. Note that this is inevitable when solving such a system: In all experiments, we had a degeneration from sparse to dense shortly before solving the overall system. To remedy this, we incorporated the fastest known, open source linear algebra package for matrices over $\mathbb{F}_2$, namely the Method of the 4 Russians Implementation (M4RI) [1]. Experimentally, we have found that matrices with less than $\approx 1/1000$ nonzero coefficients in the corresponding matrix over $\mathbb{F}_2$ should be handled by our *sparse* strategy described above and by M4RI otherwise.

Further building blocks and how this parts are used can be found in [22]. They are avoided because of space constraints.

4   Experiments

This section consists of three parts. First we consider the model and we see a saturation of variables and monomials when adding Trivium instances. Second we present on some parameter studies to further strengthen our system of equations. Finally, we use our insights to actually attack Trivium using the techniques described in the previous sections. We stress that the overall system of equations can be generated before we get the actual data for the output. This way our attack splits into an online and offline phase. All experiments were done on an AMD-Opteron-6276@2.3 GHz with 256 nodes and 1 TB of RAM. Each node had access to at most 256 GB of RAM at a time. As we do not use any parallelizing techniques we are only using one core. Furthermore we want to stress that the online phase only requires a standard computer with 16 GB of RAM.

Saturation. When adding different Trivium instances with identical key variables but different IV constants that lie in the same master cube, the overall number of variables and monomials in the quadratic monomials of the overall system tends towards a saturation point (cf. Fig. 2a–b). More specifically; we fix $80 - i$ IV bits to zero and set the remaining i bits to all possible values from $\mathbb{F}_2^i$. The first i bits proven to be optimal for our purposes.

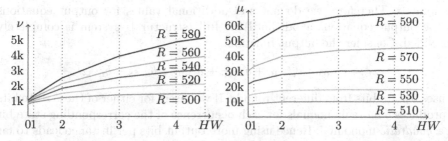

(a) Hamming Weight against number of variables ν for R rounds of Trivium in a master cube of dimension 5

(b) Hamming Weight against number of quadratic monomials μ for R rounds for Trivium in a master cube of dimension 5

Fig. 2. Saturation in the model of Trivium

We have plotted saturation for 32 instances in Fig. 2a–b, counting both the number of variables and the number of quadratic monomials needed for the system consisting of all instances of Trivium. In these figures, we have first added the IV with Hamming weight 0, then all IV with Hamming weight 1 and so forth. Note that instances with the same Hamming weight yield the same number of variables. As we can see in these graphs, the amount of variables needed to generate an instance becomes significantly *lower* if we generate instances with higher Hamming weight.

The saturation of monomials needs a lower Hamming weight of the IV, so the saturation of monomials is much flatter than the saturation of variables. Note that variables that are not in the quadratic (saturated) part of the system are only found in the linear terms. Furthermore, we stress that Fig. 2a–b are chosen only as an example. This effect also exists if the number of rounds increases (up to $R = 1152$). However, if the number of rounds grows we need to generate more instances to see this effect; it seems that we need to generate exponentially more instances to see the saturation. All in all, this points to a kind of "basis": Trivium instances for IV with small Hamming weight serve as a kind of basis for Trivium instances of higher Hamming weight. While this seems obvious when looking at the generating equations, it is still interesting to see how *strong* this effect is in practice. Unfortunately, we were unable to derive a closed formula depending on the number of instances and rounds but have to leave this as an open question.

In conclusion, saturation means that we can obtain more defining equations from many instances than we would expect from one instance alone. In a sense, this is the key observation to launch our attack.

Saturation should occur in other ciphers as well since the system of equation is generated by a repeatedly execution of an update function.

Output and Parameters. While output equations clearly help us to linearize the system, the very structure of Trivium in our model yields a lot of additional monomials. Therefore, we do not add additional values for output equations or the output equations at all until the full (structural) system is completely simplified. Consider the output function:

$$z_i := c_{i-65} + c_{i-110} + a_{i-65} + a_{i-92} + b_{i-68} + b_{i-83}.$$

It uses 6 state bits from different rounds. If we insert for either of these state bits it produces 5 more monomials for each occurrence of the corresponding state bit in a *quadratic* monomial. Hence using more output bits per instance leads to far more monomials than we can afford.

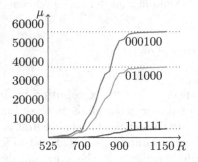
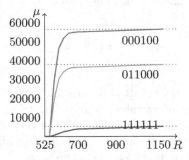

(a) Initialization rounds R against number of quadratic monomials μ for $n_o = 1$

(b) Initialization rounds against number of quadratic monomials μ for $n_o = 66$

Fig. 3. Comparing monomials with $n_o = 1$ and $n_o = 66$; Initialization rounds R against number of quadratic monomials μ; The numbers next to the lines are the significant parts of the IV (binary). The rest of the IV is zero. Consider the Hamming weight of these numbers.

Figure 3 compares systems with $T = 64$ instances and increasing number of rounds R. In the first experiment we used $n_o = 1$ and in the second $n_o = 66$. We see that many output bits do not necessarily lead to a more useful system because we get much more monomials. Even if we use two output equations we get a system with nearly double the number of monomials which we cannot solve easier. Note that Fig. 3 reflects the monomials needed for one instance while Fig. 2b shows the number of monomials needed for the whole system of equations. Therefore the saturation can also be seen in Fig. 3.

When we use $n_o = 66$ output bits the number of monomials at $R = 700$ rounds is negligibly smaller than the number of monomials for full Trivium ($R = 1152$). We choose $n_o = 66$ because for $n_o > 66$ we need to introduce new intermediate variables even for the output and that destroys the purpose of (over)defining the system for the linearization step. For $n_o = 1$ output bits, the same effect occurs for $R = 925$ rounds. Since we want to linearize our system to derive a solution we get the following conjecture.

Conjecture 1. The complexity of our attack on Trivium does not grow after $R = 925$ rounds using $n_o = 1$ output bit and after $R = 700$ rounds using $n_o = 66$ output bits. That means if we are able to break $R = 925$ rounds with one output bit we are able to break full Trivium with one output bit.

In a nutshell: Since the number of monomials does not increase, neither does the difficulty of the attack. Unfortunately, both settings are out of reach for a practical test at the moment.

Attacks. In this paragraph we describe the attacks and their complexities.

In Fig. 3 we have illustrated the number of monomials depending on the number of output bits. With this in mind we have specialized our attack to the case $n_o = 1$. Based on this, we generate a system with Trivium-625 instances.

The following table shows the number of monomials and variables needed for the full system. This also includes "dummy" variables for the output. When we add concrete data for the output bits to our system these numbers decrease rapidly. Furthermore we can see that there is a time-data trade-off when guessing variables. When we guess fewer variables we need more data to launch the attack. When we guess more variables we need less data but the time complexity increases (Table 1).

Table 1. Experimental results from the Online-phase on Trivium-625 with number of variables ν and number of monomials μ. Time is measured in Trivium computations.

R	μ	ν	#guessed variables	Data complexity	Time complexity
625	**499,741**	**15,869**	23	2^{11}	$2^{59.7}$
625	1,135,858	32,518	**0**	$2 \cdot 2^{11}$	**$2^{42.2}$**

Information of how we guess variables and how we convert the used time in seconds in trivium computation, can be found in the full version of this paper [22]. We tested 101 keys each both for correct and incorrect guesses.

When we do not guess variables, we need more data and though more instances in our symbolic system. Generating the full symbolic system becomes a challenge due to the size of the system and RAM usage in the offline-phase. Thus we generate two systems consisting of 2^{11} instances each. The two systems do not profit from each other through similar variables in the offline-phase so the number of variables and the number of monomials is more than doubled.

In the online-phase of the attack each system is reduced due to the linear output equations and similar variables. In our example in the table above we solved the system in $2^{17.1}$ seconds which leads to $2^{42.2}$ *Trivium computations* on average. Again, this experiment was conducted 101 times.

Unfortunately, we are unable to find a closed formula to predict the number of instances we would need to solve a system for a given number of rounds, as the behaviour of Trivium and the solver is quite erratic in this respect.

The real bottleneck of our attack is the generation of a symbolic system for a useful number of instances in the offline-phase. We can overcome this problem with a better implementation of the linear algebra or the ElimLin algorithm. However, we still cannot really resolve the exponential growth starting at $R = 700$ (Fig. 3) which works as a kind of barrier for our techniques used to attack Trivium. We want to encourage others to further improve or enhance the techniques used in this paper.

5   Conclusions

In this paper we have shown that algebraic attacks can be significantly improved. We achieve this by enhancing the ElimLin algorithm with a variant of eXtended Linearization and using a proper monomial ordering; in particular the last proved crucial in our experiments. Overall, we built a solver for sparse polynomial systems that can handle up to 10^6 monomials in 10^6 equations. In particular, we solved the system of equations arising from Trivium-625 with $1,135,858$ monomials and $32,518$ variables. Before our work, plain ElimLin was able to solve systems of equations with 3056 variables and 4331 monomials. While this is not quite comparable, because the systems solved are not the same, this improvement by a factor of ≈ 262 demonstrates considerable progress.

In addition, we have seen that using *many* instances of Trivium rather than only one with a long key stream significantly improves the attack. All in all, we were able to break a 625 round reduced version of Trivium in practical time ($2^{42.2}$ Trivium computations) and a data complexity of 2^{12}. Other key recovery attacks on Trivium can do better in terms of rounds with $R = 799$ but they requires a large amount of data (2^{40} bits) and time 2^{62} while guessing 62 bits. It is doubtful if this rate can be achieved in practice. An advantage of our approach is that we actually computed the full attack and did *not* make extrapolations from our results, as we do not want to make promises which are hard to keep.

Another line of research is the integration of more toolboxes into our solver, most notably SAT-solvers and more efficient sparse linear algebra packages.

While our experiments were conducted only on Trivium, we are confident that the ideas and lessons learned are also useful for the algebraic cryptanalysis of other symmetric primitives, such as block ciphers or hash functions. We want to stress that the potential of algebraic cryptanalysis can only be unleashed if equal stress is put on modelling techniques and the corresponding solver.

Acknowledgements. The first author wants to thank Wolfram Koepf (University of Kassel) for fruitful discussions and guidance. Both authors gratefully acknowledges an Emmy Noether Grant of the Deutsche Forschungsgemeinschaft (DFG).

References

1. Abbott, T., Albrecht, M., Bard, G., Bodrato, M., Brickenstein, M., Dreyer, A., Dumas, J.G., Hart, W., Harvey, D., James, J., Kirkby, D., Pernet, C., Said, W., Wood, C.: M4RI(e)–Linear Algebra over F_2 (and F_2^e). http://m4ri.sagemath.org/
2. Albrecht, M.: Algorithmic Algebraic Techniques and their Application to Block Cipher Cryptanalysis. Ph.D. thesis, Royal Holloway, University of London (2010)
3. Armknecht, F., Krause, M.: Algebraic attacks on combiners with memory. In: Boneh, D. (ed.) CRYPTO 2003. LNCS, vol. 2729, pp. 162–175. Springer, Heidelberg (2003)
4. Ars, G., Faugère, J.-C., Imai, H., Kawazoe, M., Sugita, M.: Comparison between XL and Gröbner basis algorithms. In: Lee, P.J. (ed.) ASIACRYPT 2004. LNCS, vol. 3329, pp. 338–353. Springer, Heidelberg (2004)
5. Biryukov, A., De Cannière, C.: Block ciphers and systems of quadratic equations. In: Johansson, T. (ed.) FSE 2003. LNCS, vol. 2887, pp. 274–289. Springer, Heidelberg (2003)
6. Brickenstein, M., Dreyer, A.: PolyBoRi: a framework for Groebner-basis computations with Boolean polynomials. J. Symbol. Comput. **44**(9), 1326–1345 (2009). http://dx.doi.org/10.1016/j.jsc.2008.02.017. Effective Methods in Algebraic Geometry
7. De Cannière, C., Preneel, B.: TRIVIUM. In: Robshaw, M., Billet, O. (eds.) New Stream Cipher Designs. LNCS, vol. 4986, pp. 244–266. Springer, Heidelberg (2008)
8. Courtois, N.T., Bard, G.V.: Algebraic cryptanalysis of the data encryption standard. In: Galbraith, S.D. (ed.) Cryptography and Coding 2007. LNCS, vol. 4887, pp. 152–169. Springer, Heidelberg (2007)
9. Courtois, N.T., Klimov, A.B., Patarin, J., Shamir, A.: Efficient algorithms for solving overdefined systems of multivariate polynomial equations. In: Preneel, B. (ed.) EUROCRYPT 2000. LNCS, vol. 1807, pp. 392–407. Springer, Heidelberg (2000)
10. Courtois, N.T., Meier, W.: Algebraic attacks on stream ciphers with linear feedback. In: Biham, E. (ed.) EUROCRYPT 2003. LNCS, vol. 2656, pp. 345–359. Springer, Heidelberg (2003)
11. Courtois, N.T., Pieprzyk, J.: Cryptanalysis of Block ciphers with overdefined systems of equations. In: Zheng, Y. (ed.) ASIACRYPT 2002. LNCS, vol. 2501, pp. 267–287. Springer, Heidelberg (2002)
12. Courtois, N.T., Sepehrdad, P., Sušil, P., Vaudenay, S.: ElimLin algorithm revisited. In: Canteaut, A. (ed.) FSE 2012. LNCS, vol. 7549, pp. 306–325. Springer, Heidelberg (2012)
13. Diem, C.: The XL-algorithm and a conjecture from commutative algebra. In: Lee, P.J. (ed.) ASIACRYPT 2004. LNCS, vol. 3329, pp. 323–337. Springer, Heidelberg (2004)
14. Dinur, I., Shamir, A.: Cube attacks on tweakable black box polynomials. In: Joux, A. (ed.) EUROCRYPT 2009. LNCS, vol. 5479, pp. 278–299. Springer, Heidelberg (2009)

282 F.-M. Quedenfeld and C. Wolf

15. Faugère, J.C.: A new efficient algorithm for computing gröbner bases (F4). In: Proceedings of the 2002 International Symposium on Symbolic and Algebraic Computation, ISSAC 2002, pp. 75–83. Springer (2002)
16. Faugère, J.C.: A new efficient algorithm for computing Gröbner bases without reduction to zero (F_5). In: International Symposium on Symbolic and Algebraic Computation, ISSAC 2002, pp. 75–83. ACM Press, July 2002
17. Faugère, J.C., Ars, G.: An algebraic cryptanalysis of nonlinear filter generators using Gröbner bases. Rapport de recherche 4739. www.inria.fr/rrrt/rr-4739.html
18. Faugère, J.-C., Joux, A.: Algebraic cryptanalysis of hidden field equation (HFE) cryptosystems using Gröbner bases. In: Boneh, D. (ed.) CRYPTO 2003. LNCS, vol. 2729, pp. 44–60. Springer, Heidelberg (2003)
19. Faugère, J.-C., Otmani, A., Perret, L., Tillich, J.-P.: Algebraic cryptanalysis of Mceliece variants with compact keys. In: Gilbert, H. (ed.) EUROCRYPT 2010. LNCS, vol. 6110, pp. 279–298. Springer, Heidelberg (2010)
20. Fouque, P.-A., Vannet, T.: Improving key recovery to 784 and 799 rounds of trivium using optimized cube attacks. In: Moriai, S. (ed.) FSE 2013. LNCS, vol. 8424, pp. 502–517. Springer, Heidelberg (2014)
21. Khazaei, S., Hasanzadeh, M.M., Kiaei, M.S.: Linear Sequential Circuit Approximation of Grain and Trivium Stream Ciphers. Cryptology ePrint Archive, Report 2006/141 (2006). http://eprint.iacr.org/2006/141/
22. Quedenfeld, F., Wolf, C.: Advanced Algebraic Attack on Trivium. Cryptology ePrint Archive, Report 2014/893 (2014). http://eprint.iacr.org/
23. Quedenfeld, F., Wolf, C.: Algebraic Properties of the Cube Attack. Cryptology ePrint Archive, Report 2014/800 (2014). http://eprint.iacr.org/2014/800/
24. Raddum, H.: Cryptanalytic results on Trivium (2006). http://www.ecrypt.eu.org/stream/triviump3.html
25. Schilling, T.E., Raddum, H.: Analysis of trivium using compressed right hand side equations. In: Kim, H. (ed.) ICISC 2011. LNCS, vol. 7259, pp. 18–32. Springer, Heidelberg (2012)
26. Shannon, C.E.: Communication theory of secrecy systems. Bell Syst. Techn. J. **28**, 656–715 (1949)
27. Simonetti, I., Faugère, J.C., Perret, L.: Algebraic attack against trivium. In: First International Conference on Symbolic Computation and Cryptography, SCC 2008, pp. 95–102. LMIB, Beijing (2008). http://www-polsys.lip6.fr/jcf/Papers/SCC08c.pdf
28. Sugita, M., Kawazoe, M., Perret, L., Imai, H.: Algebraic cryptanalysis of 58-Round SHA-1. In: Biryukov, A. (ed.) FSE 2007. LNCS, vol. 4593, pp. 349–365. Springer, Heidelberg (2007)
29. Teo, S., et al.: Algebraic analysis of Trivium-like ciphers (2013). http://www.eprint.iacr.org/2013/240.pdf
30. Yang, B.-Y., Chen, J.-M.: All in the XL family: theory and practice. In: Park, C., Chee, S. (eds.) ICISC 2004. LNCS, vol. 3506, pp. 67–86. Springer, Heidelberg (2005)
31. Yang, B.-Y., Chen, J.-M.: Theoretical analysis of XL over small fields. In: Wang, H., Pieprzyk, J., Varadharajan, V. (eds.) ACISP 2004. LNCS, vol. 3108, pp. 277–288. Springer, Heidelberg (2004)

Managing Massive Data

Compressing Big Data: When the Rate of Convergence to the Entropy Matters

Salvatore Aronica[1](✉), Alessio Langiu[1,2], Francesca Marzi[3],
Salvatore Mazzola[1], Filippo Mignosi[3], and Giulio Nazzicone[3]

[1] IAMC-CNR Unit of Capo Granitola, National Research Council, Trapani, Italy
{Salvatore.Aronica,Alessio.Langiu,Salvatore.Mazzola}@cnr.it
[2] King's College London, London, UK
[3] DISIM Department, University of L'Aquila, L'Aquila, Italy
{Francesca.Marzi,GiulioNazzicone}@graduate.univaq.it,
Filippo.Mignosi@univaq.it

Abstract. It is well known from a theoretical point of view that LZ78 have an asymptotic convergence to the entropy faster than LZ77. A faster rate of convergence to the theoretical compression limit should lead to a better compression ratio. In effect, early LZ78-like and LZ77-like compressors behave accordingly to the theory. On the contrary, it seems that most of the recent commercial LZ77-like compressors perform better than the other ones. Probably this is due to a strategy of optimal parsing, which is used to factorize the text and can be applied to both LZ77 and LZ78 cases, as recent results suggest. To our best knowledge there are no theoretical results concerning the rate of convergence to the entropy of both LZ77-like and LZ78-like case when a strategy of optimal parsing is used. In this paper we investigate how an optimal parsing affect the rate of convergence to the entropy of LZ78-like compressors. We discuss some experimental results on LZ78-like compressors and we consider the ratio between the speed of convergence to the entropy of a compressor with optimal parsing and the speed of convergence to the entropy of a classical LZ78-like compressor. This ratio presents a kind of *wave* effect that become bigger and bigger as the entropy of the memoryless source decreases but it seems always to slowly converge to one. These results suggest that for non-zero entropy sources the optimal parsing does not improve the speed of convergence to the entropy in the case of LZ78-like compressors.

Keywords: Lempel-Ziv compression algorithms · Text compression · Text entropy · String algorithms

1 Introduction

The most studied and used in practice dictionary compressors are the LZ77-like and the LZ78-like compressors (see for instance [19,20]). It is known that theoretically LZ78-like compressors have a faster convergence than LZ77-like

© Springer International Publishing Switzerland 2016
I.S. Kotsireas et al. (Eds.): MACIS 2015, LNCS 9582, pp. 285–289, 2016.
DOI: 10.1007/978-3-319-32859-1_24

compressors, when the texts to be compressed are generated by a memoryless source. In practice, on the contrary, it seems that the most advanced LZ77-like compressors perform better. This discrepancy between theory and practice is, for us, the main open problem in this field that clearly affects all chances of giving concrete optimization results on dictionary-based compressors.

Recent results (see [4]) show that probably the discrepancy between theory and practice between LZ78-like and LZ77-like compressors is due to the effect of using an optimal parsing strategy, which can be applied in both LZ77 and LZ78 cases, rather than to the fitting of the data to the source model or to the hidden constants in the "transient" phase.

While it is empirically shown that applying an optimal parsing to LZ77-like compressors improves the compression ratio, it is not known how an optimal parsing affect LZ78-like compression ratio.

2   Preliminaries

In [2] it is possible to find a survey on Dictionary methods and of Symbolwise methods and a description of the deep relationship among them (see also [1, 6, 15, 16]).

A dictionary compression algorithm, as noticed in [2], can be fully described by: 1. The dictionary description, i.e. a static collection of phrases or a complete algorithmic description on how the dynamic dictionary is built and updated. 2. The encoding of dictionary pointers in the compressed data. 3. The parsing method, i.e. the algorithm that splits the uncompressed data in dictionary phrases.

Compression speed and compression ratio strongly depend on the parsing method. Since we can associate a directed weighted graph $G_{A,T} = (V, E, L)$ to any dictionary compression algorithm A, any text T and any cost function $C : E \rightarrow \mathbb{R}^+$, there is a relation between parsing methods and path on such graph (see [3–5, 11]). An optimal parsing is then a path of minimal weight.

The speed of convergence to the entropy of LZ77-like and LZ78-like compressors is a still active field of researches. Many research papers appeared improving previous results or considering different settings. The interested reader can see for instance the following references [7–9, 12, 14, 17, 18].

To our best knowledge anyhow, all above researches considered only greedy parsing in both cases of LZ77-like and LZ78-like compressors and they did not considered optimal parsing strategies, whilst, as discussed at the end of [4], optimal parsing seems to be the first choice in all practical dictionary compressor.

3   Experimental Results

Early implementations of LZ78 algorithm, like *compress*, have a better compression ratio than early LZ77 implementations which use the greedy parsing. According to the experimental results reported in [4], the compression ratio of LZ77-like compressors improve when an optimal parsing is used.

Table 1. Standard deviation of the compression ratio, over 100 files generated by an i.i.d. source, achieved by an LZ78-like compressor with optimal parsing. First column reports file size. First row reports the 1's probability of the source.

Size (Byte)	1/4	1/8	1/16	1/32	1/64	1/128	1/256	1/512	1/1024
2^7	0,0257321	0,0355836	0,0453462	0,0414385	0,0377056	0,0369808	0,0254035	0,0201563	0,0143968
2^8	0,0196458	0,0252745	0,0288668	0,0266154	0,0270187	0,0212884	0,0179250	0,0160178	0,0098047
2^9	0,0167899	0,0222582	0,0183216	0,0175383	0,0161119	0,0147846	0,0104683	0,0101021	0,0061889
2^10	0,0132363	0,0146518	0,0155986	0,0137633	0,0096773	0,0096552	0,0080242	0,0060842	0,0043022
2^11	0,0096718	0,0112648	0,0099046	0,0078961	0,0085512	0,0058397	0,0050191	0,0042069	0,0029931
2^12	0,0088274	0,0076916	0,0074478	0,0066218	0,0054580	0,0040541	0,0038019	0,0024763	0,0020072
2^13	0,0054567	0,0048697	0,0050250	0,0045276	0,0034565	0,0032514	0,0027525	0,0021661	0,0013779
2^14	0,0038877	0,0033266	0,0034402	0,0030494	0,0027343	0,0018505	0,0017652	0,0015349	0,0009944
2^15	0,0029316	0,0027646	0,0025572	0,0020944	0,0017285	0,0015004	0,0012528	0,0008889	0,0007122
2^16	0,0020293	0,0017560	0,0017770	0,0014395	0,0012375	0,0010912	0,0009427	0,0005672	0,0005252
2^17	0,0011942	0,0013381	0,0010879	0,0010263	0,0008775	0,0006943	0,0005858	0,0004864	0,0003216
2^18	0,0008690	0,0009367	0,0007724	0,0008265	0,0006804	0,0004655	0,0004353	0,0003429	0,0002585
2^19	0,0004935	0,0006062	0,0005723	0,0004854	0,0004330	0,0003235	0,0002907	0,0002631	0,0001692
2^20	0,0004022	0,0004652	0,0003738	0,0003508	0,0003142	0,0002461	0,0002034	0,0001667	0,0001329

Practical performance of commercial compressors shows that LZ77-like compressors with optimal parsing achieve a better compression ratio of any known LZ78-like compressors. Therefore we can state the following conjecture which is supported by following experimental results.

Conjecture 1. The speed of convergence to the entropy of the compression ratio of LZ77-like compressors with an optimal parsing is faster than or equal to the speed of convergence to the entropy of the compression ratio of LZ78-like compressors with (and without) an optimal parsing.

We have run a series of experiments considering the compression ratio achieved by a LZW-based compressor, which is one of the LZ78-like compressors, and the compression ratio of a compressors based on LZW with an optimal parsing strategy (see [10]).

We consider LZ78-like and LZ77-like compression schemes with unbounded dictionary and *non* uniform cost of dictionary pointers, witch are common assumptions among commercial compressors. Under these assumptions, the greedy parsing, the one proposed in the LZ77 and LZ78 original papers [19, 20], and the flexible parsing [13] are not optimal.

In our experimental settings, we consider binary files of different size drowned from memoryless sources with different 1's probability. In Table 1 we report the standard deviation of the compression ratio of an LZW with optimal parsing (over 100 runs). The considered sources are i.i.d. (independent and identically distributed) with 1's probability of 1/2, 1/4, ... 1/1024. Considered file sizes are between 2^7 and 2^{24}.

Figure 1 shows the trend of the ratio of the speed of convergence of two compressors. Considering the sources having 1's probability of 1/2, 1/4, 1/8, 1/16, ... 1/1024, we have compared (i.e., computed the ratio) the speed of convergence to the entropy (i.e., the difference between compression ratio and the source entropy) of two compressors: LZW with optimal parsing and LZW without optimal parsing.

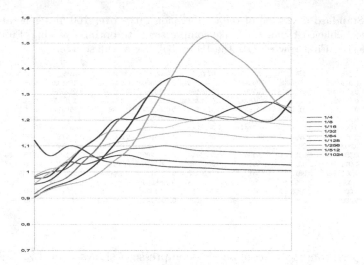

Fig. 1. Trend of the compression ratio of LZW over LZW with optimal parsing computed on file of increasing size of exponential steps stating from 2^9 up to 2^{24}.

Such ratio of LZW convergence speed over LZW with optimal parsing convergence speed has been computed on file of increasing size, by exponential steps, starting from 2^9 up to 2^{24}. This ratio has been averaged over 100 runs. Since Table 1 shows a decreasing behavior and the same happens for LZW without optimal parsing, the averaged values represented in Fig. 1 have a decreasing standard deviation, too. The curves show three phenomenons (dominant trends): 1. increasing values in the first part, 2. an oscillating trend in the middle part, and a decreasing third part. The first part is explained considering that the compression ratio achieved on such small files is affected by an overweight due to the presence of an non-optimized header in the compressed output of our compressor implementations. The middle part oscillating trend is what we have called the transient phase, and, last but not least, the third decreasing part is what suggest that the speed of the convergence to the entropy of each of the two compressors converges to one. This is our main argument supporting our conjecture.

This ratio presents a kind of *wave* effect that become bigger and bigger as the entropy of the memoryless source decreases but it seems always to slowly converge to one. According to the theory, this wave can be a *tsunami* for some families of highly compressible strings, because it has been proved that optimal parsing can improve the rate of convergence to the entropy in such cases.

Our experimental results therefore suggest that for non-zero entropy sources, an optimal parsing strategy does not improve the speed of convergence to the entropy in the case of LZ78-like compressors. At the moment this is a theoretical open problem.

It remains also open the problem of giving a proof of the fact that an optimal parsing strategy improves the speed of convergence to the entropy in the case of LZ77-like compressors.

References

1. Bell, T.C., Cleary, J.G., Witten, I.H.: Text Compression. Prentice Hall, Upper Saddle River (1990)
2. Bell, T.C., Witten, I.H.: The relationship between greedy parsing and symbolwise text compression. J. ACM **41**(4), 708–724 (1994)
3. Crochemore, M., Giambruno, L., Langiu, A., Mignosi, F., Restivo, A.: Dictionary-symbolwise flexible parsing. In: Iliopoulos, C.S., Smyth, W.F. (eds.) IWOCA 2010. LNCS, vol. 6460, pp. 390–403. Springer, Heidelberg (2011)
4. Crochemore, M., Giambruno, L., Langiu, A., Mignosi, F., Restivo, A.: Dictionary-symbolwise flexible parsing. J. Discrete Algorithms **14**, 74–90 (2012)
5. Crochemore, M., Langiu, A., Mignosi, F.: The rightmost equal-cost position problem. In: Bilgin, A., Marcellin, M.W., Serra-Sagristà, J., Storer, J.A. (eds.) DCC, pp. 421–430. IEEE, Los Alamitos (2013)
6. Crochemore, M., Lecroq, T.: Pattern-matching and text-compression algorithms. ACM Comput. Surv. **28**(1), 39–41 (1996)
7. Jacob, T., Bansal, R.K.: Almost sure optimality of sliding window Lempel-Ziv algorithm and variants revisited. IEEE Trans. Inf.Theor. **59**(8), 4977–4984 (2013)
8. Jacquet, P., Szpankowski, W.: Asymptotic behavior of the Lempel-Ziv parsing scheme and digital search trees. Theor. Comput. Sci. **144**(1&2), 161–197 (1995)
9. Jacquet, P., Szpankowski, W.: Analytic Pattern Matching. From DNA to Twitter. Cambridge University Press, Cambridge (2015)
10. Langiu, A.: Optimal Parsing for dictionary text compression. Ph.D thesis, Université Paris-Est, (2012). https://tel.archives-ouvertes.fr/tel-00804215/document
11. Langiu, A.: On parsing optimality for dictionary-based text compression - the zip case. J. Discrete Algorithms **20**, 65–70 (2013)
12. Lastras-Montano, L.A.: On certain pathwise properties of the sliding-window Lempel-Ziv algorithm. IEEE Trans. Inf. Theor. **52**(12), 5267–5283 (2006)
13. Matias, Y., Sahinalp, S.C.: On the optimality of parsing in dynamic dictionary based data compression. In: SODA, pp. 943–944 (1999)
14. Ornstein, D., Weiss, B.: Entropy and data compression schemes. IEEE Trans. Inf. Theor. **39**(1), 78–83 (1993)
15. Salomon, D.: Data compression - The Complete Reference, 4th edn. Springer, New York (2007)
16. Salomon, D.: Variable-length Codes for Data Compression. Springer-Verlag, London (2007)
17. Savari, S.A.: Redundancy of the Lempel-Ziv string matching code. IEEE Trans. Inf. Theor. **44**(2), 787–791 (1998)
18. Wyner, A.D., Wyner, A.J.: Improved redundancy of a version of the Lempel-Ziv algorithm. IEEE Trans. Inf. Theor. **41**(3), 723–731 (1995)
19. Ziv, J., Lempel, A.: A universal algorithm for sequential data compression. IEEE Trans. Inf. Theor. **23**(3), 337–343 (1977)
20. Ziv, J., Lempel, A.: Compression of individual sequences via variable-rate coding. IEEE Trans. Inf. Theor. **24**(5), 530–536 (1978)

Trends in Temporal Reasoning: Constraints, Graphs and Posets

Jacqueline W. Daykin[1,2]([✉]), Mirka Miller[3,4], and Joe Ryan[3]

[1] Department of Computer Science, Royal Holloway,
University of London, Egham, UK
[2] Department of Informatics, King's College London, London, UK
`jackie.daykin@rhul.ac.uk, jackie.daykin@kcl.ac.uk`
[3] School of Electrical Engineering and Computer Science,
University of Newcastle, New South Wales, Australia
`joe.ryan@newcastle.edu.au`
[4] Department of Mathematics, University of West Bohemia, Pilsen, Czech Republic
`mirka.miller@newcastle.edu.au`

Abstract. Temporal reasoning finds many applications in numerous fields of artificial intelligence – frameworks for representing and analyzing temporal information are therefore important. Allen's interval algebra is a calculus for temporal reasoning that was introduced in 1983. Reasoning with qualitative time in Allen's full interval algebra is NP-complete. Research since 1995 identified maximal tractable subclasses of this algebra via exhaustive computer search and also other *ad-hoc* methods. In 2003, the full classification of complexity for satisfiability problems over constraints in Allen's interval algebra was established algebraically. We review temporal reasoning concepts including a method for deciding tractability of temporal constraint satisfaction problems based on the theory of algebraic closure operators for constraints. Graph-based temporal representations such as interval and sequence graphs are discussed. We also propose novel research for scheduling algorithms based on the Fishburn-Shepp inequality for posets.

Keywords: Algebraic closure · Allen's interval algebra · Artificial intelligence · Constraint satisfaction problem · Fishburn-Shepp inequality · Graph · Poset · Qualitative temporal reasoning · Tractable satisfiability

1 Introduction

Temporal reasoning is a mature research endeavor and arises naturally in numerous diverse applications of artificial intelligence such as: planning and scheduling [A-91], natural language processing [SC-88], diagnostic expert systems [N-91], behavioural psychology [CS-73], circuit design [WH-90], software tools for comprehending the state of patients in intensive care units from their temporal information [JCMPM-10], business intelligence [KKD-08], and timegraphs, that is graphs partitioned into a set of chains supporting search which originated in the context of story comprehension [GSS-93].

© Springer International Publishing Switzerland 2016
I.S. Kotsireas et al. (Eds.): MACIS 2015, LNCS 9582, pp. 290–304, 2016.
DOI: 10.1007/978-3-319-32859-1_25

Allen [A-83] introduced an algebra of binary relations on intervals (of time), for representing and reasoning about time. These binary relations, for example *before, during, meets*, describe *qualitative* temporal information which we will be concerned with here. The problem of satisfiability for a set of interval variables with specified relations between them is that of deciding whether there exists an assignment of intervals on the real line for the interval variables, such that all of the specified relations between the intervals are satisfied. When the temporal constraints are chosen from the full Allen's algebra, this form of satisfiability problem is known to be NP-complete. However, reasoning restricted to certain fragments of Allen's algebra is generally equivalent to related well-known problems such as the interval graph and interval order recognition problems [PS-97], which in turn find application in molecular biology [GKS-94, K-93, MZ-08].

Alternative frameworks for formalizing qualitative temporal problems include the point algebra of van Beek and Cohen [VBC-90] for expressing qualitative relations between time points, the point-interval algebra of Vilain [V-82] for describing qualitative relations between time points and time intervals, as well as combinations of these. A comprehensive introduction to the concept of reasoning with qualitative temporal information is given by van Beek [VB-92].

Frameworks for handling *quantitative* temporal information have been proposed by Meiri [M-91], Kautz and Ladkin [KL-91], Gerevini *et al.* [GSS-93], Dechter *et al.* [DMP-91], Jonsson and Bäckström [JB-96], and Drakengren and Jonsson [DJ-97] who also introduced the notion of sequentiality between intervals which is relevant to reasoning about action. Many of the ideas discussed in this paper can be extended to these quantitative frameworks, however the focus here is the tractability of qualitative temporal reasoning and associated graph theory techniques. The unification of interval algebras in artificial intelligence with those of interval orders and interval graphs in combinatorics was considered by Golumbic and Shamir [GS-93], where complexity analysis led them to efficient algorithms for restrictions of the satisfiability problem.

Reasoning in these formalisms is hard, in particular for Allen's interval algebra it is NP-complete [VK-86]. Such reasoning tasks include: determining satisfiability for a set of temporal relations; finding all feasible relationships between two intervals; and deducing new relations from those that are known. Hence these important computational problems motivated the search for tractable subproblems, where reasoning can be guaranteed to be reasonably efficient.

This venture produced a number of tractable subclasses of Allen's algebra, including the *continuous endpoint subclass* of Vilain, Kautz and van Beek [VKVB-89], the *pointisable subclass* of Ladkin and Maddux [LM-88] and the point algebra of Vilain and Kautz [VK-86, VKVB-89] considered algorithmically by van Beek and Cohen [VBC-90], the *ORD-Horn subclass* of Nebel and Bürckert [NB-95], and the *starting (ending) point algebras* of Drakengren and Jonsson [DJ-97] along with further tractable classes that can express the notion of sequentiality between intervals, which is not possible in the ORD-Horn algebra [DJ-97ii].

However, in view of the large number of possible subclasses of Allen's algebra – there are in fact 2^{8192} such subclasses as explained in Sect. 2 – research focused on identifying *maximal* tractable subclasses.

Nebel and Bürckert [NB-95] identified the first maximal subclass, namely the ORD-Horn subclass, which is a strict superset of the pointisable subclass. Subsequent ones were established by Drakengren and Jonsson [DJ-97, DJ-97ii], resulting in eighteen maximal subclasses, subsuming all subclasses previously known to be tractable. This initiated the classification of the complexity of arbitrary subclasses of Allen's interval algebra.

The sheer magnitude of the problem meant that the subclasses could not even be enumerated on a computer, suggesting the need for theoretical along with brute-force computer methods. Towards the classification, Drakengren and Jonsson, for instance, gave a complete characterization of tractable inference using the notion of sequentiality in [DJ-98]; they also showed that no undiscovered tractable subalgebra could contain more than three basic relations, namely $\equiv$, ρ and ρ', where $\rho \in \{d, o, s, f\}$ – see Table 1. Hence, in order to represent complete knowledge about temporal information, the most expressive algebras were already known.

Finally, over a decade later, Krokhin, Jeavons and Jonsson [KJJ-03] completed the classification of complexity for satisfiability type problems over constraints expressed in Allen's interval algebra using only analytical techniques. They showed that this algebra contains *exactly* eighteen maximal tractable subalgebras (precisely those that had previously been identified), and reasoning in any fragment not entirely contained in one of these subalgebras is NP-complete; in other words these eighteen subalgebras include all possible tractable subsets of Allen's algebra. To obtain this powerful dichotomous result, they combined two novel elements: firstly a new uniform description of the known maximal tractable subalgebras, and secondly systematically applying a general algebraic technique for identifying maximal subalgebras with a given property, thus exploiting the existing algebraic properties of Allen's algebra.

This survey focuses on a theoretical approach to the analysis of complexity for temporal reasoning problems by considering them as a special case of the standard *constraint satisfaction problem*, and applying known links to the algebraic closure properties of the constraints they contain – we will show that tractable subclasses of Allen's interval algebra are characterized by algebraic closure properties. Graph-related techniques for temporal reasoning are also considered; we conclude with proposing research into applying the Fishburn-Shepp inequality for posets to determine heuristics for scheduling algorithms.[1]

2   Preliminaries and Definitions

2.1   The Constraint Satisfaction Problem

The fundamental mathematical structure required to describe the constraint satisfaction problem is the *relation*.

---

[1] Throughout we assume $P \neq NP$.

Definition 1. *For any set D, and any natural number n, we denote the set of all n-tuples of elements of D by D^n. A subset of D^n is called an n-ary relation over D. For any tuple $t \in D^n$, and any i in the range 1 to n, we denote the value in the ith coordinate position of t by $t[i]$, and the tuple t by $\langle t[1], t[2], \ldots, t[n] \rangle$.*

We now define the general *constraint satisfaction problem (CSP)* which has been widely studied in the artificial intelligence community.

Definition 2. *An instance of a* constraint satisfaction problem *consists of:*

- *a finite set of variables, V;*
- *a domain of values, D;*
- *a finite set of constraints $C_1, C_2, \ldots, C_q$; each constraint C_i is a pair (s_i, R_i), where:*
  - *s_i is a tuple of variables of length m_i, called the* constraint scope*;*
  - *R_i is an m_i-ary relation over D, called the* constraint relation*.*

For each constraint, (s_i, R_i), the tuples in R_i indicate the allowed combinations of simultaneous values for the variables in s_i. The length of s_i, and of the tuples in R_i, is called the *arity* of the constraint. In particular, unary constraints specify the allowed values for a single variable, and binary constraints specify the allowed combinations of values for a pair of variables – of interest here for temporal problems; moreover, any arbitrary CSP can be converted to a binary CSP.

A *graph* is an ordered pair $G = (V, E)$ comprising a set V of *vertices* together with a set E of *edges* which are 2-element subsets of V; if these subsets comprise ordered pairs of vertices then the graph is said to be *directed*. A binary CSP can be depicted by a constraint graph: each vertex represents a variable, and each directed edge represents a constraint between variables connected by an edge – see Example 1.

A *solution* to a CSP instance is a function from the variables to the domain such that the image of each constraint scope is an element of the corresponding constraint relation.

Not only is the constraint satisfaction problem known to be NP-complete in general [M-77], but this is the case even when the constraints are restricted to binary constraints. Furthermore, restricting the allowed constraint relations to some fixed subset of all the possible relations affects the complexity of this decision problem. Consider the following definition of a restricted problem class.

Definition 3. *For any set of relations, Γ, $CSP(\Gamma)$ is defined to be the class of decision problems with:*

Instance: *A constraint satisfaction problem instance, Π, in which all constraint relations are elements of Γ or binary equality relations.*

Question: *Does Π have a solution?*

If there is an algorithm which solves every problem instance in $CSP(\Gamma)$ in polynomial time, then Γ is said to be a *tractable* set of relations.

By choosing the set of relations Γ appropriately yields specialized versions of the constraint satisfaction problem corresponding to particular computational problems. For example, when Γ is the set containing only binary disequality relations, then $\mathrm{CSP}(\Gamma)$ corresponds to the well-known class of graph colouring problems, for instance the four colour theorem, which provides useful tools in modelling a wide variety of scheduling and assignment tasks.

In the following sections we shall show that by choosing an appropriate set of relations Γ we can arrange for $\mathrm{CSP}(\Gamma)$ to correspond to various forms of temporal reasoning problems expressed in Allen's interval algebra and its subclasses.

2.2   Allen's Interval Algebra

Allen's [A-83] calculus for reasoning about time is based on the concept of *time intervals* together with *binary relations* on them. In this approach, time is considered to be an infinite dense ordered set, such as the rationals $\mathbf{R}$, and a *time interval* X is an ordered pair of time points (X^-, X^+) such that $X^- < X^+$.

Given two time intervals, their relative positions can be described by exactly one of the members of the set $\mathbf{B}$ of 13 basic interval relations, which are depicted in Table 1. For instance, the relation *meets* could represent the display intervals of two traffic lights, and a main course is eaten *during* a three-course meal. These basic relations describe relations between *definite* intervals of time. On the other hand, *indefinite* intervals, whose exact relation may be uncertain, are described by a set of all the basic relations that may apply. For example, if a registration session is a prerequisite of a lecture, then these events must be scheduled according to the interval relation $\{p, m\}$.

The universe of Allen's interval algebra consists of all the binary relations on time intervals which can be expressed as disjunctions of the basic interval relations. These disjunctions are written as sets of basic relations, leading to a total of $2^{13} = 8192$ binary relations, including the *null relation* $\varnothing$ (also denoted by $\bot$) and the *universal relation* $\mathbf{B}$ (also denoted by $\top$). The set of all binary relations $2^{\mathbf{B}}$ is denoted by $\mathcal{A}$; every temporal relation in $\mathcal{A}$ can be defined by a conjunction of disjunctions of endpoint relations of the form $X < Y, X = Y$ and their negations.

The operations on the relations defined in Allen's algebra are: unary *converse* (denoted by $\smile$), binary *intersection* (denoted by $\cap$) and binary *composition* (denoted by $\circ$), which are defined as follows:

$$\forall\, X, Y: \qquad X r^{\smile} Y \;\leftrightarrow\; Y r X$$
$$\forall\, X, Y: \qquad X(r \cap s)Y \;\leftrightarrow\; X r Y \bigwedge X s Y$$
$$\forall\, X, Y: \qquad X(r \circ s)Y \;\leftrightarrow\; \exists Z : (X r Z \bigwedge Z s Y),$$

where X, Y, Z are intervals, and r, s are interval relations. Allen [A-83] gives a composition table for the basic relations.

Fundamental *reasoning problems* in Allen's framework have been studied by a number of authors, including Golumbic and Shamir [GS-92, GS-93], Ladkin and Maddux [LM-88], van Beek [VB-90, VB-92] and Vilain and Kautz [VK-86].

One such fundamental problem is the *satisfiability problem* for temporal relations on intervals, (ISAT), defined as follows: given a collection of temporal relations between a set of variables, decide whether there is an assignment of

Table 1. [NB-95] The set **B** of the thirteen basic qualitative relations defined by Allen. The relations $X^- < X^+$ and $Y^- < Y^+$ are always valid, hence omitted.

Basic interval relation	Symbol	Example	Endpoint relations, X
X precedes (before) Y	$p\,(\prec)$	xxx	$X^+ < Y^-$
Y preceded-by (after) X	$p \smile (\succ)$	$\quad yyy$	
X meets Y	m	$xxxx$	$X^+ = Y^-$
Y met-by X	$m \smile$	$\quad yyyy$	
X overlaps Y	o	$xxxx$	$X^- < Y^- < X^+ < Y^+$
Y overlapped-by X	$o \smile$	$\quad yyyy$	
X during Y	d	$\quad xxx$	$X^- > Y^-,\ X^+ < Y^+$
Y includes X	$d \smile$	$yyyyyyy$	
X starts Y	s	xxx	$X^- = Y^-,\ X^+ < Y^+$
Y started-by X	$s \smile$	$yyyyyyy$	
X finishes Y	f	$\quad\quad xxx$	$X^- > Y^-,\ X^+ = Y^+$
Y finished-by X	$f \smile$	$yyyyyyy$	
X equals Y	$\equiv$	$xxxx$	$X^- = Y^-,\ X^+ = Y^+$
		$yyyy$	

time intervals to the variables which satisfies all of the relations. Notice that any instance of ISAT corresponds to an instance of the constraint satisfaction problem in which all of the constraint relations are temporal relations. In other words, ISAT is equivalent to CSP($\mathcal{A}$), and hence techniques for the analysis of the general constraint satisfaction problem can be applied to the ISAT problem.

Another important temporal reasoning problem is the problem of determining the *strongest implied relation* between every pair of variables, (ISI). This problem is studied as a deductive *closure* problem in [VK-86], and known as the *minimal labeling* problem in [VB-89]. Note that ISAT and ISI are equivalent with respect to polynomial Turing reductions [VK-86].

A CSP is *path consistent*, if for every instantiation of two variables $v_i, v_j \in V$ that satisfies $v_i R_{ij} v_j \in C$ there exists an instantiation of every third variable $v_k \in V$ such that $v_i R_{ik} v_k \in C$ and $v_k R_{kj} v_j \in C$ are also satisfied. Path-consistency is hence computing the transitive closure of a set of relations between intervals. Classic algorithms proposed in temporal reasoning were based on the constraint reasoning algorithms PC-1 and PC-2 [M-77].

The following example illustrates a constraint graph expressing indefinite qualitative temporal information, along with reasoning problems.

Example 1. Consider the following CSP where the constraints are the relations of Allen's interval algebra, each J_i, say job to be scheduled, is a time interval of the form (X^-, X^+), and **B** indicates that there is no direct knowledge of the relationship between J_1 and J_3.

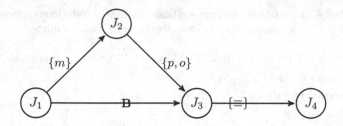

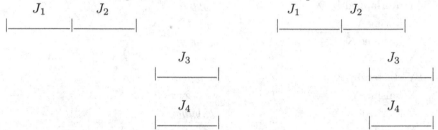

Solutions to this temporal constraint satisfaction problem are as follows:

The minimal label between J_1 and J_3 is $\{p\}$, the *precedes* relation; all other labels are already minimal.

An undirected graph $G = (V, E)$ is called an *interval graph* if its vertices can be represented by intervals on the real line, such that, two vertices are adjacent if and only if the corresponding intervals intersect. Interestingly, interval graphs were originally motivated from molecular biology, however they were later applied to establish NP-completeness of ISAT for subsets of relations, which in turn introduced the interval graph sandwich problem which arises in physical mapping of DNA material [GS-93].

A *sequence graph* [D-92] is an incomplete type of interval graph consisting of sequence chains, namely subgraphs where all constraints are sequence constraints. Sequence chains are used to reduce the number of edges in the graph while preserving the expressiveness of Allen's calculus. Sequence graphs model artificial intelligence applications where many events may occur sequentially.

2.3   Subclasses of Allen's Interval Algebra

In this section we consider restricted temporal reasoning problems in which the relations are chosen from specified subsets of the set of all temporal relations on intervals, $\mathcal{A}$. Note that there are $2^{|\mathcal{A}|}$ such subsets, that is 2^{8192}, or approximately 10^{2466} – clearly a massive combinatorial issue.

For every subset $\Gamma \subseteq \mathcal{A}$ of temporal relations, the corresponding restricted satisfiability problem ISAT(Γ) is equivalent to CSP(Γ) - hence the complexity of ISAT(Γ) can be obtained via the complexity of CSP(Γ).

We now consider some well-known tractable subclasses of Allen's algebra.

Example 2 (The continuous endpoint class, C). This class includes all temporal relations which may be defined using conjunctions of clauses of endpoint relations of the form $x = y$, $x \leq y$ and $x \neq y$, such that (1) there are only unit clauses, and (2) for each unit clause $x \neq y$, the clause form also contains a unit clause of the form $x \leq y$ or $y \leq x$.

It contains 83 relations, including $\{d, o, s\}$, $\{s^\smile, o^\smile, \equiv, f\}$, and $\{d^\smile, f^\smile, o, m, p\}$, (as well as the null relation, $\bot$). For example, the relation $\{d, o, s\}$ is defined by the following conjunction of endpoint relations (see Table 1):

$$\{(X^- \leq X^+), (X^- \neq X^+), (Y^- \leq Y^+), (Y^- \neq Y^+), (X^- \leq Y^+),$$
$$(X^- \neq Y^+), (Y^- \leq X^+), (X^+ \neq Y^-), (X^+ \leq Y^+), (X^+ \neq Y^+)\}.$$

The continuous endpoint class was first described and shown to be tractable by Vilain, Kautz and van Beek [VKVB-89], and subsequently described by Ligozat in terms of "convex relations" with respect to a lattice representation [L-97]. This subclass has the computational advantage that the path-consistency method solves ISI(C) [VB-89, VBC-90, VKVB-89].

Example 3 (The pointisable class, P). This slight generalization of class C is defined in the same way as C, but without the condition (2). It contains 188 relations, including all relations in C together with (non-convex) relations such as $\{d, o\}$, $\{d^\smile, o^\smile, f^\smile, f\}$, and $\{d^\smile, f^\smile, o, p\}$.

This class of temporal relations was first described and shown to be tractable by Ladkin and Maddux [LM-88] and studied by van Beek and Cohen [VBC-90]. Although path-consistency is not sufficient for solving ISI(P) [VB-89], it is for deciding ISAT(P) [LM-88, VK-86]. Van Beek [VB-89, VB-90] and van Beek and Cohen [VBC-90] give algorithms for solving ISI(P) in $O(n^4)$ time; van Beek specifies an algorithm for deciding ISAT(P) in $O(n^2)$ time [VB-90].

Example 4 (The ORD-Horn class, $\mathcal{H}$). This class is a strict superset of P, defined using conjunctions of *disjunctions* of the endpoint relations in P, and where each disjunction contains at most one relation which is not of the form $x \neq y$. That is, the relations permit an ORD-clause form containing only clauses with at most one positive literal. It contains 868 relations, including all those in P together with relations such as $\{f^\smile, s, o\}$, whose endpoint relations are given by the set:

$$\{(X^- \leq X^+), (X^- \neq X^+), (Y^- \leq Y^+), (Y^- \neq Y^+), (X^- \leq Y^-), (X^- \leq Y^+),$$
$$(X^- \neq Y^+), (Y^- \leq X^+), (X^+ \neq Y^-), (X^+ \leq Y^+), (X^- \neq Y^- \lor X^+ \neq Y^+)\}.$$

Nebel and Bürckert [NB-95] identified this, via machine enumeration, to be the first known maximal tractable subclass, and, the unique greatest tractable subclass amongst those that contain all 13 basic relations – comprising over 10 % of the full algebra. Further, they established that the path-consistency method is sufficient for deciding ISAT($\mathcal{H}$), implying its wider applicability [NB-95].

Ligozat [L-98] showed that any subalgebra which contains all basic relations, and a relation which is not ORD-Horn, will contain at least two of four "corner" relations: $\{d^\smile, s^\smile, o^\smile, f, d\}$, $\{o^\smile, s^\smile, d^\smile, f^\smile, o\}$ and their converses.

Example 5 (Starting point and ending point algebras). Drakengren and Jonsson [DJ-97] discovered a large family of maximal tractable subclasses, "starting point" and "ending point" algebras, denoted $S(b), S^*, E(b), E^*$ - the parameter b is chosen from specified basic relations. The six algebras $S(b)$ & $E(b)$ contain 2312 elements each, and S^* & E^* contain 1445 elements each. For brevity we only define $S(b)$; for $E(b)$, S^* and E^* see [DJ-97].

Let $r_s = \{\succ, d, o^\smile, m^\smile, f\}$, and let $r_e = \{\prec, d, o, m, s\}$. Then, for $b \in \{\succ, d, o^\smile\}$, define $S(b)$ to be the set of relations r such that:

$$\{b, b^\smile\} \subseteq r$$
$$\{b\} \subseteq r \subseteq r_s \cup \{\equiv, s, s^\smile\}$$
$$\{b^\smile\} \subseteq r \subseteq r_s^\smile \cup \{\equiv, s, s^\smile\}$$
$$r \subseteq \{\equiv, s, s^\smile\}.$$

The algebras allow for metric constraints on interval starting or ending points.

We proceed to show that tractable subclasses may be characterised as sets of relations with a particular form of algebraic invariance property.

3   Algebraic Closure Properties of Constraints

Jeavons *et al.* [JCG-97] developed a theory of *algebraic closure properties of constraint relations* to distinguish between tractable and NP-complete CSPs – we will apply this to the complexity of a subclass of Allen's interval algebra.

Definition 4 *[JCG-97]. Let R be an n-ary relation over a domain D, and let $\varphi : D^k \to D$ be a k-ary operation on D. The relation R is said to be* closed under φ *if, for all $t_1, t_2, \ldots, t_k \in R$, $\varphi(t_1, t_2, \ldots, t_k) \in R$, where $\varphi(t_1, t_2, \ldots, t_k) = \langle \varphi(t_1[1], t_2[1], \ldots, t_k[1]), \varphi(t_1[2], t_2[2], \ldots, t_k[2]), \ldots, \varphi(t_1[n], t_2[n], \ldots, t_k[n]) \rangle.$*

The algebraic approach to tractability refers to special properties of operations including: *idempotent, constant, unary, projection, semiprojection, majority, affine, or ACI (associative, commutative & idempotent)* - details in [JCG-97].

Example 6. Let the binary operator $Min : \mathbf{R}^2 \to \mathbf{R}$ be defined as:

$$Min(a, b) = a \text{ if } a \leq b; \quad Min(a, b) = b \text{ if } a > b.$$

Then Min is ACI but not essentially unary. Similarly, let $Max : \mathbf{R}^2 \to \mathbf{R}$ be defined analogously to Min using $\geq$; Max is ACI and not essentially unary. However, ternary *Median* is a majority operation – see further examples in [JCG-97].

We will apply the Min operator to illustrate closure for a temporal relation.

Example 7. The basic relation *finishes*, f, is equivalently the infinite set of four-tuples of rationals $((X^-, X^+), (Y^-, Y^+))$ which satisfy $Y^- < X^- < X^+ = Y^+$. To show that f is closed under Min: given two relations f, $((X_1^-, X_1^+), (Y_1^-, Y_1^+))$ and $((X_2^-, X_2^+), (Y_2^-, Y_2^+))$, it follows that $Min(Y_1^-, Y_2^-) < Min(X_1^-, X_2^-) < Min(X_1^+, X_2^+) = Min(Y_1^+, Y_2^+)$, which is a four-tuple belonging to f. Similarly, all 13 basic relations are closed under Min. By symmetry, all 13 basic relations are closed under Max.

The following results, which link the complexity of CSPs to their algebraic closure properties, will be applied to derive complexity in the class $\mathcal{C}$.

Theorem 1 *([JCG-97]). Any tractable set of relations Γ over a finite domain D must be closed under some not essentially unary operation (i.e. either a semi-projection, or a constant, or a majority, or an affine, or an idempotent binary operation).*

Theorem 2 *([JCG-97]). For any set of relations Γ over a finite domain D, if Γ is closed under a constant, or a majority, or an affine, or some ACI operation, then $CSP(\Gamma)$ is tractable.*

Furthermore, a set Γ, as defined in Theorem 2, is a *maximal* set of tractable relations [JCG-97]: the addition of any other relation which is not closed under the same operation changes $CSP(\Gamma)$ from a tractable into an NP-complete problem. Applying these theorems to the complexity of Allen's algebra yields:

Lemma 1. *The continuous endpoint class $\mathcal{C}$ is closed under the binary ACI operator* Min *for finite domains.*

Proof. Suppose that R is a temporal relation in $\mathcal{C}$, where R consists of a set of r basic relations. Further, let $t_i, t_j \in R$, where $t_i = \{t_{i_1}, t_{i_2}, \ldots, t_{i_r}\}$ and $t_j = \{t_{j_1}, t_{j_2}, \ldots, t_{j_r}\}$.
    Then $Min(t_i, t_j)$ is given by $Min(t_{i_k}, t_{j_k})$, for $1 \leq k \leq r$. Each operation on a pair of the same basic relations, $Min(t_{i_k}, t_{j_k})$, is given by applying Min to the four-tuples defining end-points. As shown in Example 7, any basic relation specified by the pair (t_{i_k}, t_{j_k}) is closed under Min. Hence $Min(t_i, t_j) \in R$. □
    By symmetry, the class $\mathcal{C}$ is closed under the ACI operation Max; and further, it is closed under the majority operator $Median$ (median of interval endpoints):

Lemma 2. *The continuous endpoint class $\mathcal{C}$ is closed under the ternary operator* Median *for finite domains.*

Proof. By expressing $Median$ as $Median = Max(Min(a, b), Min(Max(a, b), c))$ the result follows from Lemma 1 for Min and by symmetry for Max. □

Corollary 1. *For finite domains, the satisfiability problem for the continuous endpoint class, $ISAT(\mathcal{C})$, is tractable.*

Hence closure is an important algebraic property for tractability. Meanwhile, various computer-assisted exhaustive searches led to a classification of complexity within a large part of Allen's algebra [DJ-97, DJ-97ii, DJ-98]. For further progress, theoretical studies would be necessary, since using these methods would require dealing with more than 10^{50} individual cases. Thus with promising algebraic indications, the way was paved for further advancement theoretically.
    In 2003, Krokhin *et al.* [KJJ-03] completed the analysis of complexity for satisfiability problems expressed in Allen's algebra by showing that the known

maximal subclasses are the only forms of tractability within this interval algebra: Allen's algebra contains exactly eighteen maximal tractable subalgebras and that reasoning within any subset not included in one of these is NP-complete. First they introduced a new uniform description for all of the existing maximal tractable subalgebras by systematically "loosening" the relations in a subalgebra e.g. replacing $\{m\}$ by $\{p, m\}$. The second novel element was to exploit the algebraic properties of Allen's algebra by importing a technique from general algebra to obtain a description of maximal subalgebras with a given property. A similar algebraic approach was used for one fragment of Allen's algebra [L-98], while Krokhin *et al.* systematically applied this technique to obtain a classification for all possible fragments of Allen's algebra. Their purely analytical method breaks the proof down into a collection of six simple cases, and makes extensive use of the operations defined in the algebra (converse, intersection and composition), while exploiting the fact that tractability of a subalgebra is a pertinent hereditary property in Allen's algebra. In order to show that any fragment that is not entirely contained in one of the eighteen known maximal tractable algebras is NP-complete, they applied generic techniques: Betweenness, Derivation, and Duality. Importantly, both the result and the algebraic method can be used to classify the complexity in other temporal and spatial formalisms.

4   Posets and the Fishburn-Shepp Inequality

Following the survey of temporal reasoning and our results linking algebraic techniques with complexity of Allen's algebra, we propose here novel research directions: to specify heuristics for scheduling based on representing a collection of intervals of time with constraints as a poset, and applying the Fishburn-Shepp inequality to guide a scheduling algorithm.

Let Q be a finite *poset* (partially ordered set) with n elements and C be a chain $1 < 2 < \cdots < c$. For (Q, C), a map $\omega : Q \to C$ is *strict order-preserving* if, for all $x, y \in Q$, $x < y$ implies $\omega(x) < \omega(y)$. Let $\lambda : Q \to \{1 < 2 < \cdots < n\}$ be a *linear extension* of Q, that is, an order-preserving injection.

A poset Q is equivalently a *directed acyclic graph (DAG)*, $G = (V, E)$; for temporal reasoning, the vertices represent time intervals, and edges between vertices are labeled with relations in Allen's algebra which satisfy the partial ordering. For scheduling problems, a linear extension λ of Q (or G) can be used to schedule tasks: λ must respect interval constraints, that is relations between comparable elements.

Algorithmically, a linear extension of a DAG, G, can be determined in linear time by performing a depth-first search of G [T-76]. The set of all linear extensions, Λ, can be generated in constant amortized time, that is $O(|\Lambda|)$, while the corresponding counting problem is #P-complete; and $P(x < y)$ can be computed in time $O(n^2 + |\Lambda|)$ [PR-94].

The Fishburn-Shepp inequality [F-84, S-82] is an inequality for the number of extensions of partial orders to linear orders, expressed as follows. Suppose that x, y and z are incomparable elements of a finite poset, then

$$P(x < y)P(x < z) < P((x < y) \land (x < z))$$

where P(*) is the probability that a linear extension has the property *. By re-expressing this in terms of conditional probability, $P(x < z) < P((x < z) \mid (x < y))$, we see that $P(x < z)$ strictly increases by adding the condition $x < y$. The proposed problem is to apply the Fishburn-Shepp inequality to efficiently find a favourable schedule under specified criteria.

The poset can be represented by an adjacency matrix, while recording the number of adjacencies for each element – likely candidates for applying the Fishburn-Shepp inequality have the least adjacencies. Furthermore, this inequality can be applied to reduce the total number of linear extensions, by substituting chains for the 3-element antichains, $x < y < z$ or $x < z < y$, according to which is considered more favourable. A naive scheduling algorithm is then:

- Construct incidence matrix of the poset - record number of adjacencies
- Choose incomparable triples x, y, z; select x
- For each triple t_i replace t_i by $x < y < z$ or $x < z < y$
- Find the set L_i of linear extensions with the induced chain t_i
- Select schedule from the intersection of L_i's
- If the intersection is empty then choose the schedule from the largest set L_i.

We will illustrate this idea with an example:

Example 8. Consider a set of time intervals of jobs, $\mathcal{J} = \{J_1, J_2, J_3, J_4, J_5, J_6\}$ with a partial order $\leq$ defined by $(X^- < Y^-) \land (X^+ \leq Y^+)$, which requires scheduling. Suppose that $\mathcal{J}$ is the poset / DAG shown below, where the edges have been labeled with satisfying temporal relations, and there are two triples of incomparable elements: $\{J_2, J_3, J_5\}$ and $\{J_2, J_3, J_6\}$. Then

$P(J_2 < J_3) \land (J_2 < J_5)) = 6/30$, $P((J_3 < J_2) \land (J_3 < J_5)) = 6/30$, and $P((J_5 < J_2) \land (J_5 < J_3)) = 18/30$;

$P((J_2 < J_3) \land (J_2 < J_6)) = 12/30$, $P((J_3 < J_2) \land (J_3 < J_6)) = 12/30$, and $P((J_6 < J_2) \land (J_6 < J_3)) = 6/30$.

The largest set of linear extensions corresponding to the first triple is when $(J_5 < J_2) \land (J_5 < J_3)$ and for the second triple is (w.l.o.g.) when $(J_2 < J_3) \land (J_2 < J_6)$. Their intersection has 6 linear extensions and we arbitrarily choose $J_1 \leq J_5 \leq J_2 \leq J_6 \leq J_3 \leq J_4$ as a schedule – so, although any linear extension would suffice we are selecting one which appears more "suitable" to satisfy. A parallel solution to this temporal constraint satisfaction problem is: $J_1 = (10, 20)$, $J_5 = (5, 10)$, $J_2 = (12, 23)$, $J_6 = (20, 22)$, $J_3 = (15, 20)$ and $J_4 = (18, 23)$. However, had we induced the chain $J_5 \leq J_2 \leq J_3$ or the chain $J_2 \leq J_3 \leq J_6$ in the poset, then there would only be 9 linear extensions to compute; and only 4 extensions if we induce both chains.

Note that the partial order $(X^- < Y^-) \land (X^+ \leq Y^+)$ does not include the basic interval relation *during*, and so it is not equivalent to the ORD-Horn class $\mathcal{H}$.

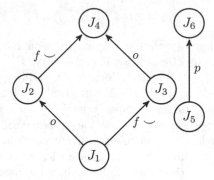

We believe this approach is worthy of fuller investigation. Questions and directions to consider: Define suitable partial orders; how/many to choose 3-element antichains and selection of the x element; iterate reducing 3-element antichains to chains; specify the scheduling criteria or heuristic; relate the partial orders to tractable and other classes of Allen's algebra; determine applications related to the posets; specify the number of relations captured by a partial order; determine limits of parallelism; and complexity issues.

5   Conclusion

We reviewed the temporal constraint satisfaction problem for reasoning in Allen's interval algebra. Jeavons *et al.* [JCG-97] established that any tractable set of relations must be closed under certain algebraic operations. Krokhin *et al.* showed that there are exactly eighteen maximal tractable subclasses [KJJ-03]. These theoretical methods improved on computer-assisted search for establishing tractability. We illustrated the algebraic method for tractability within the class $\mathcal{C}$.

Graph-based representations for temporal reasoning have been discussed: interval and sequence graphs, and DAGs. We concluded by proposing future research into representing a collection of intervals of time with constraints as a poset and applying the Fishburn-Shepp inequality to select a favourable schedule: the idea is clarified with a naive algorithm.

References

[A-83] Allen, J.F.: Maintaining knowledge about temporal intervals. Commun. ACM **26**(11), 832–843 (1983)

[A-91] Allen, J.F.: Temporal reasoning and planning. In: Allen, J.F., Kautz, H.A., Pelavin, R.N., Tenenberg, J.D. (eds.) Reasoning About Plans, Chapter 1, pp. 1–67. Morgan Kaufmann, San Mateo (1991)

[CS-73] Coombs, C.H., Smith, J.E.K.: On the detection of structures in attitudes and developmental processses. Psych. Rev. **80**, 337–351 (1973)

[DMP-91] Dechter, R., Meiri, I., Pearl, J.: Temporal constraint networks. Artif. Intell. **49**(1–3), 61–95 (1991)

[D-92] Dorn, J.: Temporal reasoning in sequence graphs. In: AAAI 1992, pp. 735–740 (1992)

[DJ-97]   Drakengren, T., Jonsson, P.: Eight maximal tractable subclasses of
            Allen's algebra with metric time. J. Artif. Intell. Res. **7**, 25–45 (1997)
[DJ-97ii]   Drakengren, T., Jonsson, P.: Twenty-one large tractable subclasses of
            Allen's algebra. Artif. Intell. **93**(1–2), 297–319 (1997)
[DJ-98]   Drakengren, T., Jonsson, P.: A complete classification of tractability in
            Allen's algebra relative to subsets of basic relations. Artif. Intell. **106**(2),
            205–219 (1998)
[F-84]   Fishburn, P.C.: A correlational inequality for linear extensions of a poset.
            Order **1**(2), 127–137 (1984)
[GSS-93]   Gerevini, A., Schubert, L.K., Schaeffer, S.: Temporal reasoning in Time-
            graph I-II. SIGART Bull. **4**(3), 21–25 (1993)
[GKS-94]   Golumbic, M.C., Kaplan, H., Shamir, R.: On the complexity of DNA
            physical mapping. Adv. Appl. Math. **15**, 251–261 (1994)
[GS-92]   Golumbic, M.C., Shamir, R.: Algorithms and complexity for reasoning
            about time. In: AAAI 1992, pp. 741–747 (1992)
[GS-93]   Golumbic, M.C., Shamir, R.: Complexity and algorithms for reasoning
            about time: a graph theoretic approach. J. ACM **40**(5), 1108–1133 (1993)
[JCG-97]   Jeavons, P., Cohen, D.A., Gyssens, M.: Closure properties of constraints.
            J. ACM **44**(4), 527–548 (1997)
[JB-96]   Jonsson, P., Bäckström, C.: A linear-programming approach to temporal
            reasoning. In: AAAI 1996, pp. 1235–1240 (1996)
[JCMPM-10] Juarez, J.M., Campos, M., Morales, A., Palma, J., Marin, R.: Applica-
            tions of temporal reasoning to intensive care units. J. Healthc. Eng. **1**(4),
            615–636 (2010)
[K-93]   Karp, R.M.: Mapping the genome: some combinatorial problems arising
            in molecular biology. In: STOC 1993, pp. 278–285 (1993)
[KL-91]   Kautz, H.A., Ladkin, P.B.: Integrating metric and qualitative temporal
            reasoning. In: AAAI 1991, pp. 241–246 (1991)
[KKD-08]   Krieger, H.-U., Kiefer, B., Declerck, T.: A framework for temporal rep-
            resentation and reasoning in business intelligence applications. In: AAAI
            Spring Symposium: AI Meets Business Rules and Process Management,
            pp. 59–70 (2008)
[KJJ-03]   Krokhin, A., Jeavons, P., Jonsson, P.: Reasoning about temporal rela-
            tions: the tractable subalgebras of Allen's interval algebra. J. ACM
            50(5), 591–640 (2003)
[LM-88]   Ladkin, P.B., Maddux, R.D.: On binary constraint networks, Technical
            report KES. U.88.8, Kestrel Institute, Palo Alto, CA (1988)
[L-97]   Ligozat, G.: Figures for thought: temporal reasoning with pictures, AAAI
            Technical report WS-97-11, pp. 31–36 (1997)
[L-98]   Ligozat, G.: "Corner" relations in Allen's algebra. Constraints **3**(2–3),
            165–177 (1998)
[M-77]   Mackworth, A.K.: Consistency in networks of relations. Artif. Intell.
            8(1), 99–118 (1977)
[MZ-08]   Mandoiu, I., Zelikovsky, A. (eds.): Bioinformatics Algorithms: Tech-
            niques and Applications. Wiley Series in Bioinformatics. Wiley Inter-
            science, Hoboken (2008)
[M-91]   Meiri, I.: Combining qualitative and quantitative constraints in temporal
            reasoning. In: AAAI 1991, pp. 260–267 (1991)
[NB-95]   Nebel, B., Bürckert, H.-J.: Reasoning about temporal relations: a maxi-
            mal tractable subclass of Allen's interval algebra. J. ACM **42**(1), 43–66
            (1995)

[N-91] Nökel, K. (ed.): Temporally Distributed Symptoms in Technical Diagnosis. LNCS (LNAI), vol. 517. Springer, Heidelberg (1991)

[PS-97] Pe'er, I., Shamir, R.: Satisfiability problems on intervals and unit intervals. Theoret. Comput. Sci. **175**(2), 349–372 (1997)

[PR-94] Pruesse, G., Ruskey, F.: Generating linear extensions fast. SIAM J. Comput. **23**(2), 373–386 (1994)

[S-82] Shepp, L.A.: The XYZ conjecture and the FKG inequality. Ann. Probab. **10**(3), 824–827 (1982)

[SC-88] Song, F., Cohen, R.: The interpretation of temporal relations in narrative. In: AAAI 1988, pp. 745–750 (1988)

[T-76] Tarjan, R.E.: Edge-disjoint spanning trees and depth-first search. Acta Inform. **6**(2), 171–185 (1976)

[VB-89] van Beek, P.: Approximation algorithms for temporal reasoning. In: IJCAI 1989, pp. 1291–1296 (1989)

[VB-90] van Beek, P.: Reasoning about qualitative temporal information. In: AAAI 1990, pp. 728–734 (1990)

[VB-92] van Beek, P.: Reasoning about qualitative temporal information. Artif. Intell. **58**(1–3), 297–326 (1992)

[VBC-90] van Beek, P., Cohen, R.: Exact and approximate reasoning about temporal relations. Comput. Intell. **6**(3), 132–144 (1990)

[V-82] Vilain, M.B.: A system for reasoning about time. In: AAAI 1982, pp. 197–201 (1982)

[VK-86] Vilain, M., Kautz, H.: Constraint propagation algorithms for temporal reasoning. In: AAAI 1986, pp. 377–382 (1986)

[VKVB-89] Vilain, M., Kautz, H., van Beek, P.: Constraint propagation algorithms for temporal reasoning: a revised report. In: Weld, D.S., de Kleer, J. (eds.) Readings in Qualitative Reasoning about Physical Systems, pp. 373–381. Morgan Kaufmann, California (1989)

[WH-90] Ward, S.A., Halstead, R.H.: Computation Structures. MIT Press, Cambridge, Mass (1990)

Reconstructing a Sparse Solution
from a Compressed Support Vector Machine

Joachim Giesen$^{(\boxtimes)}$, Sören Laue, and Jens K. Mueller

Friedrich-Schiller-Universität Jena, Jena, Germany
joachim.giesen@uni-jena.de

Abstract. A support vector machine is a means for computing a binary classifier from a set of observations. Here we assume that the observations are n feature vectors each of length m together with n binary labels, one at each observed feature vector. The feature vectors can be combined into a $n \times m$ feature matrix. The classifier is computed via an optimization problem that depends on the feature matrix. The solution of this optimization problem is a vector of dimension m from which a classifier with good generalization properties can be computed directly. Here we show that the feature matrix can be replaced by a compressed feature matrix that comprises n feature vectors of length $\ell < m$. The solution of the optimization problem for the compressed feature matrix has only dimension ℓ and can computed faster since the optimization problem is smaller. Still, the solution to the compressed problem needs to be related to the original solution. We present a simple scheme that reconstructs the original solution from a solution of the compressed problem up to a small error. For the reconstruction guarantees we assume that the solution of the original problem is sparse. We show that sparse solutions can be promoted by a feature selection approach.

1 Introduction

In a binary classification problem we are given observations

$$(x_1, y_1), \ldots, (x_n, y_n)$$

of labels $y_i \in \{-1, 1\}$ at data points x_i in some space Ω. The goal is to learn a classifier (predictor) from the observations, i.e., a mapping $\Omega \to \{-1, 1\}$ that maps the points in Ω to the label space $\{-1, 1\}$ and can be used to predict the label at any point in Ω. Typically, the points in Ω are described by features, e.g., by numerical feature functions $e_j : \Omega \to \mathbb{R}$. If m is the number of feature functions, then we can combine the features at $x \in \Omega$ into a feature vector

$$\Phi(x) := \big(e_1(x), \ldots, e_m(x)\big) \in \mathbb{R}^m.$$

The features of the observations can thus be encoded in a feature matrix $\Phi \in \mathbb{R}^{n \times m}$ whose i-th row Φ_i is given as $\Phi_i = \Phi(x_i)$, i.e., the feature vector of the i-th data point.

© Springer International Publishing Switzerland 2016
I.S. Kotsireas et al. (Eds.): MACIS 2015, LNCS 9582, pp. 305–319, 2016.
DOI: 10.1007/978-3-319-32859-1_26

A particularly simple yet powerful class of classifiers are linear classifiers (i.e., linear in the possibly non-linear feature functions)

$$\Omega \ni x \mapsto \mathrm{sign}\left(\langle \Phi(x), w \rangle\right) \in \mathbb{R},$$

with $w \in \mathbb{R}^m$. Hence, for linear classification the learning problem reduces to determining w from the feature matrix Φ of the observations. In the by now classical support vector machine approach, see [3], this is accomplished through the following Euclidean regularized optimization/learning problem,

$$\min_{w \in \mathbb{R}^m} \left(L(\Phi w, y) + c\|w\|_2^2\right)$$

where

$$L(\Phi w, y) = \frac{1}{n} \sum_{i=1}^{n} \max\{0, 1 - y_i \langle \Phi_i, w \rangle\}$$

is the hinge loss function, and $c > 0$ is a regularization parameter that controls the trade-off between the loss term and the regularization term. See also [15] for a textbook introduction into support vector machines and related topics.

The dimension of a support vector machine (optimization problem), i.e., the number of optimization variables, is the number of feature functions m. In the case that $\Omega = \mathbb{R}^m$, the feature functions are often chosen to be the coordinate functions (aka linear support vector machines)

$$e_j : x = \left(x^{(1)}, \ldots, x^{(m)}\right) \mapsto x^{(j)},$$

but m can be as large as n, as it is a common choice to have one feature function for every data point, or it can be even larger than n, for instance in the case of images, where m can be the number of pixels in an image while n is the number of images. In the following we assume that the problem is high-dimensional, i.e., the number of features m is in $\omega(\log n)$, where n is the number of data points.

The goal of compressing a support vector machine is to replace the original optimization problem by a problem with fewer optimization variables that can be solved faster. In (random projection) support vector compression the original feature vectors Φ_i are projected down to feature vectors $\bar{\Phi}_i$ using a random projection matrix $\Lambda \in \mathbb{R}^{\ell \times m}$. Hence, the compressed feature matrix $\bar{\Phi}$ whose rows are $\bar{\Phi}_i = \Phi_i \Lambda^T$ has only $\ell \cdot n$ entries, whereas the original feature matrix has $n \cdot m$ entries. The support vector machine for the compressed feature vectors, i.e., the compressed support vector machine,

$$\min_{\bar{w} \in \mathbb{R}^\ell} \left(L(\bar{\Phi}\bar{w}, y) + c\|\bar{w}\|_2^2\right).$$

has only dimension ℓ. The compressed support vector machine like the original support vector machine can be solved efficiently, namely in time linear in the size of the (compressed) feature matrix (i.e., the number of non-zero entries), using for example trust region Newton methods, see [11], or cutting plane methods, see [7]. Hence, compressing the original support vector machine only results in an asymptotically better running time if computing the compressed feature matrix takes at most linear time in the number of the non-zeros of the original feature matrix. Still, turning to a compressed support vector machine can result in a speed-up even if computing the compressed feature matrix takes super-linear time in the size of the original feature matrix, namely in the case that the feature vectors do not fit into primary memory, e.g., cache or main memory. The support vector machine problem is well known to be memory bound and not compute bound since fast support vector machine solvers need to touch each input datum only a few times. Fast support vector machine solvers [7, 11] work in iterations. If the feature vectors do not fit into the primary memory, then they need to be read again from secondary memory in every iteration. Hence, compressing the feature vectors in streaming fashion from secondary into primary memory helps to avoid accessing the slower secondary memory repeatedly. Important for the choice of the projection matrix is not only the possible speed-up, but also the ability of the resulting classifier to generalize to new data points, i.e., its statistical performance that for example can be measured in terms of a cross-validation error should not deteriorate when turning to a compressed support vector machine. The ideal case is that the optimal solution of the original support vector machine can be (approximately) reconstructed from a solution of the compressed support vector machine. In this case not only the statistical properties are well preserved, but also the resulting classifier operates on the original feature space, where the features often have a well defined meaning.

Contributions. We analyze support vector compression using standard Gaussian projection matrices Λ, and show that if ℓ, i.e., the length of the compressed feature vectors, is chosen in $\Omega(\log \max\{n, m\})$, then with high probability, the solution of the original support vector machine can be approximated arbitrarily well by $\Lambda^T \bar{w}$, where $\bar{w}$ is a solution of the compressed support vector machine. The constant that is hidden in the big-O notation for the target dimension depends on the approximation guarantee and on the sparsity of the optimal solution, which is a weaker assumption than what has been used before in the literature (see the paragraph on related work). We also show how to augment support vector machines by a feature selection approach that promotes sparse solutions.

The original feature matrix needs space in the order of $O(nm)$, whereas the compressed feature matrix only needs space of the order $O(n \log \max\{n, m\})$ if $\ell \in \Theta(\log \max\{n, m\})$. For compression and reconstruction we also need to store the projection matrix Λ whose size is of the order $O(m \log \max\{n, m\})$. Hence, by compression we can reduce the required space from $O(nm)$ to $O((m + n) \log \max\{n, m\})$. The compressed feature matrix can be computed in time in $O(mn \log \max\{n, m\})$.

The compression scheme can be utilized as follows: the feature vectors Φ_i are streamed from slow secondary memory to fast primary memory, e.g., from tape or hard drive to main memory or from main memory into a faster cache, and get compressed on the fly into the new feature vectors $\bar{\Phi}_i$ that have only $O(\log \max\{n, m\})$ entries each. The support vector machine can then be solved

---

Algorithm 1. CompressAndReconstruct

Input: feature vectors $\Phi_1, \ldots, \Phi_n \in \mathbb{R}^m$
Output: approximate solution w of the support vector machine problem

$$\min_{w \in \mathbb{R}^m} \left(L(\Phi w, y) + c\|w\|_2^2 \right)$$

compute a random projection matrix $\Lambda \in \mathbb{R}^{\ell \times m}$
for $i = 1$ **to** n **do**
   read Φ_i from secondary memory and compute $\bar{\Phi}_i = \Phi_i \Lambda^T$
end for
compute a solution $\bar{w}$ of the compressed support vector machine problem

$$\min_{\bar{w} \in \mathbb{R}^\ell} \left(L(\bar{\Phi}\bar{w}, y) + c\|\bar{w}\|_2^2 \right)$$

return $w = \Lambda^T \bar{w}$

---

for the compressed feature matrix while accessing only the fast primary memory. Finally, the solution $\bar{w}$ of the compressed support vector machine can be expanded and the expansion is an approximate solution to the original problem with approximation guarantees. This pipeline is summarized in Algorithm 1.

Related Work. Since support vector machines are among the most intensively studied and best understood machine learning techniques, there has been also quite some work on dimension reduction and/or compression for support vector machines. Here we mention only the most relevant for our work.

To the best of our knowledge, the only work on compressed support vector machines, where computing the compressed feature matrix can be accomplished in time linear in the number of non-zeros of the original feature matrix, is by Paul et al. [14] who provide generalization (margin) bounds for compressed linear support vector machines. Their work builds on results by Clarkson and Woodruff [2] and extensions by Meng and Mahoney [12] and Nelson and Nguyen [13], respectively. Clarkson and Woodruff have improved the running time for computing a low-dimensional embedding of a sparse feature matrix to a time linear in the number of non-zeros in the feature matrix. The target dimension ℓ in the scheme of Paul et al. [14] depends on the rank of the original feature matrix, whereas our target dimension depends on the sparsity of the optimal solution of the original support vector machine. Note that the low rank assumption implies that the solution is sparse in some basis which is enough for our guarantees to hold. Another difference is that we are aiming at approximating the original support vector solution with strong guarantees.

The only result that we are aware of that approximates the optimal solution of the original support vector machine from its compressed counterpart is by Zhang et al. [19] who also use Gaussian projection matrices. Their reconstruction result is again based on the assumption that the feature matrix is of low rank or can be well approximated by a low rank matrix, whereas we only need the weaker assumption that the solution of the original support vector machine is sparse. We show that sparsity of the solution can always be promoted by combining the support vector machine with some feature selection method (see Sect. 4). Feature selection is of independent interest since sparse solutions are often favored in practice because they provide predictors that are easier to interpret and faster to evaluate.

On the practical side, Yu et al. [18] address the support vector machine problem for large data that do not fit into main memory in a block minimization framework. In their framework the data are divided into blocks that fit into main memory, and at each step one block is loaded into main memory and handled by either a primal or dual support vector machine solver. The framework has been tested on data sets that are 20 times larger than the available main memory. Vedaldi and Zisserman [16] construct high-dimensional sparse feature vectors from arbitrary kernels using among others a technique called product quantization. One motivation for their work is to compress the given data—in their case large image descriptors, such that it fits into main memory. Paul et al. [14] also report on experiments for a data set (HapMap-HGDP) whose feature matrix is too large to fit into main memory. Compression turned out to be practically beneficial although the feature matrix is dense, i.e., the number of non-zeros is large.

2   Compression

In our standard compression scheme we are working with random combinations of the given feature functions. To simplify the exposition of our arguments we assume without loss of generality that the feature functions have been scaled such that all the feature vectors are contained in the m-dimensional Euclidean unit ball $\mathbb{B}^m$, e.g., by scaling the feature functions such that $e_j : \Omega \to \left(\frac{-1}{\sqrt{m}}, \frac{1}{\sqrt{m}} \right)$. From the scaled feature functions $e_1, \ldots, e_m$ we compute a smaller set of feature functions $\bar{e}_1, \ldots, \bar{e}_\ell$ with $\ell < m$ as follows

$$\bar{e}_i = \sum_{j=1}^{m} \lambda_{ij} e_j, \quad i = 1, \ldots, \ell$$

where the λ_{ij} are independent, identically distributed Gaussian random variables with expectation 0 and variance $1/m$, i.e., $\lambda_{ij} \sim \mathcal{N}\left(0, \frac{1}{m}\right)$. Using the new feature functions we replace the feature vectors

$$\Phi_i = \left(e_1(x_i), \ldots, e_m(x_i) \right)$$

by the compressed feature vectors

$$\bar{\Phi}_i = \left(\bar{e}_1(x_i), \ldots, \bar{e}_\ell(x_i) \right), \ i = 1, \ldots, n.$$

The support vector machine for the compressed feature vectors reads as

$$\min_{\bar{w} \in \mathbb{R}^\ell} \left(L(\bar{\Phi}\bar{w}, y) + c \|\bar{w}\|_2^2 \right),$$

where $\bar{\Phi} \in \mathbb{R}^{n \times \ell}$ is the matrix whose rows are the ℓ-dimensional feature vectors $\bar{\Phi}_i, i = 1, \ldots, n$. Thus, $\bar{\Phi} = \Phi \Lambda^T$, where $\Lambda = (\lambda_{ij}) \in \mathbb{R}^{\ell \times m}$ is the random Gaussian projection matrix.

3   Reconstruction

In this section we prove our main reconstruction result that is summarized in the following theorem. The theorem states that the optimal solution of the original support vector machine problem can be reconstructed by $\Lambda^T \bar{w}$ for an optimal or approximate solution $\bar{w}$ of the compressed problem up to a small error.

Theorem 1. *Assume that the optimal solution w of the original support vector machine is l-sparse, i.e., only some constant l of the m entries of w are non-zero in some fixed (that is data independent) basis. Then, for a given $\delta \in (0, 1/2)$, the solution w and the reconstruction $\Lambda^T \bar{w}$ from the optimal solution $\bar{w}$ of the compressed machine can be related as follows,*

$$(1 + (1 + 1/c)\delta)^{-2} \left(L(\Phi \Lambda^T \bar{w}, y) + c \|\Lambda^T \bar{w}\|_2^2 \right) - \delta \leq L(\Phi w, y) + c \|w\|_2^2$$
$$\leq L(\Phi \Lambda^T \bar{w}, y) + c \|\Lambda^T \bar{w}\|_2^2$$

with probability at least $1 - \frac{3}{\min\{m,n\}}$ for sufficiently large $\ell \in \Omega\left(\frac{l \log \max\{n,m\}}{\delta^2} \right)$.

We have subdivided the proof of Theorem 1 into several lemmas that exploit the structure of the projection matrix Λ and the assumption that the solution of the original support vector machine is l-sparse. The proofs combine ideas from low distortion embeddings and compressive sensing.

The first property that we want to exploit is that the projection matrix Λ is a "Johnson-Lindenstrauss"-transform, see [8], i.e., we have the following properties (see for example [17] (Page 2, Lemma 1.3) for a proof).

Lemma 1. *Let P be a finite subset of $\mathbb{R}^m$ with $n + 1$ elements, let $\delta \in (0, 1/2)$ and $\ell \geq \frac{20}{\delta^2} \log(n + 1)$. Then,*

1. *for all $p \in P$ it holds with probability at least $1 - 2(n+1)\exp(-(\delta^2 - \delta^3)\ell/4)$ that*

$$(1 - \delta)\|p\|_2^2 \leq \|\Lambda p\|_2^2 \leq (1 + \delta)\|p\|_2^2,$$

2. *for any fixed $p \in P$ and all $q \in P - \{p\}$ it holds with probability at least $1 - 4n\exp(-(\delta^2 - \delta^3)\ell/4)$ that*

$$|\langle p, q \rangle - \langle \Lambda p, \Lambda q \rangle| < \delta\|p\|\|q\|. \qquad \square$$

If we apply Lemma 1 to the feature vectors $\Phi_i \in \mathbb{B}^m$ and some fixed $w \in \mathbb{R}^m$ we get the following lemma.

Lemma 2. *For any fixed $w \in \mathbb{R}^m$, $\delta \in (0, 1)$ and $\ell \geq \frac{8}{\delta^2 - \delta^3} \log(6(n+1))$ it holds with probability at least $1 - \frac{1}{n+1}$ that*

$$|L(\bar{\Phi}\Lambda w, y) - L(\Phi w, y)| < \delta\|w\|_2.$$

Proof. Let $P = \{\Phi_1, \ldots, \Phi_n, w\}$. By Lemma 1 and taking a union bound (over the two events in Lemma 1), if we choose

$$\ell \geq \frac{8}{\delta^2 - \delta^3} \log(6(n+1)),$$

then with probability at least $1 - \frac{1}{n+1}$, the norms of all the points in P are preserved up to a factor $(1 \pm \delta)$ and the n scalar products $|\langle \Phi_i, w \rangle|$ are simultaneously preserved up to an additive error of $\delta \|w\|_2$, since by assumption the feature vectors Φ_i are contained in the m-dimensional Euclidean unit ball $\mathbb{B}^m$. Hence, we have

$$|\langle \bar{\Phi}_i, \Lambda w \rangle - \langle \Phi_i, w \rangle| < \delta \|w\|_2$$

which implies

$$|1 - y_i \langle \bar{\Phi}_i, \Lambda w \rangle - (1 - y_i \langle \Phi_i, w \rangle)| < \delta \|w\|_2,$$

and thus

$$|\max\{0, 1 - y_i \langle \bar{\Phi}_i, \Lambda w \rangle\} - \max\{0, 1 - y_i \langle \Phi_i, w \rangle\}| < \delta \|w\|_2$$

from which we derive

$$|L(\bar{\Phi} \Lambda w, y) - L(\Phi w, y)| < \delta \|w\|_2$$

with probability at least $1 - \frac{1}{n+1}$ by using the definition

$$L(\Phi w, y) = \frac{1}{n} \sum_{i=1}^{n} \max\{0, 1 - y_i \langle \Phi_i, w \rangle\}. \qquad \square$$

Next we want to exploit that the optimal solution w of the original problem is l-sparse in some fixed basis (that means, a basis independent of the data points $x_1, \ldots, x_n$), i.e., only $l < m$ of the optimal coefficients are non-zero in this basis. To exploit sparsity we use that Λ satisfies the restricted isometry property, see for example [1,5,6], with high probability. A matrix $\Lambda \in \mathbb{R}^{\ell \times m}$ satisfies the restricted isometry property (RIP) with constant $\delta_l \in (0, 1)$ if

$$(1 - \delta_l) \|p\|_2^2 \leq \|\Lambda p\|_2^2 \leq (1 + \delta_l) \|p\|_2^2$$

for all l-sparse vectors $p \in \mathbb{R}^m$, and δ_l is minimal with this property. In fact, we have the following lemma for the projection matrix Λ (see for example [6] (Page 18, Theorem 3.6) for more details).

Lemma 3. *There exists a constant $C > 0$ such that the restricted isometry constant δ_l of Λ is upper bounded by $\delta > 0$ with probability at least $1 - \varepsilon$ provided that*

$$\ell \geq \frac{C}{\delta^2} \left(l \log(m/\ell) - \log(1/\varepsilon) \right). \qquad \square$$

Note that even if w is not l-sparse in the orthonormal basis that corresponds to the feature functions $\{e_i\}$, it is by our assumption l-sparse with respect to some fixed orthonormal basis, i.e., a basis that is independent of the specific problem instance and its optimal solution w. The fixed basis can be derived from $\{e_i\}$ by applying an orthonormal transform U, i.e., Uw is l-sparse. Note that

$$\|w\|_2^2 = \|Uw\|_2^2 \quad \text{and} \quad \|Aw\|_2^2 = \|AU^TUw\|_2^2,$$

and the matrix AU^T still satisfies the restricted isometry property. Hence, we can use the following lemma for the optimal solution of the original problem.

Lemma 4. *Let $w \in \mathbb{R}^m$ be l-sparse. For the projection matrix Λ it holds that*

$$\|Aw\|_2^2 \leq (1+\delta)\|w\|_2^2$$

with probability at least $1 - 1/m$, provided $\ell \geq (C \cdot l \log m)/\delta^2$ for some constant $C > 0$.

Proof. The proof follows immediately from the RIP for the projection matrix Λ if we set $\varepsilon = 1/m$. □

Next we give an upper bound on the optimal value of the original problem.

Lemma 5. *Let $w \in \mathbb{R}^m$ be the optimal solution of the original support vector machine. For the projection matrix Λ it holds for any $\bar{w} \in \mathbb{R}^\ell$ that*

$$L(\Phi w, y) + c\|w\|_2^2 \leq L(\Phi A^T \bar{w}, y) + c\|A^T \bar{w}\|_2^2 = L(\bar{\Phi}\bar{w}, y) + c\|A^T \bar{w}\|_2^2.$$

Proof. Note that $\Lambda^T \bar{w}$ is feasible for the original problem. Hence, the optimality of w for the original problem implies that

$$L(\Phi w, y) + c\|w\|_2^2 \leq L(\Phi A^T \bar{w}, y) + c\|A^T \bar{w}\|_2^2 = L(\bar{\Phi}\bar{w}, y) + c\|A^T \bar{w}\|_2^2,$$

where the last equality follows from $\bar{\Phi} = \Phi A^T$. □

The next lemma gives an upper bound for the norm of the reconstruction $\Lambda^T \bar{w}$.

Lemma 6. *For the projection matrix Λ it holds for any $\bar{w} \in \mathbb{R}^\ell$ that*

$$\|A^T \bar{w}\|_2^2 \leq (1+\delta)\|\bar{w}\|_2^2$$

with probability at least $1 - \ell^2 \exp\left(-\frac{m\delta^2}{40\ell^2}\right)$ for large enough ℓ.

Proof. We need to bound $\|A^T \bar{w}\|_2^2 = \langle A^T \bar{w}, A^T \bar{w}\rangle$. The following holds,

$$\langle A^T \bar{w}, A^T \bar{w}\rangle = \sum_{i=1}^{m}\left(\sum_{j=1}^{\ell} \lambda_{ij}\bar{w}_j\right)^2$$

$$= \sum_{i=1}^{m}\left(\sum_{j=1}^{\ell} \lambda_{ij}^2\bar{w}_j^2 + 2\sum_{j=1}^{\ell-1}\sum_{k=j+1}^{\ell} \lambda_{ij}\lambda_{ik}\bar{w}_j\bar{w}_k\right)$$

$$= \sum_{j=1}^{\ell}\bar{w}_j^2\left(\sum_{i=1}^{m}\lambda_{ij}^2\right) + 2\sum_{j=1}^{\ell-1}\sum_{k=j+1}^{\ell}\bar{w}_j\bar{w}_k\left(\sum_{i=1}^{m}\lambda_{ij}\lambda_{ik}\right).$$

The sums $\sum_{i=1}^{m}(m\lambda_{ij}^2), j = 1,\dots,\ell$ are χ^2 distributed with m degrees of freedom. From well known concentration bounds for χ^2 distributions, see for example [17], it follows that

$$P\left[\sum_{i=1}^{m}\lambda_{ij}^2 \geq 1 + \frac{\delta}{2}\right] = P\left[\sum_{i=1}^{m}m\lambda_{ij}^2 \geq \left(1 + \frac{\delta}{2}\right)m\right] \leq \exp\left(-\frac{m\,\delta^2}{16}\left(1 - \frac{\delta}{2}\right)\right).$$

Next we bound the sums $\sum_{i=1}^{m}\lambda_{ij}\lambda_{ik}$, i.e., sums over products of independent identically normally distributed random variables. Its moment generating function is

$$M(t) = (1 - t^2/m^2)^{-m/2}.$$

Applying Chernoff's bounding method for a random variable X and $s > 0$, i.e.,

$$P[X \geq s] = P\left[\exp(tX) \geq \exp(st)\right] \leq \frac{E\left[\exp(tX)\right]}{\exp(st)} = \frac{M(t)}{\exp(st)}$$

for all $t > 0$, to $\sum_{i=1}^{m}\lambda_{ij}\lambda_{ik}$ gives

$$P\left[\sum_{i=1}^{m}\lambda_{ij}\lambda_{ik} \geq \frac{\delta}{4\ell}\right] \leq (1 - t^2/m^2)^{-m/2}\exp\left(-\frac{t\delta}{4\ell}\right).$$

Setting $t = m\sqrt{1-\tau}$ gives

$$P\left[\sum_{i=1}^{m}\lambda_{ij}\lambda_{ik} \geq \frac{\delta}{4\ell}\right] \leq \tau^{-m/2}\exp\left(-\frac{\delta m\sqrt{1-\tau}}{4\ell}\right)$$

$$= \exp\left(-\frac{m}{2}\log\tau\right)\exp\left(-\frac{\delta m\sqrt{1-\tau}}{4\ell}\right)$$

$$= \exp\left(-\frac{m}{4\ell}\left(2\ell\log\tau + \delta\sqrt{1-\tau}\right)\right).$$

A simple calculation shows that $f(\tau) = 2\ell\log\tau + \delta\sqrt{1-\tau}$ is maximized at

$$\tau = \frac{8\ell^2}{\delta^2}\left(\sqrt{1 + \frac{\delta^2}{4\ell^2}} - 1\right) = 1 - \frac{\delta^2}{16\ell^2} + \Theta\left(\frac{1}{\ell^4}\right).$$

Setting $\tau = 1 - (\delta/4\ell)^2$ gives by using $\log(1-x) = -x - \frac{x^2}{2} - \frac{x^3}{3} - \dots$ for $x < 1$,

$$f(\tau) = 2\ell\log\left(1 - \left(\frac{\delta}{4\ell}\right)^2\right) + \frac{\delta^2}{4\ell} = -\frac{\delta^2}{8\ell} - \Theta\left(\frac{1}{\ell^3}\right) + \frac{\delta^2}{4\ell} = \frac{\delta^2}{8\ell} - \Theta\left(\frac{1}{\ell^3}\right),$$

which is lower bounded by $\frac{\delta^2}{10\ell} > 0$ for large enough ℓ. Hence,

$$P\left[\sum_{i=1}^{m}\lambda_{ij}\lambda_{ik} \geq \frac{\delta}{4\ell}\right] \leq \exp\left(-\frac{m\delta^2}{40\ell^2}\right).$$

Finally, observing that $\sum_{j=1}^{\ell} \bar{w}_j^2 = \|\bar{w}\|_2^2$, and

$$\sum_{j=1}^{\ell-1} \sum_{k=j+1}^{\ell} \bar{w}_j \bar{w}_k \leq \|\bar{w}\|_1^2 \leq \ell\|\bar{w}\|_2^2$$

gives

$$\|\Lambda^T \bar{w}\|_2^2 = \langle \Lambda^T \bar{w}, \Lambda^T \bar{w} \rangle \leq \left(1 + \frac{\delta}{2}\right) \|\bar{w}\|_2^2 + 2\ell\frac{\delta}{4\ell}\|\bar{w}\|_2^2 = (1+\delta)\|\bar{w}\|_2^2$$

with probability at least $1 - \ell^2 \exp\left(-\frac{m\delta^2}{40\ell^2}\right)$ using a union bound.   □

Now we are finally prepared to prove the reconstruction theorem.

Proof [**of Theorem** 1]. The second inequality in the statement of the theorem has been shown in Lemma 5, and the first inequality follows from

$$\begin{aligned}
L(\Phi\Lambda^T \bar{w}, y) + c\|\Lambda^T \bar{w}\|_2^2 &= L(\bar{\Phi}\bar{w}, y) + c\|\Lambda^T \bar{w}\|_2^2 \\
&\leq L(\bar{\Phi}\bar{w}, y) + c(1+\delta)\|\bar{w}\|_2^2 \\
&\leq (1+\delta)\left(L(\bar{\Phi}\bar{w}, y) + c\|\bar{w}\|_2^2\right) \\
&\leq (1+\delta)\left(L(\bar{\Phi}\Lambda w, y) + c\|\Lambda w\|_2^2\right) \\
&\leq (1+\delta)\left(L(\Phi w, y) + \delta\|w\|_2 + c\|\Lambda w\|_2^2\right) \\
&\leq (1+\delta)\left(L(\Phi w, y) + \delta\|w\|_2 + c(1+\delta)\|w\|_2^2\right) \\
&\leq (1+\delta)\left(L(\Phi w, y) + (\delta + \delta\|w\|_2^2) + c(1+\delta)\|w\|_2^2\right) \\
&= (1+\delta)\left(L(\Phi w, y) + c(1 + (1+1/c)\delta)\|w\|_2^2 + \delta\right) \\
&\leq (1 + (1+1/c)\delta)(1+\delta)\left(L(\Phi w, y) + c\|w\|_2^2 + \delta\right) \\
&\leq (1 + (1+1/c)\delta)^2 \left(L(\Phi w, y) + c\|w\|_2^2 + \delta\right)
\end{aligned}$$

where the first inequality has been shown in Lemma 6, the second inequality follows from $\delta > 0$ and the non-negativity of the loss function, the third inequality follows from the optimality of $\bar{w}$ and the feasibility of Λw for the compressed problem, the fourth inequality has been shown in Lemma 2, the fifth inequality has been shown in Lemma 4, the sixth inequality is implied by $\delta\|w\|_2 \leq \delta$ if $\|w\|_2 \leq 1$ and $\delta\|w\|_2 \leq \delta\|w\|_2^2$ if $\|w\|_2 \geq 1$, and the last two inequalities are again implied by the positivity of δ and c and the non-negativity of the loss function.

Since Lemma 2 holds with probability at least $1 - \frac{1}{n+1} > 1 - \frac{1}{\min\{n,m\}}$, Lemma 4 holds with probability at least $1 - \frac{1}{m} \geq 1 - \frac{1}{\min\{n,m\}}$, and Lemma 6 holds with probability at least $1 - \ell^2 \exp\left(-\frac{m\delta^2}{40\ell^2}\right) > 1 - \frac{1}{\min\{n,m\}}$ (for sufficiently large m), a simple union bound gives that the theorem holds with probability at least $1 - \frac{3}{\min\{n,m\}}$.   □

In practice support vector machines are often not solved optimally, but only a δ'-approximation for some $\delta' > 0$ is computed. Let $\hat{w} \in \mathbb{R}^\ell$ be a δ'-approximate solution of the compressed support vector machine, i.e.,

$$\left(L(\bar{\Phi}\hat{w}, y) + c\,\|\hat{w}\|_2^2\right) - \left(L(\bar{\Phi}\bar{w}, y) + c\,\|\bar{w}\|_2^2\right) \leq \delta',$$

then it follows immediately from Theorem 1 that also the reconstruction $\Lambda^T\hat{w}$ provides an approximation of the optimal solution w for the original support vector machine whose approximation guarantee depends only on δ and δ', i.e. with high probability we have

$$(1 + (1 + 1/c)\delta)^{-2}\left(L(\Phi\Lambda^T\hat{w}, y) + c\,\|\Lambda^T\hat{w}\|_2^2\right) - \delta - \delta'$$
$$\leq L(\Phi w, y) + c\,\|w\|_2^2 \leq L(\Phi\Lambda^T\hat{w}, y) + c\,\|\Lambda^T\hat{w}\|_2^2.$$

4   Feature Selection

Our compression schemes works well when we can assume that the optimal solution of the original support vector machine is sparse. It is common practice to promote sparse solutions by L_1-regularization, i.e., by adding an L_1-regularization term $c_1\|w\|_1$ with regularization parameter c_1 to the original support vector machine problem. Unfortunately, Theorem 1 is no longer valid for the modified support vector machine since Lemmas 4 and 6 do not hold for the L_1-norm. This problem can be circumvented by the feature selection approach that we describe in the following, see also [10] for the idea. The feature selection approach is best motivated by using the adjoint problem formulation for support vector machines which is a consequence of the Representer Theorem, see [9]. The standard L_2-regularization parameter c is referred to as c_2 in the following.

Theorem 2 [Representer Theorem]. *For any loss function $L(\cdot, \cdot)$ if*

$$w = \operatorname{argmin}_{w' \in \mathbb{R}^m}\left(L(\Phi w', y) + c_2\|w'\|_2^2\right)$$

exists, then $w = \Phi^T a$ for some $a \in \mathbb{R}^n$.

It follows that we can optimize over $a \in \mathbb{R}^n$ instead of $w \in \mathbb{R}^m$, namely by substituting $w = \Phi^T a$ in the original support vector machine problem, which results in the equivalent adjoint formulation

$$\min_{a \in \mathbb{R}^n}\left(L(\Phi\Phi^T a, y) + c_2(a^T\Phi\Phi^T a)\right).$$

The matrix $\Phi\Phi^T \in \mathbb{R}^{n \times n}$ can also be written as

$$\Phi\Phi^T = \sum_{j=1}^m \Psi_j\Psi_j^T,$$

where Ψ_j is the j-th column of the matrix Φ. That is, $\Phi\Phi^T$ has been written as a sum of m rank-1 matrices $\Psi_j\Psi_j^T$ that correspond to the feature functions

$e_j, j = 1, \ldots, m$. Weighting the j-th feature function by $0 \le \mu_j \le 1$ results in new feature functions $e_j^{(\mu)} = \mu_j e_j$ and

$$\sum_{j=1}^{m} \Psi_j^{(\mu)} \left(\Psi_j^{(\mu)}\right)^T := \sum_{j=1}^{m} \mu_j^2 \Psi_j \Psi_j^T = \Phi^{(\mu)} \left(\Phi^{(\mu)}\right)^T,$$

with $\Phi^{(\mu)} := \Phi D$, where $D = D(\mu_1, \ldots, \mu_m) \in \mathbb{R}^{m \times m}$ is the diagonal matrix whose diagonal is the weight vector $\mu = (\mu_1, \ldots, \mu_m) \in [0,1]^m$. Sparse solutions can now be promoted by adding an L_1-regularization term for the weight vector μ which results in the following feature selective support vector machine

$$\min_{\mu \in [0,1]^m} \min_{w \in \mathbb{R}^m} \left(L(\Phi^{(\mu)} w, y) + c_2 \|w\|_2^2 + c_1 \|\mu\|_1\right)$$

that can be compressed as follows

$$\min_{\bar{\mu} \in [0,1]^m} \min_{\bar{w} \in \mathbb{R}^\ell} \left(L(\bar{\Phi}^{(\bar{\mu})} \bar{w}, y) + c_2 \|\bar{w}\|_2^2 + c_1 \|\bar{\mu}\|_1\right),$$

where $\bar{\Phi}^{(\bar{\mu})} \in \mathbb{R}^{n \times \ell}$ is the matrix whose rows are the ℓ-dimensional feature vectors

$$\bar{\Phi}_i^{(\bar{\mu})} = \left(\sum_{j=1}^{m} \lambda_{1j} e_j^{(\bar{\mu})}(x_i), \ldots, \sum_{j=1}^{m} \lambda_{\ell j} e_j^{(\bar{\mu})}(x_i)\right)$$

$$= \left(\sum_{j=1}^{m} \lambda_{1j} \bar{\mu}_j e_j(x_i), \ldots, \sum_{j=1}^{m} \lambda_{\ell j} \bar{\mu}_j e_j(x_i)\right)$$

for $i = 1, \ldots, n$. That is, $\bar{\Phi}^{(\bar{\mu})} = \Phi^{(\bar{\mu})} \Lambda^T$.

In the spirit of Theorem 1 we can prove an analogous theorem for the relationship between the feature selective support vector machine and its compressed counterpart.

Theorem 3. *For a given $\delta \in (0, 1/2)$, the optimal solution (w, μ) of the original feature selective support vector machine, where w is assumed to be l-sparse, and the reconstruction $(\Lambda^T \bar{w}, \bar{\mu})$ from the optimal solution $(\bar{w}, \bar{\mu})$ of the compressed feature selective support vector machine can be related as follows,*

$$(1 + (1 + 1/c_2)\delta)^{-2} \left(L(\Phi^{(\bar{\mu})} \Lambda^T \bar{w}, y) + c_2 \|\Lambda^T \bar{w}\|_2^2 + c_1 \|\bar{\mu}\|_1\right) - \delta$$

$$\le L(\Phi^{(\mu)} w, y) + c_2 \|w\|_2^2 + c_1 \|\mu\|_1$$

$$\le L(\Phi^{(\bar{\mu})} \Lambda^T \bar{w}, y) + c_2 \|\Lambda^T \bar{w}\|_2^2 + c_1 \|\bar{\mu}\|_1.$$

with probability at least $1 - \frac{3}{\min\{m,n\}}$ for sufficiently large $\ell \in \Omega\left(\frac{l \log \max\{n,m\}}{\delta^2}\right)$.

Proof. The first inequality follows from

$$L(\Phi^{(\bar\mu)}\Lambda^T\bar w, y) + c_2\,\|\Lambda^T\bar w\|_2^2 + c_1\,\|\bar\mu\|_1$$

$$= L(\bar\Phi^{(\bar\mu)}\bar w, y) + c_2\,\|\Lambda^T\bar w\|_2^2 + c_1\,\|\bar\mu\|_1$$

$$\leq L(\bar\Phi^{(\bar\mu)}\bar w, y) + c_2(1+\delta)\,\|\bar w\|_2^2 + c_1\,\|\bar\mu\|_1$$

$$\leq (1+\delta)\left(L(\bar\Phi^{(\bar\mu)}\bar w, y) + c_2\,\|\bar w\|_2^2 + c_1\,\|\bar\mu\|_1\right)$$

$$\leq (1+\delta)\left(L(\bar\Phi^{(\mu)}\Lambda w, y) + c_2\,\|\Lambda w\|_2^2 + c_1\,\|\mu\|_1\right)$$

$$\leq (1+\delta)\left(L(\Phi^{(\mu)}w, y) + \delta\|w\|_2 + c_2\,\|\Lambda w\|_2^2 + c_1\,\|\mu\|_1\right)$$

$$\leq (1+\delta)\left(L(\Phi^{(\mu)}w, y) + \delta\|w\|_2 + c_2\,(1+\delta)\|w\|_2^2 + c_1\,\|\mu\|_1\right)$$

$$\leq (1+\delta)\left(L(\Phi^{(\mu)}w, y) + (\delta + \delta\|w\|_2^2) + c_2\,(1+\delta)\|w\|_2^2 + c_1\,\|\mu\|_1\right)$$

$$= (1+\delta)\left(L(\Phi^{(\mu)}w, y) + c_2\,(1 + (1+1/c_2)\delta)\|w\|_2^2 + c_1\,\|\mu\|_1 + \delta\right)$$

$$\leq (1 + (1+1/c_2)\delta)(1+\delta)\left(L(\Phi^{(\mu)}w, y) + c_2\,\|w\|_2^2 + c_1\,\|\mu\|_1 + \delta\right)$$

$$\leq (1 + (1+1/c_2)\delta)^2\left(L(\Phi^{(\mu)}w, y) + c_2\,\|w\|_2^2 + c_1\,\|\mu\|_1 + \delta\right)$$

where the first inequality has been shown in Lemma 6, the second inequality follows from $\delta > 0$ and the non-negativity of both the loss function and $c_1\,\|\bar\mu\|_1$, the third inequality is implied by the optimality of $(\bar w, \bar\mu)$ and the feasibility of $(\Lambda w, \mu)$ for the compressed problem, the fourth inequality has been shown in Lemma 2, the fifth inequality has been shown in Lemma 4, the sixth inequality is implied by $\delta\|w\|_2 \leq \delta$ if $\|w\|_2 \leq 1$ and $\delta\|w\|_2 \leq \delta\|w\|_2^2$ if $\|w\|_2 \geq 1$, and the last two inequalities are again implied by the positivity of δ and c_2 and the non-negativity of both the loss function and $c_1\,\|\bar\mu\|_1$.

The second inequality follows from the optimality of (w, μ) and the feasibility of $(\Lambda^T\bar w, \bar\mu)$ for the original feature selective support vector machine, i.e.,

$$L(\Phi^{(\mu)}w, y) + c_2\,\|w\|_2^2 + c_1\,\|\mu\|_1 \leq L(\Phi^{(\bar\mu)}\Lambda^T\bar w, y) + c_2\,\|\Lambda^T\bar w\|_2^2 + c_1\|\bar\mu\|_1.$$

As in the proof of Theorem 1, since Lemma 2 holds with probability at least $1 - \frac{1}{n+1} > 1 - \frac{1}{\min\{n,m\}}$, Lemma 4 holds with probability at least $1 - \frac{1}{m} \geq 1 - \frac{1}{\min\{n,m\}}$, and Lemma 6 holds with probability at least $1 - \ell^2\exp\left(-\frac{m\delta^2}{40\ell^2}\right) > 1 - \frac{1}{\min\{n,m\}}$ (for sufficiently large m), a simple union bound gives that the theorem holds with probability at least $1 - \frac{3}{\min\{n,m\}}$. □

As for Theorem 1, if $(\hat w, \hat\mu)$ is only a δ'-approximate solution of the compressed feature selective support vector machine, then the reconstruction $(\Lambda^T\hat w, \hat\mu)$ is an approximation of the optimal solution (w, μ) for the original feature selective support vector machine, whose strong approximation guarantee depends only on δ and δ'.

The original and the compressed feature selective support vector machine can both be reformulated as convex-concave optimization problems that have a unique solution and can be solved efficiently.

5   Conclusions

We have shown that the optimal solution of a support vector machine can be reconstructed up to a small error from an optimal or approximate solution of a corresponding compressed support vector machine, where the compression is achieved through a standard Gaussian projection matrix. The approximation error depends on the required compression rate that can be controlled. The approximation guarantee depends only on the rather weak assumption that the solution to the original problem is sparse in some basis. This assumption is for example satisfied when the feature matrix is of low rank, an assumption that has been used before. We have also shown that our compression technique works well together with a feature selection approach that promotes sparse solutions.

Our approach is not restricted to standard support vector machines. Only Lemma 2 needs to be adapted for handling other loss functions, e.g., a logistic loss function or the Crammer Singer loss function, see [4], for multi-class classification. In future work it would be interesting to generalize the results to projection matrices that allow a faster computation of the compressed feature matrix.

Acknowledgments. This work has been carried out within the project CG Learning. The project CG Learning acknowledges the financial support of the Future and Emerging Technologies (FET) programme within the Seventh Framework Programme for Research of the European Commission, under FET-Open grant number: 255827.

References

1. Candès, E.J., Tao, T.: Near-optimal signal recovery from random projections: universal encoding strategies? IEEE Trans. Inf. Theory **52**(12), 5406–5425 (2006)
2. Clarkson, K.L., Woodruff, D.P.: Low rank approximation and regression in input sparsity time. In: Symposium on Theory of Computing Conference (STOC), pp. 81–90 (2013)
3. Cortes, C., Vapnik, V.: Support-vector networks. Mach. Learn. **20**(3), 273–297 (1995)
4. Crammer, K., Singer, Y.: On the learnability and design of output codes for multiclass problems. In: Computational Learning Theory (COLT), pp. 35–46 (2000)
5. Donoho, D.L.: Compressed sensing. IEEE Trans. Inf. Theory **52**(4), 1289–1306 (2006)
6. Fornasier, M., Rauhut, H.: Compressive Sensing, Chap. 2. Springer, New York (2011)
7. Joachims, T.: Training linear SVMs in linear time. In: ACM SIGKDD International Conference on Knowledge Discovery and Data Mining (KDD), pp. 217–226 (2006)

8. Johnson, W.B., Lindenstrauss, J.: Extensions of Lipschitz mappings into a Hilbert space. Contem. Math. **26**, 189–206 (1984)
9. Kimeldorf, G.S., Wahba, G.: A correspondence between Bayesian estimation on stochastic processes and smoothing by splines. Ann. Math. Stat. **41**, 495–502 (1970)
10. Lanckriet, G.R.G., Cristianini, N., Bartlett, P.L., El Ghaoui, L., Jordan, M.I.: Learning the kernel matrix with semi-definite programming. In: Proceedings of the Nineteenth International Conference (ICML), pp. 323–330 (2002)
11. Lin, C.-J., Weng, R.C., Keerthi, S.S.: Trust region newton method for logistic regression. J. Mach. Learn. Res. **9**, 627–650 (2008)
12. Meng, X., Mahoney, M.W.: Low-distortion subspace embeddings in input-sparsity time and applications to robust linear regression. In: Symposium on Theory of Computing Conference (STOC), pp. 91–100 (2013)
13. Nelson, J., Nguyen, H.L.: OSNAP: faster numerical linear algebra algorithms via sparser subspace embeddings. In: Annual IEEE Symposium on Foundations of Computer Science (FOCS), pp. 117–126 (2013)
14. Paul, S., Boutsidis, C., Magdon-Ismail, M., Drineas, P.: Random projections for support vector machines. In: International Conference on Artificial Intelligence (AISTATS) (2013)
15. Schölkopf, B., Smola, A.J.: Learning with Kernels: Support Vector Machines, Regularization, Optimization, and Beyond. MIT Press, Cambridge (2001)
16. Vedaldi, A., Zisserman, A.: Sparse kernel approximations for efficient classification and detection. In: IEEE Conference on Computer Vision and Pattern Recognition (ICCV) (2012)
17. Vempala, S.: The Random Projection Method. DIMACS: Series in Discrete Mathematics and Theoretical Computer Science Series. American Mathematical Society, Providence (2004)
18. Yu, H.-F., Hsieh, C.-J., Chang, K.-W., Lin, C.-J.: Large linear classification when data cannot fit in memory. In: ACM SIGKDD International Conference on Knowledge Discovery and Data Mining (KDD), pp. 833–842 (2010)
19. Zhang, L., Mahdavi, M., Jin, R., Yang, T.: Recovering optimal solution by dual random projection. In: Conference on Learning Theory (COLT) (2013)

Subquadratic-Time Algorithms for Abelian Stringology Problems

Tomasz Kociumaka, Jakub Radoszewski$^{(\boxtimes)}$, and Bartłomiej Wiśniewski

Faculty of Mathematics, Informatics and Mechanics, University of Warsaw,
Warsaw, Poland
{kociumaka,jrad}@mimuw.edu.pl, b.wisniewski@students.mimuw.edu.pl

Abstract. We propose the first subquadratic-time algorithms to a number of natural problems in abelian pattern matching (also called jumbled pattern matching) for strings over a constant-sized alphabet. Two strings are considered equivalent in this model if the numbers of occurrences of respective symbols in both of them, specified by their so-called Parikh vectors, are the same. We propose the following algorithms for a string of length n:

- Counting and finding longest/shortest abelian squares in $O(n^2/\log^2 n)$ time. Abelian squares were first considered by Erdös (1961); Cummings and Smyth (1997) proposed an $O(n^2)$-time algorithm for computing them.
- Computing all shortest (general) abelian periods in $O(n^2/\sqrt{\log n})$ time. Abelian periods were introduced by Constantinescu and Ilie (2006) and the previous, quadratic-time algorithms for their computation were given by Fici et al. (2011) for a constant-sized alphabet and by Crochemore et al. (2012) for a general alphabet.
- Finding all abelian covers in $O(n^2/\log n)$ time. Abelian covers were defined by Matsuda et al. (2014).
- Computing abelian border array in $O(n^2/\log^2 n)$ time.

This work can be viewed as a continuation of a recent very active line of research on subquadratic space and time jumbled indexing for binary and constant-sized alphabets (e.g., Moosa and Rahman, 2012). All our algorithms work in linear space.

Keywords: Jumbled pattern matching · Jumbled indexing · Abelian period · Abelian square

1 Introduction

Algorithmic abelian stringology has been extensively studied in the recent years. Abelian pattern matching (also called jumbled pattern matching) can be viewed

T. Kociumaka—Supported by Polish budget funds for science in 2013–2017 as a research project under the 'Diamond Grant' program.

J. Radoszewski—Supported by the Polish Ministry of Science and Higher Education under the 'Iuventus Plus' program in 2015–2016 grant no 0392/IP3/2015/73.

J. Radoszewski—Newton International Fellow at King's College London.

© Springer International Publishing Switzerland 2016
I.S. Kotsireas et al. (Eds.): MACIS 2015, LNCS 9582, pp. 320–334, 2016.
DOI: 10.1007/978-3-319-32859-1_27

as an approximate variant of regular pattern matching. The problem that has received most attention in this area is jumbled indexing; see [2–4,11,12,15,16]. In the binary case, initially it was known that there exists a jumbled index of linear size that can answer in constant time queries asking if there is a substring of the text that is commutatively equivalent to a pattern specified by the number of zeroes and ones (see [4]). However, the straightforward construction time of such an index was quadratic. Due to a number of works [2,15,16] $O(n^2/\log^2 n)$-time construction algorithm of such an index was obtained. This line of research eventually lead to a very recent breakthrough $O(n^{1.859})$-time algorithm by Chan and Lewenstein [3].

In this work we present the first subquadratic-time algorithms computing several basic notions of abelian stringology in the case of binary and constant-sized alphabet. This includes:

- Abelian squares, that were first considered from a combinatorial perspective by Erdös [8] and from the algorithmic perspective by Cummings and Smyth [7] (in the latter case, under the name of weak repetitions),
- Abelian periods, defined by Constantinescu and Ilie [5],
- Abelian covers, introduced by Matsuda et al. [14],
- Natural notions of abelian borders and abelian border array.

Cummings and Smyth [7] presented an $O(n^2)$-time algorithm for computing all abelian squares in a string of length n. Fici et al. [9] showed an $O(n^2)$-time algorithm computing abelian periods in a string over a constant-sized alphabet and Crochemore et al. [6] solved this problem in $O(n^2)$ time for any alphabet. Other types of abelian periods are known, including regular and full abelian periods (see [10]), for which linear-time or almost linear-time algorithms are known [13].

Matsuda et al. [14] used a slightly different definition of abelian covers than the one that we use here (that is, they considered two abelian covers different if and only if the set of starting positions of the occurrences of the covers are different) and obtained an $O(n^2)$-time algorithm for computing a representation of all such (possibly exponentially many) abelian covers.

Our Results:

- Computing the longest, the shortest and the number of all abelian squares in $O(n^2/\log^2 n)$ time and computing the distribution of all abelian squares over their center positions in $O(n^2/\log n)$ time;
- Computing the shortest (general) abelian period and all its occurrences, and an $O(n^2/\log n)$-representation of all abelian periods in $O(n^2/\sqrt{\log n})$ time;
- Computing the shortest abelian cover in $O(n^2/\log n)$ time;
- Computing the abelian border array in $O(n^2/\log^2 n)$ time and a representation of all abelian borders of prefixes of the string in $O(n^2/\log n)$ time (showing a similarity between computing abelian borders and abelian squares).

We assume constant-sized alphabet and the word-RAM model of computation. In all algorithms we apply in a fancy way the technique called *four-Russian trick*, which was first involved in abelian pattern matching by Moosa and Rahman [16] in the problem of jumbled indexing. All our algorithms work in linear space, excluding the size of the result.

Structure of the Paper: The following section includes definitions of the terms used in abelian stringology, as well as the basic notation. In Sects. 3, 4 and 5 we show our results for a binary alphabet. In Sect. 6 we show that all presented algorithms can be applied to the case of any constant-sized alphabet, preserving both time and space complexity. The Conclusions Section contains a brief discussion of possible future work.

2   Preliminaries

Let s be a string of length $n = |s|$ over an alphabet Σ of size $\sigma = |\Sigma|$. By $s[i]$, where $1 \leq i \leq n$, we denote the i-th symbol of s, and by $s[i,j]$ we denote a substring of s consisting of all symbols of s on positions from i to j inclusive. Substrings of the form $s[1,i]$ are called prefixes of s and substrings of the form $s[i,n]$ are called suffixes of s.

A *Parikh vector* of a string s, $\mathcal{P}(s)$, is a vector of size σ, where the element $\mathcal{P}(s)[l]$ stores the number of occurrences of symbol l in string s, i.e. $\mathcal{P}(s)[l] = x$ if and only if $|\{i : s[i] = l\}| = x$. The *norm* r of a Parikh vector R is the sum of its elements, $r = \Sigma_{l \in \Sigma} R[l]$. One can add or subtract Parikh vectors component-wise.

We say that two strings s and t are *abelian equivalent* (or commutatively equivalent), if s is a permutation (an anagram) of t, or equivalently $\mathcal{P}(s) = \mathcal{P}(t)$, and write $s \approx t$. We say that t is an *abelian factor* of s if and only if $\mathcal{P}(t)[l] \leq \mathcal{P}(s)[l]$ for every symbol l of the alphabet Σ. We say that a position i in a string t is an *abelian occurrence* of a string s if $s \approx t[i, i + |s| - 1]$.

We now proceed with the definitions of the main notions of abelian periodicity that we consider in the paper.

Definition 1. *An* abelian square *is a string t of length $2k$ such that $t[1,k] \approx t[k+1, 2k]$.*

Definition 2. *A pair (i,k) is a (general)* abelian period *of s if and only if there exists an index j such that $s[i, i+k-1] \approx s[i+k, i+2k-1] \approx \ldots \approx s[i+(j-1)k, i+jk-1]$ and $s[1, i-1]$ and $s[i+jk, n]$ are abelian factors of $s[i, i+k-1]$. A regular abelian period k of s is a general abelian period of s that starts at the first position of s, i.e. the general abelian period $(1,k)$ of s is a regular abelian period k of s.*

Definition 3. *An* abelian border *t of s is a prefix of s that is abelian equivalent to some suffix of s. A proper abelian border t of s is an abelian border t such that $|t| < n$. An* abelian border array *of s is a table π of length n such that $\pi[i]$ is the length of the longest proper abelian border of $s[1,i]$.*

Definition 4. *An abelian border* t *is an* abelian cover *of* s *if and only if the abelian occurrences of* t *in* s cover *the whole string* s. *That is, if* I *is a sorted sequence of positions* i *in* s *such that* $s[i, i + |t| - 1] \approx t$, *then the first element of* I *is* 1, *the last element of* I *is* $n - |t| + 1$ *and the differences between every two consecutive elements of* I *are no greater than* $|t|$.

In Sects. 3, 4 and 5 we consider strings over a binary alphabet $\Sigma = \{0, 1\}$. In this case, to check if two strings s and t of the same length are abelian equivalent it suffices to know only the number of 1 s in each of them. We denote $ones(s) = |\{i : s[i] = 1\}|$. Let us note that $ones(s[i, j])$ can be computed in constant time using an array pre-computed in linear time that stores values of $ones(s[1, i])$ for every $1 \leq i \leq n$. We assume that $ones(s[i, j]) = 0$ for $i > j$.

In the binary case we assume that $s[i, i + m - 1]$ for $m = O(\log n)$ can be stored using a constant number of integers (bit masks). An array that stores the representations of all such substrings for a given m can be computed in linear time.

3   Abelian Squares

In this section we focus on finding the longest abelian squares which have their centers in every possible index i of string s. That is, for each $1 \leq i \leq n$ we want to find the largest k such that $s[i - k, i - 1] \approx s[i, i + k - 1]$.

This problem was solved in $O(n^2)$ time by Cummings and Smyth [7]. Let us briefly restate their solution in the case of a binary alphabet. For every i we take $k = \min(i - 1, n - i + 1)$ and store in a variable r the difference $ones(s[i - k, i - 1]) - ones(s[i, i + k - 1])$. If $r = 0$ then $s[i - k, i + k - 1]$ is an abelian square. If not, then we decrease k by one, update r, and check again. We continue this way, until we find such k, for which the condition holds. One can see that this solution works in $O(n^2)$ time and $O(n)$ space. Our algorithm reduces the time complexity of this scheme by employing two optimizations.

3.1   First Optimization

For the first optimization we will use a pre-computed 3-dimensional auxiliary array A of size $(2m + 1) \times 2^m \times 2^m$; the particular value of m will be chosen later. If there exists a string w of length $2k \geq 2m$ with a m-letter prefix p, m-letter suffix q and $ones(w[1, k]) - ones(w[k + 1, 2k]) = \Delta$, and the longest abelian square centered in the middle of w has length $k - l$ for $0 \leq l \leq m$, then $A[\Delta][p][q] = l$. Otherwise (if such an abelian square does not exist or it is shorter) $A[\Delta][p][q] = -1$.

One can see that when $|\Delta| > m$, then $A[\Delta][p][q] = -1$ for every p and q, so there is no need to store this information explicitly. Each field of array A can be computed in $O(m)$ time, so the whole array A can be constructed in $O(m^2 4^m)$ time. To compute each field of the array A we obviously do not need to generate the string w.

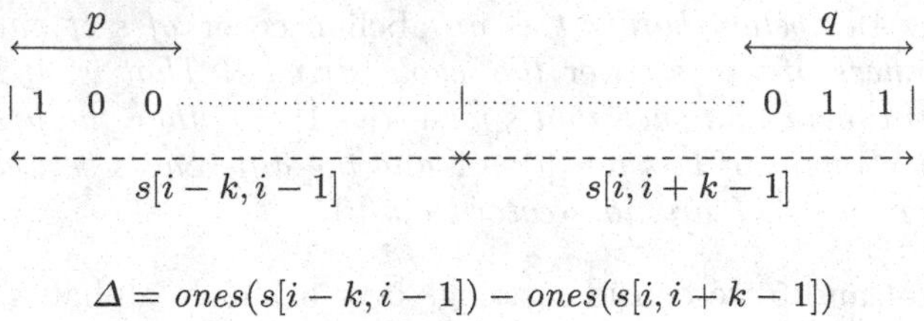

$$\Delta = ones(s[i - k, i - 1]) - ones(s[i, i + k - 1])$$

Fig. 1. How to choose p, q and Δ for given i, k and $m = 3$. If $\Delta = 1$, $A[\Delta][p][q] = 2$.

When analyzing a specific index i of string s and a specific length k we set a m-letter prefix of $s[i - k, i + k - 1]$ as p, and a m-letter suffix of $s[i - k, i + k - 1]$ as q, i.e. $p = s[i - k, i - k + m - 1]$ and $q = s[i + k - m, i + k - 1]$. Additionally let Δ be the difference in the number of 1s between the first and the second half of $s[i - k, i + k - 1]$, i.e. $\Delta = ones(s[i - k, i - 1]) - ones(s[i, i + k - 1])$; see Fig. 1.

If $A[\Delta][p][q] \neq -1$ and k is the largest possible (that is, $k = \min(i - 1, n - i + 1)$), we have found the longest abelian square with a center in i. If $A[\Delta][p][q] = -1$, we know that the length of the longest abelian square is no greater than $k - m$. We can now decrement k by m and check the array A again. We finish either when the first abelian square is found or when k drops below m; in the latter case we find the longest abelian square in $O(m)$ time. This way we can compute the length of the longest abelian square with a center in i in $O(n/m+m)$ time and lengths of longest abelian squares for every center index i of s in $O(n^2/m + nm)$ time.

Lemma 5. *The longest abelian squares for every center index i of s can be computed in $O(n^2/\log n)$ time using $O(n)$ space.*

Proof. By choosing $m = \left\lfloor \frac{\log n}{4} \right\rfloor$ the running time becomes:

$$O(n^2/m + nm + m^2 4^m) = O(n^2/\log n + n \log n + 4^{\frac{\log n}{4}} \log^2 n)$$
$$= O(n^2/\log n + \sqrt{n} \log^2 n)$$
$$= O(n^2/\log n).$$

The space complexity becomes:

$$O(n + m4^m) = O(n + 4^{\frac{\log n}{4}} \log n)$$
$$= O(n + \sqrt{n} \log n)$$
$$= O(n). \qquad \square$$

3.2 Second Optimization

We will now use a second optimization, which will allow us to further decrease the time complexity. For this purpose we will pre-compute an auxiliary array B of size $(2m - 1) \times 2^{2m-1} \times 2^{2m-1} \times 2^m \times 2^m$.

If there is a string w of length $2k + m - 1$ ($k \geq m$) that satisfies all the conditions:

- $ones(w[1, k]) - ones(w[k + 1, 2k]) = \Delta$,
- its $(2m - 1)$-letter prefix is p,
- its $(2m - 1)$-letter suffix is q,
- $c = w[k + 1, k + m - 1]$,
- b is a bit mask of length m,

then $B[\Delta][p][q][c][b]$ stores all pairs (j, l) such that $j \leq m$, $l \leq m$, $b[j] = 0$ and $k - l$ is the length of the longest abelian square with a center in $k + j$.

Every field of array B can be computed in $O(m^2)$ time with a modified algorithm for finding abelian squares by Cummings and Smyth [7] that we have already recalled at the beginning of this section. We check m center indices and m lengths of potential abelian squares. The variable r can be initialized in constant time, using Δ.

In our algorithm, instead of analyzing a specific index i, we will analyze m indices i, $i + 1$, ..., $i + m - 1$ at a time. When considering i we look at $k = \min(i - 1, n - i - m + 2)$, set the $(2m - 1)$-letter prefix of $s[i - k, i + k + m - 2]$ as p and the $(2m - 1)$-letter suffix of the same as q. We also take $\Delta = ones(s[i - k, i - 1]) - ones(s[i, i + k - 1])$ and $c = s[i, i + m - 1]$.

Additionally, we maintain a bit mask b telling us for which of these m indices we have already found the longest abelian square (we initialize all bits of b to 0). Now we check the array B with proper values for each dimension (see Fig. 2), save all pairs stored in the corresponding field of B, replace b with the updated mask and decrease k by m. We continue this way until all bits of b are 1 or until $k < m$, in which case we search for the abelian squares at positions $i, \ldots, i + m - 1$ in $O(m^2)$ time. Then we start considering $i + m$.

This solution can be implemented in two nested loops, where both the inner and the outer loop execute $O(n/m)$ runs. This way the whole algorithm works in $O(n^2/m^2 + nm^2)$ time, after constructing array B in $O(m^3 64^m)$ time. Since the array can possibly store more than one value in each field, it uses $O(m^2 64^m)$ space.

This algorithm computes the lengths of the longest abelian squares with their centers in every index i. It can be easily modified to compute the lengths of the shortest abelian squares. Instead of starting with the largest possible k and decreasing it by m in every run of the inner loop, we may start with $k = 1$

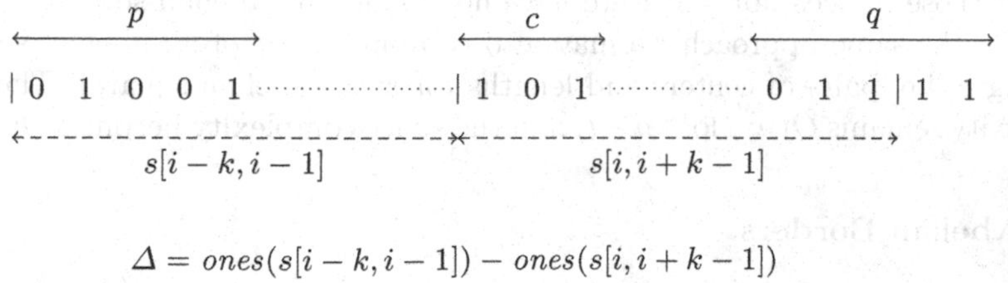

$$\Delta = ones(s[i - k, i - 1]) - ones(s[i, i + k - 1])$$

Fig. 2. How to choose p, q, c and Δ for given i, k and $m = 3$. If $\Delta = 1$ and $b = (0, 0, 1)$, then $B[\Delta][p][q][c][b] = ((1, 1), (2, 3))$.

and increase it by m in every run. The array B also needs to be modified to store information about the shortest abelian squares.

Theorem 6. *The longest (shortest) abelian squares for every center index i of s can be computed in $O(n^2/\log^2 n)$ time using $O(n)$ space.*

Proof. By choosing $m = \left\lfloor \frac{\log n}{12} \right\rfloor$ the running time becomes:

$$O(n^2/m^2 + nm^2 + m^3 64^m) = O(n^2/\log^2 n + n\log^2 n + 64^{\frac{\log n}{12}} \log^3 n)$$
$$= O(n^2/\log^2 n + \sqrt{n}\log^3 n)$$
$$= O(n^2/\log^2 n).$$

The space complexity becomes:

$$O(n + m^2 64^m) = O(n + 64^{\frac{\log n}{12}} \log^2 n)$$
$$= O(n + \sqrt{n}\log^2 n)$$
$$= O(n). \qquad \square$$

3.3   Counting All Abelian Squares

We can modify the values stored in array B so that we can use them to count the number of all abelian squares in string s. In each field of B we may store the number of all abelian squares, when the values Δ, p, q and w are specified. This can be calculated in $O(m^2)$ time, which does not affect the time complexity (nor the space complexity).

If we are interested only in computing the number of all abelian squares, we may omit maintaining the bit mask b and cut one of the dimensions of array B. This has no effect on the time and space complexity.

One can also count the number of all abelian squares, which have their center in index i for every i. To make this possible, the array B can also store a list of length m, where the j-th element of the list tells us how many abelian squares we have found for the center $i+j-1$. Since we must process m operations, the time complexity increases to $O(n^2/\log n)$. We may also say that the time complexity is $O(n^2/\log^2 n + d)$, where d is the number of all abelian squares in string s, if we omit those indices, for which we have not found any abelian square.

Using the same approach we may also compute a list of all abelian squares in string s (i.e. pairs of centers and lengths for every abelian square). The time complexity remains $O(n^2/\log^2 n + d)$ and the space complexity becomes $O(n+d)$.

3.4   Abelian Borders

It is known that a string s has an abelian border of length i if and only if it has an abelian border of length $n - i$ (see Lemma 4 in Matsuda et al. [14]). This way the longest and the shortest abelian borders are determined by each other.

We will modify the algorithm for computing abelian squares of s so that it will compute the abelian border array of s.

We still will be analyzing m indices i, $i+1$, ..., $i+m-1$ at once, but this time they stand for m ends of consecutive prefixes $s[1,i]$, $s[1,i+1]$, ..., $s[1,i+m-1]$. We will focus on finding the shortest abelian borders for these prefixes. For given i we start with $k=1$. We set Δ to the difference of numbers of 1 s in $s[1,k-1]$ and $s[i+m-k+1,i+m-1]$. We denote $(2m-1)$-letter prefix of $s[k,i+m-k]$, $p = s[k,k+2m-2]$ and $(2m-1)$-letter suffix of the same, $q = s[i-m-k+2,i+m-k]$. We fix $c = s[i,i+m-1]$ and maintain a bit mask b, where the j-th bit of b tells if we have already found the shortest abelian border of $s[1,i+j-1]$.

We will use a pre-computed auxiliary array C with the same number of dimensions and the same size of respective dimensions as array B. Given Δ, p, q, c and b the array C stores pairs (j,l) of shortest abelian borders that were found. Every field of this array can be computed in $O(m^2)$ time.

The algorithm executes two nested loops. The outer loop starts with $i=1$ and increases i by m after every run. The inner loop starts with $k=1$ and increases k by m after every run. This leads us to the following corollary.

Corollary 7. *The abelian border array of string s can be computed in $O(n^2/\log^2 n)$ time using $O(n)$ space.*

Similarly as in the case of computing a distribution of abelian squares by their centers, we can compute a representation of all abelian borders of prefixes of s in $O(n^2/\log n)$ time and $O(n)$ additional space.

4   Abelian Periods

The problem of finding all abelian periods was solved in $O(n^2)$ time for any alphabet by Crochemore et al. [6]. Let us briefly recall their solution adapted to our binary alphabet case.

One can see that k is a regular abelian period of $s[i,n]$ if and only if all the conditions below hold:

- k is a regular abelian period of $s[i+k,n]$,
- if $n \geq i+2k-1$ then $s[i+k,i+2k-1] \approx s[i,i+k-1]$,
- if $n < i+2k-1$ then $s[i+k,n]$ is an abelian factor of $s[i,i+k-1]$.

Thereby (i,k) is a (general) abelian period of s if and only if k is a regular abelian period of $s[i,n]$ and $s[1,i-1]$ is an abelian factor of $s[i,i+k-1]$.

All general abelian periods of s are of the form (i,k), where $1 \leq i,k \leq n$, thus information about all of them can be represented as an array GP of size $n \times n$, where $GP[i][k] = 1$ if (i,k) is an abelian period of s and $GP[i][k] = 0$ otherwise. We can compute this array using dynamic programming. For this purpose we use an array RP of size $n \times n$, where $RP[i][k] = 1$ if k is a regular abelian period of $s[i,n]$ and $RP[i][k] = 0$ otherwise.

We will first show how to compute RP. We set $RP[i][k] = 1$ for every i and k such that $n \leq i + k - 1$. To compute $RP[i][k]$ we need to know $RP[i+k][k]$, so we will consider all remaining pairs (i, k) in a descending order of i. Checking if two strings are abelian equivalent or if one string is an abelian factor of the other can be done in constant time, so the whole array RP can be computed in $O(n^2)$ time and space.

To compute the array GP we look at each $RP[i][k] = 1$ such that $i \leq k$ and set $GP[i][k] = 1$ if and only if $s[1, i - 1]$ is an abelian factor of $s[i, i + k - 1]$.

The length of the shortest abelian period of s, the number of all of them and the number of all abelian periods of s can be easily extracted from the GP array.

4.1 Computing RP Faster

For a better picture, let us assume that i numbers columns and k numbers rows of all arrays. Let us take $m < n$ and without loss of generality assume that m is a divisor of n (otherwise we increase n by at most $m - 1$). We can code $O(\log n)$ fields of each of the arrays GP, RP into one machine word. In our case we will pack all fields on indices $\{i, i+1, \ldots, i+m-1\} \times \{k, k+1, \ldots, k+m-1\}$ (there are exactly m^2 of them), where $m|(i-1)$ and $m|(k-1)$, into unit square arrays of size $m \times m$. If m is small enough, we can code such a unit array into one machine word, row by row. This way both GP and RP are divided regularly into n^2/m^2 unit square arrays and stored in $O(n^2/m^2)$ space.

From now on we will still reference to a single field of GP and RP on indices (i, k) or a subarray of any of them, but we are assuming that the physical representation of arrays is as described. To improve time complexity of computing RP and GP we will fill out m fields of RP at a time and m fields of GP at a time.

To compute RP we will use three auxiliary arrays: A of size $(2m+1) \times 2^{m^2} \times 2^{m-1} \times 2^{2m-2}$, E of size $2^{m^2} \times 2^{m^2} \times m$ and F of size $2^{m^2} \times 2^m \times m$.

E is an array which for given two unit arrays X and Y and a shift i, stores a unit array Z such that the first $n - i + 1$ columns of Z are the last $n - i + 1$ columns of X and the last $i - 1$ columns of Z are the first $i - 1$ columns of Y.

F is an array which for given unit array X, a column (that is, a 1-d array) C and a shift i, stores a unit array Y which is X with the i-th column replaced by C.

Finally $A[\Delta][X][p][q] = C$ if and only if there exists a string w of length at least $2k + 2m - 2$ such that all of the conditions below hold:

- $p = w[k+1, k+m-1]$,
- $q = w[2k+1, 2k+2m-2]$,
- $\Delta = ones(w[1, k]) - ones(w[k+1, 2k])$,
- $k = lm + 1$ for some integer l,
- X is a unit array containing information about regular periods on suffixes of w starting on indices from $\{k+1, k+2, \ldots, k+m\}$ and having lengths from $\{k, k+1, \ldots, k+m-1\}$,

– C is a column containing information about regular periods of w with lengths from $\{k, k+1, \ldots, k+m-1\}$.

We will use this array, to instead of calculating one value $RP[i][k]$ at a time, calculate $RP[i][k, k+m-1]$ at once. For every Δ, X, p and q, there exists exactly one C. If $|\Delta| > m$, then $C = (0, 0, \ldots, 0)$ and we do not store this explicitly, otherwise we can simply pre-compute each field of A in $O(m)$ time.

To compute RP we will first set all $RP[i][k] = 1$, where $n \leq k + i - 1$, as previously. Noticing that there are only $O(n)$ unit arrays which we will fill with 1s only partially allows us to initialize RP in $O(n^2/m^2)$ time, instead of $O(n^2)$ time.

Now for the remaining pairs (i, k), we iterate first by k, starting with $k = n - m + 1$ and decreasing it by m. For a given k we iterate by i from n to 1, decreasing it by 1.

In each operation we take $w = s[i, n]$, $X = RP[i+k, i+k+m-1][k, k+m-1]$ and p, q and Δ respectively; see Fig. 3. X can be extracted from two particular unit arrays from the n^2/m^2 unit arrays that RP consist of, by using array E. After getting a new column C from array A, that corresponds to $RP[i][k, k + m - 1]$, we can update array RP, using array F.

All the operations above are constant-time, thus the whole RP array can be computed in $O(n^2/m)$ time.

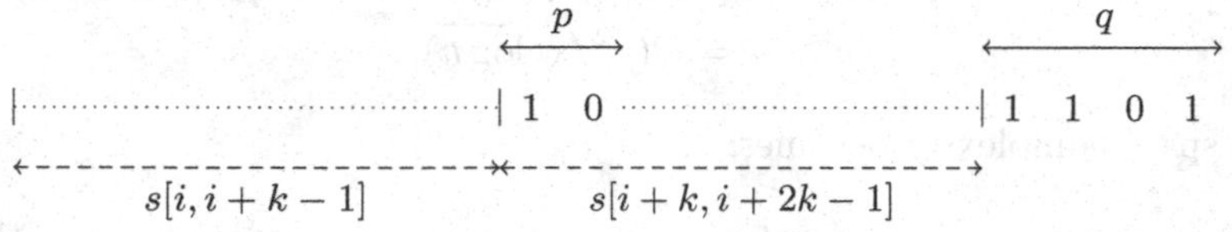

$$\Delta = ones(s[i, i+k-1]) - ones(s[i+k, i+2k-1])$$

Fig. 3. How to choose p, q and Δ for given i, k and $m = 3$. If $\Delta = 0$ and $X = \left(\begin{smallmatrix} 1 & 0 & 0 \\ 0 & 0 & 0 \\ 0 & 0 & 1 \end{smallmatrix}\right)$, then $A[\Delta][X][p][q] = \left(\begin{smallmatrix} 1 \\ 0 \\ 0 \end{smallmatrix}\right)$.

4.2 Computing GP Faster

Notice that if $RP[i][k] = 0$, then $GP[i][k] = 0$. Thus the array GP is the array RP with some fields set to 0.

To compute the array GP we will first copy the array RP into it. Then we will update $GP[i][k] = 0$ for all $i > k$. As already mentioned, that can be done in $O(n^2/m^2)$ time.

The remaining fields that have to be set to 0 are determined by all i and k for which $s[1, i-1]$ is not an abelian factor of $s[i, i+k-1]$. To detect all such i and k, for each i we compute the minimal k_i, such that $s[1, i-1]$ is an abelian factor of $s[i, i + k_i - 1]$. This can be done in linear time, since obviously $(i + 1) + k_{i+1} \geq i + k_i$; see [6].

Having i and k_i we set $GP[i][k] = 0$ for all $k < k_i$. Since all such k form an interval, it can be done in $O(n/m)$ time for specific i and k_i using the array F and in $O(n^2/m)$ time for the whole array GP.

For now the arrays RP and GP use $O(n^2/m^2)$ space. If we are interested only in counting periods of every size and storing all shortest periods, it suffices to store only m consecutive rows of both arrays, i.e. rows $k, \ldots, k+m-1$ (and all k_i's). For this only $O(n/m)$ unit arrays and $O(n)$ additional space is necessary.

In conclusion, computing arrays A, E and F costs us $O(m^2 2^{m^2} 8^m + m^3 4^{m^2})$ time and $O(m 2^{m^2} 8^m + m 4^{m^2})$ space. Counting all abelian periods costs us additionally $O(n^2/m)$ time and $O(n)$ space.

Theorem 8. *All shortest abelian periods of string s can be computed in $O(n^2/\sqrt{\log n})$ time using $O(n)$ space.*

Proof. By choosing $m = \left\lfloor \sqrt{\frac{\log n}{4}} \right\rfloor$ the running time becomes:

$$
\begin{aligned}
O(n^2/m + m^2 2^{m^2} 8^m + m^3 4^{m^2}) &= O(n^2/\sqrt{\log n} + 2^{\frac{\log n}{4}} 8^{\sqrt{(\log n)/4}} \log n \\
&\quad + 4^{\frac{\log n}{4}} \log^{\frac{3}{2}} n) \\
&= O(n^2/\sqrt{\log n} + n^{0.26} \log n + \sqrt{n} \log^{\frac{3}{2}} n) \\
&= O(n^2/\sqrt{\log n}).
\end{aligned}
$$

The space complexity becomes:

$$
\begin{aligned}
O(n + m 2^{m^2} 8^m + m 4^{m^2}) &= O(n + 2^{\frac{\log n}{4}} 8^{\sqrt{(\log n)/4}} \sqrt{\log n} + 4^{\frac{\log n}{4}} \sqrt{\log n}) \\
&= O(n + n^{0.26} \sqrt{\log n} + \sqrt{n \log n}) \\
&= O(n).
\end{aligned}
$$

$\square$

5   Abelian Covers

For given $i \leq n$ we can check if the prefix of length i of string s is an abelian cover of s. This is actually the abelian pattern matching problem, which for a constant-sized alphabet can obviously be solved in linear time.

Let us assume that we have found all occurrences of $s[1, i]$ in s and stored their (sorted) positions in a sequence I. Then $s[1, i]$ is an abelian cover of s if and only if the first element of I is 1, the last element of I is $n - i + 1$ and the differences between all pairs of consecutive elements of I are no greater than i.

Checking all $1 \leq i \leq n$, the shortest abelian cover of string s can be found in $O(n^2)$ time.

5.1   Optimizing Running Time

Assume without loss of generality that m is a divisor of n. Instead of finding one occurrence of $s[1, i]$ in s at a time, we will only find the smallest and the largest positions from each of intervals $[1, m]$, $[m + 1, 2m]$, ..., $[n - m + 1, n]$ such that $s[1, i]$ is an abelian match on s at these positions.

For this purpose we will use a pre-computed auxiliary array A of size $2^{m-1} \times (2m + 1) \times 2^{m-1}$. $A[p][\Delta][q] = (l_1, l_2)$ if and only if there are strings w of length $i + m - 1$ (text) and c of length i (potential cover) such that all of the conditions below hold:

- p is a $(m - 1)$-letter prefix of w,
- q is a $(m - 1)$-letter suffix of w,
- $\Delta = ones(w[1, i]) - ones(c)$,
- $1 \le l_1 \le l_2 \le m$,
- $-ones(p[1, l_1 - 1]) + ones(q[1, l_1 - 1]) = \Delta$, i.e. $c \approx w[l_1, l_1 + i - 1]$,
- $-ones(p[1, l_2 - 1]) + ones(p[1, l_2 - 1]) = \Delta$, i.e. $c \approx w[l_2, l_2 + i - 1]$,
- l_1 is the smallest possible,
- l_2 is the largest possible.

If no such pair exists then $A[p][\Delta][q] = (-1, -1)$. Each element of array A can be pre-computed in $O(m)$ time. We consider only $-m \le \Delta \le m$, so the whole array can be then pre-computed in $O(m^2 4^m)$ time and uses $O(m4^m)$ space.

Our algorithm for finding the shortest abelian cover works as follows. First we check if any prefix of s of length less than $2m$ is a cover of s. This can be done by the straightforward solution in $O(nm)$ time. Now we assume that the shortest cover has length at least $2m$.

We iterate with i from $2m$ to n to check if $s[1, i]$ is an abelian cover of s. For every i we start with $I = ()$. Now we iterate with j from $j = 1$ to $n - i + 1$, increasing j by m each time. We are checking if there is an abelian match on positions $[j, j+m-1]$, so we are implicitly considering substring $s[j, j+i+m-2]$.

For every i and j we denote three values:

- the $(m - 1)$-letter prefix of $s[j, j + i + m - 2]$, $p = s[j, j + m - 2]$,
- the $(m - 1)$-letter suffix of $s[j, j + i + m - 2]$, $q = s[j + i, j + i + m - 2]$ and
- the difference of 1s, $\Delta = ones(s[j, j + i - 1]) - ones(s[1, i])$.

If $|\Delta| > m$ then $s[1, i]$ does not abelian match any substring in $s[j, j + i + m - 2]$. Otherwise we can find the first and the last occurrence (in the considered interval) by referencing to the array A and, if they exist, add them to I.

After iterating with j, I contains $O(n/m)$ elements. We can then check if $s[1, i]$ is an abelian cover of s the same way, as we did it in the straightforward solution. This way the whole algorithm works in $O(nm + n^2/m)$ time.

Theorem 9. *The shortest abelian cover for string s can be computed in $O(n^2/\log n)$ time using $O(n)$ space.*

Proof. By choosing $m = \left\lfloor \frac{\log n}{4} \right\rfloor$ the running time becomes:

$$
\begin{aligned}
O(nm + n^2/m + m^2 4^m) &= O(n \log n + n^2/\log n + 4^{\frac{\log n}{4}} \log^2 n) \\
&= O(n^2/\log n + \sqrt{n} \log^2 n) \\
&= O(n^2/\log n).
\end{aligned}
$$

The space complexity becomes:

$$
\begin{aligned}
O(n + m4^m) &= O(n + 4^{\frac{\log n}{4}} \log n) \\
&= O(n + \sqrt{n} \log n) \\
&= O(n).
\end{aligned}
$$

$\square$

It is clear that we can use the same algorithm to compute lengths of all abelian covers of s. If we are interested in detecting all occurrences of all abelian covers of s, then the array A needs to store not only the pair of the smallest l_1 and the largest l_2 that hold all mentioned conditions, but all such l. This has no effect on space complexity, but the time complexity becomes $O(n^2/\log n + d)$, where d is the number of all occurrences of all abelian covers in s.

The shortest abelian cover is probably of most interest. We can use our algorithm to find it and then we can find all its occurrences in s in $O(n)$ time, using abelian pattern matching in linear time.

6   The Case of a Larger Alphabet

We have presented subquadratic-time algorithms for computing abelian squares, general abelian periods, abelian covers and abelian border array for a string over a binary alphabet. In the fundamental problem of abelian stringology—jumbled indexing—switching from binary to a larger constant-sized alphabet increases the hardness of the problem considerably [1,12]. However, all our algorithms can be applied to the case of any constant-sized alphabet, preserving both time and space complexity.

For this, in each of the presented auxiliary arrays, instead of considering Δ, that is, the difference in the number of 1 s between two substrings, we may consider a difference of Parikh vectors of the corresponding substrings. The number of possible differences we should consider now is obviously no greater than $(2m + 1)^\sigma$, where σ is the size of the alphabet. This increases the size of the auxiliary arrays and the computation time by at most a factor of $O(\log^\sigma n)$. These sizes and construction times were always $o(n^{1-\epsilon})$ for some $\epsilon > 0$. Hence, they will remain sublinear for a constant σ.

Instead of using the function *ones*, we may use an analogous constant-time function that returns a Parikh vector for a substring. In all of the algorithms we assume that a substring of length m can be stored using one machine word. For this to be still true for $\sigma > 2$, we need to divide the proposed values of m in Theorems 6 and 9 by a factor of $\log \sigma$ and in Theorem 8 by a factor of $\sqrt{\log \sigma}$.

7   Conclusions

We have presented the first subquadratic-time algorithms for computing abelian squares, general abelian periods, abelian covers and abelian border array for a string over a constant-sized alphabet. An open question is if there exist $O(n^2/\log^c n)$-time algorithms for the problems considered in this paper with a larger constant c. A further question, for all of the problems, is if there exists an $O(n^{2-\epsilon})$-time algorithm or, possibly, if there is a lower bound on the time complexity of an algorithm solving this problem. In comparison, for the seminal problem in this area, the jumbled indexing, on one hand, $O(n^{2-\epsilon})$-time algorithms are known for any constant-sized alphabet [3], but on the other hand, for sufficiently large alphabets $\sigma > 2$ conditional lower bounds are known for the query vs construction time trade-off [1] based on hardness of the 3SUM problem.

References

1. Amir, A., Chan, T.M., Lewenstein, M., Lewenstein, N.: On hardness of jumbled indexing. In: Esparza, J., Fraigniaud, P., Husfeldt, T., Koutsoupias, E. (eds.) ICALP 2014. LNCS, vol. 8572, pp. 114–125. Springer, Heidelberg (2014)
2. Burcsi, P., Cicalese, F., Fici, G., Lipták, Z.: On table arrangements, scrabble freaks, and jumbled pattern matching. In: Boldi, P., Gargano, L. (eds.) FUN 2010. LNCS, vol. 6099, pp. 89–101. Springer, Heidelberg (2010)
3. Chan, T.M., Lewenstein, M.: Clustered integer 3SUM via additive combinatorics. In: Servedio, R.A., Rubinfeld, R. (eds.) Proceedings of the Forty-Seventh Annual ACM on Symposium on Theory of Computing, STOC 2015, Portland, OR, USA, 14–17 June 2015, pp. 31–40. ACM (2015)
4. Cicalese, F., Fici, G., Lipták, Z.: Searching for jumbled patterns in strings. In: Holub, J., Žd'árek, J. (eds.) Proceedings of the Prague Stringology Conference 2009, Prague, Czech Republic, 31 August - 2 September 2009, pp. 105–117. Prague Stringology Club, Department of Computer Science and Engineering, Faculty of Electrical Engineering, Czech Technical University in Prague (2009)
5. Constantinescu, S., Ilie, L.: Fine and Wilf's theorem for abelian periods. Bull. EATCS **89**, 167–170 (2006)
6. Crochemore, M., Iliopoulos, C.S., Kociumaka, T., Kubica, M., Pachocki, J., Radoszewski, J., Rytter, W., Tyczyński, W., Waleń, T.: A note on efficient computation of all abelian periods in a string. Inf. Process. Lett. **113**(3), 74–77 (2013)
7. Cummings, L.J., Smyth, W.F.: Weak repetitions in strings. J. Comb. Math. Comb. Comput. **24**, 33–48 (1997)
8. Erdös, P.: Some unsolved problems. Hung. Acad. Sci. Mat. Kutató Intézet Közl **6**, 221–254 (1961)
9. Fici, G., Lecroq, T., Lefebvre, A., Prieur-Gaston, É.: Computing abelian periods in words. In: Holub, J., Žd'árek, J. (eds.) Proceedings of the Prague Stringology Conference 2011, pp. 184–196. Czech Technical University in Prague, Czech Republic (2011)
10. Fici, G., Lecroq, T., Lefebvre, A., Prieur-Gaston, É., Smyth, W.: Quasi-linear time computation of the abelian periods of a word. In: Holub, J., Žd'árek, J. (eds.) Proceedings of the Prague Stringology Conference 2012, pp. 103–110. Czech Technical University in Prague, Czech Republic (2012)

11. Hermelin, D., Landau, G.M., Rabinovich, Y., Weimann, O.: Binary jumbled pattern matching via all-pairs shortest paths. CoRR, abs/1401.2065 (2014)
12. Kociumaka, T., Radoszewski, J., Rytter, W.: Efficient indexes for jumbled pattern matching with constant-sized alphabet. In: Bodlaender, H.L., Italiano, G.F. (eds.) ESA 2013. LNCS, vol. 8125, pp. 625–636. Springer, Heidelberg (2013)
13. Kociumaka, T., Radoszewski, J., Rytter, W.: Fast algorithms for abelian periods in words and greatest common divisor queries. In: Portier, N., Wilke, T. (eds.) 30th International Symposium on Theoretical Aspects of Computer Science, STACS 2013, 27 February - 2 March 2013, Kiel, Germany, vol. 20 of LIPIcs, pp. 245–256. Schloss Dagstuhl - Leibniz-Zentrum fuer Informatik (2013)
14. Matsuda, S., Inenaga, S., Bannai, H., Takeda, M.: Computing abelian covers and abelian runs. In: Holub, J., Zdárek, J. (eds.) Proceedings of the Prague Stringology Conference 2014. Prague, Czech Republic, 1–3 September 2014, pp. 43–51. Department of Theoretical Computer Science, Faculty of Information Technology, Czech Technical University in Prague (2014)
15. Moosa, T.M., Rahman, M.S.: Indexing permutations for binary strings. Inf. Process. Lett. 110(18–19), 795–798 (2010)
16. Moosa, T.M., Rahman, M.S.: Sub-quadratic time and linear space data structures for permutation matching in binary strings. J. Discrete Algorithms 10, 5–9 (2012)

Using Statistical Search to Discover Semantic Relations of Political Lexica – Evidences from Bulgarian-Slovak EUROPARL 7 Corpus

Velislava Stoykova(✉)

Institute for Bulgarian Language - BAS,
52, Shipchensky Proh. Str., Bl. 17, 1113 Sofia, Bulgaria
vstoykova@yahoo.com

Abstract. The paper presents statistical approach to discover semantic relations of political lexica using parallel Bulgarian-Slovak EUROPARL 7 Corpus. It employs statistical properties incorporated in the Sketch Engine software to generate concordances, co-occurrences and colloca-tions. A comparative analysis of semantic structure of political lexica investigating synonymic, attributive and reciprocal semantic relations of most frequent key words from two parallel corpora – for both Bulgarian and Slovak languages is offered. The paper address some issue related to correct terms discovery, their translations and use in political speech. Finally, more general conclusions about semantic properties of political lexica are presented.

Keywords: Data mining · Combinatorics on words · Machine translation

1 Introduction

The Europarl parallel corpus contains proceedings of European Parliament ses-sions. It includes texts in 21 European languages belonging to different fami-lies of related and non-related languages. From its early beginning the corpus was created to process pairs of sentence aligned bilingual sub-corpora for sta-tistical machine translation [5]. The preprocessing includes identifying sentence boundaries and alignment of bilingual pairs of parallel corpora using Church and Gale algorithm [1]. In its last version EUROPARL 7 Corpus uses small number of mark-up annotations like: <CHAPTER id>, <SPEAKER id name and language> and paragraph <p>. The existing parallel corpora relate most Euro-pean languages to English including the Bulgarian-English parallel EUROPARL 7 Corpus. Further, we are going to present a research on analysis of seman-tic properties of political lexica using Bulgarian-Slovak parallel EUROPARL 7 Corpus generated on the base of EUROPARL 7.

2 Bulgarian-Slovak EUROPARL 7 Corpus

The Bulgarian-Slovak parallel EUROPARL 7 Corpus includes texts from both Bulgarian and Slovak part of EUROPARL 7 Corpus and uses the Sketch Engine

© Springer International Publishing Switzerland 2016
I.S. Kotsireas et al. (Eds.): MACIS 2015, LNCS 9582, pp. 335–339, 2016.
DOI: 10.1007/978-3-319-32859-1_28

software for processing parallel corpora, hence corpus texts are incorporated into the Sketch Engine statistical software. The corpus is not annotated for part-of-speech, not lemmatized and does not use any formal grammar. It uses annotation scheme of EUROPARL 7 Corpus and includes at about 9 215 000 words of Bulgarian language sub-corpus and 13 000 000 words of Slovak language sub-corpus allowing different types of statistical search and Corpus Query Language (CQL) regular expressions search.

3   The Sketch Engine

The Sketch Engine (SE) software [3,4] allows the use of various approaches to extract lexical semantic properties of words and most of them are with multilingual application [2]. Extracting keywords is a most common and widely used technique to define basic terms of a particular domain. The SE's software standard options for keywords extraction are based on the use of word frequency lists. However, semantic relations can be extracted by generation of related word contexts through word concordances. Concordances define context in quantitative terms and a further work is needed to be done to define semantic relations by searching for co-occurrences and collocations of related keyword.

The co-occurrences and collocations are words which are most probably to be found with a related keyword. They assign the semantic relations between the keyword and its particular collocated word which might be of similarity or of a distance. The SE approaches to evaluate corpus co-occurrences and collocations are based on defining the probability for which different criteria are evaluated. We have used techniques of $T-score$, $MI-score$ and $MI^3-score$ incorporated in the Sketch Engine for corpus processing and searching.

Basically for all, the following terms are used: N – corpus size, f_A – number of occurrences of the keyword in the whole corpus (the size of the concordance), f_B – number of occurrences of the collocated keyword in the whole corpus, f_{AB} – number of occurrences of the collocate in the concordance (number of co-occurrences). The related formulas for defining $T-score$, $MI-score$ and $MI^3-score$ are as follows:

$$\text{MI-Score } \log_2 \frac{f_{AB}N}{f_A f_B}$$

$$\text{T-Score } \frac{f_{AB} - \dfrac{f_A f_B}{N}}{\sqrt{f_{AB}}}$$

$$\text{MI}^3\text{-Score } \log_2 \frac{f_{AB}^3 N}{f_A f_B}$$

The $T-score$, $MI-score$ and $MI^3-score$ are applicable for processing parallel corpora as well.

Word list Corpus: **EUROPARL7**, Page 1 Next >		Word list Corpus: **EUROPARL7**, Page 1 Next >	
word	**Freq**	**word**	**Freq**
Европейския	31731	EÚ	42691
Комисията	31585	Európskej	29747
съюз	29533	Komisia	19208
ЕС	24063	únie	16882
председател	22153	členských	16375
Европа	21279	predsedajú	
отношение	16584	ci	16335
държавите-членки	13254	štáty	15583
		Komisie	15197
подкрепа	11605	politiky	12906
политика	11399	práva	11804
парламент	10998	členské	11683
държави	10411	Európe	10753
правата	10317	Parlament	10141
страна	10263	podporu	9723

Fig. 1. The first fourteen most frequent words from both Bulgarian ans Slovak sub-corpora.

3.1 Parallel Corpora Processing in SE

Recently, the SE has been improved with options to process parallel corpora [6]. Additionally to sentence alignment it is possible to perform various types of parallel statistical search according to common search criteria and to generate parallel aligned concordances, co-occurrences, etc.

4 Semantic Analysis

For our research we follow strategy already applied for conceptual semantic relations extraction [9] and term extraction described in [8]. We first generate parallel word frequency lists as well as keywords for both sub-corpora. We obtained similar results for which first fourteen hits are given at Fig. 1. For a more detailed semantic analysis of word contexts we generate parallel bilingual word concordances as follows:

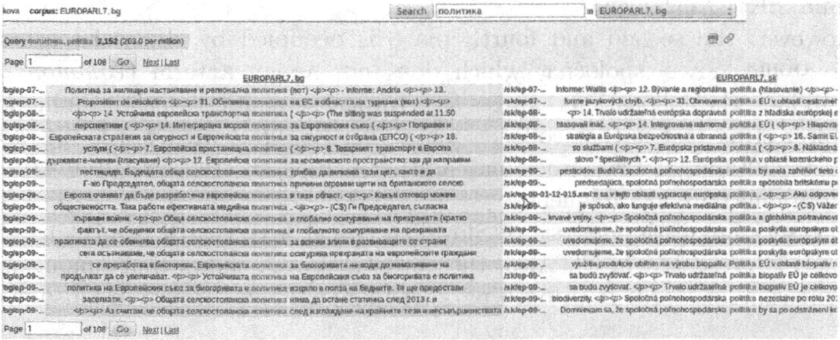

Collocation candidates

Page [1]
Next >

	Freq	T-score	MI	MI3
P | N селскостопанска	1247	35.273	9.807	30.375
P | N общата	1276	35.655	9.070	29.705
P | N външна	804	28.304	9.134	28.436
P | N обща	736	27.038	8.222	27.269
P | N сигурност	796	27.948	6.732	26.005
P | N европейската	593	24.186	7.203	25.627
P | N политика	801	27.869	6.030	25.322
P | N европейска	399	19.829	7.096	24.377
P | N сближаване	386	19.485	6.919	24.104
P | N съседство	290	16.995	8.933	25.293
P | N регионалната	280	16.705	9.216	25.474
P | N енергийна	287	16.859	7.694	24.024
P | N икономическата	308	17.396	6.834	23.368
P | N социалната	215	14.579	7.442	22.938
P | N отбрана	202	14.180	8.765	24.081

Collocation candidates

Page [1]
Next >

	Freq	T-score	MI	MI3
P | N poľnohospodárska	296	17.201	12.188	28.607
P | N spoločná	307	17.513	11.099	27.623
P | N zahraničná	130	11.399	12.071	26.116
P | N Spoločná	111	10.532	11.464	25.052
P | N regionálna	83	9.107	11.450	24.200
P | N európska	120	10.939	9.434	23.248
P | N súdržnosti	158	12.542	8.836	23.444
P | N obchodná	72	8.481	10.878	23.218
P | N politika	181	13.409	8.248	23.248
P | N energetická	68	8.238	9.907	22.082
P | N susedská	54	7.346	11.321	22.831
P | N bezpečnostná	54	7.344	10.829	22.339
P | N obranná	28	5.290	11.630	21.244
P | N sociálna	31	5.553	8.552	18.461
P | N prisťahovalecká	23	4.795	11.857	20.904

Fig. 2. The collocation candidates for the word *politics* from both Bulgarian and Slovak sub-corpora.

The most frequent content word is the adjective *European* (BG – Европейски, SK – Európskej), followed by *commission, president, member-state, politics,* etc. (BG – комисия, председател, държави-членки, политика; SK – komisia, predsedajúci, štáty, politiky). Also, the frequency of word *politics* in Bulgarian and Slovak sub-corpora (BG – политика (11399), SK – *politiky* (12906)) are very similar.

Further, we will analyze semantic properties for that keyword by generating its related parallel bilingual collocations from both sub-corpora.

4.1   Lexical Relations

Lexical relations analysis and generation of parallel bilingual collocations of the word *politics* for both sub-corpora can give possible multiword expressions for that word. The results obtained are given at Fig. 2. They present frequency lists of words which are most probable to be found with related keyword *politics*. Both lists are similar - the first word is the word *agricultural* (BG - селскостопанска, SK – poľnohospodárska), the third word is *foreign* (BG - външна, SK – zahraničná).

However, the second and fourth place is occupied by the word *common* (BG - обща, SK – spoločná) which does not assign a term (*common politics*). In fact, the collocations express a reciprocal semantics which is ambiguous. The interesting fact is that collocations of word *politics* with words like *social* (BG - социалната, SK – sociálna) and even *defence* (BG - отбранителна, SK – obranná) are less frequent.

The sociolinguistic analysis of Slovak political lexica with respect to globalization is presented in [7]. Also, the statistic search shows that among most frequent content words from both parallel bilingual sub-corpora are also words with reciprocal semantics like *regional* in sense of *national*, incorrect term translation, unknown words like *cohesion* (which is not available in the official academic multivolume dictionaries of both Bulgarian and Slovak languages) and lots of misinterpreted terms. As for the attributive collocations, we have found very frequent rate of nouns modified by more than three adjectives which also leads to ambiguity and misinterpretation.

5   Conclusion

The results of our research for analyzing semantic properties of political lexica obtained with parallel context statistical search for keywords, concordances and collocations from Bulgarian-Slovak EUROPARL 7 Corpus show surprisingly high rate frequency of keywords that express reciprocal or ambiguous semantics, incorrect translation and usage of unknown words which can be misinterpreted. Taking into account that both bilingual sub-corpora have same semantic content, and the processing methodology for both corpora was parallel, we can conclude that received results underlay specific semantic features of political lexica. Further, it will be interesting to process and compare similar results for more European languages.

References

1. Gale, W., Church, K.: A program for aligning sentences in bilingual corpora. Comput. Linguist. **19**(1), 5–102 (1993)
2. Kilgarriff, A., Reddy, S., Pomikalek, J., Avinesh, P.: A corpus factory for many languages. In: Proceedings of the LREC 2010, pp. 904–910 (2010)
3. Kilgarriff, A., Rundell, M.: Lexical profiling software and its lexicographic applications: a case study. In: Proceedings from EURALEX 2002, pp. 807–811 (2002)
4. Kilgarriff, A., Rychly, P., Smrz, P., Tugwell, D.: The sketch engine. In: Proceedings from EURALEX 2004, pp. 105–116 (2004)
5. Koehn, P.: Europarl: A parallel corpus for statistical machine translation. In: Proceedings from MT Summit, pp. 79–86 (2005)
6. Michelfeit, J.: Parallel corpora in sketch engine. In: Sketch Engine Workshop IV, Tallinn (2013) (presentation)
7. Ondrejovic, S.: Between purism and glocalism. In: Sociolinguistica Slovaca, vol. 8, pp. 25–32. VEDA (2014)
8. Stoykova, V., Petkova, E.: Automatic extraction of mathematical terms for precalculus. In: Proceedia Technology, vol. 1, pp. 464–468. Elsevier (2012)
9. Stoykova, V., Simkova, M., Majchrakova, D., Gajdosova, K.: Detecting time expressions for bulgarian and slovak language from electronic text corpora. Proc. Soc. Behav. Sci. **186**, 257–260 (2015). Elsevier

Computational Theory of Differential and Difference Equations

Simple Differential Field Extensions and Effective Bounds

James Freitag[1]([📧]) and Wei Li[2]

[1] UCLA Mathematics Department, Box 951555, Los Angeles, CA 90095-1555, USA
freitagj@gmail.com
[2] Academy of Mathematics and Systems Science, Chinese Academy of Sciences,
No. 55 Zhongguancun East Road, Beijing 100190, China
liwei@mmrc.iss.ac.cn

Abstract. We establish several variations on Kolchin's differential primitive element theorem, and conjecture a generalization of Pogudin's primitive element theorem. These results are then applied to improve the bounds for the effective Differential Lüroth theorem.

Keywords: Differential chow forms · Primitive element theorem · Model theory · Differential Lüroth theorem

1 Notation

Throughout this paper, $\mathcal{U}$ will be a fixed sufficiently large saturated differentially closed field of characteristic zero with a derivation operator δ. An element $c \in \mathcal{U}$ such that $\delta(c) = 0$ is called a constant. In this paper, all the differential fields under discussion are subfields of $\mathcal{U}$ and subscripts denote differentiation.

Let F be a differential subfield of $\mathcal{U}$ and $S \subset \mathcal{U}$. We denote respectively by $F[S]$, $F(S)$, $F\{S\}$, and $F\langle S\rangle$ the smallest subring, the smallest subfield, the smallest differential subring, and the smallest differential subfield of $\mathcal{U}$ containing F and S.

The set S is said to be *differentially dependent* over F if the set $(\delta^k a)_{a \in S, k \geq 0}$ is algebraically dependent over F, and otherwise, S is said to be *differentially independent* over F. In the case $S = \{a\}$, we also say that a is *differentially algebraic* or *differentially transcendental* over F respectively. A maximal subset Ω of S which is differentially independent over F is said to be a *differential transcendence basis* of $F\langle S\rangle$ over F. We use $\text{d.tr.deg}\, F\langle S\rangle/F$ to denote the *differential transcendence degree* of $F\langle S\rangle$ over F, which is the cardinality of Ω.

J. Freitag—Thanks Dave Marker, Omar León Sanchez, and Gabriela Jeronimo for useful conversations related to this work. JF was partially supported by NSF MSPRF 1204510.

W. Li—Partially supported by a National Key Basic Research Project of China (2011CB302400) and by grants from NSFC (60821002, 11301519) and thanks the University of California, Berkeley for providing a good research environment during her appointment as a Visiting Scholar.

© Springer International Publishing Switzerland 2016
I.S. Kotsireas et al. (Eds.): MACIS 2015, LNCS 9582, pp. 343–357, 2016.
DOI: 10.1007/978-3-319-32859-1_29

Considering F and $F\langle S\rangle$ as algebraic fields, we denote the algebraic transcendence degree of $F\langle S\rangle$ over F by $\mathrm{tr.deg}\, F\langle S\rangle/F$.

We use $F\{y_1,\ldots,y_n\}$ to denote the differential polynomial ring over F. Given a differential polynomial $f \in F\{y_1,\ldots,y_n\}$, the order of f w.r.t. y_i is the greatest number k such that $y_i^{(k)}$ appears effectively in f, which is denoted by $\mathrm{ord}(f,y_i)$. And if y_i does not appear in f, then we set $\mathrm{ord}(f,y_i) = -\infty$. The *order* of f is defined to be $\max_i \mathrm{ord}(f,y_i)$. A (resp. radical, prime) differential ideal is a (resp. radical, prime) algebraic ideal $\mathcal{I}$ of $F\{y_1,\ldots,y_n\}$ satisfying $\delta(\mathcal{I}) \subset \mathcal{I}$.

By affine space, we mean $\mathbb{A}^n = \mathcal{U}^n$. An element $\eta = (\eta_1,\ldots,\eta_n) \in \mathbb{A}^n$ is called a differential zero of $f \in F\{y_1,\ldots,y_n\}$ if $f(\eta) = 0$. The set of all differential zeros of $\Sigma \subset F\{y_1,\ldots,y_n\}$, is called a *differential variety* defined over F, denoted by $\mathbb{V}(\Sigma)$. All the differential varieties in this paper are assumed to be subsets of $\mathbb{A}^n$. For a differential variety V which is defined over F, we denote $\mathbb{I}(V)$ to be the set of all differential polynomials in $F\{y_1,\ldots,y_n\}$ that vanish at every point of V.

A differential ideal $\mathcal{I} \subset F\{y_1,\ldots,y_n\}$ is prime if and only if it has a generic point, that is, a point $\eta \in \mathbb{V}(\mathcal{I})$ such that for any $f \in F\{y_1,\ldots,y_n\}$, $f(\eta) = 0 \Leftrightarrow f \in \mathcal{I}$. Let $\mathcal{I}$ be a prime differential ideal with a generic point $(\eta_1,\ldots,\eta_n)$. Then there exist d and h such that for sufficiently large t,

$$\mathrm{tr.deg}\, F(\eta_i^{(k)} : 1 \le i \le n; k \le t)/F = d(t+1) + h.$$

The polynomial $\omega_{\mathcal{I}}(t) = d(t+1)+h$ is called the *Kolchin polynomial* of $\mathcal{I}$ and the corresponding d, h are called the *differential dimension* and *order* of $\mathcal{I}$. When $\bar{a}$ is a tuple in a differential field extension of a differential field K, then we write $\omega_{\bar{a}/K}(t)$ for $\omega_{I(\bar{a}/K)}(t)$ where $I(\bar{a}/K)$ is the differential ideal of all differential polynomials over K which vanish at $\bar{a}$.

2 Introduction

In this note, we discuss various primitive element theorems for ordinary differential field extensions. The oldest such result we consider goes back to Kolchin [7, p. 728], where in fact he proved the primitive element theorem in the more general partial differential settings. Here, we restrict to consider the ordinary differential field extensions, so for convenience, we state Kolchin's primitive element theorem in the ordinary differential case as follows:

Theorem 2.1. *Let F be a differential field containing at least one nonconstant. Let $E = F\langle a_1,\ldots,a_n\rangle$ and suppose that $\mathrm{d.tr.deg}\, E/F = 0$. Then there is some $b \in E$ such that $E = F\langle b\rangle$.*

Pogudin [12] generalized Kolchin's theorem to the case that F is a constant field, under the assumption that E contains a nonconstant. In this note, we give a mild generalization of Kolchin's theorem, and conjecture a generalization of Pogudin's theorem. We also illustrate how these generalizations are useful for improving the bounds on a problem of effective differential algebra.

Let $F\langle u\rangle$ denote the fraction field of the differential polynomial ring $F\{u\}$ in one variable. Ritt [14] proved the analog of Lüroth's theorem:

Theorem 2.2. *Let K be a differential field such that $F \subset K \subset F\langle u\rangle$. Then there is some element $g \in K$ such that $K = F\langle g\rangle$.*

Ritt's original formulation is for fields of meromorphic functions, but the general theorem follows from this case via Seidenberg's embedding theorem [18] (Kolchin first proved the general theorem in [8,9]). More recent work has focused on computational aspects of the Lüroth's theorem in the case that K is finitely generated over F. To be more precise, suppose $K = F\langle P_1(u)/Q_1(u),\ldots,P_n(u)/Q_n(u)\rangle$, then computing a Lüroth generator of K/F and giving order and degree bounds for a Lüroth generator are problems of effective differential algebra. Following Kolchin's idea, if $A(y) \in K\{y\}$ is the minimal differential polynomial of x over K w.r.t. the canonical ranking, then for any pair $(a,b) \in K^2$ of coefficients of the polynomial A satisfying that $a/b \notin F$, this a/b can serve as a Lüroth generator [9]. Thus, using the language of modern differential characteristic sets, a Lüroth generator can be computed in the following way: Given a characteristic set $Q_1(u)y_1 - P_1(u),\ldots,Q_n(u)y_n - P_n(u)$ of a prime differential ideal $\mathfrak{I} \subset F\{u,y_1,\ldots,y_n\}$ w.r.t. the elimination ranking $u < y_1 < \cdots < y_n$, compute a characteristic set $B_1(y_1,y_2),\ldots,B_{n-1}(y_1,\ldots,y_n),B_0(y_1,\ldots,y_n,u)$ of $\mathfrak{I}$ w.r.t. the elimination ranking $y_1 < \cdots < y_n < u$. Rewrite $B_0(y_1,\ldots,y_n,u) = \sum_i f_i(y_1,\ldots,y_n)\theta_i(u)$, if $\zeta = \frac{f_{i_1}(P_1(u)/Q_1(u),\ldots,P_n(u)/Q_n(u))}{f_{i_2}(P_1(u)/Q_1(u),\ldots,P_n(u)/Q_n(u))} \notin F$, then $K = F\langle\zeta\rangle$. Based on this idea, Gao and Xu [6] gave an algorithmic proof of the differential Lüroth theorem, but did not consider bounds for the degrees or order of the generator. D'Alfonso et al. [3] proved the following effective version of the theorem:

Theorem 2.3. *Let F be an ordinary differential field of characteristic 0, u differentially transcendental over F and $K = F\langle P_1(u)/Q_1(u),\ldots,P_n(u)/Q_n(u)\rangle$, where $P_j,Q_j \in F\{u\}$ are relatively prime differential polynomials of order at most $e \geq 1$ (i.e. at least one derivative of u occurs in P_j or Q_j for some j) and degree bounded by d such that each $P_j/Q_j \notin F$. Then, any Lüroth generator v of K/F can be written as the quotient of two relatively prime differential polynomials $P(u), Q(u) \in F\{u\}$ with order bounded by $\min\{\mathrm{ord}(P_j/Q_j) : 1 \leq j \leq n\}$ and total degree bounded by $\min\{(d+1)^{(e+1)n}, (nd(e+1)+1)^{2e+1}\}$.*

The connection between the differential Lüroth theorem and the primitive element theorem is related to improving the degree bounds. We should note that our manipulations are not designed to attack the problem of bounding the order, but note that as Kolchin proved in [8], any two Lüroth generators ω_1 and ω_2 are related by the formula $\omega_2 = (a\omega_1 + b)/(c\omega_1 + d)$ for some $a,b,c,d \in F$, so any two Lüroth generators should have the same order [3, see the remarks at the end of Subsect. 3.1]. Our technique is most easily employed in the case that the field F posseses a nonconstant element. In this case, the ideas of our techniques essentially derive from a mild generalization of Theorem 2.1. In the case that F is a constant differential field, our ongoing work is related to an attempt to generalize Pogudin's recent primitive element theorem [12].

In the case that F contains a nonconstant element, we improve the degree bounds as follows:

Theorem 2.4. *Let F be an ordinary differential field of characteristic 0 containing a nonconstant element. Let u be differentially transcendental over F and $K = F\langle P_1(u)/Q_1(u), \ldots, P_n(u)/Q_n(u)\rangle$, where $P_j, Q_j \in F\{u\}$ are relatively prime differential polynomials of order at most $e \geq 1$ (i.e. at least one derivative of u occurs in P_j or Q_j for some j) and degree bounded by d such that each $P_j/Q_j \notin F$ In Theorem 2.3. Then, any Lüroth generator v of K over F can be written as the quotient of two coprime differential polynomials $P(u), Q(u) \in F\{u\}$ with degree bounded by*

$$\min\{(\lceil n/2 \rceil \cdot d + 1)^{2(e+1)}, (d+1)^{n(e+1)}, (nd(e+1)+1)^{2e+1}\}.$$

In the case that the base field consists of constants, we improve the bound as follows:

Theorem 2.5. *Suppose F is a field of constants, u be differentially transcendental over F and $K = F\langle P_1(u)/Q_1(u), \ldots, P_n(u)/Q_n(u)\rangle$ with P_i, Q_i satisfying the same conditions as in Theorem 2.4. Then the total degree of a Lüroth generator of K over F is bounded by*

$$\min\{(d(n+e-1)+1)^{2(e+1)}, (nd(e+1)+1)^{2e+1}\}.$$

The paper is organized as follows. In Sect. 3, we will prove various primitive element theorems for differential fields. In Sect. 4, we will utilize our embedding results to establish the improved bounds for the differential Lüroth's Theorem.

3 Variations of the Primitive Element Theorem for Differential Field Extensions

Throughout this section, all differential fields which appear will be assumed to be subfields of $\mathcal{U}$, the fixed sufficiently large saturated differentially closed field of characteristic zero given in Sect. 1.

Lemma 3.1 *[15, p. 35]. Suppose F contains at least one nonconstant element. If $f \in F\{u\}$ is a nonzero differential polynomial with order r, then for any nonconstant $\eta \in F$, there exists an element $c_0 + c_1\eta + c_2\eta^2 + \cdots + c_r\eta^r$ which does not annul f, where $c_0, \ldots, c_r$ are constants in F.*

Remark 3.2. Note that in Lemma 3.1, we can always select the c_i from the rational number field $\mathbb{Q}$. Indeed, let $x_0, \ldots, x_r$ be arbitrary constants, i.e., the x_i are algebraically independent over F and $x_i' = 0$. Since $f(\sum_{i=0}^r c_i\eta^r) \neq 0$, $g(x_0, \ldots, x_n) = f(x_0 + x_1\eta + x_2\eta^2 + \cdots + x_r\eta^r)$ is a nonzero polynomial in $F\langle\eta\rangle[x_0, \ldots, x_r]$. Since $\mathbb{Q}$ is an infinite field, by induction on r, it is easy to show that there exists $(d_0, \ldots, d_r) \in \mathbb{Q}^{r+1}$ such that $f(d_0 + d_1\eta + d_2\eta^2 + \cdots + d_r\eta^r) \neq 0$.

Also, if $f \in F\{u_1, \ldots, u_n\}$ is a nonzero differential polynomial with order bounded by r, then for any nonconstant $\eta \in F$, there exist $c_{ij} \in \mathbb{Q} (1 \leq i \leq n; 0 \leq j \leq r)$ such that $f(\sum_{i=0}^{r} c_{0i}\eta^i, \ldots, \sum_{i=0}^{r} c_{ni}\eta^i) \neq 0$. We justify this by induction on n. The above paragraph shows that it is valid for $n = 1$. Suppose it holds for $n - 1$. Regard $f(u_1, \ldots, u_n)$ as a polynomial in $u_1, \ldots, u_{n-1}$ with coefficients in $F\{u_n\}$, then by the induction hypothesis, there exist $c_{ij} \in \mathbb{Q} (1 \leq i \leq n - 1)$ such that $g(u_n) = f(\sum_{j=0}^{r} c_{1j}\eta^j, \ldots, \sum_{j=0}^{r} c_{n-1,j}\eta^j, u_n) \neq 0$. Thus, from the case $n = 1$, there exist $c_{nj} \in \mathbb{Q}$ such that $g(\sum_{j=0}^{r} c_{nj}\eta^j) = f(\sum_{j=0}^{r} c_{1j}\eta^j, \ldots, \sum_{j=0}^{r} c_{nj}\eta^j) \neq 0$.

Lemma 3.3. *Let $F_1 \subset F$ be differential fields. Suppose that F_1 is not a field of constants. Then the Kolchin closure of F_1^n over F is $\mathbb{A}^n$.*

Proof. It suffices to show that $\mathbb{I}(F_1^n)$, the set of all differential polynomials over F which vanish at F_1^n, is the zero differential ideal. Since F contains at least a nonconstant, say η, for any nonzero $f \in F\{y_1, \ldots, y_n\}$, by Remark 3.2, there exist $c_{ij} \in \mathbb{Q} (0 \leq j \leq \mathrm{ord}(f))$ such that $f(\sum_j c_{1j}\eta^j, \ldots, \sum_j c_{nj}\eta^j) \neq 0$. Note that for each i, $\sum_j c_{ij}\eta^j \in F_1$. Thus, $\mathbb{I}(F_1^n) = [0]$. Hence, the Kolchin closure of F_1^n is $\mathbb{V}(0) = \mathbb{A}^n$. $\square$

Note that Kolchin's proof [7, p. 728] for Theorem 2.1 as well as Seidenberg's proof for [17, Theorem 1] implies the following result:

Proposition 3.4. *Let $L = F\langle \alpha_1, \ldots, \alpha_n \rangle$ and $\mathrm{d.tr.deg}\, L/F = 0$. Let F be a subfield of F such that F contains a nonconstant. Then, there exist $c_1, \ldots, c_n \in F$ such that $L = F\langle c_1\alpha_1 + \cdots + c_n\alpha_n \rangle$.*

The following result is a straightforward implication of Proposition 3.4 and here we will give a new proof from the geometric point of view.

Proposition 3.5. *Let $K = F\langle \alpha_1, \ldots, \alpha_n \rangle$ with F containing at least one nonconstant and $\mathrm{d.tr.deg}\, K/F = d > 0$. Assume without loss of generality that $\alpha_1, \ldots, \alpha_d$ is a differential transcendence basis of K over F. Then for any nonconstant subfield F_1 of F, there exist $c_{d+1}, \ldots, c_n \in F_1$ such that $K = F\langle \alpha_1, \ldots, \alpha_d, \sum_{i=d+1}^{n} c_i\alpha_i \rangle$.*

Proof. Consider the affine differential variety $V \subset \mathbb{A}^{n-d}$ given by the locus of $\alpha_{d+1}, \ldots, \alpha_n$ over $F\langle \alpha_1, \ldots, \alpha_d \rangle$. The variety V is $F\langle \alpha_1, \ldots, \alpha_d \rangle$-definable each of whose points are differentially algebraic over $F\langle \alpha_1, \ldots, \alpha_d \rangle$. Let $\bar{u} = (u_{d+1}, \ldots, u_n)$ be a tuple which are differentially independent over $F\langle \alpha_1, \ldots, \alpha_d \rangle$. Then we claim that the map $\phi_{\bar{u}} : V \to \mathbb{A}^1$ given by $\bar{x} = (x_{d+1}, \ldots, x_n) \to \sum_{i=d+1}^{n} u_i x_i$ is injective. For if not, then there are two points $\bar{a} = (a_{d+1}, \ldots, a_n)$ and $\bar{b} = (b_{d+1}, \ldots, b_n)$ such that $\sum_{i=d+1}^{n} u_i a_i = \sum_{i=d+1}^{n} u_i b_i$. But $\bar{a}$ and $\bar{b}$ are differentially algebraic over $F\langle \alpha_1, \ldots, \alpha_d \rangle$, and so $\bar{u}$ is δ-transcendental over $F\langle \alpha_1, \ldots, \alpha_d, \bar{a}, \bar{b} \rangle$. But now we have a contradiction, because $\sum_{i=d+1}^{n} u_i(a_i - b_i) = 0$, and not all of the $a_i - b_i$ can be zero, since $\bar{a} \neq \bar{b}$.

The injectivity of the map $\phi_{\bar{u}}$ on the $F\langle \alpha_1, \ldots, \alpha_d \rangle$-definable set V is a first order property (over $F\langle \alpha_1, \ldots, \alpha_d \rangle$) of the tuple $\bar{u}$. Since $\bar{u}$ is generic over $F\langle\alpha_1,\ldots,\alpha_d\rangle$, it follows by quantifier elimination that for all $\bar{v}$ in an $F\langle\alpha_1,\ldots,\alpha_d\rangle$-open subset $U \subset \mathbb{A}^{n-d}$, the map $\phi_{\bar{v}}$ is injective. There is a point $\bar{\gamma} = (\gamma_{d+1},\ldots,\gamma_n)$ of F^{n-d} in U by Lemma 3.3 applied to F relative to $F\langle\alpha_1,\ldots,\alpha_d\rangle$.

Now let $W \subset \mathbb{A}^n$ be the locus of $\alpha_1,\ldots,\alpha_n$ over F. Then the map $\pi_\gamma : \mathbb{A}^n \to \mathbb{A}^{d+1}$ given by $(x_1,\ldots,x_n) \mapsto (x_1,\ldots,x_d, \sum_{i=d+1}^n \gamma_i x_i)$ is injective on the fiber above $\alpha_1,\ldots,\alpha_d$ in W (the proper subvariety of W with $x_1 = \alpha_1,\ldots,x_d = \alpha_d$). By the genericity of $\bar{\alpha} \in W$ over F, it follows that π_γ is injective on a Kolchin open subset of $\mathbb{A}^n$. So $F\langle\bar{\alpha}\rangle \cong F\langle W\rangle \cong F\langle\pi_\gamma(W)\rangle \cong F\langle\alpha_1,\ldots,\alpha_d, \sum_{i=d+1}^n \gamma_i\alpha_i\rangle$ completing the proof. $\square$

Next, we will establish Proposition 3.5 through the use of differential Chow forms, which enables us to compute β_i effectively. Without assuming a transcendence basis beforehand, we restate the proposition as follows:

Proposition 3.6. *Assume F is a differential field with at least one nonconstant. Suppose that $K = F\langle\alpha_1,\ldots,\alpha_n\rangle$ is a finitely generated differential field extension of F such that the differential transcendence degree of K over F is d. Then for any subfield $F \subset F$ containing a nonconstant, there are $\beta_0,\ldots,\beta_d \in K$ which are F-linear combinations of $\alpha_1,\ldots,\alpha_n$ such that*

$$K = F\langle\beta_0,\ldots,\beta_d\rangle.$$

Proof. Let $\mathfrak{I} = \mathbb{I}((\alpha_1,\ldots,\alpha_n)) \subset F\{y_1,\ldots,y_n\}$. Then $\mathfrak{I}$ is of differential dimension d. Let

$$L_i = u_{i0} + u_{i1}y_1 + \cdots + u_{in}y_n \ (i = 0, 1, \ldots, d)$$

be a system of $d+1$ generic differential hyperplanes where all the u_{ij} are differentially independent over F. Denote

$$u_i = (u_{i0}, u_{i1}, \ldots, u_{in}) \, (i = 0, \ldots, d) \text{ and } u = \{u_{ij} : i = 0, \ldots, d; j \neq 0\}.$$

Let $\mathcal{P} = [\mathfrak{I}, L_0, \ldots, L_d] \subset F\{y_1, \ldots, y_n, u_0, \ldots, u_d\}$. Assume $G(u_0, \ldots, u_d)$ is the differential Chow form of $\mathfrak{I}$ and $\mathrm{ord}(G) = h$. Then by the property of the differential Chow form [5, Lemma 4.10],

$$\frac{\partial G}{\partial u_{00}^{(h)}} y_j - \frac{\partial G}{\partial u_{0j}^{(h)}} \in \mathcal{P} \ (j = 1, \ldots, n).$$

Since $\xi = (\alpha_1, \ldots, \alpha_n; -\sum_{j=1}^n u_{0j}\alpha_j, u_{01}, \ldots, u_{0n}; \ldots; -\sum_{j=1}^n u_{dj}\alpha_j, u_{d1}, \ldots, u_{dn})$ is a generic point of $\mathcal{P}$ and $\frac{\partial G}{\partial u_{00}^{(h)}} \notin \mathcal{P}$, $\frac{\partial G}{\partial u_{00}^{(h)}}(\xi) \neq 0$. Now regard $\frac{\partial G}{\partial u_{00}^{(h)}}(\xi)$ as a differential polynomial in u with coefficients in K, which is nonzero. By Lemma 3.1, there exists $a_{ij} \in F$ (for any $u_{ij} \in u$) such that

$$\frac{\partial G}{\partial u_{00}^{(h)}}(\alpha_1, \ldots, \alpha_n; -\sum_{j=1}^n a_{0j}\alpha_j, a_{01}, \ldots, a_{0n}; \ldots; -\sum_{j=1}^n a_{dj}\alpha_j, a_{d1}, \ldots, a_{dn}) \neq 0.$$

For each $k = 0, 1, \ldots, d$, let g_k be the differential polynomial in $F\{u_{00}, \ldots, u_{d0}\}$ obtained from $\frac{\partial G}{\partial u_{0k}^{(h)}}$ by replacing $u_{ij} \in u$ by a_{ij}. Then g_0 is a nonzero differential polynomial which satisfies $g_0(-\sum_{j=1}^n a_{0j}\alpha_j, \ldots, -\sum_{j=1}^n a_{dj}\alpha_j) \neq 0$.

Let $\beta_i = -\sum_{j=1}^n a_{ij}\alpha_j$ for $i = 0, \ldots, d$. We claim that $K = F\langle\beta_0, \ldots, \beta_d\rangle$. Let $\bar{L}_i = u_{i0} + a_{i1}y_1 + \cdots + a_{in}y_n \ (i = 0, \ldots, d)$ and

$$\mathcal{P}_1 = [\mathfrak{I}, \bar{L}_0, \ldots, \bar{L}_d] \subset F\{y_1, \ldots, y_n, u_{00}, \ldots, u_{d0}\}.$$

Clearly, $\mathcal{P}_1$ is a prime differential ideal with a generic point $(\alpha_1, \ldots, \alpha_n, \beta_0, \ldots, \beta_d)$. Since $\frac{\partial G}{\partial u_{00}^{(h)}}y_j - \frac{\partial G}{\partial u_{0j}^{(h)}} \in \mathcal{P}$, it is clear that $g_0 y_j - g_j \in \mathcal{P}_1$ for each $j = 1, \ldots, n$. Thus, $g_0 y_j - g_j$ vanishes at $(\alpha_1, \ldots, \alpha_n, \beta_0, \ldots, \beta_d)$, which implies

$$\alpha_j = g_j(\beta_0, \ldots, \beta_d)/g_0(\beta_0, \ldots, \beta_d).$$

Hence, $K = F\langle\beta_0, \ldots, \beta_d\rangle$. $\square$

We use the following two examples to illustrate the method given in the proof of Proposition 3.6 to compute the generators of the required forms. In the both examples, differential Chow forms can be computed using either the characteristic set method as described in [5, Remark 4.4] or the algorithms for computing differential Chow forms given in [10].

Example 3.7. Let $F = \mathbb{Q}(x)$ with derivation $\delta = \frac{d}{dx}$. Let $K = \mathbb{Q}(x)\langle\alpha_1, \alpha_2\rangle$ where α_1, α_2 are the generic solutions of $y'+1, z'$ respectively. Let $\mathfrak{I} = \mathbb{I}((\alpha_1, \alpha_2))$ $\subset F\{y, z\}$. Then $\mathfrak{I}$ is of differential dimension 0. Take $L_0 = u_0 + u_1 y + u_2 z$. Then the differential Chow form of $\mathfrak{I}$ is

$$G = (u_1 u_2' - u_2 u_1')u_0'' - 2u_1'(u_1 u_2' - u_1' u_2) - u_1''(u_0 u_2' - u_2 u_0') + u_2''(u_0 u_1' - u_0' u_1 + u_1^2).$$

So the separant of G is $S_G = u_1 u_2' - u_2 u_1'$. We can take $u_1 = -1$ and $u_2 = -x$ which does not annul S_G. Hence, $\alpha_1 + x\alpha_2$ is a primitive element of K/F, that is, $K = F\langle\alpha_1 + x\alpha_2\rangle$.

Example 3.8. Let $F = \mathbb{Q}(x)$ with $\delta = \frac{d}{dx}$. Let $K = \mathbb{Q}(x)\langle u+x, u'+x, u''\rangle$ where u is differentially transcendental over F. Clearly, the differential transcendence degree of K over F is 1. Let $u_i = (u_{i0}, u_{i1}, u_{i2}, u_{i3}) \ (i = 0, 1)$. Then we can compute the differential Chow form $G(u_0, u_1)$ of $\mathbb{I}((u+x, u'+x, u'')) \subset F\{y_1, y_2, y_3\}$, which is a differential polynomial of order 2 and differential degree 6. The separant of G is

$$\begin{aligned} S_G = u_{13}(&u_{11}u_{12}u_{03}^2 - u_{11}u_{13}u_{02}u_{03} - u_{12}u_{11}u_{03}^2 + u_{12}'u_{13}u_{01}u_{03} - u_{03}'u_{11}u_{13}u_{02} \\ &+ u_{03}'u_{12}u_{13}u_{01} + u_{13}'u_{11}u_{02}u_{03} - u_{13}'u_{12}u_{01}u_{03} + u_{02}'u_{11}u_{13}u_{03} - u_{02}'u_{13}^2u_{01} \\ &- u_{01}'u_{12}u_{13}u_{03} + u_{01}'u_{13}^2u_{02} - u_{11}^2u_{03}^2 + u_{11}u_{12}u_{02}u_{03} + 2u_{11}u_{13}u_{01}u_{03} \\ &- u_{11}u_{13}u_{02}^2 - u_{12}^2u_{01}u_{03} + u_{12}u_{13}u_{01}u_{02} - u_{13}^2u_{01}^2). \end{aligned}$$

We can take $u_{01} = 1, u_{02} = 0, u_{03} = x, u_{11} = x, u_{12} = 1, u_{13} = 1$ which does not annul S_G. Hence, $\beta_1 = (u+x) + tu'', \beta_2 = x(u+x) + (u'+x) + u''$ is a set of generators of K/F, that is, $K = F\langle\beta_1, \beta_2\rangle$.

The proofs of Lemma 3.3 and Proposition 3.5 can be generalized to the case of a differential field with finitely many commuting derivations $\delta_1, \ldots, \delta_m$, under the assumption that F contains m elements $\beta_1, \ldots, \beta_n$ whose Jacobian, $\det(\delta_i(\beta_j))$, is nonzero [7], but the proof of Proposition 3.6 is not suited for the partial case.

Proposition 3.5 does not hold in the case that the field F is the constant field:

Example 3.9. Let F be the rational number field with the trivial derivation. Let x, y be two constants in a differential extension field of F, which are algebraically independent over F. Consider $K = F\langle x, y \rangle = F(x, y)$. Then K is of differential transcendence degree 0, but the transcendence degree of K over F is 2. Clearly, there is no $a, b \in F$ such that $K = F(ax + by)$.

When F is a constant differential field, although Proposition 3.5 is not valid, we have the following similar result, which can be regarded as a consequence of Seidenberg's proof [17].

Proposition 3.10. *Assume F is a differential field of constants. Suppose that $K = F\langle \alpha_1, \ldots, \alpha_n \rangle$ is a finitely generated differential field extension of F such that the differential transcendence degree of K over F is $d > 0$. Suppose α_1 is differentially transcendental over F. Then there exist $\beta_1, \ldots, \beta_d \in K$ such that $K = F\langle \alpha_1, \beta_1, \ldots, \beta_d \rangle$ and each β_i is an F-linear combination of $\alpha_2, \ldots, \alpha_n$ and powers of α_1 bounded by h where h is the order of the differential Chow form of $\alpha_2, \ldots, \alpha_n$ over $F\langle \alpha_1 \rangle$. In particular, there exist $c_{ijk} \in \mathbb{Q}$ such that $\beta_i = \sum_{j=2}^{n} (\sum_{k=0}^{h} c_{ijk} \alpha_1^k) \alpha_j, i = 1, \ldots, d$ and $K = F\langle \alpha_1, \beta_1, \ldots, \beta_d \rangle$.*

Proof. Let $K_1 = F\langle \alpha_1 \rangle$. Then the differential transcendence degree of K over K_1 is $d - 1$. Consider the differential ideal $\mathcal{I} = \mathbb{I}((\alpha_2, \ldots, \alpha_n)) \subset K_1\{y_2, \ldots, y_n\}$. Suppose the order of $\mathcal{I}$ is equal to h. Suppose $G(u_0, \ldots, u_{d-1})$ is the differential Chow form of $\mathcal{I}$, then $\mathrm{ord}(G) = h$. Applying the similar method as in the proof of Proposition 3.6 to K/K_1, by Remark 3.2, we can find $c_{ijk} \in \mathbb{Q}$ such that for $\beta_i = \sum_{j=2}^{n} (\sum_{k=0}^{h} c_{ijk} \alpha_1^k) \alpha_j, K = F\langle \alpha_1, \beta_1, \ldots, \beta_d \rangle$. $\square$

Also, in the case that F is a constant differential field, one can establish Proposition 3.5 when making an additional assumption on the elements $\alpha_1, \ldots, \alpha_n$. The additional assumption uses terminology from model theory, which we will now introduce. Our conventions are designed to deliver the model theoretic notions in the differential algebraic setting, where some of the notions can be given significantly simpler definitions than in the general setting.

We remind the reader that $\mathcal{U}$ is a universal differential field. Let $X\mathbb{A}^n$ be a constructible set in the Kolchin topology over F; that is, X is a boolean combination of affine differential varieties over F. Then we say X is *orthogonal to the constants* if for any differential field extension K of F, any element c of the constant field $\mathcal{C}$ of $\mathcal{U}$, and any $\bar{a} \in X$, we have the equality of the Kolchin polynomials:

$$\omega_{\bar{a}/K\langle c \rangle}(t) = \omega_{\bar{a}/K}(t).$$

This implies that if c is transcendental over K, then c is transcendental over $K\langle a \rangle$.

The notion defined in the previous paragraph is a special case of the general notion defined in [1, see Ziegler's article, page 40 for additional details].

Proposition 3.11. *Suppose that $K = F\langle \alpha_1, \ldots, \alpha_n \rangle$ is a finitely generated differential field extension of F such that the differential transcendence degree of K over F is d. Assume without loss of generality that $\alpha_1, \ldots, \alpha_d$ are differentially independent over F. Suppose that $loc((\alpha_{d+1}, \ldots, \alpha_n)/F\langle \alpha_1, \ldots, \alpha_d \rangle)$ is orthogonal to the constants. Then there is $\beta_{d+1} \in K$ such that $K = F\langle \alpha_1, \ldots, \alpha_d, \beta_{d+1} \rangle$ and β_{d+1} is an $\mathbb{Q}$-linear combination of $\alpha_{d+1}, \ldots, \alpha_n$.*

Proof. First, we claim that the general result follows from the case in which $n = d+2$. This follows inductively, noting that $\mathbb{Q}$-linear combinations of $\mathbb{Q}$-linear combinations of $\alpha_{d+1}, \ldots, \alpha_n$ are again $\mathbb{Q}$-linear combinations of $\alpha_{d+1}, \ldots, \alpha_n$ and $\mathbb{Q}$-linear combinations preserve orthogonality to the constants. So, without loss of generality, assume that $n = d + 2$.

Let $X = loc_{F\langle \alpha_1, \ldots, \alpha_d \rangle}(a, b)$, the Kolchin closure of (a, b) over the ground field $F\langle \alpha_1, \ldots, \alpha_d \rangle$. Let $c_1, c_2 \in \mathcal{C}$ be independent transcendental constants over $F\langle \alpha_1, \ldots, \alpha_d \rangle$. We claim the map

$$\phi_{\bar{c}} : X \to \mathbb{A}^1$$

given by $(x, y) \mapsto c_1 x + c_2 y$ is injective on a Kolchin open subset of X. If this is not the case, there are $(x_1, y_1), (x_2, y_2) \in X$ such that (x_i, y_i) is generic on X over $F\langle \alpha_1, \ldots, \alpha_d, c_1, c_2 \rangle$ (which implies that c_1, c_2 are independent transcendentals over $F\langle \alpha_1, \ldots, \alpha_d, x_i, y_i \rangle$ for $i = 1, 2$) such that

$$c_1 x_1 + c_2 y_1 = c_1 x_2 + c_2 y_2.$$

But now taking $K = F\langle \alpha_1, \ldots, \alpha_d, c_1, x_1, y_1 \rangle$, we can see that c_2 is not transcendental over $K\langle x_2, y_2 \rangle$ as $c_2 = c_1 \cdot \frac{x_2 - x_1}{y_1 - y_2}$ is in $K\langle x_2, y_2 \rangle$. But this implies:

$$\omega_{(x_2, y_2)/K\langle c_2 \rangle}(t) \neq \omega_{(x_2, y_2)/K}(t),$$

and this contradicts the assumption that X is orthogonal to the constants.

So, there is a Kolchin open subset $U \subset X$ such that the map $\phi_{\bar{c}}|_U$ is an injective map. Injectivity is a definable property of the map $\phi_{\bar{c}}|_U$, and it holds for the generic point in $\mathcal{C}^2$, so for some Zariski open (the Kolchin open subsets of $\mathcal{C}$ are Zariski open) subset $U_1 \subset \mathcal{C}$, for all $\bar{c}' \in U_1$, the map $\phi_{\bar{c}'}|_U$ is injective. By the density of $\mathbb{Q}^2$ in $\mathcal{C}^2$, there are $\bar{q} = q_1, q_2$ for which $\phi_{\bar{q}}|_U$ is injective and thus gives an isomorphism between the differential function field of X and its image, completing the proof. $\square$

Remark 3.12. The assumption that $tp(\alpha_{d+1}, \ldots, \alpha_n/F\langle \alpha_1, \ldots, \alpha_d \rangle)$ is orthogonal to the constants is rather difficult to verify in practice. On the other hand, it is folklore of the model theory of differential fields that *most* differential equation of some order ≥ 1 and degree ≥ 2 should be strongly minimal and trivial (which implies orthogonality to the constants). For specific instances of results of this nature, see [2, 13].

There is a considerable literature devoted to verifying this condition for various specific differential equations [2, 4, 13]; proving that a given strongly minimal differential equation has trivial forking geometry (and is thus orthogonal to the constants) is also the key to proving that differential closure is not minimal [16]. As far as we can tell, only the results of [4] provide examples which are defined over a differential transcendental, which is the only case pertinent to the differential Lüroth theorem. To give the reader an idea of the hypothesis, we will give a specific example, in the case of two variables in order to keep the technicalities minimal.

So, let
$$P_1/Q_1 := S(u' + u) + R(u' + u) \cdot ((u' + u)')^2,$$

where
$$R(y) = \frac{y^2 - 1968y + 2\ 654\ 208}{2y^2(y - 1728)^2},$$

and
$$S(x) = \left(\frac{x''}{x'}\right)' - \frac{1}{2}\left(\frac{x''}{x'}\right)^2$$

is the Schwarzian derivative. Let
$$P_2/Q_2 = u' + u.$$

Then the type $tp(P_2/Q_2/F\langle P_1/Q_1\rangle)$ is the generic solution to the differential equation
$$S(x) + R(x) \cdot ((x)')^2 = P_1/Q_1.$$

By the results of [4], this type is strongly minimal. It follows that the type is nonorthogonal to the constants, since the equivalence relation of nonorthogonality refines transcedence degree on strongly minimal sets (the authors of [4] also prove that this set has trivial forking geometry).

Propositions 3.5 and 3.10 are effective in the sense that the degree of the elements which generate the differential field extension are bounded. We further conjecture the following mild strengthening of Pogudin's primitive element theorem:

Conjecture 3.13. *Assume F is a constant differential field. Suppose that $K = F\langle\alpha_1, \ldots, \alpha_n\rangle$ is a finitely generated differential field extension of F such that the differential transcendence degree of K over F is $d > 0$. Assume without loss of generality that $\alpha_1, \ldots, \alpha_d$ are a differential transcendence basis for K over F. Assume that at least one of $\alpha_{d+1}, \ldots, \alpha_n$ is a nonconstant. Then there is a polynomial $P \in \mathbb{Q}[x_{d+1}, \ldots, x_n]$ such that*
$$K = F\langle\alpha_1, \ldots, \alpha_d, P(\alpha_{d+1}, \ldots, \alpha_n)\rangle.$$

The above conjecture is a direct consequence of the following stronger conjecture.

Conjecture 3.14. *Assume F is a differential field which contains at least one nonconstant. Suppose that $K = F\langle\alpha_1, \ldots, \alpha_n\rangle$ is a finitely generated differential field extension of F such that each α_i is differentially algebraic over F and at* *least one α_i is a nonconstant. Then there is a polynomial $P \in \mathbb{Q}[x_1, \ldots, x_n]$ such that*

$$K = F\langle P(\alpha_1, \ldots, \alpha_n)\rangle.$$

The above conjecture is not true if all α_i are constants. Similar to Example 3.9, if x, y are constants which are independent algebraic indeterminates, $\mathbb{Q}(t)\langle x, y\rangle \neq \mathbb{Q}(t)\langle P(x, y)\rangle$ for any $P \in \mathbb{Q}[x, y]$.

In the current work in progress we hope to establish the conjecture in an effective form (bounding the degree of P); bounding the degree of P might then be used to improve the bounds for the degree of a Lüroth generator while working over a constant differential field.

4 Improving the Bounds in the Differential Lüroth Theorem

In this section, we explain how the results of the previous section can be applied to improve the degree bound for the differential Lüroth theorem.

4.1 The Nonconstant Case

We will work first in the case where F contains some nonconstant element, proving Theorem 2.4. The analysis is simpler in this case and the primitive element style analysis of the previous section yields improved bounds.

Suppose that

$$K = F\langle P_1(u)/Q_1(u), \ldots, P_n(u)/Q_n(u)\rangle$$

where $P_j, Q_j \in F\{u\}$ are relatively prime differential polynomials with order satisfying $e = \max\{\mathrm{ord}(P_i), \mathrm{ord}(Q_j)\} \geq 1$ and total degree bounded by d such that $P_j/Q_j \notin F$ for every $1 \leq j \leq n$.

When $x \in \mathbb{N}$, let $\lfloor x \rfloor, \lceil x \rceil$ denote the standard floor and ceiling functions, respectively. Let

$$K_1 = F\langle P_1(u)/Q_1(u), \ldots, P_{\lfloor n/2 \rfloor}(u)/Q_{\lfloor n/2 \rfloor}(u)\rangle$$

and consider the differential field extension

$$K = K_1\langle P_{\lfloor n/2 \rfloor+1}(u)/Q_{\lfloor n/2 \rfloor+1}(u), \ldots, P_n(u)/Q_n(u)\rangle.$$

Since each $P_i(u)/Q_i(u) \notin F$ and $\mathrm{d.tr.deg}\, K/F = 1$, $\mathrm{d.tr.deg}\, K/K_1 = 0$. Apply Proposition 3.4 to the extension K over K_1 with K_1 playing the role of F in Proposition 3.4, then we obtain a generator β for K over K_1 which is an F-linear combination of $P_{\lfloor n/2 \rfloor+1}(u)/Q_{\lfloor n/2 \rfloor+1}(u), \ldots, P_n(u)/Q_n(u)$. Note that this β has order at most e and degree at most $\lceil n/2 \rceil \cdot d$. Specifically, the total degree of β is bounded by the sum of the degrees of $P_{\lfloor n/2 \rfloor+1}(u)/Q_{\lfloor n/2 \rfloor+1}(u), \ldots, P_n(u)/Q_n(u)$. Here, It may happen that $K = K_1$

and in this case, the obtained β may be contained in F. If this happens, we reset $\beta = P_n(u)/Q_n(u)$.

Now we have $K = K_1\langle\beta\rangle = F\langle P_1(u)/Q_1(u),\ldots,P_{\lfloor n/2\rfloor}(u)/Q_{\lfloor n/2\rfloor}(u),\beta\rangle$. Clearly, d.tr.deg $K/F\langle\beta\rangle = 0$. Applying Proposition 3.4 to the differential field extension

$$K = F\langle P_1(u)/Q_1(u),\ldots,P_{\lfloor n/2\rfloor}(u)/Q_{\lfloor n/2\rfloor}(u),\beta\rangle$$

over $F\langle\beta\rangle$, then there is an F-linear combination of

$$P_1(u)/Q_1(u),\ldots,P_{\lfloor n/2\rfloor}(u)/Q_{\lfloor n/2\rfloor}(u)$$

which generates K over $F\langle\beta\rangle$. Call this element α and note that α has order at most e and degree at most $\lfloor n/2\rfloor\cdot d$. Specifically, the total degree of α is bounded by the sum of the total degrees of $P_1(u)/Q_1(u),\ldots,P_{\lfloor n/2\rfloor}(u)/Q_{\lfloor n/2\rfloor}(u)$.

Now, we have obtained $K = F\langle\alpha,\beta\rangle$ with $\max\{\text{ord}(\alpha),\text{ord}(\beta)\} = e_1 \leq e$ and $\deg(\alpha),\deg(\beta) \leq \lceil n/2\rceil\cdot d$. Note it may happen that $e_1 = 0$. In the following, we show that applying Theorem 2.3 to the differential field extension $F\langle\alpha,\beta\rangle$ over F, the degree of a Lüroth generator is bounded by $(\lceil n/2\rceil\cdot d + 1)^{(e+1)2}$.

Lemma 4.1. *The degree of a Lüroth generator of $F\langle\alpha,\beta\rangle$ over F is bounded by $(\lceil n/2\rceil\cdot d + 1)^{(e+1)2}$.*

Proof. If $e_1 \geq 1$, applying Theorem 2.3 directly to the differential field extension $F\langle\alpha,\beta\rangle$ over F, the bound can be obtained.

Now suppose $e_1 = 0$ and $\alpha = R_1(u)/S_1(u), \beta = R_2(u)/S_2(u) \in F(u)$. Let $u = z'$, the first derivative of a new element z. Since u is differentially transcendental over F, z is differentially transcendental over F too. Thus, $K = F\langle R_1(u)/S_1(u), R_2(u)/S_2(u)\rangle = F\langle R_1(z')/S_1(z'), R_2(z')/S_2(z')\rangle \subset F\langle z\rangle$. With respect to the new differential indeterminate z, $\max\{\text{ord}(R_i, z),\text{ord}(S_i, z)\} = 1$ which satisfying the conditions in Theorem 2.3. Thus, there exists coprime paris $(P(z), Q(z)) \in F\{z\}^2$ with $\text{ord}(P, z),\text{ord}(Q, z) \leq 1$ and $\deg(P),\deg(Q) \leq (\lceil n/2\rceil\cdot d + 1)^{(1+1)2} \leq (\lceil n/2\rceil\cdot d + 1)^{(e+1)2}$ such that $K = F\langle P(z)/Q(z)\rangle$. Since $K = F\langle R_1(u)/S_1(u), R_2(u)/S_2(u)\rangle$ has a Lüroth generator $T_1(u)/T_2(u)$ and by [8, p. 359], the two Lüroth generators $P(z)/Q(z)$ and $T_1(u)/T_2(u)$ are related by the formula $P(z)/Q(z) = (aT_1(u)/T_2(u) + b)/(cT_1(u)/T_2(u) + d)$ for some $a, b, c, d \in F$, Thus, $P(z), Q(z) \in F[z']$. Replacing z' by u in P and Q, we get a Lüroth generator $P_0(u), Q_0(u)$ satisfying $\text{ord}(P_0, u),\text{ord}(Q_0, u) = 0$ and $\deg(P_0),\deg(Q_0) \leq (\lceil n/2\rceil\cdot d + 1)^{(e+1)2}$. $\quad\square$

Combining the degree bound given in Theorem 2.3 with Lemma 4.1, we obtain the degree of a Lüroth generator of $K = F\langle P_1(u)/Q_1(u),\ldots,P_n(u)/Q_n(u)\rangle$ over F is bounded by

$$\min\{(\lceil n/2\rceil\cdot d + 1)^{(e+1)2},(d + 1)^{(e+1)n},(nd(e + 1) + 1)^{2e+1}\}.$$

This establishes Theorem 2.4.

Remark 4.2. The first quantity of the minimum taken above is $(\lceil n/2 \rceil \cdot d + 1)^{(e+1)2}$. This is almost always smaller than the second quantity $(d + 1)^{(e+1)n}$, the only pertinent exceptional case being $n = 3, e = 1$, and $d = 1$. When any of the inputs is larger, $(\lceil n/2 \rceil \cdot d + 1)^{(e+1)2}$ is less than $(d + 1)^{(e+1)n}$. It is also true that $(\lceil n/2 \rceil \cdot d + 1)^{(e+1)2}$ is very often smaller than $(nd(e + 1) + 1)^{2e+1}$, though there are infinitely many exceptional cases (essentially by picking n or d to be sufficiently large compared to e). From a practical standpoint, examples with low order, degree, and number of variables are of particular interest; when $n \leq 10$ and $d \leq 10$, $(\lceil n/2 \rceil \cdot d + 1)^{(e+1)2}$ is the smallest of the above bounds (excluding the exceptional case $n = 3, e = 1$, and $d = 1$).

4.2   The Constant Case

In this subsection, we assume F is a field of constants. Let

$$K = F\langle P_1(u)/Q_1(u), \ldots, P_n(u)/Q_n(u) \rangle$$

where each $P_i(u)/Q_i(u) \notin F$.

If $tp(P_2(u)/Q_2(u), \ldots, P_n(u)/Q_n(u)/F\langle P_1(u)/Q_1(u)\rangle)$ is orthogonal to the constants, then the analysis from the previous subsection works completely analogously with Proposition 3.11 in place of Proposition 3.5. The criterion also applies with $P_i(u)/Q_i(u)$ exchanging roles with $P_1(u)/Q_1(u)$, for any $i = 2, \ldots, n$. In the following, we do not assume such conditions on the generators P_i/Q_i.

To give the main theorem in this section, we first need several lemmas.

Lemma 4.3 *[10, Theorem 18]. Let $\mathfrak{I}$ be a prime differential ideal of differential dimension d in $F\{y_1, \ldots, y_n\}$, and $\mathcal{A} = \{A_1, \ldots, A_{n-d}\}$ a characteristic set of $\mathfrak{I}$ under an arbitrary ranking. Then $\mathrm{ord}(\mathfrak{I})$ is bounded by the Jacobi number of $\mathcal{A}$. That is,*

$$\mathrm{ord}(\mathfrak{I}) \leq \max_{\sigma} \sum_{i=1}^{n-d} \mathrm{ord}(A_i, y_{\sigma(i)}),$$

where σ runs among all injective maps from $\{1, \ldots, n - d\}$ to $\{1, \ldots, n\}$.

Lemma 4.4 *[5, Theorem 2.11]. Let $\mathfrak{I}$ be a prime differential ideal in $F\{y_1, \ldots, y_n\}$. Then $\mathrm{ord}(\mathfrak{I})$ is the maximum of all the relative orders of $\mathfrak{I}$, that is,*

$$\mathrm{ord}(\mathfrak{I}) = \max_{U} \mathrm{ord}_U(\mathfrak{I}),$$

where U is any parametric set of $\mathfrak{I}$, that is, U is a maximal subset of variables $\{y_1, \ldots, y_n\}$ such that $\mathfrak{I} \cap F\{U\} = \{0\}$.

With the above preparations, we now prove Theorem 2.5. That is, to show when F is a field of constants, then the total degree of a Lüroth generator is bounded by

$$\min\{(d(n + e - 1) + 1)^{2(e+1)}, (nd(e + 1)(n + e - 1) + 1)^{2e+1}\}.$$

Proof of Theorem 2.5. Let $\alpha_i = P_i(u)/Q_i(u)$ $(i = 1,\ldots,n)$ and $K = F\langle \alpha_1,\ldots,\alpha_n \rangle$. Clearly, the differential transcendence degree of K over F is 1. Also, by the hypothesis, each $\alpha_i \in F\langle u \rangle \backslash F$. Suppose h is the order of the prime differential ideal $\mathbb{I}((\alpha_2,\ldots,\alpha_n))$ over $F\langle \alpha_1 \rangle$. By Proposition 3.10, there exist $c_{jk} \in \mathbb{Q}$ such that for $\eta = \sum_{j=2}^{n}(\sum_{k=0}^{h} c_{jk}\alpha_1^k)\alpha_j$, $K = F\langle \alpha_1, \eta \rangle$. The problem is reduced to the case $n = 2$. The order of η is still bounded by e. The degree of η is bounded by $d(n + h - 1)$.

It suffices to give a bound for h. Consider the prime differential ideal

$$\mathbb{J} = \mathbb{I}((u, \alpha_1, \alpha_2, \ldots, \alpha_n)) \subset F\{y_1, \ldots, y_n, z\}.$$

It is easy to show that

$$\mathcal{A} := Q_1(z)y_1 - P_1(z), \ldots, Q_n(z)y_n - P_n(z)$$

is a characteristic set of $\mathbb{J}$ w.r.t. the elimination ranking $z < y_1 < \ldots < y_n$. Since the orders of P_i, Q_i is bounded by e, the order matrix $(s_{ij})_{n \times (n+1)}$ of $\mathcal{A}$ satisfies $s_{ii} = \mathrm{ord}(A_i, y_i) = 0$, $s_{ij} = -\infty$ $(i \neq j \leq n)$ and $s_{i,n+1} = \mathrm{ord}(A_i, z) \leq e$. So the Jacobi number of $\mathcal{A}$, $\max_\sigma \{\sum_{i=1}^{n} s_{i\sigma(i)}\}$ for σ running through injective maps from $\{1,\ldots,n\}$ to $\{1,\ldots,n+1\}$, is bounded by e. Thus, by Lemma 4.3, $\mathrm{ord}(\mathbb{J}) \leq e$. Let $\mathbb{J}_1 = \mathbb{I}((\alpha_1, \alpha_2, \ldots, \alpha_n)) = \mathbb{J} \cap F\{y_1, \ldots, y_n\}$. The Kolchin polynomials of $\mathbb{J}$ and $\mathbb{J}_1$ have the following relations: for sufficiently large t,

$$\omega_{\mathbb{J}}(t) = \mathrm{tr.deg}\, F\big(u^{(k)}, \alpha_i^{(k)} : k \leq t, i = 1, \ldots, n\big)/F$$
$$= (t+1) + \mathrm{ord}(\mathbb{J})$$
$$= (t+1) + \mathrm{ord}(\mathbb{J}_1)$$
$$\quad + \mathrm{tr.deg}\, F\big(u^{(k)}, \alpha_i^{(k)} : k \leq t, i = 1, \ldots, n\big)/F\big(\alpha_i^{(k)} : k \leq t, i = 1, \ldots, n\big)$$

Thus, $\mathrm{ord}(\mathbb{J}_1) \leq \mathrm{ord}(\mathbb{J}) \leq e$. Note that $h = \mathrm{tr.deg}\, F\langle \alpha_1, \alpha_2, \ldots, \alpha_n \rangle / F\langle \alpha_1 \rangle$ is also equal to the relative order of $\mathbb{J}$ w.r.t. the parametric set $\{y_1\}$, by Lemma 4.4, $h \leq \mathrm{ord}(\mathbb{J}) \leq e$.

So the degree of η is bounded by $d(n + e - 1)$. Hence, by Theorem 2.3, the degree of a Lüroth generator is bounded by

$$\min\{(d(n + e - 1) + 1)^{2(e+1)}, (2d(e + 1)(n + e - 1) + 1)^{2e+1}\}.$$

$\square$

Remark 4.5. In most of the cases, especially when either n or e is large, we have

$$(d(n + e - 1) + 1)^{2(e+1)} < (2d(e + 1)(n + e - 1) + 1)^{2e+1}$$

and

$$(d(n + e - 1) + 1)^{2(e+1)} < (nd(e + 1) + 1)^{2e+1}.$$

Hence, the degree bound given in Theorem 2.5 is smaller than that in Theorem 2.3.

As an experiment to compare the two bounds, we have computed more than 10,000 randomly generated tuples (n, d, e) simulating the pertinent cases of the bounds when each of the variables is less than 30, and our bound gives the better result approximately 94.3 % of the time.

In future work, we hope to prove Conjecture 3.13 in an effective manner and use the result to improve the bounds for the degree in the effective differential Lüroth theorem for the case that the base field is constant.

References

1. Bouscaren, E.: Proof of the Geometric Mordell-Lang Conjecture. In: Hrushovski's, E. (ed.) Model Theory and Algebraic Geometry. Lecture Notes in Mathematics, vol. 1696, pp. 177–196. Springer, Heidelberg (1998)
2. Brestovski, M.: Algebraic independence of solutions of differential equations of the second order. Pac. J. Math. **140**(1), 1–19 (1989)
3. D'Alfonso, L., Jeronimo, G., Solernó, P.: Effective differential lüroth's theorem. J. Algebra **406**, 1–19 (2014)
4. Freitag, J., Scanlon, T.: Strong minimality and the j-function. J. Eur. Math. Soc. (2015). Accepted
5. Gao, X.-S., Li, W., Yuan, C.-M.: Intersection theory in differential algebraic geometry: generic intersections and the differential chow form. Trans. Am. Math. Soc. **365**(9), 4575–4632 (2013)
6. Gao, X.-S., Tao, X.: Lüroth theorem in differential fields. J. Syst. Sci. Complex. **15**(4), 376–383 (2002)
7. Kolchin, E.R.: Extensions of differential fields, I. Ann. Math. **43**, 724–729 (1942)
8. Kolchin, E.R.: Extensions of differential fields, II. Ann. Math. **45**(2), 358–361 (1944)
9. Kolchin, E.R.: Extensions of differential fields, III. Bull. Am. Math. Soc. **53**(4), 397–401 (1947)
10. Li, W., Li, Y.: Computation of differential chow forms for ordinary prime differential ideals. Adv. Appl. Math. **72**, 77–112 (2015). doi:10.1016/j.aam.2015.09.004
11. Marker, D., Messmer, M., Pillay, A.: Model Theory of Fields. A.K. Peters/CRC Press (2005)
12. Gleb, A.: The primitive element theorem for differential fields with zero derivation on the ground field. J. Pure Appl. Algebra **219**(9), 4035–4041 (2015)
13. Pong, W.Y.: On a result of rosenlicht. Commun. Algebra **30**(12), 5933–5939 (2002)
14. Ritt, J.F.: Differential Equations from the Algebraic Standpoint, vol. 14. American Mathematical Soc., New York (1932)
15. Ritt, J.F.: Differential Algebra. Dover Publications, New York (1950)
16. Rosenlicht, M.: The nonminimality of the differential closure. Pac. J. Math. **52**, 529–537 (1974)
17. Seidenberg, A.: Some basic theorems in differential algebra (characteristic p, arbitrary). Trans. Amer. Math. Soc. **73**, 174–190 (1952)
18. Seidenberg, A.: Abstract differential algebra and the analytic case. Proc. Am. Math. Soc. **9**(1), 159–164 (1958)

A New Bound for the Existence of Differential Field Extensions

Richard Gustavson[1]($\boxtimes$) and Omar León Sánchez[2]

[1] Department of Mathematics, CUNY Graduate Center,
365 Fifth Avenue, New York, NY 10016, USA
rgustavson@gradcenter.cuny.edu
[2] Department of Mathematics and Statistics, McMaster University,
Hamilton Hall, 1280 Main Street West, Hamilton, ON L8S 4K1, Canada
oleonsan@math.mcmaster.ca

Abstract. We prove a new upper bound for the existence of a differential field extension of a differential field (K, Δ) that is *compatible* with a given field extension of K. In 2014, Pierce provided an upper bound in terms of lengths of certain antichain sequences of $\mathbb{N}^m$ equipped with the product order. This result has had several applications to effective methods in differential algebra such as the effective differential Nullstellensatz problem. Using a new approach involving Macaulay's theorem on the Hilbert function, we produce an improved upper bound.

Keywords: Algebraic theory of differential equations · Fields with several commuting derivations · Differential field extensions

1 Preliminaries

Let (K, Δ) be a differential field of characteristic zero with m commuting derivations $\Delta = \{\partial_1, \ldots, \partial_m\}$. We place two orders on $\mathbb{N}^m$, the product order $\leq$ and the degree-lexicographic order $\trianglelefteq$. Given $\xi = (u_1, \ldots, u_m) \in \mathbb{N}^m$, we let the degree of ξ be $\deg \xi = u_1 + \cdots + u_m$. For any $r \in \mathbb{N}$, we set $\Gamma(r) = \{\xi \in \mathbb{N}^m : \deg \xi \leq r\}$.

We will consider field extensions of K whose generators over K are indexed by elements of $\mathbb{N}^m$; more precisely, we consider extensions of the form $K(a^\xi : \xi \in \Gamma(r))$ for some $r \in \mathbb{N}$. A generator a^ξ is said to be a *leader* if it is algebraic over $K(a^\tau : \tau \triangleleft \xi)$, and a leader a^ξ is said to be *minimal* if there is no leader a^τ with $\tau < \xi$.

A differential field extension (M, Δ') of (K, Δ), with $\Delta' = \{D_1, \ldots, D_m\}$ and $D_i|_K = \partial_i$, is said to be *compatible* with the field extension $L := K(a^\xi : \xi \in \Gamma(r))$ if $L \subseteq M$ and whenever $\xi \in \Gamma(r-1)$, then $D_i a^\xi = a^{\xi+\mathbf{i}}$, where $\mathbf{i}$ is the tuple with a one in the i-th entry and zeros elsewhere. On the other hand, the field $L = K(a^\xi : \xi \in \Gamma(r))$ is said to satisfy the *differential condition* if there exist derivations $D_i : K(a^\xi : \xi \in \Gamma(r-1)) \to L$ extending ∂_i such that $D_i a^\xi = a^{\xi+\mathbf{i}}$ for all $\xi \in \Gamma(r-1)$. Note that the existence of a differential field extension (M, Δ') of (K, Δ) compatible with L implies that L satisfies the differential condition; however, the converse is of course not generally true.

© Springer International Publishing Switzerland 2016
I.S. Kotsireas et al. (Eds.): MACIS 2015, LNCS 9582, pp. 358–361, 2016.
DOI: 10.1007/978-3-319-32859-1_30

In [7], Pierce showed that there is an integer $s \geq r$ depending only on m and r (generally much larger than r) such that if there exists a field extension of L of the form $K(a^\xi : \xi \in \Gamma(s))$ satisfying the differential condition, then there exists a differential field extension of (K, Δ) compatible with L. However, he does not deal with the issue of determining the minimal value of s for which this property holds. In this note we denote this minimal value by T. Let us note that Pierce finds an upper bound for T in terms of maximal lengths of certain antichain sequences of $(\mathbb{N}^m, \leq)$, and in [4] this upper bound is shown to be bounded above by $2A(m + 3, 4r - 1)$ where A denotes the Ackermann function. In this note we present an improvement of this upper bound.

These types of results have been proven to be very fruitful. There have been several applications, including determining the consistency of a collection of system of polynomial (partial) differential equations [2] and bounding the number of solutions to such equations [1].

2   Results

In [7, Theorem 4.3], Pierce shows that if there is a field extension $L = K(a^\xi : \xi \in \Gamma(2r))$ of the differential field (K, Δ) such that any minimal leader a^ξ satisfies $\xi \in \Gamma(r)$, then there is a differential field extension of (K, Δ) that is compatible with L. As a corollary one gets that $T \leq 2^{\mathfrak{L}_{m,f}+1} r$, where $f : \mathbb{N} \to \mathbb{N}$ is the function $f(i) = 2^i r$ and $\mathfrak{L}_{m,f}$ denotes the maximal length of an antichain sequence of $(\mathbb{N}^m, \leq)$ of degree growth bounded by f.

Now, given an antichain $\bar{\xi} \subseteq \mathbb{N}^m$, we let

$$\gamma\left(\bar{\xi}\right) = \{LCM(\eta, \zeta) : \eta \neq \zeta \text{ with } \eta, \zeta \in \bar{\xi}\}, \tag{1}$$

where $LCM(\eta, \zeta)$ denotes the least upper bound (or least common multiple) of η and ζ. For a field extension $L = K(a^\xi : \xi \in \Gamma(r))$ of K, we let $\gamma(L)$ denote $\gamma\left(\bar{\xi}\right)$, where $\bar{\xi} = (\xi_1, \ldots, \xi_k)$ and $a^{\xi_1}, \ldots, a^{\xi_k}$ are the minimal leaders of L. Note that $\gamma(L) \subseteq \Gamma(2r)$.

We have the following improvement of [7, Theorem 4.3]:

Theorem 1. *Let* $L = K(a^\xi : \xi \in \Gamma(r))$ *be a field extension of* K *satisfying the differential condition for some integer* $r \geq 0$. *Suppose further that*

($\sharp$) *For every* $\tau \in \gamma(L) \setminus \Gamma(r)$ *and* $1 \leq i < j \leq m$ *such that* $a^{\tau-i}$ *and* $a^{\tau-j}$ *are leaders, there exists a sequence of minimal leaders* $a^{\eta_1}, \ldots, a^{\eta_s}$ *such that* $\eta_\ell \leq \tau - \mathbf{k}_\ell$, *with* $k_1 = i$, $k_s = j$ *and some* $k_2, \ldots, k_{s-1}$, *and*

$$\deg LCM(\eta_\ell, \eta_{\ell+1}) \leq r \quad \text{for } \ell = 1, \ldots, s-1. \tag{2}$$

Then, there is a differential field extension of (K, Δ) *compatible with* L.

Given $r \geq 0$, let $g : \mathbb{N} \to \mathbb{N}$ be the function $g(1) = r$, $g(2 + i) = r + i$ for $i \geq 0$. Recall that $\mathfrak{L}_{m,g}$ denotes the maximal length of an antichain sequence of $(\mathbb{N}^m, \leq)$ with degree growth bounded by g. Using the above theorem, and several arguments like Macaulay's theorem on the Hilbert function [5], we can prove:

Theorem 2. *With the above notation, if $K(a^\xi : \xi \in \Gamma(g(\mathfrak{L}_{m,g}) + 1))$ is a field extension of K satisfying the differential condition, then there is a differential field extension of (K, Δ) compatible with $K(a^\xi : \xi \in \Gamma(r))$. In particular,*

$$T \leq g(\mathfrak{L}_{m,g}) + 1. \tag{3}$$

This result is similar in nature to [7, Theorem 4.10], but the proof is very different and it yields an explicit upper bound. Moreover, the value $g(\mathfrak{L}_{m,g})$ is much smaller than the upper bound $2^{\mathfrak{L}_{m,f}+1}r$ of T computed in [4]. For instance, when $m = 1$ we have $g(\mathfrak{L}_{m,g}) + 1 = r$ (which is expected) while $2^{\mathfrak{L}_{m,f}+1}r = 4r$. Also, note that when $m = 2$ our new bound yields $T \leq 2r$ which appears to be a new result.

Remark 1. When $m = 1$, extensions of (K, ∂) of the form $K(a^\xi : \xi \in \Gamma(r))$ satisfying the differential condition have previously been studied using the language of differential kernels [3, §3]. In this language, having $T \leq r$ when $m = 1$ corresponds to the fact that every differential kernel has a realization (see [3, Proposition 3]).

As we mentioned above, an upper bound for T is given in [4] to be $T < 2A(m + 3, 4r - 1)$. Since the Ackermann function grows incredibly quickly in its first variable, even having $m = 1$ produces very large bounds. However, due to our construction, a result of Moreno Socías [6] yields the following:

Corollary 1. *For each $r \geq 1$, we have*

$$T \leq A(m, r). \tag{4}$$

While this upper bound is still in terms of the Ackermann function, the first input now depends only on m instead of $m + 3$, which means that it produces much better results for small values of m. For example, in [4] it is shown that when $m = 3$ and $r = 1$, then $T \leq 2^{71}$, and when $m = 3$ and $r = 2$, then $T \leq 2^{2^{2^{520}+520}+2^{520}+521}$. However, when $m = 3$, our results yield that when $r = 1$ then $T \leq 4$, and when $r = 2$ then $T \leq 12$.

So far we have only dealt with the case of differential field extensions (M, Δ') of (K, Δ) that are *differentially generated* by a single element, namely $a^{\mathbf{0}}$, where $\mathbf{0}$ denotes the zero tuple. Our arguments can be extended to admit n-many differential generators (the minimal value T is defined in the natural way). To do this, we consider n-copies of $\mathbb{N}^m$, that is $\mathbb{N}^m \times n$, with the following order: $(\xi, i) \leq (\eta, j)$ if and only if $i = j$ and $\xi \leq \eta$. For an element $(\xi, i) \in \mathbb{N}^m \times n$ we let $\deg(\xi, i) = \deg \xi$, and $\Gamma(r) = \{(\xi, i) \in \mathbb{N}^m \times n : \deg \xi \leq r\}$.

Fix $r \geq 0$ and let g be the function defined above. For arbitrary $n \in \mathbb{N}$ we let $\mathfrak{L}_{m,g}^n$ be the maximal length of an antichain sequence of $(\mathbb{N}^m \times n, \leq)$ with degree growth bounded by g. One can prove the following extension of Theorem 2 and Corollary 1:

Theorem 3. *If $K(a_i^\xi : (\xi, i) \in \Gamma(g(\mathcal{L}_{m,g}^n + 1))$ is a field extension of K satisfying the differential condition, then there is a differential field extension of (K, Δ) compatible with $K(a_i^\xi : (\xi, i) \in \Gamma(r))$. Consequently,*

$$T \leq A_n(m, r), \tag{5}$$

where $A_1(x, y) = A(x, y)$ and $A_{i+1}(x, y) = A(x, A_i(x, y))$ for $i > 0$.

In particular, when $m = 2$, the above theorem yields $T \leq 2^n r$.

References

1. Freitag, J., León Sánchez, O.: Effective uniform bounding in partial differential fields. Adv. Math. **288**, 308–336 (2016)
2. Gustavson, R., Kondratieva, M., Ovchinnikov, A.: New effective differential Nullstellensatz. Adv. Math. **290**, 1138–1158 (2016)
3. Lando, B.A.: Jacobi's bound for the order of systems of first order differential equations. Trans. Am. Math. Soc. **152**(1), 119–135 (1970)
4. León Sánchez, O., Ovchinnikov, A.: On bounds for the effective differential Nullstellensatz. J. Algebra **449**, 1–21 (2016)
5. Macaulay, F.S.: Some properties of enumeration in the theory of modular systems. Proc. Lond. Math. Soc. **26**(2), 531–555 (1927)
6. Moreno Socías, G.: An ackermannian polynomial ideal. In: Mattson, H.F., Mora, T., Rao, T.R.N. (eds.) Applied Algebra, Algebraic Algorithms and Error-Correcting Codes. LNCS, vol. 539, pp. 269–280. Springer, Heidelberg (1991)
7. Pierce, D.: Fields with several commuting derivations. J. Symb. Logic **79**(01), 1–19 (2014)

Dimension Polynomials of Intermediate Fields of Inversive Difference Field Extensions

Alexander Levin[✉]

The Catholic University of America, Washington, DC 20064, USA
levin@cua.edu
http://faculty.cua.edu/levin

Abstract. Let K be an inversive difference field, L a finitely generated inversive difference field extension of K, and F an intermediate inversive difference field of this extension. We prove the existence and establish properties and invariants of a numerical polynomial that describes the filtration of F induced by the natural filtration of the extension L/K associated with its generators. Then we introduce concepts of type and dimension of the extension L/K considering chains of its intermediate fields. Using properties of dimension polynomials of intermediate fields we obtain relationships between the type and dimension of L/K and difference birational invariants of this extension carried by its dimension polynomials. Finally, we present a generalization of the obtained results to the case of multivariate dimension polynomials associated with a given inversive difference field extension and a partition of the basic set of translations.

Keywords: Inversive difference field · Inversive difference module · Filtration · Dimension polynomial

1 Introduction

Dimension polynomials of inversive difference modules and inversive difference field extensions, first introduced in [9], play the same role in difference algebra, as Hilbert polynomials play in commutative algebra and algebraic geometry. (A similar role in differential algebra is played by differential dimension polynomials introduced by E. Kolchin in [6]; see also [7, Chap. 2].) Several applications of dimension polynomials to the study of inversive difference algebraic structures are based on the fact that if P is a prime reflexive difference ideal in a ring of inversive difference polynomials $R = K\{y_1, \ldots, y_s\}^*$ over an inversive difference field K, then the quotient field of R/P is an inversive difference field extension of K generated by the images of y_i in R/P. The dimension polynomial of this extension, therefore, characterizes the ideal P; assigning such polynomials to prime reflexive difference ideals has led to a number of new results on the Krull-type dimension of inversive difference algebras (see, for example, [16] and [13, Sect. 4.6]). Another important application of difference dimension polynomials

© Springer International Publishing Switzerland 2016
I.S. Kotsireas et al. (Eds.): MACIS 2015, LNCS 9582, pp. 362–376, 2016.
DOI: 10.1007/978-3-319-32859-1_31

is based on the fact that the univariate difference dimension polynomial of a system of algebraic difference equations (defined as the dimension polynomial of the inversive difference field extension associated with the system) expresses the A. Einstein's strength of this system (see [12] and [13, Chap. 7]). In this connection, the study of difference dimension polynomials and methods of their computation is of primary importance for the qualitative theory of difference equations.

In this paper we prove the existence and describe some properties of a univariate dimension polynomial associated with an intermediate inversive difference field of a finitely generated inversive difference field extension. (Note that this result implies the existence of a dimension polynomial that describes the strength of a system of difference equations with a group action in the sense of A. Einstein.) We also present a more general theorem on a multivariate dimension polynomial associated with such an intermediate field. Furthermore, using chains of intermediate fields, we introduce concepts of type and dimension of an inversive difference field extension and find relationships between these characteristics and difference birational invariants of the extension.

2   Preliminaries

Throughout the paper $\mathbb{Z}$, $\mathbb{N}$, $\mathbb{Q}$, and $\mathbb{R}$ denote the sets of all integers, all nonnegative integers, all rational numbers, and all real numbers, respectively. As usual, $\mathbb{Q}[t]$ will denote the ring of polynomials in one variable t with rational coefficients. By a ring we always mean an associative ring with a unity. Every ring homomorphism is unitary (maps unit onto unit), every subring of a ring contains the unity of the ring. Unless otherwise indicated, by the module over a ring A we mean a left A-module. Every module over a ring is unitary and every algebra over a commutative ring is unitary as well.

By a *difference ring* we mean a commutative ring R together with a finite set $\sigma = \{\alpha_1, \ldots, \alpha_n\}$ of mutually commuting injective endomorphisms of R. The set σ is called a *basic set* of R and the endomorphisms α_i are called *translations*. We also say that R is a σ-ring. A subring (ideal) S of R is called a difference (or σ-) subring (respectively, ideal) of R if $\alpha(S) \subseteq S$ for every $\alpha \in \sigma$. If all translations of R are automorphisms, we set $\sigma^* = \{\alpha_1, \ldots, \alpha_n, \alpha_1^{-1}, \ldots, \alpha_n^{-1}\}$ and say that R is an *inversive difference ring* or a σ^*-ring. If a difference (respectively, inversive difference) ring R is a field, it is called a *difference* (or σ-) *field* (respectively, an *inversive difference* (or σ^*-) *field*).

If R is an inversive difference ring with a basic set $\sigma = \{\alpha_1, \ldots, \alpha_n\}$, then Γ will denote the free commutative group of all power products of the form $\gamma = \alpha_1^{k_1} \ldots \alpha_n^{k_n}$ where $k_i \in \mathbb{Z}$ ($1 \leq i \leq n$). The *order* of such an element γ is defined as ord $\gamma = \sum_{i=1}^{n} |k_i|$; furthermore, for every $r \in \mathbb{N}$, we set $\Gamma(r) = \{\gamma \in \Gamma \mid \text{ord } \gamma \leq r\}$.

A subring (ideal) R_0 of a σ^*-ring R is said to be a σ^*-subring (respectively, σ^*-ideal) of R if R_0 is closed with respect to the action of any translation $\alpha \in \sigma^*$. If R is an inversive difference ring and a prime σ-ideal is reflexive, it is referred to as a prime σ^*-ideal of R.

If R is a σ^*-ring R and $S \subseteq R$, then the intersection of all σ^*-ideals of R containing S is denoted by $[S]^*$. Clearly, $[S]^*$ is the smallest σ^*-ideal of R containing S; as an ideal, it is generated by the set $\Gamma S = \{\gamma(a) | \gamma \in \Gamma, a \in S\}$. If $J = [S]^*$, the elements of the set S are called σ^*-generators of J; if $S = \{a_1, \ldots, a_r\}$, we write $J = [a_1, \ldots, a_r]^*$ and say that the σ^*-ideal J is finitely generated.

Let R and S be two difference (in particular, inversive difference) rings with the same basic set σ, that is, elements of the set σ act on each of the rings as mutually commuting endomorphisms. A ring homomorphism $\phi : R \to S$ is called a *difference* (or σ-) *homomorphism* if $\phi(\alpha a) = \alpha \phi(a)$ for any $\alpha \in \sigma$, $a \in R$. It is easy to see that the kernel of such a mapping is a reflexive difference ideal of R.

Let R be an inversive difference ring with a basic set σ, R_0 a σ^*-subring of R and $B \subseteq R$. The intersection of all σ^*-subrings of R containing R_0 and B is called the σ^*-*subring of R generated by the set B over R_0*, it is denoted by $R_0\{B\}^*$. (As a ring, $R_0\{B\}$ coincides with the ring $R_0[\{\gamma(b) | b \in B, \gamma \in \Gamma\}]$ obtained by adjoining the set $\{\gamma(b) | b \in B, \gamma \in \Gamma\}$ to the ring R_0.) The set B is said to be the set of σ^*-*generators* of the σ^*-ring $R_0\{B\}^*$ over R_0. If this set is finite, $B = \{b_1, \ldots, b_k\}$, we say that $R' = R_0\{B\}^*$ is a finitely generated inversive difference (or σ^*-) ring extension (or overring) of R_0 and write $R' = R_0\{b_1, \ldots, b_k\}^*$.

If R is a difference (σ-) field and R_0 a subfield of R such that $\alpha(a) \in R_0$ for any $a \in R_0, \alpha \in \sigma$, then R_0 is said to be a *difference* (or σ-) *subfield* of R; R, in turn, is called a *difference* (or σ-) *field extension* or a *difference* (or σ-) *overfield* of R_0. In this case we also say that we have a σ-field extension R/R_0. If R is inversive and its subfield R_0 is a σ^*-subring of R, then R_0 is said to be an *inversive difference* (or σ^*-) *subfield* of R while R is called an *inversive difference* (or σ^*-) *field extension* or an *inversive difference* (or σ^*-) *overfield* of R_0. (We also say that we have a σ^*-field extension R/R_0.) If $R_0 \subseteq R_1 \subseteq R$ is a chain of σ- (σ^*-) field extensions, we say that R_1/R_0 is a *difference* or σ- (respectively, *inversive difference* or σ^*-) *field subextension* of R/R_0 or that R_1 is an *intermediate difference* (σ-) or, respectively, *inversive difference* (σ^*-) *field* of R/R_0.

If R is a σ^*-field, R_0 a σ^*-subfield of R and $B \subseteq R$, then the intersection of all σ^*-subfields of R containing R_0 and B is denoted by $R_0\langle B \rangle^*$ (or $R_0\langle b_1, \ldots, b_k \rangle^*$ if $B = \{b_1, \ldots, b_k\}$ is a finite set). This is the smallest σ^*-subfield of R containing R_0 and B; it coincides with the field $R_0(\{\gamma(b) | b \in B, \gamma \in \Gamma\})$. The set B is called the set of σ^*-*generators* of the σ^*-field $R_0\langle B \rangle^*$ over R_0.

The following theorem, whose proof can be found in [8, Sect. 6.4], introduces the (univariate) dimension polynomial of a finitely generated inversive difference field extension.

Theorem 1. *Let $L = K\langle \eta_1, \ldots, \eta_s \rangle^*$ be an inversive difference field extension of an inversive difference field K generated by a finite set $\eta = \{\eta_1, \ldots, \eta_s\}$. With the above notation, there exists a polynomial $\phi_{\eta|K}(t) \in \mathbb{Q}[t]$ such that*

(i) $\phi_{\eta|K}(r) = \mathrm{trdeg}_K K(\{\gamma \eta_j | \gamma \in \Gamma(r), 1 \le j \le s\})$ *for all sufficiently large* $r \in \mathbb{Z}$;

(ii) $\deg \phi_{\eta|K} \leq n = \operatorname{Card} \sigma$ *and the polynomial* $\phi_{\eta|K}(t)$ *can be written as*

$$\phi_{\eta|K}(t) = \sum_{i=0}^{n} a_i \binom{t+i}{i} \quad \text{where } a_0, \ldots, a_n \in \mathbb{Z} \text{ and } 2^n | a_n.$$

(iii) $d = \deg \phi_{\eta|K}$, a_n *and* a_d *do not depend on the set of* σ^*-*generators* η *of* L/K *(*$a_d \neq a_n$ *if and only if* $d < n$*). Moreover,* $\dfrac{a_n}{2^n}$ *is equal to the difference (or* σ-*) transcendence degree of* L *over* K *(denoted by* σ-$\operatorname{trdeg}_K L$*), that is, to the maximal number of elements* $\xi_1, \ldots, \xi_k \in L$ *such that the family* $\{\gamma\xi_i \mid \gamma \in \Gamma, 1 \leq i \leq k\}$ *is algebraically independent over* K.

(iv) *If the elements* $\eta_1, \ldots, \eta_s$ *are* σ-*algebraically independent over* K *(that is, the set* $\{\gamma\eta_i \mid \gamma \in \Gamma, 1 \leq i \leq s\}$ *is algebraically independent over* K*), then*

$$\phi_{\eta|K}(t) = s \sum_{k=0}^{n} (-1)^{n-k} 2^k \binom{n}{k} \binom{t+k}{k}.$$

The polynomial $\phi_{\eta|K}(t)$ is called the σ^*-*dimension polynomial* of the σ^*-field extension L of K associated with the system of σ^*-generators η. The numbers $d = \deg \phi_{\eta|K}$ and a_d are called σ-*type* and *typical* σ-*transcendence degree* of L/K; they are denoted by σ-$\operatorname{type}_K L$ and σ-$t.\operatorname{trdeg}_K L$, respectively.

Methods and examples of computation of σ^*-dimension polynomials can be found in [8, Chaps. 5 and 9], [17], and [13, Sect. 7.7].

Let R be an inversive difference ring with a basic set $\sigma = \{\alpha_1, \ldots, \alpha_n\}$, Γ the free commutative group generated by σ, and $Y = \{y_1, \ldots, y_s\}$ a set of symbols. Then the polynomial ring $R = K[\{\gamma y_j | \gamma \in \Gamma, 1 \leq j \leq s\}]$ in a denumerable set of indeterminates γy_j can be treated as an inversive difference ring extension of R where $\alpha(\gamma y_j) = (\alpha\gamma)y_j$ for any $\alpha \in \sigma^*$, $\gamma \in \Gamma$, $1 \leq j \leq s$. The ring R is called a *ring of inversive difference* (or σ^*-) *polynomials* over R; it is denoted by $R\{y_1, \ldots, y_s\}^*$.

A system of the form

$$f_i(y_1, \ldots, y_s) = 0 \qquad (i \in I),$$

where $f_i(y_1, \ldots, y_s) \in K\{y_1, \ldots, y_s\}^*$, is said to be a system of algebraic difference (or σ^*-) *equations* over K. By a solution of this system we mean an s-tuple $(a_1, \ldots, a_s)$ with coordinates in some inversive difference overring of K that annuls all f_i. In other words, $f_i(y_1, \ldots, y_s)$ becomes zero if one replaces every entry γy_j in f_i by γa_j, $(\gamma \in \Gamma, 1 \leq j \leq s)$.

Let K be an inversive difference (σ^*-) field and $\mathfrak{P}$ the reflexive difference ideal generated by a set of σ^*-polynomials $\{f_i | i \in I\}$ in $K\{y_1, \ldots, y_s\}^*$. If this ideal is prime (then the system is referred to as a prime system of algebraic σ^*-equations), one can consider the corresponding field of quotients $Q(K\{y_1, \ldots, y_s\}^*/\mathfrak{P}) = K\langle\eta_1, \ldots, \eta_s\rangle^*$ where η_j is the canonical image of y_j in $K\{y_1, \ldots, y_s\}^*/\mathfrak{P}$ $(1 \leq j \leq s)$. The corresponding polynomial $\phi_{\eta|K}(t)$, whose existence is established by Theorem 1, is called the σ^*-*dimension polynomial* of the system of difference equations $f_i(y_1, \ldots, y_s) = 0$ $(i \in I)$. This polynomial has the following interpretation as the difference version of the strength of a system of differential equations defined by A. Einstein in [1].

Let

$$A_i(f_1, \ldots, f_s) = 0 \qquad (i = 1, \ldots, p) \tag{1}$$

be a system of equations in finite differences with respect to s unknown grid functions $f_1, \ldots, f_s$ in n real variables $x_1, \ldots, x_n$ with coefficients in some functional field K. Suppose that the difference grid, whose nodes form the domain of considered functions, has equal cells of dimension $h_1 \times \cdots \times h_n$ $(h_1, \ldots, h_n \in \mathbb{R})$ and fills the whole space $\mathbb{R}^n$. (As an example, one can consider a field K consisting of a zero function and fractions of the form u/v where u and v are grid functions defined almost everywhere and vanishing at a finite number of nodes.)

Let us fix some node $\mathcal{P}$ and say that *a node $\mathcal{Q}$ has order i* (with respect to $\mathcal{P}$) if the shortest path from $\mathcal{P}$ to $\mathcal{Q}$ along the edges of the grid consists of i steps (by a step we mean a path from a node of the grid to a neighbor node along the edge between these two nodes). Let us consider the values of the unknown grid functions $f_1, \ldots, f_s$ at the nodes whose order does not exceed r $(r \in \mathbb{N})$. If $f_1, \ldots, f_s$ should not satisfy any system of equations (or any other condition), their values at nodes of any order can be chosen arbitrarily. Because of the system in finite differences (and equations obtained from the equations of the system by transformations of the form $f_j(x_1, \ldots, x_s) \mapsto f_j(x_1 + k_1 h_1, \ldots, x_s + k_n h_n)$ with $k_1, \ldots, k_n \in \mathbb{Z}$, $1 \le j \le s$), the number of independent values of the functions $f_1, \ldots, f_s$ at the nodes of order $\le r$ decreases. This number, which is a function of r, is considered as a "measure of strength" of the system in finite differences (in the sense of A. Einstein). We denote it by S_r.

If the transformations α_j of the field of coefficients K defined by

$$\alpha_j f(x_1, \ldots, x_n) = f(x_1, \ldots, x_{j-1}, x_j + h_j, \ldots, x_n)$$

$(1 \le j \le n)$ are automorphisms of this field, then K can be considered as an inversive difference field with the basic set $\sigma = \{\alpha_1, \ldots, \alpha_n\}$. Furthermore, assume that the replacement of the unknown functions f_i by σ^*-indeterminates y_i $(i = 1, \ldots, s)$ in the ring $K\{y_1, \ldots, y_n\}^*$ leads to a prime system of algebraic σ^*-equations. Then the σ^*-dimension polynomial of this system is said to be the σ^*-*dimension polynomial of the given system in finite differences*. Clearly, $\phi_{\eta|K}(r) = S_r$ for any $r \in \mathbb{N}$, so the σ^*-dimension polynomial is the measure of strength of such a system. One can find methods and examples of computation of σ^*-dimension polynomials of some systems of equations in finite differences in [13, Sects. 7.7, 7.8].

3   The Main Theorem

The following result is an essential generalization of Theorem 1.

Theorem 2. *Let K be an inversive difference $(\sigma^*\text{-})$ field with $\sigma = \{\alpha_1, \ldots, \alpha_n\}$ and let $L = K\langle \eta_1, \ldots, \eta_s \rangle^*$ be an inversive difference $(\sigma^*\text{-})$ field extension of K generated by a finite set $\eta = \{\eta_1, \ldots, \eta_s\}$. Let F be an intermediate σ^*-field of the extension L/K and for any $r \in \mathbb{N}$, let $F_r = F \bigcap K(\{\gamma \eta_j | \gamma \in \Gamma(r), 1 \le j \le s\})$. Then there exists a polynomial $\phi_{K,F,\eta}(t) \in \mathbb{Q}[t]$ such that*

(i) $\phi_{K,F,\eta}(r) = \mathrm{trdeg}_K F_r$ for all sufficiently large $r \in \mathbb{Z}$;

(ii) $\deg \phi_{K,F,\eta} \leq n$ and $\phi_{K,F,\eta}(t)$ can be written as $\phi_{K,F,\eta}(t) = \displaystyle\sum_{i=0}^{n} b_i \binom{t+i}{i}$

where $b_0, \ldots, b_n \in \mathbb{Z}$.

(iii) $d = \deg \phi_{K,F,\eta}(t)$, b_n and b_d do not depend on the set of σ^*-generators η of the extension L/K. Furthermore, $2^n | b_n$ and $\dfrac{b_n}{2^n} = \sigma\text{-trdeg}_K F$.

The polynomial $\phi_{K,F,\eta}(t)$ is called a σ^*-dimension polynomial of the intermediate field F associated with the set of σ^*-generators η of L/i. The numbers $d = \deg \phi_{K,F,\eta}(t)$ and b_d are called the *relative σ-type* and *relative typical σ-transcendence degree* of the intermediate field F of L/K. These characteristics, denoted by $\sigma\text{-type}_{L/K}(F)$ and $\sigma\text{-t.}\,\mathrm{trdeg}_{L/K}(F)$, respectively, depend only on the σ^*-fields K, L, and F.

The proof of Theorem 2 is based on properties of inversive difference modules introduced in [9] and the fact that the module of Kähler differentials associated with a σ^*-field extension L/K can be equipped with a structure of an inversive difference L-module. (The idea of using modules of Kähler differentials for the study of inversive difference field extensions has come from a similar approach explored by J. Johnson in the study of differential field extensions, see [3,5]. This approach was also used by the author in the study of dimension polynomials of intermediate differential fields, see [14].) In what follows we give a brief account of this approach; more results on inversive difference (σ^*-) modules and, in particular, on σ^*-modules of Kähler differentials can be found in [13, Sects. 3.5 and 4.2].

Let K be an inversive difference field with a basic set $\sigma = \{\alpha_1, \ldots, \alpha_n\}$ and let Γ be the commutative group generated by σ. Let $\mathcal{E}$ (or $\mathcal{E}_K$ if one needs to indicate the field K) denote the set of all finite sums of the form $\sum_{\gamma \in \Gamma} a_\gamma \gamma$ where $a_\gamma \in K$ (such a sum is called a σ^*-*operator* over K; two σ^*-operators are equal if and only if their corresponding coefficients are equal).

The set $\mathcal{E}$ can be treated as a ring with respect to its natural structure of a left K-module and the relationships $\alpha a = \alpha(a)\alpha$ ($a \in K$, $\alpha \in \sigma^* = \{\alpha_1, \ldots, \alpha_n, \alpha_1^{-1}, \ldots, \alpha_n^{-1}\}$)) extended by distributivity. It is said to be a *ring of σ^*-operators* over K.

If $w = \sum_{\gamma \in \Gamma} a_\gamma \gamma \in \mathcal{E}$, then the number $\mathrm{ord}\, w = \max\{\mathrm{ord}\, \gamma \,|\, a_\gamma \neq 0\}$ is called the *order* of the σ^*-operator w. In what follows, we treat $\mathcal{E}$ as a filtered ring with the ascending filtration $(\mathcal{E}_r)_{r\in\mathbb{Z}}$ where $(\mathcal{E})_r = 0$ if $r < 0$ and $(\mathcal{E})_r = \{w \in \mathcal{E} \,|\, \mathrm{ord}\, w \leq r\}$ if $r \geq 0$.

An *inversive difference module* over K (also called a σ^*-K-module) is defined as a left $\mathcal{E}$-module M, that is, a vector K-space where elements of σ^* act as additive mutually commuting operators such that $\alpha(ax) = \alpha(a)\alpha x$ for any $\alpha \in \sigma^*, x \in M, a \in K$.

We say that M is a finitely generated σ^*-K-module if M is finitely generated as a left $\mathcal{E}$-module.

By a filtration of a σ^*-K-module M we mean an exhaustive and separated filtration of M as a $\mathcal{E}$-module, that is, an ascending chain $(M_r)_{r\in\mathbb{Z}}$ of vector K-subspaces of M such that $\mathcal{E}_r M_s \subseteq M_{r+s}$ for all $r, s \in \mathbb{Z}$, $M_r = 0$ for all sufficiently small $r \in \mathbb{Z}$, and $\bigcup_{r \in \mathbb{Z}} M_r = M$. Such a filtration is called *excellent* if every M_r $(r \in \mathbb{Z})$ is finitely generated over K and there exists $r_0 \in \mathbb{Z}$ such that $M_r = \mathcal{E}_{r-r_0} M_{r_0}$ for any $r \geq r_0$.

The proofs of the following two results can be found in [8, Sects. 6.3 and 6.7].

Theorem 3. *With the above notation, let M be a σ^*-K-module with an excellent filtration $(M_r)_{r \in \mathbb{Z}}$. Then there is a polynomial $\psi(t) \in \mathbb{Q}[t]$ such that:*

(i) $\psi(r) = \dim_K M_r$ for all sufficiently large $r \in \mathbb{Z}$.
(ii) $\deg \psi \leq n$ and $\psi(t)$ is of the form

$$\psi(t) = \sum_{i=0}^{n} a_i \binom{t+i}{i}$$

where $a_0, \ldots, a_n \in \mathbb{Z}$ and $2^n | a_n$.
(iii) $d = \deg \psi(t)$, a_n and a_d do not depend on the excellent filtration $(M_r)_{r \in \mathbb{Z}}$ of M. Furthermore, $\dfrac{a_n}{2^n}$ is equal to the σ^*-dimension of M over K (denoted by σ^*-$\dim_K M$), that is, to the maximal number of elements $x_1, \ldots, x_k \in M$ such that the family $\{\gamma x_i \mid \gamma \in \Gamma, 1 \leq i \leq k\}$ is linearly independent over K.

Theorem 4. *Let $\mu : N \to M$ be an injective homomorphism of filtered σ^*-K-modules M and N with filtrations $(M_r)_{r \in \mathbb{Z}}$ and $(N_r)_{r \in \mathbb{Z}}$, respectively. (It means that μ is a homomorphism of $\mathcal{E}$-modules and $\mu(N_r) \subseteq M_r$ for any $r \in \mathbb{Z}$.) If the filtration of M is excellent, then the filtration of N is also excellent.*

Proof of Theorem 2. As before, let $L = K\langle \eta_1, \ldots, \eta_s \rangle^*$ and let $\Omega_{L|K}$ denote the module of Kähler differentials associated with the extension L/K. Then $\Omega_{L|K}$ can be treated as a σ^*-L-module where the action of the elements of σ^* is defined in such a way that $\alpha(d\zeta) = d\alpha(\zeta)$ for any $\zeta \in L$, $\alpha \in \sigma^*$ (see [13, Lemma 4.2.8]).

Let $M = \Omega_{L|K}$ and for any $r \in \mathbb{N}$, let M_r denote the vector L-space generated by all elements $d\zeta$ where $\zeta \in K(\bigcup_{i=1}^{s} \Gamma(r)\eta_i)$. It is easy to check that $(M_r)_{r \in \mathbb{Z}}$ ($M_r = 0$ if $r < 0$) is an excellent filtration of the σ^*-L-module M.

Let F be any intermediate σ^*-field of L/K and let $F_r = F \bigcap K(\{\gamma\eta_j \mid \gamma \in \Gamma(r), 1 \leq j \leq s\})$ $(r \in \mathbb{N})$. Let $\mathcal{E}_L$ denote the ring of σ^*-operators over L and let N be the $\mathcal{E}_L$-submodule of M generated by all elements of the form $d\zeta$ where $\zeta \in F$. Furthermore, for any $r \in \mathbb{N}$, let N_r be the vector L-space generated by all elements $d\zeta$ where $\zeta \in F_r$.

It is easy to see that if one sets $N_r = 0$ for $r < 0$, then the family $(N_r)_{r \in \mathbb{Z}}$ becomes a filtration of the σ^*-L-module N, and the embedding $N \to M$ becomes a homomorphism of filtered σ^*-L-modules. Since the filtration $(M_r)_{r \in \mathbb{Z}}$ is excellent, one can apply Theorem 4 and obtain that the filtration $(N_r)_{r \in \mathbb{Z}}$ is also excellent. Therefore, by Theorem 3, there exists a polynomial $\phi_{K,F,\eta}(t) \in \mathbb{Q}[t]$ such that $\phi_{K,F,\eta}(t)(r) = \dim_K N_r$ for all sufficiently large $r \in \mathbb{Z}$.

Since a family $(\zeta_i)_{i \in I}$ of elements of F_r $(r \in \mathbb{Z})$ is algebraically independent over K if and only if the family $(d\zeta_i)_{i \in I}$ is linearly independent over L,

$\dim_K N_r = \mathrm{trdeg}_K F_r$ for all $r \in \mathbb{N}$. Applying Theorem 3 we obtain the statement of Theorem 2.

Note that if $F = L$, then Theorem 2 implies the result of Theorem 1. Furthermore, Theorem 2 shows that the Einstein's strength of a prime system of algebraic σ^*-equations, whose solution should be invariant with respect to the action of any group G commuting with basic operators α_i, is expressed by a polynomial function. (We mean that $\alpha_i G = G\alpha_i$ for $i = 1, \ldots, n$ and $g(a) = a$ for any $g \in G$, $a \in K$.) Indeed, in this case the fixed field F of the group G is an intermediate σ^*-field of the corresponding σ^*-field extension L/K, so the polynomial $\phi_{K,F,\eta}(t)$ (where $\eta = \{\eta_1, \ldots, \eta_s\}$ is a system of σ^*-generators of L/K) expresses the A. Einstein's strength of the system with the group action.

4   Type and Dimension of an Inversive Difference Field Extension

Let K be an inversive difference field with a basic set $\sigma = \{\alpha_1, \ldots, \alpha_n\}$ and $L = K\langle \eta_1, \ldots, \eta_n \rangle^*$ a σ^*-field extension of K generated by a finite set $\eta = \{\eta_1, \ldots, \eta_s\}$. (We keep the notation introduced in Sect. 2). Let $\mathfrak{U}$ denote the set of all intermediate σ^*-fields of the extension L/K and let

$$\mathfrak{B}_\mathfrak{U} = \{(F, E) \in \mathfrak{U} \times \mathfrak{U} \mid F \supseteq E\}.$$

Furthermore, let $\overline{\mathbb{Z}}$ denote the ordered set $\mathbb{Z} \bigcup \{\infty\}$ (where the natural order on $\mathbb{Z}$ is extended by the condition $a < \infty$ for any $a \in \mathbb{Z}$).

Proposition 1. *With the above notation, there exists a unique mapping $\mu_\mathfrak{U}$: $\mathfrak{B}_\mathfrak{U} \to \overline{\mathbb{Z}}$ such that*

(i) $\mu_\mathfrak{U}(F, E) \geq -1$ for any pair $(F, E) \in \mathfrak{B}_\mathfrak{U}$.
(ii) If $d \in \mathbb{N}$, then $\mu_\mathfrak{U}(F, E) \geq d$ if and only if $\mathrm{trdeg}_E F > 0$ and there exists an infinite descending chain of intermediate σ^*-fields

$$F = F_0 \supseteq F_1 \supseteq \cdots \supseteq F_r \supseteq \cdots \supseteq E \tag{2}$$

such that

$$\mu_\mathfrak{U}(F_i, F_{i+1}) \geq d - 1 \quad (i = 0, 1, \ldots). \tag{3}$$

Proof. In order to show the existence and uniqueness of the desired mapping $\mu_\mathfrak{U}$, one just needs to mimic the proof of the corresponding statement for chains of prime differential ideals presented in [4]. Namely, let us set $\mu_\mathfrak{U}(F, E) = -1$ if $F = E$ or the field extension F/E is algebraic. If $(F, E) \in \mathfrak{B}_\mathfrak{U}$, $\mathrm{trdeg}_E F > 0$ and for every $d \in \mathbb{N}$, there exists a chain of intermediate σ^*-fields (2) with condition (3), we set $\mu_\mathfrak{U}(F, E) = \infty$. Otherwise, we define $\mu_\mathfrak{U}(F, E)$ as the maximal integer d for which condition (ii) holds (that is, $\mu_\mathfrak{U}(F, E) \geq d$). It is clear that the mapping $\mu_\mathfrak{U}$ defined in this way is unique.

With the notation of the last proposition, we define the *type* of the inversive difference (σ^*-) field extension L/K as the integer

$$\text{type}_\sigma(L/K) = \sup\{\mu_{\mathfrak{U}}(F,E) \,|\, (F,E) \in \mathfrak{B}_{\mathfrak{U}}\}. \tag{4}$$

Furthermore, we define the *dimension* of the σ^*-extension L/K as the number $\dim_\sigma(L/K) = \sup\{q \in \mathbb{N} \,|\, \text{there exists a chain } F_0 \supseteq F_1 \supseteq \cdots \supseteq F_q \text{ such that } F_i \in \mathfrak{U} \text{ and}$

$$\mu_{\mathfrak{U}}(F_{i-1}, F_i) = \text{type}_\sigma(L/K) \quad (i = 1, \ldots, q). \tag{5}$$

It is easy to see that for any pair of intermediate σ^*-fields $(F, E) \in \mathfrak{B}_{\mathfrak{U}}$, $\mu_{\mathfrak{U}}(F, E) = -1$ if and only if the field extension E/F is algebraic. It is also clear that if $\text{type}_\sigma(L/K) < \infty$, then $\dim_\sigma(L/K) > 0$.

The following result provides a relationship between the introduced characteristics of a finitely generated field σ^*-extension and the invariants of its σ^*-dimension polynomials established in Theorem 1.

Theorem 5. *Let K be an inversive difference (σ^*-) field with $\sigma = \{\alpha_1, \ldots, \alpha_n\}$ and let L be a finitely generated σ^*-field extension of K. Then*

(i) $\text{type}_\sigma(L/K) \leq \sigma\text{-type}_K L \leq n$.
(ii) If $\sigma\text{-trdeg}_K L > 0$, then $\text{type}_\sigma(L/K) = n$ and $\dim_\sigma(L/K) = \sigma\text{-trdeg}_K L$.
(iii) If $\sigma\text{-trdeg}_K L = 0$, then $\text{type}_\sigma(L/K) < n$.

Proof. Let $\eta = \{\eta_1, \ldots, \eta_s\}$ be a system of σ^*-generators of L over K, let Γ denote the free commutative group generated by σ, and for every $r \in \mathbb{N}$, let $\Gamma(r) = \{\gamma \in \Gamma \,|\, \text{ord}\,\gamma \leq r\}$. Furthermore, for every $r \in \mathbb{N}$, we set $L_r = K(\{\gamma\eta_i \,|\, \gamma \in \Gamma(r), 1 \leq i \leq s\})$, and if F is any intermediate σ^*-field of the extension L/K, then F_r will denote the field $F \cap L_r$.

By Theorem 2, there is a polynomial $\phi_{K,F,\eta}(t) \in \mathbb{Q}[t]$ such that $\phi_{K,F,\eta}(r) = \text{trdeg}_K F_r$ for all sufficiently large $r \in \mathbb{N}$, $\deg \phi_{K,F,\eta} \leq n$, and $\phi_{K,F,\eta}(t)$ can be written as $\phi_{K,F,\eta}(t) = \sum_{i=0}^n b_i \binom{t+i}{i}$ where $b_0, \ldots, b_n \in \mathbb{Z}$, $2^n | b_n$ and $\dfrac{b_n}{2^n} = \sigma\text{-trdeg}_K F$. Since the field K and the system of σ^*-generators of L/K are fixed in the following considerations, we will denote the polynomial $\phi_{K,F,\eta}(t)$ by $\phi_F(t)$.

It is easy to see that if E and F are two intermediate σ^*-fields of L/K and $F \supseteq E$, then $\phi_F(t) \geq \phi_E(t)$. It means that $\phi_F(s) \geq \phi_E(s)$ for all sufficiently large $s \in \mathbb{N}$. (Note that, as it is shown in [18], the set W of all differential dimension polynomials of finitely generated differential field extensions is well ordered with respect to this ordering. At the same time, as it is proved in [8, Chap. 2], W is also a set of all difference dimension polynomials). Also, if $F \supseteq E$ and $\phi_F(t) = \phi_E(t)$, then the field extension F/E is algebraic. Indeed, if some element $x \in F$ is transcendental over E, then there exists $r_0 \in \mathbb{N}$ such that $x \in F_r$ for all $r \geq r_0$. Therefore, $\text{trdeg}_K F_r = \text{trdeg}_K E_r + \text{trdeg}_{E_r} F_r > \text{trdeg}_K E_r$ for all $r \geq r_0$ hence $\phi_F(t) > \phi_E(t)$ contrary to our assumption.

Let $\mathfrak{U}$ and $\mathfrak{B}_{\mathfrak{U}}$ be the sets introduced at the beginning of this section. The following statement is the main step of the proof of Theorem 5.

Lemma 1. *For any $d \in \mathbb{Z}$, $d \geq -1$, and for any pair $(F, E) \in \mathfrak{B}_\mathfrak{U}$, the inequality*

$$\mu_\mathfrak{U}(F, E) \geq d$$

implies the inequality

$$\deg(\phi_F(t) - \phi_E(t)) \geq d.$$

Proof. We proceed by induction on d. Since $\deg(\phi_F(t) - \phi_E(t)) \geq -1$ for any pair $(F, E) \in \mathfrak{B}_\mathfrak{U}$ and $\deg(\phi_F(t) - \phi_E(t)) \geq 0$ if $\operatorname{trdeg}_E F > 0$, our statement is true for $d = -1$ and $d = 0$. (As usual we assume that the degree of the zero polynomial is -1.)

Let $d > 0$ and let the statement be true for all nonnegative integers less than d. Let $\mu_\mathfrak{U}(F, E) \geq d$ for some pair $(F, E) \in \mathfrak{B}_\mathfrak{U}$, so that there exists a chain of intermediate σ^*-fields (2) such that $\mu_\mathfrak{U}(F_i, F_{i+1}) \geq d - 1$ $(i = 0, 1, \dots)$. If $\deg(\phi_{F_i}(t) - \phi_{F_{i+1}}(t)) \geq d$ for some $i \in \mathbb{N}$, then $\deg(\phi_F(t) - \phi_E(t)) \geq \deg(\phi_{F_i}(t) - \phi_{F_{i+1}}(t)) \geq d$, so the statement of the lemma is true.

Suppose that for every $i \geq 0$, one has $\deg(\phi_{F_i}(t) - \phi_{F_{i+1}}(t)) = d - 1$, that is,

$$\phi_{F_i}(t) - \phi_{F_{i+1}}(t) = \sum_{j=0}^{d-1} b_j^{(i)} \binom{t+j}{j}$$

where $b_0^{(i)}, \dots, b_{d-1}^{(i)} \in \mathbb{Z}$, $b_{d-1}^{(i)} > 0$. Then

$$\phi_F(t) - \phi_{F_{i+1}}(t) = \sum_{k=0}^{i}(\phi_{F_k}(t) - \phi_{F_{k+1}}(t)) = \sum_{j=0}^{d-1} c_j^{(i)} \binom{t+j}{j}$$

where $c_0^{(i)}, \dots, c_{d-1}^{(i)} \in \mathbb{Z}$ and $c_{d-1}^{(i)} = \sum_{k=0}^{i} b_{d-1}^{(k)}$. Therefore, $c_{d-1}^{(0)} < c_{d-1}^{(1)} < \dots$ and $\lim_{i\to\infty} c_{d-1}^{(i)} = \infty$. On the other hand,

$$\phi_F(t) - \phi_{F_{i+1}}(t) \leq \phi_F(t) - \phi_E(t) = \sum_{j=0}^{d-1} a_j^{(i)} \binom{t+j}{j}$$

for some $a_0^{(i)}, \dots, a_{d-1}^{(i)} \in \mathbb{Z}$. If $\deg(\phi_F(t) - \phi_E(t)) = d - 1$, then we would have $c_{d-1}^{(i)} < a_{d-1}$ for all $i \in \mathbb{N}$ contrary to the fact that $\lim_{i\to\infty} c_{d-1}^{(i)} = \infty$. Thus, $\deg(\phi_F(t) - \phi_E(t)) \geq d$, so our lemma is proved.

Completion of the Proof of Theorem 5. Since $\deg(\phi_F(t) - \phi_E(t)) \leq \sigma\text{-type}_K L \leq n$ for any pair $(F, E) \in \mathfrak{B}_\mathfrak{U}$, the last lemma implies that $\text{type}_\sigma(L/K) \leq \sigma\text{-type}_K L \leq n$. If $\sigma\text{-trdeg}_K L = 0$, then $\sigma\text{-type}_K L < n$ hence $\text{type}_\sigma(L/K) \leq \sigma\text{-type}_K L < n$. Thus, it remains to prove statement (ii) of the theorem.

Let $\sigma\text{-trdeg}_K L = k > 0$, let $x_1, \dots, x_k$ be a difference transcendence basis of L over K, and let $F = K\langle x_1 \rangle^*$. Clearly, in order to prove that $\text{type}_\sigma(L/K) = n$ it is sufficient to show that $\mu_\mathfrak{U}(F, K) \geq n$. This inequality, in turn, immediately follows from the consideration of the following n descending chains of σ^*-fields where for any positive integers $i_1, i_2, \ldots, i_n$,

$$L^{(i_1)} = K\langle(\alpha_1 - 1)^{i_1} x_1\rangle^*, L^{(i_1,i_2)} = L^{(i_1+1)}\langle(\alpha_1 - 1)^{i_1}(\alpha_2 - 1)^{i_2} x_1\rangle^*, \ldots,$$

$$L^{(i_1,\ldots,i_n)} = L^{(i_1,\ldots i_{n-2},i_{n-1}+1)}\langle(\alpha_1 - 1)^{i_1} \cdots (\alpha_{n-1} - 1)^{i_{n-1}}(\alpha_n - 1)^{i_n} x_1\rangle^* :$$

$$F = K\langle x_1\rangle^* \supset L^{(1)} \supset \cdots \supset L^{(i_1)} \supset L^{(i_1+1)} \supset \cdots \supset K,$$

$$L^{(i_1)} \supset L^{(i_1,1)} \supset \cdots \supset L^{(i_1,i_2)} \supset L^{(i_1,i_2+1)} \supset \cdots \supset L^{(i_1+1)}, \ldots,$$

$$L^{(i_1,\ldots,i_{n-1})} \supset L^{(i_1,\ldots,i_{n-1},1)} \supset L^{(i_1,\ldots,i_n)} \supset \cdots \supset L^{(i_1,\ldots,i_{n-1}+1)}.$$

These chains show that $\mu_\mathfrak{U}(L^{(i_1,\ldots,i_n)}, L^{(i_1,\ldots,i_{n-1},i_n+1)}) \geq 0$, $\mu_\mathfrak{U}(L^{(i_1,\ldots,i_{n-1})}, L^{(i_1,\ldots,i_{n-1},i_{n-1}+1)}) \geq 1, \ldots, \mu_\mathfrak{U}(F, K) \geq n$. As we have seen, the last inequality implies that $\text{type}_\sigma(L/K) = n$.

Replacing K by $K\langle x_1, \ldots, x_j\rangle$ and x_1 by x_{j+1} ($0 \leq j \leq k - 1$) in the above chains, we obtain that $\mu_\mathfrak{U}(K\langle x_1, \ldots, x_{j+1}\rangle^*, K\langle x_1, \ldots, x_j\rangle^*) = n = \text{type}_\sigma(L/K)$. Therefore, $\dim_\sigma(L/K) \geq k = \sigma\text{-trdeg}_K L$.

On the other hand, $\dim_\sigma(L/K) \leq k$. Indeed, let $F_0 \supseteq F_1 \supseteq \cdots \supseteq F_q$ be a chain of intermediate σ^*-fields of L/K such that $\mu_\mathfrak{U}(F_i, F_{i+1}) = \text{type}_\sigma(L/K) = n$ for $i = 0, \ldots, q - 1$. Clearly, in order to prove our inequality, it is sufficient to show that $q \leq k$.

For every $i = 0, \ldots, q$, the corresponding dimension polynomial $\phi_i(t) = \phi_{K,F_i,\eta}(t)$, whose existence is established by Theorem 2, can be written as $\phi_i(t) = \sum_{j=0}^n a_j^{(i)} \binom{t+j}{j}$ where $a_j^{(i)} \in \mathbb{Z}$ ($0 \leq i \leq q - 1$, $0 \leq j \leq n$) and $b_n^{(i)} = \frac{a_n^{(i)}}{2^n} \in \mathbb{Z}$. Then

$$\phi_0(t) - \phi_q(t) = \sum_{i=1}^q (\phi_{i-1}(t) - \phi_i(t)) = \sum_{i=1}^q \sum_{j=0}^n (a_j^{(i-1)} - a_j^{(i)})\binom{t+j}{j}$$

$$= (a_n^{(0)} - a_n^{(q)})\binom{t+n}{n} + o(t^n)$$

where $o(t^n)$ denote a polynomial of degree at most $n-1$. Since $\mu_\mathfrak{U}(F_i, F_{i+1}) = n$, one has $\deg(\phi_i(t) - \phi_{i+1}(t)) = n$ (see Lemma 1). Therefore, $b_n^{(0)} > b_n^{(1)} > \cdots > b_n^{(q)}$, hence

$$b_n^{(0)} - b_n^{(q)} = \sum_{i=1}^q (b_n^{(i-1)} - b_n^{(i)}) \geq q.$$

On the other hand,

$$\phi_0(t) - \phi_q(t) \leq \phi_{K,L,\eta}(t) - \phi_{K,F_q,\eta}(t) \leq \phi_{K,L,\eta}(t) = \phi_{\eta|K}(t) = \sum_{i=0}^n a_i \binom{t+i}{i}$$

where $a_0 = 2^n \sigma\text{-trdeg}_K L$ (we use the notation of Theorem 1). Therefore, $q \leq b_n^{(0)} - b_n^{(q)} \leq k = \sigma\text{-trdeg}_K L$. This completes the proof of Theorem 5.

5   Multivariate Dimension Polynomials of Intermediate σ^*-field Extensions

In this section we present a result that generalizes both Theorem 3.1 and the theorem on multivariate dimension polynomial of a finitely generated inversive difference field extension proved in [15], see Theorem 6 below.

Let K be an inversive difference field with basic set of automorphisms $\sigma = \{\alpha_1, \ldots, \alpha_n\}$. Let us fix a representation of σ as the union of p disjoint subsets $(p \geq 1)$:

$$\sigma = \sigma_1 \cup \cdots \cup \sigma_p \tag{6}$$

where

$$\sigma_1 = \{\alpha_1, \ldots, \alpha_{n_1}\}, \ \sigma_2 = \{\alpha_{n_1+1}, \ldots, \alpha_{n_1+n_2}\}, \ \ldots,$$

$$\sigma_p = \{\sigma_{n_1+\cdots+n_{p-1}+1}, \ldots, \alpha_n\} \ (n_1 + \cdots + n_p = n).$$

If $\gamma = \alpha_1^{k_1} \ldots \alpha_n^{k_n} \in \Gamma$ $(k_i \in \mathbb{Z})$ then the *order of* γ *with respect to* σ_i $(1 \leq i \leq p)$ is defined as $\sum_{\nu=n_1+\cdots+n_{i-1}+1}^{n_1+\cdots+n_i} |k_\nu|$; it is denoted by $\mathrm{ord}_i \gamma$. (If $i = 1$, the last sum is replaced by $\sum_{\nu=1}^{n_1} |k_\nu|$.) Clearly, $\sum_{i=1}^p \mathrm{ord}_i \gamma = \mathrm{ord}\, \gamma$.

In what follows, for any $r_1, \ldots, r_p \in \mathbb{N}$, we set

$$\Gamma(r_1, \ldots, r_p) = \{\gamma \in \Gamma \mid \mathrm{ord}_i \gamma \leq r_i \ (i = 1, \ldots, p)\}.$$

Furthermore, if $(j_1, \ldots, j_p)$ is any permutation of the set $\{1, \ldots, p\}$, let $<_{j_1, \ldots, j_p}$ denote the lexicographic order on $\mathbb{N}^p$ such that $(r_1, \ldots, r_p) <_{j_1, \ldots, j_p} (s_1, \ldots, s_p)$ if and only if either $r_{j_1} < s_{j_1}$ or there exists $k \in \mathbb{N}$, $1 \leq k \leq p - 1$, such that $r_{j_\nu} = s_{j_\nu}$ for $\nu = 1, \ldots, k$ and $r_{j_{k+1}} < s_{j_{k+1}}$.

If $\Sigma \subseteq \mathbb{N}^p$, then Σ' will denote the set $\{e \in \Sigma \mid e$ is a maximal element of Σ with respect to one of the $p!$ lexicographic orders $<_{j_1, \ldots, j_p}\}$. Say, if $\Sigma = \{(1,1,1), (2,3,0), (0,2,3), (2,0,5), (3,3,1), (4,1,1), (2,3,3)\} \subseteq \mathbb{N}^3$, then $\Sigma' = \{(2,0,5), (3,3,1), (4,1,1), (2,3,3)\}$.

The following result, which generalizes Theorem 1, was proved in [15].

Theorem 6. *Let* $L = K\langle \eta_1, \ldots, \eta_s \rangle^*$ *be a* σ^*-field extension generated by a set $\eta = \{\eta_1, \ldots, \eta_s\}$. *Then there exists a polynomial* $\Phi_\eta(t_1, \ldots, t_p)$ *in* p *variables* $t_1, \ldots, t_p$ *with rational coefficients such that*

(i) $\Phi_\eta(r_1, \ldots, r_p) = \mathrm{trdeg}_K K(\bigcup_{j=1}^s \Gamma(r_1, \ldots, r_p)\eta_j)$ *for all sufficiently large* $(r_1, \ldots, r_p) \in \mathbb{N}^p$ *(it means that there exist nonnegative integers* $s_1, \ldots, s_p$ *such that the last equality holds for all* $(r_1, \ldots, r_p) \in \mathbb{N}^p$ *with* $r_1 \geq s_1, \ldots, r_p \geq s_p$*);*

(ii) $\deg_{t_i} \Phi_\eta \leq n_i$ *for* $i = 1, \ldots, p$ *(hence* $\deg \Phi_\eta \leq n$*) and* $\Phi_\eta(t_1, \ldots, t_p)$ *can be represented as*

$$\Phi_\eta(t_1, \ldots, t_p) = \sum_{i_1=0}^{n_1} \cdots \sum_{i_p=0}^{n_p} a_{i_1 \ldots i_p} \binom{t_1 + i_1}{i_1} \cdots \binom{t_p + i_p}{i_p}$$

where $a_{i_1 \ldots i_p} \in \mathbf{Z}$ *and* $2^n \mid a_{n_1 \ldots n_p}$.

(iii) *Let* $E_\eta = \{(i_1, \ldots, i_p) \in \mathbb{N}^p \,|\, 0 \le i_k \le n_k \ (k = 1, \ldots, p) \text{ and } a_{i_1 \ldots i_p} \ne 0\}$.
Then $d = \deg \Phi_\eta$, $a_{n_1 \ldots n_p}$, *elements* $(k_1, \ldots, k_p) \in E'_\eta$, *the corresponding coefficients* $a_{k_1 \ldots k_p}$ *and the coefficients of the terms of total degree* d *do not depend on the choice of the system of* σ^*-*generators* η.

The polynomial $\Phi_\eta(t_1, \ldots, t_p)$ is called the *inversive difference* (or σ^*-) *dimension polynomial* of the σ^*-field extension L/K associated with the set of σ^*-generators η and partition (6) of the basic set of automorphisms.

The proof of Theorem 6 presented in [15] is based on the method of characteristic sets with respect to several term orderings for inversive difference polynomials. The idea of this method comes from [10–12] where a similar approach is used for the study of bivariate difference-differential dimension polynomials and multivariate dimension polynomials of modules over rings of Ore polynomials. An alternative way for obtaining results on dimension polynomials of differential, difference and inversive difference field extension is to prove the existence and describe properties of dimension polynomials associated with filtered modules over the corresponding rings of (differential, difference or inversive difference) operators and then explore the relationship between the transcendence degree of a field extension and the dimension of the corresponding module of Kähler differentials. (This approach is used in [3,5,8, Chaps. 5, 6] and some other works.) In what follows, we use properties of modules of Kähler differentials to prove the central result of this section:

Theorem 7. *With the notation of Theorem 6, let* F *be an intermediate* σ^*-*field of the* σ^*-*field extension* L/K *and for any* $r_1, \ldots, r_p \in \mathbb{N}^p$, *let*

$$F_{r_1, \ldots, r_p} = F \bigcap K\left(\bigcup_{j=1}^{s} \Gamma(r_1, \ldots, r_p)\eta_j \right).$$

Then there is a polynomial $\Phi_{K,F,\eta}(t_1, \ldots, t_p) \in \mathbb{Q}[t_1, \ldots, t_p]$ *in* p *variables* $t_1, \ldots, t_p$ *such that*

(i) $\Phi_{K,F,\eta}(r_1, \ldots, r_p) = \operatorname{trdeg}_K F_{r_1, \ldots, r_p}$ *for all sufficiently large* $(r_1, \ldots, r_p) \in \mathbb{N}^p$;

(ii) $\deg_{t_i} \Phi_{K,F,\eta} \le n_i$ *for* $i = 1, \ldots, p$ *(hence* $\deg \Phi_{K,F,\eta} \le n = \operatorname{Card} \sigma$) *and* $\Phi_{K,F,\eta}(t_1, \ldots, t_p)$ *can be written as*

$$\Phi_{K,F,\eta}(t_1, \ldots, t_p) = \sum_{i_1=0}^{n_1} \cdots \sum_{i_p=0}^{n_p} b_{i_1 \ldots i_p} \binom{t_1 + i_1}{i_1} \cdots \binom{t_p + i_p}{i_p}$$

where $b_{i_1 \ldots i_p} \in \mathbb{Z}$ *and* $2^n \,|\, b_{n_1 \ldots n_p}$.

(iii) *Let* $E_{K,F,\eta} = \{(i_1, \ldots, i_p) \in \mathbb{N}^p \,|\, 0 \le i_k \le n_k \ (k = 1, \ldots, p) \text{ and } b_{i_1 \ldots i_p} \ne 0\}$. *Then* $d = \deg \Phi_{K,F,\eta}$, $b_{n_1 \ldots n_p}$, *elements* $(k_1, \ldots, k_p) \in E'_{K,F,\eta}$, *the corresponding coefficients* $b_{k_1 \ldots k_p}$ *and the coefficients of the terms of total degree* d *do not depend on the choice of the system of* σ^*-*generators* η *of* L/K.

Proof. We will mimic the method of the proof of Theorem 2 using the results on multivariate dimension polynomials of inversive difference (σ^*-) modules over L. Let $\mathcal{E}$ be the ring of σ^*-operators over L considered as a filtered ring with p-dimensional filtration $\{\mathcal{E}_{r_1,\ldots,r_p} \mid (r_1,\ldots,r_p) \in \mathbb{Z}^p\}$ where for any $r_1,\ldots,r_p \in \mathbb{N}^p$, $\mathcal{E}_{r_1,\ldots,r_p}$ is the vector L-subspace of $\mathcal{E}$ generated by the set $\Gamma(r_1,\ldots,r_p)$, and $\mathcal{E}_{r_1,\ldots,r_p} = 0$ if at least one r_i is negative. If M is a σ^*-L-module (that is, a left $\mathcal{E}$-module), then a family $\{M_{r_1,\ldots,r_p} \mid (r_1,\ldots,r_p) \in \mathbb{Z}^p\}$ of vector K-subspaces of M is said to be a p-dimensional filtration of M if

(i) $M_{r_1,\ldots,r_p} \subseteq M_{s_1,\ldots,s_p}$ for any $(r_1,\ldots,r_p), (s_1,\ldots,s_p) \in \mathbb{Z}^p$ such that $r_i \leq s_i$ for $i = 1,\ldots,p$.

(ii) $\displaystyle\bigcup_{(r_1,\ldots,r_p)\in\mathbb{Z}^p} M_{r_1,\ldots,r_p} = M$.

(iii) There exists a p-tuple $(r_1^{(0)},\ldots,r_p^{(0)}) \in \mathbb{Z}^p$ such that $M_{r_1,\ldots,r_p} = 0$ if $r_i < r_i^{(0)}$ for at least one index i ($1 \leq i \leq p$).

(iv) $\mathcal{E}_{r_1,\ldots,r_p} M_{s_1,\ldots,s_p} \subseteq M_{r_1+s_1,\ldots,r_p+s_p}$ for any $(r_1,\ldots,r_p), (s_1,\ldots,s_p) \in \mathbb{Z}^p$.

If every vector L-space $M_{r_1,\ldots,r_p}$ is finite-dimensional and there exists an element $(h_1,\ldots,h_p) \in \mathbb{Z}^p$ such that $\mathcal{E}_{r_1,\ldots,r_p} M_{h_1,\ldots,h_p} = M_{r_1+h_1,\ldots,r_p+h_p}$ for any $(r_1,\ldots,r_p) \in \mathbb{N}^p$, the p-dimensional filtration $\{M_{r_1,\ldots,r_p} \mid (r_1,\ldots,r_p) \in \mathbb{Z}^p\}$ is called *excellent*.

It is easy to see that if $z_1,\ldots,z_k$ is a finite system of generators of a vector σ^*-L-module M, then $\{\sum_{i=1}^{k} \mathcal{E}_{r_1,\ldots,r_p} z_i \mid (r_1,\ldots,r_p) \in \mathbb{Z}^p\}$ is an excellent p-dimensional filtration of M.

As we have seen, the module of Kähler differentials $\Omega_{L|K}$ can be treated as a σ^*-L-module such that $\alpha(d\zeta) = d\alpha(\zeta)$ for any $\zeta \in L$, $\alpha \in \sigma^*$. Furthermore, $\Omega_{L|K} = \sum_{i=1}^{s} \mathcal{E} d\eta_i$, and if $(\Omega_{L|K})_{r_1\ldots r_p}$ ($r_1,\ldots,r_p \in \mathbb{N}$) is the vector L-subspace of $\Omega_{L|K}$ generated by the set $\{d\eta \mid \eta \in K(\{\gamma\eta_j \mid \gamma \in \Gamma(r_1,\ldots,r_p), 1 \leq j \leq s\})\}$ and $(\Omega_{L|K})_{r_1\ldots r_p} = 0$ whenever at least one r_i is negative, then $\{(\Omega_{L|K})_{r_1\ldots r_p} \mid (r_1,\ldots,r_p) \in \mathbb{Z}^p\}$ is an excellent p-dimensional filtration of $\Omega_{L|K}$.

Let N be the $\mathcal{E}$-submodule of $\Omega_{L|K}$ generated by all elements of the form $d\zeta$ where $\zeta \in F$. Furthermore, for any $r_1,\ldots,r_p \in \mathbb{N}$, let $N_{r_1,\ldots,r_p}$ be the vector L-space generated by all elements $d\zeta$ where $\zeta \in F_{r_1,\ldots,r_p}$. Setting $N_{r_1,\ldots,r_p} = 0$ if $(r_1,\ldots,r_p) \in \mathbb{Z}^p \setminus \mathbb{N}^p$, we obtain a p-dimensional filtration of the σ^*-L-module N, and the embedding $N \to \Omega_{L|K}$ becomes a homomorphism of p-filtered σ^*-L-modules. Using this fact, one can mimic the proof of Theorem 3.2.8 of [13] to show that the filtration $\{N_{r_1,\ldots,r_p} \mid (r_1,\ldots,r_p) \in \mathbb{Z}^p\}$ is excellent. Now the result of Theorem 7 immediately follows from the fact that $\dim_L N_{r_1,\ldots,r_p} = \mathrm{trdeg}_K F_{r_1,\ldots,r_p}$ for all $(r_1,\ldots,r_p) \in \mathbb{N}^p$ (as we have mentioned in the proof of Theorem 2, a family $(\zeta_i)_{i \in I}$ of elements of L (in particular, of $F_{r_1,\ldots,r_p}$) is algebraically independent over K if and only if the family $(d\zeta_i)_{i \in I}$ is linearly independent over L) and [13, Theorem 3.5.8]. (The last theorem states that under the above conditions, there exists a polynomial $\Phi_{K,F,\eta}(t_1,\ldots,t_p) \in \mathbb{Q}[t_1,\ldots,t_p]$ such that $\Phi_\eta(r_1,\ldots,r_p) = \dim_L N_{r_1,\ldots,r_p}$ for all sufficiently large $(r_1,\ldots,r_p) \in$

$\mathbb{Z}^p$ and $\Phi_{K,F,\eta}(t_1, \ldots, t_p)$ satisfies conditions (ii) of Theorem 7. Statement (iii) of Theorem 7 can be obtained in the same way as statement (iii) of Theorem 2 of [15].)

References

1. Einstein, A.: The Meaning of Relativity. Appendix II (Generalization of gravitation theory), 4th edn., pp. 133–165. Princeton (1953)
2. Kolchin, E.R.: The notion of dimension in the theory of algebraic differential equations. Bull. Am. Math. Soc. **70**, 570–573 (1964)
3. Johnson, J.L.: Kähler differentials and differential algebra. Ann. Math. **89**(2), 92–98 (1969)
4. Johnson, J.L.: A notion on Krull dimension for differential rings. Comment. Math. Helv. **44**, 207–216 (1969)
5. Johnson, J.L.: Kähler differentials and differential algebra in arbitrary characteristic. Trans. Am. Math. Soc. **192**, 201–208 (1974)
6. Kolchin, E.R.: Some problems in differential algebra. In: Proceedings of the International Congress of Mathematicians (Moscow - 1966), Moscow, pp. 269–276 (1968)
7. Kolchin, E.R.: Differential Algebra and Algebraic Groups. Academic Press, New York (1973)
8. Kondrateva, M.V., Levin, A.B., Mikhalev, A.V., Pankratev, E.V.: Differential and Difference Dimension Polynomials. Kluwer Academic Publishers, Dordrecht (1999)
9. Levin, A.B.: Characteristic polynomials of inversive difference modules and some properties of inversive difference dimension. Russ. Math. Surv. **35**(1), 217–218 (1980)
10. Levin, A.B.: Reduced Groebner bases, free difference-differential modules and difference-differential dimension polynomials. J. Symb. Comput. **30**, 357–382 (2000)
11. Levin, A.B.: Gröbner bases with respect to several orderings and multivariable dimension polynomials. J. Symb. Comput. **42**(5), 561–578 (2007)
12. Levin, A.B.: Computation of the strength of systems of difference equations via generalized Gröbner bases. In: Grobner Bases in Symbolic Analysis, pp. 43–73. Walter de Gruyter (2007)
13. Levin, A.B.: Difference Algebra. Springer, New York (2008)
14. Levin, A.B.: Dimension polynomials of intermediate fields and Krull-type dimension of finitely generated differential field extensions. Math. Comput. Sci. **4**(2–3), 143–150 (2010)
15. Levin, A.B.: Multivariate dimension polynomials of inversive difference field extensions. In: Barkatou, M., Cluzeau, T., Regensburger, G., Rosenkranz, M. (eds.) AADIOS 2012. LNCS, vol. 8372, pp. 146–163. Springer, Heidelberg (2014)
16. Levin, A.B., Mikhalev, A.V.: Type and dimension of finitely generated G-algebras. Contemp. Math. **184**, 275–280 (1995)
17. Mikhalev, A.V., Pankratev, E.V.: Computer Algebra. Calculations in Differential and Difference Algebra. Moscow State Univ. Press, Moscow (1989)
18. Sit, W.: Well-ordering of certain numerical polynomials. Trans. Am. Math. Soc. **212**, 37–45 (1975)

A "Polynomial Shifting" Trick in Differential Algebra

Gleb Pogudin[✉]

Moscow State University, Moscow, Russia
pogudin.gleb@gmail.com

Throughout the paper all fields are assumed to be of characteristic zero and "differential" means "ordinary differential".

Standard proofs of the primitive element theorem [1, V, Theorem 4.6] and the Noether normalization lemma [1, VIII, Theorem 2.1] are based on a consideration of "generic combinations" of initial generators. We propose a differential counterpart of this argument which we call a "polynomial shifting" trick. It is an important part of recent proofs of a strengthened version of Kolchin's primitive element theorem (see [3, Theorems 1 and 2]) and a differential analog of the Noether normalization lemma (see [4, Theorem 1]). This trick turned out to be quite flexible and constructive. We hope that this method will be useful dealing with problems of the same flavour.

In what follows $A\{x\}$ is an algebra of differential polynomials in x over A and $A\langle x \rangle$ is its field of fractions. Let $F \subset E$ be an extension of differential fields. Then, for $a \in E$ by $F\langle a \rangle$ we denote the differential subfield of E generated by a and F. Let $A \subset B$ be an extension of differential k-algebras. Then, for $b \in B$ by $A\{b\}$ we denote the differential subalgebra of B generated by b and A. Let k, K, E and F be differential fields.

The trick is based on the following simple lemma (see [2, p. 35] for a bit weaker version).

Lemma 1. *Let $P(x) \in K\{x\}$ be a nonzero differential polynomial and $\mathrm{ord}\, P \leqslant n$. Assume that there exists $t \in K$ such that $t' = 1$. Then, there exists a polynomial $s(t) \in \mathbb{Q}[t]$ such that $\deg_t s \leqslant n$ and $P(s(t)) \neq 0$.*

We will illustrate the method by applying it to the following lemma:

Lemma 2. *Let $P(x, y) \in k\{x, y\}\backslash k\{x\}$. Then, there exists a polynomial $q(t) \in \mathbb{Q}[t]$ such that $\frac{\partial}{\partial y}P(x + q(y), y) \neq 0$.*

Proof. Let $V_\Lambda = \{\Lambda_0, \Lambda_1, \ldots\}$ be a set of variables. We extend the derivation from $k\{x, y\}$ to $k\{x, y\}[V_\Lambda]$ by $(\Lambda_i)' = y'\Lambda_{i+1}$. A variable Λ_i should be thought of as a placeholder for the expression $q^{(i)}(y)$. Hence, the formula $(\Lambda_i)' = y'\Lambda_{i+1}$ is a rewritten chain rule for the derivative. More formally, let us fix an arbitrary polynomial $q(t) \in \mathbb{Q}[t]$. Then, the $k\{x, y\}$-linear map $\varphi_q \colon k\{x, y\}[V_\Lambda] \to k\{x, y\}$ defined by $\varphi_q(\Lambda_i) = q^{(i)}(y)$ is a homomorphism of differential k-algebras.

We see that $\frac{\partial}{\partial y}\varphi_q(\Lambda_i) = \varphi_q(\Lambda_{i+1})$. Thus, we can compute $\frac{\partial}{\partial y}\varphi_q\left(P(x + \Lambda_0, y)\right)$ in terms of Λ_i's. By n denote $\mathrm{ord}_x P$ and by S denote the separant of P with respect

© Springer International Publishing Switzerland 2016
I.S. Kotsireas et al. (Eds.): MACIS 2015, LNCS 9582, pp. 377–379, 2016.
DOI: 10.1007/978-3-319-32859-1_32

to x. Then

$$\frac{\partial}{\partial y}\varphi_q\left(P(x+\Lambda_0,y)\right) = \varphi_q\left(\Lambda_{n+1}S(x+\Lambda_0,y)+T\right),$$

where $T \in k\{x,y\}[\Lambda_0,\ldots,\Lambda_n]$. Since T does not depend on Λ_{n+1}, $R = \Lambda_{n+1}S(x+\Lambda_0,y)+T$ is a nonzero polynomial in $\Lambda_0,\ldots,\Lambda_{n+1}$ over $k\langle x,y\rangle$. Let us define a derivation on $k\langle x,y\rangle[V_\Lambda]$ by $D(z) = \frac{z'}{y'}$. By the definition, $D(y) = 1$ and $D(\Lambda_i) = \Lambda_{i+1}$. Thus, R can be considered as a differential polynomial in Λ_0 over $k\langle x,y\rangle$. Due to Lemma 1, there exists a polynomial $q(t) \in \mathbb{Q}[t]$ such that $\varphi_q(R) \neq 0$. Hence, $\frac{\partial}{\partial y}P(x+q(y),y) \neq 0$.

To sum up, the trick is a way to treat a "generic polynomial" in some differential variable y. The formula $(\Lambda_i)' = y'\Lambda_{i+1}$ carries the information that Λ_i is a polynomial in y, so we need no longer keep this in mind. Moreover, Λ_i is a "homogeneous" object, i.e. depends solely on y. This allows us to compute partial derivative of an expression involving Λ_i's with respect to y and its derivations.

The proof of the following theorem is based on a more sophisticated version of the same argument.

Theorem 1 ([3], **Theorem 1**). *Let $E = F\langle a,b\rangle$, $\mathrm{trdeg}_F E < \infty$, and $b' \neq 0$. Then, there exists $p(x) \in \mathbb{Q}[x]$ such that $\mathrm{trdeg}_F F\langle a+p(b)\rangle = \mathrm{trdeg}_F F\langle a,b\rangle$.*

Proof (Sketch of the proof). Again, let $V_\Lambda = \{\Lambda_0, \Lambda_1, \ldots\}$ be a set of variables. We extend the derivation from E to $E[V_\Lambda]$ by $(\Lambda_i)' = b'\Lambda_{i+1}$. For $p(t) \in \mathbb{Q}[t]$ by φ_p we denote the homomorphism of differential E-algebras $E[V_\Lambda] \to E$ defined by $\varphi_p(\Lambda_i) = p^{(i)}(b)$.

Let $c = a + \Lambda_0$ and $n = \mathrm{trdeg}_F E$. Then, $\mathrm{trdeg}_{F(V_\Lambda)} E(V_\Lambda) = n$. Hence, $c,\ldots,c^{(n)}$ are algebraically dependent over $F(V_\Lambda)$. Let $R(x_0,\ldots,x_n) \in F[V_\Lambda][c,\ldots,c^{(n)}]$ be a "minimal dependence" between $c,\ldots,c^{(n)}$ over $F(V_\Lambda)$. Then, $\varphi_p(R)$ is an algebraic dependence between $b, \varphi_p(c),\ldots,\varphi_p(c^{(n)})$ over F. All we need is a polynomial $p(t) \in \mathbb{Q}[t]$ such that $\varphi_p(R)$ is a nonzero polynomial with respect to b. Or, equivalently, we need $p(t) \in \mathbb{Q}[t]$ such that $\frac{\partial}{\partial b}\varphi_p(R) \neq 0$. This can be done in the same way as in Lemma 2.

The above theorem is a first step of the proof of the following strengthened version of Kolchin's primitive element theorem.

Theorem 2 ([3], **Theorem 2**). *Let $E = F\langle a_1,\ldots,a_m\rangle$, $\mathrm{trdeg}_F E < \infty$, and E contains a nonconstant. Then, there exists $a \in E$ such that $E = F\langle a\rangle$.*

In [4] the following analog of the Noether normalization lemma was proved using two-stage "polynomial shifting".

Theorem 3 ([4], **Theorem 1**). *Let B be a differentially finitely generated integral differential k-algebra of differential transcendence degree d. Then, there exist differentially independent elements $b_1,\ldots,b_d \in B$ such that for every prime differential ideal $\mathfrak{p} \subset A = k\{b_1,\ldots,b_d\}$ there exists a prime differential ideal $\mathfrak{q} \subset B$ such that $\mathfrak{q} \cap A = \mathfrak{p}$.*

We leave the proof of the following lemma as an exercise to an interested reader.

Lemma 3 *Let $E = F\langle c \rangle$, $\operatorname{trdeg}_F E < \infty$ and $c' \neq 0$. Let $P(x) \in F\{x\}$ be a nonconstant differential polynomial. Then, there exists a polynomial $s(t) \in \mathbb{Q}[t]$ such that $\operatorname{trdeg}_F F\langle P(s(c)) \rangle = \operatorname{trdeg}_F E$.*

References

1. Lang, S.: Algebra. Graduate Texts in Mathematics. Springer, New York (2002)
2. Ritt, J.F.: Differential Algebra, vol. 33. Colloquium publications of AMS, New York (1948)
3. Pogudin, G.A.: The primitive element theorem for differential fields with zero derivation on the base field. J. Pure Appl. Algebra **219**(9), 4035–4041 (2015)
4. Pogudin G.A., A differential analog of the Noether normalization lemma, preprint, submitted to ArXiv

Data and Knowledge Exploration

Searching for Geometric Theorems Using Features Retrieved from Diagrams

Wenya An, Xiaoyu Chen[✉], and Dongming Wang

LMIB – SKLSDE – School of Mathematics and Systems Science,
Beihang University, Beijing 100191, China
chenxiaoyu@buaa.edu.cn

Abstract. Searching for knowledge objects from knowledge bases is a basic problem that need be investigated in the context of knowledge management. For geometric knowledge objects such as theorems, natural language representations may not exactly reveal the features and structures of geometric entities, and that is why keyword-based searching is often unsatisfactory. To obtain high-quality results of searching for theorems in plane Euclidean geometry with images of diagrams as input, we propose a method using geometric features retrieved from the images. The method consists of four main steps: (1) retrieve geometric features, with formal representations, from an input image of a diagram D using pattern recognition and numerical verification; (2) construct a graph G corresponding to D from the retrieved features and weaken G to match graphs produced from formal representations of theorems in OpenGeo, an open geometric knowledge base; (3) calculate the degree of relevance between G and the graph for each theorem found from OpenGeo; (4) rank the resulting theorems according to their degrees of relevance. This method, based on graph matching, takes into account the structures of diagrams and works effectively. It is capable of finding out theorems of higher degree of relevance and may have potential applications in geometric knowledge management and education.

Keywords: Theorem searching · Graph matching · Degree of relevance · Knowledge management

1 Introduction

Searching for information, a fundamental activity of human beings in all eras, has never become so complex, so challenging, and so demanding as nowadays when the world is flooded with data and information. There are many sophisticated information retrieval (IR) systems which have been developed to retrieve desired and valuable information from diverse sources of data in efficient ways. In order to speak about information being *desired* or *valuable*, one has to introduce suitable mechanisms to specify desired information and to use quantitative estimates to measure the value or usefulness of each piece of information. For example, some of the IR systems choose to rank documents according to their

© Springer International Publishing Switzerland 2016
I.S. Kotsireas et al. (Eds.): MACIS 2015, LNCS 9582, pp. 383–397, 2016.
DOI: 10.1007/978-3-319-32859-1_33

numerical scores labeled on the basis of the vector-space model and the probabilistic model [16].

We are concerned with searching for information (more concretely, for geometric knowledge objects) from knowledge bases for which pattern matching is essential (see, e.g., [20]). For knowledge base queries, there are mainly two kinds: factual and conceptual [19]. The former identifies pieces of information relevant to the input through expansion of the query terms and expansion of themes, while the latter identifies potential existence of information in particular areas by specifying terminologies. In addition, ontology-based retrieval models or algorithms (e.g., the one described in [14]) may be used to support semantic search and knowledge-base exploitation.

Mathematics is constituted by multi-layer knowledge with various kinds of representations. Most of mathematical objects may be represented using functions and relations in which complex structures may be not obvious. It is rather difficult to retrieve mathematical objects merely based on keywords. Mathematical information retrieval is accomplished in general with query construction, normalization, indexing, and matching [21]. Currently available systems for searching mathematical expressions are based mostly on tree indexing (see, for instance, [6,10–13]). The retrieval of mathematical information from natural language text is another challenging issue, involving the use of techniques and tools from computational linguistics and artificial intelligence. The reader is referred to [8] for the system *mArachna*, which is capable of retrieving certain mathematical information from scientific books in German.

Figures are widely used to represent or illustrate mathematical knowledge in general, and geometric knowledge in particular. How to retrieve geometric information and how to discover geometric knowledge from figures are both interesting questions that may be asked. For existing work related to these questions, we may mention [9], in which declarative, procedural, and analytic approaches are used to describe geometric figures and a search mechanism based on a graph database is developed, and [2,17], in which it is shown how geometric theorems can be discovered automatically from images of diagrams.

In this paper, we present an inexact query method for searching geometric theorems stored in the open geometric knowledge base OpenGeo [18]. Our method is different from the query method presented in [9]. We construct undirected graphs, which are easier to be built, instead of directed graphs for queries and theorems stored in the database. We use types to describe nodes and edges. It is not necessary to consider directed edges when types of nodes and edges are fixed. For example, if two nodes' types are `point` and `line` respectively and the type of the edge between these two nodes is `incident`, then there is only one possibility that the point is incident to the line. On the other hand, weights are used to measure importance of nodes in an undirected graph. Considering an inexact query, we will weaken the query graph according to the weights of nodes to find out theorems. So it is possible to emphasize different parts of the query graph in the weakening process. If one wants to keep the nodes that have perpendicular relations, higher weights can be given to these nodes than to others. Different allocations of weights lead to different searching results, satisfying various application requests. Our method allows users to allocate weights by their will-ingness. In addition, to reduce searching space to save time, irrelevant theorems which have no possibilities to satisfy the query are filtered out before hand.

In detail, for a given image of a diagram, the method works by first retriev-ing geometric features (including geometric objects and their relations) implied in the image using pattern-recognition methods and numerical verification tech-niques [2, 17] and then constructing a graph G corresponding to the diagram in the input image from the retrieved features. The graph G is simplified and weak-ened to match graphs produced from theorems in OpenGeo and the degree of rele-vance between G and the graph of each theorem found from OpenGeo is calculated and used to rank the resulting theorems. This inexact query method is capable of figuring out theorems of high degree of relevance with the diagram in the input image and may be used to explore properties of similar diagrams, to find relevant theorems for illustration, and to seek for analogous techniques of theorem prov-ing. After a short review of formal representation of geometric theorems and the structure and implementation of OpenGeo in Sect. 2, we will outline the process of searching for theorems from OpenGeo using features retrieved from images of diagrams in Sect. 3 and define the degree of relevance for ranking searching results in Sect. 4. Experimental results with a preliminary implementation of our search-ing method will be reported in Sect. 5. Conclusions drawn from this work will be discussed briefly and together with future work in Sect. 6.

2   OpenGeo: A Formalized Geometric Knowledge Base

OpenGeo [18] is an open online geometric knowledge base, containing typical geometric knowledge objects (such as definitions, theorems, and proofs) and web-based interfaces and tools to support users to manage the knowledge objects stored in OpenGeo.

2.1   Representation of Geometric Knowledge Objects in OpenGeo

Geometric knowledge objects in OpenGeo are categorized into specific classes, including definition, axiom, theorem, proof, problem, and algorithm (which are interconnected according to the structure of geometric knowl-edge objects [3]). Each class may contain several data items. For example, the class of theorem contains knowledgeName (identifiers for the theorem), formalRepresentation (processable by other tools for automated reason-ing, computation, transformation, etc.), naturalRepresentation (for human users to read), algebraicRepresentation (algebraic expressions for the theo-rem), diagramInstruction (instructions for drawing diagrams of the theorem), nondegeneracyCondition (constraints to make the theorem rigorous and unam-biguous), figure (images or diagrams constructed by using dynamic geometry software for the theorem), and keyWords.

The data stored in the formalRepresentation item is represented in GDL, a formal Geometry Description Language [1] which is readable and processable, and can be easily transformed into or from natural languages. For example, the theorem named after Robert Simson, illustrated by the figure in Fig. 1 and stated in English as "the feet of the perpendiculars from a point to the sides of a triangle are collinear if and only if the point lies on the circumcircle of the triangle," may be represented in GDL as

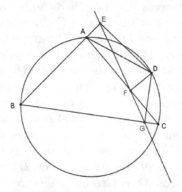

Fig. 1. The figure of Simson's theorem

Theorem(Simson, Theorem,
assume($\{A, B, C, D\} :=$ {point(), point(), point(), point()},
    incident(D, circumcircle(triangle(A, B, C)))),
show(collinear(foot(D, line(A, B)), foot(D, line(A, C)), foot(D,
    line(B, C)))))).

This formal representation of the theorem can be used for automated proving of the theorem and automated generation of diagrams illustrating the theorem [1]. We will use the data stored in the formalRepresentation item of theorem class to construct graphs of theorems for searching in Sect. 3.

2.2 OpenGeo Extension

To search for theorems in OpenGeo with an image of diagram as query input, it is effective to firstly retrieve geometric features from both the input image and the theorems in OpenGeo and then contrast the retrieved features to determine the degrees of relevance between the diagram in the input image and the theorems. By geometric features of a diagram or a theorem, we mean geometric entities and their relations which are involved in the diagram or the theorem. They can be represented as a graph in the way that an entity is mapped into a node, while a relation is mapped into an undirected edge. Formally, a pair (ID, Type) is used to represent a node, where ID is an integer automatically assigned to the node and Type indicates the type of the entity (see Table 1); a triple (FirstNodeID, SecondNodeID, Type) is used to represent an edge, where FirstNodeID and SecondNodeID are the ID's of the two nodes that the edge connects and Type indicates the type of the relation (see Table 2). Different types of nodes are related to different types of edges. For example, D (in Table 1) means the distance

Table 1. Types of nodes

Type	Entity
P	a point
L	a line, a segment, or a halfline
C	a circle
TRI	a triangle
QUAD	a quadrilateral
D	a distance

Table 2. Types of edges

Type	Relation
inc	a point is incident to a line or a circle
ind	endpoints of a segment
perp	two lines are perpendicular
para	two lines are parallel
equ	equivalence
tri	vertices of a triangle
quad	vertices of a quadrilateral

between two points, no matter whether the line segment between the two points exists or not in the diagram. But if L (in Table 1) appears, there must be a line appearing in the diagram. Relations between entities of type L could be perpendicular relation (of type perp in Table 2), parallel relation (of type para in Table 2) while quantities of type D have the equivalence relation (of type equ in Table 2).

For instance, the graph for the geometric features of Simson theorem may be represented as follows.

− Nodes:

$$(0, L), \quad (1, L), \quad (2, L), \quad (3, C), \quad (4, P), \quad (5, L),$$
$$(6, P), \quad (7, L), \quad (8, P), \quad (9, L), \quad (10, L), \quad (11, TRI),$$
$$(12, P), \quad (13, P), \quad (14, P), \quad (15, P).$$

− Edges:

$$(0, 5, perp), (1, 7, perp), (2, 9, perp), (12, 11, tri), (13, 11, tri), (14, 11, tri),$$
$$(12, 0, inc), (13, 0, inc), (12, 1, inc), (14, 1, inc), (13, 2, inc), (14, 2, inc),$$
$$(4, 5, inc), (4, 0, inc), (15, 5, inc), (6, 7, inc), (6, 1, inc), (15, 7, inc),$$
$$(8, 9, inc), (8, 2, inc), (15, 9, inc), (4, 10, inc), (6, 10, inc), (8, 10, inc),$$
$$(15, 3, inc), (12, 3, inc), (13, 3, inc), (14, 3, inc).$$

To facilitate the fetching of geometric features of theorems, OpenGeo has been extended by adding a data item, named feature, to store graphs generated automatically from formal representations of theorems.

3   Searching for Geometric Theorems in OpenGeo

The process of searching for geometric theorems in OpenGeo consists of three main steps: (1) retrieving geometric features from an input image of diagram; (2) filtering out irrelevant theorems using the retrieved features; (3) matching the features of each remaining theorem with the features of the diagram in the input image to obtain theorems of high relevance. We detail these steps in the following three subsections.

3.1   Retrieving Geometric Features from Diagrams

Chen, Song, and Wang [2,17] proposed a method to detect basic geometric entities (such as points, lines, and circles), to recognize labels of basic geometric
entities, and to mine basic geometric relations (such as incidence, parallelism,
perpendicularity, and equivalence) from images of diagrams by using techniques
and tools of pattern recognition and numerical verification. The retrieved geometric features, represented in GDL, can be transformed into graph representations. For example, the following graph representation may be produced for the
diagram shown in Fig. 2.

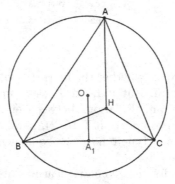

Fig. 2. An image of diagram as query input

- Nodes:

$(0,\mathtt{P})$, $(1,\mathtt{L})$, $(2,\mathtt{D})$, $(3,\mathtt{D})$, $(4,\mathtt{P})$, $(5,\mathtt{L})$, $(6,\mathtt{L})$,
$(7,\mathtt{L})$, $(8,\mathtt{C})$, $(9,\mathtt{L})$, $(10,\mathtt{L})$, $(11,\mathtt{L})$, $(12,\mathtt{P})$, $(13,\mathtt{D})$,
$(14,\mathtt{D})$, $(15,\mathtt{D})$, $(16,\mathtt{TRI})$, $(17,\mathtt{L})$, $(18,\mathtt{L})$, $(19,\mathtt{L})$, $(20,\mathtt{P})$,
$(21,\mathtt{P})$, $(22,\mathtt{P})$.

- Edges:

$(0,2,\mathtt{ind})$, $(21,2,\mathtt{ind})$, $(0,3,\mathtt{ind})$, $(22,3,\mathtt{ind})$, $(4,13,\mathtt{ind})$, $(20,13,\mathtt{ind})$,
$(4,14,\mathtt{ind})$, $(21,14,\mathtt{ind})$, $(4,15,\mathtt{ind})$, $(22,15,\mathtt{ind})$, $(1,9,\mathtt{perp})$, $(5,10,\mathtt{perp})$,
$(6,11,\mathtt{perp})$, $(0,1,\mathtt{inc})$, $(21,1,\mathtt{inc})$, $(22,1,\mathtt{inc})$, $(22,5,\mathtt{inc})$, $(20,5,\mathtt{inc})$,
$(20,6,\mathtt{inc})$, $(21,6,\mathtt{inc})$, $(4,7,\mathtt{inc})$, $(0,7,\mathtt{inc})$, $(20,9,\mathtt{inc})$, $(21,10,\mathtt{inc})$,
$(22,11,\mathtt{inc})$, $(9,12,\mathtt{inc})$, $(10,12,\mathtt{inc})$, $(11,12,\mathtt{inc})$, $(4,17,\mathtt{inc})$, $(20,17,\mathtt{inc})$,
$(4,18,\mathtt{inc})$, $(21,18,\mathtt{inc})$, $(4,19,\mathtt{inc})$, $(22,19,\mathtt{inc})$, $(20,8,\mathtt{inc})$, $(21,8,\mathtt{inc})$,
$(22,8,\mathtt{inc})$, $(20,16,\mathtt{tri})$, $(21,16,\mathtt{tri})$, $(22,16,\mathtt{tri})$, $(2,3,\mathtt{equ})$, $(13,15,\mathtt{equ})$,
$(14,15,\mathtt{equ})$.

Geometric information retrieved from images of diagrams may contain entities and relations irrelevant to the geometric features that the diagrams are
expected to illustrate. To obtain geometric features for the purpose of searching,
the following simple rules may be applied to remove redundant points, lines, and
circles.

– If a point is involved in no more than two relations and types of these relations are `inc`, then remove this point and the corresponding relations. For example, if a point in a diagram is just the intersection of two lines, then it does not show any important geometric features of the diagram. Therefore, this point and the two relations of `inc` can be removed.
– If a line or a circle is not involved in any relations, then remove it. In other words, a line or a circle without relations in a diagram does not show any important geometric features of the diagram, so it can be removed.

3.2 Filtering Out Irrelevant Theorems Using Features

For efficient searching in OpenGeo, it is necessary to filter out irrelevant theorems before starting the process of feature matching. In view of the importance of geometric relations in the construction of diagrams, we adopt the following rules to filter out some irrelevant theorems.

For any graph s produced for a theorem in OpenGeo and a graph q produced for the diagram in the input image, let the types of edges of s and q be collected in sets $\mathbf{C}_s$ and $\mathbf{C}_q$ and the numbers of edges of the same types of s and q be collected in sets $\mathbf{N}_s$ and $\mathbf{N}_q$, respectively.

1. If $\mathbf{C}_q \setminus \mathbf{C}_s \neq \emptyset$, then the theorem with graph s is considered as irrelevant.
2. If $\mathbf{C}_q \subset \mathbf{C}_s$ and there exists a c in $\mathbf{C}_q$ such that the number of edges of type c in $\mathbf{N}_q$ is greater than that in $\mathbf{N}_s$, then the theorem with graph s is considered as irrelevant.

If either of the two conditions is satisfied, it is impossible that q is a subgraph of s. So the theorem with graph s is irrelevant to the query.

3.3 Matching Geometric Objects and Relations

By means of representing geometric features using graphs (see Sect. 2.2), the problem of feature matching can be converted to that of graph matching. For the latter there is a universal method, called *GraphGrep* and introduced by Giugno and Shasha [7]. This method proceeds by first creating a database, then parsing the query graph and filtering the database, and finally finding subgraphs matching the query graph. The resulting graphs produced by GraphGrep contain the query graph as a subgraph. Using such exact matching, it is hardly possible to find out theorems which are relevant with the query diagram only to some degree.

What we actually want is inexact matching. To achieve this, we add a weakening process before using GraphGrep, that is, first weakening the query graph by eliminating certain nodes and edges and then using GraphGrep to find graphs for theorems in OpenGeo that match the weakened query graph exactly. The following steps can be used to weaken the query graph.

1. *Compute weights of nodes.* Let $\mathbf{R}$ be the set {`inc`, `ind`, `perp`, `para`, `equ`, `tri`, `quad`} of types of edges, where the weight for each $\mathrm{T} \in \mathbf{R}$ is pre-given

and denoted by w_T. Let $\mathbf{W_R} = [w_{\text{inc}}, w_{\text{ind}}, w_{\text{perp}}, w_{\text{para}}, w_{\text{equ}}, w_{\text{tri}}, w_{\text{quad}}]$, $\mathbf{V} = \{v_1, v_2, \ldots, v_p\}$ be the set of nodes, and $\mathbf{E} = \{e_1, e_2, \ldots, e_q\}$ be the set of edges of the query graph. The weight of an edge e_i, denoted by w_i^e, is defined to be w_{T_i}, where T_i is the type of e_i, and the weight of a node v_i connected by $e_{i_1}, e_{i_2}, \ldots, e_{i_n}$ $(1 \le i_1, i_2, \ldots, i_n \le q)$ is defined to be $w_{T_{i_1}} + w_{T_{i_2}} + \cdots + w_{T_{i_n}}$. Let the weight of v_i be denoted by w_i^v and $\mathbf{W_V} = [w_1^v, w_2^v, \ldots, w_p^v]$.[1]

2. *Sort nodes with respect to a specific order.* Let the types of nodes be ordered as $\mathtt{D} \prec \mathtt{P} \prec \mathtt{L} \prec \mathtt{C} \prec \mathtt{TRI} \prec \mathtt{QUAD}$. Sort the nodes in $\mathbf{V}$ with respect to the order $\prec$, introduced according to the following rules:
   (a) if $w_i^v < w_j^v$, then $v_i \prec v_j$; if $w_i^v = w_j^v$, then go to (b);
   (b) if the number of edges connected to v_i is greater than that of edges connected to v_j, then $v_i \prec v_j$; if the two numbers are equal, then go to (c);
   (c) if the type of $v_i \prec$ the type of v_j, then $v_i \prec v_j$; if the types are identical, then go to (d);
   (d) if the ID of v_i is less than that of v_j, then $v_i \prec v_j$.

3. *Remove a node and the edges connected to the node.* Let the nodes of the query graph be ordered as $v_{s_1} \prec v_{s_2} \prec \cdots \prec v_{s_p}$ and denote by $\mathbf{E}_{v_{s_i}} (1 \le i \le p)$ the set of edges that are connected to v_{s_i}. Then $\mathbf{V} \setminus \{v_{s_i}\}$ and $\mathbf{E} \setminus \mathbf{E}_{v_{s_i}}$ are respectively the set of nodes and the set of edges of the weakened graph, obtained from the query graph by removing the node v_{s_i} from $\mathbf{V}$ and all the edges connected to v_{s_i} from $\mathbf{E}$.

4   Processing Results of Searching

Using the method of inexact matching presented in the preceding section, one can find a set of theorems in OpenGeo whose graphs match the query graph of the diagram in the given image. It remains to rank the found theorems, so that those which are most relevant to what the diagram may illustrate are placed on the top.

4.1   Computing Degrees of Relevance

Given the image of a diagram D as query input, we want to define, for each theorem T whose graph matches the graph of D, a quantity rel_T^D, ranging from 0% to 100% and called the *degree of relevance* between D and T, to measure how relevant T is to D. For two theorems T_1 and T_2, if $\text{rel}_{T_1}^D < \text{rel}_{T_2}^D$, then theorem T_2 is said to be more relevant with D than theorem T_1. The degree of relevance should meet the following three requirements.

- **Complete.** Let (V, E) and (V_T, E_T) be the graph representations for D and T, respectively. If $V = V_T$ and $E = E_T$, then $\text{rel}_T^D = 100\%$; if $V \cap V_T = \emptyset$ and $E \cap E_T = \emptyset$, then $\text{rel}_T^D = 0\%$.

---

[1] For example, let $\mathbf{V} = \{(0, \mathtt{P}), (1, \mathtt{L}), (2, \mathtt{L})\}$ and $\mathbf{E} = \{(0, 1, \mathtt{inc}), (0, 2, \mathtt{inc}), (1, 2, \mathtt{perp})\}$. If $\mathbf{W_R} = [1, 1, 2, 2, 1, 1, 1]$, then $\mathbf{W_V} = [2, 3, 3]$.

- **Intuitive.** Let (V, E), (V_{T_1}, E_{T_1}), and (V_{T_2}, E_{T_2}) be the graph representations for diagram D and theorems T_1 and T_2, respectively, and let $m_k = \frac{|V \cap V_{T_k}| + |E \cap E_{T_k}|}{|V_{T_k}| + |E_{T_k}|}$ for $k = 1, 2$.[2] If $m_{k_1} < m_{k_2}$, then $\text{rel}^D_{T_{k_1}} < \text{rel}^D_{T_{k_2}}$ $(k_1, k_2 \in \{1, 2\})$.

- **Orderly.** Let D_n be the diagram for which the graph is obtained by weakening the query graph n times $(n = 1, 2, \ldots, |V|)$. Suppose that theorems T_a and T_b match D_a and D_b, respectively. If $a < b$, then $\text{rel}^D_{T_a} > \text{rel}^D_{T_b}$.

The degree of relevance may be defined in different ways to meet the above requirements. In what follows, we provide one definition and show its soundness. Similarity of graphs has been studied in the past. Maximum common edge subgraphs are used for calculation of graph similarity in [15] and Dehmer and others [5] use generalized trees which are directed and hierarchical graphs to measure structural similarity of graphs. Most of the methods focus on general graphs. Our method is based on weighted and undirected graphs and takes into account geometric characteristics.

Let
$$\mathbf{G}_{D_g} = (\{v_{r_{g,1}}, v_{r_{g,2}}, \ldots, v_{r_{g,m_g}}\}, \{e_{r_{g,1}}, e_{r_{g,2}}, \ldots, e_{r_{g,n_g}}\})$$

be the representation of the graph resulting from $\mathbf{G}_D$ after being weakened g times $(g = 0, 1, \ldots, m_0 - 1)$,[3] and

$$\mathbf{G}_{T_g} = (\{v_{t_{g,1}}, v_{t_{g,2}}, \ldots, v_{t_{g,l_g}}\}, \{e_{t_{g,1}}, e_{t_{g,2}}, \ldots, e_{t_{g,h_g}}\})$$

be the graph representation for a theorem T_g whose graph matches the query graph of D_g exactly. Let the set $\mathbf{W_R}$ of weights be given, the set of weights of edges in the graph of D be $\{w^e_{r_{0,1}}, w^e_{r_{0,2}}, \ldots, w^e_{r_{0,n_0}}\}$, and the set of weights of edges in the graph of D_g be $\{w^e_{r_{g,1}}, w^e_{r_{g,2}}, \ldots, w^e_{r_{g,n_g}}\}$. Then the degree of relevance between T_g and D is defined as

$$\text{rel}^D_{T_g} = \text{mat}_g \cdot (\text{mtr}_g - \text{mtr}_{g+1}) + \text{mtr}_{g+1}, \tag{1}$$

where

$$\text{mat}_g = \frac{1}{2} \cdot \left(\frac{m_g}{l_g} + \frac{n_g}{h_g} \right), \tag{2}$$

$$\text{mtr}_k = \frac{1}{2} \cdot \left(\frac{m_k}{m_0} + \frac{\sum_{j=1}^{n_k} w^e_{r_{k,j}}}{\sum_{j=1}^{n_0} w^e_{r_{0,j}}} \right), \quad k = g, g+1, \tag{3}$$

and $g = 0, 1, \ldots, m_0 - 1$. In the above definition, mat_g and mtr_g measure the degree of matching between $\mathbf{G}_{D_g}$ and $\mathbf{G}_{T_g}$ and the degree of matching between $\mathbf{G}_{D_g}$ and $\mathbf{G}_D$, respectively.

Assertion. The degree of relevance defined above is complete, intuitive, and orderly.

The correctness of this assertion can be seen from the following arguments.

---

[2] $|S|$ denotes the number of elements in set S.
[3] When $g = 0$, $\mathbf{G}_{D_0} = \mathbf{G}_D$ is the graph representation for the query diagram D.

1. Complete. If $\mathbf{G}_D$ and $\mathbf{G}_{T_0}$ are equivalent, then $\mathbf{G}_D, \mathbf{G}_{D_0}$, and $\mathbf{G}_{T_0}$ are all the same. Therefore, $\mathrm{mtr}_0 = 1$, $\mathrm{mat}_0 = 1$, and thus $\mathrm{rel}_{T_0}^D = 1$, which means that the degree of relevance is $100\,\%$. If for any theorem T_0, there is neither node nor edge of $\mathbf{G}_D$ which matches the nodes or edges of $\mathbf{G}_{T_0}$, then $\mathrm{mat}_0 = 0$, $\mathrm{mtr}_1 = 0$, and thus $\mathrm{rel}_{T_0}^D = 0$, which means that the degree of relevance is $0\,\%$.

2. Intuitive. According to the definition, when mtr_g and mtr_{g+1} are fixed, $\mathrm{mtr}_g - \mathrm{mtr}_{g+1} > 0$ holds. Therefore, the larger mat_g is, the higher $\mathrm{rel}_{T_g}^D$ is.

3. Orderly. From the formulae in the definition, it is easy to deduce that $\mathrm{rel}_{T_g}^D > \mathrm{mtr}_{g+1}$ and $\mathrm{rel}_{T_{g+1}}^D < \mathrm{mtr}_{g+1}$. Therefore, $\mathrm{rel}_{T_g}^D > \mathrm{rel}_{T_{g+1}}^D$.

4.2 Ranking the Results

Retrieved theorems can be ranked according to the degrees of their relevance with the query diagram. For example, five theorems $T_1, \ldots, T_5$ found of degrees $85\,\%$, $90\,\%$, $45\,\%$, $92\,\%$, and $79\,\%$ of relevance, respectively, with the query diagram may be ranked top-down in the order of T_4, T_2, T_1, T_5, T_3. From the ranking, it is easy to see which theorems are most relevant to the query input.

5 Implementation and Experimental Results

Now we explain how the searching method presented in the previous sections has been implemented using Python and provide some experimental results to show the performance of the method with our preliminary implementation.

5.1 Implementation Issues

The searching procedure contains five modules: parsing, filtering, exact matching (GraphGrep), similarity measuring, and reducing. Through the parsing module, both the input image of a query diagram and the formal representations of theorems in OpenGeo are parsed to yield graph representations of geometric features. By comparing the numbers of entities and relations in the query graph with those in the graph of each theorem in OpenGeo, the filtering module serves to reduce search space and produce a set of candidate graphs. Then GraphGrep is used to determine which candidate graphs match the query graph exactly. For each resulting graph after exact matching, the degree of relevance between this graph and the query graph is calculated in the module of similarity measuring. While the degrees of relevance are higher than a pre-specified percentage (threshold) and the given number of weakening operations is not reached, the query graph is (further) weakened in the reducing module and the procedure repeats with the weakened query graph instead of the query graph.

5.2 Examples and Experiments

To see how well the searching procedure performs, let us take the image of the diagram shown in Fig. 2 as an example. With this image as query input, the

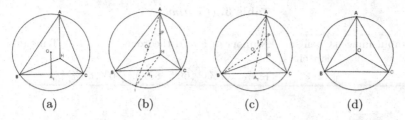

(a)   (b)   (c)   (d)

Fig. 3. (a) Searching result of exact matching; (b)–(c) Searching results after weakening the query graph once; (d) Searching result after weakening the query graph twice.

procedure can find one theorem in OpenGeo, which is illustrated by the diagram shown in Fig. 3(a). The degree of relevance between this found theorem and the

Table 3. Selected experimental results

Input image	OpenGeo capacity	Number of found theorems			Time (s)
		(80%, 100%]	(60%, 80%]	(50%, 60%]	
	189	2	0	1	0.146
	189	3	1	0	0.153
	189	1	3	11	0.206
	189	8	17	4	0.177
	189	1	0	0	0.079
	189	0	3	4	0.265
	189	0	3	16	0.267

(*Continued*)

Table 3. (*Continued*)

Input image	OpenGeo capacity	Number of found theorems			Time (s)
		(80%, 100%]	(60%, 80%]	(50%, 60%]	
	189	3	4	2	0.126
	189	0	1	1	0.269
	189	1	2	9	0.189
	189	1	4	0	0.195
	189	2	3	1	0.149
	189	2	0	1	0.161
	189	3	1	0	0.172
	189	1	3	11	0.163
	189	0	0	0	–
	189	0	0	0	–

query diagram is 99.61 %. If the query graph is weakened once, then two other theorems from OpenGeo, illustrated by the diagrams shown in Fig. 3(b) and (c), can be found and the degrees of relevance between these two theorems and the query diagram are 81.97 % and 81.04 %, respectively. If the query graph is weakened twice, then another theorem, illustrated by the diagram shown in Fig. 3(d), can be found and the degree of relevance between this theorem and the query diagram is 71.95 %. These searching results indicate that the procedure we have implemented is capable of finding geometric theorems with images of diagrams as query input and the measure we have introduced for the degrees of relevance between found theorems and query diagrams is sound.

We have made experiments on more than 40 images of diagrams (scanned from the book [4]) to test our searching procedure. Selected experimental results are given in Table 3. The first column shows the input query images and the second column presents the number of theorems stored in OpenGeo for searching. With each input query image, the procedure may find some theorems, the number of which is recoded in the three sub-columns of the third column: the first sub-column counts the number of found theorems whose degrees of relevance with the query diagram belong to the interval (80 %, 100 %] and the second and the third sub-column count the numbers of found theorems whose degrees of relevance belong to the intervals (60 %, 80 %] and (50 %, 60 %], respectively. The last column of Table 3 shows the running time of the searching procedure. The experimental results demonstrate that our searching procedure works effectively for most input images as query. For the input images in the last two rows of Table 3, the procedure can only find theorems whose degrees of relevance are less than 50 %.

6 Conclusion and Future Work

We have proposed a method to tackle the problem of searching for geometric theorems with images of diagrams as query input. The method uses geometric features retrieved from the images of diagrams and is based on graph matching. This method also treats weakened query graphs as a bridge to find out relevant theorems from the query graph. It is capable not only of finding theorems which the query diagrams likely illustrate, but also of ranking the found theorems according to their degrees of relevance with query diagrams. Preliminary experiments show that our method as well as its implementation works effectively for theorem searching in OpenGeo. We will improve and generalize the method, e.g., by including more types of geometric entities and relations, will extend our implementation for searching theorems in other geometric knowledge bases, e.g., graph databases, will try various methods to calculate degrees of relevance, and will develop a user-friendly interface for geometric theorem processing.

Acknowledgements. This work has been supported by the project SKLSDE-2015ZX-18 and the NSFC project 11371047.

References

1. Chen, X.: Representation and automated transformation of geometric statements. J. Syst. Sci. Complexity **27**(2), 382–412 (2014)
2. Chen, X., Song, D., Wang, D.: Automated generation of geometric theorems from images of diagrams. Ann. Math. Artif. Intell. **74**(3–4), 333–358 (2015)
3. Chen, X., Wang, D.: Formalization and specification of geometric knowledge objects. Math. Comput. Sci. **7**(4), 439–454 (2013)
4. Chou, S.-C.: Mechanical Geometry Theorem Proving. D. Reidel, Dordrecht (1988)
5. Dehmer, M., Emmert-Streib, F., Kilian, J.: A similarity measure for graphs with low computational complexity. Appl. Math. Comput. **182**, 447–459 (2006)
6. Einwohner, T.H., Fateman, R.J.: Searching techniques for integral tables. In: 1995 International Symposium on Symbolic and Algebraic Computation, pp. 133–139. ACM (1995)
7. Giugno, R., Shasha, D.: GraphGrep: a fast and universal method for querying graphs. In: 16th International Conference in Pattern Recognition, pp. 112–115. IEEE (2002)
8. Grottke, S., Jeschke, S., Natho, N., Seiler, R.: mArachna: a classification scheme for semantic retrieval in elearning environments in mathematics. Recent Research Developments in Learning Technologies (2005)
9. Haralambous, Y., Quaresma, P.: Querying geometric figures using a controlled language, ontological graphs and dependency lattices. In: Watt, S.M., Davenport, J.H., Sexton, A.P., Sojka, P., Urban, J. (eds.) Intelligent Computer Mathematics (CICM 2014). LNAI, vol. 8543, pp. 298–311. Springer, Heidelberg (2014)
10. Hashimoto, H., Hijikata, Y., Nishida, S.: Incorporating breadth first search for indexing MathML objects. In: The International Conference on Systems, Man and Cybernetics, pp. 3519–3523. IEEE (2008)
11. Kamali, S., Tompa, F.W.: Improving mathematics retrieval. In: Towards a Digital Mathematics Library, pp. 37–48 (2009)
12. Kamali, S., Tompa, F.W.: A new mathematics retrieval system. In: 19th ACM International Conference on Information and Knowledge Management, pp. 1413–1416. ACM (2010)
13. Kohlhase, M., Sucan, I.: A search engine for mathematical formulae. In: Calmet, J., Ida, T., Wang, D. (eds.) Artificial Intelligence and Symbolic Computation (AISC 2006). LNAI, vol. 4120, pp. 241–253. Springer, Heidelberg (2006)
14. Kumar, S., Manjeet, S., Avik, D.: OWL-based ontology indexing and retrieving algorithms for Semantic Search Engine. In: The 7th International Conference of Computing and Convergence Technology, pp. 1135–1140. IEEE (2012)
15. Raymond, J.W., Gardiner, E.J., Willett, P.: RASCAL: calculation of graph similarity using maximum common edge subgraphs. Comput. J. **45**, 631–644 (2002)
16. Singhal, A.: Modern information retrieval: a brief overview. IEEE Data Eng. Bull. **24**, 35–43 (2001)
17. Song, D., Wang, D., Chen, X.: Discovering geometric theorems from scanned and photographed images of diagrams. In: Botana, F., Quaresma, P. (eds.) Automated Deduction in Geometry (ADG 2014). LNAI, vol. 9201, pp. 149–165. Springer, Heidelberg (2015)
18. Wang, D., Chen, X., An, W., et al.: OpenGeo: an open geometric knowledge base. In: Hong, H., Yap, C. (eds.) Mathematical Software (ICMS 2014). LNCS, vol. 8592, pp. 240–245. Springer, Heidelberg (2014)

19. Wical, K.: Concept knowledge base search and retrieval system. Patent 6,038,560. U.S (2000)
20. Woods, W.A.: Knowledge base retrieval. In: Brodie, M.L., Mylopoulos, J. (eds.) On Knowledge Base Management Systems, pp. 179–195. Springer, New York (1986)
21. Zanibbi, R., Blostein, D.: Recognition and retrieval of mathematical expressions. Int. J. Doc. Anal. Recogn. 15, 331–357 (2012)

New Method for Instance Feature Selection Using Redundant Features for Biological Data

Waad Bouaguel[1]([⊠]), Emna Mouelhi[2], and Ghazi Bel Mufti[3]

[1] LARODEC, ISG, University of Tunis, Tunis, Tunisia
bouaguelwaad@mailpost.tn
[2] ISG, University of Tunis, Tunis, Tunisia
Mouelhi.emmna@yahoo.fr
[3] LARIME, ESSEC, University of Tunis, Tunis, Tunisia
belmufti@yahoo.com

Abstract. Biological data bases are characterized by a very large number of features and a few instances which make classification more difficult and time consuming. This problem can be solved using feature selection approach. The Filter feature selection method ranks features according to their significance level. Then it selects the most significant features and discards the rest. The discarded features may provide some useful information and could be useful to further consideration. Hence, we propose a new feature selection method that uses these eliminated features in order to increase the classification performance and avoid the curse of dimensionality. The new approach is based on the idea of transforming the value of the similar features into new instances for the retained features. We aim to reduce the feature space by performing features selection and increasing the learning space in creating new instances using the redundant features.

Keywords: Curse of dimensionality · Relief · Feature selection · Filter

1 Introduction

The rapid progress and growth of biological data require a deep analysis of massive data collected by the pharmaceutical studies and cancer therapy at the genomic and proteomic level. Different kinds of information are stored in biological databases such as the data about sequences of macromolecules, chromosomes description, protein structures, etc. This huge quantity of data becomes the main focus of information systems that represent store and return them appropriately to the user. These data give information about the new sequences of the gene or the protein discovered, their location in the cell and the description on these data [6].

The incredible number of genes makes the treatments of biological data more challenging. As there are thousands of gene expressions and only few dozens of observations in a typical gene expression of a data set, the number of genes d is usually of 1000 to 10000 order, while the number of biological observations

© Springer International Publishing Switzerland 2016
I.S. Kotsireas et al. (Eds.): MACIS 2015, LNCS 9582, pp. 398–405, 2016.
DOI: 10.1007/978-3-319-32859-1_34

n is somewhere between 10 and 100. This phenomenon is called the curse of dimensionality [2, 13].

The curse of dimensionality is a term introduced by [1] to describe the problem caused by the exponential increase in volume, associated with adding extra dimensions to Euclidean space, such a condition makes the application of many classification or clustering methods a hard task to achieve [5]. In fact, if the number of features increases, the classifiers performance increases until the optimal number of features is reached. Further increase of this number, without increasing the number of training samples results in a decrease in classifier performance.

This problem can be solved using the dimensionality reduction process. Dimensionality reduction is the process of converting data of very high dimensionality into data of much lower dimensionality such that, each of the lower dimensions conveys more information for efficient data processing. The field of dimensionality reduction is divided into two categories of methods: feature extraction and feature selection. According to [8, 10], the first method consists of finding a transformation to a lower dimensional space by creating a new smaller feature set combining the original features. while feature extraction tends to change the features, the second method tends to maintain the meaning of the features. The feature selection process meaning reduces the size of the feature space to a manageable size in order to consider a future work [8, 12]. Usually, feature selection is used more than feature extraction before classification, since it preserves all information on the importance of each feature, whereas in the extraction function the obtained features are generally not interpretable [4].

According to [13], there are two categories of feature selection algorithms: filter and wrapper feature selection methods. Filter methods select the best features by evaluating the fundamental properties of data, making them fast and simple to implement. Wrapper methods consider the selection of a set of features as a search problem, where the aim is to find the best features according to the classifiers accuracy, making results well-matched to the predetermined classification algorithm. Although effective wrapper methods are time consuming as they are turned to the specific interaction with the classifier, which make filter methods more adopted to big data [3]. Filter methods are characterized by several benefits, in fact they select features without implying any learning algorithm and they are generally faster and tend to be computationally less expensive than wrapper methods. Filter methods, rank in terms of effectiveness. This means that it select feature 1, then feature 2, and so on. Assume that we have selected up to d = 5 features, so we use features $\{1, 2, 3, 4, 5\}$ and discard the rest. Among the selected features some may be redundant and should be discarded. The discarded features provide some useful information. For example features 1 and 2 are correlated, so feature 1 may be retained and feature 2 is eliminated. Although feature 1 is kept as the most pertinent, feature 2 might not be much worse than feature 1, and so could be useful to consider. Hence, the disadvantage of filter methods is that some features that may seem less important and redundant, and are thus discarded, may bear valuable information. It seems a bit of a waste to throw away such information, that could possibly in some way contribute to improving model performance. Hence the idea is to find a way to reduce the dimensionality of the features and in the same use the additional information. A possible solution is to perform a feature selection in order to reduce the feature space and then use the eliminated features in order to create new instances and increase the learning space and avoid the curse of dimensionality. This paper is organized as follows: Sect. 2 presents a new method for instance feature selection based on two stages. Then, Sect. 3 describes the used datasets and the performance metrics and summarize the obtained results and conclusions are drawn in Sect. 4.

2   New Method for Instance Feature Selection

As discussed before in the previous section our new approach aim to reduce the number of feature and in the same time create new instances using the redundant features. We start by a simple feature selection in which we choose the relevant features and eliminate the rest. The relevant features are then divided into two subsets: the most relevant features and relevant and redundant features. Then, the feature with high similarity to the target feature are retained and the redundant one are transformed into new instances before the classification. More precisely, this approach entails two principals steps: the first step is feature ranking and pertinence study and the second step is similarity study and instances creation. Figure 1 summarizes this two steps of our approach.

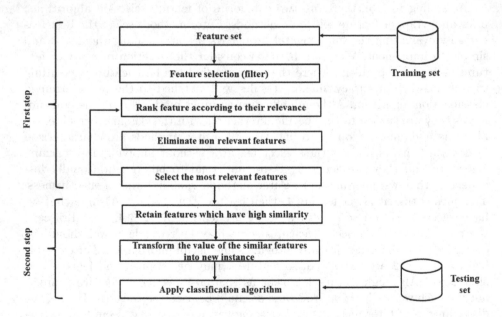

Fig. 1. Flowchart of transformation process.

2.1   Feature Ranking and Pertinence Study

As a first step, we start by performing a feature selection using Relief algorithm in order to rank the features according to their pertinence level to the target feature. Relief is one of the most famous feature selection method based on distance measures [9]. The fundamental idea of Relief is to estimate the relevance of features according to how well their values separate the instances of the same and different classes that are near each other [14]. For a dataset with n instances and d features the complexity of relief is in order of $O(nd)$, which makes it very practical to data sets with large number of instances and features, such as biological datasets. Relief returns a list of sorted features from the most important feature to the least one.

The obtained list is divided into two groups: the first one represents the most relevant features the second one represents the irrelevant features. Irrelevant features are those that can never contribute to improve the predictive accuracy of classification model. Removing such features reduces the dimension of the search space and speeds up the learning algorithm. Hence, the second one is eliminated and the first one is retained.

2.2   Similarity Study and Instance Creation

Although simple Relief doesn't remove redundant features. If the feature weights are superior to a particular threshold, these features will be selected even though many of them are highly correlated to each other [9]. Therefore, the first set of features may contain redundant features. They basically bring similar information as other features. For example, a dataset may include two features which provide similar information as date of birth and age.

Keeping these redundant features may confuse the learning algorithm. However, these kinds of features may bear valuable information, that could possibly in some way contribute to improve model performance. Lets take an illustrative example, consider we have ten features $X = (x^1; x^6; x^5; x^4; x^3; x^2; x^8; x^7; x^9; x^{10})$, we apply Relief to rank features in terms of effectiveness. We obtain this list $X = (x^2; x^4; x^6; x^3; x^5; x^1; x^8; x^7; x^9; x^{10})$. Assume that we have selected up to six features (pertinent features) and discard the rest (irrelevant features). Therefore, we obtain this list $X = (x^2; x^4; x^6; x^3; x^5; x^6)$. In the second step, we apply the correlation function in order to detect the redundant features. Therefore, the correlation results give x^1 redundant and similar to x^2, x^3 redundant and similar to x^4. The redundant features x^1 and x^3 could be eliminated and transformed into new instance for x^2 and x^4. Thus, we propose to create new instances based on the value of the redundant features. So, x^1 will add four new instances to x^2 and x^3 will add four new instances to x^4. More details are given in Fig. 2.

Hence, the most significant features are retained and the redundant ones are used to create new instance. In this way, we increase the size of the learning sample, which will improve the classification performance and no information is discarded. All the important steps of our proposed approach are given in Fig. 1.

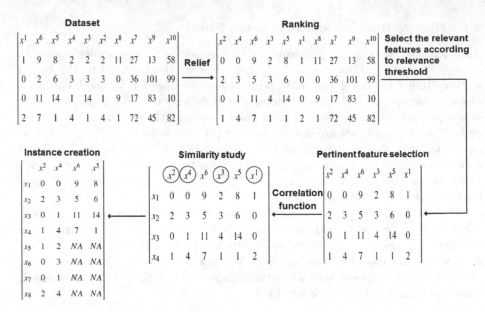

Fig. 2. Illustrative example of transforming features into new instances.

3 Experimental Investigations

The experiments were conducted on Central Nervous System (CNS), a large data set concerned with the prediction of central nervous system embryonal tumor outcome based on gene expression. This data set includes 60 samples containing 39 medulloblastoma survivors and 21 treatment failures. These samples are described by 7129 genes [11]. We consider also the Leukemia microarry gene expression dataset that consists of 72 samples which are all acute leukemia patients, either acute lymphoblastic leukemia (47 ALL) or acute myelogenous leukemia (25 AML). The total number of genes to be tested is 7129 [7].

In the literature, when we have a huge number of features (superior to 1000 features) we can select only 20 % of the datasets in order to facilitate the handling of data. Consequently, we have randomly chosen about 1426 features from CNS and Leukemia datasets. Then, in order to obtain a reliable model, we split the dataset into training and testing samples using a 10-cross-validation. The result of the proposed approach may contain missing values which should be handled before beginning the classification process. We apply the mean substitution for missing data, is to replace missing values with mean or median values over all instances.

Feature ranking and pertinence study step, is based on relief algorithm in order to rank features according to rates of pertinence, and to select the most relevant one. If the relevance weight of features is greater than a particular threshold, then these features are considered as pertinent. In this step of experimental setting, we have used the empirical studies to define the correctly threshold for

Relief algorithm. In fact, we tested four thresholds $\{0.2; 0.4; 0.6; 0.8\}$ for each datasets in order to choose the best results. Table 1 gives an example of the empirical study of the similarity and correlation thresholds for CNS dataset with DT.

Table 1. Thresholds related to CNS dataset (DT)

Pertinence rate for Relief algorithm				
Correlation rates	0.2*	0.4	0.6	0.8
0.4	0.9	0.925	0.892	0.852
0.5*	0.926*	0.909	0.917	0.869
0.6	0.9	0.926	0.893	0.844
0.7	0.833	0.926	0.918	0.86
0.8	0.883	0.926	0.909	0.629

In similarity study and instance creation step we use correlation function in order to distinguish the relevant and redundant features from the set of pertinent feature. In the literature, the similarity study can be more efficient when the threshold interval varies between $[0.4; 0.8]$. Hence, we tested five thresholds $\{0.4; 0.5; 0.6; 0.7; 0.8\}$ in order to choose the best results.

The performance of our proposed method is evaluated using the standard information retrieval performance measures: precision and recall metrics on support vector machine (SVM) with a polynomial basis function kernel and J48 decision tree (DT) classifiers. The experiments were repeated 10 times.

Table 2. Performances for CNS dataset.

	Number of attributes	Precision (%)	Recall (%)
DT			
Instance based feature selection	702	83(%)	91(%)
Relief algorithm	1620	73(%)	80(%)
With all features	7129	48(%)	51(%)
SVM			
Instance based feature selection	702	87(%)	87(%)
Relief algorithm	1620	74(%)	77(%)
With all features	7129	52(%)	47(%)

Table 2 compares the performance of DT and SVM for the CNS dataset, when the algorithms are trained over all the features or on the set obtained by the new approach and Relief algorithm. While comparing Relief algorithm

Table 3. Performances for Leukemia dataset.

	Number of attributes	Precision (%)	Recall (%)
DT			
Instance based feature selection	659	76(%)	80(%)
Reliefalgorithm	1502	63(%)	68(%)
With all features	7129	47(%)	53(%)
SVM			
Instance bascd feature selection	659	81(%)	83(%)
Relief algorithm	1502	67(%)	70(%)
With all features	7129	42(%)	48(%)

with the new instance feature selection approach based on the simple SVM and DT classification rule, it is evident that the later performs significantly much better. It can be further verified from Table 3 that the new algorithm performs significantly much better than the Relief. This performance is due to the fact that redundant features are eliminated and in the same time we have more data to train our algorithm. Over all the obtained results show that using feature selection before classification improves the classification performance. However feature a reduced set of features is more useful in presence of a suitable learning sample size.

The computed values or scores of recall, precision, and the F-measures are used to measure the performance of the feature selection methods. The differences between any two features selection methods may be due to chance or there is a significant difference between them. To rule out the possibility that the difference is due to chance and to confirm our conclusions, statistical hypothesis testing is used.

Here, we are interested in determining whether the mean values of a given performance measure significantly differ accordingly with the used feature selection method and classification method. A two-way Analysis of variance (ANOVA) is performed to test the difference between different features selection methods and classification methods. The first factor represent the different feature selection methods and the second represent the different classification methods. Using the ANOVA results we find that there are statistically significant differences between the different feature selection methods (p-value = 0.007).

4   Conclusion

In this paper, we studied the effect of feature selection on biological data sets and the links between the feature space and the size of the learning space. We proposed a new feature selection algorithm that performs a dimensionality reduction using Relief algorithm, which allows to rank the features in ascending order. After we apply the correlation function to extract redundant features and use this latest to increase the size of the learning space. Therefore, classification becomes easy and efficient.

References

1. Bellman, R.: Processus Adaptive Control: A Guided Tour. Princeton University Press, Princeton (1961)
2. Brahim, A.B., Bouaguel, W., Limam, M.: Combining feature selection and data classification using ensemble approaches: application to cancer diagnosis and credit scoring, ch. 24, pp. 517–532. Taylor and Francis (2014)
3. Bouaguel, W.: On Feature Selection Methods for Credit Scoring. Ph.D. thesis, Institut Superieur de Gestion de Tunis (2015)
4. Bouaguel, W., Mufti, G.B.: An improvement direction for filter selection techniques using information theory measures and quadratic optimization. Int. J. Adv. Res. Artif. Intell. 1(5), 7–11 (2012)
5. For Biotechnology Information, N. C.: Genbank growth (2008)
6. Froidevaux, C., Boulakia, S.C.: Intégration de sources de données génomiques du web
7. Golub, T.R., Slonim, D.K., Tamayo, P., Huard, C., Gaasenbeek, M., Mesirov, J.P., Coller, H., Loh, M.L., Downing, J.R., Caligiuri, M.A., Bloomfield, C.D.: Molecular classification of cancer: class discovery and class prediction by gene expression monitoring. Science 286, 531–537 (1999)
8. Guerif, S.: Rduction de dimension en apprentissage numrique non supervise. Ph.D. thesis, Universit Paris 13 (2006)
9. Kira, K., Rendell, L.A.: A practical approach to feature selection. In: Proceedings of the Ninth International Workshop on Machine Learning, pp. 249–256. Morgan Kaufmann Publishers Inc., San Francisco (1992)
10. Kurzynski, M.W., Rewak, A.: The GA-based bayes-optimal feature extraction procedure applied to the supervised pattern recognition. In: Rutkowski, L., Tadeusiewicz, R., Zadeh, L.A., Zurada, J.M. (eds.) ICAISC 2008. LNCS (LNAI), vol. 5097, pp. 620–631. Springer, Heidelberg (2008)
11. Pomeroy, S.L., Tamayo, P., Gaasenbeek, M., Sturla, L.M., Angelo, M., McLaughlin, M.E., Kim, J.Y.H., Goumnerova, L.C., Black, P.M., Lau, C., Allen, J.C., Zagzag, D., Olson, J.M., Curran, T., Wetmore, C., Biegel, J.A., Poggio, T., Mukherjee, S., Rifkin, R., Califano, A., Stolovitzky, G., Louis, D.N., Mesirov, J.P., Lander, E.S., Golub, T.R.: Prediction of central nervous system embryonal tumour outcome based on gene expression. Nature 415(6870), 436–442 (2002)
12. Richard, J., Qiang, S.: Computational Intelligence and Feature Selection: Rough and Fuzzy Approaches. John Wiley and Sons, Canada (2008)
13. Salvador, G., Julin, L., Francisco, H.: Data preprocessing in Data Mining. Janusz Kacprzyk, Polish Academy of Sciences, Warsaw, Poland. Springer (2015)
14. Yu, L., Liu, H.: Feature selection for high-dimensional data: A fast correlation-based filter solution. In: Proceedings of the International Conference on Machine Learning (ICML), pp. 856–863 (2003)

Faceted Search for Mathematics

Radu Hambasan[✉] and Michael Kohlhase

Computer Science, Jacobs University Bremen, Bremen, Germany
radu.hambasan@gmail.com

Abstract. Faceted search is one of the most practical ways to browse a large corpus of information. Information is categorized automatically for a given query and the user is given the opportunity to further refine his/her query. Many search engines offer a powerful faceted search engine, but only on the textual level. Faceted Search in the context of Math Search is still unexplored territory.

In this paper, we describe one way of solving the faceted search problem in mathematics: by extracting recognizable formula schemata from a given set of formulae and using these schemata to divide the initial set into formula classes. Also, we provide a direct application by integrating this solution with existing services.

1 Introduction

Search engines have become the prevalent tool for exploring the ever growing trove of digital data on the Internet. Although text search engines (e.g. Google or DuckDuckGo) are sufficient for most uses, they are limited when it comes to finding scientific content. STEM documents (Science, Technology, Engineering and Mathematics) contain mathematical formulae which cannot be properly indexed by a text search engine as they are structured expressions of operators (fractions, square-roots, subscripts and superscripts) and tokens.

A good math search engine is needed by several user groups. One user group would be an airline manufacturer, searching for formulae in their engineering whitepapers. In the case of research centers, like CERN, valuable time would be saved if scientists would have a fast, reliable and powerful math search engine to analyse previous related work. Still another user group is represented by university students who would be empowered by search engines, when their textbooks are not limited to text, but also include formulae. For all these applications, we first need a strong math search engine and second, a large corpus of math to index. Correspondingly, Math Information Retrieval is a small but vibrant research topic, we refer the reader to the recent Math-2 Task [1] at the NTCIR-11 IR Evaluation Campaign for an overview.

The Cornell e-Print Archive arXiv [3], is an example of such a corpus, containing over a million STEM documents from various scientific fields (Physics, Mathematics, Computer Science, Quantitative Biology, Quantitative Finance and Statistics). Zentralblatt Math (ZBMath) [17] has abstracts/reviews of all published papers (currently 3 .5 million) since 1859. A search engine for these

© Springer International Publishing Switzerland 2016
I.S. Kotsireas et al. (Eds.): MACIS 2015, LNCS 9582, pp. 406–420, 2016.
DOI: 10.1007/978-3-319-32859-1_35

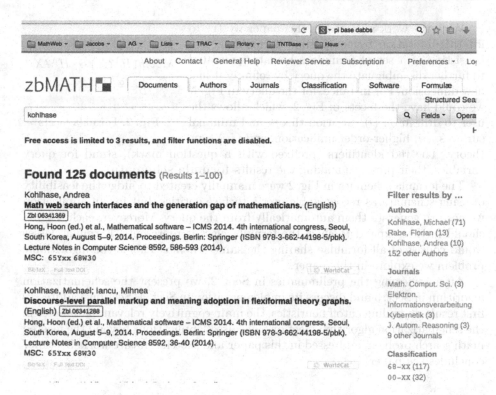

Fig. 1. Faceted search in ZBMath

must provide an expressive query language and query-refining options to be able to retrieve useful information.

Zentralblatt Math also provides a powerful search engine called "structured search". This engine is also capable of faceted search[1]. Figure 1 shows a typical situation: a user searched for a keyword (here an author name) and the faceted search generated links for search refinements (the **facets**) on the right. Currently, facets for the primary search dimensions are generated – authors, journals, MSC (Mathematics Subject Classification) [13]. This allows the user to further explore the result space, even without knowing in advance the specifics of what he/she is looking for. Unfortunately these facets are all metadata-driven and not specific to mathematics – the MSC facet is an exception, but it is rather vague because it will only provide information about the field of mathematics to which an article belongs. If the authors use formulae from another field in their paper, the results will suffer a drop in relevance.

---

[1] Faceted search originates from image search [16] and the results showed that users prefer a category-based approach to searching, even if the interface is initially unfamiliar.

With the work reported in this paper we try to lift this limitation by computing "formula facets" consisting of a set of formula schemata generated to further disambiguate the query by refining it in a new dimension. For instance, for the query in Fig. 1, we could have the facets in Fig. 2, which allows the user to drill in on (*i*) variation theory and minimal surfaces, (*ii*) higher-order unification, and (*iii*) type theory. The red identifiers (prefixed with a question mark), stand for query variables, their presence making the results **formula schemata**.

$$\int_{?M} ?\Phi(d_p?f)dvol$$

$$\lambda?X.h(H^1?X)\cdots H^n?X$$

$$\frac{?\Gamma \vdash ?A \gg ?\alpha}{?D}$$

Fig. 2. Formula facets

The formula schemata in Fig. 2 were manually created to judge the feasibility of using schemata as recognizable user interface entities, but for an application we need to generate them automatically from the query. Moreover, each schema should further expand to show the formula class it represents. Formula classes would consist of all formulae sharing the same schema. This is the algorithmic problem we explore in this paper.

After reviewing the preliminaries in Sect. 2, we present the schematization algorithm in Sect. 3 and discuss its implementation in Sect. 4. Section 5 addresses first results in finding cutoff heuristics, the main cognitively relevant parameter in the schematization algorithm. Section 6 discusses applications beyond the faceted math search problem addressed in this paper and sketches future work. Section 7 concludes the paper.

2 Preliminaries

We will now present the systems on which our work is based:

- **MathWebSearch** provides the necessary index structure for schema search.
- **Elasticsearch** provides hits in response to text queries, as well as run aggregations on the hits. These hits represent formulae to be schematized.
- **arXiv** provides a large corpus of mathematical documents that we can index and run our system on.
- **LaTeXML** converts LaTeX expressions to MathML.

As discussed in Sect. 1, the goal of this project is to develop a scalable formula schematization engine, capable of dividing a set of query hits into classes, according to the generated formula schemata.

We have set the following end-user requirements for our system:

R1. it should be able to generate formula schemata from a given set of formulae and the resulting schemata should be easily recognizable by the user.
R2. it should be able to classify the given set of formulae according to the generated schemata.
R3. the system should be massively scalable, i.e. capable of answering queries with hundreds of thousands of formulae in a matter of seconds.

At its core, the MathWebSearch [8] system (MWS) is a content-based search engine for mathematical formulae. It indexes MathML [12] formulae, using a technique derived from automated theorem proving: Substitution Tree Indexing [7]. Recently, it was augmented with full-text search capabilities, combining keyword queries with unification-based formula search. The engine serving text queries is Elasticsearch (below). From now on, in order to avoid confusion, we will refer to the core system (providing just formula query capability) as MWS and to the complete service (MWS + Elasticsearch) as TeMaSearch (Text + Math Search).

Internal to MWS, each mathematical expression is encoded as a set of substitutions based on a depth-first traversal of its Content MathML tree. Furthermore, each tag from the Content MathML tree is encoded as a TokenID, to lower the size of the resulting index. The (bijective) mapping is also stored together with the index and is needed to reconstruct the original formula. The index itself is an in-memory trie of substitution paths.

To facilitate fast retrieval, MWS stores FormulaIDs in the leaves of the substitution tree. These are integers uniquely associated with formulae, and they are used to store the context in which the respective expressions occurred. These identifiers are stored in a separate LevelDB [11] database.

MathWebSearch exposes a RESTful HTTP API which accepts XML queries. A valid query must obey the Content MathML format, potentially augmented with *qvar* variables which match any subterms. A *qvar* is a wildcard in a query, with the restriction that if two *qvars* have the same name, they must be substituted in the same way.

Elasticsearch [5] is a powerful and efficient full text search and analytics engine, built on top of Lucene [2]. It can scale massively, because it partitions data in shards and is also fault tolerant, because it replicates data. It indexes schema-free JSON documents and the search engine exposes a RESTful web interface. The query is also structured as JSON and supports a multitude of features via its domain specific language: nested queries, filters, ranking, scoring, searching using wildcards/ranges and faceted search.

arXiv is a repository of over one million publicly accessible scientific papers in STEM fields. For the NTCIR-11 challenge [8], MWS indexed over 8.3 million paragraphs (totaling 176 GB) from arXiv. We will base our queries on this large index, because it provides a rich database of highly relevant formulae. Moreover, Elasticsearch will have more formulae on which it can run aggregations, also leading to more relevant results.

An overwhelming majority of the digital scientific content is written using LaTeX or TeX [10], due to its usability and popularity among STEM researchers. However, formulae in these formats are not good candidates for searching because they do not display the mathematical structure of the underlying idea. For this purpose, conversion engines have been developed to convert LaTeX expressions to more organized formats such as MathML.

An open source example of such a conversion engine is LaTeXML [14]. The MathWebSearch project relies heavily on it, to convert arXiv documents from LaTeX to XHTML which is later indexed by MWS. It exposes a powerful API,

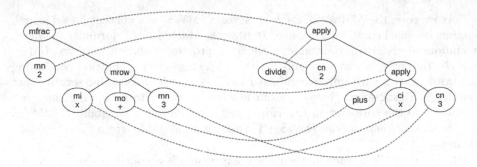

Fig. 3. The CMML/PMML parallel markup

accepting custom definition files which relate T_EX elements to corresponding XML fragments that should be generated. For the scope of this project, we are more interested in another feature of L^AT_EXML: cross-referencing between Presentation MathML and Content MathML. While converting T_EX entities to Presentation MathML trees, L^AT_EXML assigns each PMML element a unique identifier which is later referenced from the corresponding Content MathML element. In this manner, we can modify the Content MathML tree and reflect the changes in the Presentation MathML tree which can be displayed to the user.

Figure 3 illustrates the parallel markup for $\frac{2}{x+3}$. On the left side we have Presentation MathML and on the right side, Content MathML. As we can see, every Content element has a Presentation correspondent, except the divide operator, whose meaning is reflected in the structure of the displayed formula.

3   Schematization of Formula Sets

In this section, we provide a theoretical approach to the problem of generating formula schemata, by formalizing the problem and describing an efficient algorithm to solve it. First we formulate the problem at hand more carefully.

Definition 1. *Given a set $\mathcal{D}$ of documents (fragments) – e.g. generated by a search query, a **coverage** $0 < r \leq 1$, and a **width** n, the **Formula Schemata Generation** (FSG) problem requires generating a set $\mathcal{F}$ of at most n formula schemata (content MathML expressions with **qvar** elements for query variables), such that $\mathcal{F}$ covers $\mathcal{D}$ with coverage r.*

Definition 2. *We say that a set $\mathcal{F}$ of formula schemata **covers** a set $\mathcal{D}$ of document fragments, with **coverage** r, iff at least $r \cdot |\mathcal{D}|$ formulae from $\mathcal{D}$ are an instance $\sigma(f)$ of some $f \in \mathcal{F}$ for a substitution σ.*

The algorithm that we present requires a MWS index of a corpus. Given such an index, and a set $\mathcal{D}$ of formulae (as CMML expressions), we can find the set $\mathcal{F}$ in the following way:

1. Parse the given CMML expressions similarly to MWS queries, to obtain their encoded DFS representations.
2. Choose a reasonable cutoff heuristic, see below.
3. Unify each expression with the index, up to a given threshold (given by the above heuristic).
4. Keep a counter for every index path associated with the unifications. Since we only match up to a threshold, some formulae will be associated with the same path (excluding the leaves). We increase the counter each time we find a path already associated with a counter.
5. Sort these path-counter pairs by counter in descending order and take the first n (n being the width required by the FSG).
6. If the threshold depth was smaller than a formula's expression depth, the path associated with it will have missing components. We replace the missing components with qvars to generate the schema and return the result set.

Figure 4a shows a MWS index with encoded expressions which were simplified at depth 1. This means that the Content MathML representation of the formulae was truncated at depth 1 and then encoded in the index, resulting in a "simplified index".

The formulae's paths represent their depth-first traversal. Every formula can be reconstructed given its path in the index. The circles represent index nodes and the number inside represents the token's ID. When we reach a leaf node, we completely described a formula. This is encoded in the leaf node by an ID, which can be used to retrieve the formula from the database. The length of the arrows symbolizes the depth of the omitted subterms (for higher depths, we have longer

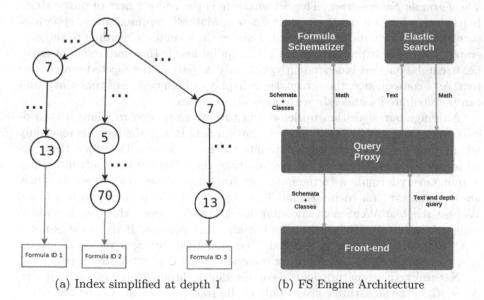

(a) Index simplified at depth 1   (b) FS Engine Architecture

Fig. 4. Aspects of the formula search system

arrows). Notice how both formula with ID 1 and formula with ID 3 show the same "path" when ignoring subterms below a cutoff depth (the simplification depth), which in this case is 1.

4   Implementation

In this section, we explain the key details of the formula classifier's implementation, the overall system architecture, as well as the challenges and trade-offs associated with the taken design decisions.

Design Overview. The full faceted search system comprises of the following components: the Formula Schematizer 4, Elasticsearch, a proxy to mediate communication between the Schematizer and Elasticsearch and a Web front-end. The architecture of the system is shown in Fig. 4b.

Once the user enters a query (which consists of keywords and a depth), the front-end forwards the request to a back-end proxy. The proxy sends the text component of the query to Elasticsearch and receives back math contained in matching documents. Afterwards, it sends the retrieved math and the depth parameter (from the original query) to the Schematizer. The Schematizer will respond with a classification of the math in formula classes, as well as the corresponding schema for each class. Finally, the proxy forwards the result to the front-end which displays it to the user.

In the following sections, we will explain the core components of the system in detail and describe the challenges faced during implementation.

The Formula Schematizer. The Schematizer is the central part of our system. It receives a set of formulae in their Content MathML representation, generates corresponding formula schemata and classifies the formulae according to the generated schemata. It provides an HTTP endpoint and is therefore self-contained, i.e. it can be queried independently, not only as part of the faceted search system. As a consequence, the Schematizer displays a high degree of versatility, and can be integrated seamlessly with other applications.

Although our algorithm works well in theory, we needed to adapt it considering various MathWebSearch implementation details, e.g. the index is read-only (therefore we cannot store extra data into the index nodes). Therefore, the overall idea/theory is the same, but now we take the following shortcut: instead of unifying every formula with the index, we just pretend we do and instead generate a "signature" for each formula. This signature is the path shown in Fig. 4a. We use the MathWebSearch encoding for MathML nodes, where each node is assigned an integer ID based on its tag and text content. If the node is not a leaf, then only the tag is considered. The signature will be a vector of integer IDs, corresponding to the pre-order traversal of the Content MathML tree.

Naturally, the signature depends on the depth chosen for the cutoff heuristic. At depth 0, the signature consists only of the root token of the Content MathML expression. At full depth (the maximum depth of the expression), the signature is the same as the depth-first traversal of the Content MathML tree.

Based on these computed signatures, we divide the input set of formulae into formula classes, i.e. all formulae with the same signature belong to the same class. For this operation we keep an in-memory hash table, where the keys are given by the signatures and the values are sets of formulae which have the signature key. After filling the hash table, we sort it according to the number of formulae in a given class, since the signatures which cover the most formulae should come at the beginning of the reported result.

The Schematizer caller can place an optional limit on the maximum number of schemata to be returned. If such a limit was specified, we apply it to our sorted list of signatures and take only the top ones.

As a last step, we need to construct Content MathML trees from the signatures, to be able to show the schemata as formulae to the user. We are able to do this because we know the arity of each token and the depth used for cutoff. The tree obtained after the reconstruction might be incomplete, so we insert query variables in place of missing subtrees. We finally return these Content MathML trees with query variables (the formula schemata), together with the formulae which they cover.

Presentation by Replacement. After obtaining the schemata and formula classes, we need to be able to display the result to the user. One possibility would be to have the Schematizer return Content MathML expressions for the schemata and use an XSL stylesheet [15] to convert them to Presentation MathML. This approach would unfortunately generate unrecognizable schemata due to the inherent ambiguity of CMML. For instance, a csymbol element can be represented in several different ways depending on the notation being used. Additionally, we cannot reliably foresee all possible rules that should be implemented in the stylesheet and as a consequence some formulae will be wrongly converted.

Since the XSL conversion is unreliable, we will make use of the cross-reference system provided by LATEXML, as discussed before. Instead of returning Content MathML expressions, the Schematizer will use the first formula in each class as a template and "punch holes into it", effectively returning the ID of the nodes that are to be substituted with query variables. We will use this IDs to replace the referenced PMML nodes with `<mi>` nodes representing the qvars.

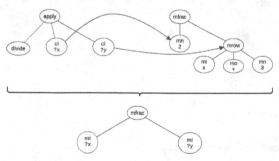

Figure 5 shows the presentation by replacement technique for a given schema. The Schematizer returned a schema which was checked against the first formula in its class ($\frac{2}{x+3}$) to generate two substitutions, marked with red on the left side. Due to the cross-reference system provided by LATEXML, we are able to find the corresponding PMML elements and substitute them with `<mi>` tokens. The result will be displayed to the user as $\frac{?x}{?y}$.

Fig. 5. Presentation by replacement

Performance. We designed the Schematizer to be a very lightweight daemon, both as memory requirements and as CPU usage. To test if we achieved this goal, we benchmarked it on a server running Linux 3.2.0, with 10 cores (Intel Xeon CPU E5-2650 2.00 GHz) and 80 GB of RAM.

We obtained the 1123 expressions to be schematized by querying Elasticsearch with the keyword "Fermat". While the overall time taken by the faceted search engine was around 5 s, less than a second was spent in the Schematizer. Also, the CPU utilized by the Schematizer never rose higher than 15 % (as indicated by the top utility). Asymptotically, the algorithm would run in $O(N)$ time, where N is the number of input formulae. We are able to reach linear time performance, because each formula is processed exactly once and the signature is stored in a hash table, as discussed in Sect. 4.

Due to its design principles, the Schematizer is almost indefinitely scalable, because it does not require shared state between formulae and can therefore be implemented as a MapReduce [4] job, where mappers compute the signature of assigned formulae and reducers assemble the signature hash table. However, the unification algorithm currently used by MathWebSearch is linear is in the number

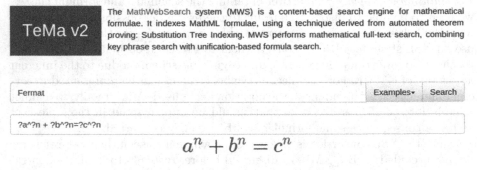

Fig. 6. TeMa v2 query results

of nodes of the Content MathML expression that is being unified. Therefore, the current search engine implementation would pose challenges for scalability, although the Schematizer itself will be able to extend easily.

The Front-End. We have integrated the Schematizer into a Math Search Engine which is capable of mathematical faceted search.

The TemaV2 front-end extends TeMaSearch to be able to perform mathematical faceted search. It is intended for users who want to filter query results based on a given facet (formula schema in this case). The look and feel is similar to the previous version of TeMaSearch, as shown in Fig. 6, where the first input field is used to specify keywords and the second one is used to specify LATEX-style formulae for the query. When returning results, a "Math Facets" menu will be presented to the user. Figure 6 shows the results of a query for "Fermat" and $?a^{?n} + ?b^{?n} = ?c^{?n}$. Besides the regular TeMaSearch results, the user is also presented with a "Math Facets" section.

When the "Math Facets" section is expanded the user can see the top 10 schemata (ranked with respect to their coverage), as shown in Fig. 7. We have also implemented a "search-on-click" functionality that allows the user to do a fresh search using the clicked schema and the initial keyword, which effectively filters the current results.

5   Finding a Cutoff Heuristic

To generate formula schemata, we must define a "cutoff heuristic", which tells the program when two formulae belong to the same schema class. If there is no heuristic, two formulae would belong to the same class, only if they were identical. However, we want formulae that have something in common to be grouped together, even if they are not perfectly identical. The cutoff heuristic is the parameter in the schematization algorithm that determines the suitability of schemata for the various information access tasks at hand. As this is essentially a user-driven, cognitive task it is not a priori clear what cutoff heuristics will perform best.

To explore the space of heuristics, we have implemented a special front-end and used that to evaluate heuristics for the math search task discussed above. As a proper user-level evaluation was beyond the scope of this paper, we have implemented various heuristics and discussed them with the ZBMath group in the context of the ZBMath corpus, this led to the development of the dynamic cutoff heuristic presented at the end of this section.

A Schema Evaluation Front-End. The SchemaSearch front-end provides just a textual search input field. It is intended for users who want an overview of the formulae contained in a corpus. As shown in Fig. 8a, the user can enter a set of keywords for the query, as well as a schema depth, which defaults to 3. The maximum result size is not accessible to the user, to prevent abuses and reduce server load. There is also an "R" checkbox which specifies if the cutoff

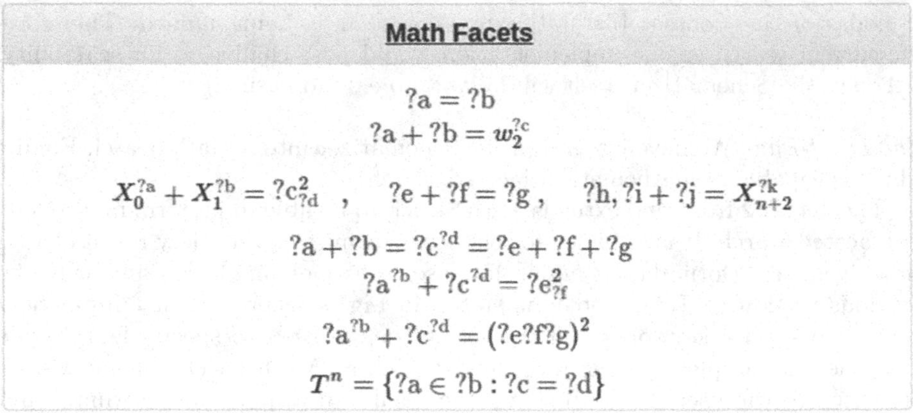

Math Facets

$$?a = ?b$$
$$?a + ?b = w_2^{?c}$$

$$X_0^{?a} + X_1^{?b} = ?c_{?d}^2 \ , \quad ?e + ?f = ?g \ , \quad ?h, ?i + ?j = X_{n+2}^{?k}$$

$$?a + ?b = ?c^{?d} = ?e + ?f + ?g$$
$$?a^{?b} + ?c^{?d} = ?e_{?f}^2$$

$$?a^{?b} + ?c^{?d} = (?e?f?g)^2$$

$$T^n = \{?a \in ?b : ?c = ?d\}$$

Fig. 7. Math facets in TeMa v2

depth should be absolute or relative. If relative, the depth should be given in percentages.

Figure 8a shows the formula schemata at depth 3, over the arXiv corpus, for a query containing the keyword "Kohlhase". By default, the top 40 schemata are shown, but the results are truncated for brevity. The bold number on the left side of each result item indicates how many formulae are present in each formula class. For instance, the third schema represents a formula class containing 10 formulae. The entities marked in blue are query variables (qvars).

Figure 8b shows the expansion of a formula class. There are 22 formulae in the class given by this particular math schema, as indicated by the count on the left upper side, out of which only ten are shown. We can see 2 unnamed query variables marked with blue as $?a$ and $?b$. By seeing the schema, the user can form an impression about the general structure of the formulae from that class. After expanding the class, the listing of concrete formulae appears. If the user clicks on one of them, he is redirected to the source document from which that expression was extracted.

Another class expansion which showcases the schematization can be seen in Fig. 8c. By seeing this schema, the user can abstract away the complexity of the formulae and obtain a "summary" of the meaning behind it. Also, by expanding the class he can explore several related formulae easily, because they are grouped together.

Dynamic Cutoff. We have experimented with several possibilities for the heuristic and found out that a *dynamic cutoff* which preserves the operators leads to more intuitive results. We can identify the operators by looking at the first child of the `apply` token in the CMML tree. The user is given the option to have an absolute (fixed) or relative (depending on the depth of the CMML tree) cutoff for the operands.

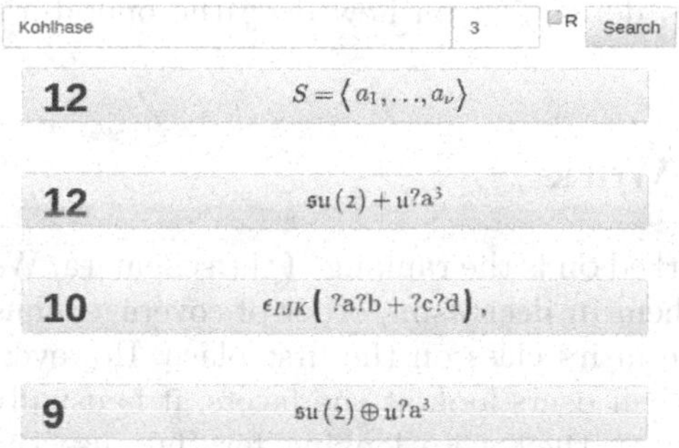

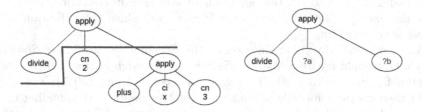

(a) Faceted Results at depth 3

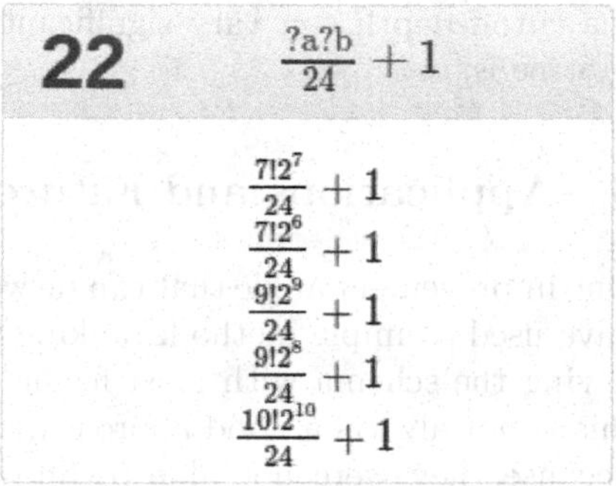

(b) Expansion of a Formula Class 1

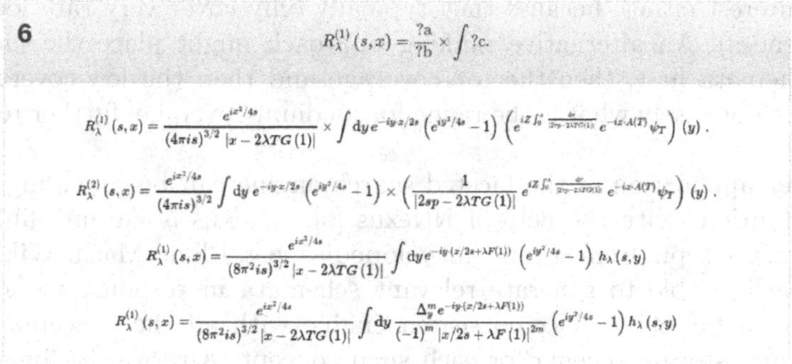

(c) Expansion of a Formula Class 2

Fig. 8. Generated schemata (Color figure online)

Fig. 9. Dynamic cutoff

Figure 9 illustrates this heuristic at depth 1. The divide element was kept, because it was the first child of apply, while the other children were removed. If we were to use a depth of 2, the plus element would also be included in the schema.

This heuristic is not simply keeping another tree node uncut. If the current node is an operator, it can also have multiple levels of children and therefore we need to keep that entire subtree from being cut. What this means, is that the cutoff depth can vary significantly, depending on how deep the operator's subtree is.

6   Applications and Future Work

One improvement angle that can be worked on is the ranking of the schemata. We have used a simple method, ranking them in decreasing order of coverage, thus having the schema with most formulae in its class on the first place. However, this is not always a good approach. When users look at the facets, it is usually because they were not able to find what they were looking for (because the result set is too large). The first schemata cover most of the formulae users have already looked at, so they are not of interest. However, the last schemata are not of interest either, because they typically only cover very rare formulae (1-2 occurrences). An alternative ranking approach might place the *medium-coverage* schemata first, then the top-coverage and then the low-coverage. In order to define precisely what is the range for medium-coverage, further research is required.

One other application of the faceted search engine can be providing mathematical definitions with the help of NNexus [6]. NNexus is an auto-linker for mathematical concepts from several encyclopedias, e.g. PlanetMath, Wikipedia. Assuming we are able to generate relevant schemata in response to keyword queries, we can target the faceted search engine with all the concepts stored by NNexus and store a schema for each such concept. Afterwards, for a given query, we can obtain the schema and check it against our stored set of schemata. If we find it, we can link the given expression to its mathematical definition. Given a large number of stored concepts and a high schemata relevance, the user should be able to see the definition of any encountered formulae on the Web. For example, hovering over $a^2 + b^2 = c^2$ will show the definition of the Pythagorean theorem.

Another, more direct, application of the Schematizer would be *Similarity Search*. One could create a MathWebSearch based search engine, which accepts an input formula and a similarity degree (between 0 % and 100 %). The engine would then create a formula schema at a relative depth corresponding to the similarity degree and use this schema to search the corpus. This approach defines the similarity between two formulae as the percentage of the CMML tree depth that they share.

Last, but not least, we will need to invest in a full user-level evaluation of the utility of formula facets, and the influence of various cutoff heuristics on that.

7   Conclusion

We have presented the design and implementation of a system capable of mathematical faceted search. Moreover, we have described a general-purpose scalable Schematizer which can generate intuitive and recognizable formula schemata and divide expressions into formula classes according to said schemata. Consequently, we have successfully addressed all challenges outlined in Sect. 2.

Although the Schematizer provides recognizable formulae, some queries to SchemaSearch (e.g. using an author as keyword) provide hits with a very low relevance. This is because we cannot distinguish between the work of the author and work where the author is cited at the textual level. As a consequence, searching for "Fermat" would also show formulae from papers where Fermat was cited and if these papers are numerous, as it happens with known authors, would provide the user with misleading results. This suggests that a better source of mathematical expressions might be required for the SchemaSearch demo.

The implementation of the Schematizer presented here is licensed under GPL v3.0 and code is available at http://github.com/KWARC/mws/.

Acknowledgements. This work has been supported by the Leibniz Association under grant SAW-2012-FIZ_KA-2 (Project MathSearch). The authors gratefully acknowledge fruitful discussions with Fabian Müller, Wolfram Sperber, and Olaf Teschke in the MathSearch Project, which led to this research (the ZBMath information service uses faceted search on the non-formula dimensions very successfully) and clarified the requirements from an application point of view.

References

1. Aizawa, A., et al.: NTCIR-11 Math-2 task overview. In: Kando, N., Joho, H., Kishida, K. (eds.) NTCIR Workshop 11 Meeting, pp. 88–98. NII, Tokyo (2014). http://research.nii.ac.jp/ntcir/workshop/OnlineProceedings11/pdf/NTCIR/OVERVIEW/01-NTCIR11-OV-MATH-AizawaA.pdf
2. Apache Lucene. https://lucene.apache.org/. Accessed 4 Oct 2015
3. arxiv.org e-Print archive. http://www.arxiv.org. Accessed 12 June 2012
4. Dean, J., Ghemawat, S.: MapReduce: simplified dataprocessing on large clusters (2004)
5. Elastic Search. http://www.elasticsearch.org/. Accessed 7 Dec 2014
6. Ginev, D., Corneli, J.: NNexus reloaded. In: Watt, S.M., Davenport, J.H., Sexton, A.P., Sojka, P., Urban, J. (eds.) CICM 2014. LNCS, vol. 8543, pp. 423–426. Springer, Heidelberg (2014)
7. Graf, P.: Substitution tree indexing. In: Hsiang, J. (ed.) Rewriting Techniques and Applications. LNCS, pp. 117–131. Springer, Heidelberg (1994)
8. Hambasan, R., Kohlhase, M., Prodescu, C.: MathWeb-search at NTCIR-11. In: Kando, N., Joho, H., Kishida, K. (eds.) NTCIR Workshop 11 Meeting, pp. 114–119. NII, Tokyo (2014). http://research.nii.ac.jp/ntcir/workshop/OnlineProceedings11/pdf/NTCIR/Math-2/05-NTCIR11-MATH-HambasanR.pdf
9. Kando, N., Joho, H., Kishida, K. (eds.) NTCIR Workshop 11 Meeting. NII, Tokyo, Japan (2014)
10. LaTeX - A document preparation system. https://www.latex-project.org/. Accessed 4 Oct 2015
11. LevelDB. http://leveldb.org/. Accessed 21 Dec 2014
12. Mathematical Markup Language. http://www.w3.org/TR/MathML3/
13. Mathematics Subject Classification (MSC) SKOS (2012). http://msc2010.org/resources/MSC/2010/info/. Accessed 31 Aug 2012

14. Miller, B.: LaTeXML: A $\LaTeX$ to XML Converter. http://dlmf.nist.gov/LaTeXML/. Accessed 12 Mar 2013

15. XSLT for Presentation MathML in a Browser. 20 December 2000. http://dpcarlisle.blogspot.de/2009/12/xslt-for-presentation-mathml-in-browser.html#uds-search-results. Accessed 4 Apr 2015

16. Yee, K.-P., et al.: Faceted metadata for image search, browsing. In: Proceedings of the SIGCHI Conference on Human Factors in Computing Systems. ACM Press, pp. 401–408 (2003). http://dl.acm.org/citation.cfm?id=642681

17. Zentralblatt Math Website. http://zbmath.org/. Accessed 7 Dec 2014

Evaluation of a Predictive Algorithm for Converting Linear Strings to Mathematical Formulae for an Input Method

Shizuka Shirai$^{(\boxtimes)}$ and Tetsuo Fukui

Mukogawa Women's University, Nishinomiya, Japan
shirai@mukogawa-u.ac.jp, fukui@mukogawa-u.ac.jp

Abstract. Recently, some computer-aided assessment (CAA) systems are able to assess learner's answers using mathematical expressions. However, the standard input method for mathematics is cumbersome for novice learners. In 2011, the last author Fukui proposed a new mathematical input method similar to the ones used for inputting Japanese characters in many systems. This method allows users to input mathematical expressions using colloquial-style mathematical string. However, users must convert each element contained in the colloquial-style mathematical string. In this study, we propose a predictive algorithm for converting the whole mathematical formulae.

Keywords: Math input method · Predictive conversion · Machine learning

1 Introduction

Recently, computer-aided assessment (CAA) systems have been utilized in mathematics education, with some CAA systems able to assess learner's answers using mathematical expressions. However, the standard input method for mathematics is cumbersome for novice learners [1].

In 2011, Professor Tetsuo Fukui of Mukogawa Women's University proposed a new mathematical input method [2,3] that allows users to input mathematical expressions using colloquial-style linear strings. This is converted to the desired two-dimensional mathematical expression in a similar way to how Japanese characters are created in many operating systems. This method enables users to input almost any mathematical expression without learning new command syntax. However, users must convert each element contained in the colloquial-style mathematical string in order, from left to right.

To improve this situation, we have proposed a predictive algorithm that uses a structured perceptron mainly for natural language processing [4]. We have examined the prediction accuracy using two parameter sets on an evaluation dataset containing 800 mathematical formulae from a mathematics textbook [5]. Our algorithm achieved an accuracy of up to 95.0 % in terms of the top ten ranking.

© Springer International Publishing Switzerland 2016
I.S. Kotsireas et al. (Eds.): MACIS 2015, LNCS 9582, pp. 421–425, 2016.
DOI: 10.1007/978-3-319-32859-1_36

However, we also found that the score parameter continues to rise linearly during the learning process.

The present study addresses this shortcoming by improving the efficiency of mathematical input. Our proposed system uses an intelligent system to convert from linear strings to mathematical formulae with a predictive algorithm that has stable score parameters.

2   Predictive Conversion [4]

2.1   Linear String Rules

The linear string rule for a mathematical expression can be described as follows: Set the key letters (or words) corresponding to the elements of a mathematical expression linearly in the order of colloquial (or reading) style, without considering two-dimensional placement and delimiters.

In other words, a key letter (or word) consists of the ASCII code(s) corresponding to the initial or clipped form (such as the LATEX-form) of the objective mathematical symbol. Therefore, one key often supports many mathematical symbols. For example, when a user wants to input β^2, the linear string is denoted by "b2". Here, "b" denotes the "beta" symbol, and there is no need to include a power sign (such as the caret symbol (^)). In the case of $\frac{1}{\beta^2+4}$, the linear string is denoted by "1/b2+4", where it is not necessary to surround the denominator (which is generally the operand of an operator) by parentheses, because they are never printed.

2.2   Design of Intelligently Predictive Conversion

We have proposed a predictive algorithm that converts a linear string s into the most suitable mathematical expression y_p. For prediction purposes, we devised a method by which each candidate for selection was ranked in terms of its suitability. Our method uses a function $\text{Score}(y)$ to allocate a score that is proportional to the probability of a mathematical expression y occurring. This enables us to predict the candidate y_p using Eq. (1) as being the most suitable expression with the maximum score. Here, $Y(s)$ in Eq. (1) is the totality of all possible mathematical expressions converted from s, which is calculated by three procedures: key separation, structure analysis, and applying all the possible mathematical elements.

$$y_p \text{ s.t. } \text{Score}(y_p) = \max\{\text{Score}(y)|y \in Y(s)\} \tag{1}$$

A mathematical expression consists of mathematical symbols, namely numbers, variables, or *operators*[1], together with the operating relations between a specific operator and part of the expression. Therefore, we represent a mathematical expression by a tree structure consisting of nodes and edges, which

---

[1] In this article, "operator" is used in the sense of operating on, i.e., performing actions on elements in terms of their arrangements for two-dimensional mathematical notation.

correspond to the symbols and operating relations, respectively. In other words, any math expression y is characterized by all the nodes and edges included in y. We identify each node or edge as a mathematical element in a formula.

2.3 Predictive Algorithm

Let us assume that the probability of a certain math element occurring is proportional to its frequency of use. Then, the probability of occurrence of a mathematical expression y, which is possibly converted from a given string s, is estimated from the total score of all the math elements included in y. Given the numbering of each element from 1 to F_{total}, which is the total number of all elements, let θ_f be the score of the $f(=1, \cdots, F_{total})$-th element, and let $x_f(y)$ be the number of times the f-th element is included in y. Then, Score(y) in Eq. (1) is estimated by Eq. (2), where the score vector $\boldsymbol{\theta}^T = (\theta_1, \cdots, \theta_{F_{total}})$ and the F_{total}-dimensional vector $\boldsymbol{X} = (x_f(y))$, $f = 1, \cdots, F_{total}$.

$$h_\theta\left(\boldsymbol{X}(y)\right) = \boldsymbol{\theta}^T \cdot \boldsymbol{X}(y) = \sum_{f=1}^{F_{total}} \theta_f x_f(y) \tag{2}$$

Equation (2) is in agreement with the hypothesis function of linear regression, and $\boldsymbol{X}(y)$ is referred to as the characteristic vector of y. To solve our linear regression problem and predict the probability of a mathematical expression occurring, we conduct supervised machine learning on m elements of a training dataset $\{(s_1, y_1), (s_2, y_2), \cdots, (s_m, y_m)\}$. Our learning algorithm for the optimized score vector is performed by the following four-step procedure:

Step 1. Initialization: $\boldsymbol{\theta} = \boldsymbol{0}$, $i = 1$
Step 2. Decision of a candidate: y_p s.t. $h_\theta\left(\boldsymbol{X}(y_p)\right) = \max\{h_\theta\left(\boldsymbol{X}(y)\right)|y \in Y(s_i)\}$
Step 3. Training parameter: if$(y_p \neq y_i)$ {

$$\begin{aligned}\theta_f &:= \theta_f + 1 \quad \text{for } \{f \leq F_{total}|x_f(y_i) > 0\} \\ \theta_{\bar{f}} &:= \theta_{\bar{f}} - 1 \quad \text{for } \{\bar{f} \leq F_{total}|x_{\bar{f}}(y_p) > 0\}\end{aligned} \tag{3}$$

}
Step 4. if$(i < m)$ { i=i+1; go to **Step 2** for repetition.}
    else {Output θ and end.}

This learning algorithm is very simple, and is similar to machine learning using a structured perceptron for natural language processing.

3 Main Algorithm

We have examined the prediction accuracy using two score learning parameter sets on an evaluation dataset containing 800 mathematical formulae from a mathematics textbook [5]. The two parameter sets of θ for scoring were trained by the following two algorithms, which were programed in Java on a desktop computer (MacOS 10.9, 3.2 GHz Intel core i3, 8 GB memory):

Algorithm 1. Step 1–Step 4, using Eq. (3).
Algorithm 2. Step 1–Step 4, with **Step 3** using

$$\begin{aligned}
\theta_f &:= \theta_f + 2 \quad \text{for } \{f \leq F_{total} | x_f(y_i) > 0\} \\
\theta_{\bar{f}} &:= \theta_{\bar{f}} - 1 \quad \text{for } \{\bar{f} \leq F_{total} | x_{\bar{f}}(y_p) > 0\}
\end{aligned} \tag{4}$$

in place of Eq. (3).

In the experimental evaluation, we measured the ratio of correct predictions from among 100 test datasets after learning the parameters with Algorithms 1 and 2 using a training dataset consisting of another 700 formulae.

The results show that the prediction accuracy of "Best 10", namely the top ten ranking, was about 89.2 % with Algorithm 1. However, Algorithm 2 achieved an average prediction accuracy of approximately 95.0 % for the top ten ranking. This accuracy is sufficient for a mathematical input interface. However, the score parameter continues to rise while Algorithm 2 is undergoing learning [4].

In this study, we propose the following Algorithm 2' to overcome the problem suffered by Algorithm 2.

Algorithm 2'. Step 1–Step 4, where **Step 3** uses Eq. (5).

$$\begin{aligned}
\text{if}(\theta_f < S_{\max})\{\theta_f &:= \theta_f + 2 \quad \text{for } \{f \leq F_{total} | x_f(y_i) > 0\}\} \\
\theta_{\bar{f}} &:= \theta_{\bar{f}} - 1 \quad \text{for } \{\bar{f} \leq F_{total} | x_{\bar{f}}(y_p) > 0\}
\end{aligned} \tag{5}$$

Here, $S_{\max}$ in Eq. (5) is a suitable upper bound for any mathematical element score.

The machine learning results with Algorithm 2 and Algorithm 2' in the case that $S_{\max} = 20$ are given in Table 1 for various sizes of the training dataset. It can be seen that the accuracy of the "Best 1" with Algorithm 2' was approximately 68.3 % after being trained 700 times. This algorithm achieved 90.5 % accuracy for the top three ranking, and 96.2 % for the top ten ranking. With a training set of size 700, there is no statistically significant difference (at the 5 % level) between the results for Algorithm 2 and those for Algorithm 2' in the "Best 1", "Best 3", or "Best 10" cases. Additionally, the learning curves for both algorithms

Table 1. Prediction accuracy using Algorithms 2 and 2'

Training number	Best 1 (%)		Best 3 (%)		Best 10 (%)		Correct Score	
	Algo. 2	Algo. 2'	Algo. 2	Algo. 2'	Algo. 2	Algo. 2'	Algo. 2	Algo. 2'
0	25.9(3.8)	25.9(3.8)	41.3(4.4)	41.3(4.4)	52.3(4.3)	52.3(4.3)	2.8(0.1)	2.8(0.1)
100	53.3(14.6)	54.2(13.8)	82.5(6.4)	82.6(6.4)	88.5(4.3)	88.7(4.3)	307.0(58.1)	283.0(74.0)
200	60.3(5.0)	65.7(6.8)	86.1(4.2)	87.7(3.4)	91.7(3.2)	93.0(2.7)	568.9(99.4)	428.6(110.6)
300	64.1(5.1)	69.5(6.1)	89.1(3.2)	88.3(3.1)	93.8(2.9)	94.0(3.0)	964.3(186.8)	494.1(134.7)
400	67.7(5.7)	67.9(6.3)	90.1(3.1)	88.8(2.4)	94.1(3.1)	94.3(2.8)	1103.4(75.2)	536.7(148.8)
500	67.6(5.7)	69.2(5.6)	90.6(2.9)	89.8(3.0)	94.5(2.8)	95.2(2.7)	1290.6(99.8)	566.2(162.7)
600	69.1(4.6)	70.6(5.2)	90.8(2.7)	90.9(2.7)	94.3(2.5)	95.9(2.5)	1492.2(106.5)	590.0(169.4)
700	68.5(6.0)	68.3(6.1)	91.1(2.5)	90.5(2.8)	95.0(2.5)	96.2(2.3)	1692.9(114.7)	608.0(180.0)

Numbers in parentheses denote *SD*.

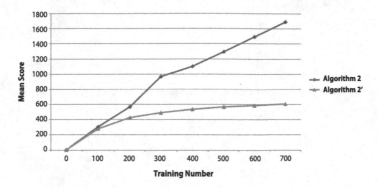

Fig. 1. Change in score parameter given by Algorithms 2 and 2'

change at the same skill rate for each of these cases. The mean scores among the correct expressions ("correct score" for short) in the test dataset for each training number are presented in the fifth column of Table 1 and illustrated in Fig. 1. Whereas the correct score with Algorithm 2 increases in proportion to the training number n (decision coefficient: $R^2 = 0.98$), the correct score with Algorithm 2' only increases at a rate of $\log n$ ($R^2 = 0.96$).

4   Conclusion and Future Work

We have proposed a predictive algorithm with a prediction accuracy of 96.2 % for the top ten ranking by improving a previous algorithm in terms of a structured perceptron for stable score parameter learning. The mean CPU time for predicting each mathematical expression was 0.44 s (SD = 6.75).

Finally, the most important avenues for future research are to shorten the time for prediction and develop an intelligent mathematical input interface by implementing our proposed predictive algorithm.

This work was supported by JSPS KAKENHI Grant Number 26330413.

References

1. Sangwin, C.J.: Computer Aided Assessment of Mathematics Using STACK. In: Proceedings of ICME, vol.12 (2012)
2. Fukui, T.: An intelligent method of interactive user interface for digitalized mathematical expressions (in Japanese). In: RIMS Kokyuroku, vol. 1780, pp. 160–171 (2012)
3. Shirai, S., Fukui, T.: Development and evaluation of a web-based drill system to master basic math formulae using a new interactive math input method. In: Hong, H., Yap, C. (eds.) ICMS 2014. LNCS, vol. 8592, pp. 621–628. Springer, Heidelberg (2014)
4. Fukui, T., Shirai, S.: Predictive algorithm from linear string to mathematical formulae for math input method. In: Proceedings of 21st Conference on Applications of Computer Algebra 2015 in Kalamata, Greece, pp. 17–22 (2015)
5. Iidaka, S., Matsumoto, Y., et al.: Mathematics I, 001, TOKYO SHOSEKI (2012)

Algorithm Engineering
in Geometric Computing

Linear Programs and Convex Hulls Over Fields of Puiseux Fractions

Michael Joswig[✉], Georg Loho[✉], Benjamin Lorenz[✉],
and Benjamin Schröter[✉]

Institut Für Mathematik, TU Berlin, MA 6-2, Str. des 17. Juni 136,
10623 Berlin, Germany
{joswig,loho,lorenz,schroeter}@math.tu-berlin.de

Abstract. We describe the implementation of a subfield of the field of formal Puiseux series in `polymake`. This is employed for solving linear programs and computing convex hulls depending on a real parameter. Moreover, this approach is also useful for computations in tropical geometry.

Keywords: Linear programming over ordered fields · Convex hull computation over ordered fields · Rational functions · Puiseux series · Tropical convex hull computation

1 Introduction

It is well known and not difficult to see that the standard concepts from linear programming (LP), e.g., the Farkas Lemma and LP duality, carry over to an arbitrary ordered field; e.g., see [7, Sect. 2] or [16, Sect. 2.1]. Traces of this can already be found in Dantzig's monograph [8, Chapter 22]. This entails that any algorithm whose correctness rests on these LP corner stones is valid over any ordered field. In particular, this holds for the simplex method and usual convex hull algorithms. A classical construction, due to Hilbert, turns a field of rational functions, e.g., with real coefficients, into an ordered field; see [30, Sect. 147]. In [16] Jeroslow discussed these fields in the context of linear programming in order to provide a rigorous foundation of the so-called "big M method". The purpose of this note is to describe the implementation of the simplex method and of a convex hull algorithm over fields of this kind in the open source software system `polymake` [14].

Hilbert's ordered field of rational functions is a subfield of the field of formal Puiseux series $\mathbb{R}\{t\}$ with real coefficients. The latter field is real-closed by the Artin–Schreier Theorem [27, Theorem 12.10]; by Tarski's Principle (cf. [28]) this implies that $\mathbb{R}\{t\}$ has the same first order properties as the reals. The study of

M. Joswig—Partially supported by Einstein Foundation Berlin and Deutsche Forschungsgemeinschaft (DFG) within the Priority Program 1489 "Experimental Methods in Algebra, Geometry and Number Theory".

© Springer International Publishing Switzerland 2016
I.S. Kotsireas et al. (Eds.): MACIS 2015, LNCS 9582, pp. 429–445, 2016.
DOI: 10.1007/978-3-319-32859-1_37

polyhedra over $\mathbb{R}\{\!\{t\}\!\}$ is motivated by tropical geometry [9], especially tropical linear programming [2]. The connection of the latter with classical linear programming has recently lead to a counter-example [1] to a "continuous analogue of the Hirsch conjecture" by Deza, Terlaky and Zinchenko [10]. In terms of parameterized linear optimization (and similarly for the convex hull computations) our approach amounts to computing with sufficiently large (or, dually, sufficiently small) positive real numbers. Here we do *not* consider the more general algorithmic problem of stratifying the parameter space to describe all optimal solutions of a linear program for *all* choices of parameters; see, e.g., [17] for work into that direction.

This paper is organized as follows. We start out with summarizing known facts on ordered fields. Then we describe a specific field, $\mathbb{Q}\{t\}$, which is the field of rational functions with rational coefficients and rational exponents. This is a subfield of $\mathbb{Q}\{\!\{t\}\!\}$, which we call the field of *Puiseux fractions*. It is our opinion that this is a subfield of the formal Puiseux series which is particularly well suited for exact computations with (some) Puiseux series; see [22] for an entirely different approach. In the context of tropical geometry Markwig [23] constructed a much larger field, which contains the classical Puiseux series as a proper subfield. For our applications it is relevant to study the evaluation of Puiseux fractions at sufficiently large rational numbers. In Sect. 3 we develop what this yields for comparing convex polyhedra over $\mathbb{R}\{\!\{t\}\!\}$ with ordinary convex polyhedra over the reals. The tropical geometry point of view enters the picture in Sect. 4. We give an algorithm for solving the dual tropical convex hull problem, i.e., the computation of generators of a tropical cone from an exterior description. Allamigeon, Gaubert and Goubault gave a combinatorial algorithm for this in [4], while we use a classical (dual) convex hull algorithm and apply the valuation map. The benefit of our approach is more geometric than in terms of computational complexity: in this way we will be able to study the fibers of the tropicalization map for classical versus tropical cones for specific examples. Section 5 sketches the `polymake` implementation of the Puiseux fraction arithmetic and the LP and convex hull algorithms. The LP solver is a dual simplex algorithm with steepest edge pivoting, and the convex hull algorithm is the classical beneath-and-beyond method [11,18]. An overview with computational results is given in Sect. 6.

2   Ordered Fields and Rational Functions

A field $\mathbb{F}$ is *ordered* if there is a total ordering $\leq$ on the set $\mathbb{F}$ such that for all $a, b, c \in \mathbb{F}$ the following conditions hold:

(i) if $a \leq b$ then $a + c \leq b + c$,
(ii) if $0 \leq a$ and $0 \leq b$ then $0 \leq a \cdot b$.

Any ordered field necessarily has characteristic zero. Examples include the rational numbers $\mathbb{Q}$, the reals $\mathbb{R}$ and any subfield in between.

Given an ordered field $\mathbb{F}$ we can look at the ring of univariate polynomials $\mathbb{F}[t]$ and its quotient field $\mathbb{F}(t)$, the field of rational functions in the indeterminate

t with coefficients in $\mathbb{F}$. On the ring $\mathbb{F}[t]$ we obtain a total ordering by declaring $p < q$ whenever the leading coefficient of $q - p$ is a positive element in $\mathbb{F}$. Extending this ordering to the quotient field by letting

$$\frac{u}{v} < \frac{p}{q} \; :\Longleftrightarrow \; uq < vp,$$

where the denominators v and q are assumed positive, turns $\mathbb{F}(t)$ into an ordered field; see, e.g., [30, Sect. 147]. This ordered field is called the "Hilbert field" by Jeroslow [16].

By definition, the exponents of the polynomials in $\mathbb{F}[t]$ are natural numbers. However, conceptually, there is no harm in also taking negative integers or even arbitrary rational numbers as exponents into account, as this can be reduced to the former by clearing denominators and subsequent substitution. For example,

$$\frac{2t^{3/2} - t^{-1}}{1 + 3t^{-1/3}} = \frac{2t^{5/2} - 1}{t + 3t^{2/3}} = \frac{2s^{15} - 1}{s^6 + 3s^4}, \tag{1}$$

where $s = t^{1/6}$. In this way that fraction is written as an element in the field $\mathbb{Q}(t^{1/6})$ of rational functions in the indeterminate $s = t^{1/6}$ with rational coefficients. Further, if $p \in \mathbb{F}(t^{1/\alpha})$ and $q \in \mathbb{F}(t^{1/\beta})$, for natural numbers α and β, then the sum $p + q$ and the product $p \cdot q$ are contained in $\mathbb{F}(t^{1/\gcd(\alpha,\beta)})$. This shows that the union

$$\mathbb{F}\{t\} = \bigcup_{\nu \geq 1} \mathbb{F}(t^{1/\nu}) \tag{2}$$

is again an ordered field. We call its elements *Puiseux fractions*. The field $\mathbb{F}\{t\}$ is a subfield of the field $\mathbb{F}\{\!\{t\}\!\}$ of *formal Puiseux series*, i.e., the formal power series with rational exponents of common denominator. For an algorithmic approach to general Puiseux series see [22].

The map val which sends the rational function p/q, where $p, q \in \mathbb{F}[t^{1/\nu}]$, to the number $\deg_t p - \deg_t q$ defines a non-Archimedean valuation on $\mathbb{F}(t)$. Here we let $\mathrm{val}(0) = \infty$. As usual the *degree* is the largest occurring exponent. The valuation map extends to Puiseux series. More precisely, for $f, g \in \mathbb{F}\{t\}$ we have the following:

(i) $\mathrm{val}(f \cdot g) = \mathrm{val}(f) + \mathrm{val}(g)$,
(ii) $\mathrm{val}(f + g) \leq \max(\mathrm{val}(f), \mathrm{val}(g))$.

If $\mathbb{F} = \mathbb{R}$ is the field of real numbers we can evaluate a Puiseux fraction $f \in \mathbb{R}\{t\}$ at a real number τ to obtain the real number $f(\tau)$. This map is defined for all $\tau > 0$ except for the finitely many poles, i.e., zeros of the denominator. Restricting the evaluation to positive numbers is necessary since we are allowing rational exponents. The valuation map satisfies the equation

$$\lim_{\tau \to \infty} \log_\tau |f(\tau)| = \mathrm{val}(f). \tag{3}$$

That is, seen on a logarithmic scale, taking the valuation of f corresponds to interpreting t like an infinitesimally large number. Reading the valuation map in terms of the limit (3) is known as *Maslov dequantization*, see [24].

Occasionally, it is also useful to be able to interpret t as a *small* infinitesimal. To this end, one can define the *dual degree* $\deg^*$, which is the smallest occurring exponent. This gives rise to the *dual valuation* map $\mathrm{val}^*(p/q) = \deg_t^* p - \deg_t^* q$ which yields

$$\mathrm{val}^*(f+g) \;\geq\; \min(\mathrm{val}^*(f), \mathrm{val}^*(g)) \quad \text{and} \quad \lim_{\tau \to 0} \log_\tau |f(\tau)| \;=\; \mathrm{val}^*(f).$$

Changing from the primal to the dual valuation is tantamount to substituting t by t^{-1}.

Remark 1. The valuation theory literature often employs the dual definition of a valuation. The Eq. (3) is the reason why we usually prefer to work with the primal.

Up to isomorphism of valuated fields the valuation on the field $\mathbb{F}(t)$ of rational functions is unique, e.g., see [30, Sect. 147]. As a consequence the valuation on the slightly larger field of Puiseux fractions is unique, too.

To close this section let us look at the algorithmically most relevant case $\mathbb{F} = \mathbb{Q}$. Then, in general, the evaluation map sends positive rationals to not necessarily rational numbers, again due to fractional exponents. By clearing denominators in the exponents one can see that evaluating at $\sigma > 0$ ends up in the totally real number field $\mathbb{Q}(\sqrt[\nu]{\sigma})$ for some positive integer ν. For instance, evaluating the Puiseux fraction from Eq. (1) would give an element of $\mathbb{Q}(\sqrt[6]{\sigma})$.

3   Parameterized Polyhedra

Consider a matrix $A \in \mathbb{F}\{t\}^{m \times (d+1)}$. Then the set

$$C := \left\{ x \in \mathbb{F}\{t\}^{d+1} \mid A \cdot x \geq 0 \right\}$$

is a polyhedral cone in the vector space $\mathbb{F}\{t\}^{d+1}$. Equivalently, C is the set of feasible solutions of a linear program with $d+1$ variables over the ordered field $\mathbb{F}\{t\}$ with m homogeneous constraints, the rows of A. The Farkas–Minkowski–Weyl Theorem establishes that each polyhedral cone is finitely generated. A proof for this result on polyhedral cones over the reals can be found in [31, Sects. 1.3 and 1.4] under the name "Main theorem for cones". It is immediate to verify that the arguments given hold over any ordered field. Therefore, there is a matrix $B \in \mathbb{F}\{t\}^{(d+1) \times n}$, for some $n \in \mathbb{N}$, such that

$$C = \left\{ B \cdot a \mid a \in \mathbb{F}\{t\}^n, \, a \geq 0 \right\}. \tag{4}$$

The columns of B are points and the cone C is the non-negative linear span of those.

Let L be the *lineality space* of C, i.e., L is the unique maximal linear subspace of $\mathbb{F}\{t\}^{d+1}$ which is contained in C. If $\dim L = 0$ the cone C is *pointed*. Otherwise, the set C/L is a pointed polyhedral cone in the quotient space $\mathbb{F}\{t\}^{d+1}/L$. A *face* of C is the intersection of C with a supporting hyperplane. The faces are partially ordered by inclusion. Each face contains the lineality space. Adding the entire cone C as an additional top element we obtain a lattice, the *face lattice* of C. The maximal proper faces are the *facets* which form the co-atoms in the face lattice. The *combinatorial type* of C is the isomorphism class of the face lattice (e.g., as a partially ordered set). Notice that our definition says that each cone is combinatorially equivalent to its quotient modulo its lineality space.

Picking a positive element τ yields matrices $A(\tau) \in \mathbb{F}^{m \times (d+1)}$ and $B(\tau) \in \mathbb{F}^{(d+1) \times n}$ as well as a polyhedral cone $C(\tau) = \{x \in \mathbb{F}^{d+1} \mid A(\tau) \cdot x \geq 0\}$ by evaluating the Puiseux fractions at the parameter τ. Here and below we will assume that τ avoids the at most finitely many poles of the $(m+n) \cdot (d+1)$ coefficients of A and B.

Theorem 1. *There is a positive element $\tau_0 \in \mathbb{F}$ so that for every $\tau > \tau_0$ we have*

$$C(\tau) \;=\; \{ B(\tau) \cdot \alpha \mid \alpha \in \mathbb{F}^n, \, \alpha \geq 0 \},$$

and evaluating at τ maps the lineality space of C to the lineality space of $C(\tau)$. Moreover, the polyhedral cones C and $C(\tau)$ over $\mathbb{F}\{t\}$ and $\mathbb{F}$, respectively, share the same combinatorial type.

Proof. First we show that an orthogonal basis of the lineality space L evaluates to an orthogonal basis of the lineality space of $C(\tau)$. For this, consider two vectors $x, y \in \mathbb{F}\{t\}^{d+1}$ and pick τ large enough to avoid their poles and zeros. Then, the scalar product of x and y vanishes if and only if the scalar product of $x(\tau)$ and $y(\tau)$ does. Hence, the claim follows.

Now we can assume that the polyhedral cone C is pointed, i.e., it does not contain any linear subspace of positive dimension. If this is not the case the subsequent argument applies to the quotient C/L.

Employing orthogonal bases, as for the lineality spaces above, shows that the evaluation maps the linear hull of C to the linear hull of $C(\tau)$, preserving the dimension. So we may assume that C is full-dimensional, as otherwise the arguments below hold in the linear hull of C.

Let $\ell \leq \binom{m}{d}$ be the number of d-element sets of linearly independent rows of the matrix A. For each such set of rows the set of solutions to the corresponding homogeneous system of linear equations is a one-dimensional subspace of $\mathbb{F}\{t\}^{(d+1)}$. For each such system of homogeneous linear equations pick two non-zero solutions, which are negatives of each other. We arrive at 2ℓ vectors in $\mathbb{F}\{t\}^{(d+1)}$ which we use to form the columns of the matrix $Z \in \mathbb{F}\{t\}^{(d+1) \times 2\ell}$.

By the Farkas–Minkowski–Weyl theorem, we may assume that the columns of B from (4) only consist of the rays of C and that the rays of C form a subset of the columns of Z. In particular, the columns of B occur in Z. Since the cone C is pointed, the matrix B contains at most one vector from each opposite pair of the columns of Z. This entails that B has at most ℓ columns.

Further, the real matrix $Z(\tau)$ contains all rays of $C(\tau)$ for each τ that avoids the poles of A and Z. In the following, we want to show that those columns of $Z(\tau)$ which form the rays of $C(\tau)$ are exactly the columns of $B(\tau)$.

We define $s(j, k) \in \mathbb{F}\{t\}$ to be the scalar product of the jth row of A and the kth column of Z. The $m \cdot 2\ell$ signs of the scalar products $s(j, k)$, for $j \in [m]$ and $k \in [2\ell]$, form the *chirotope* of the linear hyperplane arrangement defined by the rows of A (in fact, due to taking two solutions for each homogenous system of linear equations, we duplicate the information of the chirotope). For almost all $\tau \in \mathbb{F}$ evaluating the Puiseux fractions $s(j, k)$ at τ yields an element of $\mathbb{F}$. For sufficiently large τ the sign of $s(j, k)$ agrees with its evaluation. This follows from the definition of the ordering on $\mathbb{F}\{t\}$, cf. [16, Proposition, Sect. 1.3].

Let $\tau_0 \in \mathbb{F}$ be larger than all the at most finitely many poles of A and Z. Further, let τ_0 be large enough such that the chirotope of $A(\tau)$ agrees with the chirotope of A for all $\tau > \tau_0$.

By construction the rays of C correspond to the non-negative columns of the chirotope whose support, given by the 0 entries, is maximal with respect to inclusion; these are exactly the columns of B. The corresponding columns of the chirotope of $A(\tau)$, for $\tau > \tau_0$, yield the rays of $C(\tau)$, which, hence, are the columns of $B(\tau)$.

The same holds for the facets of C and $C(\tau)$. The facets of C correspond to the non-negative rows of the chirotope whose support, given by the 0 entries, is maximal with respect to inclusion.

Now the claim follows since the face lattice of a polyhedral cone is determined by the incidences between the facets and the rays.   □

A statement related to Theorem 1 occurs in Benchimol's PhD thesis [5]. The Proposition 5.12 in [5] discusses the combinatorial structure of tropical polyhedra (arising as the feasible regions of tropical linear programs). Yet here we consider the relationship between the combinatorial structure of Puiseux polyhedra and their evaluations over the reals. As in the proof of [5, Proposition 5.12] we could derive an explicit upper bound on the optimal τ_0. To this end one can estimate the coefficients of the Puiseux fractions in Z, which are given by determinantal expressions arising from submatrices of A. Their poles and zeros are bounded by Cauchy bounds (e.g., see [26, Theorem 8.1.3]) depending on those coefficients. We leave the details to the reader.

A *convex polyhedron* is the intersection of finitely many affine halfspaces. It is a called a *polytope* if it is bounded. Restricting to cones allows a simple description in terms of homogeneous linear inequalities. Yet this encompasses arbitrary polytopes and polyhedra, as they can equivalently be studied through their homogenizations. In fact, all implementations in `polymake` are based on this principle. For further reading we refer to [31, Sect. 1.5]. We visualize Theorem 1 with a very simple example.

Example 1.   Consider the polytope P in $\mathbb{R}\{t\}^2$ for large t defined by the four inequalities

$$x_1, x_2 \geq 0, \qquad x_1 + x_2 \leq 3, \qquad x_1 - x_2 \leq t.$$

The evaluations at $\tau \in \{0, 1, 3\}$ are depicted in Fig. 1. For $\tau = 0$ we obtain a triangle, for $\tau = 1$ a quadrangle and for $\tau \geq 3$ a triangle again. The latter is

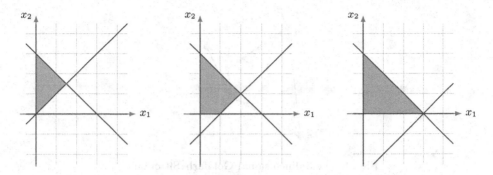

Fig. 1. Polygon depending on a real parameter as defined in Example 1

the combinatorial type of the polytope P over the field of Puiseux fractions with real coefficients.

Corollary 1. *The set of combinatorial types of polyhedral cones which can be realized over $\mathbb{F}\{t\}$ is the same as over $\mathbb{F}$.*

Proof. One inclusion is trivial since $\mathbb{F}$ is a subfield of $\mathbb{F}\{t\}$. The other inclusion follows from the preceding result.   □

For $A \in \mathbb{F}\{t\}^{m \times d}$, $b \in \mathbb{F}\{t\}^m$ and $c \in \mathbb{F}\{t\}^d$ we consider the linear program $\mathrm{LP}(A, b, c)$ over $\mathbb{F}\{t\}$ which reads as

$$\begin{aligned}\text{maximize} \quad & c^\top \cdot x \\ \text{subject to} \quad & A \cdot x = b, \ x \geq 0.\end{aligned} \tag{5}$$

For each positive $\tau \in \mathbb{F}$ (which avoids the poles of the Puiseux fractions which arise as coefficients) we obtain a linear program $\mathrm{LP}(A(\tau), b(\tau), c(\tau))$ over $\mathbb{F}$. Theorem 1 now has the following consequence for parametric linear programming.

Corollary 2. *Let $x^* \in \mathbb{F}\{t\}^d$ be an optimal solution to the LP (5) with optimal value $v \in \mathbb{F}\{t\}$. Then there is a positive element $\tau_0 \in \mathbb{F}$ so that for every $\tau > \tau_0$ the vector $x^*(\tau)$ is an optimal solution for $\mathrm{LP}(A(\tau), b(\tau), c(\tau))$ with optimal value $v(\tau)$.*

The above corollary was proved by Jeroslow [16, Sect. 2.3]. His argument, based on controlling signs of determinants, is essentially a local version of our Theorem 1. Moreover, determining all the rays of a polyhedral cone can be reduced to solving sufficiently many LPs. This could also be exploited to derive another proof of Theorem 1 from Corollary 2.

Remark 2. It is worth to mention the special case of a linear program over the field $\mathbb{F}\{t\}$, where the coordinates of the linear constraints, in fact, are elements of the field $\mathbb{F}$ of coefficients, but the coordinates of the linear objective function

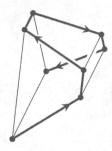

Fig. 2. The 3-dimensional Goldfarb–Sit cube.

are arbitrary elements in $\mathbb{F}\{t\}$. That is, the feasible domain is a polyhedron, P, over $\mathbb{F}$. Evaluating the objective function at some $\tau \in \mathbb{F}$ makes one of the vertices of P optimal. Solving for all values of τ, in general, amounts to computing the entire normal fan of the polyhedron P. This is equivalent to solving the dual convex hull problem over $\mathbb{F}$ for the given inequality description of P; see also [17]. Here we restrict our attention to solving parametric linear programs via Corollary 2.

The next example is a slight variation of a construction of Goldfarb and Sit [15]. This is a class of linear optimization problems on which certain versions of the simplex method perform poorly.

Example 2. We fix $d > 1$ and pick a positive $\delta \leq \frac{1}{2}$ as well as a positive $\varepsilon < \frac{\delta}{2}$. Consider the linear program

$$\text{maximize} \quad \sum_{i=1}^{d} \delta^{d-i} x_i$$
$$\text{subject to} \quad 0 \leq x_1 \leq \varepsilon^{d-1}$$
$$x_{j-1} \leq \delta x_j \leq \varepsilon^{d-j} \delta - x_{j-1} \qquad \text{for } 2 \leq j \leq d.$$

The feasible region is combinatorially equivalent to the d-dimensional cube. Applying the simplex method with the "steepest edge" pivoting strategy to this linear program with the origin as the start vertex visits all the 2^d vertices. Moreover, the vertex-edge graph with the orientation induced by the objective function is isomorphic to (the oriented vertex-edge graph of) the Klee–Minty cube [20]. See Fig. 2 for a visualization of the 3-dimensional case.

We may interpret this linear program over the reals or over $(\mathbb{R}\{\delta\})\{\varepsilon\}$, the field of Puiseux fractions in the indeterminate ε with coefficients in the field $\mathbb{R}\{\delta\}$. This depends on whether we want to view δ and ε as indeterminates or as real numbers. Here we consider the ordering induced by the dual valuation val*, i.e., δ and ε are *small* infinitesimals, where $\varepsilon \ll \delta$. Two more choices arise from considering ε a constant in $\mathbb{R}\{\delta\}$ or, conversely, δ a constant in $\mathbb{R}\{\varepsilon\}$. Note that our constraints on δ and ε are feasible in all four cases.

Our third and last example is a class of linear programs occurring in [1]. For these the central path of the interior point method with a logarithmic barrier function has a total curvature which is exponential as a function of the dimension.

Example 3. Given a positive integer r, we define a linear program over the field $\mathbb{Q}\{t\}$ (with the primal valuation) in the $2r + 2$ variables $u_0, v_0, u_1, v_1, \ldots, u_r, v_r$ as follows:

$$
\begin{aligned}
\text{minimize} \quad & v_0 \\
\text{subject to} \quad & u_0 \leq t, \; v_0 \leq t^2 \\
& \left. \begin{aligned} u_i &\leq t u_{i-1}, \; u_i \leq t v_{i-1} \\ v_i &\leq t^{1-\frac{1}{2^i}} (u_{i-1} + v_{i-1}) \end{aligned} \right\} \quad \text{for } 1 \leq i \leq r \\
& u_r \geq 0, \; v_r \geq 0.
\end{aligned}
$$

Here it would be interesting to know the exact value for the optimal τ_0 in Theorem 1, as a function of r. Experimentally, based on the method described below, we found $\tau_0 = 1$ for $r = 1$ and $\tau_0 = 2^{2^{r-1}}$ for r at most 5. We conjecture the latter to be the true bound in general.

To find the optimal bound for a given constraint matrix A we can use the following method. One can solve the dual convex hull problem for the cone C, which is the feasible region in homogenized form, to obtain a matrix B whose columns are the rays of C. This also yields a submatrix of A corresponding to the rows which define facets of C. Without loss of generality we may assume that submatrix is A itself. Let τ_0 be the largest zero or pole of any (Puiseux fraction) entry of the matrix $A \cdot B$. Then for every value $\tau > \tau_0$ the sign patterns of $(A \cdot B)(\tau)$ and $A \cdot B$ coincide, and so do the combinatorial types of C and $C(\tau)$. Determining the zeros and poles of a Puiseux fraction amounts to factorizing univariate polynomials.

4   Tropical Dual Convex Hulls

Tropical geometry is the study of the piecewise linear images of algebraic varieties, defined over a field with a non-Archimedean valuation, under the valuation map; see [21] for an overview. The motivation for research in this area comes from at least two different directions. First, tropical varieties still retain a lot of interesting information about their classical counterparts. Therefore, passing to the tropical limit opens up a path for combinatorial algorithms to be applied to topics in algebraic geometry. Second, the algebraic geometry perspective offers opportunities for optimization and computational geometry. Here we will discuss how classical convex hull algorithms over fields of Puiseux fractions can be applied to compute tropical convex hulls; see [19] for a survey on the subject; a standard algorithm is the tropical double description method of [3].

The *tropical semiring* $\mathbb{T}$ consists of the set $\mathbb{R} \cup \{-\infty\}$ together with $u \oplus v = \max(u, v)$ as the addition and $u \odot v = u + v$ as the multiplication. Extending these operations to vectors turns $\mathbb{T}^{d+1}$ into a semimodule. A *tropical cone* is the sub-semimodule

$$
\text{tcone}(G) = \{\lambda_1 \odot g_1 \oplus \cdots \oplus \lambda_n \odot g_n \mid \lambda_1, \ldots, \lambda_n \in \mathbb{T}\}
$$

generated from the columns $g_1, \ldots, g_n$ of the matrix $G \in \mathbb{T}^{(d+1) \times n}$. Similar to classical cones, tropical cones admit an exterior description [13]. It is known that every tropical cone is the image of a classical cone under the valuation map val: $\mathbb{R}\{\!\{t\}\!\} \to \mathbb{T}$; see [9]. Based on this idea, we present an algorithm for computing generators of a tropical cone from a description in terms of tropical linear inequalities; see Algorithm 1 below.

Before we can start to describe that algorithm we first need to discuss matters of general position in the tropical setting. The *tropical determinant* of a square matrix $U \in \mathbb{T}^{\ell \times \ell}$ is given by

$$\text{tdet}(U) = \bigoplus_{\sigma \in S_\ell} u_{1\pi(1)} \odot \cdots \odot u_{\ell\pi(\ell)}. \tag{6}$$

Here S_ℓ is the symmetric group of degree ℓ; computing the tropical determinant is the same as solving a linear assignment optimization problem. Consider a pair of matrices $H^+, H^- \in \mathbb{T}^{m \times (d+1)}$ which serve as an exterior description of the tropical cone

$$Q = \left\{ z \in \mathbb{T}^{(d+1)} \;\middle|\; H^+ \odot z \geq H^- \odot z \right\}. \tag{7}$$

In contrast to the classical situation we have to take two matrices into account. This is due to the lack of an additive inverse operation. We will assume that $\min(H_{ij}^+, H_{ij}^-) = -\infty$ for any pair $(i,j) \in [m] \times [d+1]$, i.e., for each coordinate position at most one of the corresponding entries in the two matrices is finite. Then we can define

$$\chi(i,j) := \begin{cases} 1 & \text{if } H_{ij}^+ \neq -\infty \\ -1 & \text{if } H_{ij}^- \neq -\infty \\ 0 & \text{otherwise.} \end{cases}$$

For each term $u_{1\pi(1)} \odot \cdots \odot u_{\ell\pi(\ell)}$ in (6) we define its *sign* as

$$\text{sgn}(\pi) \cdot \chi(1, \pi(1)) \cdots \chi(\ell, \pi(\ell)),$$

where $\text{sgn}(\pi)$ is the sign of the permutation π. Now the exterior description (7) of the tropical cone Q is *tropically sign-generic* if for each square submatrix U of $H^+ \oplus H^-$ we have $\text{tdet}(U) \neq -\infty$ and, moreover, the signs of all terms $u_{1\pi(1)} \odot \cdots \odot u_{\ell\pi(\ell)}$ which attain the maximum in (6) agree. By looking at 1×1-submatrices U we see that in this case all coefficients of the matrix $H^+ \oplus H^-$ are finite and thus $\chi(i,j)$ is never 0.

Proof (Correctness of Algorithm 1). The main lemma of tropical linear programming [2, Theorem 16] says the following. In the tropically sign-generic case, an exterior description of a tropical cone can be obtained from an exterior description of a classical cone over Puiseux series by applying the valuation map to the constraint matrix coefficient-wise. This statement assumes that the classical cone is contained in the non-negative orthant. We infer that

---

Algorithm 1. A dual tropical convex hull algorithm

Input: pair of matrices $H^+, H^- \in \mathbb{T}^{m \times (d+1)}$ which provide a tropically
       sign-generic exterior description of the tropical cone Q from (7)
Output: generators for Q
pick two matrices $A^+, A^- \in \mathbb{R}\{\!\{t\}\!\}^{m \times (d+1)}$ with strictly positive entries such
that $\mathrm{val}(A^+) = H^+$ and $\mathrm{val}(A^-) = H^-$;
apply a classical dual convex hull algorithm to determine a matrix
$B \in \mathbb{R}\{\!\{t\}\!\}^{(d+1) \times n}$ such that
$$\{ B \cdot a \mid a \in \mathbb{R}\{\!\{t\}\!\}^n, a \geq 0\} = \left\{ x \in \mathbb{R}\{\!\{t\}\!\}^{(d+1)} \;\middle|\; (A^+ - A^-) \cdot x \geq 0, x \geq 0 \right\} ;$$
return $\mathrm{val}(B)$;

$$Q = \left\{ z \in \mathbb{T}^{m \times (d+1)} \;\middle|\; H^+ \odot z \geq H^- \odot z \right\}$$
$$= \mathrm{val}\left(\left\{ x \in \mathbb{R}\{\!\{t\}\!\}^{m \times (d+1)} \;\middle|\; A^+ \cdot x \geq A^- \cdot x, x \geq 0 \right\} \right)$$
$$= \mathrm{val}\left(\{ B \cdot a \mid a \in \mathbb{R}\{\!\{t\}\!\}^n, a \geq 0 \} \right).$$

Now [9, Proposition 2.1] yields $Q = \mathrm{val}(\{ B \cdot a \mid a \in \mathbb{R}\{\!\{t\}\!\}^n, a \geq 0 \}) = \mathrm{tcone}(\mathrm{val}(B))$. This ends the proof.   □

The correctness of our algorithm is not guaranteed if the genericity condition is not satisfied. The crucial properties of the lifted matrices A^+, A^- are not necessarily fulfilled. It is an open question of how an exterior description over $\mathbb{T}$ is related to an exterior description over $\mathbb{R}\{\!\{t\}\!\}$ in the general setting. We are even lacking a convincing concept for the "facets" of a general tropical cone.

5   Implementation

As a key feature the `polymake` system for discrete geometry is designed as a `Perl/C++` hybrid, that is, both programming languages are used in the implementation and also both programming languages can be employed by the user to write further code. One main advantage of Perl is the fact that it is interpreted; this makes it suitable as the main front end for the user. Further, Perl has its strengths in the manipulation of strings and file processing. C++ on the other hand is a compiled language with a powerful template mechanism which allows to write very abstract code which, nonetheless, is executed very fast. Our implementation, in C++, makes extensive use of these features. The implementation of the dual steepest edge simplex method, contributed by Thomas Opfer, and the beneath-beyond method for computing convex hulls (see [11] and [18]) are templated. Therefore `polymake` can handle both computations for arbitrary number field types which encode elements in an ordered field.

Based on this mechanism we implemented the type `RationalFunction` which depends on two generic template types for coefficients and exponents. Note that the field of coefficients here does not have to be ordered. Our proof-of-concept implementation employs the classical Euclidean GCD algorithm for normalization. Currently the numerator and the denominator are chosen coprime such that the denominator is normalized with leading coefficient one. For the most interesting case $\mathbb{F} = \mathbb{Q}$ it is known that the coefficients of the intermediate polynomials can grow quite badly, e.g., see [12, Example 1]. Therefore, as expected, this is the bottleneck of our implementation. In a number field or in a field with a non-Archimedean valuation the most natural choice for a normalization is to pick the elements of the ring of integers as coefficients. The reason for our choice is that this more generic design does not make any assumption on the field of coefficients. This makes it very versatile, and it fits the overall programming style in `polymake`. A fast specialization to the rational coefficient case could be based on [12, Algorithm 11.4]. This is left for a future version.

The `polymake` implementation of Puiseux fractions $\mathbb{F}\{t\}$ closely follows the construction described in Sect. 2. The new number type is derived from `RationalFunction` with overloaded comparison operators and new features such as evaluating and converting into `TropicalNumber`. An extra template parameter `MinMax` allows to choose whether the indeterminate t is a small or a large infinitesimal.

There are other implementations of Puiseux series arithmetic, e.g., in `Magma` [6] or `MATLAB` [29]. However, they seem to work with finite truncations of Puiseux series and floating-point coefficients. This does not allow for exact computations of the kind we are interested in.

6   Computations

We briefly show how our `polymake` implementation can be used. Further, we report on timings for our LP solver, tested on the Goldfarb–Sit cubes from Example 2, and for our (dual) convex hull code, tested on the polytopes with a "long and winded" central path from Example 3.

6.1   Using `polymake`

The following code defines a 3-dimensional Goldfarb–Sit cube over the field $\mathbb{Q}\{t\}$, see Example 2. We use the parameters $\varepsilon = t$ and $\delta = \frac{1}{2}$. The template parameter `Min` indicates that the ordering is induced by the dual valuation val*, and hence the indeterminate t plays the role of a small infinitesimal.

```
polytope > $monomial=new UniMonomial<Rational,Rational>(1);
polytope > $t=new PuiseuxFraction<Min>($monomial);
polytope > $p=goldfarb_sit(3,2*$t,1/2);
```

The polytope object, stored in the variable `$p`, is generated with a facet description from which further properties will be derived below. It is already equipped with a `LinearProgram` subobject encoding the objective function from Example 2. The following lines show the maximal value and corresponding vertex of this linear program as well as the vertices derived from the outer description. Below, we present timings for such calculations.

```
polytope > print $p->LP->MAXIMAL_VALUE;
(1)
polytope > print $p->LP->MAXIMAL_VERTEX;
(1) (0) (0) (1)
polytope > print $p->VERTICES;
(1) (0) (0) (0)
(1) (t^2) (2*t^2) (4*t^2)
(1) (0) (t) (2*t)
(1) (t^2) (t -2*t^2) (2*t -4*t^2)
(1) (0) (0) (1)
(1) (t^2) (2*t^2) (1 -4*t^2)
(1) (0) (t) (1 -2*t)
(1) (t^2) (t -2*t^2) (1 -2*t + 4*t^2)
```

As an additional benefit of our implementation we get numerous other properties for free. For instance, we can compute the parameterized volume, which is a polynomial in t.

```
polytope > print $p->VOLUME;
(t^3 -4*t^4 + 4*t^5)
```

That polynomial, as an element of the field of Puiseux fractions, has a valuation, and we can evaluate it at the rational number $\frac{1}{12}$.

```
polytope > print $p->VOLUME->val;
3
polytope > print $p->VOLUME->evaluate(1/12);
25/62208
```

6.2 Linear Programs

We have tested our implementation by computing the linear program of Example 2 with polyhedra defined over Puiseux fractions.

The simplex method in polymake is an implementation of a (dual) simplex with a (dual) steepest edge pricing. We set up the experiment to make sure our Goldfarb–Sit cube LPs behave as badly as possible. That is, we force our implementation to visit all $n = 2^d$ vertices, when d is the dimension of the input. Table 1 illustrates the expected exponential growth of the execution time of the linear program. In three of our four experiments we choose δ as $\frac{1}{2}$. The computation over $\mathbb{Q}\{\varepsilon\}$ costs a factor of about 80 in time, compared with the rational cubes for a modest $\varepsilon = \frac{1}{6}$. However, taking a small ε whose binary encoding takes more than 18,000 bits is substantially more expensive than the computations over the field $\mathbb{Q}\{\varepsilon\}$ of Puiseux fractions. Taking δ as a second small infinitesimal is possible but prohibitively expensive for dimensions larger than twelve.

6.3 Convex Hulls

We have also tested our implementation by computing the vertices of the polytope from Example 3. For this we used the client long_and_winding which

Table 1. Timings (in s) for the Goldfarb–Sit cubes of dimension d with $\delta = \frac{1}{2}$. For ε we tried a small infinitesimal as well as two rational numbers, one with a short binary encoding and another one whose encoding is fairly large. For comparison we also tried both parameters as indeterminates.

d	m	n	$\mathbb{Q}\{\varepsilon\}$ ε	$\mathbb{Q}$ $\varepsilon = \frac{1}{6}$	$\mathbb{Q}$ $\varepsilon = \frac{2}{174500}$	$(\mathbb{Q}\{\delta\})\{\varepsilon\}$ $\varepsilon \ll \delta$
3	6	8	0.010	0.003	0.005	0.101
4	8	16	0.026	0.001	0.017	0.353
5	10	32	0.064	0.002	0.065	1.034
6	12	64	0.157	0.007	0.253	2.877
7	14	128	0.368	0.006	0.829	7.588
8	16	256	0.843	0.016	2.643	19.226
9	18	512	1.906	0.039	7.703	47.806
10	20	1024	4.258	0.090	21.908	118.106
11	22	2048	9.383	0.191	59.981	287.249
12	24	4096	20.583	0.418	160.894	687.052

creates the $d = (2r+2)$-dimensional polytope given by $m = 3r+4$ facet-defining inequalities. Over the rationals we evaluated the inequalities at 2^{2^r} which probably gives the correct combinatorics; see the discussion at the end of Example 3. This very choice forces the coordinates of the defining inequalities to be integral, such that the polytope is rational. The number of vertices n is derived from that rational polytope. The running times grow quite dramatically for the parametric input (Table 2). This overhead could be reduced via a better implementation of the Puiseux fraction arithmetic.

6.4   Experimental Setup

Everything was calculated on the same Linux machine with `polymake` perpetual beta version 2.15-beta3 which includes the new number type, the templated simplex algorithm and the templated beneath-and-beyond convex hull algorithm. All timings were measured in CPU seconds and averaged over ten iterations. The simplex algorithm was set to use only one thread.

All tests were done on `openSUSE` 13.1 (x86_64), with Linux kernel 3.11.10-25, `clang` 3.3 and `perl` 5.18.1. The rational numbers use a `C++`-wrapper around the `GMP` library version 5.1.2. As memory allocator `polymake` uses the `pool_allocator` from `libstdc++`, which was version 4.8.1 for the experiments.

The hardware for all tests was:

Intel(R) Core(TM) i7-3930K CPU @ 3.20 GHz
bogomips: 6400.21
MemTotal: 32928276 kB

Table 2. Timings (in s) for convex hull computation of the feasibility set from Example 3. All timings represent an average over ten iterations. If any test exceeded a one hour time limit this and all larger instances of the experiment were skipped and marked $-$.

r	d	m	n	$\mathbb{Q}\{t\}$	$\mathbb{Q}$
1	4	7	11	0.018	0.000
2	6	10	28	0.111	0.000
3	8	13	71	0.754	0.010
4	10	16	182	15.445	0.036
5	12	19	471	1603.051	0.150
6	14	22	1226	-	0.737
7	16	25	3201	-	4.001
8	18	28	8370	-	25.093
9	20	31	21901	-	223.240
10	22	34	57324	-	1891.133

Acknowledgments. We thank Thomas Opfer for contributing to and maintaining within the polymake project his implementation of the dual simplex method, originally written for his Master's Thesis [25].

References

1. Allamigeon, X., Benchimol, P., Gaubert, S., Joswig, M.: Long and winding central paths, preprint (2014). arXiv:1405.4161
2. Allamigeon, X., Benchimol, P., Gaubert, S., Joswig, M.: Tropicalizing the simplex algorithm. SIAM J. Discrete Math. **29**(2), 751–795 (2015). http://dx.doi.org/10.1137/130936464
3. Allamigeon, X., Gaubert, S., Goubault, É.: The tropical double descriptionmethod. In: STACS 2010: 27th International Symposium on TheoreticalAspects of Computer Science, LIPIcs. Leibniz International Proceedings in Informatics,vol. **5**, pp. 47–58. Schloss Dagstuhl. Leibniz-Zent. Inform., Wadern (2010)
4. Allamigeon, X., Gaubert, S., Goubault, É.: Computing the vertices of tropical polyhedra using directed hypergraphs. Discrete Comput. Geom. **49**(2), 247–279 (2013). http://dx.doi.org/10.1007/s00454-012-9469-6
5. Benchimol, P.: Tropical aspects of linear programming. Theses, École Polytechnique, December 2014. https://hal-polytechnique.archives-ouvertes.fr/tel-01198482
6. Bosma, W., Cannon, J., Playoust, C.: The MAGMA algebra system I. The user language. J. Symbolic Comput. **24**(3–4), 235–265 (1997). http://dx.doi.org/10.1006/jsco.1996.0125
7. Charnes, A., Kortanek, K.O.: On classes of convex and preemptive nuclei for n-person games. In: Proceedings of the Princeton Symposium on Mathematical Programming (Princeton University, 1967). pp. 377–390. Princeton University Press, Princeton (1970)

8. Dantzig, G.B.: Linear Programming and Extensions. Princeton University Press, Princeton, N.J. (1963)
9. Develin, M., Yu, J.: Tropical polytopes and cellular resolutions. Experiment. Math. **16**(3), 277–291 (2007). http://projecteuclid.org/euclid.em/1204928529
10. Deza, A., Terlaky, T., Zinchenko, Y.: Central path curvature and iteration-complexity for redundant Klee-Minty cubes. In: Gao, D.Y., Sherali, H.D. (eds.) Advances in Applied Mathematics and Global Optimization. Adv. Mech. Math., vol. 17, pp. 223–256. Springer, New York (2009). http://dx.doi.org/10.1007/978-0-387-75714-8_7
11. Edelsbrunner, H.: Algorithms in Combinatorial Geometry. EATCS Monographs on Theoretical Computer Science, vol. 10. Springer-Verlag, Berlin (1987). http://dx.doi.org/10.1007/978-3-642-61568-9
12. von zur Gathen, J., Gerhard, J.: Modern Computer Algebra, 2nd edn. Cambridge University Press, Cambridge (2003)
13. Gaubert, S., Katz, R.D.: Minimal half-spaces and external representation of tropical polyhedra. J. Algebraic Combin. **33**(3), 325–348 (2011). http://dx.doi.org/10.1007/s10801-010-0246-4
14. Gawrilow, E., Joswig, M.: polymake: a framework for analyzing convex polytopes. In: Kalai, G., Ziegler, G.M. (eds.) Polytopes–combinatorics and computation (Oberwolfach, 1997). DMV Sem., pp. 43–73. Birkhäuser, Basel (2000)
15. Goldfarb, D., Sit, W.Y.: Worst case behavior of the steepest edge simplex method. Discrete Appl. Math. **1**(4), 277–285 (1979). http://dx.doi.org/10.1016/0166-218X(79)90004-0
16. Jeroslow, R.G.: Asymptotic linear programming. Oper. Res. **21**(5), 1128–1141 (1973). http://dx.doi.org/10.1287/opre.21.5.1128
17. Jones, C.N., Kerrigan, E.C., Maciejowski, J.M.: On polyhedral projection and parametric programming. J. Optim. Theor. Appl. **138**(2), 207–220 (2008). http://dx.doi.org/10.1007/s10957-008-9384-4
18. Joswig, M.: Beneath-and-beyond revisited. In: Joswig, M., Takayama, N. (eds.) Algebra, Geometry, and Software Systems, pp. 1–21. Springer, Berlin (2003)
19. Joswig, M.: Tropical convex hull computations. In: Litvinov, G.L., Sergeev, S.N. (eds.) Tropical and Idempotent Mathematics, Contemporary Mathematics, vol. 495, pp. 193–212. American Mathematical Society, Providence (2009)
20. Klee, V., Minty, G.J.: How good is the simplex algorithm? In: Inequalities, III (Proceedings of Third Symposium, University of California, Los Angeles, California, 1969; dedicated to the memory of Theodore S. Motzkin), pp. 159–175. Academic Press, New York (1972)
21. Maclagan, D., Sturmfels, B.: Introduction to Tropical Geometry, Graduate Studies in Mathematics, vol. 161. American Mathematical Society, Providence, RI (2015)
22. Mannaa, B., Coquand, T.: Dynamic Newton-Puiseux theorem. J. Log. Anal. **5**, 22 (2013). Paper 5
23. Markwig, T.: A field of generalised Puiseux series for tropical geometry. Rend. Semin. Mat. Univ. Politec. Torino **68**(1), 79–92 (2010)
24. Maslov, V.P.: On a new superposition principle for optimization problem. In: Séminaire sur les équations aux dérivées partielles, pp. 1985–1986, Exp. No. XXIV, 14. École Polytech., Palaiseau (1986)
25. Opfer, T.: Entwicklung eines exakten rationalen dualen Simplex-Lösers. Master's thesis, TU Darmstadt (2011)
26. Rahman, Q.I., Schmeisser, G.: Analytic Theory of Polynomials, London Mathematical Society Monographs. New Series, vol. 26. The University Press, Clarendon Press, Oxford (2002)

27. Salzmann, H., Grundhöfer, T., Hähl, H., Löwen, R.: The classical fields, Encyclopedia of Mathematics and its Applications, vol. 112. Cambridge University Press, Cambridge (2007). http://dx.org/10.1017/CBO9780511721502
28. Tarski, A.: A Decision Method for Elementary Algebra and Geometry. RAND Corporation, Santa Monica, California (1948)
29. The MathWorks Inc.: MATLAB, version 8.4.0.150421 (R2014b). Natick, Massachusetts (2014)
30. van der Waerden, B.L.: Algebra II. Unter Benutzung von Vorlesungen von E. Artin und E. Noether, 6th edn. Springer, Berlin (1993). Mit einem Geleitwort von Jürgen Neukirch
31. Ziegler, G.M.: Lectures on Polytopes. Graduate Texts in Mathematics, vol. 152. Springer-Verlag, New York (1995)

Another Classroom Example of Robustness Problems in Planar Convex Hull Computation

Marc Mörig[✉]

Faculty of Computer Science, Department of Simulation and Graphics,
Otto-von-Guericke University of Magdeburg, Universitätsplatz 2,
39106 Magdeburg, Germany
moerig@isg.cs.unimagdeburg.de

Abstract. Algorithms in computational geometry are designed under the assumption of exact real arithmetic. Indiscriminately replacing exact real arithmetic by hardware floating-point arithmetic almost inevitably leads to robustness problems. Kettner et al. provide examples where rounding errors let such straightforward implementations of incremental convex hull computation crash, loop forever, or silently compute garbage. We complement their work by providing problematic examples for another planar convex hull algorithm.

Keywords: Implementation · Numerical robustness problems · Floating-point geometry

1 Introduction

Algorithms in computational geometry are designed under the assumption of exact real arithmetic at unit cost [6]. Simply using hardware floating-point arithmetic as an indiscriminate substitute in implementations almost inevitably leads to robustness problems [7,9]. However, there are only few examples documented in the literature: The LEDA book [4] gives some instructive examples in Sect. 9.6 and reports on experiments that fail for floating-point based implementations in Sects. 10.7 and 10.8. Shewchuk [8] discusses an example where the computation of a 2D Delaunay triangulation by divide and conquer fails. Moreover, it is folklore that straightforward implementations of Jarvis' march for computing planar convex hulls can loop forever for certain nearly degenerate input data. Computing the convex hull of a set of points in the plane is one of the best studied problems in computational geometry. Kettner et al. [3] show how to create input data that lets an implementation of incremental planar convex hull computation fail in various ways. After fixing an implementation, they examine the decisions the implementation will make for input points in a certain critical range of space. This allows them to select input points where these decisions are incorrect and lead to failures. Failures generated include not only computing output that is not valid, but also infinite looping, or the program crashing. However, the planar convex hull algorithm they study does not have worst-case

© Springer International Publishing Switzerland 2016
I.S. Kotsireas et al. (Eds.): MACIS 2015, LNCS 9582, pp. 446–450, 2016.
DOI: 10.1007/978-3-319-32859-1_38

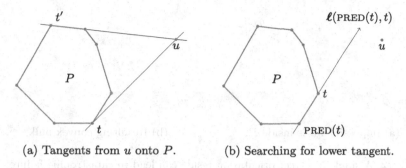

(a) Tangents from u onto P.   (b) Searching for lower tangent.

Fig. 1. Basic steps in a 2D convex hull computation.

optimal running time. The same holds for Jarvis' march. Both have quadratic worst-case complexity.

We complement the work by Kettner et al. by providing examples for the plane sweep variant of Graham's scan [1], a convex hull algorithm with worst-case optimal running time $O(n \log n)$. At our companion website [5] we provide an implementation of the algorithm and further material illustrating our examples.

2   Short Description of Algorithm and Predicates

Let p, q, and r be three points in the plane, and let $\ell(p, q)$ be the oriented line passing first through p and then through q. The 2D orientation predicate determines the position of r relative to $\ell(p, q)$. If $p = (p_x, p_y)$, $q = (q_x, q_y)$, and $r = (r_x, r_y)$, then the predicate is tantamount to computing the sign of the determinant

$$D_{O2} = \begin{vmatrix} p_x & p_y & 1 \\ q_x & q_y & 1 \\ r_x & r_y & 1 \end{vmatrix} = \begin{vmatrix} q_x - p_x & q_y - p_y \\ r_x - p_x & r_y - p_y \end{vmatrix}. \tag{1}$$

The three points p, q and r are collinear if and only if D_{O2} is zero. Otherwise, r is to the left of $\ell(p, q)$, if D_{O2} is greater than zero and r is to the right of $\ell(p, q)$, if D_{O2} is smaller than zero. Besides coordinate comparison this is the only predicate used in the algorithm.

The algorithm itself is a variant of Graham's scan [1] and proceeds as follows. We process the points one by one, in xy-lexicographical order. We maintain the convex hull P of already processed points as the circular sequence of vertices in counterclockwise order along its boundary. Thus, when we arrive at a new point u, we have to update P. Since we process points from left to right, u is not contained in P. Consider the two tangents from u onto P, see Fig. 1a. Each tangent touches exactly one or two vertices of P. Let t' be the vertex of P furthest from u that the upper tangent touches, and let t be the vertex of P furthest from u that the lower tangent touches. Note that $t = t'$ is possible in case P is a segment. In the sequence of vertices of P, we replace all vertices between t and t' with u.

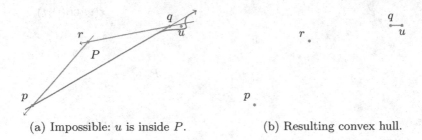

(a) Impossible: u is inside P.   (b) Resulting convex hull.

Fig. 2. A single incorrect orientation result can lead to catastrophic failure.

How do we find t and t'? Let q be the point that was processed in the previous update step. Then q is the rightmost vertex of P. We start with $t = q$ and check the position of u relative to $\ell(\text{PRED}(t), t)$ with the 2D orientation predicate, see Fig. 1b. As long as u is not on the left side, we advance t to $\text{PRED}(t)$ and check the position of u again. The vertex t' can be found analogously. There is one exceptional case: if P is a segment we have to stop the search for t and t' after at most one step.

As for the 2D orientation predicate we use the straightforward implementation. We compute an approximation D'_{O2} of D_{O2} as

$$D'_{O2} = (q_x \ominus p_x) \otimes (r_y \ominus p_y) \ominus (r_x \ominus p_x) \otimes (q_y \ominus p_y), \tag{2}$$

where $\otimes$ and $\ominus$ are floating-point multiplication and subtraction. The point p is distinguished in the computation of D'_{O2} and is called the pivot point. Many familiar mathematical properties like associativity or distributivity do not hold for floating-point operations. Therefore, while permuting p, q, and r will lead to sign changes only in D_{O2}, this is in general not the case for D'_{O2}. Our results can be reproduced using floating point arithmetic compliant to the IEEE 754 standard [2].

3   How It Fails

What can go wrong with this algorithm when a floating-point based orientation predicate is used? The search for t and t' may stop to soon, in which case the resulting polygon is not convex anymore. This may lead to incorrect output or more problems in later steps, since the correctness of the search for t and t' depends on P being convex! The search may also stop too late, in that case vertices of P are cut away and may not be contained in the end result.

Carrying the first case to the extreme, the search for both t' and t may stop at q. This can never occur geometrically, since for full dimensional P the upper and lower tangent touch P in different vertices. But it can occur due to a single incorrect result from the 2D orientation predicate, as illustrated in Fig. 2a. Here the current hull polygon P has vertices p, q and r and is about to be updated

Let

$$p = (-10.04094770362331879, \quad -7.506293383338360492)$$
$$q = (\quad 1.056089924324703055, \quad -0.9655180522057801307)$$
$$r = (\quad -5.5608992432470305545, \quad -2.255180522057801307)$$
$$u = (\quad 1.0560899243247048318, \quad -0.9655180522057800196)$$
$$u' = (\quad 1.0560899243247043877, \quad -0.9655180522057800196)$$

In the drawings below, we classify each point with floating-point coordinates near q if it is to the right (-), to the left (+), or on $(\bullet)$ line ℓ drawn solid. Correctly classified points are drawn in grey, while misclassified points are colored $(\text{-}, \text{+}, \bullet)$..

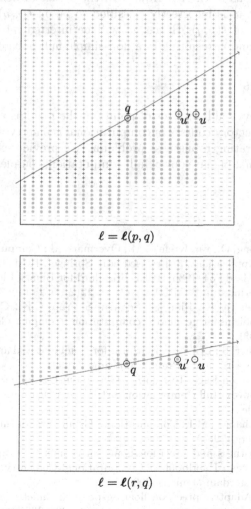

$$\ell = \boldsymbol{\ell}(p, q)$$

$$\ell = \boldsymbol{\ell}(r, q)$$

Fig. 3. Zoom in on points q and u in Fig. 2.

with point u. Figure 3 shows the lattice of points with floating-point coordinates close to q, among them u, and how the 2D orientation predicate classifies their position relative to $\ell(p, q)$ and $\ell(r, q)$. Point u is correctly classified to be right of $\ell(r, q)$, but incorrectly classified to be left of $\ell(p, q)$. Since we process points in sorted order, u must be outside P, but the incorrect classification moves u inside P, at least from the view of the algorithm. We have $t = t' = q$ and all vertices between t and t' are to be replaced by u. The implementation may now traverse the sequence of hull vertices, starting at t, and remove vertices until t' is reached. This will remove all vertices except q. Then u is inserted, resulting in the hull polygon shown in Fig. 2b.

For another example, suppose we are updating the current hull P with the point u', two positions to the left from u in the lattice of floating-point points in Fig. 3. The point u' is incorrectly classified to be on $\ell(r, q)$, and incorrectly classified to be left of $\ell(p, q)$. Hence, u' will be inserted into the sequence of vertices between q and r, resulting in a non-simple hull polygon.

4   Conclusions

Computing the convex hull of a set of points in the plane is a problem studied in almost all introductory courses in computational geometry and hence a prime candidate for illustrating robustness problems. We provide examples of failure for worst-case optimal planar convex hull computation, complementing the work of Kettner et al.

References

1. de Berg, M., Cheong, O., van Krefeld, M., Overmars, M.: Computational Geometry: Algorithms and Applications, 3rd revised edn. Springer, Heidelberg (2008)
2. ANSI, IEEE Standard 754–1985 : IEEE Standard for Binary Floating-Point Arithmetic (1985). Reprinted in SIGPLAN Notices 22(2), 9–25 (1987)
3. Kettner, L., Mehlhorn, K., Pion, S., Schirra, S., Yap, C.K.: Classroom examples of robustness problems in geometric computation. Comput. Geom. Theor. Appl. 40(1), 61–78 (2008)
4. Mehlhorn, K., Näher, S.: LEDA: A Platform for Combinatorial and Geometric Computing. Cambridge University Press, Cambridge (1999)
5. Mörig, M.: Companion Pages to Another Classroom Example of Robustness Problems in Planar Convex Hull Computation. http://wwwisg.cs.uni-magdeburg.de/ag/ClassroomExample/
6. Preparata, F.P., Shamos, M.I.: Computational Geometry: An Introduction, 1st edn. Springer, Heidelberg (1985)
7. Schirra, S.: Robustness and precision issues in geometric computation. In: Sack, J.R., Urrutia, J. (eds.) Handbook of Computational Geometry, chap. 14, pp. 597–632. Elsevier, Amsterdam, January 2000
8. Shewchuk, J.R.: Adaptive precision floating-point arithmetic and fast robust geometric predicates. Discrete Comput. Geom. 18(3), 305–363 (1997)
9. Yap, C.K.: Robust geometric computation. In: Handbook of Discrete and Computational Geometry, chap. 41, pp. 927–952, 2nd edn. CRC (2004)

Precision-Driven Computation in the Evaluation of Expression-Dags with Common Subexpressions: Problems and Solutions

Marc Mörig[(✉)] and Stefan Schirra[(✉)]

Department of Simulation and Graphics, Faculty of Computer Science,
Otto-von-Guericke University of Magdeburg, Universitätsplatz 2,
39106 Magdeburg, Germany
stschirr@ovgu.de

Abstract. Precision-driven computation is a recursive scheme for the approximate evaluation of arithmetic expression-dags that allows for specifying the accuracy of evaluation results in advance. We illustrate and explain how current implementations of precision driven arithmetic may negate advantages from sharing common subexpressions by re-evaluating these subexpressions many times. Since the number of re-evaluations depends on seemingly minor details of expression structure and evaluation strategy, significant performance differences may arise between otherwise competitive implementations of precision driven arithmetic for the same user code and then again between otherwise equivalent user codes for the same evaluation strategy as well. We present a new evaluation strategy that separates precision propagation from expression evaluation and thereby avoids multiple evaluations of common subexpressions completely.

Keywords: Precision-driven computation · Exact geometric computation · Expression-dag-based number types · Verified numerical computing

1 Introduction

Algorithms in computational geometry are designed under the assumption of exact real arithmetic at unit cost [13]. Simply using hardware floating-point arithmetic as an indiscriminate substitute in implementations almost inevitably leads to robustness problems: implementations may crash, loop forever, or silently compute garbage [5], due to inconsistent decisions caused by rounding errors [14,19].

An effective approach to overcome these robustness issues is the exact geometric computation paradigm [17] which calls for correct decisions by geometric predicates. Correctly computing the signs of arithmetic expressions in geometric predicates trivially ensures correct decisions and hence consistency. Regarding control flow, an implementation behaves as its theoretical counterpart.

© Springer International Publishing Switzerland 2016
I.S. Kotsireas et al. (Eds.): MACIS 2015, LNCS 9582, pp. 451–465, 2016.
DOI: 10.1007/978-3-319-32859-1_39

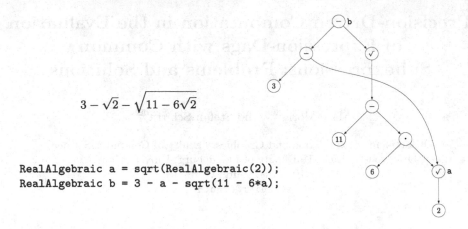

$$3 - \sqrt{2} - \sqrt{11 - 6\sqrt{2}}$$

```
RealAlgebraic a = sqrt(RealAlgebraic(2));
RealAlgebraic b = 3 - a - sqrt(11 - 6*a);
```

Fig. 1. An expression, a corresponding dag and code leading to this dag.

Thus, the exact geometric computation paradigm ensures topological and combinatorial correctness. However, since we now detect degeneracies correctly, an implementation has to handle them appropriately. Note that the exact geometric computation paradigm does not ask for exact numerical values. All we need are sufficiently accurate approximations that allow us to determine the requested signs correctly.

Recording the computation history of numerical values in expression-dags, i.e., expression-"trees" that may share common subexpressions, allows one to (re)compute an approximation of the value of the expression at any time at any accuracy. Figure 1 shows an expression-dag for a simple expression with square root operations. Using the expression-dag we can adaptively compute better and better bigfloat approximations and defer high precision computations until we really need them. We use constructive zero separation bounds [3,6,9,12, 15] to resolve the cases where the actual value of an expression is zero. This way, we can adaptively compute the sign of an expression correctly. Since all sign computations and hence all decisions in geometric predicates are exact, inconsistencies caused by numerical imprecision are abandoned.

Precision-driven computation [20] is a key technique for the efficient evaluation of expression-dags. In order to improve accuracy, we could simply re-evaluate the dag with higher precision starting from the leaves and determine the resulting accuracies along the way. Precision-driven computation, however, starts at the root and recursively specifies accuracy of evaluation results in advance. We present more details of precision-driven computation in Sect. 3.

2   Expression-Dag-Based Number Types

Number types CORE::Expr [4,21] and leda::real [2,7] encapsulate precision-driven adaptive evaluation of expression-dags in C++ classes. Thanks to the wrapping a user need not know anything about the details of the implementation in order to get correct decisions. More recently, `RealAlgebraic`, another expression-dag-based number type has been designed and implemented [10,11]. Like the most recent version of `CORE::Expr` this number type is a C++ class template. This makes `RealAlgebraic` fairly adjustable. For instance, `RealAlgebraic` allows one to exchange the bigfloat arithmetic used to compute approximations, to select a floating-point filter, to use different strategies for deferring dag-construction, e.g., by using error-free floating-point transformations or adding tests that check whether the result of a floating-point computation is exact. `RealAlgebraic` provides a default variant. Like `CORE::Expr` and `leda::real` it supports a subset of the real algebraic numbers that includes the rational numbers and is closed under the basic arithmetic operations $\pm$, $-$, $\cdot$, $/$ and $\sqrt[d]{\ }$. For a discussion of the use of such number types in geometric computing we refer the reader to [8,16].

3   Precision-Driven Computation

The purpose of expression-dag-based number-types is computing the sign of expressions correctly. To compute the sign of the expression represented by a dag node v, current expression-dag-based number types compute increasingly accurate bigfloat approximations and maintain error bounds. For example, in `RealAlgebraic` each node v stores an approximation $\hat{v}$ and an error $\mathbf{e}_v$ such that

$$|\hat{v} - v| \le \mathbf{e}_v.$$

Here and in the sequel we use v to denote both a node and the exact value of the associated expression. The sign of the node is known, once $|\hat{v}| > \mathbf{e}_v$ or $|\hat{v}| + \mathbf{e}_v$ is smaller than the computed zero separation bound for the expression represented by v. In the latter case, we have $v = 0$.

To compute or improve the approximation of a dag node, straightforward interval arithmetic may be used: for example, let $z = x \cdot y$ be a multiplication node. Then

$$|\hat{z} - z| \le |\hat{z} - \hat{x}\hat{y}| + |\hat{x}(\hat{y} - y)| + |y(\hat{x} - x)|$$
$$\le |\hat{z} - \hat{x}\hat{y}| + |\hat{x}|\mathbf{e}_y + (|\hat{y}| + \mathbf{e}_y)\mathbf{e}_x. \tag{1}$$

Thus, if we compute $\hat{z} \leftarrow \hat{x} \odot_p \hat{y}$ using bigfloat arithmetic with a relative error of 2^{-p}, we can set $\mathbf{e}_z \leftarrow 2^{-p}|\hat{z}| + |\hat{x}|\mathbf{e}_y + (|\hat{y}| + \mathbf{e}_y)\mathbf{e}_x$, where the right hand side is computed using low precision bigfloat arithmetic with rounding away from zero. Similar error estimates exist for the remaining operations $\pm$, $/$, $\sqrt[d]{\ }$. We may therefore improve the approximation of all dag nodes by increasing the precision p and recomputing all dag nodes, reporting the resulting error from bottom to top. There is a clear disadvantage to this method: the accuracy of the final result becomes known only after the computation of the final approximation.

Precision-driven computation on the other hand specifies the accuracy before recomputing an approximation. Current implementations work in a recursive fashion: to compute an approximation of a node they first compute approximations for its children. We illustrate this by describing how multiplication is handled by `RealAlgebraic`. The starting point is an error estimate quite similar to Eq. (1):

$$\begin{aligned}
|\hat{z} - z| &\leq |\hat{z} - \hat{x}\hat{y}| + |\hat{x}(\hat{y} - y)| + |y(\hat{x} - x)| \\
&\leq 2^{-p}|\hat{x}\hat{y}| + (|x| + e_x)e_y + |y|e_x \\
&\leq 2^{-p}|\hat{x}\hat{y}| + |x|e_y + |y|e_x + e_x e_y.
\end{aligned} \tag{2}$$

Assume we want to compute $\hat{z}, e_z$ such that $e_z < \tilde{e}_z$. First we select $\tilde{e}_x$, and $\tilde{e}_y$ such that

$$\tilde{e}_x \leq \frac{\tilde{e}_z}{4|y|}, \qquad \tilde{e}_y \leq \frac{\tilde{e}_z}{4|x|}, \qquad \tilde{e}_x \tilde{e}_y \leq \frac{\tilde{e}_z}{4},$$

and recursively recompute $\hat{x}$ and $\hat{y}$, requesting $e_x \leq \tilde{e}_x$ and $e_y \leq \tilde{e}_y$. Then we select a precision p such that

$$2^{-p} \leq \frac{\tilde{e}_z}{4|\hat{x}\hat{y}|}$$

and recompute $\hat{z} \leftarrow \hat{x} \odot_p \hat{y}$ with a relative error of 2^{-p}. Finally, by Eq. (2), we can set $e_z \leftarrow \tilde{e}_z$. In order to perform precision-driven computation, we need error estimates like Eq. (2) for all involved operations. Yap [18] provides such estimates for $\pm, \cdot, /, \sqrt[d]{\ }$, and a few elementary functions, both for relative and absolute error.

Other expression-dag-based number types perform precision-driven computation in a slightly different way. `leda::real` uses the error estimate

$$|\hat{z} - z| \leq 2^{-p}|\hat{x}\hat{y}| + |\hat{x}|e_y + |y|e_x$$

instead of Eq. (2), see [1]. It has only three terms and allows to select a slightly smaller precision (i.e., $2^{-p} \leq \tilde{e}_z/(2|\hat{x}\hat{y}|)$). However, the estimate forces us to recompute $\hat{x}$ before we can select $\tilde{e}_y$, since no upper bound on the future value of $\hat{x}$ can be determined easily. The same holds for division operations: one child node is recomputed before the accuracy requirement for the other child is determined.

`CORE::Expr` [18] ignores the second order error term $e_x e_y$ in Eq. (2). This is accounted for by recomputing the complete approximation using interval arithmetic once the precision p has been determined. Prior to version 2, `CORE::Expr` performs ring operations $\pm$ and $\cdot$ exactly by adjusting the bigfloat precision, which allows one to omit the error term $2^{-p}|\hat{x}\hat{y}|$ in Eq. (2).

Since accuracies are chosen recursively, the name precision-driven is a bit deceptive. Accuracy-driven might have been more self-explanatory.

4   Common Problems with Common Subexpressions

A strength of adaptive exact decisions number types based on expression-dags is their user-friendliness. Ideally, you can use these number-types like any other number type without knowing anything about the internals. However, as a user you probably expect these number types to show roughly the same performance when exchanging operands in addition, multiplication, or equality testing, since these operations are commutative. Unfortunately, with current implementations of precision-driven computation number types `leda::real`, `CORE::Expr`, and `RealAlgebraic` do not necessarily behave like this in the presence of common subexpressions. In principle, having common subexpressions instead of several copies of the same expression is advantageous. In particular, it allows us to compute better separation bounds. The main advantage, however, is the potential reduction of evaluation cost. However, current precision-driven evaluation strategies may void this advantage.

In order to illustrate the problem we pick an example from the test suite of `RealAlgebraic`: verification of the formula

$$\sum_{i=0}^{n-1} r^i = \frac{1 - r^n}{1 - r} \qquad \text{for } r \neq 1, \tag{3}$$

for geometric series. Figure 2 shows a generic implementation parameterized by a number type which verifies Eq. (3) for any r and n. Note how the code reuses r^{i-1} for the computation of r^i.

```
template <class NumberType>
bool geom_series(const NumberType r, const int n)
{ NumberType s=0, p=1;
  for(int i=0;i<n;i++)
  { s = s + p;
    p = p * r;
  }
  NumberType t = ( 1 - p )/( 1 - r );
  return (t==s);
}
```

Fig. 2. Verification code for Eq. (3).

When running the code in Fig. 2 for the default variant of `RealAlgebra-ic` with $r = \sqrt{13}$ and $n = 2048$ on our test platform this takes 0.12 seconds. However, if we replace (`t==s`) in the last line of Fig. 2 by (`s==t`) and repeat the experiment, suddenly this takes 61.49 seconds. The running time explodes. A small code change leads to a huge difference in running times. This is not user-friendly at all.

Let us have a closer look. Bigfloat arithmetic is the main cost factor in expression-dag evaluation, so we repeat both experiments for $n = 128$ and check the amount of bigfloat arithmetic used. It turns out that there is a large difference in the usage of multiplication, while the difference for $\pm, /, \sqrt{\ }$ is negligible.

Figure 3 shows how often a bigfloat multiplication of a certain precision is performed when running the code in Fig. 2 and its modification. Actually, we conflate precisions corresponding to the same number of limbs, i.e., the number of computer words used in bigfloat arithmetic. It is clearly visible that the problem is not due to higher precision in bigfloat operations. However, many more operations of roughly the same precision are performed by the (s==t) variant. The y-axis is logarithmic, so the number of bigfloat multiplications roughly squares.

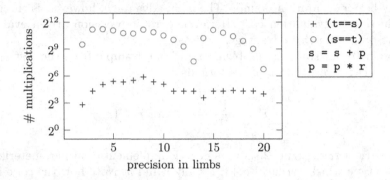

Fig. 3. Number of bigfloat multiplications for $n = 128$ for `RealAlgebraic`.

With (t==s) we get an expression-dag like that shown in Fig. 4a while we get an expression-dag like that shown in Fig. 4b if we replace (t==s) by (s==t). Since in both cases the sign of the root node is zero, the value at the root node has to be approximated repeatedly until we reach the separation bound. Using $r = \sqrt{13}$, we also make sure that no node becomes known exactly at some point. Let us consider a single step of precision-driven computation at the root node. For binary operations, the precision driven computation of `RealAlgebraic` first recursively improves the first operand, then the second operand and finally recomputes the value at the node itself.

Due to the order of traversal, in the case of (t==s), we visit each multiplication node coming from the right side first, where left and right corresponds to the drawings in Fig. 4. Later we arrive at this node again, coming from the left side. Upon the second arrival, we again request an approximation of certain accuracy from this node. If we are lucky, we need not recompute the approximation. This is what happens in case (t==s). In case (s==t) we arrive from the left side first. Later we arrive at a node again, coming from the right side. This time the present accuracy is insufficient and we must recompute the approximation of this node and much worse, all nodes below. Thus, we have to perform quadratically many bigfloat multiplications, while in case (t==s) the value at each node is recomputed only once.

In case (s==t), we approximate each multiplication node several times, with increasing accuracy, until finally reaching the necessary maximal accuracy. This increase in precision is not visible in the coarse statistics in Fig. 3 which conflates

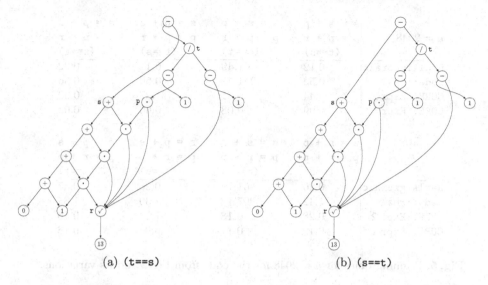

(a) (t==s)   (b) (s==t)

Fig. 4. Expression-dags generated for $r = \sqrt{13}$ and $n = 4$.

precisions corresponding to the same number of limbs. We only see that more operations with the same number of limbs are performed. Thus the already available approximation can be slightly insufficient only, making the difference invisible.

We can reproduce such behavior with more or less significant impact on running time for all existent expression-dag-based C++ number types, not just RealAlgebraic. All our experiments were run on an Intel Core i5-660 processor with 3.33 GHz running Ubuntu. The code is compiled using g++ 4.6.3 with optimization flags -O3. We use the precompiled leda::real coming with LEDA 6.4 and CORE::Expr version 1.8 as shipped with CGAL 4.6. Furthermore, we use CORE::Expr version 2.1.1 in addition because of the redesign of CORE::Expr starting with version 2.0 [21].

Besides the equality test in the last line, there are two more commutative operations, s = s + p and p = p * r, in the code in Fig. 2. We study swapping operands in these operations in our experiments as well. Figure 5 shows running times for $n = 2048$, $r = \sqrt{13}$, and all combinations of swaps for RealAlgebraic, leda::real, CORE::Expr 1, and CORE::Expr 2. Since leda::real uses an evaluation strategy very similar to RealAlgebraic, it shows the same behavior on exactly the same examples. CORE::Expr 2 slows down significantly if we swap the operands in the addition operation from s = s + p to s = p + s. Again, a small code change leads to a big change in running time.

Regarding, CORE::Expr 1, the impact of operand swapping is much less severe. CORE::Expr 1 slows down if we use (s==t) and swap the operands in the multiplication operation from p = p * r to p = r * p. Expression-dags illustrating the bad cases for CORE::Expr are shown in Fig. 6. The precision-driven computation in CORE::Expr 1 performs ring operations $\pm$, $\cdot$ exactly and

$n = 2048$	s = s + p p = p * r (t==s)	s = s + p p = p * r (s==t)	s = p + s p = p * r (t==s)	s = p + s p = p * r (s==t)
RealAlgebraic	0.12	61.49	0.12	0.22
leda::real	0.52	291.45	0.50	0.96
CORE::Expr 2	0.44	0.31	37.27	0.32
CORE::Expr 1	0.09	0.08	0.06	0.08

	s = s + p p = r * p (t==s)	s = s + p p = r * p (s==t)	s = p + s p = r * p (t==s)	s = p + s p = r * p (s==t)
RealAlgebraic	0.26	61.78	0.26	0.37
leda::real	1.42	297.13	1.39	1.84
CORE::Expr 2	0.28	0.18	37.18	0.16
CORE::Expr 1	0.12	0.17	0.06	0.16

Fig. 5. Running times for $n = 2048$ for the code from Fig. 2 and its variations.

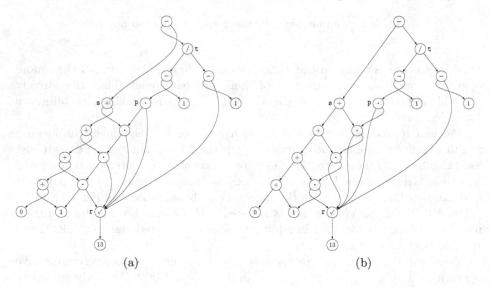

(a)   (b)

Fig. 6. Bad expression-dags for CORE::Expr 2 and CORE::Expr 1.

actually computes the error bound for a node, instead of setting it to the requested value. Therefore, the actual error stored in a node may be smaller than originally requested, which helps to reduce the number of re-evaluations in this particular scenario. However, this does not always help, as the accuracy requirements in repeated visits may differ much more than can be bridged by a slightly better error bound computation. With CORE::Expr 1, the slow-down is largely due to repeated evaluations of $\sqrt{13}$. If we start the experiments such that a sufficiently accurate approximation at the $\sqrt{13}$ node is already available, the measured running times for CORE::Expr 1 are the same for all code variations.

In very bad cases, the evaluation strategy totally negates the advantage of sharing common subexpressions. The dag is evaluated as if it were an expression-tree, storing copies of an identical subexpression for each reference to it. Any dag sharing common subexpressions is prone to such problems. If a node is referenced several times, there is a good chance that all parents have different accuracy requirements on this node. If we arrive from a parent with low requirements first, we must recompute the node upon a later arrival. Since the accuracy requirements propagated by precision-driven computation are usually tight, this will almost always trigger a recomputation of all descendent nodes. Even if the problem does not appear in cascaded form, we might easily loose a factor of two by going wrong just once near the root node. This happens for example for `RealAlgebraic` when we use $s = p + s$, $p = p * r$, and swap the operands in the equality test, see columns 3 and 4 in the top table in Fig. 5, as well as the top part of Fig. 8.

5   Improved Evaluation Strategies

How can we avoid such unexpected dependence on the order of operands in commutative operations? The error estimates used in `leda::real` require to recompute the approximation of one child node before the accuracy requirement for the other child can be computed. `CORE::Expr` and `RealAlgebraic` do not have such a requirement. Their error estimates allow us to recur to any child first. It is however by no means clear how to choose an order to avoid recomputation for all nodes in the dag globally. However, we can choose at random. If we do this for `RealAlgebraic`, we get the running times shown in Fig. 7 in rows labeled randomized. The corresponding usage of bigfloat arithmetic for $n = 128$ is shown in the middle part of Fig. 8, labeled randomized.

$n = 2048$	s = s + p p = p * r (t==s)	s = s + p p = p * r (s==t)	s = p + s p = p * r (t==s)	s = p + s p = p * r (s==t)
RealAlgebraic	0.12	61.49	0.12	0.22
randomized	0.31	0.35	0.22	0.38
topsorted	0.12	0.12	0.12	0.12

	s = s + p p = r * p (t==s)	s = s + p p = r * p (s==t)	s = p + s p = r * p (t==s)	s = p + s p = r * p (s==t)
RealAlgebraic	0.26	61.78	0.26	0.37
randomized	0.23	0.26	0.27	0.24
topsorted	0.12	0.12	0.12	0.12

Fig. 7. Running times for $n = 2048$ for the code from Fig. 2 and its variations.

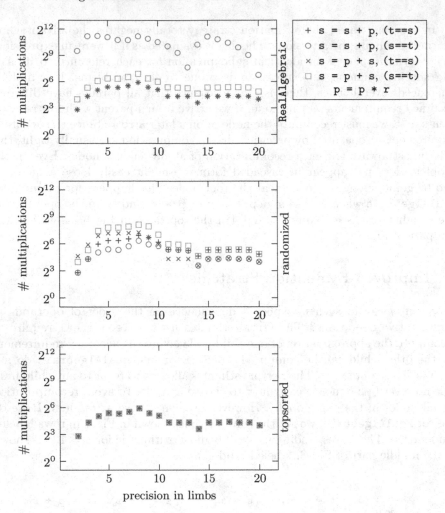

Fig. 8. Number of bigfloat multiplications for $n = 128$ for `RealAlgebraic`, the randomized variant, and the topologically sorting variant for some variations of the code in Fig. 2.

With the randomized strategy, both running time and number of bigfloat multiplications are significantly lower in the (`s==t`) case compared to the default version of `RealAlgebraic`, but somewhat higher in the (`t==s`) case. By randomizing dag traversal, the expected number of node re-evaluations in our example becomes linear: It suffices to consider the dag rooted at `s` for the analysis. Let $M(i)$ be the number of bigfloat multiplications triggered by the addition node with distance i to node `s`. Then

$$M(i) = \begin{cases} i - 1 & \text{if we traverse to the right first,} \\ i - 1 + M(i - 1) & \text{otherwise.} \end{cases}$$

Thus the expected number of multiplications is bounded by $2(i-1)$. On average, we perform at most twice the optimal number of bigfloat multiplications, which our experiments nicely confirm. Randomization helps, but the average number of bigfloat operations in a randomized evaluation strategy is still larger than the number of bigfloat multiplications of the default version of `RealAlgebraic` in a favorable case. Furthermore, randomization makes the behavior less predictable. There still might be significant performance differences even without making any changes in the code.

The key observation for improvement is that the error estimate in Eq. (2) and like estimates for other operations do not require to recompute approximations of child nodes at all. A node can wait until all its parents have registered their accuracy requirements and all its children have recomputed their approximation. Only then we have to compute a new approximation. To this end, we store the requested error $\tilde{e}_v$ in each dag node v. This allows us to improve the approximation of a node using the following algorithm after sorting nodes topologically.

Algorithm 1. (Topological Precision Driven Arithmetic).
Let u be a dag node and $e > 0$. Then TOPSORTEDPRECISIONDRIVEN *computes an approximation $\hat{u}$ such that*

$$|u - \hat{u}| \leq e.$$

1: **procedure** TOPSORTEDPRECISIONDRIVEN (u, e)
2:     $\tilde{e}_u \leftarrow \min\{e, \tilde{e}_u\}$
3:     *let D be the dag rooted at u*
4:     **for** *all nodes $v \in D$ in topological order* **do**
5:         **if** $\tilde{e}_v < e_v$ **then**
6:             **for** *all children w of v* **do**
7:                 *compute error bound r to be requested from w*
8:                 $\tilde{e}_w \leftarrow \min\{r, \tilde{e}_w\}$
9:     **for** *all nodes $v \in D$ in reverse topological order* **do**
10:         **if** $\tilde{e}_v < e_v$ **then**
11:             *compute necessary precision p and recompute $\hat{v}$*
12:         $e_v \leftarrow \tilde{e}_v$

We need to initialize $\tilde{e}_v$ only once when v is created, since it never becomes larger than e_v. By processing nodes in topological order in the first stage, we ensure that upon arriving at a node v, all its parents have registered their accuracy requirements. At this point we have all the data necessary to compute the requirements from v to its children. In the second stage, processing nodes in reverse topological order ensures that the children of a node have already been re-evaluated with the necessary accuracy.

With the new strategy, we get the running times shown in Fig. 7 in rows labeled topsorted. The corresponding usage of bigfloat arithmetic for $n = 128$ is shown in the bottom part of Fig. 8, labeled topsorted. Sorting the nodes of a dag topologically takes linear time in the size of the dag, so there is no asymptotic disadvantage compared to other traversal methods. In our experiments, there is no running time overhead for topological sorting observable.

While the geometric series example magnifies the effect, problems with the evaluation of common subexpressions are also present in other scenarios, but often much less noticeable. We use computation of the convex hull of intersection points of line segments to illustrate this in the context of cascaded geometric computation. The segments we generate have random positive integer endpoints less than 280. We then compute all intersection points of these segments by brute-force. The coordinates of the input points for the convex hull algorithm are rational numbers. We use intersection points as input for the convex hull computation in order to make the problem arithmetically more demanding. Due to the construction there are many collinearities among the points, see Fig. 9, so some degenerate orientation tests might arise in the subsequent convex hull computation. We measure the time only for the convex hull computation, after all segment intersection points have been computed and the expression-dags for their coordinates have been constructed.

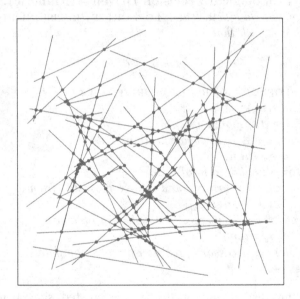

Fig. 9. Intersection points for convex hull computation.

Figure 10 plots running times for the default version of `RealAlgebraic` and the new variant that uses topological sorting. It shows that the new variant roughly achieves the same running time, sometimes it is even faster. Orientation tests in the convex hull computation involve rational coordinates of three points each. In the implementation of the orientation test, the coordinates of the point chosen as pivot element appear twice. Hence, common subexpressions arise. Furthermore, for each point, its rational coordinates have a common denominator, so there are common subexpressions in the expression-dags

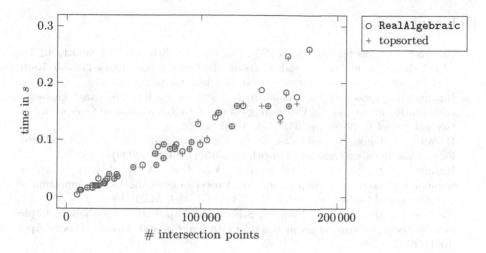

Fig. 10. Running times for convex hull of intersection points.

of the input data already. Apparently, in many cases, the new evaluation strategy saves re-evaluation time for common subexpression thereby compensating for the additional cost of topological sorting.

6   Conclusions

As illustrated by our experiments, the performance of the currently available expression-dag-based number types `leda::real` and `CORE::Expr` can depend on the order of operands in commutative operations. The same holds for the default version of `RealAlgebraic`. Clearly it is undesirable that the running time is sensitive to such small code changes. A user may rewrite a part of her code and end up with significantly worse performance. Furthermore, the fact that different number types may favor different operand orders makes their experimental comparison quite difficult. One of them might seem inefficient only because expressions are created in a way favoring the other number type(s).

We present a new evaluation strategy which uses topological sorting to determine an evaluation order prior to precision-driven evaluation. Our new precision-driven evaluation strategy nicely eliminates the user-unfriendly dependence on operand order at marginal additional cost, if any. Another advantage of the algorithm is that it may be parallelized to the amount allowed by the expression-dag. Any two nodes not connected by a directed path can be processed in parallel in both stages. This is especially interesting for the second stage, where expensive bigfloat operations are performed.

References

1. Burnikel, C., Fleischer, R., Funke, S., Mehlhorn, K., Schirra, S., Schmitt, S.: The LEDA class real number - extended version. Technical report, ECG-TR-363110-01, Max-Planck-Institut für Informatik, Saarbrücken, Germany (2005)
2. Burnikel, C., Fleischer, R., Mehlhorn, K., Schirra, S.: Efficient exact geometric computation made easy. In: Proceedings of the 15th Symposium on Computational Geometry (SoCG 1999), pp. 341–350. ACM (1999)
3. Burnikel, C., Funke, S., Mehlhorn, K., Schirra, S., Schmitt, S.: A separation bound for real algebraic expressions. Algorithmica **55**(1), 14–28 (2009)
4. Karamcheti, V., Li, C., Pechtchanski, I., Yap, C.K.: A core library for robust numeric and geometric computation. In: Proceedings of the 15th Symposium on Computational Geometry (SoCG 1999), pp. 351–359. ACM (1999)
5. Kettner, L., Mehlhorn, K., Pion, S., Schirra, S., Yap, C.K.: Classroom examples of robustness problems in geometric computation. Comput. Geom.: Theory Appl. **40**(1), 61–78 (2008)
6. Li, C., Yap, C.K.: A new constructive root bound for algebraic expressions. In: Proceedings of the 12th ACM-SIAM Symposium on Discrete Algorithms (SODA 2001), pp. 496–505. SIAM (2001)
7. Mehlhorn, K., Näher, S.: LEDA: A Platform for Combinatorial and Geometric Computing. Cambridge University Press, Cambridge (1999)
8. Mehlhorn, K., Schirra, S.: Exact computation with leda_real - theory and geometric applications. In: Alefeld, G., Rohn, J., Rump, S.M., Yamamoto, T. (eds.) Symbolic Algebraic Methods and Verification Methods. Springer Mathematics, pp. 163–172. Springer, Wien, Austria (2001)
9. Mignotte, M.: Identification of algebraic numbers. J. Algorithms **3**, 197–204 (1982)
10. Mörig, M.: Algorithm Engineering for Expression Dag Based Number Types. Ph.D. thesis, Otto-von-Guericke-Universität Magdeburg (2015)
11. Mörig, M., Rössling, I., Schirra, S.: On the design and implementation of a generic number type for real algebraic number computations based on expression dags. Math. Comput. Sci. **4**(4), 539–556 (2010)
12. Pion, S., Yap, C.K.: Constructive root bound for k-ary rational input numbers. Theor. Comput. Sci. **369**(1–3), 361–376 (2006)
13. Preparata, F.P., Shamos, M.I.: Computational Geometry: An Introduction. Monographs in Computer Science, 1st edn. Springer, New York (1985)
14. Schirra, S.: Robustness and precision issues in geometric computation. In: Sack, J.R., Urrutia, J. (eds.) Handbook of Computational Geometry, chap. 14, pp. 597–632. Elsevier, Amsterdam, The Netherlands (2000)
15. Schirra, S.: Much ado about zero. In: Albers, S., Alt, H., Näher, S. (eds.) Efficient Algorithms. LNCS, vol. 5760, pp. 408–421. Springer, Heidelberg (2009)
16. Schirra, S.: On the use of adaptive, exact decisions number types based on expression-dags in geometric computing. In: 26th Canadian Conference on Computational Geometry (CCCG 2014) (2014)
17. Yap, C.K.: Towards exact geometric computation. Comput. Geom.: Theory Appl. **7**(1–2), 3–23 (1997)
18. Yap, C.K.: On guaranteed accuracy computation. In: Geometric Computation, pp. 322–373. World Scientific (2004)
19. Yap, C.K.: Robust geometric computation. In: Handbook of Discrete and Computational Geometry, 2nd edn., chap. 41, pp. 927–952. CRC (2004)

20. Yap, C.K., Dubé, T.: The exact computation paradigm. In: Computing in Euclidean Geometry, 2nd edn., pp. 452–486. World Scientific (1995)
21. Yu, J., Yap, C., Du, Z., Pion, S., Brönnimann, H.: The design of Core 2: a library for exact numeric computation in geometry and algebra. In: Fukuda, K., Hoeven, J., Joswig, M., Takayama, N. (eds.) ICMS 2010. LNCS, vol. 6327, pp. 121–141. Springer, Heidelberg (2010)

Real Complexity: Theory and Practice

Rigorous Numerical Computation of Polynomial Differential Equations Over Unbounded Domains

Olivier Bournez[1], Daniel S. Graça[2,3]([✉]), and Amaury Pouly[1,2]

[1] LIX, Ecole Polytechnique, 91128 Palaiseau Cedex, France
[2] CEDMES/FCT, Universidade do Algarve, C. Gambelas, 8005-139 Faro, Portugal
dgraca@ualg.pt
[3] SQIG/Instituto de Telecomunicações, Lisbon, Portugal

Abstract. In this abstract we present a rigorous numerical algorithm which solves initial-value problems (IVPs) defined with polynomial differential equations (i.e. IVPs of the type $y' = p(t,y)$, $y(t_0) = y_0$, where p is a vector of polynomials) for any value of t. The inputs of the algorithm are the data defining the initial-value problem, the time T at which we want to compute the solution of the IVP, and the maximum allowable error $\varepsilon > 0$. Using these inputs, the algorithm will output a value $\tilde{y}_T$ such that $\|\tilde{y}_T - y(T)\| \leq \varepsilon$ in time polynomial in T, $-\log \varepsilon$, and in several quantities related to the polynomial IVP.

1 Introduction

With the appearance of fast and cheap digital computing devices in the last decades, digital computers have become increasingly important as a simulation tool in many fields, ranging from weather forecast to finance. The idea underlying such simulations is simple: pick some system which we want to study and simulate it on a computer using some numerical method. Quite often we can obtain in this manner information about the system which we could not collect otherwise. Think, for example, about the case of weather forecast.

However, this poses a fundamental question: how reliable are these simulations? The truth is that, although historically such simulations have already given fundamental insights (like suggesting that dynamical systems can have *strange attractors* [6]), due to phenomena like sensitive dependence on initial conditions, in general not much is known about the overall error committed in such simulations.

It therefore seems to make sense to develop numerical methods with the property that we can rigorously tell which is the error done when we apply such methods. This is in contrast to what happens usually in numerical analysis where, at best, only estimates of the error are presented. On the other side, to obtain rigorous bounds on the error, we need to use more complicated methods, which are more amenable for analysis, and which are usually slower or might even be unfeasible for practical implementation. In general, it is not trivial to devise numerical methods which are practical to use and for which the error can be rigorously determined.

© Springer International Publishing Switzerland 2016
I.S. Kotsireas et al. (Eds.): MACIS 2015, LNCS 9582, pp. 469–473, 2016.
DOI: 10.1007/978-3-319-32859-1_40

To achieve a balance between these contradicting requirements, it makes sense to consider restricted classes of problems, which are nonetheless general enough to be of practical importance.

In this paper, we consider initial-value problems (IVPs) defined with polynomial ordinary differential equations (ODEs)

$$\begin{cases} y'(t) = p(y) \\ y(t_0) = y_0 \end{cases} \tag{1}$$

where p is a vector of polynomials. We consider, without loss of generality that the system is autonomous since the independent variable t can always be written as an extra variable y_{n+1} satisfying $y'_{n+1} = 1$. We note that almost any IVP written with the "usual" functions of Analysis (trigonometric functions, exponentials, their inverses and compositions, etc.) can always be rewritten as a polynomial ODE (see e.g. [3,9]).

Therefore IVPs with the format (1) are sufficiently broad to include a wide range of IVPs of practical interest. Moreover, since the right-hand side consists of relatively simple functions (polynomials), we are able to rigorously analyze the error committed when we solve numerically (1) by using properties of polynomials.

2   Solving IVPs Over Unbounded Domains

It is standard practice to analyze numerical methods which solve IVPs only over a compact time interval $[0, T]$. This is both true in the Numerical Analysis literature (see e.g. [1]) as it is in the Theoretical Computer Science literature (see e.g. [5]).

However, in practice, people seldom set a valid time interval $[0, T]$ before implementing a numerical procedure, be it for the simple reason that they sometimes do not even know which might be the relevant value for T before doing some numerical simulations.

Therefore, it seems desirable to devise numerical methods which make no prior assumptions on the values which T might take. Of course, the time needed to execute the algorithm (the computational complexity) depends on T: in general, the higher T, the more time the algorithm will take to execute, but it seems to be a non-trivial task to determine which is the dependence of the execution time of the algorithm with respect to T.

There is a "conventional wisdom" that the unbounded time domain case can be reduced to the bounded time one, for which many results exist (see e.g. [4,5]). However, this is not true, since in the bounded case many parameters which are important for the (unbounded) complexity are hidden in the constant of the "big-O" notation. A very simple example illustrates this problem. Assume that $y : I \to \mathbb{R}^d$ is the solution of

$$\begin{cases} y_1(0) = 1 \\ y_2(0) = 1 \\ \dots \\ y_n(0) = 1 \end{cases} \qquad \begin{cases} y'_1(t) = y_1(t) \\ y'_2(t) = y_1(t)y_2(t) \\ \dots \\ y'_d(t) = y_1(t)\cdots y_n(t) \end{cases}$$

It follows from [7] that for any fixed, compact I, y is polynomial time computable. On the other hand, this system can be solved explicitly and yields:

$$y_1(t) = e^t \qquad y_{n+1}(t) = e^{y_n(t)-1} \qquad y_d(t) = e^{e^{\cdot^{\cdot^{\cdot^{e^{e^t}-1}}}}-1}$$

One immediately sees that, since y is a tower of exponentials, y cannot be polynomial time computable over $\mathbb{R}$.

Note that this discrepancy arises because, in the bounded time case, the size of compact I is not taken as a parameter of the problem (because it is fixed). Also note that the dimension d of the system is hardly ever taken into account, although it has a huge influence on the resulting complexity. More precisely, if I is bounded then the complexity of computing $y(t)$ can be seen to be polynomial in t, but more than exponential in the length of the interval I and on d: this part is usually hidden in the "big-O" part of the constants.

3   Contributions

The main contribution of this abstract is to show that there is a numerical method, which we denote as SolvePIVPEx (see Sect. 4 for more details), which can rigorously solve IVPs (1) over unbounded domains.

Theorem 1 (Complexity and Correctness of SolvePIVPEx**).** *Let* $t \in \mathbb{R}$, $\varepsilon > 0$, *and assume that* y *satisfies* (1) *over* $[t_0, t]$. *Let*

$$x = \text{SolvePIVPEx}(t_0, y_0, p, t, \varepsilon)$$

where SolvePIVPEx *is a numerical method described in Sect. 4. Then*

- $\| x - y(t) \| \leqslant \varepsilon$
- *the arithmetic complexity of the algorithm is bounded by*

$$\text{poly}(k^d, \text{Len}(t_0, t), \log \| y_0 \|, -\log \varepsilon)$$

- *the bit complexity of the algorithm is bounded by*

$$\text{poly}(k, \text{Len}(t_0, t), \log \| y_0 \|, \log \Sigma p, -\log \varepsilon)^d$$

where k *is the maximum degree of the components of* p, d *is the number of components of* p, Σp *is the sum of the absolute values of the coefficients of* p, *and* $\text{Len}(t_0, t)$ *is a bound on the length of the curve* $y(\cdot)$ *from the point* $(t_0, y(t_0))$ *to the point* $(t, y(t))$.

4 The Numerical Method SolvePIVPEx and Sketch of the Proof of Theorem 1

For reasons of space, we will not present the algorithm defining the numerical method SolvePIVPEx nor the detailed proof of Theorem 1 (see [8] for more details). However, in this section, we briefly sketch the ideas underlying SolvePIVPEx and the proof of Theorem 1.

The numerical method SolvePIVPEx is based on a generic adaptive Taylor meta-algorithm which numerically solves (1). This is a meta-algorithm in the sense that, in a first approach, we leave open the question of how we choose some of the parameters of the algorithm. The goal of this meta-algorithm is, given as input $t \in \mathbb{Q}$ and $0 < \varepsilon < 1$ and the initial condition of (1), to compute $x \in \mathbb{Q}^d$ such that $\| x - y(t) \| < \varepsilon$.

We assume that the meta-algorithm uses the following values:

- $n \in \mathbb{N}$ is the number of steps of the algorithm
- $t_0 < t_1 < \ldots < t_n = t$ are the intermediate times
- $\delta t_i = t_{i+1} - t_i \in \mathbb{Q}$ are the time steps
- for $i \in \{0, \ldots, n-1\}$, $\omega_i \in \mathbb{N}$ is the order at time t_i and $\mu_i > 0$ is the rounding error at time t_i
- $\tilde{y}_i \in \mathbb{Q}^d$ is the approximation of y at time t_i.

This meta-algorithm works by solving the ODE (1) with initial condition $y(t_i) = \tilde{y}_i$ over a small time interval $[t_i, t_{i+1}]$, yielding as a result the approximation $\tilde{y}_{i+1}$ of $y(t_{i+1})$. This approximation over this small time interval is obtained using a Taylor approximation of order ω_i (we also do not fix, in a first approach, the value ω_i to analyze its influence on the error and on the time complexity of the algorithm. After this analysis is done, we can choose appropriate values for ω_i) using the polynomial algorithm given in [2]. This procedure is repeated recursively over $[t_0, t_1], [t_1, t_2], \ldots, [t_i, t_{i+1}], \ldots$ until we reach the desire time $t_n = t$. This introduces three potential sources of errors: (i) a global error due to the fact that, on the interval $[t_i, t_{i+1}]$ we do not solve $y' = p(y)$ with the initial value $y(t_i)$ but instead with the initial value $\tilde{y}_i$; (ii) a truncation error over $[t_i, t_{i+1}]$ because we only compute a truncated Taylor series of the solution instead of the full Taylor series; (iii) a rounding error because we might only have a finite number of bits to store partial results.

Using the crucial fact that the right-hand side of (1) consists of polynomials, at each time step t_i, one can present an argument based on Cauchy majorants to establish a lower bound on the local radius of convergence. We can choose the step length $t_{i+1} - t_i$ to be a constant fraction of the estimated radius of convergence, and the majorants can also be used to select a suitable truncation order ω_i. One can also show, using Gronwalls Lemma, that the propagation of errors from one step to the next can be controlled, and depends on a bound on the length of the curve $y(\cdot)$ over the domain under consideration. This last parameter needs to be fed to the algorithm, but we can automatically determine a suitable value for it, since we can decide if a (rational) value is large enough to be fed as a bound to the length of the curve. By using some (arbitrary, say the value 1) initial guess and by restarting the method with a larger guess if needed, we can continue this procedure until we decide that we have obtained a high enough value which can be used as a bound for the length of the curve.

Proceeding in this manner we end up fixing the parameters of the meta-algorithm (length of time steps, order of the Taylor approximation of each step, etc.) and we end up with an algorithm SolvePIVPEx which satisfies the conditions of Theorem 1.

Acknowledgments. D. Graça was partially supported by *Fundação para a Ciência e a Tecnologia* and EU FEDER POCTI/POCI via SQIG - Instituto de Telecomunicações through the FCT project UID/EEA/50008/2013.

References

1. Atkinson, K.E.: An Introduction to Numerical Analysis, 2nd edn. Wiley, New York (1989)
2. Bostan, A., Chyzak, F., Ollivier, F., Salvy, B., Schost, É., Sedoglavic, A.: Fast computation of power series solutions of systems of differential equations. In: SODA 2007, pp. 1012–1021, January 2007
3. Graça, D.S., Campagnolo, M.L., Buescu, J.: Computability with polynomial differential equations. Adv. Appl. Math. **40**(3), 330–349 (2008)
4. Kawamura, A.: Lipschitz continuous ordinary differential equations are polynomial-space complete. Comput. Complex. **19**(2), 305–332 (2010)
5. Ko, K.I.: Computational Complexity of Real Functions. Birkhäuser, Basel (1991)
6. Lorenz, E.N.: Deterministic non-periodic flow. J. Atmos. Sci. **20**, 130–141 (1963)
7. Müller, N., Moiske, B.: Solving initial value problems in polynomial time. In: Proceedings of 22 JAIIO - PANEL 1993, Part 2, pp. 283–293 (1993)
8. Pouly, A.: Continuous Models of Computation: From Computability to Complexity. Ph.D. thesis, Ecole Polytechnique/Universidade do Algarve (2015)
9. Warne, P.G., Warne, D.P., Sochacki, J.S., Parker, G.E., Carothers, D.C.: Explicit a-priori error bounds and adaptive error control for approximation of nonlinear initial value differential systems. Comput. Math. Appl. **52**(12), 1695–1710 (2006). http://dx.doi.org/10.1016/j.camwa.2005.12.004

Using Taylor Models in Exact Real Arithmetic

Franz Brauße[1], Margarita Korovina[2], and Norbert Müller[1(✉)]

[1] Abteilung Informatikwissenschaften, Universität Trier, Trier, Germany
mueller@uni-trier.de, brausse@informatik.uni-trier.de
[2] A.P. Ershov Institute of Informatics Systems, Novosibirsk, Russia
rita.korovina@gmail.com

Abstract. Software libraries for Exact Real Arithmetic implement the
theory of computability on non-denumerable sets. Usually they are based
on interval arithmetic. We discuss enhancements where the interval arithmetic is augmented by versions of Taylor models. Although this has no
effect on the abstract notion of computability, the efficiency of implementations can be improved dramatically.

1 Introduction

This paper deals with the theoretical background for recent improvements on
the iRRAM software [12] for exact real arithmetic (ERA). ERA can be viewed
as implemented version of the theory of computability on non-denumerable
sets, often called Type-2-Theory of Effectivity (TTE, [3,16]). The techniques
discussed could also easily be applicable to other software for exact real computations like [1,6], as those software packages also are based on interval arithmetic.

A serious problem common to all interval algorithms is that they suffer from
wrapping effects, i.e. unnecessary growth of approximation intervals during a
computation. Reducing such wrapping effects is an important issue in interval
arithmetic [13, p. 15ff]. The Taylor models proposed by Makino/Berz [11] are
the most prominent way to deal with such effects. They are widely applied in
software for interval arithmetic based on double precision numbers. Basically,
Taylor models are higher-dimensional polynomials with real (or rather double
precision) coefficients enhanced with an error interval. A further approach is
affine arithmetic [5] which can be interpreted as a version of the Taylor models
restricted to polynomials of degree 1. One of the oldest approaches is 'generalized
interval arithmetic' [8,15] which is similar to affine arithmetic, but now with
interval coefficients.

Although interval methods are also the basis for the representations most
commonly used in TTE, Taylor models have not yet been considered in this
area. There are only few papers like [2,6,9,14] pointing in this direction.

The research leading to these results has received funding from the People Programme (Marie Curie Actions) of the European Union's Seventh Framework
Programme FP7/2007-2013/ under REA grant agreement no. PIRSES-GA-2011-
294962-COMPUTAL and from the DFG/RFBR grant CAVER BE 1267/14-1 and
14-01-91334.

I.S. Kotsireas et al. (Eds.): MACIS 2015, LNCS 9582, pp. 474–488, 2016.
DOI: 10.1007/978-3-319-32859-1_41

We assume that the reader has some basic knowledge about computability with Turing machines and relativized computations via numberings, which can be found e.g. in [7,16]. As special notation we will use $f\colon \subseteq A \to B$ to indicate that f is a partial function from A to B. With $[A \to B]$ we denote the set of all partial functions from A to B. In the following boldface symbols denote vectors. For any vector $\boldsymbol{A}$ let A_i be the i-th component of $\boldsymbol{A}$. For a set $\boldsymbol{A}$ of real vectors let $\operatorname{diam}(\boldsymbol{A}) = \sup_{\boldsymbol{x},\boldsymbol{y} \in A} \|\boldsymbol{x} - \boldsymbol{y}\|$ be the diameter of $\boldsymbol{A}$ and let $\operatorname{int}(\boldsymbol{A})$ be the interior of $\boldsymbol{A}$.

2 Computability Using Wrapping Families

When dealing with non-denumerable sets, computability is usually defined with one of the following extensions of the well-known Turing machines: *oracle Turing machines* (with function oracles) for functions from $[([\Sigma^* \to \Sigma^*] \times \Sigma^*) \to \Sigma^*]$ or *Type-2-Turing machines* (with infinite input and output streams) for functions from $[\Sigma^\infty \to \Sigma^\infty]$, where in both cases Σ is some finite alphabet. In [10] the author uses oracle Turing machines, while in [16] the author prefers Type-2-Turing machines. Both approaches are equivalent with respect to questions of computability and both use *representations* to transfer the resulting computability notion to a notion of computability that is valid for real numbers (or related structures).

To simplify notations, we do not use these basic definitions of computability on $[\Sigma^* \to \Sigma^*]$ or Σ^∞ directly but use a derived computability notion for sequences $[\mathbb{N} \to S]$ instead, for suitable countable sets S. To this end, we assume that these sets S are equipped with numberings $\nu_S\colon \subseteq \mathbb{N} \to S$. A canonical way to translate the basic computability definitions into computability on $[\mathbb{N} \to S]$ can be found in [16].

In the following we will consider representations of $\mathbb{R}^d$, where for simplicity d is an arbitrary fixed dimension. Using sequences $[\mathbb{N} \to S]$ instead of $[\Sigma^* \to \Sigma^*]$ or Σ^∞ as the set of names, a representation of these real vectors is just a mapping of the form $[[\mathbb{N} \to S] \to \mathbb{R}^d]$. Most often S is chosen as a suitable countable basis for a topology on $\mathbb{R}^d$. One example is the representation ϱ^d of real vectors in [16], where S is a countable set of open d-dimensional boxes. More precisely $S = \mathbb{I}^d$ where $\mathbb{I}$ is the set of open intervals with dyadic endpoints:

$$\mathbb{I} := \left\{ \left(\frac{m_1}{2^n}, \frac{m_2}{2^n} \right) \mid m_1, m_2 \in \mathbb{Z}, n \in \mathbb{N}, m_1 < m_2 \right\}$$

In that setting, a sequence $p : \mathbb{N} \to \mathbb{I}^d$ represents a real vector $\boldsymbol{x} = \varrho^d(p)$ iff $\lim_{n \in \mathbb{N}} \operatorname{diam}(p(n)) = 0$ and $\bigcap_{n \in \mathbb{N}} p(n) = \{\boldsymbol{x}\}$.

Computability of a function $f\colon \subseteq \mathbb{R}^d \to \mathbb{R}^d$ is defined via a transformation process of boxes $O \in \mathbb{I}^d$ into new boxes $O' \in \mathbb{I}^d$ with $f(O) \subseteq O'$. However, the image $f(O)$ of $O \in \mathbb{I}^d$ is usually not a box. So O' has to be an overestimation of $f(O)$. Using O' implies that $f(O)$ is '*wrapped*' in $\partial O'$ at the expense of a bigger volume of O' compared to $f(O)$. As the definition of ϱ^d demands convergence to a point, this overestimation is not a critical issue in theoretical approaches.

In this paper we consider modifications of the set $\mathbb{I}^d$ to improve ERA.

Definition 1. *A countable family* $\mathcal{A} = \{A_n \mid n \in \mathbb{N}\}$ *of sets* $A_n \subseteq \mathbb{R}^d$ *is called* wrapping, *if*

$$\forall \boldsymbol{x} \in \mathbb{R}^d \; \forall \varepsilon \in \mathbb{R}_{>0} \; \exists A \in \mathcal{A} : \mathrm{diam}(A) \leq \varepsilon \wedge \boldsymbol{x} \in \mathrm{int}(A)$$

The corresponding representation $\tau_{\mathcal{A}} \colon \subseteq [\mathbb{N} \to \mathcal{A}] \to \mathbb{R}^d$ *is defined as follows:*

$$\tau_{\mathcal{A}}(p) := \boldsymbol{x} \text{ iff the sequence } p : \mathbb{N} \to \mathcal{A} \text{ satisfies}$$
$$\lim_{n \in \mathbb{N}} \mathrm{diam}(p(n)) = 0 \quad \wedge \quad \bigcap_{n \in \mathbb{N}} p(n) = \{\boldsymbol{x}\}$$

Such a sequence p *is then called a* $\tau_{\mathcal{A}}$-*name of* $\boldsymbol{x}$.

Obviously, $\mathbb{I}^d$ is an example of a wrapping family, and the representation ϱ^d is identical to $\tau_{\mathbb{I}^d}$. However, there are many different ways to define a wrapping family: $\mathcal{A}$ could be based on closed intervals instead of the open intervals $\mathbb{I}$, we could additionally allow point intervals consisting of a single number, or half-open intervals, or the union of finitely/infinitely many intervals etc. Also computation diagrams or symbolic notations could be used to define wrapping families. In the next section, we will elaborate in more detail how Taylor models fit into this approach.

All these wrapping families give rise to possible representations of real vectors. In order to compare those, we use the notion of reducibility in TTE [16], which is similar to the one used in the theory of numberings [7]. Suppose δ_1, δ_2 are representations. If the identity function id is (δ_1, δ_2)-computable, then δ_1 is called computably reducible to δ_2 ($\delta_1 \leq_c \delta_2$). For representations a weaker and topologically motivated reducibility is also important [16]: If id is (δ_1, δ_2)-continuous, then δ_1 is called topologically reducible to δ_2 ($\delta_1 \leq_t \delta_2$). Reducibility in both directions ($\delta_1 \leq_c \delta_2 \leq_c \delta_1$ or $\delta_1 \leq_t \delta_2 \leq_t \delta_1$) is called 'equivalence' and denoted by '$\equiv_c$' and '$\equiv_t$', respectively.

Lemma 1. *If* $\mathcal{A}$ *and* $\mathcal{B}$ *are wrapping, then* $\tau_{\mathcal{A}}$ *and* $\tau_{\mathcal{B}}$ *are topologically equivalent.*

Proof. We will only show $\tau_{\mathcal{A}} \leq_t \tau_{\mathcal{B}}$. Because of symmetry, we also get the inverse, so $\tau_{\mathcal{A}} \equiv_t \tau_{\mathcal{B}}$.

W.l.o.g. we first assume that $\mathbb{R}^d \in \mathcal{B}$. (Later we will remove this assumption). In consequence, for any $A \in \mathcal{A}$ the set $\{B \in \mathcal{B} \mid A \subseteq B\}$ is non-empty and $\inf\{\mathrm{diam}(B) \mid B \in \mathcal{B} \wedge A \subseteq B\}$ is well defined.

So there exists a function $w_{\mathcal{A}}^{\mathcal{B}} : \mathcal{A} \times \mathbb{N} \to \mathcal{B}$ connecting the wrapping families with following properties for any $A \in \mathcal{A}$ and any $n \in \mathbb{N}$

$$A \subseteq w_{\mathcal{A}}^{\mathcal{B}}(A, n)$$
$$\mathrm{diam}(w_{\mathcal{A}}^{\mathcal{B}}(A, n)) \leq 2^{-n} + \inf\{\mathrm{diam}(B) \mid B \in \mathcal{B} \wedge A \subseteq B\}$$

The function $w_{\mathcal{A}}^{\mathcal{B}}$ results in a 'good' wrapping of the first parameter A in a member of family $\mathcal{B}$ with additional overestimation of not more than 2^{-n}. Please note that we do not require $w_{\mathcal{A}}^{\mathcal{B}}$ to be computable.

Now we extend $w_{\mathcal{A}}^{\mathcal{B}}$ to an operator $W_{\mathcal{A}}^{\mathcal{B}} : [\mathbb{N} \to \mathcal{A}] \to [\mathbb{N} \to \mathcal{B}]$ as follows: For any $p : \mathbb{N} \to \mathcal{A}$ we let

$$W_{\mathcal{A}}^{\mathcal{B}}(p) := \left(w_{\mathcal{A}}^{\mathcal{B}}(p(n), n)\right)_{n \in \mathbb{N}}$$

As $W_{\mathcal{A}}^{\mathcal{B}}$ is defined just using initial sequences of its argument p, $W_{\mathcal{A}}^{\mathcal{B}}$ is continuous.

To prove $\tau_{\mathcal{A}} \leq_t \tau_{\mathcal{B}}$ consider $p \in \text{dom}(\tau_{\mathcal{A}})$ and let $\boldsymbol{x} := \tau_{\mathcal{A}}(p)$, $q := W_{\mathcal{A}}^{\mathcal{B}}(p)$. We have to show $q \in \text{dom}(\tau_{\mathcal{B}})$ and $\tau_{\mathcal{A}}(p) = \tau_{\mathcal{B}}(q)$.

For any $m \in \mathbb{N}$ there exists $B_m \in \mathcal{B}$ with $\text{diam}(B_m) \leq 2^{-m}$ and $\boldsymbol{x} \in \text{int}(B_m)$. So there is a $k \in \mathbb{N}$, $k \geq m$, with $\{\boldsymbol{y} \in \mathbb{R}^d : \|\boldsymbol{y} - \boldsymbol{x}\| < 2^{-k-1}\} \subseteq B_m$. Now for every $n \geq k$ with $\text{diam}(p(n)) \leq 2^{-k-2}$ we have $p(n) \subseteq B_m$ and hence $\text{diam}(q(n)) \leq 2^{-n} + \text{diam}(B_m) \leq 2^{-n} + 2^{-m} \leq 2^{1-m}$. This proves that $\lim_{n \in \mathbb{N}} \text{diam}(q(n)) = 0$. By construction we also have $\boldsymbol{x} \in p(n) \subseteq q(n)$, so additionally $\bigcap_{n \in \mathbb{N}} q(n) = \{\boldsymbol{x}\}$.

In case $\mathbb{R}^d \notin \mathcal{B}$ we proceed as follows: We use $\mathcal{B}' = \mathcal{B} \cup \{\mathbb{R}^d\}$ and construct $W_{\mathcal{A}}^{\mathcal{B}'}$ as above. Now for $p \in \text{dom}(\tau_{\mathcal{A}})$, $q' := W_{\mathcal{A}}^{\mathcal{B}'}(p)$ contains only a finite number of occurrences of $\mathbb{R}^d$. Then we apply a second continuous operator $T : [\mathbb{N} \to \mathcal{B}'] \to [\mathbb{N} \to \mathcal{B}]$ that simply eliminates all occurrences of $\mathbb{R}^d$ from its argument. Thus $T(q')$ is an infinite subsequence of q' as $p \in \text{dom}(\tau_{\mathcal{A}})$. Therefore $W_{\mathcal{A}}^{\mathcal{B}} := T \circ W_{\mathcal{A}}^{\mathcal{B}'}$ is a continuous operator with the desired property. $\square$

Computational equivalence $\equiv_c$, however, depends on additional computational properties of $\mathcal{A}$ and $\mathcal{B}$. Recall that both families must be countable, so implicitly there are numberings $\nu_{\mathcal{A}}$ and $\nu_{\mathcal{B}}$ giving rise to a notion of computability between $\mathcal{A}$ and $\mathcal{B}$. These numberings might stem from an implementation of corresponding data structures in programming languages. A sufficient condition for computational equivalence using these underlying numberings is as follows:

Lemma 2. *Suppose $\mathcal{A}$ and $\mathcal{B}$ are wrapping families and there is a $((\nu_{\mathcal{A}}, \nu_{\mathbb{N}}), \nu_{\mathcal{B}})$-computable multivalued function $w_{\mathcal{A}}^{\mathcal{B}} : \mathcal{A} \times \mathbb{N} \rightrightarrows \mathcal{B}$ such that for all $A \in \mathcal{A}$, $n \in \mathbb{N}$,*

$$A \subseteq w_{\mathcal{A}}^{\mathcal{B}}(A, n) \ \wedge \ \text{diam}(w_{\mathcal{A}}^{\mathcal{B}}(A, n)) \leq 2^{-n} + \inf\{\text{diam}(B) \mid B \in \mathcal{B} \wedge A \subseteq B\}.$$

Then $\tau_{\mathcal{A}} \leq_c \tau_{\mathcal{B}}$, i.e. $\tau_{\mathcal{A}}$ is computably reducible to $\tau_{\mathcal{B}}$.

Proof. We can reuse almost all of the previous proof: Now $w_{\mathcal{A}}^{\mathcal{B}}$ is given as a computable multivalued function, and we again are able to define:

$$W_{\mathcal{A}}^{\mathcal{B}}(p) := \left(w_{\mathcal{A}}^{\mathcal{B}}(p(n), n)\right)_{n \in \mathbb{N}}$$

Then $W_{\mathcal{A}}^{\mathcal{B}}$ is a computable (multivalued) realizer for the reduction $\tau_{\mathcal{A}} \leq_c \tau_{\mathcal{B}}$. $\square$

Please note that multivaluedness of $w_{\mathcal{A}}^{\mathcal{B}}$ is of advantage here as it allows different names u, v to deliver different results in $\mathcal{B}$ although they denote the same set $A = \nu_{\mathcal{A}}(u) = \nu_{\mathcal{A}}(v)$. In an implementation this means that we are not forced to 'normalize' names prior to applying the transformation $w_{\mathcal{A}}^{\mathcal{B}}$.

In Lemma 2 the overestimation for the diameter is bounded by 2^{-n}. Instead of this we could as well have used any function $g : \mathbb{N} \to \mathbb{R}_{>0}$ with $\lim_{n \to \infty} g(n) = 0$.

3   Computability Using Taylor Models

Taylor models can be viewed as a special way to define wrapping families. To see this, we first introduce the usual definition of these Taylor models and then generalize it in a way to allow a classification in TTE.

In [11] a Taylor model is given as a pair $T = (g, I)$, where $g(\boldsymbol{\lambda}) = \sum_n c_n \boldsymbol{\lambda}^n$ is a polynomial in a vector $\boldsymbol{\lambda} = (\lambda_1, \ldots, \lambda_k)$ of variables called 'error symbols', for arbitrary arity $k \in \mathbb{N}$, and I is an interval enclosing any approximation errors (called *interval remainder* of the Taylor model). The error symbols λ_i denote unknown values within the interval $\mathbb{U} := [-1, 1]$ and allow to express functional dependencies between different Taylor models that share those error symbols, i.e. within a vector of Taylor models.

As a first generalization we prefer to use closed intervals $c'_n = [\tilde{c}_n \pm \varepsilon_n] \subseteq \mathbb{R}$ (with center $\tilde{c}_n$ and radius ε_n) as coefficients instead of the point coefficients c_n used in [11]. However, we may use point intervals $c'_n = [\tilde{c}_n \pm 0]$ to implement points. Additionally we combine the point c_0 and the interval remainder I into an interval coefficient c'_0, hence we do not need to specify I any longer and are able to join g and I into a single polynomial. So a Taylor model T is just a polynomial in k variables, where the coefficients c'_n are closed intervals.

Most implementations restrict the coefficients (points as well as intervals) to be based on double precision numbers. As this is just a finite set of numbers, it is close at hand that for ERA infinite data types like dyadic (or rational) numbers should be used instead. But also arbitrary computable numbers might be used to define the coefficients.

In the following let us fix $\mathbb{K}$ to be some countable set of interval coefficients. We will not explicitly define $\mathbb{K}$ but just assume that $\mathbb{K}$ is a wrapping family itself (for $\mathbb{R}$). Additionally we assume that $\nu_{\mathbb{K}}$ is a numbering of $\mathbb{K}$ such that there are computable multivalued reductions $w_{\mathbb{K}}^{\mathbb{I}}$ and $w_{\mathbb{I}}^{\mathbb{K}}$ as in Lemma 2 which prove that the representation $\tau_{\mathbb{K}}$ is computationally equivalent to ϱ.

In order to derive wrapping families for $\mathbb{R}^d$ from Taylor models, we consider d-dimensional vectors of models. As the error symbols $\boldsymbol{\lambda}$ are used to express functional dependencies, we need a careful specification of the joint image of such a vector of Taylor models.

Definition 2. *For vectors $\boldsymbol{T}$ of Taylor models define* $image(\boldsymbol{T}) \subseteq \mathbb{R}^d$ *as follows:*

- *For each single Taylor model T in k variables with interval coefficients c'_n and each value $\boldsymbol{\lambda} = (\lambda_1, \ldots, \lambda_k) \in \mathbb{U}^k$ the polynomial $T(\boldsymbol{\lambda}) = \sum_n c'_n \cdot \boldsymbol{\lambda}^n$ is evaluated in the following way:*
  - *For an index $\boldsymbol{n} = (n_1, \ldots, n_k)$ let $\boldsymbol{\lambda}^n := \lambda_1^{n_1} \cdot \ldots \cdot \lambda_k^{n_k}$. This corresponds to the usual multi-index notation. Each $\boldsymbol{\lambda}^n$ thus is a scalar value in $\mathbb{U}$.*
  - *Each $c'_n \cdot \boldsymbol{\lambda}^n$ denotes the usual product of an interval with a scalar value.*
  - *Finally $\sum_n c'_n \cdot \boldsymbol{\lambda}^n$ is evaluated as a sum of (independent) intervals.*
- *For vectors $\boldsymbol{T} = (T_1, \ldots, T_d)$ of Taylor models and $\boldsymbol{\lambda} \in \mathbb{U}^k$ let $\boldsymbol{T}(\boldsymbol{\lambda})$ be the d-dimensional Cartesian product $\boldsymbol{T}(\boldsymbol{\lambda}) := \times_{i=1}^d T_i(\boldsymbol{\lambda})$ of the intervals $T_i(\boldsymbol{\lambda})$.*
- *Finally let* $image(\boldsymbol{T}) := \bigcup_{\boldsymbol{\lambda} \in \mathbb{U}^k} \boldsymbol{T}(\boldsymbol{\lambda})$ *be the set denoted by $\boldsymbol{T}$.*

A simple example of Taylor models of order 1 in $k = 2$ variables and for $d = 2$ is given in Fig. 1.

Using that $\mathbb{K}$ is a wrapping family, we are able to construct vectors $\boldsymbol{T}$ such that $image(\boldsymbol{T})$ has arbitrary small diameter. Thus we may use $image(\boldsymbol{T})$ as basis for a definition of wrapping families.

$$T(\lambda_1, \lambda_2) = \begin{pmatrix} [0,0] + [2,4] \cdot \lambda_1 + [10,10] \cdot \lambda_2 \\ [0,0] + [0,4] \cdot \lambda_1 + [-1,2] \cdot \lambda_2 \end{pmatrix}$$

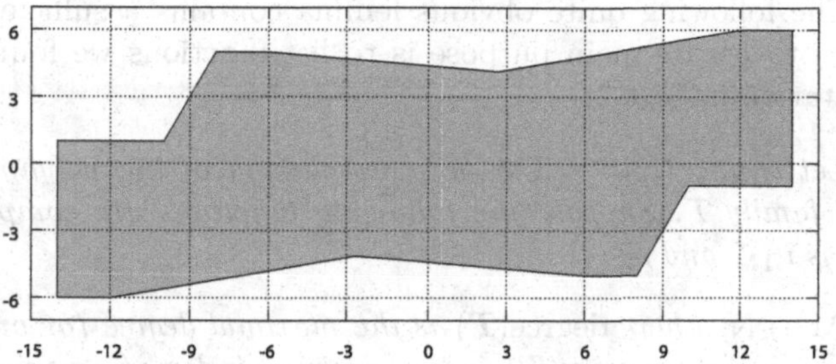

Fig. 1. Example of a vector T of Taylor models and the denoted set image(T) $\subseteq \mathbb{R}^2$.

Definition 3. *For $k \in \mathbb{N}$ let TM_k be the set of d-dimensional vectors of polynomials of arbitrary degrees in k variables and with coefficients from $\mathbb{K}$. Additionally let $\mathrm{TM} := \bigcup_{k \in \mathbb{N}} \mathrm{TM}_k$.*

The Taylor model wrapping families $\mathcal{T}_k$ and $\mathcal{T}$ are defined as follows:

$$\mathcal{T}_k := \{\mathrm{image}(T) \mid T \in \mathrm{TM}_k\}$$
$$\mathcal{T} := \bigcup_{k \in \mathbb{N}} \mathcal{T}_k$$

For a sequence $s : \mathbb{N} \to \mathrm{TM}$ of Taylor model vectors let image(s) be the sequence image(s) : $j \mapsto \mathrm{image}(s(j))$ of the denoted sets.

Using further restrictions on the Taylor models, we are able to identify at least the following slightly different subfamilies that have been used in literature. (Remember that we allow coefficients from $\mathbb{K}$ and not just intervals with double precision endpoints.)

1. **Affine arithmetic** [5]: Polynomials of order ≤ 1, only coefficients c_0' are non-point intervals.
2. **Generalized interval arithmetic** [8]: Polynomials of order ≤ 1, all coefficients c_n' are arbitrary intervals.
3. **Classical Taylor models** [11]: Polynomials of arbitrary order, only coefficients c_0' are non-point intervals.

The example in Fig. 1 actually fits to generalized interval arithmetic. Please note that due to the interval coefficients of λ_1 and λ_2, the denoted set is not convex, in contrast to the linear models in affine arithmetic.

As we restrict $\mathbb{K}$ to be a (countable) wrapping family, all possible k-ary Taylor models build a countable set. Additionally already the Taylor models of order 0 induce wrapping families. By Lemma 1 the deduced representations for all (sub-)families of Taylor models are topologically equivalent as soon as they contain arbitrary polynomials of order 0.

To examine computational equivalence, we need to have a closer look at the numbering of the Taylor models. Although they are defined via polynomials, this does not immediately imply that their numbering admits access to the single coefficients. The following quite obvious lemma contains a sufficient condition for equivalence to ϱ^d. Its main purpose is to list functions we found useful for the implementation in Sect. 5:

Lemma 3. *Let* $\nu_{\mathrm{TM}} : \subseteq \mathbb{N} \to \mathrm{TM}$ *be a numbering of the polynomials in the Taylor model family* $\mathcal{T}$. *Suppose the following functions are computable w.r.t. the numberings* ν_{TM} *and* $\nu_{\mathbb{K}}$:

1. degree : $\mathrm{TM} \to \mathbb{N}$, *where* degree($\boldsymbol{T}$) *is the maximal degree (or order) of* $\boldsymbol{T}$.
2. arity : $\mathrm{TM} \to \mathbb{N}$, *where* arity($\boldsymbol{T}$) *is the maximal index of a variable in* $\boldsymbol{T}$.
3. coeff : $\mathrm{TM} \times [1, \ldots, d] \times \mathbb{N}^* \to (\mathbb{K} \cup \{\bot\})$, *where* coeff($\boldsymbol{T}, i, \boldsymbol{m}$) *is the coefficient* $c_{\boldsymbol{m}}$ *of* $\boldsymbol{\lambda}^{\boldsymbol{m}}$ *in* $T_i(\boldsymbol{\lambda}) = \sum_n c_n \boldsymbol{\lambda}^n$, *if it exists ($\bot$ otherwise), for* $i = 1, \ldots, d$.
4. init : $\mathbb{K}^d \to \mathrm{TM}$, *where* init($\boldsymbol{c}$) *is the vector* $\boldsymbol{T} = (T_i)_i$ *of constant polynomials* $T_i \equiv c_i$, *for* $i = 1, \ldots, d$.

Then the representations ϱ^d, $\tau_{\mathcal{T}}$ *and* $\tau_{\mathcal{T}_k}$ *are computationally equivalent, for arbitrary* $k \in \mathbb{N}$.

Proof. Using the first 3 functions, all coefficients can be extracted from $\boldsymbol{T}$. Using Lemma 2 for $\mathbb{K}$ and $\mathbb{I}$ and the mapping $w_{\mathbb{K}}^{\mathbb{I}}$, these coefficients can be transformed to enclosing open intervals. Then the polynomials T_i, $i = 1, \ldots, d$, can be evaluated on these intervals. This already proves $\tau_{\mathcal{T}} \leq_c \varrho^d$.

For the inverse direction, we only need to transform a vector of open intervals to $\mathbb{K}^d$ using $w_{\mathbb{I}}^{\mathbb{K}}$ and then apply the fourth function, thus proving $\varrho^d \leq_c \tau_{\mathcal{T}_0}$. Finally, $\tau_{\mathcal{T}_0} \leq_c \tau_{\mathcal{T}_k} \leq_c \tau_{\mathcal{T}}$ is trivial.   □

As long as reasonable implementations are used (in the sense of the previous lemma and with computable $w_{\mathbb{K}}^{\mathbb{I}}$, $w_{\mathbb{I}}^{\mathbb{K}}$), Taylor models offer no new aspects concerning questions of computability. For efficient implementations however, the situation is far more complex and all variants of Taylor models are very helpful.

4   Arithmetic on Taylor Models in ERA

In ERA, a computation on real numbers is performed quite similar to ordinary computations on double precision numbers, just using a different data type. So as long as any input is already represented in the initial state of the computation, we have to deal with a (usually finite) sequence $x_0 \rightsquigarrow x_1 \rightsquigarrow x_2 \rightsquigarrow \ldots \rightsquigarrow x_n$ of real vectors during the computation, where each vector $x_i \in \mathbb{R}^d$ represents the complete data space after i computational steps. Each such step itself consists of the application of a (computable) function $f_i : \subseteq \mathbb{R}^d \to \mathbb{R}^d$ on x_i, so $x_{i+1} = f_i(x_i)$. At the very end some property $P(x_n)$ of x_n is produced as output. An example for P could be printing a component of x_n with a certain number of significant decimal digits.

Each vector x_i itself corresponds to a sequence of sets $O_{i,j}$ in the underlying representation. These sets might be open boxes (for ϱ^d), or elements of a wrapping family, like d-dimensional Taylor models. Essentially we deal with double sequences as follows:

$$x_0 \overset{f_0}{\longmapsto} x_1 \overset{f_1}{\longmapsto} x_2 \overset{f_2}{\longmapsto} \dots \overset{f_{n-1}}{\longmapsto} x_n$$

$$\left.\begin{matrix} \vdots \\ O_{0,2} \\ O_{0,1} \\ O_{0,0} \end{matrix}\right\} \longmapsto \left\{\begin{matrix} \vdots \\ O_{1,2} \\ O_{1,1} \\ O_{1,0} \end{matrix}\right. \longmapsto \left\{\begin{matrix} \vdots \\ O_{2,2} \\ O_{2,1} \\ O_{2,0} \end{matrix}\right\} \longmapsto \dots \longmapsto \left\{\begin{matrix} \vdots \\ O_{n,2} \\ O_{n,1} \\ O_{n,0} \end{matrix}\right.$$

The initial sequence $(O_{0,j})_{j\in\mathbb{N}}$ identifying x_0 should be computable, furthermore any $O_{i+1,j}$ should depend computably on (only finitely many) values $O_{i,k}$. The task of ERA implementations now is to find some $O_{n,j}$ that allows to deduce $P(x_n)$. This can be achieved in different ways.

The iRRAM library uses an approach where each horizontal line is treated independently. So it generates $O_{0,j} \mapsto O_{1,j} \mapsto O_{2,j} \mapsto \dots \mapsto O_{n,j}$, each with the inclusion property $f_i(O_{i,j}) \subseteq O_{i+1,j}$ and starting with $j = 0$ as a 'control parameter'. As long as the resulting $O_{n,j}$ is not precise enough to deduce $P(x_n)$, the index j is increased to some larger value j' of the control parameter and a new computation $O_{0,j'} \mapsto O_{1,j'} \mapsto \dots$ starts. Details on how the control parameter influences precisions can be found in [12].

An advantage of this approach is that the memory footprint is quite small. Additionally it is quite easy to enhance the underlying data structures for real numbers (previously only using closed intervals) by Taylor models, as the process of incrementing j is independent from the single computations $O_{i,j} \mapsto O_{i+1,j}$.

To enhance iRRAM by Taylor models, in parallel to the already implemented interval operations we only have to implement Taylor model versions of these operations. As the Taylor models are based on polynomials with interval coefficients, this turns out to be quite simple. Basic arithmetic operations like addition, subtraction and multiplication lead from polynomials again to polynomials. For other holomorphic operations, the image of a polynomial is not necessarily a polynomial any longer. In that case truncated Taylor series can be used. The truncation error is simply added to the basic coefficient c'_0 in the image. Here we will not go into details but refer to the original paper [11].

In the rest of this section we would like to address two aspects where the ERA implementation is different from ordinary Taylor model implementations: *sweeping* and *polishing*. Both are important to keep the data structures small and efficient.

Apart from addition and subtraction, operations on Taylor models tend to increase the order of the used polynomials significantly. This might easily lead to exponentially growing data structures. So it is very important to apply *sweeping*: whenever appropriate, any monomial $c'_n \cdot \lambda^n$ may be replaced by the monomial $c'_n \cdot \mathbb{U} \cdot \lambda^k$ for arbitrary $k < n$. This way, the order may be decreased at will.

As each error symbol λ_i is a variable over $\mathbb{U}$, this sweeping retains the inclusion property mentioned above, as it usually enlarges the denoted set. Thus sweeping does not violate the correctness of computations. The disadvantage is that sweeping removes information about functional dependencies to some extent and thus adds overestimation to the Taylor model, but it also helps to control the size of the Taylor model.

Please note that for ordinary Taylor models, only the basic coefficient c'_0 is allowed to be a non-point interval. Therefore, for this family, sweeping can only use $k = 0$ and must replace $c'_n \cdot \lambda^n$ by $c'_n \cdot \mathbb{U}$. For ERA we propose to use $k \neq 0$ when sweeping to retain at least linear functional dependencies.

Sweeping reduces degrees and usually keeps the set of error variables unchanged. Please note that we can use sweeping to eliminate all occurrences of some error variable.

During longer computations, the radii ε_n of all coefficient intervals $c'_n = [\tilde{c}_n \pm \varepsilon_n]$ slowly grow, due to sweeping or just due to overestimations necessitated by applying operations. Without counteraction, a Taylor model computation would eventually degenerate and suffer from wrapping effects as ordinary interval computations.

In order to keep the radii reasonably small, at any time any monomial $[\tilde{c}_n \pm \varepsilon_n] \cdot \lambda^n$ may be split into two monomials $[\tilde{c}_n \pm 0] \cdot \lambda^n + [\varepsilon_n \pm 0] \cdot \lambda_{\mathrm{new}} \cdot \lambda^n$, where λ_{new} is an error symbol that is not being used elsewhere.

This splitting increases the number of error symbols as well as the order of monomials. However it keeps the denoted set unchanged, therefore the inclusion property still holds.

We combine sweeping and splitting in a single operation called *polishing* as follows: First sweeping is used to reduce the degree of monomials with big radii and to reduce the set of error variables. Afterwards the big radii are reduced by splitting. Please note that there exist many possible variants and combinations of sweeping and splitting. In our implementation we use a heuristic described in more detail in Sect. 5.

The implementation of the polishing operation in the iRRAM is influenced by the control parameter mentioned before: The result of a polishing may change with increased precision. We only need that asymptotically the images of polished Taylor models converge to the same points as the unpolished versions. Thus the necessary properties of polishing can formally be defined as follows.

Definition 4. *A polishing is a multivalued function* $\pi : \mathrm{TM} \times \mathbb{N} \rightrightarrows \mathrm{TM}$ *satisfying the following properties:*

1. *Inclusion:* $\forall T \in \mathrm{TM} \ \forall j \in \mathbb{N} : \mathrm{image}(T) \subseteq \mathrm{image}(\pi(T, j))$
2. *Consistency: For a sequence* $s : \mathbb{N} \to \mathrm{TM}$ *of Taylor model vectors let* $\pi(s) : j \mapsto \pi(s(j), j)$ *be the sequence of the polished vectors. Whenever* $\tau_T(\mathrm{image}(s))$ *is defined, then also* $\tau_T(\mathrm{image}(\pi(s)))$ *must be defined with* $\tau_T(\mathrm{image}(s)) = \tau_T(\mathrm{image}(\pi(s)))$.

Based on Lemma 3, a polishing operator for a vector T of Taylor models could be implemented by overestimating $\mathrm{image}(T)$ through bounding boxes, thus reducing the number of error symbols in the resulting Taylor model to zero. This would still be good enough for both the inclusion and the consistency property. However, the notion of polishing also enables us to keep an arbitrary number of error symbols and thus retain some functional dependencies leading to better enclosures; an example is provided by Algorithm 1. The number of remaining error variables after applying polishing greatly influences the efficiency of single arithmetic operations.

5 Taylor Models in iRRAM

The iRRAM [12] is a software library implemented in C++. It is licensed under LGPL and it is freely available at https://github.com/norbert-mueller/iRRAM. The central structure is a class REAL, whose objects behave like real numbers in TTE [16]. Internally, simplified interval arithmetic based on multiple-precision floating-point numbers is used. Intervals take the form $[c \pm \varepsilon]$ where the center c is implemented as an arbitrarily long multiple-precision number using the MPFR library [17]. Two 32-bit integers m, e are used to represent the mantissa and the exponent of the radius $\varepsilon = m \cdot 2^e$. In consequence, computations with these intervals are only slightly more expensive than with isolated multiple-precision numbers. This interval type can also be used easily as the coefficient type $\mathbb{K}$ for Taylor models. (Of course, here we have to ignore that center c and exponent e still are finite structures.)

In the iRRAM implementation we currently have two data types for reals: REAL is based on ϱ, the other one (named TM) uses generalized interval arithmetic similar to [8]. Although this implemented version is just linear at the moment, the use of non-point intervals for the first order coefficients gives a much better behaviour than linear Taylor models with points (or point intervals), see e.g. [4].

So the user may choose from two different data types for $\mathbb{R}$. By Lemma 3 these are computationally equivalent and there are conversion methods within the constructors for the data types. Please note that, according to the previous section and the horizontal working principle of the iRRAM, internally algorithms only have to be applied to intervals or to Taylor models, but not to the sequences defining the representations.

In ERA details about the internal structure of the implementation must be concealed from the user and it is not possible to get access to any details about the actual data representing a real number. In an implementation based on Taylor models information required to be hidden would be the order of the polynomials, the number of error symbols, or the diameter of interval coefficients, for example.

In consequence, the user can not check whether operations like sweeping or polishing should be done, but she only is allowed to call methods that test for the necessity of these housekeeping routines and, if appropriate, perform the internal cleaning of the data structures. An example is given in Algorithm 2.

In the following we present a short overview on the main methods applicable to these data types on the user level and their internal implementation in iRRAM.

1. Mutual conversions between the data types REAL and TM are implemented by two constructors REAL(TM) and TM(REAL). Both constructors are applied to single real numbers. On the user level they both realize reductions between the representations, i.e. they are appear to be just the identity function on $\mathbb{R}$.
   – Using X = REAL(T) constructs a name for a real number given in representation ϱ^d from a name for the same number given in representation τ_T.
   – Vice versa, T = TM(X) constructs a name in representation τ_T from a name in representation ϱ^d.
   Due to the 'horizontal' mode of operation of the iRRAM described in Sect. 4. the constructor X = REAL(T) only has to compute a single interval I (for X) from a single Taylor model T (for T). This can be implemented as an evaluation of $I = \text{image}(T)$ as in Definition 2 for the one-dimensional case. The control parameter j is only used implicitly influencing the precision of the single interval operations.

   The constructor T = TM(X) simply takes the current interval for X and converts it to a Taylor model of arity zero.
2. Arithmetic operations '∘' on real numbers are available for variables of type TM using operator overloading. On the user level, these overloaded operators behave exactly like the denoted operations '∘'.

   Internally, an operation T3=T1∘T2 takes the current Taylor models T_1 for T1 and T_2 for T2 and constructs a new Taylor model T_3 as result, which then is stored in T3. During the construction of T_3, any arising monomials of degree ≥ 2 are immediately reduced to linear with sweeping.
3. The sweeping operations are included in the arithmetic operations and in the polishing below. Therefore there is no direct access to sweeping operations on the user level.
4. Polishing must be triggered manually. This is done via a call polish(TM&T) for a single TM variable or as polish(vector<TM>&T) for vectors. In both cases, the arguments of the calls are replaced by 'polished' versions thereof. According to Lemma 3, this does not change the represented values. Thus on the user level polishing has no visuable effects.

   A simplified implementation of such a polishing is given in Algorithm 1, where it is applied to a vector of d variables of type TM. Part (b) of the algorithm simplifies each single involved Taylor model by sweeping away all of its error symbols with exception of the largest, which is determined in Part (a). The results of (b)ii. and (b)iv. in this algorithm implicitly depend on the control parameter, which is omitted here for simplicity. Parts (c) and (d) split the resulting basic coefficient into two monomials using one new error symbol.
5. Currently, error symbols can only created by polishing.

Note that applying Algorithm 1 for polishing increases the number of error symbols by d. The algorithm does ensure on the other side that each component of the result T' uses at most 2 of these error symbols: one old error symbol (with the biggest coefficient) and one new error symbol. Two of the three coefficients even are point intervals.

Although the polishing operation can be applied on individual variables, it should rather be applied to many (or all) existing variables of type TM at the

---

Algorithm 1. Simplified version of polishing implemented in iRRAM.

Input: Linear $\boldsymbol{T} \in \text{TM}_k$: for each T_i index set S_i &coefficients $c_n = [\tilde{c}_n \pm \varepsilon_n]$, $\boldsymbol{n} \in S_i$.

Output: Linear $\boldsymbol{T}' \in \text{TM}_{k+d}$ of input's structure s.t. image($\boldsymbol{T}$) $\subseteq$ image($\boldsymbol{T}'$).

1. **for all** $i \in \{1, \ldots, d\}$ **do**
   - (a) $\hat{n} \leftarrow \arg\max_{n \in S_i \setminus \{0\}} |\tilde{c}_n|$ *(if $S_i \setminus \{0\} = \emptyset$, skip polishing this component: $T_i' \leftarrow T_i$)*
   - (b) **if** $|\tilde{c}_{\hat{n}}| > 2^{30} \cdot \varepsilon_{\hat{n}}$ **then**   *(error of the monomial with highest effect is small enough)*
     - i. $c_{\hat{n}}' \leftarrow c_{\hat{n}}$ *(retain these dependencies on the computation's input)*
     - ii. $[\tilde{s} \pm \varepsilon] \leftarrow \sum_{n \in S_i \setminus \{\hat{n}\}} c_n \cdot \mathbb{U}$ *(merge remaining coefficients into one interval)*
     - **else**   *(otherwise $c_{\hat{n}}$ already accumulated too much wrapping)*
     - iii. $c_{\hat{n}}' \leftarrow 0$ *(and is not retained)*
     - iv. $[\tilde{s} \pm \varepsilon] \leftarrow \sum_{n \in S_i} c_n \cdot \mathbb{U}$ *(merge all coefficients and keep no dependencies)*
   - (c) $c_0' \leftarrow [\tilde{s} \pm 0]$ *(represent center and radius as point-interval coefficients, each)*
   - (d) $c_{e_{k+i}}' \leftarrow [\varepsilon \pm 0]$, where e_{k+i} is the $(k+i)$-th unit vector *(new error symbol λ_{k+i})*
   - (e) $T_i'(\boldsymbol{\lambda}) \leftarrow c_0' + c_{\hat{n}}' \boldsymbol{\lambda}^{\hat{n}} + c_{e_{k+i}}' \lambda_{k+i}$ *(polished TM has 2 to 3 monomials)*
2. **return** $\boldsymbol{T}' = (T_1', \ldots, T_d')$

---

same time. In this case, the number of error symbols in use is restricted to at most $2 \cdot d$. Thus the computational overhead per arithmetic operation with data type TM compared to data type REAL is kept at a constant level (but depending on d). It is however just one way to handle an ever-growing number of error variables which are necessary to keep the interval coefficients themselves from suffering from wrapping effects.

6   Experimental Results

As an application example we consider the well known logistic map. This map is a discrete dynamical system which exhibits chaotic behaviour for many parameters. It has been used many times in computable analysis, e.g. in [2,14], and can be regarded as a reference problem and benchmark. It is defined as follows for initial values $x_0 \in [0, 1]$:

$$x_{i+1} := c \cdot x_i \cdot (1 - x_i)$$

The parameter c is usually taken from the closed interval $[3, 4]$. In consequence the function $c \cdot x \cdot (1 - x)$ is non-negative on $x \in [0, 1]$ and takes its maximal value $c/4 \leq 1$ at $x = 0.5$. So all values of the sequence $(x_i)_{i \in \mathbb{N}}$ will be in the interval $[0, 1]$.

Two implementations of this sequence in the iRRAM package for the initial values $x = 0.125$ are given in Algorithm 2. One version is based on the new Taylor model data type, the second one uses REAL, but only lines 3, 10 and 12 differ. The parameter c as well as the index n of the value x_n to be computed are given as inputs. The output of the program is a list of the decimal representation of all the values x_i with 12 significant decimals, for $0 \leq i \leq n$.

The following table gives a few results for specific values of c and n. The given timings were computed on an Intel CPU (i5-460M @ 2.53GHz). The value for

---

Algorithm 2. Logistic map using Taylor models in iRRAM

```
1  // Version for Taylor models        1  // Version for REAL datatype
2  #include "iRRAM.h"                   2  #include "iRRAM.h"
3  #include "TaylorModel.h"             3
4  using namespace iRRAM;               4  using namespace iRRAM;
5  void compute() {                     5  void compute() {
6     REAL c;                           6     REAL c;
7     int n;                            7     int n;
8     cin >> c;                         8     cin >> c;
9     cin >> n;                         9     cin >> n;
10    TM x = REAL(0.125);              10    REAL x = REAL(0.125);
11    for (int i=0; i<=n; i++) {       11    for (int i=0; i<=n; i++) {
12       TM::polish(x);                12
13       cout << REAL(x) << "\n";      13       cout << REAL(x) << "\n";
14       x = x*c*(REAL(1)-x);          14       x = x*c*(REAL(1)-x);
15    }                                15    }
16 }                                   16 }
```

'precision' is the highest number of bits the underlying multiple precision software MPFR [17] had to be called with. This 'precision' essentially corresponds to the control parameter j in the double sequence $O_{i,j}$ from the beginning of Sect. 4. For some of the values, already the parameter $j = 0$ was sufficient, where the iRRAM uses double precision numbers.

	Data type TM				Data type REAL			
c	$n = 10000$		$n = 100000$		$n = 10000$		$n = 100000$	
	time [s]	precision [bits]	time [s]	precision [bits]	time [s]	precision [bits]	time [s]	precision [bits]
3.125	0.09	double	0.90	double	1.08	18581	266	175466
3.56982421875	0.09	double	0.94	double	0.85	18581	363	219405
3.75	0.64	5894	115	57301	1.60	23299	400	219405
3.82	0.75	7440	148	71699	1.38	23299	340	219405
3.830078125	0.09	double	0.92	double	1.40	23299	337	219405
3.84	0.09	136	0.89	136	1.46	23299	354	219405

The number of bits necessary for the computations with type TM closely match the average bit loss reported in [2] for these values of c. The value $c = 3.56982421875 = 7311 \cdot 2^{-11}$ has also been recommended there (leading to an orbit of length 64), as well as $c = 3.830078125 = 1961 \cdot 2^{-9}$ (having an orbit of length 3 in an area of chaotic behaviour).

From these experimental results it is obvious that the Taylor models may lead to significant improvements in an ERA implementation.

7   Conclusions and Future Work

In this paper we presented a view on the theoretical background for enhancing ERA by Taylor models. The approach using wrapping families opens up the way to many generalizations.

In typical implementations using double precision numbers, Taylor models allow verified computations that could not be performed at all with ordinary interval arithmetic. The theoretical point of view taken in this paper shows that for the notion of computability on real numbers Taylor models do not increase the set of computable functions.

Although Taylor models increase the complexity of single operations significantly, their excellent behaviour concerning error propagation can result in great improvements in efficiency.

We expect that a rigorous investigation of this improved efficiency will also lead to refinements of complexity theoretical results in areas where the use of Taylor series is close at hand, for example in analytic continuation and for ODE solving.

Further work is necessary to finalize the integration of operations on Taylor models into the iRRAM library, also optimizations of the internal data structures and the used algorithms are planned in order to improve the efficiency for higher-dimensional systems. Higher-order Taylor models have to be implemented in the library. We also need closer investigations to optimize the heuristics for polishing.

References

1. Bauer, A., Kavkler, I.: Implementing real numbers with RZ. Electr. Notes Theor. Comput. Sci. **202**, 365–384 (2008)
2. Blanck, J.: Efficient exact computation of iterated maps. J. Log. Algebr. Program. **64**(1), 41–59 (2005)
3. Brattka, V., Hertling, P., Weihrauch, K.: A tutorial on computable analysis. New Computational Paradigms: Changing Conceptions of What is Computable, pp. 425–491. Springer, New York (2008)
4. Brauße, F., Korovina, M., Müller, N.T.: Towards using exact real arithmetic for initial value problems. In: PSI: 10th Ershov Informatics Conference, 25–27, Innopolis, Kazan, Russia, to appear in Lecture Notes in Computer Science (2015)
5. De Figueiredo, L.H., Stolfi, J.: Affine arithmetic: concepts and applications. Numerical Algorithms **37**(1–4), 147–158 (2004)
6. Duracz, J., Farjudian, A., Konečný, M., Taha, W.: Function Interval Arithmetic. In: Hong, H., Yap, C. (eds.) ICMS 2014. LNCS, vol. 8592, pp. 677–684. Springer, Heidelberg (2014)
7. Ershov, Y.: Handbook of Computability Theory, chapter Theory of numberings, pp. 473–503. North-Holland, Amsterdam (1999)
8. Hansen, E.R.: A generalized interval arithmetic. In: Interval Mathemantics: Proceedings of the International Symposium, Karlsruhe, West Germany, May 20–24, pp. 7–18 (1975)
9. Khanh, T.V., Ogawa, M.: rasat: SMT for polynomial inequality. Technical Report Research Report IS-RR–003, JAIST (2013)

10. Ko, K.-I.: Complexity theory of real functions. Birkhauser Boston Inc., Cambridge, MA, USA (1991)
11. Makino, K., Berz, M.: Higher order verified inclusions of multidimensional systems by taylor models. Nonlinear Analysis: Theory, Methods & Applications, 47(5), 3503–3514, Proceedings of the Third World Congress of Nonlinear Analysts (2001)
12. Müller, N.T.: The iRRAM: Exact arithmetic in C++. Lecture notes in computer science, 2991:222–252 (2001)
13. Nedialkov, N.S., Jackson, K.R., Corliss, G.F.: Validated solutions of initial value problems for ordinary differential equations. Appl. Math. Comput. 105(1), 21–68 (1999)
14. Spandl, C.: Computational complexity of iterated maps on the interval (extended abstract). In: Proceedings Seventh International Conference on Computability and Complexity in Analysis, CCA 2010, Zhenjiang, China, 21–25th , pp. 139–150, 2010 June 2010
15. Tupper, J.A.: Graphing equations with generalized interval arithmetic. Master's thesis, University of Toronto (1996)
16. Weihrauch, K.: Computable analysis: an introduction. Springer-Verlag New York Inc, Secaucus, NJ, USA (2000)
17. Zimmermann, P.: Reliable Computing with GNU MPFR. In: Fukuda, K., Hoeven, J., Joswig, M., Takayama, N. (eds.) ICMS 2010. LNCS, vol. 6327, pp. 42–45. Springer, Heidelberg (2010)

On the Computational Complexity of Positive Linear Functionals on $\mathcal{C}[0;1]$

Hugo Férée[1,2,3,4] and Martin Ziegler[1,5(✉)]

[1] Department of Mathematics, TU Darmstadt, Schlossgartenstr. 7,
64289 Darmstadt, Germany
[2] Université de Lorraine, LORIA, UMR 7503, 54506 Vandœuvre-lès-Nancy, France
[3] Inria, 54600 Villers-lès-Nancy, France
[4] CNRS, LORIA, UMR 7503, 54506 Vandœuvre-lès-Nancy, France
[5] KAIST, School of Computing, 291 Daehak-ro, 34141 Daejeon, Republic of Korea
ziegler@cs.kaist.ac.kr

Abstract. The Lebesgue integration has been related to polynomial counting complexity in several ways, even when restricted to smooth functions. We prove analogue results for the integration operator associated with the Cantor measure as well as a more general second-order **#P**-hardness criterion for such operators. We also give a simple criterion for relative polynomial time complexity and obtain a better understanding of the complexity of integration operators using the Lebesgue decomposition theorem.

1 Motivation and Introduction

Devising a complexity theory of higher-type computation is an ongoing endeavour since at least 25 years [Cook91, KaCo96, IBR01, KaCo10, FGH13, FeHo13, KSZ15]. Perhaps more modestly, we are interested in classifying the continuous linear functionals Ψ on the space $\mathcal{C}[0;1]$ of continuous functions on the real unit interval: first non-uniformly, that is, investigate the computational complexity of the real number $\Psi(f)$ for arbitrary but fixed polynomial-time computable $f \in \mathcal{C}[0;1]$; and then uniformly with (approximations, in some sense, to) f 'given' by means of oracle access, yet still for fixed Ψ.

According to the *Riesz–Markov–Kakutani Representation Theorem* precisely every positive (i.e. monotone) linear functional $\Psi : \mathcal{C}[0;1] \to \mathbb{R}$ is of the form $\Psi(f) = \int_0^1 f(t)\, d\nu(t)$ for some regular Borel measure ν on $[0;1]$. *Lebesgue's Decomposition Theorem* in turn asserts each such ν to admit a (unique) decomposition $\nu = \nu_d + \nu_c + \nu_s$, where

(i) ν_d is discrete,
(ii) ν_c is absolutely continuous w.r.t. the canonical (*i.e.* Lebesgue) measure λ,

Supported in part by the *Marie Curie International Research Staff Exchange Scheme Fellowship* 294962 within the 7th European Community Framework Programme and by the *German Research Foundation* (DFG) with project Zi 1009/4-1. A SHORT version of this work was presented at CCA 2015.

© Springer International Publishing Switzerland 2016
I.S. Kotsireas et al. (Eds.): MACIS 2015, LNCS 9582, pp. 489–504, 2016.
DOI: 10.1007/978-3-319-32859-1_42

(iii) and ν_s is singular continuous.

This theorem can be useful in the study of the computational complexity of an integration operator, by determining the complexity of each component.

Remark 1.

(i) The prototype of a discrete measure is Dirac's family δ_z with $\delta_z([a;b]) = 1$ if $z \in [a;b]$ and $\delta_z([a;b]) = 0$ otherwise. The induced positive linear functional is simply evaluation at z — and polynomial-time computable uniformly in z, essentially by definition.
   More generally every discrete measure on $[0;1]$ has the form $\nu_d = \sum_{j \in \mathbb{N}} \delta_{z_j} \cdot w_j$ for two sequences $(z_j) \subseteq [0;1]$ and $(w_j) \subseteq [0;\infty)$ with $\sum_j w_j < \infty$.

(ii) The prototype of an absolutely continuous measure on $[0;1]$ is thus λ defined by $\lambda([a;b]) = b - a$; and the complexity of its induced positive linear functional on $\mathcal{C}[0;1]$, namely of definite Riemann integration, has been characterized as #P_1 (Fact 3a+b); and indefinite Riemann integration as #P (Fact 3c+d); cmp. [Frie84]. Moreover restricting to continuously differentiable argument does not reduce the worst-case complexity.
   In general, according to the classical Radon–Nikodym Theorem, to every absolutely continuous measure ν_c on $[0;1]$ there exists some measurable $\varphi : [0;1] \rightarrow [0;\infty)$ such that $\int_0^x f(t) \, d\nu_c(t) = \int_0^x f(t)\varphi(t) \, dt$ holds for all $f \in \mathcal{C}[0;1]$ and $0 \leq x \leq 1$.

(iii) The prototype of a singular continuous measure is Cantor's, that is, given by the Devil's Staircase or Cantor–Lebesgue–Vitali function $S : [0;1] \rightarrow [0;1]$ as cumulative distribution and inducing as functional the parametric Riemann-Stieltjes integral $(f, x) \mapsto \int_0^x f(t) \, dS(t)$.

1.1   Recap of Discrete, Real, and Second-Order Complexity Theory

We presume familiarity with discrete complexity theory and only briefly recall the classes

- P of decision problems $L \subseteq \Sigma^*$ to which membership "$\vec{u} \in L$?" is decidable within a number of steps polynomial in the input length $|\vec{u}|$;
- EXP of decision problems decidable in time bounded by some exponential polynomial in the input length;
- PSPACE of decision problems to which membership is decidable using an at most polynomial amount of memory;
- FP of total function problems $f : \Sigma^* \rightarrow \mathbb{N}$ computable in a number of steps polynomial in the input length with output encoded in binary;
- NP of decision problems $L \subseteq \Sigma^*$ of the following form for some $V \in$ P and some integer polynomial p: $L = \{\vec{u} \mid \exists \vec{v} : |\vec{v}| \leq p(|\vec{u}|), \langle \vec{u}, \vec{v} \rangle \in V\}$.
- #P of counting (i.e. function) problems of the form

$$\psi : \Sigma^* \ni \vec{u} \mapsto \#(\{\vec{v} \mid |\vec{v}| \leq p(|\vec{u}|), \langle \vec{u}, \vec{v} \rangle \in V\}) \in \mathbb{N}$$

- #P_1 of *unary* counting problems of the form

$$\psi_1 : \mathbb{N} \ni n \mapsto \#\big(\{\vec{v} \mid |\vec{v}| \leq p(n), \langle 1^n, \vec{v} \rangle \in V\}\big) \in \mathbb{N}$$

with hierarchy $\mathsf{P} \subseteq \dfrac{\mathsf{P}^{\#\mathsf{P}_1}}{\mathsf{NP}} \subseteq \mathsf{P}^{\#\mathsf{P}} \subseteq \mathsf{PSPACE} \subseteq \mathsf{EXP}$.

In particular recall that $\#\mathsf{P}$ may not be closed even under simple functions [HeOg02, Sect. 5.2]. Here Σ denotes some fixed finite alphabet containing at least symbols 0 and 1; and

$$\Sigma^* \times \Sigma^* \ni (\vec{v}, \vec{w}) \mapsto \langle \vec{v}, \vec{w} \rangle \in \Sigma^*$$

an injective polynomial-time computable string pairing function having polynomial-time decidable image and polynomial-time computable partial inverse.

Concerning the complexity of real functions we refer to [Weih00, Example 7.2.14] and [Ko91, COROLLARY 2.21]: Computing $f : [0;1] \to [0;1]$ within time $t(n)$ means (or rather is equivalent) to compute in the discrete sense of complexity $t(n)$ some function $\tilde{f} : \{0,1\}^* \to \{0,1\}^*$ such that it holds for $\mu(n) := t(n+1) + 1$, for all $n \in \mathbb{N}$, and for all $\vec{u} \in \{0,1\}^n$:

$$\forall x \in [0;1]: \quad |x - \mathrm{bin}(\vec{u})/2^{\mu(n)}| \leq 2^{-\mu(n)} \quad \Rightarrow \quad |f(x) - \mathrm{bin}(\tilde{f}(\vec{u}))/2^n| \leq 2^{-n}. \quad (1)$$

Here, for $\vec{v} \in \{0,1\}^n$, $\mathrm{bin}(\vec{v})/2^{|\vec{v}|} := \sum_{j=0}^{n-1} v_j 2^{j-n} \in [0;1)$ is a dyadic rational. Again $\{0,1\}$ can be replaced by any other at least binary alphabets without affecting the complexity more than polynomially. In the sequel for the Cantor distribution ternary rational approximations $\mathrm{tri}(\vec{u})/3^{|\vec{u}|} = \sum_{j=0}^{m-1} u_j 3^{j-m}$ for $\vec{u} \in \{0,1,2\}^m$ will often turn out as convenient.

Recall that $\mu : \mathbb{N} \to \mathbb{N}$ satisfying $d(x, x') \leq 2^{-\mu(n)} \Rightarrow e\big(f(x), f(x')\big) \leq 2^{-n}$ is called a *modulus of continuity* of $f : X \to Y$ with metric spaces (X, d) and (Y, e). Concerning a uniform complexity of operators in analysis, we follow [KaCo10, Sect. 3] in letting $\mathsf{Pred} := \{0,1\}^{(\{0,1\}^*)}$ denote the set of all predicates on finite binary strings; $\mathsf{Reg} \subseteq \{0,1\}^{**} := (\{0,1\}^*)^{(\{0,1\}^*)}$ the family of all (total) mapping $\varphi : \{0,1\}^* \to \{0,1\}^*$ that are length-monotonous in the sense of satisfying $|\varphi(\vec{u})| \leq |\varphi(\vec{v})|$ whenever $|\vec{u}| \leq |\vec{v}|$. In this case, the size function $|\varphi| : \mathbb{N} \ni |\vec{v}| \mapsto |\varphi(\vec{v})| \in \mathbb{N}$ of φ is well-defined. A *second-order polynomial* P is a term over $+, \times, \mathbb{N}$ and first-order variable n as well as second-order variable ℓ. An oracle Turing machine $\mathcal{M}^?$ *computes* a partial function $F :\subseteq \mathsf{Reg} \to \mathsf{Reg}$ when producing $F(\varphi)(\vec{v})$, given $\vec{v} \in \{0,1\}^*$ and oracle access to $\varphi \in \mathrm{dom}(F)$. $\mathcal{M}^?$ runs in *second-order polynomial time* if it makes a number steps bounded by $P(|\varphi|, |\vec{v}|)$ for some second-order polynomial P. The following notions and bold-face complexity classes are from [Kawa11, DEFINITIONS 2.10–2.13]:

Definition 2.

(a) **P** *is the class of total* $F : \mathsf{Reg} \to \mathsf{Pred}$ *computable in second-order polynomial time.*

(b) **FP** *is the class of total* $F : \mathsf{Reg} \to \mathsf{Reg}$ *computable in second-order polynomial time.*

(c) **NP** *is the class of total* $G : \text{Reg} \to \text{Pred}$ *of the form*

$$G(\varphi)(\vec{v}) = 1 \;\Leftrightarrow\; \exists \vec{w} \in \{0,1\}^{P(|\varphi|,|\vec{v}|)} : F(\varphi)(\vec{v},\vec{w}) = 1$$

for some $F \in \mathbf{P}$ *and some second-order polynomial* P.

(d) **#P** *is the class of total* $G : \text{Reg} \to \text{Reg}$ *of the form*

$$G(\varphi)(\vec{v}) = \text{bin}\left(\#\{\vec{w} \in \{0,1\}^{P(|\varphi|,|\vec{v}|)} : F(\varphi)(\vec{v},\vec{w}) = 1\}\right)$$

for some $F \in \mathbf{P}$ *and some second-order polynomial* P.

(e) *For* $F, G :\subseteq \text{Reg} \to \text{Reg}$, *a second-order polynomial-time (Weihrauch-) reduction from* F *to* G *is a triple* (U,V,W) *with* $U,V,W \in \mathbf{FP}$ *such that* $U(\varphi) \in \text{dom}(G)$ *for every* $\varphi \in \text{dom}(F)$ *and*

$$\forall \vec{v} \in \{0,1\}^* : \quad F(\varphi)(\vec{v}) \;=\; W\big(\varphi\big)\Big\langle G\big(U(\varphi)\big)\big((V(\varphi))(\vec{v})\big), \vec{v}\Big\rangle$$

(f) *Some* $\varphi \in \text{Reg}$ *encodes* $f \in C([0;1],[0;1])$ *if it is of the form*

$$\varphi : \{0,1\}^* \;\ni\; \vec{u} \;\mapsto\; 1^{\mu(n)}\, 0\, \text{bin}\big(\tilde{f}(\vec{u})\big)$$

for some modulus of continuity μ *of* f *and* $\tilde{f}$ *according to Eq. (1).*

(g) $F :\subseteq \text{Reg} \to \text{Reg}$ *represents some operator* $\Lambda :\subseteq C([0;1],[0;1]) \to C([0;1],[0;1])$ *if it maps every encoding* $\varphi \in \text{Reg}$ *of some* $f \in \text{dom}(\Lambda)$ *to some encoding* $F(\varphi)$ *of* $\Lambda(f)$.

(h) *We may identify such an operator with the functional*

$$\Lambda :\subseteq C([0;1],[0;1]) \times [0;1] \;\ni\; (f,x) \;\mapsto\; \Lambda(f)(x) \in [0;1] \;.$$

(j) $\Lambda :\subseteq C([0;1],[0;1]) \to C([0;1],[0;1])$ *is computable in second-order polynomial-time if it admits a representative* $F :\subseteq \text{Reg} \to \text{Reg}$ *computable in second-order polynomial-time.*

Compare [BrGh11,HiPa13] for a computable version of Item (e). We record that closure under composition of second-order polynomial-time computability yields transitivity of second-order polynomial-time reducibility. As opposed to the first-order complexity classes with the P/NP Millennium Prize and related open problems, the second-order versions are generally known distinct.

Fact 3.

(a) If $f : [0;1] \to [0;1]$ is polynomial-time computable and $\#\mathbf{P}_1 \subseteq \mathbf{FP}_1$, then $\int_0^1 f(t)\,dt$ is again polynomial-time computable.

(b) There exists a polynomial-time computable smooth (i.e. C^∞) $f : [0;1] \to [0;1]$ such that polynomial-time computability of $\int_0^1 f(t)\,dt$ implies $\#\mathbf{P}_1 \subseteq \mathbf{FP}_1$.

(c) If $f : [0;1] \to [0;1]$ is polynomial-time computable and $\#\mathbf{P} \subseteq \mathbf{FP}$, then $[0;1] \ni x \mapsto \int_0^x f(t)\,dt$ is again polynomial-time computable.

(d) There exists a polynomial-time computable smooth $f : [0;1] \to [0;1]$ such that polynomial-time computability of $[0;1] \ni x \mapsto \int_0^x f(t)\,dt$ implies $\#\mathbf{P} \subseteq \mathbf{FP}$.

(e) For any $G :\subseteq \mathsf{Reg} \to \mathsf{Reg}$ representing (in the sense of Definition 2g) indefinite integration

$$\int : C([0;1],[0;1]) \ni f \mapsto \left([0;1]^2 \ni (x,y) \mapsto \int_x^y f(t)\,dt\right) \in C^1([0;1];[0;1]) \quad (2)$$

there exists a $\Gamma \in \#\mathsf{P}$ and a second-order polynomial-time reduction from G to F.

(f) For every $F \in \#\mathsf{P}$ and every $G :\subseteq \mathsf{Reg} \to \mathsf{Reg}$ representing the restriction $\int\big|_{C^\infty([0;1],[0;1])}$ there exists a second-order polynomial-time reduction from F to G.

For the first four items see [Ko91, THEOREMS 5.32+5.33]. They are non-uniform in that f is considered fixed and not as input. In other words, they only consider the image of a complexity class by the operator. This contrasts with the uniform Items (e) and (f), essentially [Kawa11, THEOREM 4.21], where the complexity of the operator itself is considered.

1.2   Overview, Techniques, and Related Work

The present work investigates the non-uniform and uniform computational complexity of other (types of) positive linear functionals on $C[0;1]$ and $C^\infty[0;1]$. Similarly to Fact 3, Sects. (2) and (3) relate Cantor integration non-uniformly and uniformly equivalent to $\#\mathsf{P}_1$, $\#\mathsf{P}$, and $\#\mathsf{P}$. Perhaps surprisingly, it is thus as hard as ordinary/absolutely continuous Riemann integration. (Along the way we prove the Devil's Staircase S to be computable in polynomial time.)

On the other hand, Example 16 constructs singular continuous measures that does render integration polynomial-time computable — after Subsect. 4.1 identifying classes of measures for which integration is $\#\mathsf{P}$-hard. Conversely, Example 14 constructs a discrete measure rendering integration $\#\mathsf{P}$-hard — based on Subsect. 4.2 devising classes of measures for which integration is polynomial-time computable.

Proof techniques are essentially refinements and variations of those employed in [Ko91, Sect. 5.4] and [Kawa11]: On the one hand encoding a polynomial-time decidable verifier $V \subseteq \{0,1\}^* \times \{0,1\}^*$ as polynomial-time computable (smooth) real function f_V consisting of infinitely many 'steps' such that the hard discrete counting problem $\#\{\vec{y} : (\vec{x},\vec{y}) \in V\}$ can be recovered from approximations of the continuous integral over f_V w.r.t. the measure under consideration; and on the other hand expressing approximations to said integral as discrete counting problem with polynomial-time decidable verifier; and uniformly analyzing the 'reduction' $V \mapsto f_V$ as well as its converse in terms of second-order polynomial-time complexity theory. In fact jointly scaling the steps in x-direction and y-direction is a delicate trade-off: such as to (i) recover discrete arguments $\vec{x} \in \{0,1\}^*$ from approximations to real arguments $x \in [0;1]$ as well as (ii) recover discrete results $\#\{\vec{y} : (\vec{x},\vec{y}) \in V\}$ from approximations to the real values $\int f_V(t)\,dS(t)$ while (iii) maintaining continuity, smoothness, and polynomial-time computability of f_V; cmp. Remark 18(b).

Regarding more general but qualitative computability investigations of measures the reader may refer for instance to [Schr07, HRW12, MTY14, Coll14].

2　Smooth Cantor Integration Is at Least as Hard as Continuous Riemann Integration

Proposition 4.

(a) Cantor's Function $S : [0;1] \to [0;1]$ is Hölder-continuous with exponent $\alpha = \ln(2)/\ln(3)$ and computable within polynomial time.

(b) For every interval $I = [a;b] \subseteq \mathbb{R}$ and every non-decreasing continuous $g : I \to \mathbb{R}$ it holds $\int_{g(a)}^{g(b)} f(t)\,dt = \int_I f(g(s))\,dg(s)$.

Proof.

(a) Recall that S is the uniform limit of a sequence of piecewise linear functions defined inductively by $S_0 := \mathrm{id} : [0;1] \to [0;1]$ and $S_{n+1}(t) :=$
$$\begin{cases} S_n(3t)/2 & \text{if } t \le \frac{1}{3} \\ 1/2 & \text{if } \frac{1}{3} \le t \le \frac{2}{3} \\ S_n(3t-2)/2 + 1/2 & \text{if } \frac{2}{3} \le t \end{cases}$$
More precisely $\|S_{n+1} - S_n\|_\infty \le \|S_n - S_{n-1}\|_\infty/2$, hence $\|S_n - S\|_\infty \le 2^{-n}$. Note that S_n has $\sum_{j=1}^{n} 2^j = 2^{n+1} - 2$ breakpoints at certain triadic rational points $t \in \mathbb{T}_n := \mathbb{Z}/3^n$. Moreover the restriction $S_n|_{\mathbb{T}_n} \to \mathbb{D}_n := \mathbb{Z}/2^n$ is well-defined and uniformly computable in time polynomial in n. According to (a minor adaptation of) [Ko91, THEOREM 2.22], S is therefore computable in polynomial time.

(b) Consider the generalized Darboux sums

$$U\big((s_j), f \circ g, g\big) = \sum_j \sup_{s \in [s_j, s_{j+1}]} f\big(g(s)\big) \cdot \big(g(s_{j+1}) - g(s_j)\big),$$

$$L\big((s_j), f \circ g, g\big) = \sum_j \inf_{s \in [s_j, s_{j+1}]} f\big(g(s)\big) \cdot \big(g(s_{j+1}) - g(s_j)\big)$$

by hypothesis both converging (from above and below, respectively) to $\int_I f(g(s))\,dg(s)$, where (s_j) denotes a partition of I. Substituting $t_j := g(s_j)$ thus yields a partition of $g(I)$ with classical Darboux sums $U\big((t_j), f, \mathrm{id}\big) = U\big((s_j), f \circ g, g\big)$ and $L\big((t_j), f, \mathrm{id}\big) = L\big((s_j), f \circ g, g\big)$ converging to $\int_{g(I)} f(t)\,dt$; and vice versa. $\qquad\square$

It follows that $\mathcal{C}[0;1] \ni f \mapsto f \circ S \in \mathcal{C}[0;1]$ and $\mathcal{C}^1[0;1] \ni f \mapsto f \circ S \in C^{0,\alpha}[0;1]$ are well-defined reductions from Riemann to Cantor integration computable within second-order polynomial time. Applied to Friedman and Ko's polynomial-time computable $f \in C^\infty([0;1],[0;1])$ with #P_1-'complete' integral, one obtains a polynomial-time computable Hölder-continuous $h : [0;1] \to [0;1]$ such that $\int_0^1 h(t)\,dS(t)$ is not computable in polynomial time unless #$P_1 \subseteq \mathsf{FP}$.

Note that $f \circ S$ is not differentiable in general. Moreover the reduction seems restricted to definite integration (and thus only achieves $\#P_1$-hardness rather than $\#P$) since the Cantor integration bounds a and b cannot computably be recovered from the Riemann ones $S(a)$ and $S(b)$ even in the multivalued sense.

Cantor integration is as hard as Riemann integration in the non-uniform and the uniform senses.

Theorem 5.

(a) There exists a polynomial-time computable smooth (i.e. infinitely often differentiable) $h : [0;1] \rightarrow [0;1]$ such that $\int_0^1 h(t)\,dS(t)$ is not computable in polynomial time unless $\#P_1 \subseteq FP$.

(b) There exists a polynomial-time computable smooth $h : [0;1] \rightarrow [0;1]$ such that $[0;1]^2 \ni (a,b) \mapsto \int_{\min\{a,b\}}^{\max\{a,b\}} h(t)\,dS(t)$ is not computable in polynomial time unless $\#P \subseteq FP$.

(c) For every $F \in \#P$ and every $G :\subseteq \text{Reg} \rightarrow \text{Reg}$ representing the indefinite Cantor integration operator on smooth arguments, that is, the mapping

$$C([0;1],[0;1]) \ni f \mapsto \left([0;1]^2 \ni (x,y) \mapsto \int_{\min\{x,y\}}^{\max\{x,y\}} f(t)\,dS(t) \in [0;1]\right) \in C^{0,\alpha}([0;1]^2,[0;1])$$

there exists a second-order polynomial-time reduction from F to G.

Proof. The proofs are inspired from the proof of the hardness of the Riemann integration in [Kawa11] and omitted due to page constraints.

3   Continuous Cantor Integration is at Most as Hard as Smooth Riemann Integration

Reducing the problem of approximating $\int_0^s f(t)\,dS(t)$ up to error 2^{-n} to that of approximating $\int_0^s g_n(t)\,dt$ for some smooth g_n is easy: Since the Cantor measure concentrates all weight to 2^n subintervals $I_{n,k}$ of $[0;1]$ while neglecting the complementing ones, define f_n to be zero on the latter and otherwise equal to f cut and 'squeezed' into the $I_{n,k}$. This f_n is only piecewise continuous but can be approximated up to L^1-error 2^{-n} by a smooth one — depending on n. Based on the following result, Corollary 17 will yield some g independent of n — at the expense of certain 'post-processing' the integral's value.

Theorem 6.

(a) Let $f : [0;1] \rightarrow [0;1]$ be computable in polynomial time and suppose $\#P_1 \subseteq FP$. Then the definite Cantor integral over f, that is the real number $\int_0^1 f(t)\,dS(t)$, is again computable in polynomial time.

(b) Let $f : [0;1] \rightarrow [0;1]$ be computable in polynomial time and suppose $\#P \subseteq FP$. Then the indefinite Cantor integral over f, that is the mapping $[0;1]^2 \ni (x,y) \mapsto \int_{\min\{x,y\}}^{\max\{x,y\}} f(t)\,dS(t)$, is again computable in polynomial time.

(c) *For every $G :\subseteq \text{Reg} \to \text{Reg}$ representing the mapping from Theorem 5(c) there is a second-order polynomial-time reduction from G to some $F \in$ #P.*

In-/definite Cantor integration is thus at most as hard as Riemann integration (and, equivalently, #P).

Proof.

(a) Let $\mu : \mathbb{N} \to \mathbb{N}$ be a polynomial modulus of continuity of f and, modifying Eq. (1) as indicated, $\tilde{f} : \{0,1,2\}^* \to \{0,1\}^*$ computable in polynomial time such that

$$x \in [0;1] \land \left|x - \text{tri}(\vec{u})/3^{\mu(n)}\right| \leq 3^{-\mu(n)} \;\Rightarrow\; \left|f(x) - \text{bin}(\vec{v})/2^n\right| \leq 2^{-n}\;.$$

Then the following function $\psi_1 : \{1\}^* \to \mathbb{N}$ belongs to #P$_1$:

$$\psi_1(1^n) := \#\{(\vec{w},\vec{v}) \in \{0,2\}^{\mu(n)} \times \{0,1\}^n : \text{bin}\left(\tilde{f}(\vec{w})\right) \geq \text{bin}(\vec{v})\}$$

The Cantor distribution assigns weight $1/2^m$ to each interval $\left[\frac{\text{tri}(\vec{w})}{3^m}; \frac{\text{tri}(\vec{w})+1}{3^m}\right]$, $\vec{w} \in \{0,2\}^m$. Moreover f varies by at most 2^{-n} on each such interval for $m := \mu(n)$. Therefore $\psi_1(1^n)/2^{n+\mu(n)}$ is a Darboux sum approximating $\int_0^1 f(t)\,dS(t)$ up to error 2^{1-n}.

(b) Similarly to (a), but now take into account triadic approximations $\text{tri}(\vec{a})/3^{\mu(n)}$ to $\min\{x,y\}$ and $\text{tri}(\vec{b})/3^{\mu(n)}$ to $\max\{x,y\}$ in the #P-function

$$\psi(1^n,\vec{a},\vec{b}) := \#\{(\vec{w},\vec{v}) \in \{0,2\}^{\mu(n)} \times \{0,1\}^n : \text{tri}(\vec{a}) \leq \text{tri}(\vec{w}) \leq \text{tri}(\vec{b}),\, \text{bin}\left(\tilde{f}(\vec{w})\right) \geq \text{bin}(\vec{v})\}$$

(c) Consider $H : \text{Reg} \to \text{Reg}$, defined by $H(\varphi)(\langle 1^n,\vec{a},\vec{b}\rangle,\langle\vec{w},\vec{v}\rangle) := 1$ if

$$\vec{v} \in \{0,1\}^n,\; \vec{a},\vec{b} \in \{0,1,2\}^{\mu(n)},\; \vec{w} \in \{0,2\}^{\mu(n)},$$

$$\text{tri}(\vec{a}) \leq \text{tri}(\vec{w}) \leq \text{tri}(\vec{b}),\, \text{bin}\left(\tilde{f}(\vec{w})\right) \geq \text{bin}(\vec{v})$$

for $1^{\mu(n)}\,0\,\text{bin}\left(\tilde{f}(\vec{w})\right) := \varphi(\vec{w})$, and $H(\varphi)(\langle 1^n,\vec{a},\vec{b}\rangle,\langle\vec{w},\vec{v}\rangle) := 0$ otherwise. Then obviously $H \in$ P holds, and hence $F \in$ #P for

$$F(\varphi)(\langle 1^n,\vec{a},\vec{b}\rangle) = \text{bin}\left(\#\{(\vec{w},\vec{v}) \in \{0,2\}^{\mu(n)} \times \{0,1\}^n :\right.$$

$$\left. \text{tri}(\vec{a}) \leq \text{tri}(\vec{w}) \leq \text{tri}(\vec{b}),\, \text{bin}\left(\tilde{f}(\vec{w})\right) \geq \text{bin}(\vec{v}),\, 1^{\mu(n)}\,0\,\text{bin}\left(\tilde{f}(\vec{w})\right) := \varphi(\vec{w})\}\right)$$

satisfying $\left|F(\varphi)(\langle 1^n,\vec{a},\vec{b}\rangle)/2^{n+\mu(n)} - \displaystyle\int_{\text{tri}(\vec{a})/3^{|\vec{a}|}}^{\text{tri}(\vec{b})/3^{|\vec{b}|}} f(t)\,dS(t)\right| \leq 2^{-n}$ by (b). □

4   Generalized Hardness and Tractability Conditions

4.1   Hardness

The analysis of the similarities between the proofs of uniform #P-hardness of Lebesgue and Cantor integrations gives a list of simple criteria, which can be applied to more cases.

Theorem 7. *Let μ be a measure over $[0;1]$ such that there for every second-order polynomial P, there are rational nonempty open intervals I_w^f and $I_{w,w'}^f$ computable in uniform second-order polynomial time, where $f \in \mathbb{N} \to \mathbb{N}$, $w, w' \in \Sigma*$, and $|w'| \leq P(|w|)$, such that:*

(a) $w_1 \neq w_2 \implies I_{w_1}^f \cap I_{w_2}^f = \emptyset$

(b) $w_1' \neq w_2' \implies I_{w,w_1'}^f \cap I_{w,w_2'}^f = \emptyset$

(c) $I_{w,w'}^f \subseteq I_w^f$

(d) The function $(\mathbb{N} \to \mathbb{N}) \times \mathbb{D} \ni f, d \mapsto \begin{cases} \langle w, w' \rangle & \text{if } d \in I_{w,w'}^f \\ \varepsilon & \text{otherwise} \end{cases}$

  is second-order polynomial time computable.

(e) There exists m_w^f polynomial time computable with respect to $|f|$ and $|w|$ s.t.

$$1 \leq m_w^f \cdot \int_{I_{w,w'}^f} s_{I_{w,w'}^f} \, d\mu \leq 1 + 2^{-P(f,|w|)}$$

  where $s \in C[0;1]$ is any polynomial time computable function s.t. $s(0) = s(1) = 0$.

*Then for every **#P** function F, there exists a second-order polynomial time reduction from F to some $G :\subseteq \text{Reg} \to \text{Reg}$ representing the definite μ integration. In addition, if s is smooth and vanishes at 0 and 1, then this can be restricted to integration of smooth functions.*

Proof. Let $F \in$ **#P** and F_0 be the second-order counting function associated with the second-order polynomial P_0. Given an input oracle φ, we define a continuous function $U(\varphi)$ this way:

$$U(\varphi)(x) = \begin{cases} |I_{w,w'}^f|^{|w|} s_{I_{w,w'}^f}(x) & \text{if } x \in I_{w,w'}^f \text{ and } F_0(\varphi, w, w') = 1 \\ 0 & \text{otherwise.} \end{cases}$$

where $s_{(a;b)}(x) = \frac{s(\frac{x-a}{b-a})}{b-a}$ and $f = P_0(|\varphi|) + 1$.

First, U is well-defined, since the intervals $I_{w,w'}^f$ are pairewise disjoint. It also has a polynomial time computable rational approximation function: given a rational q and a precision n, decide in polynomial time (using d) in which interval q is (and output 0 if it is in none). Then, $|I_{w,w'}^f|^{|w|} s_{I_{w,w'}^f}(x)$ can be computed in polynomial time, since s and the endpoints of $I_{w,w'}^f$ are. It is also easy to see that the modulus of continuity of the function $x \mapsto |I_{w,w'}^f|^{|w|} s_{I_{w,w'}^f}(x)$ is the same as the one of s, and thus $U(\varphi)$ has a polynomial modulus of continuity (even independent from φ). Altogether, this proves that U is polynomial time computable.

Secondly, if s is smooth, then so is $U(\varphi)$. Indeed, the kth derivative of $U(\varphi)$ at $x \in I_{w,w'}^f$ is equal to $|I_{w,w'}^f|^{|w|-(k+1)} \cdot s^{(k)}(\frac{x-a}{b-a})$ where $I_{w,w'}^f = (a;b)$. Now, if (x_n) converges to $x \in [0;1]$, then either

- x_n is infinitely many times in a given interval $I^f_{w,w'}$. Since it is open, it is eventually in this interval, in which case $U(\varphi)^{(k)}(x_n)$ converges to $U(\varphi)^{(k)}(x_n)$ by continuity of $U(\varphi)^{(k)}$ on $I^f_{w,w'}$ (by smoothness of s);
- otherwise x_n can not be infinitely many times in more that two such intervals since they do not intersect. In this case, x is one of the endpoints of such intervals and $U(\varphi)(x_n)$ converges to $0 = U(\varphi)(x)$ (since $s(0) = s(1) = 0$);
- otherwise, x_n is eventually outside the union of the intervals (i.e. in a closed set, so where x also belongs) and $U(\varphi)(x_n) = 0 = U(\varphi)(x)$;
- finally, x_n can be decomposed into a sequence outside any interval (whose image by $U(\varphi)^{(k)}$ converges to 0), or in an interval $I^f_{w_n,w'_n}$ occurring only finitely many times. This implies that the sequence w_n diverges to $+\infty$ (since there are only a finite number of w' for a given w) and thus $U(\varphi)^{(k)}$ also converges to 0 on this subsequence (since $s^{(k)}$ is bounded, and $|I^f_{w,w'}|^{|w|-(k+1)}$ converges to 0). This is indeed equal to $U(\varphi)^{(k)}(x)$, otherwise we would be in the first case.

Finally, $F(\varphi, w)$ can be indeed computed in polynomial time from the μ-integral of U. Indeed, according to hypothesis e), $F(\varphi, w) =$

$$\sum_{|w'|\leq P_0(|\varphi|,|w|)} 1 \leq \sum_{|w'|\leq P_0(|\varphi|,|w|)} m_w^{|\varphi|} \cdot \int_{I^{P_0(|\varphi|,|w|)}_{w,w'}} s_{I^{P_0(|\varphi|,|w|)}_{w,w'}} d\mu \leq F(\varphi, w) \cdot (1 + 2^{-(P_0(|\varphi|,|w|)+1)}),$$

and since $F(\varphi, w) \leq 2^{P_0(|\varphi|,|w|)}$, and that the sum of the integrals is equal to the integral over I^f_w, we obtain:

$$F(\varphi, w) \leq m_w^{|\varphi|} \cdot \int_{I^{P_0(|\varphi|,|w|)}_w} U(\varphi) d\mu \leq F(\varphi, w) + \frac{1}{2}.$$

In other words we can define a second-order polynomial time computable function $W(\varphi, g, w)$ which computes a $\frac{1}{2}$-approximation of $m_w^{|\varphi|}$ multiplied by the real number represented by g and outputs the closest integer. In this case, we obtain:

$$F(\varphi, w) = W(\varphi)(G(U(\varphi))(I^{P_0(|\varphi|,|w|)}_w), w),$$

if G represents the definite μ-integration. $\square$

4.2   Tractability

Conversely, there is some simple sufficient condition for a positive linear operator to be polynomial time computable with respect to an oracle. For this, we will use the main result of [FGH13], where the authors define the sets of relevant points $(R_n)_{n\in\mathbb{N}}$ of a real norm on $\mathcal{C}[0; 1]$. Roughly speaking, R_n is the set of points of $[0; 1]$ where it is sufficient to know an input 1-Lipschitz function $f \in \mathcal{C}[0; 1]$ in order to determine its norm with precision 2^{-n} (see the original article for a precise definition). The theorem states that polynomial time (relatively to an oracle) computable real norms are exactly those which depend on a 'small' set of points in this sense:

Definition 8. *A set A can be polynomially covered, if there exists a polynomial P such that for all $n \in \mathbb{N}$, A can be covered by $P(n)$ balls of radius 2^{-n}. In other words, A has metric entropy $\log \circ P$.*

Fact 9 ([FGH13]). *A real norm can be computed in polynomial time relatively to an oracle if and only if its sets of relevant points $(R_n)_{n \in \mathbb{N}}$ can be polynomially covered uniformly in n.*

Even if an integration operator is not a norm, it can be completed into one, so that we can apply a weak form of the previous theorem.

Theorem 10. *If the support of a measure ν can be polynomially covered, then the corresponding indefinite integration operator $(x, y, g) \mapsto \int_x^y g \, d\nu$ is computable in polynomial time with respect to an oracle.*

Proof. Let ν be such a measure, with support S. There exists a polynomial time computable norm F over $\mathcal{C}[0;1]$. For $x \leq y$ in $[0;1]$, the operators $G_{x,y}^+(f) = F(f) + \int_x^y f^+ d\nu$ and $G_{x,y}^-(f) = F(f) + \int_x^y f^- d\nu$ (where f^+ and f^- are the positive and negative parts of f) are norms. We need to separate the positive and negative parts in order to make these operators always positive.

The set of relevant points R_n of ν-integration is included in the closure of S. Indeed, the integral of any 1-Lipschitz function defined on a neighborhood of $x \notin \bar{S}$ is equal to zero as soon as this neighborhood does not intersect S, by definition of the support of a measure.

Thus, for all n, the set of relevant points of $G_{x,y}^+$ and $G_{x,y}^-$ are included in $\bar{S} \cup R_n^F$, where $(R_n^F)_n$ are the relevant sets of F. By application of Fact 9 to F, $(R_n^F)_n$ can be polynomially covered uniformly in n, and since it is also the case for S, and thus for its closure, it is true for the union.

By application of the other implication of Fact 9, these norms are polynomial time computable with respect to an oracle. In fact, this is also true for the corresponding operators G^+ and G^-, uniformly in x and y. Since the indefinite ν-integration operator is equal to $G^+ - G^-$, it is also polynomial time computable with respect to an oracle, and allows us to conclude.   □

Even though allowing an arbitrary oracle may seem powerful, such operators are still weaker than Lebesgue or Cantor integrals.

Corollary 11. *The indefinite integration operator associated with such a measure is not #P-hard.*

Indeed, such an operator would allow to compute Lebesgue or Cantor integration operator in relative polynomial time, which is impossible (in particular by an application of Fact 9); cmp. also [KaPa14].

4.3   Applications and Examples

First, let us focus on the absolutely continuous case, *i.e.* where the integral of a function f is equal to the Lebesgue integral of $f \cdot g$, where g is a measurable function.

The simplest non-trivial example is the Lebesgue integration (with $g = 1$) is already **#P**-hard, which makes us believe that this is the case in general. It is already the case if g is polynomial time computable.

Proposition 12. *Let $g \in C[0; 1]$ be polynomial time computable and not identically zero. Then $(x, y, f) \mapsto \int_x^y f(t) \cdot g(t) dt$ is **#P**-hard.*

Proof. Omitted due to page limitations.

However, we don't know if this still holds for functions g with higher complexity. Intuitively, it seems that we need g to be polynomial time computable in order to retrieve some information about f from its integral (see Hypothesis e of Theorem 7).

Now we can have a look at the case of discrete measures, *i.e.* corresponding to posive linear operators F of the form: $F(f) = \sum_{n \in \mathbb{N}} \alpha_n \cdot f(\beta_n)$, where $\alpha_n > 0$ and $\beta_n \in [0; 1]$. It is not surprising that when this sum is finite, then the integration operator F is polynomial time computable with respect to an oracle (where an appropriate oracle encodes the α_i's and β_i's). But Theorem 10 even gives a more general result.

Proposition 13. *If F is a discrete integration operator of the form $F(f) = \sum_{n \in \mathbb{N}} \alpha_n \cdot f(\beta_n)$, such that the set $B = \{\beta_i \mid i \in \mathbb{N}\}$ can be polynomially covered, then F is computable in relative polynomial time.*

Proof. In this case, the support of F is contained in the closure of B, which is thus can also be polynomially covered, and Theorem 10 applies.   □

Conversely, there are discrete measures defining **#P**-hard integration operators.

Example 14. Let $F(f) = \sum_{w \in \{0,1\}^*} 2^{-2|w|} f(\overline{0.w.1}^2)$, where $\bar{w}^2$ is the real number with binary expansion w. Its sequence of scaling factors decreases exponentially slowly, whereas its set of evaluation points covers all the dyadic rational numbers of the open interval $(0, 1)$. This discrete positive linear operator is not computable in (relative) polynomial time. Moreover, we can apply Theorem 7 and deduce that it is **#P**-hard.

The conditions of the two theorems are not always necessary and there are cases where none of them apply. But it seems that most of the time, a result can still be obtained using the general shape of discrete measures.

Example 15. Let $F(f) = \sum_n f(d_n)$, where (d_n) is the standard enumeration of the dyadic rational numbers of $[0; 1]$. It is an integration operator relative to a discrete measure whose support is $[0; 1]$. Since an interval can't be polynomially covered, we can not apply Theorem 10. However, a direct application of [FGH13] or straightforward analysis allows us to prove that it is polynomial time computable. Indeed, to compute $F(f)$, it is sufficient to compute $f(d_0), \ldots, f(d_{\mu(n)})$, if f has modulus of continuity μ.

Finally, the last case is the one of singular continuous measures. It is the hardest one, since there is no simple characterization of such measures.

We have already seen with the Cantor measure that an integration operator for such a measure can be **#P**-hard. But a similar measure can also be polynomial time computable.

Example 16. The Cantor set is defined by the intersection of sets $(C_n)_{n\in\mathbb{N}}$, where C_n is the union of 2^n disjoint intervals of size 3^{-n}. If we define a Cantor-like set $C' = \bigcap_{n\in\mathbb{N}} C'_n$, where C'_n is the intersection of 2^{-n} intervals of size $3^{2^{-n}}$ (*i.e.* exponentially smaller), then the corresponding measure has support C', which can be polynomially covered. By a direct application of Theorem 10, the associated positive linear operator is polynomial time computable with respect to an oracle. If in addition the endpoints of these intervals are polynomial time computable uniformly in n, then it is even simply polynomial time computable.

5 Conclusion and Perspectives

We have completed the complexity-theoretic classification of the three 'proto-types' of positive linear functionals on $\mathcal{C}[0;1]$: evaluation (discrete) is polynomial-time computable whereas both Riemann (absolutely continuous) and Cantor (singular continuous) integration both correspond to the discrete complexity class $\#P_1$. More precisely they are uniformly second-order polynomial-time equivalent in the following sense:

Corollary 17.

(a) There exists a second-order polynomial-time computable operator $U :$ $C([0;1],[0;1]) \to C^\infty([0;1];[0;1])$, and second-order polynomial-time computable functionals $V_1, V_2, W : C([0;1],[0;1]) \times [-1;1] \to [0;1]$ such that the following holds:

$$\forall f \in C([0;1],[0;1]) \;\; \forall 0 \le a \le b \le 1 : \int_a^b f(t)\,dS(t) \;=\; W\!\left(f, \int_{V_1(f,a)}^{V_2(f,b)} U(f)(t)\,dt\right)$$

(b) There exists a second-order polynomial-time computable operator $U :$ $C([0;1],[0;1]) \to C^\infty([0;1];[0;1])$ and second-order polynomial-time computable functionals $V_1, V_2, W : C([0;1],[0;1]) \times [0;1] \to [0;1]$ such that the following holds:

$$\forall f \in C([0;1],[0;1]) \;\; \forall 0 \le a \le b \le 1 : \int_a^b f(t)\,dt \;=\; W\!\left(f, \int_{V_1(f,a)}^{V_2(f,b)} U(f)(t)\,dS(t)\right)$$

Proof. Combine the second-order polynomial-time reductions of (smooth) Cantor integration to and from **#P** according to Theorems 5(c) and 6(c) with the known second-order polynomial-time reductions of (smooth) Riemann integration to and from **#P**. □

Remark 18.

(a) [Kawa11, THEOREMS 4.18+4.21] originally have asserted maximization and integration of bivariate functions

$$C([0;1]^2) \ni f \mapsto \big([0;1] \ni s \mapsto \max\{f(s,t) : 0 \le t \le 1\}\big) \in C[0;1] \quad (3)$$

$$C([0;1]^2) \ni g \mapsto \Big([0;1] \ni s \mapsto \int_0^1 g(s,t)\,dt\Big) \in C[0;1] \quad (4)$$

to be **NP**-complete and **#P**-complete, respectively. Since **NP** trivially reduces to **#P**, transitivity yields the existence of a second-order polynomial-time reduction from maximization (3) to integration (4): which seems quite surprising.

(b) Fact 3(e+f) refers to a univariate variant of Eq. (4) with varying lower and upper integration bounds. In fact maximization in Eq. (3) even remains **NP**-complete when only varying the upper bound, that is, the operator $C([0;1]) \ni f \mapsto \big([0;1] \ni s \mapsto \max\{f(t) : 0 \le t \le s\}\big) \in C[0;1]$. Since such a reduction according to Definition 2(e) is permitted only one invocation of the integration operator, we wonder whether also integration remains **#P**-complete with only upper bound varying: $C([0;1]) \ni g \mapsto$

$$\Big([0;1] \ni y \mapsto \int_0^y f(t)\,dt\Big) \in C([0;1]).$$

5.1   Prototype Vs. the General Case

Our investigation of the complexity of positive linear functionals Ψ on $\mathcal{C}[0;1]$ has focused on prototypical examples of each of the three basic types according to Riesz–Markov–Kakutani. For instance the d-dimensional Poisson problem has been shown [KSZ14] to boil down to absolutely continuous integration (ii). It is, however, easy to find Ψ that are harder than these prototypes: for example evaluation (i) at some **EXP**-complete point $z \in [0;1]$. This leads to

Question 19. Is there an integrable $g : [0;1] \to (0;\infty)$ such that $\int_0^1 f(t) \cdot g(t)\,dt$ is polynomial-time computable for every polynomial-time computable $f \in \mathcal{C}[0;1]$ even in case $\mathbf{P} \ne \mathbf{NP} \ne \mathbf{P}^{\#\mathbf{P}} \ne \mathbf{PSPACE}$?

Restricted to continuous g the answer is negative: Each such admits distinct rational (and in particular polynomial-time computable) $a,b \in [0;1]$ and $k \in \mathbb{N}$ with $1/k \le |g| \le k$ on $[a;b]$; w.l.o.g. $g = |g|$ there. Record that the polynomial-time computable smooth $h : [0;1] \to [0;1]$ with $\#\mathbf{P}_1$-'complete' $\int_0^1 h(t)\,dt$ can be achieved to vanish (with all derivatives) on $(-\infty;0] \cup [1;\infty)$; cmp. [Ko91, THEOREM 5.32d]. Scaling $f(t) := h\big(\frac{t-a}{b-a}\big)$ thus is still smooth on $[0;1]$ and polynomial-time computable; yet approximating

$$(b-a) \cdot k \cdot \int_0^1 f(t) \cdot g(t)\,dt \in \Big[\int_0^1 h(t)\,dt \;;\; k^2 \cdot \int_0^1 h(t)\,dt\Big]$$

up to error $2^{-n}/k^2$ recovers $\int_0^1 h(t)\,dt$ up to error 2^{-n}. This leaves it to look for integrable, nowhere essentially bounded g; cmp. http://math.stackexchange.com/questions/620959.

Conjecture 20 (Added in proof). The computational complexity of integration $C[0;1] \ni f \mapsto \int_0^1 f(t)\,d\nu$ is related to the *joint Kolmogorov-Shannon Entropy* of ν, defined http://math.stackexchange.com/questions/111260 as mapping

$$\mathbb{N} \ni n \mapsto \inf\left\{ \sum_{j=1}^J \mu(S_j) \cdot \log 1/\mu(S_j) : [0;1] \subseteq \bigcup_{j=1}^J S_j,\ \mathrm{diam}(S_j) < 2^{-n} \right\} \in \mathbb{N}.$$

References

[BrGh11] Brattka, V., Gherardi, G.: Weihrauch degrees, omniscience principles and weak computability. J. Symb. Log. **76**(1), 143–176 (2011)

[Coll14] Collins, P.: Computable Stochastic Processes (2014). arXiv:1409.4667

[Cook91] Cook, S.A.: Computability and complexity of higher type functions. In: Moschovakis, Y.N. (ed.) Logic from Computer Science. Mathematical Sciences Research Institute Publications, pp. 51–72. Springer, Heidelberg (1991)

[FeHo13] Férée, H., Hoyrup, M.: Higher-order complexity in analysis. In: Proceedings 10th International Conference on Computability and Complexity in Analysis (CCA 2013)

[FGH13] Férée, H., Gomaa, W., Hoyrup, M.: Analytical properties of resource-bounded real functionals. J. Complex. **30**(5), 647–671 (2014)

[Frie84] Friedman, H.: The computational complexity of maximization and integration. Adv. Math. **53**, 80–98 (1984)

[HeOg02] Hemaspaandra, L.A., Ogihara, M.: The Complexity Theory Companion. Springer, Heidelberg (2002)

[HiPa13] Higuchi, K., Pauly, A.: The degree structure of Weihrauch-reducibility. Log. Methods Comput. Sci. **9**(2), 1–17 (2013)

[HRW12] Hoyrup, M., Rojas, C., Weihrauch, K.: Computability of the Radon-Nikodym derivative. Computability **1**, 1–11 (2012)

[IBR01] Irwin, R., Kapron, B., Royer, J.: On characterizations of the basic feasible functionals part I. J. Funct. Program. **11**, 117–153 (2001)

[KaCo96] Kapron, B.M., Cook, S.A.: A new characterization of type-2 feasibility. SIAM J. Comput. **25**(1), 117–132 (1996)

[KaCo10] Kawamura, A., Cook, S.A.: "Complexity theory for operators in analysis. In: Proceedings of 42nd Annual ACM Symposium on Theory of Computing (STOC 2010), pp. 495–502 (2012). (full version in ACM Transactions in Computation Theory, vol. 4:2 , article 5.)

[KaPa14] Kawamura, A., Pauly, A.: Function spaces for second-order polynomial time. In: Beckmann, A., Csuhaj-Varjú, E., Meer, K. (eds.) CiE 2014. LNCS, vol. 8493, pp. 245–254. Springer, Heidelberg (2014)

[Kawa11] Kawamura, A.: Computational complexity in analysis and geometry, Dissertation, University of Toronto (2011)

[Ko91] Ko, K.-I.: Computational Complexity of Real Functions. Birkhäuser, Boston (1991)

[KSZ14] Kawamura, A., Steinberg, F., Ziegler, M.: Complexity of Laplace's and Poisson's Equation, abstract. Bull. Symb. Log. **20**(2), 231 (2014). Full version to appear in Logical Methods in Computer Science

[KSZ15] Kawamura, A., Steinberg, F., Ziegler, M.: Computational Complexity Theory for classes of integrable functions. In: JAIST Logic Workshop Series (2015)

[MTY14] Mori, T., Tsujii, Y., Yasugi, M.: Computability of probability distributions and characteristic functions. Log. Methods Comput. Sci. **9**, 3 (2013)

[Schr07] Schröder, M.: Admissible representations of probability measures. Electron. Notes Theoret. Comput. Sci. **167**, 61–78 (2007)

[Weih00] Weihrauch, K.: Computable Analysis. Springer, Heidelberg (2000)

Average-Case Bit-Complexity
Theory of Real Functions

Matthias Schröder[1], Florian Steinberg[1], and Martin Ziegler[1,2]($\boxtimes$)

[1] TU Darmstadt, Darmstadt, Germany
m@zie.de
[2] KAIST, Daejeon, South Korea

Abstract. We introduce, and initiate the study of, average-case bit-complexity theory over the reals: Like in the discrete case a first, naïve notion of polynomial average runtime turns out to lack robustness and is thus refined. Standard examples of explicit continuous functions with increasingly high worst-case complexity are shown to be in fact easy in the mean; while a further example is constructed with both worst and average complexity exponential: for topological/metric reasons, i.e., oracles do not help. The notions are then generalized from the reals to represented spaces; and, in the real case, related to randomized computation.

1 Introduction and Motivation

In worst-case analyses of algorithms, rare instances may dominate the cost of a typically easy problem. For example Hoare's (original) `QuickSort` in some cases takes time quadratic in the number N of items but, on average over all possible input permutations $\pi : [N] \to [N]$, only $\mathcal{O}(N \log N)$ steps — in agreement with practical experience of usually being a highly efficient method! Moreover this quasi-linear average-time easily translates into a randomized algorithm with similar expected worst-case runtime: simply apply a random permutation first.

In general the asymptotic average-case cost of an algorithm is defined with respect to a fixed probability distribution $\mathbb{P}_n$ on all inputs of length $n \to \infty$. The choice of $\mathbb{P}_n$ of course matters and is generally subject to discussion, though. For instance regarding floating-point calculations, which distribution would you consider natural on the rational unit cube $[0;1)^n \cap \mathbb{Q}^n$, already in case $n = 1$? The real cube/torus $[0;1)^n$ on the other hand — and more generally every compact topological group — does come with a canonical probability measure.

Now Ko and Friedman [KoFr82] have applied discrete complexity theory to real functions computed by approximation up to absolute output error $1/2^n$,

---

Supported in part by the *Marie Curie International Research Staff Exchange Scheme Fellowship* 294962 within the 7th European Community Framework Programme, by the *German Research Foundation* (DFG) with project Zi 1009/4-1, and by the International Research Training Group 1529. A preliminary version of this work was presented at CCA 2015. Note added in proof: Theorem 9 is conditional to the assertion of Lemma 11(c).

I.S. Kotsireas et al. (Eds.): MACIS 2015, LNCS 9582, pp. 505–519, 2016.
DOI: 10.1007/978-3-319-32859-1_43

roughly corresponding to n correct binary output digits after the radix point. This worst-case setting renders purportedly easy numerical operators unusually hard [Ko91, KORZ14]. We thus suggest applying average-case complexity theory to computation over the reals in the bit model and its generalization over represented spaces [Weih03, Schr04]. The present work introduces the formal notions. Section 2 reveals the 'standard' examples of explicit (but admittedly artificial) $f : [0; 1] \to \mathbb{R}$ with worst-case complexity an exponential tower to be in fact polynomial-time computable on average; and constructs a function whose complexity is exponential both on average and in the worst-case. This complements investigations in the unit-cost model of real computation [Ritt00]. Section 3 then generalizes the relevant concepts from $[0; 1]$ to represented space in the sense of Weihrauch's *Type-2 Theory of Effectivity* (TTE); and establishes a connection between average-case and randomized expected time complexity. Our hope, expressed in Sect. 4, is for a rigorous explanation why theoretically proven hard numerical problems are regularly solved efficiently in practice.

1.1 Real Worst-Case and Average-Case Complexity

In the approximate/analytic model of computing, both real arguments x and results $f(x)$ cannot be input/output exactly but are approximated by sequences of floating point numbers with increasing absolute precision. In the following definition, Item (a) carefully refines [Ko91, DEFINITION 2.1A], [Weih00, EXAMPLE 7.2.14.3], and [KMRZ15, DEFINITION 2.1a].

Definition 1.

(a) *A (ρ-) name of a real number x is any integer sequence $(a_m)_m$ with $\frac{a_m - 1}{2^m} \leq x < \frac{a_m + 1}{2^m}$. Abbreviate with $\rho^{-1}[x] := \prod_m \{\lfloor x \cdot 2^m \rfloor, \lfloor 1 + x \cdot 2^m \rfloor\}$ the collection of all such names, considered as subset of the Baire space $\mathbb{Z}^\omega$.*

(b) *A Type-2 Machine $\mathcal{M}^O$ (i.e. a Turing machine with read-only input and one-way output tape and access to the — usually empty — oracle $O \subseteq \{0, 1\}^*$) is said to compute the partial function $f : \subseteq \mathbb{R} \to \mathbb{R}$ if it, upon input of any name $\bar{a} = (a_m)_m$ (encoded in binary using delimiters while quelling leading zeros) of some $x \in \text{dom}(f)$, produces a (similarly encoded) name $\bar{b} = (b_n)_n$ of $f(x)$.*

(c) *Refining (b), let $T_{\mathcal{M}^O}(\bar{a}, n) \in \mathbb{N} := \{0, 1, 2, \ldots\}$ denote the number of steps $\mathcal{M}^O$ makes on input $\bar{a}$ before producing b_n according to (a). Define, for fixed $x \in \text{dom}(f)$, the local worst-case running time over all names of x by*

$$T_{\mathcal{M}^O}(x, n) := \sup\{T_{\mathcal{M}^O}(\bar{a}, n) \mid \bar{a} \in \rho^{-1}[x]\} \in \bar{\mathbb{N}} := \mathbb{N} \cup \{\infty\}.$$

(d) *The (global) worst-case running time of $\mathcal{M}^O$ computing f is $T_{\mathcal{M}^O, f}(n) := \sup\{T_{\mathcal{M}^O}(x, n) : x \in \text{dom}(f)\}$. It is (worst-case) polynomial if $\forall n : T_{\mathcal{M}^O, f}(n) \leq P(n)$ holds for some polynomial $P \in \mathbb{N}[N]$; equivalently: iff $\mathbb{N} \ni n \mapsto \frac{1}{n} \cdot \sup_{x \in \text{dom}(f)} \left(T_{\mathcal{M}^O}(x, n)\right)^\varepsilon$ is bounded for some $\varepsilon > 0$.*

(e) Let (X, Σ, λ) denote a probability measure space and $T : X \times \mathbb{N} \to [0, \infty]$ a measurable function. Abbreviate with $\bar{T}(n) := \int_X T(x, n)\, d\lambda(x) \in [0; \infty]$ the associated average; and call T naïvely polynomial on average if $\bar{T}(n) \leq P(n)$ for some $P \in \mathbb{N}[N]$. On the other hand we say that T is (non-naïvely) polynomial on average if there exists some $\varepsilon > 0$ such that $\mathbb{N} \ni n \mapsto \frac{1}{n} \int_{\mathrm{dom}(f)} \big(T(x, n)\big)^{\varepsilon}\, dx$ is bounded.

(f) For metric spaces (X, d) and (Y, e), a local modulus of continuity of $f :$ $X \to Y$ is a mapping $\mu : X \times \mathbb{Z} \to \bar{\mathbb{Z}} := \mathbb{Z} \cup \{-\infty\}$ satisfying

$$\forall x, x' \in X : \quad d(x, x') < 2^{-\mu(x,n)} \quad \Rightarrow \quad e\big(f(x), f(x')\big) < 2^{-n}.$$

A (global) modulus of (uniform) continuity does not dependent on x.

(g) If (X, Σ, λ) is also a probability measure space, call f Lipschitz on average if it admits a measurable local modulus of continuity μ such that, for some constant C and all n, it holds $\int_X \mu(x, n)\, d\lambda(x) \leq n + C$.

Note how naïvely and non-naïvely polynomial average running times in (e) arise from the equivalent notions of polynomial worst-case bounds in (d). However naïve polynomial averages are not robust under polynomial slowdown; see [Gold97, Sect. A] or Remark 2(c). Hence we focus in the sequel on the non-naïve notion — which also underlies Leonid Levin's structural average-case complexity theory [BoTr06, Sect. 2.2.1]. Definition 1(g) is motivated by Remark 2(d) below.

Remark 2.

(a) For every compact $K \subseteq \mathbb{R}$, and in particular for the singleton $K := \{x\}$, the set $\rho^{-1}[K] := \bigcup_{x \in K} \rho^{-1}[x]$ is compact in $\mathbb{Z}^{\omega}$. The mapping $\rho^{-1}[K] \ni \bar{a} \mapsto T_{\mathcal{M}\circ}(\bar{a}, n) \in \mathbb{N}$ is locally constant, i.e., continuous. For compact $\mathrm{dom}(f)$ the supremum in Definition 1(d) is thus a maximum and in particular finite.

(b) The mapping $x \mapsto T_{\mathcal{M}\circ}(x, n)$ is in general discontinuous: component $\lfloor x \cdot 2^m \rfloor$ of $\rho^{-1}[x]$ 'jumps' at dyadic arguments $x \in \mathbb{D}_m := \{a/2^m : a \in \mathbb{Z}\}$. On the other hand $x \mapsto T_{\mathcal{M}\circ}(x, n)$ is locally constant on non-dyadic $x \notin \mathbb{D} := \bigcup_m \mathbb{D}_m$: For $m \in \mathbb{N}$ and $\delta := \min\{|x - x'| : x' \in \mathbb{D}_m\}$, the first m components of $\rho^{-1}[x]$ and $\rho^{-1}[x']$ agree whenever $|x - x'| < \delta$; and by continuity of $\bar{a} \mapsto T_{\mathcal{M}}^O(\bar{a}, n)$ so do $T_{\mathcal{M}\circ}(x, n)$ and $T_{\mathcal{M}\circ}(x', n)$ for appropriate $m = m(n)$. The preimage $\{x : T_{\mathcal{M}}^O(x, n) = N\}$ is thus an open subset of $\mathrm{dom}(f) \cap \mathbb{D}$ plus some subset of countable $\mathbb{D}$ and in particular Σ_2–measurable, so Definition 1(e) indeed applies to $T_{\mathcal{M}\circ}$.

(c) Let (X, Σ, λ) be a probability measure space, $f : X \to [0, \infty]$ a measurable map, and $\varepsilon > 0$. Then $\left(\int_X f(x)^{\varepsilon}\, d\lambda(x) \right)^{1/\varepsilon} \leq \int_X f(x)\, d\lambda(x)$ holds according to the Reverse Hölder Inequality $\|f \cdot 1\|_1 \geq \|f\|_{\varepsilon} \cdot \|1\|_{-1/(1/\varepsilon - 1)}$. In particular naïve polynomial on average implies polynomial on average — but not vice versa:

$$T(x, n) := \exp(n) \text{ for } 0 \leq x \leq 1/2^n, \quad T(x, n) := 0 \text{ for } 1/2^n < x \leq 1$$

has $\int_0^1 T(x, n)\, dx = (e/2)^n$ exceeding polynomial growth, but for $\varepsilon := 1/2$ it holds $\frac{1}{n} \int_0^1 \sqrt{T(x, n)}\, dx = \frac{1}{n}(\sqrt{e}/2)^n \to 0$.

According to [BoTr06, Proposition 5], $T : X \times \mathbb{N} \to [0; \infty]$ is polynomial on average iff

$$\exists \delta > 0 \; \exists P \in \mathbb{N}[N] \; \forall n \in \mathbb{N} \; \forall t > 0 : \quad \lambda(\{x : T(x, n) \geq t\}) \leq P(n)/t^{\delta}.$$

(d) Continuous $f : X \to Y$ is bounded iff it admits a local modulus of continuity attaining the value $-\infty$; it is Lipschitz-continuous iff it admits a global modulus of continuity of the form $n \mapsto n + C$ for some $C \in \mathbb{N}$.

If X is convex, $n \mapsto \mu(n+1) - 1$ is again a modulus of continuity: Whenever $d(x, y) < 2^{1-\mu(n+1)}$, there exists $z \in X$ with $d(x, z), d(z, y) < 2^{-\mu(n+1)}$; hence $e(f(x), f(y)) \leq e(f(x), f(z)) = e(f(z), f(y)) < 2^{-(n+1)} + 2^{-(n+1)}$. So w.l.o.g. $\mu(n) \leq \mu(n+1) - 1$, and it suffices to consider the restriction $\mu : \mathbb{N} \to \bar{\mathbb{Z}}$.

Suppose machine $\mathcal{M}^O$ computes $f :\subseteq \mathbb{R} \to \mathbb{R}$. Then $\mu(x, n) := T_{\mathcal{M}^O}(x, n + 1) + 1$ is a local modulus of continuity of f; cmp. [KMRZ15, Fact 3e].

(e) Suppose $f : (0; 1) \to \mathbb{R}$ is continuously differentiable with $|f'|$ decreasing. Then it has $\mu(x, n) := \lceil \max\{n + \mathrm{lb}\, |f'(x/2)|, \ln(2/x)\} \rceil$ as local modulus of continuity, where $\mathrm{lb}(x) := \ln(x)/\ln(2)$ denotes the binary logarithm:

By the *Mean Value Theorem*, to $0 < x < 1$ and $0 < h < x$, there exists some $\xi \in (x - h, x)$ such that $|f(x-h) - f(x)| = |f'(\xi)| \cdot h \leq |f'(x-h)| \cdot h < 1/2^n$ and similarly $|f(x) - f(x+h)| \leq |f'(x-h)| \cdot h < 1/2^n$: provided $h < \min\{x/2, 2^{-n}/|f'(x/2)|\} \geq 2^{-\mu(x,n)}$.

Average-case complexity gauges the expected cost (mostly runtime, but also memory etc.) of deterministic computations on random inputs — as opposed to probabilistic/randomized computations; cmp. Definition 7(c,d) below.

2   Average Versus Worst-Case Complexity

Due to its topological/metric aspect, real (as opposed to discrete) complexity theory permits explicit constructions of provably 'hard' functions [KMRZ15, Fact 3g]. However, these turn out as 'easy' on average. In fact, iterating them yields an arbitrarily large gap between worst-case and average-case complexity; see Items (a) and (b) in the following

Theorem 3.

(a) The function

$$f : [0; 1] \to [0; 1] \quad \text{with} \quad f(0) = 0 \quad \text{and} \quad f(x) = 1/\ln(e/x) = \tfrac{1}{1 - \ln x} \text{ else}$$

is computable (without oracle) in exponential time but, admitting no subexponential modulus of uniform continuity, not faster, even relative to any oracle — in the worst case: Its (both naïve and non-naïve) average time complexity is polynomial. Moreover, f is Lipschitz on average.

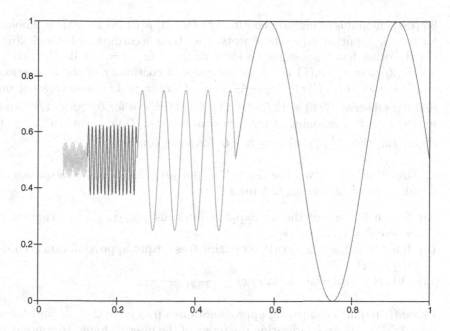

Fig. 1. Graph of the function analyzed in Theorem 3(c)

(b) *More generally the d-fold iterate $f^{(d)} = f^{(d-1)} \circ f : [0;1] \to [0;1]$ is computable without oracle in time an exponential tower of height d but not faster in the worst case, even relative to any oracle — while still having (both naïve and non-naïve) polynomial average time complexity and being Lipschitz on average.*

(c) *The function $g : [0;1] \to [-1;1]$ depicted in Fig. 1 and defined by*

$$0 \mapsto 0, \quad (2^{-j}; 2^{-j+1}] \ni t \mapsto \sin\left(\pi \cdot 2^{2^{j \cdot j}} \cdot (t \cdot 2^j - 1)\right)/2^j, \quad j = 1, 2, \ldots$$

is computable (without oracle) in exponential worst-case time but not in time polynomial on average (regardless of oracle access).

(d) *The function $F : (0;1] \times (1;\infty) \to (1;\infty)$ defined by $F(x,n) := x^{-1+1/n}$ is polynomial on average, but the composition $(x, n) \mapsto F\left(x, F(x, n)\right)$ is not.*

Item (c) thus complements (a) and (b) with an instance that is equally hard in the worst and average case. It can be regarded as an explicit, real, and relativizing variant of [LiVi97, Theorem 4.6.1]. The proofs are essentially worst-case and average-case analyses of moduli of continuity, based on Remark 2(d) and collected in the following subsection.

2.1 Proof of Theorem 3

(a) First let us record

$$f(x) = |f(0) - f(x)| < 1/2^n \Leftrightarrow x < \exp(1 - 2^n) = 2^{(1-2^n)/\ln 2} =: x_n \quad (1)$$

So μ is a modulus of uniform continuity of f iff $\mu(n) \geq (1 - 2^n)/\ln 2$ holds for $n \in \mathbb{N}$, requiring exponential worst-case time according to Remark 2(d). On the other hand, according to Remark 2(e), $\mu(x, n) = n + \text{lb}\,|f'(x/2)| \leq n + \text{lb}(2/x) = n - \ln(x)$ is a local modulus of continuity of the restriction $f|_{(0;1/e]}$ since $0 < f'(x) = \frac{1}{x \cdot (1 - \ln x)^2} = f^2(x)/x \leq 1/x$ is decreasing on $(0; 1/e]$: Observe $f''(x) = (2f(x) - 1) \cdot f^2(x)/x^2 < 0$ for $0 < x < 1/e$. And $\mu(x, n) = n$ is a modulus of uniform continuity of $f|_{[1/e;1]}$ since $f'(x) \leq 1$ there: Together $\int_0^1 \mu(x, n)\,dx \leq n + C$ for some constant C.

Algorithmically, given the desired output precision n and (a sequence of dyadic approximations $a_m/2^m$ to) $x \in [0; 1]$,

(i) Search for one of the overlapping intervals $[x_{k+1}; x_{k-1}]$ to contain x where $k = 1, \ldots, n$

(ii) If not found, $x \leq x_n$ justifies printing 0 as output approximation according to (1).

(iii) Otherwise calculate $1/\ln(e/x) = \frac{1}{1 + \text{lb}(1/x) \cdot \ln 2}$.

Phase (i) requires comparing approximations up to error $2^{-2^{\mathcal{O}(k)}}$ and takes time $2^{\mathcal{O}(k)}$ to do so; k indicating the index of the interval found to contain x. The reciprocal $y := 1/x \leq 1/x_{k+1}$ in Phase (iii) has $2^{\mathcal{O}(k)}$ digits before the radix point (integer part); hence approximating it up to absolute error 2^{-n} takes time polynomial in $2^{\mathcal{O}(k)} + n$. Now $\text{lb}(y)$ is the length of the integer part plus the binary logarithm of the remainder; it thus reduces the magnitude of the intermediate result, allowing the remaining calculations to be performed in time polynomial in $k + n$. Arguments x in the interval $[x_k; x_{k-1})$ of length $x_{k-1} - x_k \leq x_{k-1}$ thus suffice with a number of steps polynomial in $2^{\mathcal{O}(k)} + n$ to process, leading to an average running time of

$$\bar{T}(n) \leq \sum_k \text{poly}\left(2^{\mathcal{O}(k)} + n\right) \cdot x_k$$

bounded by a polynomial in n since the x_k decay superexponentially fast.

(b) Consider first the case $d = 2$ and record that

$$f\big(f(x)\big) \leq 2^{-n} \iff x \leq \exp\left(1 - \exp(2^n - 1)\right) =: x_n^{(2)},$$

so any modulus $\mu^{(2)}$ of uniform continuity of $f^{(2)}$ must be an exponential tower of height two as lower bound to the worst-case running time.

Conversely $(f \circ f)'(x) = f'\big(f(x)\big) \cdot f'(x) = \big(f(f(x))\big)^2 \cdot f(x)/x \in (0; 1/x)$ has $(f \circ f)''(x) = \big(f(f(x))\big)^2 \cdot f(x) \cdot \big(f(x) - 1 + 2f(x) \cdot f(f(x))\big)/x^2 < 0$ for all $x \in (0; \delta_2)$ where $\delta_2 := 0.32083$. Remark 2(e) thus asserts $\mu^{(2)}(x, n) = n + \text{lb}\, f'(x/2) \leq n - \ln(x)$ to be an integrable local modulus of continuity of $f^{(2)}|_{(0;\delta_2]}$; and $f^{(2)}|_{[\delta_2;1]}$ is even continuously differentiable: together $\int_0^1 \mu(x, n) = n + C$ for some constant C.

Replacing in (a) the x_k with $x_k^{(2)}$ yields an algorithm with running time an exponential tower of height two. More precisely Phase (iii) then takes a number of steps polynomial in $2^{2^{\mathcal{O}(k)}} + n$, leading to an average running time $\bar{T}^{(2)}(n) \leq \sum_k \text{poly}\left(2^{2^{\mathcal{O}(k)}} + n\right) \cdot x_k^{(2)}$ again bounded by polynomial in n.

For general d observe that

$$\left(f^{(d)}\right)'(x) = f'\left(f^{(d)}(x)\right) \cdot \left(f^{(d-1)}\right)'(x)$$

$$= \frac{\left(f^{(d-1)}(x)\right)^2}{f^{(d-1)}(x)} \cdot \left(f^{(d-1)}(x)\right)^2 \cdot f^{(d-2)}(x) \cdots f^{(2)}(x) \cdot f(x)/x$$

by induction hypothesis takes values in $(0, 1/x]$ and is strictly decreasing on $(0; \delta_d)$ for some $\delta_d > 0$. So similarly define $x_n^{(d)}$ as exponential tower of height $d + 1$ in order to extend the above considerations.

(c) Let us first record that, for $a, b \in \mathbb{Z}$ and $f :\subseteq \mathbb{R} \to \mathbb{R}$ and $z \in \text{dom}(f)$, $\mu : \text{dom}(f) \times \mathbb{Z} \to \mathbb{Z}$ is a local modulus of continuity of f iff $\nu : (y, n) \mapsto \mu(y \cdot 2^a + z, n+b) + a$ is a local modulus of continuity of $\tilde{f}(y) := f(y \cdot 2^a + z) \cdot 2^b$. Moreover its fractional moments satisfy $\int \nu(y, n)^\varepsilon \, dy = \int \left(\mu(x, n + b) + a\right)^\varepsilon dx / 2^a$. Now consider an optimal local modulus of continuity of $s \mapsto \sin(\pi s)/2$; in particular $\mu(\cdot, n)$ is 1-periodic and satisfies $\forall n < 0 : \mu(\cdot, n) \equiv -\infty$ as well as $\forall n > 1 : \mu(\cdot, n) \geq 1$. Then $g_j := g\big|_{(2^{-j}; 2^{-j+1}]}$ has optimal modulus

$\mu_j : (t, n) \mapsto \mu\left(s_j(t), n-j\right) + j + 2^{j^2}$ for $s_j(t) := 2^{2^{j \cdot j}} \cdot (t \cdot 2^j - 1) \in (0; 2^{2^{j \cdot j}}]$; and any oracle machine $\mathcal{M}^O$ computing g_j satisfies $T_{\mathcal{M}^O}(t, n) \geq \mu_j(t, n - 1) - 1$ according to Remark 2(d). We can therefore bound the fractional moments of its running time as follows: $\int_0^1 \left(T_{\mathcal{M}^O}(t, n)\right)^\varepsilon dt =$

$$= \sum_{j=1}^\infty \int_{2^{-j}}^{2^{-j+1}} \left(T_{\mathcal{M}^O}(t, n)\right)^\varepsilon dt \geq \int_{2^{-j}}^{2^{-j+1}} \left(\mu_j(t, n-1) - 1\right)^\varepsilon dt \Big|_{j=n-3}$$

$$= \int_0^{2^{2^{j \cdot j}}} \left(\mu(s, n - 1 - j) - 1 + j + 2^{j^2} - 1\right)^\varepsilon / 2^{j+2^{j \cdot j}} \, ds \Big|_{j=n-3}$$

$$= \int_0^1 \left(\mu(s, n - 1 - j) - 1 + j + 2^{j^2}\right)^\varepsilon ds / 2^j \Big|_{j=n-3} \geq (2^{j^2})^\varepsilon / 2^j \Big|_{j=n-3}$$

since $\mu(s, n - 1 - j) - 1 \geq 0$. On the other hand $g(t)$ is trivially approximated by 0 up to error $1/2^n$ for $t < 1/2^n$; and suffices with time polynomial in 2^{n^2} to compute $g_j(t)$ in case $n \geq j$ similarly to a).

(d) Note $\frac{1}{n} \int F(x, n) \, dx = x^{1/n}\big|_0^1 = 1$ for all $n \geq 1$. On the other hand, abbreviating $y := 1 - 1/n \in (0; 1)$, $F\left(x, F(x, n)\right) = \frac{1}{x} \cdot x^{x^y} \geq \exp(-1/ey)/x$ is not even integrable. Indeed, $0 = \frac{d}{dx} x^{x^y} = x^{y-1} \cdot (y \cdot \ln x - 1) \cdot x^{x^y}$ shows the minimum of x^{x^y} to be attained at $x = \exp(-1/y)$ of value $\exp(-1/ey)$. $\square$

3   Average-Case and Randomized Expected Polynomial-Time Type-2 Computation

We have so-far focused on average-case complexity of mappings on the certainly most intuitive domain $[0; 1]$. Recall that algorithms in the bit-cost model operate not on real numbers as entities but on sequences of approximations formalized as ρ–names (Definition 1(a)); and each $x \in [0; 1]$ has infinitely many of them: necessarily so, cmp. the proof of [Weih00, Theorem 4.1.15]. Now the above algorithms and arguments about moduli of continuity did not depend on which of its many ρ–names a real argument x is given by but it is easy to construct algorithms that do so, at least artificially. According to Definition 1(c) their average-case complexity is the mean over all $x \in [0; 1]$ in the worst-case over all ρ–names of x. However a full Average-Complexity Theory should weigh in all names of all arguments, preferably also for spaces other than $[0; 1]$.

The *Type-2 Theory of Effectivity* [Weih00, Sect. 3] considers computability on topological spaces X whose elements are 'encoded' by means of infinite binary sequences according to a fixed partial surjection $\alpha :\subseteq \{0, 1\}^\omega \to X$ called a *representation*. Definition 7 in Subsect. 3.2 thus considers average-case complexity on X w.r.t. a given probability distribution on $\mathrm{dom}(\alpha)$.

3.1   Probability Distributions on the Represented Space $[0; 1]$

Let X denote some space with representation $\alpha :\subseteq \{0, 1\}^\omega \to X$. Often a distribution on X is specified or otherwise canonical; which raises the problem of suitably refining that to a 'natural' one on $\mathrm{dom}(\alpha)$, that is, assign sub-weights to each $\bar{\sigma} \in \alpha^{-1}(x)$. This may in general be difficult, regarding that already $\mathrm{dom}(\alpha)$ usually has measure zero w.r.t. fair coinflipping on $\{0, 1\}^\omega$; compare the *Borel-Kolmogorov Paradox*. On $X = [0; 1]$ we now present two such distributions.

The first one is based on [ScSi06, Definitions 3 + 5 and Theorems 8 + 14], describing probability distributions on $\mathrm{dom}(\alpha)$ by means of certain stochastic processes on $\{0, 1\}^*$ in the spirit of [Weih00, Theorem 2.3.7.2]:

Definition 4. *Let Σ denote a finite alphabet.*

(a) *A probabilistic process on Σ^ω is a function $\pi : \Sigma^\omega \to [0; 1]$ mapping the empty string () to 1 and satisfying $\pi(\vec{w}) = \sum_{b \in \Sigma} \pi(\vec{w}\, b)$ for all $\vec{w} \in \Sigma^*$.*

(b) *A probabilistic name for a continuous representation $\alpha :\subseteq \Sigma^\omega \to X$ of a topological space X is a probabilistic process π on Σ^ω satisfying, for all open subsets $U \subseteq X$ and all open $V_1, V_2 \subseteq \Sigma^\omega$ with $V_1 \cap \mathrm{dom}(\alpha) = \alpha^{-1}[U] = V_2 \cap \mathrm{dom}(\alpha)$, $\hat{\pi}(V_1) = \hat{\pi}(V_2)$.*

(c) *Here $\hat{\pi}$ denotes the outer regular Borel measure on Cantor space Σ^ω defined by $\hat{\pi}(W\, \Sigma^\omega) := \sum_{\vec{w} \in W} \pi(\vec{w})$ for all prefix-free $W \subseteq \Sigma^\omega$.*

We illustrate this approach for the signed digit representation ρ_{sd} on $[0; 1]$, known [Weih00, Example 7.2.14] uniformly quadratic-time equivalent to the representation ρ from Definition 1.

Example 5. Consider alphabets $\Sigma := \{0, 1, \bar{1}\}$ and $\bar{\Sigma} := \{0, 1, \bar{1}, .\}$ as well as the mapping

$$\rho_{\mathrm{sd}} : 0.0^* 1 \Sigma^\omega + 1.0^* \bar{1} \Sigma^\omega \ni b_0 . b_1 b_2 \ldots \mapsto \sum\nolimits_{n \in \mathbb{N}} b_n 2^{-n}.$$

Moreover consider the surjective mapping $\pi : \bar{\Sigma}^* \to [0; 1]$ with $() \mapsto 1$,

$$0 \mapsto \tfrac{1}{2}, \quad 0. \mapsto \tfrac{1}{2}, \quad 0.0^n \mapsto 2^{-n-1}, \quad 0.0^n 1 \mapsto 2^{-n-2}, \quad 0.0^n 1 \Sigma^m \mapsto 2^{-n-2} \cdot 3^{-m}$$

$$1 \mapsto \tfrac{1}{2}, \quad 1. \mapsto \tfrac{1}{2}, \quad 1.0^n \mapsto 2^{-n-1}, \quad 1.0^n \bar{1} \mapsto 2^{-n-2}, \quad 1.0^n \bar{1} \Sigma^m \mapsto 2^{-n-2} \cdot 3^{-m}$$

and 0 otherwise. Then π is non-zero precisely on initial segments of $\mathrm{dom}(\rho_{\mathrm{sd}})$: $\pi^{-1}(0; 1] = \{\vec{w} \in \bar{\Sigma}^* : \vec{w} \Sigma^\omega \cap \mathrm{dom}(\rho_{\mathrm{sd}}) \neq \emptyset\}$. And $\pi(\vec{w}) = \sum_{b \in \Sigma} \pi(\vec{w} b)$ holds, hence π is a *probabilistic name* for $\rho_{\mathrm{sd}} : \mathrm{dom}(\rho_{\mathrm{sd}}) \to [0; 1]$ in the sense of [ScSi06, Definition 5]. Moreover the Borel measure it induces on $[0; 1]$ according to [ScSi06, Theorem 8] coincides with the canonical one. $\qquad\square$

Also note that $\pi^{-1}(0; 1]$ is a regular language; while, for the representation ρ from Definition 1(a), $\rho^{-1}[0; 1)$ seems more complicated. However, since $\rho^{-1}[x]$ is essentially a complete infinite binary tree for every x, we still can translate the canonical probability distribution on $[0; 1)$ to a 'uniform' one on $\rho^{-1}[0; 1) \subseteq \mathbb{N}^\omega$, implicitly identified by some suitable encoding with a subset of $\{0, 1\}^\omega$.

Example 6. Let $\rho :\subseteq \mathbb{Z}^\omega : (a_m)_m \mapsto \lim_m a_m / 2^m$ and consider the mapping

$$[0; 1) \times \{0, 1\}^\omega \ni \left(x, (s_m)_m\right) \mapsto \left(\lfloor s_m + x \cdot 2^m \rfloor\right)_m \in \rho^{-1}[0; 1) \subsetneq \prod_m \{0, 1, \ldots 2^m\}$$

with inverse

$$\left(\rho \otimes \tilde{\rho}\right) : \rho^{-1}[0; 1) \ni (a_m)_m \mapsto \left(\rho\left((a_m)_m\right), \begin{pmatrix} 0 : & \rho\left((a_m)_m\right) \geq a_n / 2^n \\ 1 : & \rho\left((a_m)_m\right) < a_n / 2^n \end{pmatrix}_n \right),$$

thus constituting a bijection which is Σ_2–measurable: recall Remark 2(b) and its proof. So we can well-define a Borel probability measure $\hat{\lambda}$ on $\rho^{-1}[0; 1)$ as the pushforward $\hat{\lambda}(A) := (\lambda \otimes \tilde{\lambda})((\rho \otimes \tilde{\rho})[A])$, where $\lambda \otimes \tilde{\lambda}$ denotes the product measure on the Borel subsets of $[0; 1) \times \{0, 1\}^\omega$ induced jointly by Lebesgues' λ on $[0; 1)$ and by the canonical/uniform/fair-coin-flip measure $\tilde{\lambda}$ on $\{0, 1\}^\omega$, defined by $\vec{s} \circ \{0, 1\}^\omega \mapsto 2^{-|\vec{s}|}$; equivalently: $\lambda \otimes \tilde{\lambda}$ is the Haar measure on $[0; 1) \times \{0, 1\}^\omega$ considered as compact topological group w.r.t. $\left(x, (s_m)_m, y, (r_m)_m\right) \mapsto \left(x + y \bmod 1, (s_m + r_m \bmod 2)_m\right)$. $\qquad\square$

3.2 Local Deterministic Average Versus Randomized Expected Worst-Case

We are now ready to formalize average-case complexity theory on represented spaces — and randomized type-2 computation:

Definition 7.

(a) *Fix a finite alphabet Σ and a Type-2 Machine $\mathcal{M}^O$ computing some function $F :\subseteq \Sigma^\omega \to \Sigma^\omega$. Let $T_{\mathcal{M}^O}(\bar{s}, n)$ denote the number of steps $\mathcal{M}^O$ makes on input $\bar{s}$ before producing the n-th symbol of $F(\bar{s})$, and fix a Borel probability measure λ on $\mathrm{dom}(F)$. $\mathcal{M}^O$ computes F in λ–average polynomial time iff there exists some $\varepsilon > 0$ such that $\mathbb{N} \ni n \mapsto \frac{1}{n} \int_{\mathrm{dom}(F)} \left(T_{\mathcal{M}^O}(\bar{s}, n)\right)^\varepsilon d\lambda(\bar{s})$ is bounded.*

(b) *Let X be a compact topological space with continuous representation $\alpha :\subseteq \Sigma^\omega \to X$ and compact $\mathrm{dom}(\alpha) \subseteq \Sigma^\omega$ equipped with Borel probability measure λ. Let $\beta :\subseteq \Sigma^\omega \to Y$ denote another represented space and consider a partial multivalued mapping $f :\subseteq X \rightrightarrows Y$. It is (α, β)–computable in λ–average polynomial time (relative to oracle O) iff there exists a Type-2 machine $\mathcal{M}^O$ computing in λ–average polynomial time according to a) some (α, β)–realizer $F : \alpha^{-1}[\mathrm{dom}(f)] \to \mathrm{dom}(\beta)$ of f in the sense that $\beta(F(\bar{s})) \in f(\alpha(\bar{s}))$ for every $\bar{s} \in \mathrm{dom}(\alpha)$ with $f(\alpha(\bar{s})) \neq \emptyset$.*

(c) *We say that $F :\subseteq \Sigma^\omega \to \Sigma^\omega$ is computed by a randomized Type-2 Machine $\mathcal{M}$ if $\mathcal{M}$ is an ordinary Type-2 machine computing the mapping $\mathrm{dom}(F) \times \{0,1\}^\omega \ni (\bar{s}, \bar{r}) \mapsto F(\bar{s})$. Its expected local runtime is $\mathbb{E}_{\bar{r}}\left[T_{\mathcal{M}}(\bar{s}, \bar{r}, n)\right]$, where $\mathbb{E}_{\bar{r}}[T(\bar{r})] := \int_{\{0,1\}^\omega} T(\bar{r}) d\lambda(\bar{r})$ denotes the expected value of measurable $T : \{0,1\}^\omega \to [0; \infty]$.*

(d) *Similarly, $f :\subseteq X \rightrightarrows Y$ is (α, β)–computable in (worst-case) expected polynomial time iff there exists $\varepsilon > 0$ and a randomized Type-2 machine $\mathcal{M}$ computing some (α, β)–realizer $F : \alpha^{-1}[\mathrm{dom}(f)] \to \mathrm{dom}(\beta)$ of f such that $\alpha^{-1}[\mathrm{dom}(f)] \times \mathbb{N} \ni (\bar{s}, n) \mapsto \frac{1}{n}\left(\bar{T}_{\mathcal{M}}(\bar{s}, n)\right)^\varepsilon$ is bounded.*

Randomized computation is ubiquitous in the discrete realm [HMRR98]. In the type-2 setting it has been considered for instance in [Boss08, BHG13]; see also [Ko91, Definition 5.27]. Our conception here corresponds in the discrete realm to the complexity class $\mathcal{ZPP}$, namely total and always correct *Las Vegas* algorithms with expected polynomial runtime almost surely in the worst-case w.r.t. input $\bar{s}$ on average w.r.t. the additional input $\bar{r}$ that does not affect correctness. Bets are generally inclined towards $\mathcal{ZPP} = \mathcal{P}$.

Recall that Definition 1(d) considered the maximum runtime over all possible names of every fixed real argument x, on average w.r.t. $x \in \mathrm{dom}(f)$; whereas Definition 7(b) takes the average runtime over all names of elements in $\mathrm{dom}(f)$. And as already indicated an algorithm may well be fast on some names of an argument x yet slow on other names of the same x. This raises the question of whether such a dependence can always be removed. We show that in the real case it probably (pun) can: Similarly to QuickSort mentioned initially — although by more involved methods and analyses — randomization turns average into expected worst-case behaviour in the sense of the following tailored

Definition 8. *Suppose $\mathcal{M}$ computes $f :\subseteq \mathbb{R} \to \mathbb{R}$. Let, for fixed $x \in \mathrm{dom}(f)$, denote by*

$$\bar{T}_{\mathcal{M}}(x, n) := \int_{\bar{s} \in \{0,1\}^\omega} T_{\mathcal{M}}\left((\rho \otimes \tilde{\rho})^{-1}(x, \bar{s}), n\right) d\tilde{\lambda}(\bar{s})$$

its local average running time *over all (ρ-) names $\bar{a}$ of x, where $\tilde{\lambda}$ denotes the canonical measure on $\{0,1\}^\omega$. W.l.o.g. $\bar{T}_\mathcal{M}(x,n) \geq 1$ for all x,n.*

Theorem 9. *Suppose $\mathcal{M}$ computes $f : [0;1) \to \mathbb{R}$ in polynomial average local average time in the sense that there exists $\varepsilon > 0$ rendering $\mathbb{N} \ni n \mapsto \frac{1}{n} \cdot \int_0^1 (\bar{T}_\mathcal{M}(x,n))^\varepsilon \, dx$ bounded.*

Then f is also computable by a randomized Type-2 Machine in polynomial average expected worst-case time in the following sense: There exists some $\delta > 0$ and a Type-2 Machine $\mathcal{M}'$ producing, given any ρ-name $\bar{a} \in \mathbb{N}^\omega$ of some $x \in [0;1)$ and any infinite binary string $\bar{r} \in \{0,1\}^\omega$, a ρ-name $\bar{b} \in \mathbb{Z}^\omega$ of $f(x)$ such that the following map is well-defined and bounded:

$$ \mathbb{N} \ni n \mapsto \frac{1}{n} \int_{[0;1)} \mathbb{E}_{\bar{r}} \left[T_{\mathcal{M}'}(x, \bar{r}, n) \right]^\delta dx, $$

recalling the local worst-case runtime $T_{\mathcal{M}'}(x, \bar{r}, n) = \sup \{ T_{\mathcal{M}'}(\bar{a}, \bar{r}, n) \mid \bar{a} \in \rho^{-1}[x] \}$.

Note that both hypothesis and conclusion are concerned with double integrals $\int_0^1 \int_{\{0,1\}^\omega} \cdots d\tilde{\lambda} \, dx$. However the bijection $\rho \otimes \tilde{\rho}$ between $\rho^{-1}[0;1)$ and $[0;1) \times \{0,1\}^\omega$ employed in Definition 8 is far from effective, not to mention efficient!

3.3   Proof of Theorem 9

Intuitively, a machine $\mathcal{M}$ computing $f : [0;1) \to \mathbb{R}$ in polynomial average local average time will be efficient in evaluating $x \mapsto f(x)$ for 'most' of the ρ–names $\bar{a}' = (a'_m)_m \in \mathbb{N}^\omega$ of 'most' x; but could take very long time for 'few bad' names $\bar{a}$ that is, have large $T(x,n)$ but small $\bar{T}(x,n)$ for some (and possibly not so few) x. So the idea is to have an algorithm first replace every bad input $\bar{a}$ with with a good name $\bar{a}'$ of the same x — but of course we merely know such $\bar{a}'$ to exist and be abundant; so let us exploit randomization to guess, given $\bar{a}$, an equivalent but good $\bar{a}'$ with sufficiently high probability. Now according to Example 6 the set $\rho^{-1}[x]$ of names $\bar{a}'$ of x is an infinite binary tree; so it seems promising to create a random name by traversing this tree and branching according to the random coin flip sequence $\bar{r}$. Unfortunately the above bijection between $\rho^{-1}[0;1) \ni \bar{a}$ and $[0;1) \times \{0,1\}^\omega$ is non-effective, not to mention efficient.

Observation 10. Consider $0 < m \in \mathbb{N}$, $a \in \mathbb{Z}$, and the interval $I_{a,m} := [\frac{a-1}{2^m}; \frac{a+1}{2^m})$ according to Definition 1(a). Let $k = k(a) \in \mathbb{N}$ be maximal such that 2^k divides a, written $2^k \mid a$. Then there exist precisely 2^{m-k} pairs (m', a') with $\mathbb{N} \ni m' < m$ and $a' \in \mathbb{Z}$ and $I_{a',m'} \supseteq I_{a,m}$.

Proof. Suppose a is odd (i.e. $k = 0$). Then $I_{\lfloor a/2 \rfloor, m-1}, I_{\lceil a/2 \rceil, m-1}$ are two subintervals of the claimed form containing $I_{a,m}$. If a is even, however, then $I_{b,m-1} \subseteq I_{a,m}$ iff $b = a/2$: only one. Inductively, $I_{\lfloor b/2 \rfloor, m-1}$ and $I_{\lceil b/2 \rceil, m-1}$ are the unique subintervals containing $I_{b,m}$ or $I_{b+1,m}$s for even b.   □

So consider the machine $\mathcal{M}'$ which, given a ρ–name $\bar{a}$ of x, behaves at phase $m \in \mathbb{N}$ as follows: Read a_m and produce, for $k_m = k(a_m) \in \mathbb{N}$, one of the 2^{m-k_m} finite sequences $(a_0', a_1', \ldots, a_{m-1}')$ satisfying $I_{a_0',1}, \ldots, I_{a_{m-1}',m-1} \supseteq I_{a_m,m}$ according to Observation 10 uniformly at random; then simulate $\mathcal{M}$ on this modified input — until the latter requests to read the $(m+1)$-st data: in which case $\mathcal{M}'$ restarts and enters phase $m+1$ (this time of course quelling any output b_n of $\mathcal{M}$ already printed during phase m).

Note that, since $I_{a_j',j} \subseteq I_{a_m,m}$ for all $j < m$, the partial name $(a_0', a_1', \ldots a_{m-1}')$ extends to a total one $\bar{a}'$ of the same x; hence $\mathcal{M}'$ indeed produces a name $\bar{b}$ of $f(x)$. In fact, according to Observation 10, the name $\bar{a}'$ of x that phase m simulates $\mathcal{M}$ on is chosen uniformly at random among $2^{m-k(a_m)}$ of the 2^m possible ones according to Definition 1(a). And by Markov's Inequality this reduced sample $A_m(\bar{a})$ still yields an expected local runtime that does not exceed the average $\bar{T}_\mathcal{M}(x, n)$ by too much — at least for sufficiently many $x \in [0; 1)$, based on the following

Lemma 11.

(a) For $x \in [0; 1)$ and $m \in \mathbb{N}$ abbreviate $k(x, m) := \max\left(k(\lfloor x \cdot 2^m \rfloor), k(\lfloor 1 + x \cdot 2^m \rfloor)\right)$. Then $\int_0^1 c^{k(x,m)}\, dx < \frac{4-c}{2-c}$ whenever $0 \le c < 2$.

(b) For $\bar{a} \in \rho^{-1}[0; 1)$ and $\bar{r} \in \{0, 1\}^\omega$ let $M = M(\bar{a}, \bar{r}, n) \in \mathbb{N}$ denote the index of the phase in which $\mathcal{M}'$ simulating $\mathcal{M}$ first achieves guaranteed output precision $1/2^n$. Then it holds

$$\mathbb{E}_{\bar{r}}\left[T_{\mathcal{M}'}(\bar{a}, \bar{r}, n) \mid M = m \right] \le \mathrm{poly}\left(2^{k(\rho(\bar{a}),m)} \cdot \bar{T}_\mathcal{M}(\rho(\bar{a}), n) \right),$$

where $\mathbb{E}[\, X \mid C\,]$ generically denotes the expected value of random variable $X : \Omega \to \mathbb{R}$ conditional to measurable $C \subseteq \Omega$ and "$Y = y$" is short for $\{\omega : Y(\omega) = y\}$.

(c) To every $q \ge 1$ there exists a $C_q \in \mathbb{N}$ such that every $w_m \ge 0$ and integrable $g_m : [0; 1] \to [0; \infty]$ $(m \in \mathbb{N})$ satisfies

$$\int_0^1 \left(\sum_m w_m \cdot g_m^q(x) \right)^{1/q} dx \le C_q \cdot \left(\sum_m w_m \right)^{1/q} \cdot \sup_m \int_0^1 g_m(x)\, dx.$$

Making $\mathcal{M}'$ double, instead of incrementing, m in each phase can improve the polynomial bound in (b) to a linear one but is beyond our present purpose.

Proof.

(a) First consider $k'(x, m) := \max\{k \le m : 2^k \mid \lfloor x \cdot 2^m \rfloor\}$. Then $k'(x, m) = \min\{k(a), m\}$ for all $a \in A_m := \{0, 1, \ldots, 2^m - 1\}$ and $\frac{a}{2^m} \le x < \frac{a+1}{2^m}$. Moreover precisely half of the (i.e. 2^{m-1} many) $a \in A_m$ are odd, that is, have $k(a) = 0$; 2^{m-2} of $a \in A_m$ have $k(a) = 1$, 2^{m-3} with $k = 2$, and so on until $2^{m-(m-1)} = 2$ with $k = m-2$ (namely $a = 2^{m-2}$ and $a = 3 \cdot 2^{m-2}$), one (namely $a = 2^{m-1}$) with $k = m - 1$, and one (namely $a = 0$) with $k = m$). Hence $\int_0^1 c^{k'(x,m)}\, dx = \sum_{a=0}^{2^m - 1} c^{\min\{k(a), m\}}/2^m =$

$$\underbrace{c^0/2 + c^1/4 + c^2/8 + \cdots + c^{m-2}/2^{m-1} + c^{m-1}/2^m + (c^m/2^{m+1}}_{} + c^m/2^{m+1})$$

is a geometric sum bounded by $\frac{1}{2} \cdot \frac{1}{1-c/2} + \frac{1}{2} = \frac{4-c}{4-2c}$. Similarly, $\int_{1/2^m}^{1+1/2^m} c^{k'(x,m)}\, dx < \frac{4-c}{4-2c}$. Finally observe $k(x,m) = \max\{k'(x,m), k'(x + 1/2^m, m)\}$.

(b) Let $A_m(\bar{a}) \subseteq \rho^{-1}[\rho(\bar{a})]$ denote the set of ρ–names $\bar{a}'$ which $\mathcal{M}'$ on input $\bar{a}$ samples from in phase m according to Observation 10, that is, with $\tilde{\lambda}(\tilde{\rho}[A_m(\bar{a})]) = 2^{-k(a_m)}$. By Markov's Inequality for non-negative $T_\mathcal{M}$, the expected runtime of $\mathcal{M}$ restricted to A_m is not too much larger than the average runtime over entire $\rho^{-1}[\rho(\bar{a})]$: By abuse of names,

$$\mathbb{E}_{\bar{a}'}\big[\, T_\mathcal{M}(\bar{a}',n) \mid A_m \,\big] \cdot \tilde{\lambda}(A_m) \leq \mathbb{E}_{\bar{a}'}\big[\, T_\mathcal{M}(\bar{a}',n) \mid \rho^{-1}[x] \,\big] = \bar{T}_\mathcal{M}(x,n).$$

Moreover, $\mathcal{M}$ can in each step read at most one data item; hence the number m of phases performed by $\mathcal{M}'$, each taking time polynomial in $T_\mathcal{M}$, satisfies $m \leq T_\mathcal{M}$. The claim thus follows from stochastic independence of the coin flips during distinct phases. □

We can conclude the average expected runtime analysis: According to Lemma 11(b) there exists some $d \in \mathbb{N}$ with $\mathbb{E}_{\bar{r}}\big[\, T_{\mathcal{M}'}(x,\bar{r},n) \mid M = m \,\big] \leq d \cdot 2^{d \cdot k(x,m)} \cdot \bar{T}_\mathcal{M}^d(x,n)$ since the right-hand side only depends on $\rho(\bar{a})$ and not on $\bar{a}$ itself. Let $\delta := \min(\varepsilon/2, 1/4)/d$. Then, by virtue of the Cauchy-Schwarz inequality in (*),

$$\int_0^1 \mathbb{E}_{\bar{r}}\big[T_{\mathcal{M}'}(x,\bar{r},n)\big]^\delta\, dx/n =$$

$$= \frac{1}{n} \int_0^1 \Big(\sum_m \mathbb{P}[M = m] \cdot \mathbb{E}\big[\, T_{\mathcal{M}'}(x,\bar{r},n) \mid M = m \,\big]\Big)^\delta dx$$

$$\leq \frac{1}{n} \int_0^1 d^\delta \cdot \bar{T}_\mathcal{M}^{d\cdot\delta}(x,n) \cdot \Big(\sum_m \mathbb{P}[M = m] \cdot 2^{d \cdot k(x,m)}\Big)^\delta dx$$

$$\overset{(*)}{\leq} \frac{d^\delta}{n} \sqrt{\int_0^1 \underbrace{\bar{T}_\mathcal{M}^{2d\cdot\delta}(x,n)}_{\leq \bar{T}_\mathcal{M}^\varepsilon}\, dx} \cdot \sqrt{\int_0^1 \Big(\sum_m \underbrace{\mathbb{P}[M=m]}_{=:w_m} \cdot \underbrace{2^{d\cdot k(x,m)}}_{\leq g_m^q(x)}\Big)^{2\delta} dx}$$

$$\leq \frac{d^\delta}{n} \cdot \sqrt{\mathcal{O}(n)} \cdot \sqrt{\Big(\sum_m \mathbb{P}[M=m]\Big)^{1/q} \cdot \sup_m \int_0^1 g_m(x)\, dx} \leq \mathcal{O}(1)$$

by hypothesis and Lemma 11(a,c) for $g_m(x) := \sqrt{2}^{k(x,m)}$ and $1/q := 2\delta$.   □

4   Conclusion and Perspectives

We have transferred average-case complexity from the discrete realm to functions on the reals with its natural probability distribution; and we have demonstrated (the 'standard' and generalized) examples of high worst-case complexity to have in fact polynomial average-case complexity; while a more involved one has both average and worst-case complexity exponential in the output precision: for topological/metric reasons, i.e., oracles do not help. We have generalized these notions to represented spaces equipped with a probability distribution; and described and analyzed a randomized algorithm replacing worst-case inputs with equivalent, 'typical' ones, thus turning average into expected runtime.

In fact average-case complexity theory explains for, and bridges, the apparent gap between many problems being hard in theory while admitting apparently efficient solutions in practice — in the discrete case: cmp., e.g., [COKJ10].

A similar effect has been observed, but lacks explanation, in real complexity theory: [Ko91, Theorems 3.7 + 5.33] construct polynomial-time computable smooth functions $f : [0; 1] \to [0; 1]$ whose running maximum or Riemann integral are again polynomial-time computable iff $\mathcal{P} = \mathcal{NP}$ or $\mathcal{FP} = \#\mathcal{P}$, respectively; however on 'typical/practical' instances f standard numerical methods seem to work very well, hinting that the hard ones may in some sense be rare — and maximization/integration perhaps in fact easy 'on average' w.r.t. to an appropriate probability distribution on a suitable space of continuous real functions, such as the Wiener measure.

References

[BHG13]   Brattka, V., Hölzl, R., Gherardi, G.: Probabilistic computability and choice. Inf. Comput. **242**(C), 249–286 (2015)

[Boss08]   Bosserhoff, V.: Notions of probabilistic computability on represented spaces. J. Univ. Comput. Sci. **146**(6), 956–995 (2008)

[BoTr06]   Bogdanov, A., Trevisan, L.: Average-case complexity. Found. Trends Theor. Comput. Sci. 2(1), 1–106 (2006). arXiv:cs/0606037

[COKJ10]   Coja-Oghlan, A., Krivelevich, M., Vilenchik, D.: Why almost all k-colorable graphs are easy to color. Theor. Comput. Syst. **46**(3), 523–565 (2010)

[Gold97]   Goldreich, O.: Notes on Levin's theory of average-case complexity. In: Goldreich, O. (ed.) Studies in Complexity and Cryptography. LNCS, vol. 6650, pp. 233–247. Springer, Heidelberg (2011)

[HMRR98]   Habib, M., McDiarmid, C., Ramirez-Alfonsin, J., Reed, B.: Probabilistic Methods for Algorithmic Discrete Mathematics. Springer, Heidelberg (1998)

[KMRZ15]   Kawamura, A., Müller, N., Rösnick, C., Ziegler, M.: Computational benefit of smoothness: parameterized bit-complexity of numerical operators on analytic functions and Gevrey's hierarchy. J. Complex. **31**(5), 689–714 (2015)

[KoFr82]   Ko, K.-I., Friedman, H.: Computational complexity of real functions. Theor. Comput. Sci. **20**, 323–352 (1982)

[Ko91]   Ko, K.-I.: Computational Complexity of Real Functions. Birkhäuser, Boston (1991)

[KORZ14]   Kawamura, A., Ota, H., Rösnick, C., Ziegler, M.: Computational complexity of smooth differential equations. Log. Methods Comput. Sci. **10**, 1 (2014)

[LiVi97] Li, M., Vitányi, P.: An Introduction to Kolmogorov Complexity and Its Applications. Springer, Heidelberg (1997)

[Ritt00] Ritter, K. (ed.): Average-Case Analysis of Numerical Problems. Lecture Notes in Mathematics, vol. 1733. Springer, Heidelberg (2000)

[Schr04] Schröder, M.: Spaces allowing type-2 complexity theory revisited. Math. Log. Q. **50**, 443–459 (2004)

[ScSi06] Schröder, M., Simpson, A.: Representing probability measures using probabilistic processes. J. Complex. **22**, 768–782 (2006)

[Weih00] Weihrauch, K.: Computable Analysis. Springer, Heidelberg (2000)

[Weih03] Weihrauch, K.: Computational complexity on computable metric spaces. Math. Log. Q. **49**(1), 3–21 (2003)

Certifying Trajectories of Dynamical Systems

Joris van der Hoeven$^{(\boxtimes)}$

Laboratoire d'informatique, UMR 7161 CNRS,
Campus de l'École polytechnique, 1, rue Honoré d'Estienne d'Orves,
Bâtiment Alan Turing, CS35003, 91120 Palaiseau, France
vdhoeven@texmacs.org

Abstract. This paper concerns the reliable integration of dynamical systems with a focus on the computation of one specific trajectory for a given initial condition at high precision. We describe several algorithmic tricks which allow for faster parallel computations and better error estimates. We also introduce "Lagrange models". These serve a similar purpose as the more classical Taylor models, but we will show that they allow for larger step sizes, especially when the truncation orders get large.

Keywords: Reliable computation · Dynamical systems · Certified integration · Ball arithmetic · Taylor models · Multiple precision computations

A.M.S. subject classification: 65G20, 37-04

1 Introduction

Description of the problem and background. Let $\Phi \in \mathbb{C}[F_1, \ldots, F_d]^d$ be a polynomial vector field and consider the dynamical system

$$f' = \Phi(f). \tag{1}$$

Given an initial condition $f(u) = I \in \mathbb{C}^d$ at $u \in \mathbb{R}$, a target point $z > u$ such that f is analytic on $[u, z]$, the topic of this paper is to compute $f(z)$.

On actual computers, this problem can only be solved at finite precisions, although the user might request the precision to be as large as needed. One high level way to formalize this is to assume that numbers in the input (i.e. the coefficients of Φ and I, as well as u and z) are *computable* [26,27] and to request $f(z)$ again to be a vector of computable complex numbers.

From a practical point of view, it is customary to perform the computations using *interval arithmetic* [1,9,11,17–19,24]. In our complex setting, we prefer to use a variant, called *ball arithmetic* or *midpoint-radius arithmetic*. In our main problem, this means that we replace our input coefficients by complex balls, and that the output again to be a vector of balls. Throughout this paper, we assume that the reader is familiar with interval and ball arithmetic. We refer to [5,7] for basic details on ball arithmetic.

I.S. Kotsireas et al. (Eds.): MACIS 2015, LNCS 9582, pp. 520–532, 2016.
DOI: 10.1007/978-3-319-32859-1_44

It will be convenient to denote balls using a bold font, e.g. $\boldsymbol{f}(z) \in \mathbb{C}^d$. The explicit compact ball with center c and radius r will be denoted by $\mathcal{B}(c, r)$. Vector notation will also be used systematically. For instance, if $c \in \mathbb{C}^d$ and $r \in (\mathbb{R}^>)^d$ with $\mathbb{R}^> = \{x \in \mathbb{R} : x > 0\}$, then $\mathcal{B}(c, r) = (\mathcal{B}(c_1, r_1), \ldots, \mathcal{B}(c_d, r_d))$.

Sometimes, it is useful to obtain further information about the dependence of the value $f(z)$ on the initial conditions; this means that we are interested in the *flow* $f(z, I)$, which satisfies the same differential Eq. (1) and the initial condition $f(u, I) = I$. In particular, the *first variation* $V = \partial f / \partial I$ is an important quantity, since it measures the sensitivity of the output on the initial conditions. If κ denotes the condition number of V, then it will typically be necessary to compute with a precision of at least $\log_2 \kappa$ bits in order to obtain any useful output.

There is a vast literature on the reliable integration of dynamical systems [3,10,12–17,20–22]. Until recently, most work focused on low precision, allowing for efficient implementations using machine arithmetic. For practical purposes, it is also useful to have higher order information about the flow. *Taylor models* are currently the most efficient device for performing this kind of computations [14,15].

Main strategy for certification of high precision trajectories. In this paper, we are interested in the time complexity of reliable integration of dynamical systems. We take a more theoretical perspective in which the working precision might become high. We are interested in certifying one particular trajectory, so we do not request any information about the flow beyond the first variation.

From this complexity point of view it is important to stress that there is a tradeoff between efficiency and quality: faster algorithms can be designed if we allow for larger radii in the output. Whenever one of these radii becomes infinite, then we say that the integration method *breaks down*: a ball with infinite radius no longer provides any useful information. Now some "radius swell" occurs structurally, as soon as the condition number of V becomes large. But high quality integration methods should limit all other sources of precision loss.

The outline of this paper is as follows:

1. For problems from reliable analysis it is usually best to perform certifications at the outermost level. In our case, this means that we first compute the entire numeric trajectory with a sufficient precision, and only perform the certification at a second stage. We will see that this numeric computation is the only part of the method which is essentially sequential.
2. The problem of certifying a complete trajectory contains a global and a local part. From the global point of view, we need to cut the trajectory in smaller pieces that can be certified by local means, and then devise a method to recombine the local certificates into a global one.
3. For the local certification, we will introduce *Lagrange models*. As in the case of Taylor models, this approach is based on Taylor series expansions, but the more precise error estimates allow for larger time steps.

The first idea has been applied to many problems from reliable computation (it is for instance known as Hansen's method in the case of matrix inversion).

Nevertheless, we think that progress is often possible by applying this simple idea even more systematically.

Outline of the paper. In Sect. 2, we start with a quick survey of the most significant facts about numerical integration schemes for the Eq. (1). We also present a new parallel and dichotomic algorithm for increasing the precision of an already computed trajectory. In Sect. 3, we will see that a similar strategy can be used for certifying trajectories.

In Sect. 3 we present a method for reducing the problem of certifying a global trajectory to the local problem of certifying sufficiently short trajectories. It is interesting to compare this approach with the more classical stepwise certification scheme, along with the numerical integration itself. The problem with stepwise schemes is that they are essentially sequential and thereby give rise to a linear precision loss in the number of steps. The global approach reduces this to a logarithmic precision loss only. The global strategy already pays off in the linear case [3] and it is possible to reorganize the computations in such a way that they can be re-incorporated into an iterative scheme. The global approach was first generalized to the non linear case in [5]. Our current presentation has the merit of conceptual simplicity and ease of implementation.

The main contribution of this paper concerns the introduction of "Lagrange models" and the way that such models allow for larger time steps. This material is presented in Sect. 4; an earlier (non refereed) version appeared in the lecture notes [7]. Classical Taylor models approximate analytic functions f on the compact unit disk (say) by a polynomial $P \in \mathbb{C}[z]$ of degree $< n$ and an error $\varepsilon \geqslant 0$ with the property that $|f(z) - P(z)| \leqslant \varepsilon$ for all $|z| \leqslant 1$. The idea behind Lagrange models is to give a more precise meaning to the "big Oh" in $f(z) = f_0 + \cdots + f_{n-1}z^{n-1} + O(z^n)$. More precisely, it consists of a polynomial $P \in \mathbb{C}[z]$ of degree $< n$ with ball coefficients and an $\varepsilon \geqslant 0$ such that $f(z) \in P(z) + \mathcal{B}(0, \varepsilon)z^n$ for all $|z| \leqslant 1$. The advantage comes from the fact that the integration operator has norm 1 for general analytic functions on the unit disk but only norm $1/(n+1)$ for analytic functions that are divisible by z^n. Although Lagrange model arithmetic can be a constant times more expensive than Taylor model arithmetic, the more precise error estimates allow us to increase the time step.

2 Fast Numerical Integration

2.1 Classical Algorithms

From a high level perspective, the integration problem between two times $u < z$ can be decomposed into two parts: finding suitable intermediate points $u = z_0 < z_1 < \cdots < z_{s-1} < z_s = z$ and the actual computation of $f(z_k)$ as a function of $f(z_{k-1})$ for $k = 1, \ldots, s$ (or as a function of $f(z_{k-1}), \ldots, f(z_{k-i})$ for some schemes).

The optimal choice of intermediate points is determined by the distance to the closest singularities in the complex plane as well as the efficiency of the scheme that computes $f(z_k)$ as a function of $f(z_{k-1})$. For $t \in [u, z]$, let $\varrho(t)$ be the convergence radius of the function f at t. High order integration schemes will enable us to take $|z_k - z_{k-1}| \geqslant c\varrho(z_{k-1})$ for some fixed constant $c > 0$. Lower order schemes may force us to take smaller steps, but perform individual steps more efficiently. In some cases (e.g. when f admits many singularities just above the real axis, but none below), it may also be interesting to allow for intermediate points $z_1, \ldots, z_{s-1}$ in $\mathbb{C}$ that keep a larger distance with the singularities of f.

For small working precisions, Runge-Kutta methods [23] provide the most efficient schemes for numerical integration. For instance, the best Runge-Kutta method of order 8 requires 11 evaluations of f for each step. For somewhat larger working precisions (e.g. quadruple precision), higher order methods may be required in order to produce accurate results. One first alternative is to use relaxed power series computations [4,8] which involve an overhead n^2 for small orders n and $n \log^2 n$ for large orders. For very large orders, a power series analogue of Newton's method provides the most efficient numerical integration method [2,6,25]. This method actually computes the first variation of the solution along with the solution itself, which is very useful for the purpose of this paper.

2.2   Parallelism

Another interesting question concerns the amount of computations that can be done in parallel. In principle, the integration process is essentially sequential (apart from some parallelism which may be possible at each individual step). Nevertheless, given a full numeric solution at a sufficiently large precision p, we claim that a solution at a roughly doubled precision can be computed in parallel.

More precisely, for each z, u and I, let $f(z, u, I)$ be the solution of (1) with $f(u, u, I) = I$, and denote $V(z, u, I) = (\partial f / \partial I)(z, u, I)$. We regard $f(z, u, I)$ as the "transitional flow" between u and z, assuming the initial condition I at u. Notice that $V(u, u, I) = \mathrm{Id}$ and, for $u < v < z$,

$$f(z, u, I) = f(z, v, f(v, u, I))$$
$$V(z, u, I) = V(z, v, f(v, u, I))V(v, u, I).$$

Now assume that we are given $f_{k,0;p} \approx f(z_k)$ for $k = 1, \ldots, s$ and at precision p. Then we may compute $V_{k,k-1;p} \approx V(z_k, z_{k-1}, f(z_{k-1}))$ in parallel at precision p. Using a dichotomic procedure of depth $\lceil \log_2 s \rceil$, we will compute $f_{k,0;2p} \approx f(z_k)$ at precision $2p$ in parallel for $k = 1, \ldots, s$, together with $V_{k,0;p} \approx V(z_k, z_0, f(z_0))$ at precision p.

More precisely, assume that $s \geqslant 2$ and let $m = \lceil s/2 \rceil$. We start with the recursive computation of $f_{k,0;2p} \approx f(z_k)$ and $f_{m+k,m;2p} \approx f(z_{m+k}, z_m, f_{m;p})$ at precision $2p$ for $k = 1, \ldots, m$ (resp. $k = 1, \ldots, s - m$), together with $V_{k,0;p} \approx V(z_k, z_0, f(z_0))$ and $V_{m+k,m;p} \approx V(z_{m+k}, z_m, f(z_m))$ at precision p. Setting $\delta = f_{m,0;2p} - f_{m,0;p}$, we take

$$f_{m+k,0;2p} := f_{m+k,m;2p} + V_{m+k,m;p}\delta$$
$$V_{m+k,0;p} := V_{m+k,m;p}V_{m,0;p}$$

for $k = 1, \ldots, s - m$. These formulas are justified by the facts that

$$f(z_{m+k}) \approx f(z_{m+k}, z_m, f_{m,0;p} + \delta)$$
$$\approx f_{m+k,m;2p} + V_{m+k,m;p}\delta$$

at precision $2p$ and $V(z_{m+k}, z_0, f(z_0)) \approx V_{m+k,m;p}V_{m,0;p}$ at precision p.

The above algorithm suggests an interesting practical strategy for the integration of dynamical systems on massively parallel computers: the fastest processor(s) in the system plays the rôle of a "spearhead" and performs a low precision integration at top speed. The remaining processors are used for enhancing the precision as soon as a rough initial guess of the trajectory is known. The spearhead occasionally may have to redo some computations whenever the initial guess drifts too far away from the actual solution. The remaining processors might also compute other types of "enhancements", such as the first and higher order variations, or certifications of the trajectory. Nevertheless, the main bottleneck on a massively parallel computer seems to be the spearhead.

3   Global Certification

3.1   From Local to Global Certification

A *certified integrator* of the dynamical system (1) can be defined to be a ball function

$$f : (z, u, I) \mapsto f(z, u, I)$$

with the property that $f(z, u, I) \in f(z, u, I)$ for any $u < z$ and $I \in I$. An *extended certified integrator* additionally requires a ball function

$$V : (z, u, I) \mapsto V(z, u, I)$$

with the property that $V(z, u, I) \in V(z, u, I)$ for any $u < z$ and $I \in I$.

A *local certified integrator* of (1) is a special kind of certified integrator which only produces meaningful results if z and u are sufficiently close (and in particular $|z - u| < \varrho(u)$). In other words, we allow the radii of the entries of $f(z, u, I)$ to become infinite whenever this is not the case. Extended local certified integrators are defined similarly.

One interesting problem is how to produce global (extended) certified integrators out of local ones. The most naive strategy for doing this goes as follows. Assume that we are given a local certified integrator f^{loc}, as well as $u < z$ and I. If the radii of the entries of $f^{\mathrm{loc}}(z, u, I)$ are "sufficiently small" (finite, for instance, but we might require more precise answers), then we define $f^{\mathrm{glob}}(z, u, I) := f^{\mathrm{loc}}(z, u, I)$. Otherwise, we take $v = (z + u)/2$ and define $f^{\mathrm{glob}}(z, u, I) := f^{\mathrm{glob}}(z, v, f^{\mathrm{glob}}(v, u, I))$. One may refine the strategy by including additional exception handling for breakdown situations. It is well known that, unfortunately, this naive strategy produces error estimates of extremely poor quality (due to the wrapping effect, and for several other reasons).

3.2   Certifying a Numerical Trajectory

A better strategy is to first compute a numerical solution to (1) together with its first variation and to certify this "extended solution" at a second stage. So assume that we are given a subdivision $u = z_0 < \cdots < z_s = z$ and approximate values $f_0 \approx f(z_0), \ldots, f_s \approx f(z_s)$, as well as $V_0 \approx V(z_0), \ldots, V_s \approx V(z_s)$. We proceed as follows:

Stage 1. We first produce reasonable candidate enclosures $\boldsymbol{f}_1 = \boldsymbol{f}(z_1, z_0, \boldsymbol{I}), \ldots,$ $\boldsymbol{f}_s = \boldsymbol{f}(z_s, z_0, \boldsymbol{I})$ with $f(z_k, z_0, I) \in \boldsymbol{f}_k$ for all $k = 1, \ldots, s$ and $I \in \boldsymbol{I}$. Let f_0 denote the center of $\boldsymbol{f}_0 = \boldsymbol{I}$, ρ its radius, and let p be the current working precision. For some large constant $K \gg 1$, a good typical ansatz would be to take

$$\boldsymbol{f}_k = f_k + 2V_k\boldsymbol{\delta},$$

where

$$\boldsymbol{\delta} = \mathcal{B}(0, \rho + K2^{-p}|f_0|).$$

At the very end, we will have to prove the correctness of the ansatz, thereby producing a certificate for the numerical trajectory.

Stage 2. We compute $\boldsymbol{V}_{k,k-1} = \boldsymbol{V}^{\mathrm{loc}}(z_k, z_{k-1}, \boldsymbol{f}_{k-1})$ for $k = 1, \ldots, s$ using an extended local integrator. Given $0 \leqslant j < k \leqslant s$, and assuming correctness of the ansatz enclosures $\boldsymbol{f}_j, \ldots, \boldsymbol{f}_{k-1}$, this provides us with a certified enclosure

$$\boldsymbol{V}_{k,j} = \boldsymbol{V}_{k,k-1} \cdots \boldsymbol{V}_{j+1,j} \tag{2}$$

for $V(z_k, z_j, \boldsymbol{f}_j)$.

Stage 3. We compute $\boldsymbol{\varphi}_{k,k-1} = \boldsymbol{f}^{\mathrm{loc}}(z_k, z_{k-1}, \mathcal{B}(f_{k-1}, 0))$ for $k = 1, \ldots, s$ using a local integrator. Given $0 \leqslant j < k \leqslant s$, and assuming correctness of the ansatz enclosures $\boldsymbol{f}_j, \ldots, \boldsymbol{f}_{k-1}$, this provides us with certified enclosures

$$\boldsymbol{\varphi}_{k,j} = f_k + \sum_{i=j+1}^{k} \boldsymbol{V}_{k,i}(\boldsymbol{f}_{i,i-1} - f_i) \tag{3}$$

$$\boldsymbol{f}_{k,j} = \boldsymbol{\varphi}_{k,j} + \boldsymbol{V}_{k,j}(\boldsymbol{f}_j - f_j) \tag{4}$$

for $f(z_k, z_j, \mathcal{B}(f_j, 0))$ and $f(z_k, z_j, \boldsymbol{f}_j)$.

Stage 4. We finally check whether $\boldsymbol{f}_{k,0} \subseteq \boldsymbol{f}_k$ for $k = 1, \ldots, s$. If this is the case, then the correctness of the ansatz $\boldsymbol{f}_k$ follows by induction over k. Otherwise, for each index k with $\boldsymbol{f}_{k,0} \not\subseteq \boldsymbol{f}_k$ we replace our ansatz $\boldsymbol{f}_k$ by a new one $\tilde{\boldsymbol{f}}_k$ as follows: we consider $\boldsymbol{\delta}_{k,0} := \boldsymbol{f}_{k,0} - f_k$, $\boldsymbol{\delta}_k := \boldsymbol{f}_k - f_k$, and take $\tilde{\boldsymbol{f}}_k := f_k + 2\sup(\boldsymbol{\delta}_{k,0}, \boldsymbol{\delta}_k)$. We next return to Stage 2 with this new ansatz. We return an error if no certificate is obtained after a suitable and fixed number of such iterations.

Remark 1. If we want to certify our trajectory with a high precision p, then we clearly have to compute the enclosures f_k with precision p. On the other hand, the enclosures $V_{k,j}$ are only needed during the auxiliary computations (3) and (4) and it actually suffices to compute them with a much lower precision (which remains bounded if we let $p \to \infty$). For the initial ansatz, we essentially need this precision to be sufficiently large such that $V_k \delta \subseteq 2V_k \delta$ for $k = 1, \ldots, s$. In general, we rather must have $f_k + V_k \delta \subseteq f_k$.

3.3   Algorithmic Considerations and Parallelism

The next issue concerns the efficient and high quality evaluation of the formulas (2) and (3). The main potential problem already occurs in the case when Φ is constant, and (2) essentially reduces to the computation of the k-th power V^k of a ball matrix V. Assuming standard ball matrix arithmetic, the naive iterative method

$$V^k = VV^{k-1}$$

may lead to an exponential growth of the relative error as a function of k. Here we understand the relative error of a ball matrix to be the norm of the matrix of radii divided by the norm of the matrix of centers. The bad exponential growth occurs for instance for the matrix

$$V = \begin{pmatrix} 1 & 1 \\ -1 & 1 \end{pmatrix},$$

which corresponds to the complex number $1 + \mathrm{i}$. The growth remains polynomial in k in the case of triangular matrices V. When using binary powering

$$V^{2k} = V^k V^k$$
$$V^{2k+1} = VV^k V^k,$$

the growth of the relative error is systematically kept down to a polynomial in k.

For this reason, it is recommended to evaluate (2) and (3) using a similar dichotomic algorithm as in Sect. 2.2. More precisely, we will compute $\varphi_{k,0}$ and $V_{k,0}$ using a parallel dichotomic algorithm for $k = 1, \ldots, s$. Assuming that $s \geqslant 2$, let $m = \lceil s/2 \rceil$. We start with the recursive computation of $\varphi_{k,0}$ and $V_{k,0}$ for $k = 1, \ldots, m$, as well as $\varphi_{m+k,m}$ and $V_{m+k,m}$ for $k = 1, \ldots, s - m$. Then we compute

$$V_{m+k,0} := V_{m+k,m}V_{m,0}$$
$$\varphi_{m+k,0} := \varphi_{m+k,m} + V_{m+k,m}(\varphi_{m,0} - f_m)$$

for $k = 1, \ldots, s - m$.

The depth of this dichotomic method is $O(\log s)$. Given the initial numerical trajectory, it follows that the cost of the certification grows only with $\log s$ on sufficiently parallel computers. It is also interesting to notice that the parallel dichotomic technique that we used to double the precision uses very similar ingredients as the above way to certify trajectories. We found this analogy to apply on several other occasions, such as the computation of eigenvalues of matrices. This is probably due to the similarity between ball arithmetic and arithmetic in jet spaces of order one.

4 Lagrange Models

4.1 Taylor Models

Let $\mathcal{D} = \mathcal{B}(0, r)$ be the compact disk of center zero and radius r. A *Taylor model* of order $n \in \mathbb{N}$ on $\mathcal{D}$ consists of a polynomial $P \in \mathbb{C}[z]$ of degree $< n$ together with an error $\varepsilon \in \mathbb{R}^{>}$. We will denote such a Taylor model by $P + \mathcal{B}_{\mathcal{D}}(\varepsilon)$ and consider it as a ball of functions: given an analytic function f on $\mathcal{D}$ and $\boldsymbol{f} = P + \mathcal{B}_{\mathcal{D}}(\varepsilon)$, we write $f \in \boldsymbol{f}$ if $\|f - P\|_{\mathcal{D}} = \sup_{z \in \mathcal{D}} |f(z) - P(z)| \leqslant \varepsilon$.

Basic arithmetic on Taylor models works in a similar way as ball arithmetic. The ring operations are defined as follows:

$$(P + \mathcal{B}_{\mathcal{D}}(\delta)) \pm (Q + \mathcal{B}_{\mathcal{D}}(\varepsilon)) = (P \pm Q) + \mathcal{B}_{\mathcal{D}}(\delta + \varepsilon)$$
$$(P + \mathcal{B}_{\mathcal{D}}(\delta)) \cdot (Q + \mathcal{B}_{\mathcal{D}}(\varepsilon)) = (P \cdot Q)_{<n} +$$
$$\mathcal{B}_{\mathcal{D}} \left(\lceil P \rceil_{\mathcal{D}} \varepsilon + \lceil Q \rceil_{\mathcal{D}} \delta + \varepsilon \delta + \lceil (P \cdot Q)_{\geqslant n} \rceil_{\mathcal{D}} \right).$$

Given a polynomial $A \in \mathbb{C}[z]$ (or an analytic function A), the product formula uses the notations

$$A_{<n} = A_0 + \cdots + A_{n-1} z^{n-1}$$
$$A_{\geqslant n} = A_n z^n + A_{n+1} z^{n+1} + \cdots,$$

and $\lceil A \rceil_{\mathcal{D}}$ denotes any upper bound for $\|A\|_{\mathcal{D}}$ that is easy to compute. One may for instance take

$$\lceil A \rceil_{\mathcal{D}} = |A_0| + \cdots + |A_{\deg A}|.$$

Now consider the operation $\int$ of integration from zero $\left(\int f \right)(z) = \int_0^z f(u) \mathrm{d}u$. The integral of a Taylor model may computed using

$$\int (P + \mathcal{B}_{\mathcal{D}}(\delta)) = \int P_{<n-1} + \mathcal{B}_{\mathcal{D}}(r |P_{n-1}|/n + r \delta).$$

This formula is justified by the mean value theorem.

In practice, the numerical computations at a given working precision involve additional rounding errors. Bounds for these rounding errors have to be added to the errors in the above formulas. It is also easy to define Taylor models on disks $\mathcal{B}(c, r)$ with general centers as being given by a Taylor model on $\mathcal{D}$ in the variable $z' = z - c$. For more details, we refer to [14, 15].

4.2   Lagrange Models

A *Lagrange model* of order $n \in \mathbb{N}$ on $\mathcal{D}$ is a functional "ball" of the form $\boldsymbol{P} + \mathcal{B}_{\mathcal{D}}(\varepsilon)z^n$, where $\boldsymbol{P} \in \mathbb{C}[z]$ is a ball polynomial of degree $< n$ and $\varepsilon \in \mathbb{R}^>$ the so called *tail bound*. Given an analytic function f on $\mathcal{D}$ and $\boldsymbol{f} = \boldsymbol{P} + \mathcal{B}_{\mathcal{D}}(\varepsilon)z^n$, we write $f \in \boldsymbol{f}$ if $f_k \in \boldsymbol{P}_k$ for all $k < n$ and $\|f_{\geqslant n}z^{-n}\|_{\mathcal{D}} \leqslant \varepsilon$. The name "Lagrange model" is motivated by Taylor–Lagrange's formula, which provides a careful estimate for the truncation error of a Taylor series expansion. We may also regard Lagrange models as a way to substantiate the "big Oh" term in the expansion $f = f_0 + \cdots + f_{n-1}z^{n-1} + O(z^n)$.

Basic arithmetic on Lagrange models works in a similar way as in the case of Taylor models:

$$(\boldsymbol{P} + \mathcal{B}_{\mathcal{D}}(\delta)z^n) \pm (\boldsymbol{Q} + \mathcal{B}_{\mathcal{D}}(\varepsilon)z^n) = (\boldsymbol{P} \pm \boldsymbol{Q}) + \mathcal{B}_{\mathcal{D}}(\delta + \varepsilon)z^n$$
$$(\boldsymbol{P} + \mathcal{B}_{\mathcal{D}}(\delta)z^n) \cdot (\boldsymbol{Q} + \mathcal{B}_{\mathcal{D}}(\varepsilon)z^n) = (\boldsymbol{P} \cdot \boldsymbol{Q})_{<n} +$$
$$\mathcal{B}_{\mathcal{D}}\left(\llbracket \boldsymbol{P} \rrbracket_{\mathcal{D}}\varepsilon + \llbracket \boldsymbol{Q} \rrbracket_{\mathcal{D}}\delta + \varepsilon\delta + \llbracket (\boldsymbol{P} \cdot \boldsymbol{Q})_{\geqslant n} \rrbracket_{\mathcal{D}}\right)z^n.$$

This time, we may take

$$\llbracket \boldsymbol{A} \rrbracket_{\mathcal{D}} = \lceil \boldsymbol{A}_0 \rceil + \cdots + \lceil \boldsymbol{A}_{\deg \boldsymbol{A}} \rceil.$$
$$\lceil \mathcal{B}(c, \rho) \rceil = |c| + \rho$$

as the "easy to compute" upper bound of a ball polynomial $\boldsymbol{A} \in \mathbb{C}[z]$. The main advantage of Lagrange models with respect to Taylor models is that they allow for more precise tail bounds for integrals:

$$\int (\boldsymbol{P} + \mathcal{B}_{\mathcal{D}}(\delta)z^n) = \int \boldsymbol{P}_{<n-1} + \mathcal{B}_{\mathcal{D}}(r\lceil \boldsymbol{P}_{n-1} \rceil/n + r\delta/(n+1))z^n.$$

Indeed, for any function f on $\mathcal{D}$, integration on a straight line segment from 0 to any $z \in \mathcal{D}$ yields

$$\left| \int_0^z f(u)u^n \mathrm{d}u \right| = \left| \int_0^{z^{n+1}} \frac{f\left(\sqrt[n+1]{v}\right)}{n+1} \mathrm{d}v \right| \leqslant \frac{r\|f\|_{\mathcal{D}}}{n+1},$$

whence $\left\| \int(fz^n) \right\|_{\mathcal{D}} \leqslant r\|f\|_{\mathcal{D}}/(n+1)$.

The main disadvantage of Lagrange models with respect to Taylor models is that they require more data (individual error bounds for the coefficients of the polynomial) and that basic arithmetic is slightly more expensive. Indeed, arithmetic on ball polynomials is a constant time more expensive than ordinary polynomial arithmetic. Nevertheless, this constant tends to one if either n or the working precision p gets large. This makes Lagrange models particularly well suited for high precision computations, where they cause negligible overhead, while improving the quality of tail bounds for integrals. For efficient implementations of basic arithmetic on ball polynomials, we refer to [5].

4.3 Reliable Integration of Dynamical Systems

Let us now return to the dynamical system (1). We already mentioned relaxed power series computations and Newton's method as two efficient techniques for the numerical computation of power series solutions to differential equations. These methods can still be used for ball coefficients, modulo some preconditioning or suitable tweaking of the basic arithmetic on ball polynomials; see [5] for more details. In order to obtain a local certified integrator in the sense of Sect. 3.1, it remains to be shown how to compute tail bounds for truncated power solutions at order n.

From now on, we will be interested in finding local certified solutions of (1) at the origin. We may rewrite the system (1) together with the initial condition $f(0) = I$ as a fixed point equation

$$f = I + \int \Phi(f). \tag{5}$$

Now assume that a Lagrange model f satisfies

$$I + \int \Phi(f) \subseteq f. \tag{6}$$

Then we claim that for any $I \in I$, there is an analytic function $f \in f$ that satisfies (5). Indeed, the analytic functions f with $f \in f$ form a compact set, so the operator $f \in f \mapsto I + \int \Phi(f) \in f$ admits a fixed point for any $I \in I$. This fixed point is unique, since its coefficients can be computed uniquely by induction.

Using ball power series computations we already know how to compute a ball polynomial P of degree $< n$ such that

$$I + \int \Phi(P) \subseteq P + O(z^n). $$

Taking $f = P + \mathcal{B}_D(\varepsilon)z^n$, it remains to be shown how to compute $\varepsilon \in (\mathbb{R}^>)^d$ in such a way that (6) is satisfied. Now denoting by J the Jacobian matrix of Φ, and putting $\varepsilon = \mathcal{B}_D(\varepsilon)z^n$, we have

$$\Phi(f) \subseteq \Phi(P) + J(f)\varepsilon.$$

Writing $Q + \mathcal{B}_D(\delta)z^n$ for $I + \int \Phi(P)$ and $\delta = \mathcal{B}_D(\delta)z^n$, we thus have $Q \subseteq P$ and it suffices to find ε such that

$$\delta + \int J(f)\varepsilon \subseteq \varepsilon. \tag{7}$$

Assuming that all eigenvalues of $J(f)$ are strictly bounded by $(n+1)/r$, it suffices to "take"

$$\varepsilon = \left\| \left(1 - \tfrac{r}{n+1}J(f)\right)^{-1} \right\|_D \delta. \tag{8}$$

We have to be a little bit careful here, since $J(\boldsymbol{f})$ depends on ε. Nevertheless, the formula (8) provides a good ansatz: starting with $\varepsilon^{[0]} = 0$, we may define

$$\varepsilon^{[i+1]} := \left[\!\!\left[\left(1 - \tfrac{r}{n+1} J(\boldsymbol{P} + \mathcal{B}_{\mathcal{D}}(\varepsilon^{[i]}) z^n) \right)^{-1} \right]\!\!\right]_{\mathcal{D}} \delta \tag{9}$$

for all i. If r was chosen small enough, then this sequence quickly stabilizes. Assuming that $\varepsilon^{[l+1]} \approx \varepsilon^{[l]}$, we set $\varepsilon = \varepsilon^{[l]} + 2(\varepsilon^{[l+1]} - \varepsilon^{[l]})$, and check whether (7) holds. If so, then we have obtained the required Lagrange model solution of (6). Otherwise, we will need to decrease r, or increase n and the working precision.

4.4 Discussion

Several remarks are in place about the method from the previous subsection. Let us first consider the important case when $\boldsymbol{I} = \mathcal{B}(I, 0)$ is given exactly, and let R denote the convergence radius of the unique solution f of (5). For large working precisions p and expansion orders n, we can make δ arbitrarily small. Assuming that the eigenvalues of $J(f)$ are strictly bounded by $(n+1)/r$, this also implies that $\varepsilon^{[1]}, \varepsilon^{[2]}, \dots$ become arbitrarily small, and that $\varepsilon = \varepsilon^{[1]} + 2(\varepsilon^{[2]} - \varepsilon^{[1]})$ satisfies (7). In other words, for any $r < R$, there exists a sufficiently large n (and working precision p) for which the method succeeds.

Let us now investigate what happens if we apply the same method with Taylor models instead of Lagrange models. In that case, the Eq. (8) becomes

$$\varepsilon = [\![(1 - r J(\boldsymbol{f}))^{-1}]\!]_{\mathcal{D}} \delta.$$

On the one hand this implies that the method will break down as soon as $1/r$ reaches the largest eigenvalue of $J(f)$, which may happen for $r \ll R$. Even if J is constant (i.e. $f' = \Phi(f)$ reduces to the differential equation $f' = Jf$ for a constant matrix J), the step size cannot exceed the inverse of the maximal eigenvalue of J. On the other hand, and still in the case when J is constant, we see that Lagrange models allow us to take a step size which is $n+1$ times as large. In general, the gain will be smaller since J usually admits larger eigenvalues on larger disks. Nevertheless, Lagrange models will systematically allow for larger step sizes.

Notice that the matrices that we need to invert in (8) and (9) admit Lagrange model entries, which should really be regarded as functions. Ideally speaking, we would like to compute a uniform bound for the inverses of the evaluations of these matrices at all points in $\mathcal{D}$. However, this may be computationally expensive. Usually, it is preferable to replace each Taylor model entry $\boldsymbol{g} + \mathcal{B}_{\mathcal{D}}(\eta) z^n$ of the matrix to be inverted by a ball enclosure $\boldsymbol{g}_0 + \cdots + \boldsymbol{g}_{n-1} \mathcal{B}(0, r^{n-1}) + \mathcal{B}(0, \eta r^n)$. The resulting ball matrix can be inverted much faster, although the resulting error bounds may be of inferior quality.

We finally stress (once more) that bound computations usually require a far smaller accuracy than the working precision. This makes it interesting to consider the problem of computing tail bounds for Lagrange (and Taylor) models independently from the problem of evaluating them: for accurate evaluations, we need the working precision p and the expansion order n to be approximately proportional. But the tail bound computations could be done at a much smaller precision $p' \ll n$. In particular, certifying the convergence of a solution to (1) on a compact disk can often be done using low precision computations only.

References

1. Alefeld, G., Herzberger, J.: Introduction to Interval Analysis. Academic Press, New York (1983)
2. Brent, R.P., Kung, H.T.: Fast algorithms for manipulating formal power series. J. ACM **25**, 581–595 (1978)
3. Gambill, T.N., Skeel, R.D.: Logarithmic reduction of the wrapping effect with application to ordinary differential equations. SIAM J. Numer. Anal. **25**(1), 153–162 (1988)
4. van der Hoeven, J.: Relax, but don't be too lazy. JSC **34**, 479–542 (2002)
5. van der Hoeven, J.: Ball arithmetic. In: Beckmann, A., Gaßner, C., Löwe, B. (eds.) International Workshop on Logical approaches to Barriers in Computing and Complexity, no. 6 in Preprint-Reihe Mathematik, pp. 179–208. Ernst-Moritz-Arndt-Universität Greifswald, February 2010
6. van der Hoeven, J.: Newton's method and FFT trading. JSC **45**(8), 857–878 (2010)
7. van der Hoeven, J.: Calcul analytique. In: Journées Nationales de Calcul Formel (2011), vol. 2. Les cours du CIRM. CEDRAM 2011. Exp. No. 4, p. 85 (2011). http://ccirm.cedram.org/ccirm-bin/fitem?id=CCIRM_2011_2_1_A4_0
8. van der Hoeven, J.: Faster relaxed multiplication. In: Proceedings of the ISSAC 2014, pp. 405–412, Kobe, Japan, July 2014
9. Jaulin, L., Kieffer, M., Didrit, O., Walter, E.: Applied Interval Analysis. Springer, London (2001)
10. Kühn, W.: Rigorously computed orbits of dynamical systems without the wrapping effect. Computing **61**, 47–67 (1998)
11. Kulisch, U.W.: Computer Arithmetic and Validity. Theory, Implementation, and Applications. Studies in Mathematics, vol. 33. de Gruyter, Berlin (2008)
12. Lohner, R.: Einschließung der Lösung gewöhnlicher Anfangs- und Randwertaufgaben und Anwendugen. Ph.D. thesis, Universität Karlsruhe (1988)
13. Lohner, R.: On the ubiquity of the wrapping effect in the computation of error bounds. In: Kulisch, U., Lohner, R., Facius, A. (eds.) Perspectives on Enclosure Methods, pp. 201–217. Springer, New York (2001)
14. Makino, K., Berz, M.: Remainder differential algebras and their applications. In: Berz, M., Bischof, C., Corliss, G., Griewank, A. (eds.) Computational Differentiation: Techniques, Applications and Tools, pp. 63–74. SIAM, Philadelphia (1996)
15. Makino, K., Berz, M.: Suppression of the wrapping effect by Taylor model-based validated integrators. Technical report MSU report MSUHEP 40910, Michigan State University (2004)
16. Moore, R.E.: Automatic local coordinate transformations to reduce the growth of error bounds in interval computation of solutions to ordinary differential equation. In: Rall, L.B. (ed.) Error in Digital Computation, vol. 2, pp. 103–140. Wiley, Hoboken (1965)
17. Moore, R.E.: Interval Analysis. Prentice Hall, Englewood Cliffs (1966)
18. Moore, R.E., Kearfott, R.B., Cloud, M.J.: Introduction to Interval Analysis. SIAM Press, Philadelphia (2009)

19. Neumaier, A.: Interval Methods for Systems of Equations. Cambridge University Press, Cambridge (1990)
20. Neumaier, A.: The wrapping effect, ellipsoid arithmetic, stability and confedence regions. In: Albrecht, R., Alefeld, G., Stetter, H.J. (eds.) Validation Numerics. Computing Supplementum, vol. 9, pp. 175–190. Springer, Heidelberg (1993)
21. Neumaier, A.: Taylor forms - use and limits. Reliable Comput. **9**, 43–79 (2002)
22. Nickel, K.: How to fight the wrapping effect. In: Nickel, K. (ed.) Interval Mathematics 1985. LNCS, vol. 212, pp. 121–132. Springer, Heidelberg (1985)
23. Press, W.H., Teukolsky, S.A., Vetterling, W.T., Flannery, B.P.: Numerical recipes, the art of scientific computing, 3rd edn. Cambridge University Press, Cambridge (2007)
24. Rump, S.M.: Verification methods: rigorous results using floating-point arithmetic. Acta Numerica **19**, 287–449 (2010)
25. Sedoglavic, A.: Méthodes seminumériques en algèbre différentielle; applications à l'étude des propriétés structurelles de systèmes différentiels algébriques en automatique. Ph.D. thesis, École polytechnique (2001)
26. Turing, A.: On computable numbers, with an application to the Entscheidungsproblem. Proc. London Maths. Soc. **2**(42), 230–265 (1936)
27. Weihrauch, K.: Computable Analysis. Springer, Heidelberg (2000)

Global Optimization

A New Matrix Splitting Based Relaxation for the Quadratic Assignment Problem

Marko Lange[(✉)]

Institute for Reliable Computing, Hamburg University of Technology,
Hamburg, Germany
marko.lange@tuhh.de

Abstract. Nowadays, the quadratic assignment problem (QAP) is widely considered as one of the hardest of the NP-hard problems. One of the main reasons for this consideration can be found in the enormous difficulty of computing good quality bounds for branch-and-bound algorithms. The practice shows that even with the power of modern computers QAPs of size $n > 30$ are typically recognized as huge computational problems. In this work, we are concerned with the design of a new low-dimensional semidefinite programming relaxation for the computation of lower bounds of the QAP. We discuss ways to improve the bounding program upon its semidefinite relaxation base and give numerical examples to demonstrate its applicability.

Keywords: Quadratic assignment problem · Semidefinite programming · Relaxation

1 Introduction

The quadratic assignment problem (QAP) was introduced by Koopmans and Beckmann [11] in 1957 as a mathematical model for problems in the allocation of indivisible resources. The class of QAPs entails a great number of applications from different scenarios in the topic of combinatorial optimization. This includes problems arising in location theory, facility layout, VLSI design, communications and various other fields. For extensive lists of applications of QAPs, we refer to the survey works by Pardalos et al. [17], Burkard et al. [4], Çela [5], Loiola et al. [13] and most recently Burkard et al. [3].

In this work, we are concerned with the computation of lower bounds for QAPs which can be formulated in Koopmans-Beckmann trace formulation [8]:

$$\inf_{X \in \Pi^n} \quad \mathrm{tr}(AXBX^T + CX^T), \qquad \text{(KBQAP)}$$

where $A, B, C \in \mathbb{R}^{n \times n}$ are the parameter matrices of the QAP, Π^n denotes the set of $n \times n$ permutation matrices, and tr() terms the trace function. More precisely, our concern is a new technique for the construction of a low-dimensional semidefinite programming (SDP) relaxation for (KBQAP).

© Springer International Publishing Switzerland 2016
I.S. Kotsireas et al. (Eds.): MACIS 2015, LNCS 9582, pp. 535–549, 2016.
DOI: 10.1007/978-3-319-32859-1_45

Our main contribution is the introduction of a new relaxation approach based on interrelated matrix splitting. The derivation of the corresponding framework can be found in Subsect. 2.2. Subsequently, we discuss additional cuts which are based on techniques introduced by Mittelmann and Peng in [15]. In Subsect. 3.1, we propose a way to tighten the respective constraints by exploiting a degree of freedom that is present in the original versions of these cuts.

1.1 Notation and Preliminaries

Unless otherwise stated, we assume that both matrices A and B are symmetric. Furthermore, without loss of generality, it is assumed that the diagonal elements of A and B are equal to zero. If this is not the case, the corresponding costs can be shifted to the linear term by setting $C_{\text{new}} := C + \text{diag}(A)\,\text{diag}(B)^T$, where $\text{diag}(A)$ denotes a column vector formed of the diagonal elements of A. Throughout this paper, $B = \sum_{i=1}^{n} \lambda_i q_i q_i^T$ shall denote the eigenvalue decomposition of B.

If not stated otherwise, $\|\cdot\|$ is used for the spectral norm. The trace inner product of two real matrices G, H is denoted by $\langle G, H \rangle := \text{tr}(G^T H)$. Furthermore, we write $H^\dagger$ for the Moore-Penrose pseudoinverse of H [16,20]. If H is an operator, $\mathcal{R}(H)$ denotes its range in the sense of its image. In the case that H is a matrix, we use the same notation referring to its column space.

The cone of symmetric positive semidefinite matrices is of major importance for every discussion about SDP problems. We denote the space of $n \times n$ symmetric matrices by $\mathbb{S}^n$ and its positive semidefinite subset by $\mathbb{S}^n_+$. In this context, we also utilize the relation sign '$\succeq$' to denote a Loewner's partial ordering, i.e. $H \succeq G$ is used to note the positive semidefiniteness of $H - G$. In addition to the already mentioned sets, we consider the space of $m \times n$ matrices $\mathbb{M}^{m,n}$ and the set of $n \times n$ double stochastic matrices $\mathbb{D}^n$. By e we denote the n dimensional column vector of all ones and $I := [e_1, \ldots, e_n]$ is used for the $n \times n$ identity matrix. Generally, we spare redundant informations on matrix dimensions. For instance, we write $\mathbb{M}^m$ instead of $\mathbb{M}^{m,m}$. Moreover, in cases where the dimension is evident from the context, the accompanying indicators may be discarded completely.

Complementary to the diag-operator, $\text{off}(H)$ denotes a column vector that contains all off-diagonal elements of the matrix H. This vector is obtained by vertical concatenation of the columns of H, but without its diagonal elements. Another considered linear transformation is the triangular vectorization of a matrix; $\text{tri}(H)$ denotes the vector obtained from the vertical concatenation of the columns of H taking solely its lower triangular elements (without matrix diagonal) into account. These operators may also be used in combination with relations, for instance $\{=_{\text{off}}, \geq_{\text{off}}, \leq_{\text{off}} \ldots\}$. In case of the subscript $_{\text{off}}$, the respective relations apply only to the off-diagonal elements of the corresponding matrices, hence $A \geq_{\text{off}} B$ is the short form for $\text{off}(A) \geq \text{off}(B)$.

2   QAP Relaxations Based on Matrix Splitting

Relaxation is a fundamental approach for the computation of lower or upper bounds of intractable programming problems. It can be used directly as an approximation of the original problem, for bound computations in branch-&-bound and branch-&-cut approaches, or as a tool to measure the quality of other bounding algorithms. In regard to the form of the given optimization problem the first step of a relaxation process requires the reformulation of the original problem. The second step comprises the removal or replacement of constraints that are the cause for intractability.

One of the most popular relaxation approaches for quadratic programming problems is based on vector lifting. A good source for relaxations of this kind is given by Zhao et al. in [24]. Compared to newer low-dimensional SDP relaxations for the QAP, relaxation frameworks based on vector lifting have their strength in the computation of tighter bounds. Their major drawbacks are the large number of $\mathcal{O}(n^4)$ variables and the accompanying computational costs.

There are some efforts to reduce the computational costs of these high dimensional SDP relaxations, see for instance [2,9,21,23]. Nevertheless, regarding QAP instances of size $n > 30$ and with little symmetry, the computational costs for solving SDP relaxation frameworks based on vector lifting remain too high for practical usage.

2.1   Non-redundant Positive Semidefinite Matrix Splitting

For a special class of QAPs - instances which are associated with Hamming and Manhatten distances - Mittelmann and Peng [15] pursued the idea of another low-dimensional SDP relaxation framework. The presented bounds not only involve a less expensive computational process, they are also provably tighter than the ones proposed in [6] by Ding and Wolkowicz. In [18] and [19], Peng et al. generalized the matrix splitting approach for other classes of the QAP.

If the parameter matrix B is positive semidefinite, the equality $Y = XBX^T$ can be relaxed to the convex semidefinite relation $Y \succeq XBX^T$. The implementation of the latter is usually realized by utilization of the Schur complement inequality [1], here

$$\begin{bmatrix} B & BX^T \\ XB & Y \end{bmatrix} = \begin{bmatrix} B^{\frac{1}{2}} \\ XB^{\frac{1}{2}} \end{bmatrix} \begin{bmatrix} B^{\frac{1}{2}} & B^{\frac{1}{2}}X^T \end{bmatrix} \in \mathbb{S}_+^{2n}. \tag{1}$$

In general, however, B does not satisfy any definiteness property. Peng et al. [18,19] dealt with this case by applying a non-redundant positive semidefinite matrix splitting scheme.

Definition 1. *For a given matrix B a matrix pair (B_1, B_2) is called a positive semidefinite matrix splitting of B if it satisfies*

$$B = B_1 - B_2, \quad B_1, B_2 \in \mathbb{S}_+. \tag{2}$$

The splitting is said to be redundant if there exists a nonzero positive semidefinite matrix R, such that

$$B_1 - R \in \mathbb{S}_+, \quad B_2 - R \in \mathbb{S}_+. \tag{3}$$

If $R \equiv 0$ is the only feasible matrix that is positive semidefinite and satisfies (3), we say that the splitting is non-redundant.

For the relaxation framework *F-SVD* introduced in [18], the authors used the following non-redundant splitting:

$$B_+ = \sum_{i:\lambda_i>0} \lambda_i q_i q_i^T \quad \text{and} \quad B_- = \sum_{i:\ \lambda_i<0} -\lambda_i q_i q_i^T. \tag{4}$$

Together with (1) and the observations that

$$\forall X \in \Pi^n, B \in \mathbb{M}^n: \quad \mathrm{diag}(XBX^T) = X\,\mathrm{diag}(B), \quad XBX^T e = XBe, \tag{5}$$

we derive the SDP basis of their framework, here referred to as *B-SVD*:

$$\inf_{X\in\mathbb{D}^n,\ Y_+,Y_-\in\mathbb{S}^n} \quad \langle A, Y_+ - Y_- \rangle + \langle C, X \rangle \tag{6a}$$

$$\text{s.t.} \quad \begin{bmatrix} B_+ & B_+X^T \\ XB_+ & Y_+ \end{bmatrix} \in \mathbb{S}_+, \quad \begin{bmatrix} B_- & B_-X^T \\ XB_- & Y_- \end{bmatrix} \in \mathbb{S}_+, \tag{6b}$$

$$\mathrm{diag}(Y_+) = X\,\mathrm{diag}(B_+), \quad \mathrm{diag}(Y_-) = X\,\mathrm{diag}(B_-), \tag{6c}$$

$$Y_+e = XB_+e, \quad Y_-e = XB_-e, \tag{6d}$$

where the variables Y_+ and Y_- are used to relax the quadratic terms XB_+X^T and XB_-X^T, respectively.

In regard to a matrix splitting based SDP relaxation such as (6), Peng et al. demonstrated the general advantage of non-redundant matrix splittings over redundant ones, see [19, Theorem 1]. Roughly speaking the theorem states that for any redundant positive semidefinite matrix splitting there exists a non-redundant splitting which leads to a tighter relaxation. Even though additional constraints on the respective variables may change this circumstance, the absence of redundancies in the positive semidefinite matrix splitting is a good indicator for a beneficial splitting scheme.

2.2   Interrelated Matrix Splitting

A particularly beautiful property of the positive semidefinite matrix splitting defined in (4) is that the ranges of the matrices B_+, B_- are orthogonal to each other, such that $\mathcal{R}(B_+) \cap \mathcal{R}(B_-) = \{0\}$ and $B_+B_- = B_-B_+ \equiv 0$. As an immediate consequence of this circumstance, B_+ and B_- are simultaneously diagonalizable. It would be a great advantage if we could make use of these inter-relations in the actual relaxation. Unfortunately, it is quite difficult to exploit the corresponding properties in form of beneficial SDP constraints. For the design of new relaxation strategies, we need a different kind of interrelation. In this subsection, we say goodbye to the idea of redundancy-free positive semidefinite matrix splitting pairs (B_+, B_-) and present a new splitting scheme:

$$B = B_\triangle - B_\triangledown \quad \text{with additional conditions on } (B_\triangle, B_\triangledown). \tag{7}$$

By the introduction of specific redundancies, we induce the presence of artificial correlations between the respective splitting parts. These interrelations shall be used to construct new types of constraints which are applicable in the corresponding QAP relaxations.

A beneficial interrelation property for the relaxation of QAP is the semidefinite inverse relation

$$B_\triangle \succeq B_\triangledown^{-1} \succeq 0. \tag{8}$$

The existence of the inverse $B_\triangledown^{-1}$ implies the regularity of $B_\triangledown$ and thereby also the regularity of $B_\triangle$. By the matrix equality

$$B_\triangledown - B_\triangle^{-1} = B_\triangle^{-1} \underbrace{(B_\triangle - B_\triangledown^{-1})}_{\succeq 0} B_\triangle^{-1} + (I - B_\triangledown^{-1} B_\triangle^{-1})^T \underbrace{B_\triangledown}_{\succeq 0} (I - B_\triangledown^{-1} B_\triangle^{-1}),$$

it is furthermore apparent that (8) implies the validity of

$$B_\triangledown \succeq B_\triangle^{-1} \succeq 0 \tag{9}$$

Indeed, it is straightforward to show that the conditions (8) and (9) are equivalent.

The discussed interrelation property can be exploited by transferring the same to the relaxation variables for the quadratic terms $Y_\triangle = X B_\triangle X^T$ and $Y_\triangledown = X B_\triangledown X^T$. The orthogonality of permutation matrices $X \in \Pi$ gives

$$X B_\triangledown^{-1} X^T = (X B_\triangledown X^T)^{-1}.$$

Relation (8) therefore requires $X B_\triangle X^T \succeq (X B_\triangledown X^T)^{-1} \succeq 0$ providing the basis for the constraint $Y_\triangle \succeq Y_\triangledown^{-1} \succeq 0$. The latter condition can be realized by utilization of the Schur complement inequality [1]:

$$\begin{bmatrix} Y_\triangle & I \\ I & Y_\triangledown \end{bmatrix} \in \mathbb{S}_+^{2n}. \tag{10}$$

For the attainment of tight SDP conditions, we are looking for matrices $B_\triangle$ and $B_\triangledown$ with minimal traces. This is the case for the splitting that satisfies the identity $B_\triangle = B_\triangledown^{-1}$.

Theorem 1. *Let $B \in \mathbb{S}^n$ be given and consider the minimization problem*

$$\begin{aligned} \inf_{B_\triangle, B_\triangledown \in \mathbb{S}^n} \quad & \text{tr}(B_\triangle) + \text{tr}(B_\triangledown) \\ \text{s.t.} \quad & B_\triangle \succeq B_\triangledown^{-1} \succeq 0, \\ & B_\triangle - B_\triangledown = B. \end{aligned} \tag{11}$$

A solution to this program is given by the matrix pair $(B_\triangle, B_\triangledown)$ *defined as*

$$B_\triangle := \frac{1}{2}\left(B + \sqrt{B^2 + 4I}\right), \qquad B_\triangledown := B_\triangle - B. \tag{12}$$

This pair satisfies the identity $B_\triangle = B_\triangledown^{-1}$.

Proof. The multiplication of the matrices defined in (12) gives

$$B_\triangle B_\triangledown = \frac{1}{2}\left(B + \sqrt{B^2 + 4I}\right)\frac{1}{2}\left(\sqrt{B^2 + 4I} - B\right) = \frac{1}{4}\left(B^2 + 4I - B^2\right) = I$$

and proves $B_\triangle = B_\triangledown^{-1}$. It is also straightforward to check that $(B_\triangle, B_\triangledown)$ satisfies the constraints of problem (11), hence states a feasible point. For now, let us assume that there is some solution $(\hat{B}_\triangle, \hat{B}_\triangledown)$ that accompanies a smaller objective value than the matrix pair from (12), thus $\mathrm{tr}(\hat{B}_\triangle) < \mathrm{tr}(B_\triangle)$. By definition, the matrices B, $B_\triangle$ and $B_\triangledown$ are all three simultaneously diagonalizable. Let $\{q_1, \ldots, q_n\}$ denote the set of the corresponding orthonormal eigenvectors, then

$$\sum_{i=1}^{n} q_i^T \hat{B}_\triangle q_i = \mathrm{tr}(\hat{B}_\triangle) < \mathrm{tr}(B_\triangle) = \sum_{i=1}^{n} q_i^T B_\triangle q_i$$

and therefore

$$\exists k \in \{1, \ldots, n\}: \quad q_k^T \hat{B}_\triangle q_k < q_k^T B_\triangle q_k.$$

Since $B_\triangle - \hat{B}_\triangle = B_\triangledown - \hat{B}_\triangledown$, this also means that $q_k^T \hat{B}_\triangledown q_k < q_k^T B_\triangledown q_k$, such that

$$q_k^T \hat{B}_\triangle q_k < q_k^T B_\triangle q_k = \lambda_k(B_\triangle) = \lambda_k(B_\triangledown)^{-1} = (q_k^T B_\triangledown q_k)^{-1} < (q_k^T \hat{B}_\triangledown q_k)^{-1}.$$

Moreover, the positive semidefinitenes of

$$\begin{bmatrix} q_k^T \hat{B}_\triangledown^{-1} q_k & 1 \\ 1 & q_k^T \hat{B}_\triangledown q_k \end{bmatrix} = \begin{bmatrix} q_k^T \hat{B}_\triangledown^{-\frac{1}{2}} \\ q_k^T \hat{B}_\triangledown^{\frac{1}{2}} \end{bmatrix} \begin{bmatrix} q_k^T \hat{B}_\triangledown^{-\frac{1}{2}} \\ q_k^T \hat{B}_\triangledown^{\frac{1}{2}} \end{bmatrix}^T \in \mathbb{S}_+^2$$

implies a nonnegative determinant of this matrix, which in turn requires that $(q_k^T \hat{B}_\triangledown^{-1} q_k)(q_k^T \hat{B}_\triangledown q_k) \geqslant 1$. Taken together, we obtain the inequality

$$q_k^T \hat{B}_\triangle q_k < (q_k^T \hat{B}_\triangledown q_k)^{-1} \leqslant q_k^T \hat{B}_\triangledown^{-1} q_k,$$

which violates the positive semidefinite condition $\hat{B}_\triangle \succeq \hat{B}_\triangledown^{-1}$, thereby contradicts our assumption and finishes the proof.

The efficiency of constraint (10) depends to a significant amount on the scaling of B. For QAP instances where $\|B\|$ is much greater than 1, the formulas in (4) and (12) give

$$B_\triangle = \frac{1}{2}(B + \sqrt{B^2 + 4I}) \approx \frac{1}{2}(B + \sqrt{B^2}) = B_+, \quad B_\triangledown \approx B_-.$$

Hence, the splitting differs only slightly from the PSD splitting based on the spectral value decomposition, and the effect of the inverse interrelation on the corresponding feasible set is hardly noticeable. On the other hand, if $\|B\| \ll 1$, the validity of (8) is purchased by introducing a relatively large redundancy:

$$\hat{B}_\Delta = \frac{1}{2}(B + \sqrt{B^2 + 4I}) \approx I, \quad \hat{B}_\nabla \approx I.$$

To counteract this behavior, we utilize a linear homogeneous function $\tau \colon \mathbb{M}^n \to \mathbb{R}$ and replace (8) with

$$B_\Delta \succeq \tau(B)^2 B_\nabla^{-1} \succeq 0. \tag{13}$$

For any positive real scaling factor α, the condition

$$\alpha B_\Delta \succeq \tau(\alpha B)^2 (\alpha B_\nabla)^{-1} \succeq 0$$

is equivalent to (13). This circumstance is easily apparent from the linearity of τ and demonstrates scaling invariance of this relation. By numerical tests, we discovered that the trace norm of a projection of the respective matrix is a suitable base for τ. In the actual implementation, we use the renormalization function τ defined as

$$\tau(B) := \frac{1}{4n} \sum_{i=1}^n \sigma_i(PBP), \tag{14}$$

where the orthogonal projection matrix P is defined as $P := I - \frac{1}{n}ee^T$, and $\sigma_i(\cdot)$ denotes the i-th singular value of the corresponding matrix. Among the tested matrix norms and various scalings of these, the choice given in (14) worked best for a large range of problems.

For QAP instances with low-rank parameter matrices B, it is possible to strengthen the semidefinite constraint by replacing the inverse property in (8) with the pseudoinverse relations

$$B_\Delta \succeq B_\nabla^\dagger \succeq 0 \quad \text{and} \quad B_\nabla \succeq B_\Delta^\dagger \succeq 0. \tag{15}$$

Any matrix pair (B_Δ, B_∇) that complies with these two conditions necessarily satisfies

$$\mathcal{R}(B_\Delta) \supseteq \mathcal{R}(B_\nabla^\dagger) = \mathcal{R}(B_\nabla) \supseteq \mathcal{R}(B_\Delta^\dagger) = \mathcal{R}(B_\Delta),$$

such that $\mathcal{R}(B_\Delta) = \mathcal{R}(B_\nabla)$. This in turn demonstrates the equivalence of (15) and the semidefinite condition

$$\begin{bmatrix} B_\Delta & G \\ G & B_\nabla \end{bmatrix} \in \mathbb{S}_+^{2n},$$

where G is the orthogonal projection matrix for the space $\mathcal{R}(B_\Delta) \cup \mathcal{R}(B_\nabla)$.

In the actual implementation, we take the approach one step further by incorporating the renormalization function τ and weighting the utilization of the inverse interrelation property against the introduced redundancy. In order to achieve these objectives, we apply the following program:

$$\inf_{B_\Delta, B_\nabla, G \in \mathbb{S}^n} \quad \mathrm{tr}(B_\Delta) + \mathrm{tr}(B_\nabla) - \xi \, \mathrm{tr}(G)$$

$$\text{s.t.} \quad \begin{bmatrix} B_\Delta & G \\ G & B_\nabla \end{bmatrix} \in \mathbb{S}_+^{2n},$$

$$B_\Delta - B_\nabla = B, \tag{16}$$

$$\|G\| \leqslant \tau(B),$$

where ξ is a nonnegative real value that serves as a threshold for the introduced redundancy.

The choice of ξ influences the effectiveness of the generalized inverse interrelation. For the extreme $\xi = 0$ the result is equivalent to the pure non-redundant matrix splitting utilized in relaxation (6), hence $(B_\Delta, B_\nabla, G) = (B_+, B_-, 0_{(n,n)})$. On the other hand, for $\xi > 2$ the attained splitting corresponds to the normalized version of the original inverse property given in (13). By no means, however, ξ is used as a trade-off between speed and quality of the respective relaxations. The best bounding results are obtained for values in between these extremes. For the numerical examples in the last section, we use $\xi = \frac{3}{2}$ as this value works well for a large range of problems.

The last piece in the puzzle of designing a new matrix splitting based SDP relaxation for the QAP is the construction of the corresponding quadratic semidefinite constraints. For the optimal matrix triple (B_Δ, B_∇, G) to problem (16), we have $G = B_\Delta^{\frac{1}{2}} B_\nabla^{\frac{1}{2}} = B_\nabla^{\frac{1}{2}} B_\Delta^{\frac{1}{2}}$. In the following relaxation framework, we implement the relation

$$\begin{bmatrix} X B_\Delta X^T & X G X^T \\ X G X^T & X B_\nabla X^T \end{bmatrix} \succeq \begin{bmatrix} X B_\Delta^{\frac{1}{2}} \\ X B_\nabla^{\frac{1}{2}} \end{bmatrix} \begin{bmatrix} B_\Delta^{\frac{1}{2}} X^T & B_\nabla^{\frac{1}{2}} X^T \end{bmatrix}$$

via utilization of the respective Schur complement inequality. Finally, we are in the position to present the SDP basis of the inverse interrelated matrix splitting relaxation, here referred to as *B-IIMS*:

$$\inf_{X \in \mathbb{D}^n, \ G, Y_\Delta, Y_\nabla \in \mathbb{S}^n} \quad \langle A, Y_\Delta - Y_\nabla \rangle + \langle C, X \rangle \tag{17a}$$

$$\text{s.t.} \quad \begin{bmatrix} I & B_\Delta^{\frac{1}{2}} X^T & B_\nabla^{\frac{1}{2}} X^T \\ X B_\Delta^{\frac{1}{2}} & Y_\Delta & G \\ X B_\nabla^{\frac{1}{2}} & G & Y_\nabla \end{bmatrix} \in \mathbb{S}_+^{3n}, \tag{17b}$$

$$\begin{bmatrix} \left(\tau(B)I - B_\Delta^{\frac{1}{2}} B_\nabla^{\frac{1}{2}} \right)^\dagger & U X^T \\ X U & \tau(B)I - G \end{bmatrix} \in \mathbb{S}_+^{2n}, \tag{17c}$$

$$\mathrm{diag}(Y_\Delta) = X \, \mathrm{diag}(B_\Delta), \quad \mathrm{diag}(Y_\nabla) = X \, \mathrm{diag}(B_\nabla),$$
$$\mathrm{diag}(G) = X \, \mathrm{diag}(B_\Delta^{\frac{1}{2}} B_\nabla^{\frac{1}{2}}), \tag{17d}$$

$$Y_\Delta e = X B_\Delta e, \quad Y_\nabla e = X B_\nabla e, \tag{17e}$$

where U denotes the orthogonal projection matrix to the column space of $\tau(B)I - B_\triangle^{\frac{1}{2}} B_\triangledown^{\frac{1}{2}}$, that is

$$U := \left(\tau(B)I - B_\triangle^{\frac{1}{2}} B_\triangledown^{\frac{1}{2}} \right)^{\dagger} \left(\tau(B)I - B_\triangle^{\frac{1}{2}} B_\triangledown^{\frac{1}{2}} \right).$$

3 Additional Cuts Based on Symmetric Functions

For many QAPs, it is possible to attain a significant improvement of the respective SDP relaxations by applying additional bounds to its optimization variables. In [15] and [18], Mittelmann, Peng and Li introduced new inequality constraints based on symmetric functions [14].

Definition 2. *A function* $f(v) \colon \mathbb{R}^n \to \mathbb{R}$ *is said to be symmetric if for any permutation matrix* $X \in \Pi^n$, *the relation* $f(v) = f(Xv)$ *holds.*

One of these functions, namely the additive function $f(v) = \langle e, v \rangle$, has already been used for the constraints (6d) and (17e). Other symmetric functions, that are useful for the construction of valid constraints, are the minimum and the maximum function as well as p-norms:

$$\forall v \in \mathbb{R}^n : \quad \min(v) = \min_{1 \leqslant i \leqslant n} v_i, \quad \max(v) = \max_{1 \leqslant i \leqslant n} v_i, \quad \mathcal{L}_p(v) = \left(\sum_{i=1}^n |v_i|^p \right)^{\frac{1}{p}}.$$

If applied to a matrix $M \in \mathbb{M}^{m,n}$, these operators act along the rows of the matrix, i.e.

$$\min(M) = [\min(e_1^T M), \ldots, \min(e_m^T M)]^T.$$

In [15, 18, 19] and also [10], the minimum and maximum functions are used to obtain linear bounds for several optimization variables.

$$e_i^T X \min(B) \leqslant (Y_+ - Y_-)_{ij} \leqslant e_i^T X \max(B) \quad \text{for} \quad 1 \leqslant i, j \leqslant n. \tag{18}$$

The same authors used constraints based on p-norm conditions for a further tightening of their relaxation frameworks:

$$\mathcal{L}_p(Y_+ - Y_-) \leqslant X \mathcal{L}_p(B_+ - B_-). \tag{19}$$

In [18], Peng et al. extended this approach by applying the same kind of constraint to each matrix variable Y_+ and Y_- as well as their sum.

3.1 Further Improvements

The linear inequalities given in (18) can be presented in the form of so-called sum-matrix inequalities. In accordance to [19], a sum-matrix is defined as:

Definition 3. *A matrix $M \in \mathbb{M}^n$ is called a sum-matrix if M is representable as*

$$M = ve^T + ew^T \tag{20}$$

for some $v, w \in \mathbb{R}^n$. In the symmetric case it is $v = w$.

Let $v_{\min} := \min(B)$ and $v_{\max} := \max(B)$ denote the vectors consisting of the minimal and maximal row elements of B, respectively. Condition (18) may then be rewritten as

$$X v_{\min} e^T \leqslant Y_+ - Y_- \leqslant X v_{\max} e^T.$$

Indeed, by the nonnegativity of X, it is straightforward to show that $v_{\min} e^T \leqslant B \leqslant v_{\max} e^T$ implies

$$X v_{\min} e^T = X v_{\min} e^T X^T \leqslant X B X^T \leqslant X v_{\max} e^T X^T = X v_{\max} e^T$$

and thus yields (18). The last observation motivates a further exploitation of sum-matrix inequalities for the attainment of tighter constraints. Define for instance

$$w_{\min} := \min(B - e v_{\min}^T) \quad \text{and} \quad w_{\max} := \max(B - e v_{\max}^T).$$

It obviously is $v_{\min} e^T + e w_{\min}^T \leqslant B \leqslant v_{\max} e^T + e w_{\max}^T$, which in turn gives the inequality constraints

$$X v_{\min} e^T + e w_{\min}^T X^T \leqslant Y_+ - Y_- \leqslant X v_{\max} e^T + e w_{\max}^T X^T.$$

By $w_{\min} \geqslant 0$ and $w_{\max} \leqslant 0$, it is apparent that these bounds are at least as good as the ones in (18).

For the linear inequalities based on the minimum respectively maximum function, Mittelmann and Peng [15] pointed out that - since the diagonal elements of Y_+ and Y_- are already described by the corresponding equality constraints - it is sufficient to account solely the off-diagonal variables. We further observe that, due to symmetry of B, the symmetric parts of the respective sum-matrices satisfy the same bounding conditions, i.e.

$$ve^T + ew^T \leqslant_{\text{off}} B \quad \implies \quad \tfrac{1}{2}(v+w)e^T + \tfrac{1}{2}e(v+w)^T \leqslant_{\text{off}} B. \tag{21}$$

Let the gap between a sum-matrix $ve^T + ew^T = (v_i + w_j)$ and an arbitrary real matrix $B = (b_{ij})$ of the same dimension be defined as

$$\delta_{\text{gap}}(B, v, w) := \sum_{\substack{i,j \\ i \neq j}} |b_{ij} - v_i - w_j|. \tag{22}$$

A suitable approach for the attainment of tight sum-matrix inequalities is the minimization of the respective gaps.

By $\delta_{\text{gap}}(B, v, w) = \delta_{\text{gap}}(B, \tfrac{1}{2}(v+w), \tfrac{1}{2}(v+w))$ and the implication in (21), it is apparently sufficient to concentrate on the lower respectively upper triangular elements of symmetric sum-matrices. The following linear programming problem can be used to compute lower and upper symmetric sum-matrix bounds for B that accompany minimal gaps:

$$\inf_{v_l, v_u \in \mathbb{R}^n} \quad \langle e, v_u - v_l \rangle$$
$$\text{s.t.} \quad v_l e^T + e v_l^T \leqslant_{\text{tri}} B \leqslant_{\text{tri}} v_u e^T + e v_u^T. \tag{23}$$

The solution vectors to this problem are used to implement the following linear inequality conditions

$$X v_l e^T + e v_l^T X^T \leqslant_{\text{tri}} Y_+ - Y_- \leqslant_{\text{tri}} X v_u e^T + e v_u^T X^T. \tag{24}$$

Suitable approaches for a further tightening of these bounds are the application of multiple varying sum-matrix inequalities and the construction of the same type of bounds for linear combinations of the respective matrix variables.

In a very similar way, it is possible to apply the sum-matrix reformulation technique from above for a tightening of the respective p-norm based constraints. However, numerical tests have shown that the effect of these extensions is relatively small. For this reason, we avoid the necessity of further computations for the determination of suitable sum-matrix updates.

4 Numerical Results

In the last section of this paper, we want to discuss the practical applicability of the presented relaxation strategy on the basis of numerical tests. For this purpose, we compare our own frameworks with one of the best performing low-dimensional SDP relaxations for the QAP, namely F-SVD which was introduced in [18].

The actual implementation of the SDP problems is realized via Yalmip [12] in Octave [7]. The used solver is SDPT3 [22]. For the presentation of the respective bounds, we follow the style in [18] and use the relative gap defined as

$$R_{\text{gap}} = 1 - \frac{\text{Lower bound from relaxation}}{\text{Optimal or best known feasible objective value}}.$$

The corresponding computation times are listed in seconds under the 'CPU' columns. Since the discussed relaxation frameworks are not designed for a specific class of QAPs, we chose the instances for our numerical tests arbitrarily from the quadratic assignment problem library [4]. The names in the column 'prob.' consists of three or four letters which indicate the names of their authors or contributors, and a number that gives their dimension. If the authors provided multiple problem instances for the same dimension, the respective instance is indicated by another letter at the end of the name. For more information on the naming scheme and the individual applications, see [4].

Prior to the comparison of the full frameworks with all additional constraints being applied, in Table 1, we compare the pure SDP relaxation bases presented in Sect. 2.

Table 1. Selected bounds for comparison of base relaxations

Prob.	B-SVD		B-IIMS	
	$R_{gap}(\%)$	CPU	$R_{gap}(\%)$	CPU
Esc16b	17.34	2	17.09	3
Had20	5.34	4	3.61	6
Kra32	42.64	13	32.27	36
LiPa40a	4.88	28	3.31	63
Nug30	12.39	11	9.93	22
Scr20	60.02	5	45.35	7
Ste36a	57.54	25	44.97	64
Tai30b	15.82	17	15.34	41
Tai50a	39.03	103	28.37	244
Tho40	14.94	40	13.06	91

The results presented in Table 1 reveal the significant differences between the considered relaxation approaches. As expected, the new relaxation program *B-IIMS* is more expensive than *B-SVD*. On the other hand, for many problem instances the additional computational costs pay off by resulting in significantly improved lower bounds.

For the attainment of the full relaxation frameworks, we extend the problems (6) and (17) by adding the constraints (24) together with the 2-norm conditions of the form (19) which are also present in *F-SVD*. We denote the full version of problem (6) by *F-SVD2*, since the only difference two the framework *F-SVD* from [18] is the utilization of different inequality constraints. Instead of the $8n^2 - 8n$ minimum and maximum bound inequalities applied in *F-SVD*, here we solely use the $n^2 - n$ constraints from (24). The full version of problem (17) applies the respective adaptations of the same constraints. The integration is realized simply by replacing the term $Y_+ - Y_-$ with $Y_\triangle - Y_\triangledown$. We follow the general naming scheme and denote this program by *F-IIMS*.

The results in Table 2 demonstrate the efficiency of the constraints in (24) compared to the significantly greater number of linear inequalities used in *F-SVD*. The difference between the bounds computed with *F-SVD* and *F-SVD2* is generally really small whereas the computation times of *F-SVD2* are noticeable shorter. Nevertheless, the results in Table 2 also reveal that the combined effect of the additional linear bounds applied in *F-SVD* is superior to the improvement of a single sum-matrix bound. The sheer number of additional constraints is difficult to beat.

The second observation from the results given in Table 2 is that the presence of the additional cuts diminishes the effect of the incorporation of the artificial inverse interrelation property. Among the tested QAP instances there is even an instance for which the application of the inverse interrelated matrix splitting approach is disadvantageous. Overall, the computational costs as well as the

Table 2. Selected bounds for comparison of full QAP relaxations

Prob.	F-SVD		F-SVD2		F-IIMS	
	R_{gap}(%)	CPU	R_{gap}(%)	CPU	R_{gap}(%)	CPU
Esc16b	5.82	3	6.73	3	6.56	4
Had20	2.53	8	2.67	6	2.32	8
Kra32	18.77	34	18.93	24	18.67	36
LiPa40a	0.11	74	0.24	42	0.23	74
Nug30	8.05	26	8.12	18	7.88	33
Scr20	16.18	10	16.24	7	16.01	8
Ste36a	19.06	55	19.35	40	18.55	55
Tai30b	12.69	32	12.88	24	13.50	48
Tai50a	21.43	250	21.58	197	21.49	240
Tho40	12.61	76	12.76	54	12.13	102

bounding quality of the frameworks *F-SVD* and *F-IIMS* are very similar. The latter relaxation, however, has a greater potential for even stronger bounds, for instance via the utilization of so-called QAP reformulations or the incorporation of a similar number of linear inequalities as used in *F-SVD*. An elaborate investigation of these possibilities is left for subsequent studies.

Acknowledgments. The author thanks the anonymous referees for their helpful remarks that lead to a better structure of the paper.

References

1. Albert, A.: Conditions for positive and nonnegative definiteness in terms of pseudoinverses. SIAM J. Appl. Math. **17**(2), 434–440 (1969). http://dx.doi.org/10.1137/0117041
2. Burer, S., Monteiro, R.D.: Local minima and convergence in low-rank semidefinite programming. Math. Program. **103**(3), 427–444 (2004). http://dx.doi.org/10.1007/s10107-004-0564-1
3. Burkard, R.E., Dell'Amico, M., Martello, S.: Assignment Problems. SIAM Philadelphia (2012). http://dx.doi.org/10.1137/1.9781611972238
4. Burkard, R.E., Karisch, S.E., Rendl, F.: QAPLIB - a quadratic assignment problem library. J. Global Optim. **10**(4), 391–403 (1997). http://dx.doi.org/10.1023/A:1008293323270
5. Çela, E.: The Quadratic Assignment Problem: Theory and Algorithms. Combinatorial Optimization, vol. 1. Springer, New York (1998). http://dx.doi.org/10.1007/978-1-4757-2787-6
6. Ding, Y., Wolkowicz, H.: A low-dimensional semidefinite relaxation for the quadratic assignment problem. Math. Oper. Res. **34**(4), 1008–1022 (2009). http://dx.doi.org/10.1287/moor.1090.0419

7. Eaton, J.W., Bateman, D., Hauberg, S.: GNU Octave version 3.0.1 manual: a high-level interactive language for numerical computations. CreateSpace Independent Publishing Platform (2009). http://www.gnu.org/software/octave/doc/interpreter

8. Edwards, C.S.: A branch and bound algorithm for the Koopmans-Beckmann quadratic assignment problem. Combinatorial Optimization II. Mathematical Programming, vol. 13, pp. 35–52. Springer, New York (1980). http://dx.doi.org/10.1007/BFb0120905

9. de Klerk, E., Sotirov, R.: Exploiting group symmetry in semidefinite programming relaxations of the quadratic assignment problem. Math. Program. **122**(2), 225–246 (2010). http://dx.doi.org/10.1007/s10107-008-0246-5

10. de Klerk, E., Sotirov, R., Truetsch, U.: A new semidefinite programming relaxation for the quadratic assignment problem and its computational perspectives. INFORMS J. Comput. **27**(2), 378–391 (2015). http://dx.doi.org/10.1287/ijoc.2014.0634

11. Koopmans, T.C., Beckmann, M.: Assignment problems and the location of economic activities. Econometrica **25**(1), 53–76 (1957). http://dx.doi.org/10.2307/1907742

12. Löfberg, J.: YALMIP: a toolbox for modeling and optimization in MATLAB. In: Proceedings of the CACSD Conference (2004). http://users.isy.liu.se/johanl/yalmip

13. Loiola, E.M., de Abreu, N.M.M., Boaventura-Netto, P.O., Hahn, P., Querido, T.: A survey for the quadratic assignment problem. Eur. J. Oper. Res. **176**(2), 657–690 (2007). http://dx.doi.org/10.1016/j.ejor.2005.09.032

14. Macdonald, I.G.: Symmetric Functions and Hall Polynomials, 2nd edn. The Clarendon Press, Oxford University Press, New York (1995)

15. Mittelmann, H., Peng, J.: Estimating bounds for quadratic assignment problems associated with Hamming and Manhattan distance matrices based on semidefinite programming. SIAM J. Optim. **20**(6), 3408–3426 (2010). http://dx.doi.org/10.1137/090748834

16. Moore, E.H.: On the reciprocal of the general matrix. Bull. Am. Math. Soc. **26**, 394–395 (1920)

17. Pardalos, P.M., Rendl, F., Wolkowicz, H.: The quadratic assignment problem: a survey and recent developments. In: Quadratic Assignment and Related Problems. DIMACS Series in Discrete Mathematics and Theoretical Computer Science, vol. 16, pp. 1–42. American Mathematical Society (AMS) (1994)

18. Peng, J., Mittelmann, H., Li, X.: A new relaxation framework for quadratic assignment problems based on matrix splitting. Math. Program. Comput. **2**(1), 59–77 (2010). http://dx.doi.org/10.1007/s12532-010-0012-6

19. Peng, J., Zhu, T., Luo, H., Toh, K.C.: Semi-definite programming relaxation of quadratic assignment problems based on nonredundant matrix splitting. Comput. Optim. Appl. (2014). http://dx.doi.org/10.1007/s10589-014-9663-y

20. Penrose, R., Todd, J.A.: A generalized inverse for matrices. Math. Proc. Camb. Phil. Soc. **51**(03), 406–413 (1955). http://dx.doi.org/10.1017/S0305004100030401

21. Rendl, F., Sotirov, R.: Bounds for the quadratic assignment problem using the bundle method. Math. Program. **109**(2–3), 505–524 (2007). http://dx.doi.org/10.1007/s10107-006-0038-8

22. Tütüncü, R.H., Toh, K.C., Todd, M.J.: Solving semidefinite-quadratic-linear programs using SDPT3. Math. Program. **95**(2), 189–217 (2003). http://dx.doi.org/10.1007/s10107-002-0347-5

23. Yang, L., Sun, D., Toh, K.C.: SDPnal+: a majorized semismooth Newton-CG augmented Lagrangian method for semidefinite programming with nonnegative constraints. Math. Program. Comput. (2015). http://dx.doi.org/10.1007/s12532-015-0082-6
24. Zhao, Q., Karisch, S.E., Rendl, F., Wolkowicz, H.: Semidefinite programming relaxations for the quadratic assignment problem. J. Comb. Optim. **2**(1), 71–109 (1998). http://dx.doi.org/10.1023/A:1009795911987

Global Optimization of H_∞ Problems: Application to Robust Control Synthesis Under Structural Constraints

Dominique Monnet$^{(\boxtimes)}$, Jordan Ninin, and Benoit Clement

Lab-STICC, IHSEV Team, ENSTA-Bretagne,
2 rue Francois Verny, 29806 Brest, France
dominique.monnet@ensta-bretagne.org,
{jordan.ninin,benoit.clement}@ensta-bretagne.fr

Abstract. In this paper, a new technique to compute a synthesis structured Robust Control Law is developed. This technique is based on global optimization methods using a Branch-and-Bound algorithm. The original problem is reformulated as a min/max problem with non-convex constraint. Our approach uses interval arithmetic to compute bounds and accelerate the convergence.

1 Context

Controlling an autonomous vehicle or a robot requires the synthesis of control laws for steering and guiding. To generate efficient control laws, a lot of specifications, constraints and requirements have been translated into norm constraints and then into a constraint feasibility problem. This problem has been solved, sometimes with relaxations, using numerical methods based on LMI (Linear Matrix Inequalities) or SDP (Semi Definite Program) [2,3]. The main limitation of these approaches is the complexity of the controller for implementation in an embedded system. However, if a physical structure is imposed on the control law in order to make the implementation easier, the synthesis of this robust control law is much more complex. And this complexity has been identified as a key issue for several years. An efficient first approach based on local non-smooth optimization was given by Apkarian and Noll [1].

In this talk, we will present a new approach based on **global optimization** in order to generate **robust control laws**.

2 H_∞ Control Synthesis Under Structural Constraints

We illustrate our approach with an example of the control of a periodic second order system G with a PID controller K subjected to two frequency constraints on the error e and on the command u of the closed-loop system, see Fig. 1. The objective is to find $\boldsymbol{k} = (k_p, k_i, k_d)$ to stabilize the closed-loop system while minimizing the H_∞ norm of the controlled system to ensure robustness.

© Springer International Publishing Switzerland 2016
I.S. Kotsireas et al. (Eds.): MACIS 2015, LNCS 9582, pp. 550–554, 2016.
DOI: 10.1007/978-3-319-32859-1_46

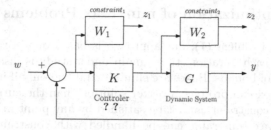

Fig. 1. 2-blocks H_∞ problem

The H_∞ norm of a dynamic system P is defined as follows:

$$||P||_\infty = \sup_\omega(\sigma_{\max}(P(j\omega))),$$

with $\sigma_{\max}$ the greatest singular value of the transfer function P and j the imaginary unit.

In our particular case, the closed-loop system can be interpreted as two SISO systems (Single In Single Out). The H_∞ norm of a SISO system is the maximum of the absolute value of the transfer function. Indeed, to minimize the H_∞ norm of our example, we need to solve the following min/max problem:

$$
\begin{cases}
\min_k \max \left(\sup_\omega \left| \dfrac{W_1(j\omega)}{1 + G(j\omega)K(j\omega)} \right|, \sup_\omega \left| \dfrac{W_2(j\omega)K(j\omega)}{1 + G(j\omega)K(j\omega)} \right| \right), \\
\\
s.t. \qquad \text{The closed-loop system must be stable.}
\end{cases}
\tag{1}
$$

The stability constraint of a closed-loop is well-known: the roots of denominator part of the transfer function $\frac{1}{1+G(s)K(s)}$ must have a non-positive real part [4]. Using Routh-Hurwitz stability criterion [5], this constraint can be reformulated as a set of non-convex constraints.

Proposition 1. *Let us consider a polynomial $Q(s) = a_n s^n + a_{n-1} s^{n-1} + \cdots + a_1 s + a_0$. The real parts of its roots are negative if the entries in the first column of the following table are positive:*

$v_{1,1} = a_n$	$v_{1,2} = a_{n-2}$	$v_{1,3} = a_{n-4}$	$v_{1,4} = a_{n-6}$ ...
$v_{2,1} = a_{n-1}$	$v_{2,2} = a_{n-3}$	$v_{2,3} = a_{n-5}$	$v_{2,4} = a_{n-7}$ ...
$v_{3,1} = \frac{-1}{v_{2,1}} \begin{vmatrix} v_{1,1} & v_{1,2} \\ v_{2,1} & v_{2,2} \end{vmatrix}$	$v_{3,2} = \frac{-1}{v_{2,1}} \begin{vmatrix} v_{1,1} & v_{1,3} \\ v_{2,1} & v_{2,3} \end{vmatrix}$	$v_{3,3} = \frac{-1}{v_{2,1}} \begin{vmatrix} v_{1,1} & v_{1,4} \\ v_{2,1} & v_{2,4} \end{vmatrix}$	...
$v_{4,1} = \frac{-1}{v_{3,1}} \begin{vmatrix} v_{2,1} & v_{2,2} \\ v_{3,1} & v_{3,2} \end{vmatrix}$	$v_{4,2} = \frac{-1}{v_{3,1}} \begin{vmatrix} v_{2,1} & v_{2,3} \\ v_{3,1} & v_{3,3} \end{vmatrix}$	...	...
$v_{5,1} = \frac{-1}{v_{4,1}} \begin{vmatrix} v_{3,1} & v_{3,2} \\ v_{4,1} & v_{4,2} \end{vmatrix}$	...	...	...
$\vdots$	$\cdot\cdot$	$\cdot\cdot\cdot$	$\cdot\cdot$

Indeed, applying Proposition 1 with $Q(s) = 1 + G(s)K(s)$, the H_∞ control synthesis under structural constraint is reformulated as a min/max problem with non-convex constraints.

3   Global Optimization of min/max Problems

In order to solve Problem (1), our approach is based on a Branch-and-Bound technique [7]. At each iteration, the domain under study is bisected to improve the computation of bounds. Boxes are eliminated if and only if it is certified that no point in the box can produce a better solution than the current best one, or that at least one constraint cannot be satisfied by any point in such a box.

The non-convex constraint can be handled with constraint programming techniques. In our approach, we use the ACID algorithm [8] which reduces the width of the boxes and so accelerates the convergence of the branch-and-bound.

The key point of our approach concerns the computation of the bounds of the objective function. In our example, the objective function can be reformulated as the following expression, with $x = (k_p, k_i, k_d)$:

$$f(x) = \sup_{\omega \in [\omega_{\min}, \omega_{\max}]} g(x, \omega). \tag{2}$$

At each iteration, Algorithm 1 is used to compute a lower bound of this function over a box $[x]$. This algorithm is also a branch-and-bound algorithm based on Interval Arithmetic. But, for not wasting time, we limit the maximum number of iterations for computing faster lower bounds. Each element $([\omega], ub_\omega)$ stored in $\mathcal{L}$ is composed of: (i) $[\omega]$ a sub-interval of $[\omega_{\min}, \omega_{\max}]$ and (ii) ub_ω an upper bound of g over $[x] \times [\omega]$.

Algorithm 1. Computation of bounds of f over a box $[x]$

Require: g: the function under study (see Equation 2); x: an initial box; $\mathcal{L}$: the list of boxes; *nbIter*: the maximal number of iterations.

1: Initialization: $(lb_{out}, ub_{out}) = (-\infty, \infty)$.
2: **for** nb := 1 **to** nbIter **do**
3:      Extract an element (ω, ub_ω) from $\mathcal{L}$.
4:      Bisect ω into two sub-boxes ω_1 and ω_2.
5:      **for** i:=1 **to** 2 **do**
6:          Compute lb_{ω_i} and ub_{ω_i} a lower and an upper bound of $g(x, \omega)$ over $[x] \times [\omega_i]$ using Interval Arithmetic techniques [6].
7:          **if** $lb_{\omega_i} > lb_{out}$ **then**
8:             $lb_{out} := lb_{\omega_i}$, {Update the best lower bound}
9:             Remove from $\mathcal{L}$ all the elements j such as $ub_{\omega_j} < lb_{out}$,
10:          **end if**
11:          **if** $ub_{\omega_i} > lb_{out}$ **then**
12:             Add $(\omega_i, ub_{\omega_i})$ in $\mathcal{L}$,
13:          **end if**
14:      **end for**
15: **end for**
16: $ub_{out} := \max_{(\omega_i, ub_{\omega_i}) \in \mathcal{L}} ub_{\omega_i}$
17: **return** (lb_{out}, ub_{out}): a lower and an upper bound of f over x.

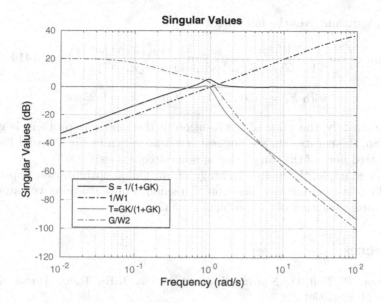

Fig. 2. Weighting functions and singular values of the solution (Color figure online)

Thanks to Interval Analysis, at the end of Algorithm 1, we can ensure that the value of the maximum of f over $[x]$ is included in $[lb_{out}, ub_{out}]$.

4   Application

In our example, we consider a second-order system and weighting functions W_1 and W_2 penalizing the error signal and control signal respectively:

$$G(s) = \frac{1}{s^2 + 1.4s + 1}, \ K(s) = k_p + \frac{k_i}{s} + \frac{k_d s}{1 + s}.$$

$$W_1(s) = \frac{s + 100}{100s + 1}, \qquad W_2(s) = \frac{10s + 1}{s + 10}.$$

We want to find k_p, k_i and k_d the coefficients of the structured controller K such that the closed-loop system respects the constraints:

$$\max \left(\|\frac{W_1(j\omega)}{1 + G(j\omega)K(j\omega)}\|_\infty, \|\frac{W_2(j\omega)K(j\omega)}{1 + G(j\omega)K(j\omega)}\|_\infty \right) \leq 1$$

The control is bounded by $[-2, 2]$, and we limit the interval of ω to $[10^{-2}, 10^2]$.

Our algorithm gives the following result:

$$\max\left(\sup_{\omega}\left|\frac{W_1(j\omega)}{1 + G(j\omega)K(j\omega)}\right|, \sup_{\omega}\left|\frac{W_2(j\omega)K(j\omega)}{1 + G(j\omega)K(j\omega)}\right|\right) = 2.1414$$

with $k_p = -0.0425$, $k_i = 0.4619$, $k_d = 0.2566$.

Unfortunately, the value of the solution of the min/max problem is greater than 1. So, the constraints are not respected as shown in Fig. 2 (solid lines are above dotted lines of the same color at some frequencies).

In this example, the main advantage of our global optimization approach is that unlike classical method based on non-smooth optimization, we can certify that no robust solution of our problem exists.

References

1. Apkarian, P., Noll, D.: Nonsmooth H_∞ synthesis. IEEE Trans. Autom. Control **51**(1), 71–86 (2006)
2. Arzelier, D., Clement, B., Peaucelle, D.: Multi-objective H_2/H_∞/impulse-to-peak control of a space launch vehicle. Eur. J. Control **12**(1), 57–70 (2006)
3. Boyd, S., El Ghaoui, L., Feron, E., Balakrishnan, V.: Linear Matrix Inequalities in System and Control Theory. SIAM, vol. 15. SIAM, Philadelphia (1994)
4. Petersen, I.R., Tempo, R.: Robust control of uncertain systems: classical results and recent developments. Automatica **50**(5), 1315–1335 (2014)
5. Hurwitz, A.: Ueber die Bedingungen, unter welchen eine Gleichung nur Wurzeln mit negativen reellen Theilen besitzt. Math. Ann. **46**(2), 273–284 (1895)
6. Jaulin, L., Kieffer, M., Didrit, O., Walter, E.: Applied Interval Analysis, with Examples in Parameter and State Estimation Robust Control and Robotics. Springer, London (2001)
7. Ninin, J.: Optimisation Globale basée sur l'Analyse d'Intervalles: Relaxation Affine et Limitation de la Mémoire. PhD thesis, Institut National Polytechnique de Toulouse, Toulouse (2010)
8. Trombettoni, G., Chabert, G.: Constructive interval disjunction. In: Principles and Practice of Constraint Programming–CP 2007, pp. 635–650. Springer (2007)

Global Optimization Based on Contractor Programming: An Overview of the *IBEX* Library

Jordan Ninin[✉]

Lab-STICC, IHSEV Team, ENSTA-Bretagne,
2 rue Francois Verny, 29806 Brest, France
jordan.ninin@ensta-bretagne.fr

Abstract. *IBEX* is an open-source C++ library for constraint processing over real numbers. It provides reliable algorithms for handling nonlinear constraints. In particular, roundoff errors are also taken into
account. It is based on interval arithmetic and affine arithmetic. The
main feature of *IBEX* is its ability to build strategies declaratively
through the contractor programming paradigm. It can also be used as a
black-box solver or with an AMPL interface. Two emblematic problems
that can be addressed are: (i) **System solving**: A guaranteed enclosure
for each solution of a system of (nonlinear) equations is calculated; (ii)
Global optimization: A global minimizer of some function under nonlinear constraints is calculated with guaranteed and reliable bounds on
the objective minimum.

1 Kernel of *IBEX*

Considering sets in place of single points is not a common point of view in the
Mathematical Programming communities. Unlike classical optimization tools,
IBEX library relies on set-membership approach [1]. These methods and algorithms do not consider single numerical values, or floating-point numbers, but
manipulate sets. The interval arithmetic offers a solid theoretical basis to represent and to calculate with subsets of $\mathbb{R}^n$.

1.1 Interval Arithmetic

An interval is a closed connected subset of $\mathbb{R}$. A non-empty interval $[x]$ can be
represented by its endpoints: $[x] = [\underline{x}, \overline{x}] = \{x : \underline{x} \leq x \leq \overline{x}\}$ where $\underline{x} \in \mathbb{R} \cup \{-\infty\}$,
$\overline{x} \in \mathbb{R} \cup \{+\infty\}$ and $\underline{x} \leq \overline{x}$. The set of intervals is denoted by $\mathbb{IR}$.

In *IBEX*, three external implementations of the Interval Arithmetic can be
linked: *Filib++* [12], *Profil-Bias* [10], *Gaol* [9]. To improve the portability and
the compatibility of *IBEX*, a homemade interval arithmetic is available without
using low-level functionality which can be dependent on the architecture of the
CPU. All the arithmetic has been patched to comply with the new IEEE 1788-
2015 Standard for Interval Arithmetic [2].

© Springer International Publishing Switzerland 2016
I.S. Kotsireas et al. (Eds.): MACIS 2015, LNCS 9582, pp. 555–559, 2016.
DOI: 10.1007/978-3-319-32859-1_47

1.2   Affine Arithmetic

The affine arithmetic is a technique to compute lower and upper bounds of functions over an interval. It is based on the same principle as interval arithmetic excepted that the quantities are represented by an affine form, see [17].

As in interval arithmetic, the usual operations and functions are extended to deal with affine forms. For the non-affine operations and some transcendental functions, such as the square root, the logarithm, the inverse and the exponential, several algorithms exist depending on the use: if you focus on the computation of the bounds, or on performance, or on linear approximation, or on the reliability, etc. Indeed, seven different versions are available to satisfy all the needs of the user [16].

1.3   Contractor Programming

The concept of *contractor* is directly inspired by the ubiquitous concept of filtering algorithms in constraint programming [7]. The strength of *IBEX* lies mainly in this concept. Every algorithm in *IBEX* is included as a *Contractor*.

Definition 1. *Let* $\mathbb{X} \subseteq \mathbb{R}^n$ *be a feasible region.*
The operator $\mathcal{C}_{\mathbb{X}} : \mathbb{IR}^n \to \mathbb{IR}^n$ *is a contractor for* $\mathbb{X}$ *if:*

$$\forall [\boldsymbol{x}] \in \mathbb{IR}^n, \begin{cases} \mathcal{C}_{\mathbb{X}}([\boldsymbol{x}]) \subseteq [\boldsymbol{x}], & \textit{(contraction)} \\ \mathcal{C}_{\mathbb{X}}([\boldsymbol{x}]) \cap \mathbb{X} \supseteq [\boldsymbol{x}] \cap \mathbb{X}. & \textit{(completeness)} \end{cases}$$

This definition means that: (i) Filtering gives a subdomain of the input domain $[\boldsymbol{x}]$; (ii) the resulting subdomain $\mathcal{C}_{\mathbb{X}}([\boldsymbol{x}])$ contains all feasible points. No solution is *lost*. A contractor is defined by a feasible region $\mathbb{X}$, and its purpose is to eliminate a part of a domain which is not in $\mathbb{X}$.

All set operators can be extended to contractors. For example, the intersection of two contractors creates a contractor for the intersection of these two sets. In the same way, the hull of two contractors creates a contractor for the disjunction of these constraints.

Definition 2. *Let* $\mathbb{X}$ *and* $\mathbb{Y} \subseteq \mathbb{R}^n$ *be two feasible regions.*

$$\begin{aligned} CtcCompo: \quad & (\mathcal{C}_{\mathbb{X}} \cap \mathcal{C}_{\mathbb{Y}})([\boldsymbol{x}]) = \mathcal{C}_{\mathbb{X}}([\boldsymbol{x}]) \cap \mathcal{C}_{\mathbb{Y}}([\boldsymbol{x}]) \\ CtcUnion: \quad & (\mathcal{C}_{\mathbb{X}} \cup \mathcal{C}_{\mathbb{Y}})([\boldsymbol{x}]) = \mathcal{C}_{\mathbb{X}}([\boldsymbol{x}]) \cup \mathcal{C}_{\mathbb{Y}}([\boldsymbol{x}]) \\ CtcFixPoint: \quad & \mathcal{C}^{\infty} = \mathcal{C} \circ \mathcal{C} \circ \mathcal{C} \circ \dots \end{aligned}$$

Using these properties, interacting, combining and merging heterogeneous techniques becomes simple.

2   Lists of Contractors

In this section, a small part of all contractors available in *IBEX* is described:

CtcFwdBwd and CtcHC4 : the *atomic* contractors

Forward-backward is a classical algorithm in constraint programming for contracting quickly with respect to one equality or inequality constraint. See, e.g., [4,8]. However, the more occurrences of variables in the expression of the (in)equality, the less accurate the contraction. Hence, this contractor is often used as an "atomic" contractor embedded in a higher-level operator like Propagation or Shaving.

HC4 is another classical algorithm of constraint programming. It allows to contract according to a system of constraints. The basic idea is to calculate the fix point of a set of n contractors $C_1, \ldots, C_n$, i.e. $(C \circ \cdots \circ C)^\infty$, without calling a contractor when it is unnecessary.

Ctc3BCid and CtcAcid: the shaving contractors

Ctc3BCID is a shaving operator. It is an implementation of the 3BCID algorithm defined in [18]. The shaving operator applies a contractor C on sub-parts (*slices*) of the input box. If a slice is entirely eliminated by C, the input box can be contracted by removing the slice from the box. This operator can be viewed as a generalization of the SAC algorithm in discrete domains [5]. This concept with continuous constraint was first introduced in [11] with the "3B" algorithm. In [11], the sub-contractor C was *CtcHC4*. In *IBEX*, the idea was extended and Ctc3BCID can be combined with every contractor.

CtcAcid is an adaptive version of the 3BCID contractor. The *handled* number of variables for which a shaving will be performed is adaptively tuned. The ACID algorithm alternates: (i) small tuning phases (during e.g. 50 nodes) where the shaving is called on a number of variables double of the last tuned value (all variables in the first tuning phase); statistics are computed in order to determine an average number of interesting calls, The number of variables to be handled in the next running phase is computed at the end of the tuning phase; (ii) and large running phases (during e.g. 950 nodes) where 3BCID is called with the number of variables determined during the last tuning phase.

CtcPolytopeHull:Contractors based on Linear relaxation

Considering a system of linear inequalities, CtcPolytopeHull gives the possibility to contract a box to the hull of the polytope (the set of feasible points). This contractor calls a linear solver linked with *IBEX* (CPLEX, Soplex or CLP) to calculate for each variable x_i, the following bounds: $\min\limits_{Ax \leq b \wedge x \in [x]} x_i$ and $\max\limits_{Ax \leq b \wedge x \in [x]} x_i$, where $[x]$ is the box to be contracted.

If some constraints are nonlinear, Linearization procedures can automatically linearize the non-linear constraints. There exists some built-in linearization techniques in *IBEX*:

(i) LinearRelaxXTaylor: a corner-based Taylor relaxation [3];
(ii) LinearRelaxAffine2: a relaxation based on affine arithmetic [15];
(iii) LinearRelaxCombo: a combination of the two previous techniques (the polytope is basically the intersection of the polytopes calculated by each technique).

CtcQInter: the q-relaxed intersection

If a set of constraints is based on physical data, it is not uncommon that some of this data is wrong. In this situation, the q-relaxed intersection of contractors can be applied to this problem.

The q-relaxed intersection of m subsets $\mathbb{X}_1, \ldots, \mathbb{X}_m$ of $\mathbb{R}^n$ is the set of all $x \in \mathbb{R}^n$ which belong to at least $(m - q)$ $\mathbb{X}_i$. We denote it by $\mathbb{X}^{\{q\}} = \bigcap^{\{q\}} \mathbb{X}_i$.

Since the q-relaxed intersection is a set operator, we have extended this notion to contractors: $\left(\bigcap^{\{q\}} \mathcal{C}_{\mathbb{X}_i}\right)([\boldsymbol{x}]) = \bigcap^{\{q\}} \left(\mathcal{C}_{\mathbb{X}_i}([\boldsymbol{x}])\right)$. This contractor allows modeling the possibility of invalid constraints: it can also be used for robust optimization.

In [6], Carbonnel et al. found an algorithm with a complexity $\theta(nm^2)$ to compute a box which contains the q-relaxed intersection of m boxes of $\mathbb{R}^n$.

CtcExist and CtcForAll: the contractors with quantifiers

Another possibility is to project a subset of $\mathbb{R}^n$ over one or more dimensions. For example, if a constraint needs to be satisfied for all values of a parameter in a given set, such as $\{x \in \mathbb{R}^n : \forall t \in \mathbb{X} \subseteq \mathbb{R}^m, g(x,t) \leq 0\}$, few solvers are available to deal with it. Another example is when a constraint needs to be satisfied for at least one value of the parameter, such as $\{x \in \mathbb{R}^n : \exists t \in \mathbb{X} \subseteq \mathbb{R}^m, g(x,t) \leq 0\}$.

Two operators are defined as contractors. The first one is CtcForAll and the second one is CtcExist. CtcForAll contracts each part of $[\boldsymbol{x}]$ which is contracted by $\mathcal{C}([\boldsymbol{x}] \times \{y\})$ for any $y \in \mathbb{Y}$. Indeed, each part $[\boldsymbol{a}]$ of $[\boldsymbol{x}]$, such as $\exists y \in \mathbb{Y}, ([\boldsymbol{a}], y) \notin \mathbb{Z}$, can be removed. Thus, each part $[\boldsymbol{b}]$ of $[\boldsymbol{x}]$, such as $\forall y \in \mathbb{Y}, ([\boldsymbol{b}], y) \in \mathbb{Z}$, is kept. A similar algorithm is used in CtcExist.

3   Optimization Strategies

To find the global optimum of a problem in a reliable way, *IBEX* included several global optimization strategies. The principle of these algorithms is based on a branch-and-bound technique [13]. At each iteration, the domain under study is bisected to improve the computation of bounds. Boxes are eliminated if and only if it is certified that no point in the box can produce a better solution than the current best one, or that at least one constraint cannot be satisfied by any point in such a box. To accelerate convergence, contractors are used at each iteration to prune the width of boxes.

The default optimization strategy is based on a mathematical model of the optimization problem that needs to be solved. This model can be constructed directly using the symbolic kernel of *IBEX* or using the AMPL interface. The default contractor inside is the following: CtcAcid $\cap$ (CtcPolytopeHull $\cap$ CtcHC4)$^\infty$.

The performance of the default optimizer is comparable to the global optimizer *BARON*. However, our approach is completely reliable, and can deal with more general problems (with trigonometric function).

Moreover, a general pattern is also available [14]. This pattern only requires a contractor defined on the feasible set X of the problem and another contractor on the unfeasible set $\overline{X}$. Indeed, *IBEX* can address more complex real-life problems with disjunction constraint, quantifiers, outliers, non-linearity, trigonometric functions, etc.

References

1. IBEX: a C++ numerical library based on interval arithmetic and constraint programming. http://www.ibex-lib.org
2. IEEE SA - 1788-2015 - IEEE Standard for Interval Arithmetic
3. Araya, I., Trombettoni, G., Neveu, B.: A contractor based on convex interval Taylor. In: Beldiceanu, N., Jussien, N., Pinson, É. (eds.) CPAIOR 2012. LNCS, vol. 7298, pp. 1–16. Springer, Heidelberg (2012)
4. Benhamou, F., Granvilliers, L.: Continuous and interval constraints. Handb. Constraint Program. **2**, 571–603 (2006)
5. Bessiere, C., Debruyne, R.: Theoretical analysis of singleton arc consistency. In: Proceedings of ECAI-04 workshop on Modeling and Solving Problems with Constraints (2004)
6. Carbonnel, C., Trombettoni, G., Vismara, P., Chabert, G.: Q-intersection algorithms for constraint-based robust parameter estimation. In AAAI'14-Twenty-Eighth Conference on Artificial Intelligence, pp. 26–30 (2014)
7. Chabert, G., Jaulin, L.: Contractor programming. Artif. Intell. **173**(11), 1079–1100 (2009)
8. Collavizza, H., Delobel, F., Rueher, M.: A note on partial consistencies over continuous domains. In: Maher, M.J., Puget, J.-F. (eds.) CP 1998. LNCS, vol. 1520, pp. 147–161. Springer, Heidelberg (1998)
9. Goualard, F.: Gaol: NOT Just Another Interval Library. University of Nantes, France (2005)
10. Knuppel, O.: PROFIL/BIAS-a fast interval library. Computing **53**(3–4), 277–287 (1994)
11. Lhomme, O.: Consistency techniques for numeric CSPs. In: IJCAI, vol. 93, pp. 232–238. Citeseer (1993)
12. Nehmeier, M., von Gudenberg, J.W.: FILIB++, expression templates, the coming interval standard. Reliable Comput. **15**(4), 312–320 (2011)
13. Ninin, J.: Optimisation Globale basée sur l'Analyse d'Intervalles: Relaxation Affine et Limitation de la Mémoire. Ph.D. thesis, Institut National Polytechnique de Toulouse, Toulouse (2010)
14. Ninin, J., Chabert, G.: Global optimization based on contractor programming. In: XII GLOBAL OPTIMIZATION WORKSHOP, pp. 77–80 (2014)
15. Ninin, J., Messine, F., Hansen, P.: A reliable affine relaxation method for global optimization. 4OR **13**(3), 247–277 (2014)
16. Stolfi, J., de Figueiredo, L.: Self-validated numerical methods and applications. In: Monograph for 21st Brazilian Mathematics Colloquium (1997)
17. Stolfi, J., de Figueiredo, L.H.: Self-validated numerical methods and applications. In: Monograph for 21st Brazilian Mathematics Colloquium. IMPA/CNPq, Rio de Janeiro, Brazil (1997)
18. Trombettoni, G., Chabert, G.: Constructive interval disjunction. In: Bessière, Christian (ed.) CP 2007. LNCS, vol. 4741, pp. 635–650. Springer, Heidelberg (2007)

The Bernstein Branch-and-Prune Algorithm for Constrained Global Optimization of Multivariate Polynomial MINLPs

Bhagyesh V. Patil[(✉)]

Cambridge Centre for Advanced Research in Energy Efficiency in Singapore,
50 Nanyang Avenue, Singapore 639798, Singapore
bhagyesh.patil@gmail.com

Abstract. This paper address the global optimization problem of polynomial mixed-integer nonlinear programs (MINLPs). A improved branch-and-prune algorithm based on the Bernstein form is proposed to solve such MINLPs. The algorithm use a new pruning feature based on the Bernstein form, called the Bernstein box and Bernstein hull consistency. The proposed algorithm is tested on a set of 16 MINLPs chosen from the literature. The efficacy of the proposed algorithm is brought out via numerical studies with the previously reported Bernstein algorithms and several state-of-the-art MINLP solvers.

1 Introduction

Optimizing a MINLP is a challenging task and has been a point of attraction to many researchers from academia as well as industry. This work present a new solution procedure for such MINLPs. Typically, this work addresses MINLPs of the following form:

$$\min_{x} \quad f(x)$$

subject to

$$g_i(x) \leq 0, \quad i = 1, 2, \ldots, m$$
$$h_j(x) = 0, \quad j = 1, 2, \ldots, n \tag{1}$$
$$x_k \in \mathbf{x} \subseteq \mathbb{R}, \quad k = 1, 2, \ldots, l_d$$
$$x_k \in \{0, 1, \ldots, q\} \subset \mathbb{Z}, \quad k = l_d + 1, \ldots, l$$

where $f : \mathbb{R}^l \mapsto \mathbb{R}$ is the (possibly nonlinear polynomial) objective function, $g_i : \mathbb{R}^l \mapsto \mathbb{R}(i = 1, 2, \ldots, m)$, and $h_j : \mathbb{R}^l \mapsto \mathbb{R}(j = 1, 2, \ldots, n)$ are the (possibly nonlinear polynomial) inequality and equality constraint functions. Further, $\mathbf{x} := [\underline{x}, \overline{x}]$ is an interval in $\mathbb{R}$, $x_k(k = 1, 2, \ldots, l_d)$ are continuous decision variables, and the rest of $x_k(k = l_d + 1, \ldots, l)$ are integer decision variables with values 0 to q, $q \in \mathbb{Z}$.

Several techniques exist in literature to solve MINLP problems. Most of these techniques either decompose and reformulate the original problem (1) into a

© Springer International Publishing Switzerland 2016
I.S. Kotsireas et al. (Eds.): MACIS 2015, LNCS 9582, pp. 560–575, 2016.
DOI: 10.1007/978-3-319-32859-1_48

series of mixed-integer linear programs (MILPs) and nonlinear programs (NLPs), or they attempt to solve a NLP relaxation in a branch-and-bound framework. The interested reader can refer [2] and references therein for more specific details about these techniques.

Recently, global optimization algorithms based on the Bernstein polynomial approach has been proposed (see [11,12]), and found to be very effective in solving small to medium dimensional polynomial MINLPs of the form (1). The current scope of the work involve systematic extension of the above proposed Bernstein global optimization algorithms to form a new improved algorithm. The improved algorithm is of a branch-and-prune type and use consistency techniques (constraint propagation) based on the Bernstein form. The consistency techniques prune regions from a solution search space that surely do not contain the global minimizer(s) [5,6], hence this improved algorithm is defined as the Bernstein branch-and-prune algorithm for the MINLPs (that is, BBPMINLP). The algorithm BBPMINLP has some new features: (a) the consistency techniques are framed in a context of the Bernstein form, namely Bernstein box consistency and Bernstein hull consistency. (b) a new form of domain contraction step based on the application of Bernstein box and Bernstein hull consistency to a constraint $f(x) \leq \tilde{f}$ is introduced. The main feature of the algorithm BBPMINLP is, all operations (branching and pruning) are done using the Bernstein coefficients.

The performance of the algorithm BBPMINLP is compared with the earlier reported Bernstein algorithms BMIO [12] and IBBBC [11], as well as with several state-of-the-art MINLP solvers on a collection of 16 test problems chosen from the literature. The performance comparison is made on the basis of the number of boxes processed (between the algorithms BMIO, IBBBC, and BBPMINLP), and ability to locate a correct global minimum (between state-of-the-art MINLP solvers and the algorithm BBPMINLP). The findings are reported at the end of the paper.

The rest of the paper is organized as follows. In Sect. 2, the reader is introduced to some background of the Bernstein form. In Sect. 3, the consistency techniques are introduced. In sequel, Bernstein box and Bernstein hull consistency techniques are also presented. In Sect. 4, the main global optimization algorithm BBPMINLP to solve the MINLP problems is presented. Finally, some conclusions based on the present work are presented in the Sect. 5.

2   Background

This section briefly presents some notions about the Bernstein form. Due to the space limitation, a simple univariate Bernstein form is introduced. A comprehensive background and mathematical treatment for a multivariate case can be found in [12].

We can write a univariate l-degree polynomial p over an interval $\mathbf{x}$ in the form

$$p(x) = \sum_{i=0}^{l} a_i x^i, \quad a_i \in \mathbb{R}. \tag{2}$$

Now the polynomial p can be expanded into the Bernstein polynomials of the same degree as below

$$p(x) = \sum_{i=0}^{l} b_i(\mathbf{x}) B_i^l(x).$$ (3)

where B_i^l are the Bernstein basis polynomials and $b_i(\mathbf{x})$ are the Bernstein coefficients give as below

$$B_i^l = \binom{l}{i} x^i (1-x)^{1-i}.$$ (4)

$$b_i(\mathbf{x}) = \sum_{j=0}^{i} \frac{\binom{i}{j}}{\binom{l}{j}} a_j, \quad i = 0, \ldots, l.$$ (5)

Equation (3) is referred as the Bernstein form of a polynomial.

Theorem 1 *(Range enclosure property). Let p be a polynomial of degree l, and let $\overline{p}(\mathbf{x})$ denote the range of p on a given interval $\mathbf{x}$. Then,*

$$\overline{p}(\mathbf{x}) \subseteq B(\mathbf{x}) := [\min(b_i(\mathbf{x})), \max(b_i(\mathbf{x}))].$$ (6)

Proof: See [4].

Remark 1. The above theorem says that the minimum and maximum coefficients of the array $(b_i(\mathbf{x}))$ provide lower and upper bounds for the range. This forms the Bernstein range enclosure, defined by $B(\mathbf{x})$ in Eq. (6). The Bernstein range enclosure can successively be sharpened by the continuous domain subdivision procedure [4].

3 Consistency Techniques

The consistency techniques are used for pruning (deleting) unwanted regions that surely do not contain the global minimizer(s) from the solution search space. This pruning is achieved by assessing consistency of the algebraic equations (in our case inequality and equality constraints) over a given box $\mathbf{x}$.

This section now describe algorithms based on the consistency ideas borrowed from [5], and expanded in context of the Bernstein form. Henceforth, these algorithms are called as Bernstein box consistency (BBC) and Bernstein hull consistency (BHC). These algorithms work as a pruning operator in the main global optimization algorithm BBPMINLP (reported in the Sect. 4).

3.1 Bernstein Box Consistency

A Bernstein box consistency (BBC) technique is used to contract the bounds on a variable domain. The implementation of a BBC involve the application of a one-dimensional Bernstein Newton contractor [9] to solve a single equation for a single variable.

Consider an equality constraint polynomial $g(\mathbf{x}) = 0$, and let $(b(\mathbf{x}))$ be the Bernstein coefficients array of $g(\mathbf{x})$. Consider any component direction, say the first, with $\mathbf{x}_1 = [a, b]$. In the BBC technique, typically an attempt is made to increase the value of a and decrease the value of b, thus effectively reducing the width of $\mathbf{x}_1$.

To increase the value of a, first find all those Bernstein coefficients of $(b(\mathbf{x}))$ corresponding to $x_1 = a$. The minimum to maximum of these coefficients gives an interval denoted by $\mathbf{g}(a)$. If $0 \notin \mathbf{g}(a)$, then the constraint is infeasible at this endpoint a, and we search starting from a, along $x_1 = [a, b]$ for the first point at which constraint becomes just feasible, that is, we try to find a zero of $g(x)$. Let us denote this zero as a'. Clearly, $g(x)$ is infeasible over $[a, a')$, and so it can discarded to get a contracted interval $[a', b]$. On the other hand, if $0 \in \mathbf{g}(a)$ then we abandon the process to increase a and instead switch over to the other endpoint b and make an attempt to decrease it in the same way as we did to increase a.

To find a zero of $\mathbf{g}$ in $[a, b]$, one iteration of the univariate version of the Bernstein Newton contractor given in [9] is used. It is as follows

$$\mathbf{N}(\mathbf{x}_1) = a - (\mathbf{g}(a)/\mathbf{g}'_{x_1}),$$
$$\mathbf{x}'_1 = \mathbf{x}_1 \cap \mathbf{N}(\mathbf{x}_1),$$

where, $\mathbf{g}(a)$ is the minimum to maximum of the Bernstein coefficients array $(b(\mathbf{x}))$ at $x_1 = a$, $\mathbf{g}'_{x_1}$ denotes an interval enclosure for the derivative of $\mathbf{g}$ on $\mathbf{x}_1$, and $\mathbf{x}'_1$ gives a new contracted interval. A similar process is carried out from the other endpoint b, and if desired, the whole process can be repeated over all other component directions to a get contracted box $\mathbf{x}'$

The algorithm for the BBC which can be applied to both equality and inequality constraints is as follows.

Algorithm Bernstein Box Consistency: $\mathbf{x}' = \mathrm{BBC}\,((b_g(\mathbf{x})), \mathbf{x}, r, x_{status,r},$ $eq\_type)$.

Inputs: The Bernstein coefficient array $(b_g(\mathbf{x}))$ of a given constraint polynomial $g(x)$, the l-dimensional box $\mathbf{x}$, the direction r (decision variable) for which the bounds are to be contracted, flag $x_{status,r}$ to indicate whether r^{th} direction (decision variable) is continuous ($x_{status,r} = 0$) or integer ($x_{status,r} = 1$), and flag eq_type to indicate whether $g(x)$ is equality constraint ($eq\_type = 0$) or inequality constraint ($eq\_type = 1$).

Outputs: A box $\mathbf{x}'$ that is contracted using Bernstein box consistency technique for a given constraint polynomial $g(x)$.

BEGIN Algorithm

1. Set $a = \inf \mathbf{x}_r$, $b = \sup \mathbf{x}_r$.
2. From the Bernstein coefficient array $(b_g(\mathbf{x}))$, compute the derivative enclosure $\mathbf{g}'_{x_r}$ in the direction x_r.

3. (Consider left endpoint of $\mathbf{x}_r$). Obtain the Bernstein range enclosure $\mathbf{g}(a)$ as the minimum to maximum from the Bernstein coefficient array of $(b_g(\mathbf{x}))$ for $x_r = a$.
4. If $eq\_type = 1$, then modify $\mathbf{g}(a)$ as $\mathbf{g}(a) = [\min \mathbf{g}(a), \inf]$.
5. If $0 \in \mathbf{g}(a)$, then we cannot increase a. Go to step 8 and try from the right endpoint b of the interval $\mathbf{x}_r$.
6. Do one iteration of the univariate Bernstein Newton contractor

$$\mathbf{N}(\mathbf{x}_r) = a - (\mathbf{g}(a)/\mathbf{g}'_{x_r}).$$
$$\mathbf{x}'_{r_a} = \mathbf{x}_r \cap \mathbf{N}(\mathbf{x}_r).$$

7. If $\mathbf{x}'_{r_a} = \emptyset$, then there is no zero of $\mathbf{g}$ on entire interval $\mathbf{x}_r$ and hence the constraint g is infeasible over box $\mathbf{x}$. EXIT the algorithm in this case with $\mathbf{x}' = \emptyset$.
8. (Consider right endpoint of $\mathbf{x}_r$). Obtain the Bernstein range enclosure $\mathbf{g}(b)$ as the minimum to maximum from the Bernstein coefficient array of $(b(\mathbf{x}))$ for $x_r = b$.
9. If $eq\_type = 1$, then modify $\mathbf{g}(b)$ as $\mathbf{g}(b) = [\min \mathbf{g}(b), \inf]$.
10. If $0 \in \mathbf{g}(b)$, then we cannot decrease b. Go to step 13
11. Do one iteration of the univariate Bernstein Newton contractor

$$\mathbf{N}(\mathbf{x}_r) = b - (\mathbf{g}(b)/\mathbf{g}'_{x_r}).$$
$$\mathbf{x}'_{r_b} = \mathbf{x}_r \cap \mathbf{N}(\mathbf{x}_r).$$

12. If $\mathbf{x}'_{r_b} = \emptyset$, EXIT the algorithm with $\mathbf{x}' = \emptyset$.
13. Compute $\mathbf{x}'_r$ as follows:
    (a) $\mathbf{x}'_r = \mathbf{x}'_{r_a} \cap \mathbf{x}'_{r_b}$, if both $\mathbf{x}'_{r_a}$ and $\mathbf{x}'_{r_b}$ are computed.
    (b) $\mathbf{x}'_r = \mathbf{x}'_{r_a}$ or $\mathbf{x}'_{r_b}$, which ever is computed.
    (c) $\mathbf{x}'_r = \mathbf{x}_r$ (both $\mathbf{x}'_{r_a}$ and $\mathbf{x}'_{r_b}$ are not computed).
14. for $k = 1, 2$ if $x_{status,r} = 1$ then
    (a) if $\mathbf{x}(r, k)$ and $\mathbf{x}'_r(r, k)$ are equal then go to substep (e).
    (b) Set $t_a = \mathbf{x}(r, k)$, and $t_b = \mathbf{x}'_r(r, k)$.
    (c) if $t_a > t_b$ then set $\mathbf{x}'_r(r, k) = \lfloor \mathbf{x}'_r(r, k) \rfloor$.
    (d) if $t_a < t_b$ then set $\mathbf{x}'_r(r, k) = \lceil \mathbf{x}'_r(r, k) \rceil$.
    (e) end (of k-loop).
15. Return $\mathbf{x}' = \mathbf{x}'_r$.

END Algorithm

3.2   Algorithm Bernstein Box Consistency for a Set of Constraints

A single application of the proposed algorithm BBC in the Sect. 3.1 can contract only one variable domain. For a multivariate constraint, in turn, we can apply BBC to each variable separately. Below algorithm, called as BBC2SET applies BBC to all the variables present in a constraint, and if there are multiple constraints, BBC2SET applies BBC to all of them simultaneously.

Algorithm BBC for a Set of Constraints: $\mathbf{x}' = \text{BBC2SET}(B, k, C, \mathbf{x}, x_{status})$.

Inputs: A cell structure B containing Bernstein coefficient arrays of all the constraint polynomials with first k Bernstein coefficient arrays are for the equality constraints, the total number of constraints C, the l–dimensional box $\mathbf{x}$, and a column vector x_{status} describing the status (continuous or integer) of the each variable x_i ($i = 1, 2, \ldots, l$).

Outputs: A contracted box $\mathbf{x}'$.

BEGIN Algorithm

1. Set $r = 0$.
2. (a) for $i = 1, 2, \ldots, l$
   (b) for $j = 1, 2, \ldots, C$
       (i) Set $r = r + 1$, and $x_{status,r} = x_{status}(r)$. if $r > l$ then $r = 1$.
       (ii) if $j < k$ then $\mathbf{x}_1 = \text{BBC}(B\{j\}, \mathbf{x}, r, x_{status,r}, 0)$.
       (iii) if $j > k$ then $\mathbf{x}_1 = \text{BBC}(B\{j\}, \mathbf{x}, r, x_{status,r}, 1)$.
       (iv) Update $\mathbf{x} = \mathbf{x} \cap \mathbf{x}_1$.
       (v) if $\mathbf{x} = \emptyset$, then set $\mathbf{x}' = \emptyset$ and EXIT the algorithm.
   (c) end (of i–loop).
   (d) end (of j–loop).
3. Return $\mathbf{x}' = \mathbf{x}$.

END Algorithm

3.3 Bernstein Hull Consistency

Similar to a BBC, a Bernstein hull consistency (BHC) technique contract bounds on a variable domain. The typical BHC procedure is as below.

Consider a multivariate equality constraint $h(x) = 0$. To apply BHC to a selected term of $h(x) = 0$, we need to keep the selected term on the left hand side and remaining all other terms need to be taken on the right hand side, that is, we write the constraint in the form $a_I x^I = h_1(x)$ where, $x = (x_1, x_2, \ldots, x_l)$ and $I = (i_1, i_2, \ldots, i_l)$. The new contracted interval for the variable $\mathbf{x}_r$ (in r^{th} direction) can be obtained as

$$\mathbf{x}'_r = \left(\frac{\mathbf{h}'}{a_I \prod \mathbf{x}_k^{i_k}} \right)^{1/i_r} \bigcap \mathbf{x}_r, \quad r = 1, 2, \ldots, l. \tag{7}$$

Here to compute $\mathbf{h}'$ we compute the Bernstein coefficients of the monomial term $a_I x^I$ and from them subtract the Bernstein coefficients of the constraint polynomial $h(x)$. The minimum to maximum of this subtracted Bernstein coefficients will give $\mathbf{h}'$. For a given constraint all the terms can be solved or only selected terms can be solved.

The algorithm for the BHC that can be applied for both equality and inequality constraints is as follows.

Algorithm Bernstein Hull Consistency: $\mathbf{x}' = \text{BHC}((b_g(\mathbf{x})), a_I, I, \mathbf{x}, x_{status}, eq\_type)$.

Inputs: The Bernstein coefficient array $(b_g(\mathbf{x}))$ of a given constraint polynomial $g(x)$, coefficient a_I of the selected term t, power I of the each variable in term t, the l–dimensional box $\mathbf{x}$, a column vector $x_{status,r}$ describing the status (if continuous, then $x_{status,r} = 0$; if integer, then $x_{status,r} = 1$) of the each variable x_r ($r = 1, 2, \ldots, l$), and flag eq_type to indicate whether $g(x)$ is equality constraint ($eq\_type = 0$) or inequality constraint($eq\_type = 1$).

Outputs: A box $\mathbf{x}'$ that is contracted using Bernstein hull consistency technique applied to a given constraint polynomial $g(x)$ and selected term t.

BEGIN Algorithm

1. Compute the Bernstein coefficient array of the selected term t as $(b_t(\mathbf{x}))$.
2. Obtain the Bernstein coefficients of the constraint inverse polynomial by subtracting $(b_g(\mathbf{x}))$ from $(b_t(\mathbf{x}))$, and then obtain its Bernstein range enclosure as the minimum to maximum of these Bernstein coefficients. Denote it as $\mathbf{h}'$.
3. if $eq\_type = 1$ then
   (a) Compute an interval $\mathbf{y}$ as $\mathbf{y} = [-\infty, 0] \cap [\min(b_g(x)), \max(b_g(x))]$.
   (b) if $\mathbf{y} = \emptyset$ then set $\mathbf{x}' = \emptyset$, and EXIT the algorithm. Else modify $\mathbf{h}'$ as $\mathbf{h}' = \mathbf{h}' + \mathbf{y}$.
4. (a) for $r = 1, 2, \ldots, l$
   (b) Compute $\mathbf{x}'_r = \left(\dfrac{\mathbf{h}'}{a_I \prod \mathbf{x}_k^{i_k}} \right)^{1/i_r} \cap \mathbf{x}_r$
   (c) for $k = 1, 2$ if $x_{status}(r) = 1$ then
       (i) if $\mathbf{x}(r, k)$ and $\mathbf{x}'_r(r, k)$ are equal then go to substep (v).
       (ii) Set $t_a = \mathbf{x}(r, k)$ and $t_b = \mathbf{x}'_r(r, k)$.
       (iii) if $t_a > t_b$ then set $\mathbf{x}'_r(r, k) = \lfloor \mathbf{x}'_r(r, k) \rfloor$.
       (iv) if $t_a < t_b$ then set $\mathbf{x}'_r(r, k) = \lceil \mathbf{x}'_r(r, k) \rceil$.
       (v) end (of k–loop).
   (d) end (of r–loop).
5. Return $\mathbf{x}'$.

END Algorithm

3.4   Algorithm Bernstein Hull Consistency for a Set of Constraints

A single application of BHC algorithm can be made only to a single term of the selected constraint. However, in practice, we may want to apply BHC to more terms, or if there is more than one constraint, we may want to call BHC several times.

Below algorithm BHC2SET applies BHC to the multiple terms and to the multiple constraints. This algorithm will call BHC several times. Our criteria for term selection is as follows. In a given constraint, if a term contains maximum power for any of the variable, then it is selected. If the term contains maximum power for two variables, then it is solved two times and so on. This criteria is inspired from the ideas about interval hull consistency reported in [5].

Algorithm BHC for a Set of Constraints: $\mathbf{x}' = \text{BHC2SET}(A, B, k, C, \mathbf{x}, x_{status})$.

Inputs: The cell structure A containing the coefficient arrays of all constraint polynomials with first k coefficient arrays are for the equality constraints, a cell structure B containing Bernstein coefficient arrays of all the constraint polynomials, where first k Bernstein coefficient arrays are for the equality constraints, the total number of constraints C, the l–dimensional box $\mathbf{x}$, and a column vector $x_{status,r}$ describing the status (if continuous, then $x_{status,r} = 0$; if integer, then $x_{status,r} = 1$) of the each variable x_r $(r = 1, 2, \ldots, l)$.

Outputs: A contracted box $\mathbf{x}'$.

BEGIN Algorithm

1. Set $r = 0$.
2. (a) for $i = 1, 2, \ldots, l$
   (b) for $j = 1, 2, \ldots, C$
      (i) Set $r = r + 1$. if $r > l$ then $r = 1$
      (ii) Select the term having the maximum power for r in the constraint j, and obtain the coefficient a_I of the selected term and I containing the power of each variable in the selected term (this shall be obtained from A).
      (iii) if $j < k$ then $\mathbf{x}_1 = \text{BHC}(B\{j\}, a_I, I, \mathbf{x}, x_{status}, 0)$.
      (iv) if $j > k$ then $\mathbf{x}_1 = \text{BHC}(B\{j\}, a_I, I, \mathbf{x}, x_{status}, 1)$.
      (v) Update $\mathbf{x} = \mathbf{x} \cap \mathbf{x}_1$.
      (vi) if $\mathbf{x} = \emptyset$ then set $\mathbf{x}' = \emptyset$, and EXIT the algorithm.
   (c) end (of j–loop).
   (d) end (of i–loop).
3. Return $\mathbf{x}' = \mathbf{x}$.

END Algorithm

4   Main Algorithm BBPMINLP

This section presents the main algorithm for constrained global optimization of the MINLPs of a form (1). The working of the algorithm is similar to a interval branch-and-bound procedure, but with following enhancements.

- This algorithm use the Bernstein form as a inclusion function for the global optimization.
- Unlike classical subdivision procedure, the algorithm use a modified subdivision procedure from [11].
- Similarly, this algorithm use a efficient cut-off test, called as a vectorized Bernstein cut-off test (VBCT) from [11].
- Further, this algorithm use the efficient Bernstein box and Bernstein hull consistency techniques. These techniques serve as a pruning operator in the algorithm, thereby speeding up the convergence of the algorithm.

Algorithm Bernstein Branch-and-Prune Constrained Optimization: $[\widetilde{y}, \widetilde{p}, U] = \text{BBPMINLP}(N, a_I, \mathbf{x}, x_{status}, \epsilon_p, \epsilon_x, \epsilon_{zero})$.

Inputs: Degree N of the variables occurring in the objective and constraint polynomials, the coefficients a_I of the objective and constraint polynomials in the power form, the initial search domain $\mathbf{x}$, a column vector $x_{status,r}$ describing the status (if continuous, then $x_{status,r} = 0$; if integer, then $x_{status,r} = 1$) of a each variable x_r $(r = 1, 2, \ldots, l)$, the tolerance parameters ϵ_p and ϵ_x on the global minimum and global minimizer(s), and the tolerance parameter ϵ_{zero} to which the equality constraints are to be satisfied.

Outputs: A lower bound $\widetilde{y}$ and an upper bound $\widetilde{p}$ on the global minimum f^*, along with a set U containing all the global minimizer(s) $\mathbf{x}^{(i)}$.

BEGIN Algorithm

1. Set $\mathbf{y} := \mathbf{x}$ and $y_{status,r} := x_{status,r}$.
2. From a_I, compute the Bernstein coefficient arrays of the objective and constraint polynomials on the box $\mathbf{y}$ respectively as $(b_o(\mathbf{y})), (b_{gi}(\mathbf{y})), (b_{hj}(\mathbf{y}))$, $i = 1, 2, \ldots, m$, $j = 1, 2, \ldots, n$.
3. Set $\widetilde{p} := \infty$ and $y := \min(b_o(\mathbf{y}))$.
4. Set $R = (R_1, \ldots, R_m, R_{m+1}, \ldots, R_{m+n}) := (0, \ldots, 0)$.
5. Initialize list $\mathcal{L} := \{(\mathbf{y}, y, R, (b_o(\mathbf{y})), (b_{gi}(\mathbf{y})), (b_{hj}(\mathbf{y})))\}$, $\mathcal{L}^{sol} := \{\}$.
6. If $\mathcal{L}$ is empty then go to step 22. Otherwise, pick the first item $(\mathbf{y}, y, R, (b_o(\mathbf{y})), (b_{gi}(\mathbf{y})), (b_{hj}(\mathbf{y})))$ from $\mathcal{L}$, and delete its entry from $\mathcal{L}$.
7. Apply the Bernstein hull consistency algorithm to the relation $f(\mathbf{y}) \leq \widetilde{p}$. If the result is empty, then delete item $(\mathbf{y}, y, R, (b_o(\mathbf{y})), (b_{gi}(\mathbf{y})), (b_{hj}(\mathbf{y})))$ and go to step 6.
$$\mathbf{y}' = \text{BHC}((b_o(\mathbf{y})), a_I, I, \mathbf{y}, y_{status,r}, 1)$$
8. Set $\mathbf{y} := \mathbf{y}'$ and compute the Bernstein coefficient arrays of the objective and constraint polynomials on the box $\mathbf{y}$, respectively as $(b_o(\mathbf{y})), (b_{gi}(\mathbf{y})), (b_{hj}(\mathbf{y}))$, $i = 1, 2, \ldots, m$, $j = 1, 2, \ldots, n$. Also set $y := \min(b_o(\mathbf{y}))$.
9. Apply the Bernstein box consistency algorithm to the $f(\mathbf{y}) \leq \widetilde{p}$. If the result is empty, then delete item $(\mathbf{y}, y, R, (b_o(\mathbf{y})), (b_{gi}(\mathbf{y})), (b_{hj}(\mathbf{y})))$ and go to step 6.
$$\mathbf{y}' = \text{BBC}((b_o(\mathbf{y})), \mathbf{y}, r, y_{status,r}, 1)$$
where bound contraction will be applied in the rth direction.

10. Set $\mathbf{y} := \mathbf{y}'$ and compute the Bernstein coefficient arrays of the objective and constraint polynomials on the box $\mathbf{y}$, respectively as $(b_o(\mathbf{y})), (b_{gi}(\mathbf{y})), (b_{hj}(\mathbf{y}))$, $i = 1, 2, \ldots, m$, $j = 1, 2, \ldots, n$. Also set $y := \min (b_o(\mathbf{y}))$.

11. {Contract domain box by applying Bernstein hull consistency to all the constraints} Apply the algorithm BHC2SET to all the constraints

$$\mathbf{y}' = \text{BHC2SET}(A_c, B_c, k, C, \mathbf{y}, y_{status}, r)$$

Here A_c is a cell structure containing the coefficient arrays of the all constraints, where the first k coefficient arrays are for the equality constraints, B_c is a cell structure containing the Bernstein coefficient arrays of the all constraints, where the first k Bernstein coefficient arrays are for the equality constraints, C is the total number of constraints, $\mathbf{y}$ is a domain box, and $\mathbf{y}'$ is the new contracted box.

12. Set $\mathbf{y} := \mathbf{y}'$ and compute the Bernstein coefficient arrays of the objective and constraint polynomials on the box $\mathbf{y}$, respectively as $(b_o(\mathbf{y})), (b_{gi}(\mathbf{y})), (b_{hj}(\mathbf{y}))$, $i = 1, 2, \ldots, m$, $j = 1, 2, \ldots, n$. Also set $y := \min (b_o(\mathbf{y}))$.

13. {Contract domain box by applying Bernstein box consistency to all the constraints} Apply the algorithm BBC2SET to all the constraints

$$\mathbf{y}' = \text{BBC2SET}(B_c, k, C, \mathbf{y}, y_{status}, r)$$

Here B_c is a cell structure containing the Bernstein coefficient arrays of all the constraints, where the first k Bernstein coefficient arrays are for the equality constraints, C is the total number of constraints, $\mathbf{y}$ is a domain box, and $\mathbf{y}'$ is a new contracted box.

14. Set $\mathbf{y} := \mathbf{y}'$ and compute the Bernstein coefficient arrays of the objective and constraint polynomials on the box $\mathbf{y}$, respectively as $(b_o(\mathbf{y})), (b_{gi}(\mathbf{y})), (b_{hj}(\mathbf{y}))$, $i = 1, 2, \ldots, m$, $j = 1, 2, \ldots, n$. Also set $y := \min (b_o(\mathbf{y}))$.

15. {Branching}
    (a) If $w(\mathbf{y}_i) = 0$ for all $i = l_d + 1, \ldots, l$ (that is, all the integer variables has been fixed to some integer values from there respective domains) then go to substep (c).
    (b) Choose a coordinate direction λ parallel to which $\mathbf{y}_{l_d+1} \times \cdots \times \mathbf{y}_l$ has an edge of maximum length, that is $\lambda \in \{i : w(\mathbf{y}) := w(\mathbf{y}_i), i = l_d+1, \ldots, l\}$. Go to step 16.
    (c) Choose a coordinate direction λ parallel to which $\mathbf{y}_1 \times \cdots \times \mathbf{y}_{l_d}$ has an edge of maximum length, that is $\lambda \in \{i : w(\mathbf{y}) := w(\mathbf{y}_i), i = 1, \ldots, l_d\}$.

16. Bisect $\mathbf{y}$ normal to direction λ, getting boxes $\mathbf{v}_1, \mathbf{v}_2$ such that $\mathbf{y} = \mathbf{v}_1 \cup \mathbf{v}_2$. The modified subdivision procedure from [11] is used.

17. for $k = 1, 2$
    (a) Set $R^k = (R_1^k, \ldots, R_m^k, R_{m+1}^k, \ldots, R_{m+n}^k) := R$.
    (b) Find the Bernstein coefficient array and the corresponding Bernstein range enclosure of the objective function (f) over $\mathbf{v}_k$ as $(b_0(\mathbf{v}_k))$ and $B_0(\mathbf{v}_k)$, respectively.

(c) Set $d_k := \min B_o(\mathbf{v}_k)$.

(d) If $\widetilde{p} < d_k$ then go to substep (j).

(e) for $i = 1, 2, \ldots, m$ if $R_i = 0$ then

   (i) Find the Bernstein coefficient array and the corresponding Bernstein range enclosure of the inequality constraint polynomial (g_i) over $\mathbf{v}_k$ as $(b_{gi}(\mathbf{v}_k))$ and $B_{gi}(\mathbf{v}_k)$, respectively.

   (ii) If $B_{gi}(\mathbf{v}_k) > 0$ then go to substep (j).

   (iii) If $B_{gi}(\mathbf{v}_k) \leq 0$ then set $R_i^k := 1$.

(f) for $j = 1, 2, \ldots, n$ if $R_{m+j} = 0$ then

   (i) Find the Bernstein coefficient array and the corresponding Bernstein range enclosure of the equality constraint polynomial (h_j) over $\mathbf{v}_k$ as $(b_{hj}(\mathbf{v}_k))$ and $B_{hj}(\mathbf{v}_k)$, respectively.

   (ii) If $0 \notin B_{hj}(\mathbf{v}_k)$ then go to substep (j).

   (iii) If $B_{hj}(\mathbf{v}_k) \subseteq [-\epsilon_{zero}, \epsilon_{zero}]$ then set $R_{m+j}^k := 1$.

(g) If $R^k = (1, \ldots, 1)$ then set $\widetilde{p} := \min(\widetilde{p}, \max B_o(\mathbf{v}_k))$.

(h) Enter $(\mathbf{v}_k, d_k, R^k)$ into the list $\mathcal{L}$ such that the second members of all items of the list do not decrease.

(j) end (of k–loop).

18. {Cut-off test} Discard all items $(\mathbf{z}, z, R, (b_o(\mathbf{z})), (b_{gi}(\mathbf{z})), (b_{hj}(\mathbf{z})))$ in the list $\mathcal{L}$ that satisfy $\widetilde{p} < z$. For the remaining items in the list $\mathcal{L}$ apply the vectorized Bernstein cut-off test from [11], and update the current minimum estimate $\widetilde{p}$.

19. Denote the first item of the list $\mathcal{L}$ by $(\mathbf{y}, y, R, (b_o(\mathbf{y})), (b_{gi}(\mathbf{y})), (b_{hj}(\mathbf{y})))$.

20. If $(w(\mathbf{y}) < \epsilon_x) \& (\max B_o(\mathbf{y}) - \min B_o(\mathbf{y})) < \epsilon_p$ then remove the item from the list $\mathcal{L}$ and enter it into the solution list $\mathcal{L}^{sol}$.

21. Go to step 6.

22. {Compute the global minimum} Set the global minimum $\widetilde{y}$ to the minimum of the second entries over all the items in $\mathcal{L}^{sol}$.

23. {Compute the global minimizers} Find all those items in $\mathcal{L}^{sol}$ for which the second entries are equal to $\widetilde{y}$. The first entries of these items contain the global minimizer(s) $\mathbf{x}^{(i)}$.

24. Return the lower bound $\widetilde{y}$ and upper bound $\widetilde{p}$ on the global minimum f^*, along with the set U containing all the global minimizer(s) $\mathbf{x}^{(i)}$.

END Algorithm

5   Numerical Studies

This section reports a numerical experimentation with the algorithm BBP-MINLP on a set of 16 test problems. These test problems were chosen from [3,8,13]. At the outset, the performance of the algorithm BBPMINLP was compared with the Bernstein algorithms BMIO in [12] and IBBBC in [11]. Further, the algorithm BBPMINLP was compared with the four state-of-the-art MINLP solvers, namely AlphaECP, BARON, Bonmin, DICOPT, whose GAMS interface is available through the NEOS server [10], and one MATLAB based open-source solver BNB20 [7].

Table 1. Description of symbols for Table 3.

Symbol	Description
l	Total number of the decision variables (binary, integer and continuous)
f^*	Bold values in this row indicates local minimum obtained
*	Indicates that the solver failed giving the message "relaxed NLP is unbounded"
**	Indicates that the solver searched one hour for the solution, still could not find the solution and therefore was terminated
***	Indicates that the solver returned the message "terminated by the solver"
****	Indicates that the solver failed giving the message "infeasible row with only small Jacobian elements"

For all computations, a desktop PC with Pentium IV 2.40 GHz processor with 2 GB RAM was used. The algorithm BBPMINLP was implemented in the MATLAB [1] with an accuracy $\epsilon = 10^{-6}$ for computing the global minimum and global minimizer(s), and a maximum limit on the number of subdivisions to be 500.

Table 1 describes the list of symbols for Table 3. Table 2 reports for the 16 test problems, the total number of boxes processed and the computational time taken in seconds to locate a correct global minimum by the Bernstein algorithms BMIO, IBBBC, and the algorithm BBPMINLP reported in this work. The algorithm BBPMINLP was compared using three different flags described as below:

- **A**: Application of the Bernstein hull consistency to the inequality and equality constraints, that is algorithm BHC2SET (see Sect. 3.4) is applied to these constraints.
- **B**: Application of the Bernstein box consistency to the inequality and equality constraints, that is algorithm BBC2SET (see Sect. 3.2) is applied to these constraints.
- **C**: Application of the Bernstein hull and box consistencies to the constraint $f(x) \leq \widetilde{f}$ ($\widetilde{f}$ is the current global minimum estimate). This serves to delete a subbox that bounds a nonoptimal point of $f(x)$.

The findings are as below. It was observed that the algorithm BMIO failed to solve for the four test problems (wester, hmittelman, sep1, tln5) and the algorithm IBBBC is unable to solve one test problem sep1. Similarly, the algorithm BBPMINLP with flags A and C is unable to solve one test problem (sep1). This is perhaps the Bernstein hull consistency in this problem was unable to sufficiently prune the search region, and hence may take more time to find the solution. However, for one test problem (tln5) we found the algorithm BBPMINLP with flag A to be more efficient than the others. In contrast, the algorithm BBPMINLP with flag B was able to successfully solve all the test problems. Moreover, it was observed for two test problems (wester, hmittelman) algorithm with flag B performed exceptionally well than the others. Overall, the performance of the

Table 2. Comparison of the number of boxes processed and computational time (in seconds) taken by the earlier Bernstein algorithms BMIO, IBBBC and the algorithm BBPMINLP.

Example	l	Statistics	BMIO	IBBBC	BBPMINLP		
					A	B	C
floudas1	2	Boxes	1003	33	29	10	31
		Time	0.45	0.08	0.3	0.10	0.18
zhu1	2	Boxes	1166	173	63	61	81
		Time	1.05	0.14	0.5	0.40	0.59
st_testph4	3	Boxes	1870	47	20	15	29
		Time	2.21	0.18	0.15	0.10	0.44
nvs21	3	Boxes	1149	785	125	67	615
		Time	0.81	0.10	0.23	0.31	1.17
gbd	4	Boxes	2201	23	23	5	15
		Time	1.40	0.09	0.11	0.02	0.28
st_e27	4	Boxes	572	21	5	5	13
		Time	0.40	0.08	0.06	0.07	0.21
zhu2	5	Boxes	2571	700	84	81	173
		Time	2.71	1.40	3.35	2.30	4.13
st_test2	6	Boxes	2987	107	17	16	5
		Time	1.63	0.18	0.30	0.12	0.11
wester	6	Boxes	*	1621	1500	4	6003
		Time		5.25	300	0.07	39.83
alan	8	Boxes	4015	1	1	1	1
		Time	3.03	0.01	0.01	0.02	0.01
ex1225	8	Boxes	6869	385	343	85	261
		Time	6.60	0.15	0.7	0.40	3.17
st_test6	10	Boxes	3003	111	18	18	91
		Time	3.57	2.68	1.25	0.70	11.51
st_test3	13	Boxes	3960	340	119	21	261
		Time	48.50	4.32	5.61	4.31	75.40
hmittelman	16	Boxes	*	431	5000	3	191
		Time		61.52	1561	1.35	316.44
sep1	29	Boxes	*	**	**	1034	**
		Time				5.96	
tln5	35	Boxes	*	>10,000	1003	2972	>8003
		Time			68.28	18.96	

* Indicates that the algorithm returned "out of memory error".
** Indicates that the algorithm did not give the result even after one hour and is therefore terminated.

Table 3. Comparison of the global minimum obtained and computational time (in seconds) taken by the algorithm BBPMINLP with state-of-the-art MINLP solvers.

Example	l	Statistics	Solver/Algorithm					
			AlphaECP	BARON	Bonmin	BNB20	DICOPT	BBPMINLP
floudas1	2	f^*	−8.5	−8.5	−8.5	**−5**	**−4**	−8.5
		Time	1.04	0.25	0.14	0.01	0.21	0.1
zhu1	2	f^*	−3.9374E +10	−3.9374E +10	−3.9374E +10	−3.9374E +10	*	−3.9374E+10
		Time	1.36	0.25	0.16	0.07		0.40
st_testph4	3	f^*	−80.5	−80.5	−80.5	−80.5	−80.5	−80.5
		Time	0.89	0.26	0.26	0.22	0.47	0.10
nvs21	3	f^*	−5.68	−5.68	−5.68	−5.68	−5.68	−5.68
		Time	15.54	1.06	0.16	0.29	0.23	0.31
gbd	4	f^*	2.2	2.2	2.2	2.2	2.2	2.2
		Time	0.5	0.25	0.22	0.03	0.22	0.02
st_e27	4	f^*	2	2	2	2	2	2
		Time	0.71	0.26	0.13	0.01	0.22	0.07
zhu2	5	f^*	**0**	−51,568	**0**	**−42,585**	**0**	−51,568
		Time	1.94	0.25	0.16	1.38	0.23	2.30
st_test2	6	f^*	−9.25	−9.25	−9.25	−9.25	−9.25	−9.25
		Time	1.83	0.32	0.31	0.29	0.84	0.12
wester	6	f^*	112,235	112,235	112,235	**	1,12,235	112,235
		Time	6.66	0.37	0.08		0.82	0.07
alan	8	f^*	2.92	2.92	2.92	2.92	2.92	2.92
		Time	0.61	0.23	0.20	0.14	1.02	0.02
ex1225	8	f^*	31	31	31	31	31	31
		Time	0.72	0.26	0.28	0.28	0.47	0.40
st_test6	10	f^*	471	471	471	471	471	471
		Time	3.56	1.42	1.17	0.82	1.42	0.70
st_test3	13	f^*	−7	−7	−7	**	−7	−7
		Time	0.94	0.27	0.61		0.98	4.31
hmittelman	16	f^*	***	13	13	19	****	13
		Time		0.42	2.62	0.09		1.35
sep1	29	f^*	−510.08	−510.08	−510.08	**−50**	−510.08	−510.08
		Time	7.91	0.06	0.04	0.14	0.001	5.96
tln5	35	f^*	**10.6**	10.3	**10.6**	**	**13.7**	10.3
		Time	12.23	0.53	52.74		0.002	18.96

algorithm BBPMINLP with flag B was seen to be the best in terms of both the number of boxes processed and the computational time it took to found a global minimum.

Table 3 reports for the 16 test problems the quality of the global minimum obtained with the algorithm BBPMINLP and the state-of-the-art MINLP solvers[1]. The bold values in the table indicate the local minimum value. For these test problems the performance of the state-of-the-art solvers was as follows:

- AlphaECP found the local minimum for two test problem (zhu2, tln5), and failed to solve one test problem (hmittelman).
- Bonmin found the local minimum for two test problems (zhu2, tln5).

---

[1] All the solver were executed in their default options for the 16 test problems considered.

– BNB20 found the local minimum for four test problems (floudas1, zhu2, hmittelman, sep1), and failed to solve three test problems (wester, st_test3, tln5).
– DICOPT found the local minimum for three test problems (floudas1, zhu2, tln5), and failed solve two test problems (zhu1, hmittelman).

However, the algorithm BBPMINLP was able to found the correct the global minimum value for all the test problems, and compares well with the state-of-the-art solvers in terms of the computational time.

6   Conclusions

In this work the Bernstein algorithm (BBPMINLP) was proposed to solve the polynomial type of MINLPs. This algorithm was composed with the two new solution search space pruning operators, namely the Bernstein box and Bernstein hull consistency. Further, the proposed algorithm also used another pruning operator based on the application of the Bernstein box and hull consistency to a constraint based on the objective function $f(x)$ and a current minimum estimate $\widetilde{f}$. This step along with a cut-off test improves the convergence of the algorithm. The performance of the proposed algorithm BBPMINLP was tested on a collection of 16 test problems. The test problems had dimensions ranging from 2 to 35 and number of constraints varying from 1 to 31. At the outset, the effectiveness of the algorithm BBPMINLP was demonstrated over the previously reported Bernstein algorithms BMIO and IBBBC. The algorithm BBPMINLP was found to be more efficient in the number of boxes processed, resulting an average reduction of 96–99 % compared to BMIO and 42–88 % compared to IBBBC. Similarly, from the computational perspective BBPMINLP was found to be well competent with the algorithms BMIO and IBBBC.

Lastly, the performance of the algorithm BBPMINLP was compared with the existing state-of-the-art MINLP solvers, such as AlphaECP, BARON, Bonmin, BNB20, and DICOPT. Test results showed the superiority of the proposed algorithm BBPMINLP over state-of-the-art MINLP solvers in terms of the solution quality obtained. Specifically, all solvers (except BARON) located local solution or failed for atleast one problem from a set of 16 test problems considered. On the otherhand, the algorithm BBPMINLP could locate correct global minimum for all the test problems. In terms of the computational time, BBPMINLP was some order of magnitudes slower than the considered MINLP solvers. However, this could be due to the difference in the computing platforms used for the algorithm implementation and testing.

Acknowledgement. This work was funded by the Singapore National Research Foundation (NRF) under its Campus for Research Excellence And Technological Enterprise (CREATE) programme and the Cambridge Centre for Advanced Research in Energy Efficiency in Singapore (CARES).

References

1. The Mathworks Inc., MATLAB version 7.1 (R14), Natick, MA (2005)
2. D'Ambrosio, C., Lodi, A.: Mixed integer nonlinear programming tools: an updated practical overview. Annals of Operations Research **204**(1), 301–320 (2013)
3. Floudas, C.A.: Nonlinear and Mixed-Integer Optimization: Fundamentals and Applications. Oxford University Press, New York (1995)
4. Garloff, J.: The Bernstein algorithm. Interval Computations **2**, 154–168 (1993)
5. Hansen, E.R., Walster, G.W.: Global Optimization Using Interval Analysis, 2nd edn. Marcel Dekker, New York (2005)
6. Hooker, J.: Logic-Based Methods for Optimization: Combining Optimization and Constraint Satisfaction. Wiley, New York (2000)
7. Kuipers, K.: Branch-and-bound solver for mixed-integer nonlinear optimization problems. MATLAB Central for File Exchange. Accessed 18 Dec. 2009
8. GAMS Minlp Model Library: http://www.gamsworld.org/minlp/minlplib/minlpstat.htm. Accessed 20 March 2015
9. Nataraj, P.S.V., Arounassalame, M.: An interval Newton method based on the Bernstein form for bounding the zeros of polynomial systems. Reliable Comput. **15**(2), 109–119 (2011)
10. NEOS server for optimization.: http://www.neos-server.org/neos/solvers/index.html. Accessed 20 March 2015
11. Patil, B.V., Nataraj, P.S.V.: An improved Bernstein global optimization algorithm for MINLP problems with application in process industry. Math. Comput. Sci. **8**(3–4), 357–377 (2014)
12. Patil, B.V., Nataraj, P.S.V., Bhartiya, S.: Global optimization of mixed-integer nonlinear (polynomial) programming problems: the Bernstein polynomial approach. Computing **94**(2–4), 325–343 (2012)
13. Zhu, W.: A provable better branch and bound method for a nonconvex integer quadratic programming problem. J. Comput. Syst. Sci. **70**(1), 107–117 (2005)

General Session

Maximum Likelihood Estimates for Gaussian Mixtures Are Transcendental

Carlos Améndola[1(✉)], Mathias Drton[2], and Bernd Sturmfels[1,3]

[1] Technische Universität, Berlin, Germany
amendola@math.tu-berlin.de
[2] University of Washington, Seattle, USA
[3] University of California, Berkeley, USA

Abstract. Gaussian mixture models are central to classical statistics, widely used in the information sciences, and have a rich mathematical structure. We examine their maximum likelihood estimates through the lens of algebraic statistics. The MLE is not an algebraic function of the data, so there is no notion of ML degree for these models. The critical points of the likelihood function are transcendental, and there is no bound on their number, even for mixtures of two univariate Gaussians.

Keywords: Algebraic statistics · Expectation maximization · Maximum likelihood · Mixture model · Normal distribution · Transcendence theory

1 Introduction

The primary purpose of this paper is to demonstrate the result stated in the title:

Theorem 1. *The maximum likelihood estimators of Gaussian mixture models are transcendental functions. More precisely, there exist rational samples $x_1, x_2, \ldots, x_N$ in $\mathbb{Q}^n$ whose maximum likelihood parameters for the mixture of two n-dimensional Gaussians are not algebraic numbers over $\mathbb{Q}$.*

The principle of maximum likelihood (ML) is central to statistical inference. Most implementations of ML estimation employ iterative hill-climbing methods, such as expectation maximization (EM). These can rarely certify that a globally optimal solution has been reached. An alternative paradigm, advanced by algebraic statistics [8], is to find the ML estimator (MLE) by solving the likelihood equations. This is only feasible for small models, but it has the benefit of being exact and certifiable. An important notion in this approach is the *ML degree*, which is defined as the algebraic degree of the MLE as a function of the data. This rests on the premise that the likelihood equations are given by polynomials.

Many models used in practice, such as exponential families for discrete or Gaussian observations, can be represented by polynomials. Hence, they have an ML degree that serves as an upper bound for the number of isolated local maxima of the likelihood function, independently of the sample size and the data. The ML

© Springer International Publishing Switzerland 2016
I.S. Kotsireas et al. (Eds.): MACIS 2015, LNCS 9582, pp. 579–590, 2016.
DOI: 10.1007/978-3-319-32859-1_49

degree is an intrinsic invariant of a statistical model, with interesting geometric and topological properties [13]. The notion has proven useful for characterization of when MLEs admit a 'closed form' [19]. When the ML degree is moderate, these exact tools are guaranteed to find the optimal solution to the ML problem [5,12].

However, the ML degree of a statistical model is only defined when the MLE is an algebraic function of the data. Theorem 1 means that there is no ML degree for Gaussian mixtures. It also highlights a fundamental difference between likelihood inference and the method of moments [3,10,15]. The latter is a computational paradigm within algebraic geometry, that is, it is based on solving polynomial equations. ML estimation being transcendental means that likelihood inference in Gaussian mixtures is outside the scope of algebraic geometry.

The proof of Theorem 1 will appear in Sect. 2. In Sect. 3 we shed further light on the transcendental nature of Gaussian mixture models. We focus on mixtures of two univariate Gaussians, the model given in (1) below, and we present a family of data points on the real line such that the number of critical points of the corresponding log-likelihood function (2) exceeds any bound.

While the MLE for Gaussian mixtures is transcendental, this does not mean that exact methods are not available. Quite to the contrary. Work of Yap and his collaborators in computational geometry [6,7] convincingly demonstrates this. Using root bounds from transcendental number theory, they provide certified answers to geometric optimization problems whose solutions are known to be transcendental. Theorem 1 opens up the possibility of transferring these techniques to statistical inference. In our view, Gaussian mixtures are an excellent domain of application for certified computation in numerical analytic geometry.

2   Reaching Transcendence

Transcendental number theory [2,11] is a field that furnishes tools for deciding whether a given real number τ is a root of a nonzero polynomial in $\mathbb{Q}[t]$. If this holds then τ is algebraic; otherwise τ is transcendental. For instance, $\sqrt{2}+\sqrt{7} = 4.059964873...$ is algebraic, and so are the parameter estimates computed by Pearson in his 1894 study of crab data [15]. By contrast, the famous constants $\pi = 3.141592653...$ and $e = 2.718281828...$ are transcendental. Our proof will be based on the following classical result. A textbook reference is [2, Theorem 1.4]:

Theorem 2 (Lindemann-Weierstrass). *If $u_1, \ldots, u_r$ are distinct algebraic numbers then $e^{u_1}, \ldots, e^{u_r}$ are linearly independent over the algebraic numbers.*

For now, consider the case of $n = 1$, that is, mixtures of two univariate Gaussians. We allow mixtures with arbitrary means and variances. Our model then consists of all probability distributions on the real line $\mathbb{R}$ with density

$$f_{\alpha,\mu,\sigma}(x) = \frac{1}{\sqrt{2\pi}} \cdot \left[\frac{\alpha}{\sigma_1} \exp\left(-\frac{(x-\mu_1)^2}{2\sigma_1^2}\right) + \frac{1-\alpha}{\sigma_2} \exp\left(-\frac{(x-\mu_2)^2}{2\sigma_2^2}\right) \right]. \quad (1)$$

It has five unknown parameters, namely, the means $\mu_1, \mu_2 \in \mathbb{R}$, the standard deviations $\sigma_1, \sigma_2 > 0$, and the mixture weight $\alpha \in [0, 1]$. The aim is to estimate the five model parameters from a collection of data points $x_1, x_2, \ldots, x_N \in \mathbb{R}$.

The *log-likelihood function* of the model (1) is

$$\ell(\alpha, \mu_1, \mu_2, \sigma_1, \sigma_2) = \sum_{i=1}^{N} \log f_{\alpha,\mu,\sigma}(x_i). \tag{2}$$

This is a function of the five parameters, while $x_1, \ldots, x_N$ are fixed constants.

The principle of maximum likelihood suggests to find estimates by maximizing the function ℓ over the five-dimensional parameter space $\Theta = [0, 1] \times \mathbb{R}^2 \times \mathbb{R}^2_{>0}$.

Remark 1. The log-likelihood function ℓ in (2) is never bounded above. To see this, we argue as in [4, Sect. 9.2.1]. Set $N = 2$, fix arbitrary values $\alpha_0 \in [0, 1]$, $\mu_{20} \in \mathbb{R}$ and $\sigma_{20} > 0$, and match the first mean to the first data point $\mu_1 = x_1$. The remaining function of one unknown σ_1 equals

$$\ell(\alpha_0, x_1, \mu_{20}, \sigma_1, \sigma_{20}) \geq \log\left[\frac{\alpha_0}{\sigma_1} + \frac{1-\alpha_0}{\sigma_{20}} \exp\left(-\frac{(x_1 - \mu_{20})^2}{2\sigma_{20}^2}\right)\right] + \text{const.}$$

The lower bound tends to ∞ as $\sigma_1 \to 0$.

Remark 1 means that there is no global solution to the MLE problem. This is remedied by restricting to a subset of the parameter space Θ. In practice, maximum likelihood for Gaussian mixtures means computing local maxima of the function ℓ. These are found numerically by a hill climbing method, such as the EM algorithm, with particular choices of starting values. See Sect. 3. This method is implemented, for instance, in the R package MCLUST [9]. In order for Theorem 1 to cover such local maxima, we prove the following statement:

There exist samples $x_1, \ldots, x_N \in \mathbb{Q}$ such that every non-trivial critical point $(\hat{\alpha}, \hat{\mu}_1, \hat{\mu}_2, \hat{\sigma}_1, \hat{\sigma}_2)$ of the log-likelihood function ℓ in the domain Θ has at least one transcendental coordinate.

Here, a critical point is *non-trivial* if it yields an honest mixture, i.e. a distribution that is not Gaussian. By the identifiability results of [20], this happens if and only if the estimate $(\hat{\alpha}, \hat{\mu}_1, \hat{\mu}_2, \hat{\sigma}_1, \hat{\sigma}_2)$ satisfies $0 < \hat{\alpha} < 1$ and $(\hat{\mu}_1, \hat{\sigma}_1) \neq (\hat{\mu}_2, \hat{\sigma}_2)$.

Remark 2. The log-likelihood function always has some algebraic critical points, for any $x_1, \ldots, x_N \in \mathbb{Q}$. Indeed, if we define the empirical mean and variance as

$$\bar{x} = \frac{1}{N} \sum_{i=1}^{N} x_i, \qquad s^2 = \frac{1}{N} \sum_{i=1}^{N} (x_i - \bar{x})^2,$$

then any point $(\hat{\alpha}, \hat{\mu}_1, \hat{\mu}_2, \hat{\sigma}_1, \hat{\sigma}_2)$ with $\hat{\mu}_1 = \hat{\mu}_2 = \bar{x}$ and $\hat{\sigma}_1 = \hat{\sigma}_2 = s$ is critical. This gives a Gaussian distribution with mean $\bar{x}$ and variance s^2, so it is trivial.

Proof (of Theorem 1). First, we treat the univariate case. Consider the partial derivative of (2) with respect to the mixture weight α:

$$\frac{\partial \ell}{\partial \alpha} = \sum_{i=1}^{N} \frac{1}{f_{\alpha,\mu,\sigma}(x_i)} \cdot \frac{1}{\sqrt{2\pi}} \left[\frac{1}{\sigma_1} \exp\left(-\frac{(x_i - \mu_1)^2}{2\sigma_1^2}\right) - \frac{1}{\sigma_2} \exp\left(-\frac{(x_i - \mu_2)^2}{2\sigma_2^2}\right) \right]. \quad (3)$$

Clearing the common denominator

$$\sqrt{2\pi} \cdot \prod_{i=1}^{N} f_{\alpha,\mu,\sigma}(x_i),$$

we see that $\partial \ell / \partial \alpha = 0$ if and only if

$$\sum_{i=1}^{N} \left[\frac{1}{\sigma_1} \exp\left(-\frac{(x_i - \mu_1)^2}{2\sigma_1^2}\right) - \frac{1}{\sigma_2} \exp\left(-\frac{(x_i - \mu_2)^2}{2\sigma_2^2}\right) \right]$$
$$\times \prod_{j \neq i} \left[\frac{\alpha}{\sigma_1} \exp\left(-\frac{(x_j - \mu_1)^2}{2\sigma_1^2}\right) + \frac{1-\alpha}{\sigma_2} \exp\left(-\frac{(x_j - \mu_2)^2}{2\sigma_2^2}\right) \right] = 0. \quad (4)$$

Letting $\alpha_1 = \alpha$ and $\alpha_2 = 1 - \alpha$, we may rewrite the left-hand side of (4) as

$$\sum_{i=1}^{N} \left[\sum_{k_i=1}^{2} \frac{(-1)^{k_i-1}}{\sigma_{k_i}} \exp\left(-\frac{(x_i - \mu_{k_i})^2}{2\sigma_{k_i}^2}\right) \right] \prod_{j \neq i} \left[\sum_{k_j=1}^{2} \frac{\alpha_{k_j}}{\sigma_{k_j}} \exp\left(-\frac{(x_j - \mu_{k_j})^2}{2\sigma_{k_j}^2}\right) \right]. \quad (5)$$

We expand the products, collect terms, and set $N_i(k) = |\{j : k_j = i\}|$. With this, the partial derivative $\partial \ell / \partial \alpha$ is zero if and only if the following vanishes:

$$\sum_{i=1}^{N} \sum_{k \in \{1,2\}^N} \exp\left(-\sum_{j=1}^{N} \frac{(x_j - \mu_{k_j})^2}{2\sigma_{k_j}^2}\right) (-1)^{k_i-1} \alpha^{|\{j \neq i: k_j=1\}|} (1-\alpha)^{|\{j \neq i: k_j=2\}|} \left(\prod_{j=1}^{N} \frac{1}{\sigma_{k_j}}\right)$$

$$= \sum_{k \in \{1,2\}^N} \exp\left(-\sum_{j=1}^{N} \frac{(x_j - \mu_{k_j})^2}{2\sigma_{k_j}^2}\right) \left(\prod_{j=1}^{N} \frac{1}{\sigma_{k_j}}\right) \sum_{i=1}^{N} (-1)^{k_i-1} \alpha^{|\{j \neq i: k_j=1\}|} (1-\alpha)^{|\{j \neq i: k_j=2\}|}$$

$$= \sum_{k \in \{1,2\}^N} \exp\left(-\sum_{j=1}^{N} \frac{(x_j - \mu_{k_j})^2}{2\sigma_{k_j}^2}\right) \left(\prod_{j=1}^{N} \frac{1}{\sigma_{k_j}}\right) \alpha^{N_1(k)-1} (1-\alpha)^{N_2(k)-1} \left[\begin{array}{c} N_1(k)(1-\alpha) \\ +N_2(k)(-\alpha) \end{array}\right]$$

$$= \sum_{k \in \{1,2\}^N} \exp\left(-\sum_{j=1}^{N} \frac{(x_j - \mu_{k_j})^2}{2\sigma_{k_j}^2}\right) \left(\prod_{j=1}^{N} \frac{1}{\sigma_{k_j}}\right) \alpha^{N_1(k)-1} (1-\alpha)^{N-N_1(k)-1} (N_1(k) - N\alpha).$$

Let $(\hat{\alpha}, \hat{\mu}_1, \hat{\mu}_2, \hat{\sigma}_1, \hat{\sigma}_2)$ be a non-trivial isolated critical point of the likelihood function. This means that $0 < \hat{\alpha} < 1$ and $(\hat{\mu}_1, \hat{\sigma}_1) \neq (\hat{\mu}_2, \hat{\sigma}_2)$. This point depends continuously on the choice of the data $x_1, x_2, \ldots, x_N$. By moving the vector with these coordinates along a general line in $\mathbb{R}^N$, the mixture parameter $\hat{\alpha}$ moves continuously in the critical equation $\partial \ell / \partial \alpha = 0$ above. By the Implicit Function Theorem, it takes on all values in some open interval of $\mathbb{R}$, and we can thus choose our data points x_i general enough so that $\hat{\alpha}$ is not an integer multiple of $1/N$. We can further ensure that the last sum above is a $\mathbb{Q}(\alpha)$-linear combination of exponentials with nonzero coefficients.

Suppose that $(\hat{\alpha}, \hat{\mu}_1, \hat{\mu}_2, \hat{\sigma}_1, \hat{\sigma}_2)$ is algebraic. The Lindemann-Weierstrass Theorem implies that the arguments of exp are all the same. Then the 2^N numbers

$$\sum_{j=1}^{N} \frac{(x_j - \hat{\mu}_{k_j})^2}{2\hat{\sigma}_{k_j}^2}, \qquad k \in \{1,2\}^N,$$

are all identical. However, for $N \geq 3$, and for general choice of data $x_1, \ldots, x_N$ as above, this can only happen if $(\hat{\mu}_1, \hat{\sigma}_1) = (\hat{\mu}_2, \hat{\sigma}_2)$. This contradicts our hypothesis that the critical point is non-trivial. We conclude that all non-trivial critical points of the log-likelihood function (2) are transcendental.

In the multivariate case, the model parameters comprise the mixture weight $\alpha \in [0,1]$, mean vectors $\mu_1, \mu_2 \in \mathbb{R}^n$ and positive definite covariance matrices $\Sigma_1, \Sigma_2 \in \mathbb{R}^n$. Arguing as above, if a non-trivial critical $(\hat{\alpha}, \hat{\mu}_1, \hat{\mu}_2, \hat{\Sigma}_1, \hat{\Sigma}_2)$ is algebraic, then the Lindemann-Weierstrass Theorem implies that the numbers

$$\sum_{j=1}^{N} (x_j - \hat{\mu}_{k_j})^T \hat{\Sigma}_{k_j}^{-1} (x_j - \hat{\mu}_{k_j}), \qquad k \in \{1,2\}^N,$$

are all identical. For N sufficiently large and a general choice of $x_1, \ldots, x_N$ in $\mathbb{R}^n$, the 2^N numbers are identical only if $(\hat{\mu}_1, \hat{\Sigma}_1) = (\hat{\mu}_2, \hat{\Sigma}_2)$. Again, this constitutes a contradiction to the hypothesis that $(\hat{\alpha}, \hat{\mu}_1, \hat{\mu}_2, \hat{\Sigma}_1, \hat{\Sigma}_2)$ is non-trivial.   □

Many variations and specializations of the Gaussian mixture model are used in applications. In the case $n = 1$, the variances are sometimes assumed equal, so $\sigma_1 = \sigma_2$ for the above two-mixture. This avoids the issue of an unbounded likelihood function (as long as $N \geq 3$). Our proof of Theorem 1 applies to this setting. In higher dimensions ($n \geq 2$), the covariance matrices are sometimes assumed arbitrary and distinct, sometimes arbitrary and equal, but often also have special structure such as being diagonal. Various default choices are discussed in the paper [9] that introduces the R package MCLUST. Our results imply that maximum likelihood estimation is transcendental for all these MCLUST models.

Example 1. We illustrate Theorem 1 for a specialization of (1) obtained by fixing three parameters: $\mu_2 = 0$ and $\sigma_1 = \sigma_2 = 1/\sqrt{2}$. The remaining two free parameters are α and $\mu = \mu_1$. We take only $N = 2$ data points, namely $x_1 = 0$ and $x_2 = x > 0$. Omitting an additive constant, our log-likelihood function equals

$$\ell(\alpha, \mu) = \log\left(\alpha \cdot e^{-\mu^2} + (1-\alpha)\right) + \log\left(\alpha \cdot e^{-(\mu-x)^2} + (1-\alpha) \cdot e^{-x^2}\right). \quad (6)$$

For a concrete example take $x = 2$. The graph of (6) for this choice is shown in Fig. 1. By maximizing $\ell(\alpha, \mu)$ numerically, we find the parameter estimates

$$\hat{\alpha} = 0.50173262959803874\ldots \quad \text{and} \quad \hat{\mu} = 1.95742494230308167\ldots \quad (7)$$

Our technique can be applied to prove that $\hat{\alpha}$ and $\hat{\mu}$ are transcendental over $\mathbb{Q}$. We illustrate this for $\hat{\mu}$.

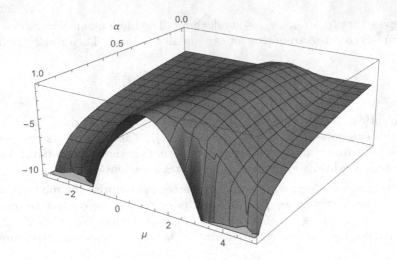

Fig. 1. Graph of the log-likelihood function for two data points $x_1 = 0$ and $x_2 = 2$.

For any $x \in \mathbb{R}$, the function $\ell(\alpha, \mu)$ is bounded from above and achieves its maximum on $[0, 1] \times \mathbb{R}$. If $x > 0$ is large, then any global maximum $(\hat{\alpha}, \hat{\mu})$ of ℓ is in the interior of $[0, 1] \times \mathbb{R}$ and satisfies $0 < \hat{\mu} \leq x$. According to a `Mathematica` computation, the choice $x \geq 1.56125...$ suffices for this. Assume that this holds. Setting the two partial derivatives equal to zero and eliminating the unknown α in a further `Mathematica` computation, the critical equation for μ is found to be

$$(x - \mu)e^{\mu^2} - x + \mu e^{-\mu(2x - \mu)} = 0. \tag{8}$$

Suppose for contradiction that both x and $\hat{\mu}$ are algebraic numbers over $\mathbb{Q}$. Since $0 < \hat{\mu} \leq x$, we have $-\hat{\mu}(2x - \hat{\mu}) < 0 < \hat{\mu}^2$. Hence $u_1 = \hat{\mu}^2$, $u_2 = 0$ and $u_3 = -\hat{\mu}(2x - \hat{\mu})$ are distinct algebraic numbers. The Lindemann-Weierstrass Theorem implies that e^{u_1}, e^{u_2} and e^{u_3} are linearly independent over the field of algebraic numbers. However, from (8) we know that

$$(x - \hat{\mu}) \cdot e^{u_1} - x \cdot e^{u_2} + \hat{\mu} \cdot e^{u_3} = 0.$$

This is a contradiction. We conclude that the number $\hat{\mu}$ is transcendental over $\mathbb{Q}$.

3   Many Critical Points

Theorem 1 shows that Gaussian mixtures do not admit an ML degree. This raises the question of how to find any bound for the number of critical points.

Problem 1. Does there exist a universal bound on the number of non-trivial critical points for the log-likelihood function of the mixture of two univariate Gaussians? Or, can we find a sequence of samples on the real line such that the number of non-trivial critical points increases beyond any bound?

Table 1. Seven critical points of the log-likelihood function in Theorem 3 with $K = 7$.

k	α	μ_1	μ_2	σ_1	σ_2	Log-likelihood
1	0.1311958	1.098998	4.553174	0.09999497	1.746049	-27.2918782147578
2	0.1032031	2.097836	4.330408	0.09997658	1.988948	-28.6397463805501
3	0.07883084	3.097929	4.185754	0.09997856	2.06374	-29.1550277534757
4	0.06897294	4.1	4.1	0.1	2.07517	-29.2858981551065
5	0.07883084	5.102071	4.014246	0.09997856	2.06374	-29.1550277534757
6	0.1032031	6.102164	3.869592	0.09997658	1.988948	-28.6397463805501
7	0.1311958	7.101002	3.646826	0.09999497	1.746049	-27.2918782147578

We shall resolve this problem by answering the second question affirmatively. The idea behind our solution is to choose a sample consisting of many well-separated clusters of size 2. Then each cluster gives rise to a distinct non-trivial critical point $(\hat{\alpha}, \hat{\mu}_1, \hat{\mu}_2, \hat{\sigma}_1, \hat{\sigma}_2)$ of the log-likelihood function ℓ from (2). We propose one particular choice of data, but many others would work too.

Theorem 3. *Fix sample size $N = 2K$ for $K \geq 2$, and take the ordered sample $(x_1, \ldots, x_{2K}) = (1, 1.2, 2, 2.2, \ldots, K, K+0.2)$. Then, for each $k \in \{1, \ldots, K\}$, the log-likelihood function ℓ from (2) has a non-trivial critical point with $k < \hat{\mu}_1 < k + 0.2$. Hence, there are at least K non-trivial critical points.*

Before turning to the proof, we offer a numerical illustration.

Example 2. For $K = 7$, we have $N = 14$ data points in the interval $[1, 7.2]$. Running the EM algorithm (as explained in the proof of Theorem 3 below) yields the non-trivial critical points reported in Table 1. Their μ_1 coordinates are seen to be close to the cluster midpoints $k + 0.1$ for all k. The observed symmetry under reversing the order of the rows also holds for all larger K.

Our proof of Theorem 3 will be based on the *EM algorithm*. We first recall this algorithm. Let $f_{\alpha,\mu,\sigma}$ be the mixture density from (1), and let

$$f_j(x) = \frac{1}{\sqrt{2\pi}\,\sigma_j} \exp\left(-\frac{(x - \mu_j)^2}{2\sigma_j^2}\right), \quad j = 1, 2,$$

be the two Gaussian component densities. Define

$$\gamma_i = \frac{\alpha \cdot f_1(x_i)}{f_{\alpha,\mu,\sigma}(x_i)}, \tag{9}$$

which can be interpreted as the conditional probability that data point x_i belongs to the first mixture component. Further, define $N_1 = \sum_{i=1}^{N} \gamma_i$ and $N_2 = N - N_1$, which are expected cluster sizes. Following [4, Sect. 9.2.2], the likelihood equations for our model can be written in the following fixed-point form:

$$\alpha = \frac{N_1}{N},\tag{10}$$

$$\mu_1 = \frac{1}{N_1} \sum_{i=1}^{N} \gamma_i x_i, \qquad\qquad \mu_2 = \frac{1}{N_2} \sum_{i=1}^{N} (1 - \gamma_i) x_i,\tag{11}$$

$$\sigma_1 = \frac{1}{N_1} \sum_{i=1}^{N} \gamma_i (x_i - \mu_1)^2, \qquad \sigma_2 = \frac{1}{N_2} \sum_{i=1}^{N} (1 - \gamma_i)(x_i - \mu_2)^2.\tag{12}$$

In the present context, the EM algorithm amounts to solving these equations iteratively. More precisely, consider any starting point $(\alpha, \mu_1, \mu_2, \sigma_1, \sigma_2)$. Then the E-step ("expectation") computes the estimated frequencies γ_i via (9). In the subsequent M-step ("maximization"), one obtains a new parameter vector $(\alpha, \mu_1, \mu_2, \sigma_1, \sigma_2)$ by evaluating the right-hand sides of the Eqs. (10)–(12). The two steps are repeated until a fixed point is reached, up to the desired numerical accuracy. The updates never decrease the log-likelihood. For our problem it can be shown that the algorithm will converge to a critical point; see e.g. [16].

Proof (of Theorem 3). Fix $k \in \{1, \ldots, K\}$. We choose starting parameter values to suggest that the pair $(x_{2k-1}, x_{2k}) = (k, k + 0.2)$ belongs to the first mixture component, while the rest of the sample belongs to the second. Explicitly, we set

$$\alpha = \frac{2}{N} = \frac{1}{K},$$

$$\mu_1 = k + 0.1, \qquad \mu_2 = \frac{K^2 + 1.2K - 2k - 0.2}{2(K - 1)},$$

$$\sigma_1 = 0.1, \qquad \sigma_2 = \frac{\sqrt{\frac{1}{12}K^4 - \frac{1}{3}K^3 + (k - \frac{43}{75})K^2 - (k^2 - k + \frac{14}{75})K + 0.01}}{K - 1}.$$

We shall argue that, when running the EM algorithm, the parameters will always stay close to these starting values. Specifically, we claim that throughout all EM iterations, the parameter values satisfy the inequalities

$$\frac{1}{4K} \le \alpha \le \frac{1}{K}, \qquad 0.09 \le \mu_1 - k \le 0.11, \qquad 0.099 \le \sigma_1 \le 0.105,\tag{13}$$

$$\frac{K}{2} + 0.1 \le \mu_2 \le \frac{K}{2} + 1.1,\tag{14}$$

$$\sqrt{\frac{K^2}{12} - \frac{K}{6} + 0.01} \le \sigma_2 \le \sqrt{\frac{K^2}{12} + \frac{K}{12} + 0.01}.\tag{15}$$

The starting values proposed above obviously satisfy the inequalities in (13), and it is not difficult to check that (14) and (15) are satisfied as well. To prove the theorem, it remains to show that (13)–(15) continue to hold after an EM update.

In the remainder, we assume that $K > 22$. For smaller values of K the claim of the theorem can be checked by running the EM algorithm. In particular, for $K > 3$, the second standard deviation satisfies the simpler bounds

$$\frac{K}{\sqrt{12}} - \frac{\sqrt{3}}{5} \leq \sigma_2 \leq \frac{K}{\sqrt{12}} + \frac{\sqrt{3}}{12}. \tag{16}$$

A key property is that the quantity γ_i, computed in the E-step, is always very close to zero for $i \neq 2k - 1, 2k$. To see why, rewrite (9) as

$$\gamma_i = \frac{1}{1 + \frac{1-\alpha}{\alpha} \frac{f_2(x_i)}{f_1(x_i)}} = \frac{1}{1 + \frac{1-\alpha}{\alpha} \frac{\sigma_1}{\sigma_2} \exp\left\{\frac{1}{2}\left(\left(\frac{x_i - \mu_1}{\sigma_1}\right)^2 - \left(\frac{x_i - \mu_2}{\sigma_2}\right)^2\right)\right\}}.$$

Since $\alpha \leq 1/K$, we have $\frac{1-\alpha}{\alpha} \geq K - 1$. On the other hand, $\frac{\sigma_1}{\sigma_2} \geq \frac{0.099}{K/\sqrt{12} + \sqrt{3}/12}$. Using that $K > 22$, their product is thus bounded below by 0.3209. Turning to the exponential term, the second inequality in (13) implies that $|x_i - \mu_1| \geq 0.89$ for $i = 2k - 2$ or $i = 2k + 1$, which index the data points closest to the kth pair. Using (16), we obtain

$$\left(\frac{x_i - \mu_1}{\sigma_1}\right)^2 - \left(\frac{x_i - \mu_2}{\sigma_2}\right)^2 \geq \left(\frac{0.89}{0.105}\right)^2 - \left(\frac{K/2 + 0.1}{K/\sqrt{12} - \sqrt{3}/5}\right)^2 \geq 67.86.$$

From $e^{33.93} > 5.4 \cdot 10^{14}$, we deduce that $\gamma_i < 10^{-14}$. The exponential term becomes only smaller as the considered data point x_i move away from the kth pair. As $|i - (2k - 1/2)|$ increases, γ_i decreases and can be bounded above by a geometric progression starting at 10^{-14} and with ratio 10^{-54}. This makes γ_i with $i \neq 2k, 2k - 1$ negligible. Indeed, from the limit of geometric series, we have

$$s_1 = \sum_{i \neq 2k-1, 2k} \gamma_i < 10^{-13}, \tag{17}$$

and similarly, $s_2 = \sum_{i \neq 2k-1, 2k} \gamma_i (x_i - k)$ satisfies

$$|s_2| = |\gamma_{2k-2}(-0.8) + \gamma_{2k+1}(1) + \gamma_{2k-3}(-1) + \gamma_{2k+2}(1.2) + \ldots| < 10^{-13}. \tag{18}$$

The two sums s_1 and s_2 are relevant for the M-step.

The probabilities γ_{2k-1} and γ_{2k} give the main contribution to the averages that are evaluated in the M-step. They satisfy $0.2621 \leq \gamma_{2k-1}, \gamma_{2k} \leq 0.9219$. Moreover, we may show that the values of γ_{2k-1} and γ_{2k} are similar, namely:

$$0.8298 \leq \frac{\gamma_{2k-1}}{\gamma_{2k}} \leq 1.2213, \tag{19}$$

which we prove by writing

$$\frac{\gamma_{2k-1}}{\gamma_{2k}} = \frac{1 + y \exp(z/2)}{1 + y},$$

and using $K > 22$ to bound

$$y = \frac{1-\alpha}{\alpha} \frac{\sigma_1}{\sigma_2} \exp\left\{\frac{1}{2}\left(\left(\frac{k - \mu_1}{\sigma_1}\right)^2 - \left(\frac{k - \mu_2}{\sigma_2}\right)^2\right)\right\},$$

$$z = \frac{0.4(k - \mu_1) + 0.04}{\sigma_1^2} - \frac{0.4(k - \mu_2) + 0.04}{\sigma_2^2}.$$

Bringing it all together, we have

$$\mu_1 = \frac{1}{N_1} \sum_{i=1}^{N} \gamma_i x_i = k + \frac{0.2\gamma_{2k} + s_2}{\gamma_{2k-1} + \gamma_{2k} + s_1}.$$

Using $\gamma_{2k} + \gamma_{2k-1} > 0.5$ and (18), as well as the lower bound in (19), we find

$$\mu_1 - k \le \frac{0.2\gamma_{2k}}{\gamma_{2k-1} + \gamma_{2k}} + \frac{s_2}{\gamma_{2k-1} + \gamma_{2k}} \le \frac{0.2\gamma_{2k}}{0.8298\gamma_{2k} + \gamma_{2k}} + 10^{-12} \le 0.11.$$

Using the upper bound in (19), we also have $0.09 \le \mu_1 - k$. Hence, the second inequality in (13) holds.

The inequalities for the other parameters are verified similarly. For instance,

$$\frac{1}{4K} < \frac{0.2621 + 0.2621}{2K} \le \frac{\gamma_{2k-1} + \gamma_{2k} + s_1}{2K} \le \frac{0.9219 + 0.9219 + 10^{-13}}{2K} < \frac{1}{K}$$

holds for $\alpha = \frac{N_1}{N}$. Therefore, the first inequality in (13) continues to be true.

We conclude that running the EM algorithm from the chosen starting values yields a sequence of parameter vectors that satisfy the inequalities (13)–(16). The sequence has at least one limit point, which must be a non-trivial critical point of the log-likelihood function. Therefore, for every $k = 1, \ldots, K$, the log-likelihood function has a non-trivial critical point with $\mu_1 \in (k, k + 0.2)$.   □

4   Conclusion

We showed that the maximum likelihood estimator (MLE) in Gaussian mixture models is not an algebraic function of the data, and that the log-likelihood function may have arbitrarily many critical points. Hence, in contrast to the models studied so far in algebraic statistics [5,8,12,19], there is no notion of an ML degree for Gaussian mixtures. However, certified likelihood inference may still be possible, via transcendental root separation bounds, as in [6,7].

Remark 3. The *Cauchy-location model*, treated in [17], is an example where the ML estimation is algebraic but the ML degree, and also the maximum number of local maxima, depends on the sample size and increases beyond any bound.

Remark 4. The ML estimation problem admits a population/infinite-sample version. Here the maximization of the likelihood function is replaced by *minimization of the Kullback-Leibler divergence* between a given data-generating distribution and the distributions in the model. The question of whether this population problem is subject to local but not global maxima was raised in [18]—in the context of Gaussian mixtures with known and equal variances. It is known that the Kullback-Leibler divergence for such Gaussian mixtures is not an analytic function [21, Sect. 7.8]. Readers of Japanese should be able to find details in [22].

As previously mentioned, Theorem 1 shows that likelihood inference is in a fundamental way more complicated than the classical *method of moments* [15]. The latter involves only the solution of polynomial equation systems. This was recognized also in the computer science literature on learning Gaussian mixtures [3,10,14], where most of the recent progress is based on variants of the method of moments rather than likelihood inference. We refer to [1] for a study of the method of moments from an algebraic perspective. Section 5 in that paper illustrates the behavior of Pearson's method for the sample used in Theorem 3.

Acknowledgements. CA and BS were supported by the Einstein Foundation Berlin. MD and BS also thank the US National Science Foundation (DMS-1305154 and DMS-1419018).

References

1. Améndola, C., Faugère, J.-C., Sturmfels, B.: Moment varieties of Gaussian mixtures. J. Algebraic Stat. arXiv:1510.04654
2. Baker, A.: Transcendental Number Theory. Cambridge University Press, London (1975)
3. Belkin, M., Sinha, K.: Polynomial learning of distribution families. SIAM J. Comput. **44**(4), 889–911 (2015)
4. Bishop, C.M.: Pattern Recognition and Machine Learning. Information Science and Statistics. Springer, New York (2006)
5. Buot, M., Hoşten, S., Richards, D.: Counting and locating the solutions of polynomial systems of maximum likelihood equations. II. The Behrens-Fisher problem. Stat. Sin. **17**(4), 1343–1354 (2007)
6. Chang, E.-C., Choi, S.W., Kwon, D., Park, H., Yap, C.: Shortest paths for disc obstacles is computable. Int. J. Comput. Geom. Appl. **16**, 567–590 (2006)
7. Choi, S.W., Pae, S., Park, H., Yap, C.: Decidability of collision between a helical motion and an algebraic motion. In: Hanrot, G., Zimmermann, P. (eds.) 7th Conference on Real Numbers and Computers, pp. 69–82. LORIA, Nancy (2006)
8. Drton, M., Sturmfels, B., Sullivant, S.: Lectures on Algebraic Statistics. Oberwolfach Seminars, vol. 39. Birkhäuser, Basel (2009)
9. Fraley, C., Raftery, A.E.: Enhanced model-based clustering, density estimation, and discriminant analysis software: MCLUST. J. Classif. **20**, 263–286 (2003)
10. Ge, R., Huang, Q., Kakade, S.: Learning mixtures of Gaussians in high dimensions. In: STOC 2015, Proceedings of the Forty-Seventh Annual ACM on Symposium on Theory of Computing, pp. 761–770 (2015)
11. Gelfond, A.O.: Transcendental and Algebraic Numbers. Translated by Leo F. Boron, Dover Publications, New York (1960)
12. Gross, E., Drton, M., Petrović, S.: Maximum likelihood degree of variance component models. Electron. J. Stat. **6**, 993–1016 (2012)
13. Huh, J., Sturmfels, B.: Likelihood geometry. In: Conca, A., et al. (eds.) Combinatorial Algebraic Geometry. Lecture Notes in Math., vol. 2108, pp. 63–117. Springer, Heidelberg (2014)
14. Moitra, A., Valiant, G.: Settling the polynomial learnability of mixtures of Gaussians. In: IEEE 51st Annual Symposium on Foundations of Computer Science, pp. 93–102 (2010)

15. Pearson, K.: Contributions to the mathematical theory of evolution. Philos. Trans. R. Soc. Lond. A **185**, 71–110 (1894)
16. Redner, R.A., Walker, H.F.: Mixture densities, maximum likelihood and the EM algorithm. SIAM Rev. **26**, 195–239 (1984)
17. Reeds, J.A.: Asymptotic number of roots of Cauchy location likelihood equations. Ann. Statist. **13**(2), 775–784 (1985)
18. Srebro, N.: Are there local maxima in the infinite-sample likelihood of Gaussian mixture estimation? In: Bshouty, N.H., Gentile, C. (eds.) COLT. LNCS (LNAI), vol. 4539, pp. 628–629. Springer, Heidelberg (2007)
19. Sturmfels, B., Uhler, C.: Multivariate Gaussian, semidefinite matrix completion, and convex algebraic geometry. Ann. Inst. Statist. Math. **62**(4), 603–638 (2010)
20. Teicher, H.: Identifiability of finite mixtures. Ann. Math. Stat. **34**, 1265–1269 (1963)
21. Watanabe, S.: Algebraic Geometry and Statistical Learning Theory. Monographs on Applied and Computational Mathematics, vol. 25. Cambridge University Press, Cambridge (2009)
22. Watanabe, S., Yamazaki, K., Aoyagi, M.: Kullback information of normal mixture is not an analytic function. IEICE Technical report, NC2004-50 (2004)

On the Quality of Some Root-Bounds

Prashant Batra(✉)

Institute for Reliable Computing, Hamburg University of Technology,
Schwarzenbergstraße 95, 21071 Hamburg, Germany
batra@tuhh.de

Abstract. Bounds for the maximum modulus of all positive (or all complex) roots of a polynomial are a fundamental building block of algorithms involving algebraic equations. We apply known results to show which are the salient features of the Lagrange (real) root-bound as well as the related bound by Fujiwara. For a polynomial of degree n, we construct a bound of relative overestimation at most $1.72n$ which overestimates the Cauchy root by a factor of two at most. This can be carried over to the bounds by Kioustelidis and Hong. Giving a very short variant of a recent proof presented by Collins, we sketch a way to further definite, measurable improvement.

Keywords: Maximum modulus of polynomial roots · Maximum overestimation · Improvements of Lagrange's bound

I.   It is well-known that the quality of root-bounds substantially influences the performance of root-finding, root-approximation and root-isolation. Numerical methods, like Newton's [14], as well as algebraic methods, e.g., isolation methods based on continued fraction expansions [3,5,6,17,18], vary in convergence speed and computational effort with the starting point determined by the root-bound.

Recently, Collins [4] emphasized the usefulness of Lagrange's real root-bound [11]: Let p be a real polynomial with Taylor expansion

$$p(x) = x^n + \sum_{i=0}^{n-1} a_i x^i \quad \in \mathbb{R}[x], \tag{1}$$

and sort the $|a_{n-i}|^{1/i}$ in non-decreasing order. Denoting the sum of the last two of those elements by $L(p)$, we have then

$$\max\{x_i \in \mathbb{R} : p(x_i) = 0\} \leq L(p).$$

After giving a proof of Lagrange's bound, Collins [4] points out that the bound $L(p)$ is no weaker than a bound (called here and in the following) $FM(p)$:

$$FM(p) := 2 \max_{i=1,\dots,n} \{|a_{n-i}|^{1/i}\}.$$

Obviously, $L(p) \leq FM(p)$, but what else is known about the numerical quality of $L(p)$? How does it compare to the best possible values, and the myriads

© Springer International Publishing Switzerland 2016
I.S. Kotsireas et al. (Eds.): MACIS 2015, LNCS 9582, pp. 591–595, 2016.
DOI: 10.1007/978-3-319-32859-1_50

of other root-bounds, for samples *cf., e.g.,* [7,12,19]? To answer these questions, let us recall first the taxonomy for root-bounds proposed by van der Sluis [20]. After recalling this frame-work and some results, we give estimates for the overestimation of $L(p)$.

For a root-bound functional applied to $f \in \mathbb{C}[x]$, $f(x) = \sum_{i=0}^{n} a_i x^i$, we may value the following qualities:

(i) computational complexity, and

(ii) overestimation with respect to $\mu(f) := \max\{|x_i| : f(x_i) = 0\}$, as well as to
$$\varrho(f) := \max\left\{\lambda \geq 0 : \lambda^n - \sum_{i=1}^{n} |a_{n-i}|\lambda^{n-i} = 0\right\}.$$

When no special structure of the considered polynomial can be used, we might like to add

(iii) homogeneity (the root-bound for f scales for the polynomial $\gamma^n f(x/\gamma)$).

For complex polynomials, the requirement that the root-bound depends only on the moduli $|a_{n-i}|$ of the coefficients is very common, and the respective requirement is

(iv) absoluteness.

Finally, the technical requirement that the root-bound functional is a non-decreasing function of the absolute values $|a_i|$ is important, and denoted as

(v) normality.

Requirements (iii) – (v) are met by the following root-bound due to Fujiwara [8]:

$$F(p) := 2 \max\left\{|a_0/2|^{1/n} ; |a_1|^{1/n-1}, \ldots, |a_{n-2}|^{1/2}, |a_{n-1}|\right\}.$$

Please note that this is not exactly the bound $FM(p)$ attributed by Collins [4] to Fujiwara. Regarding property (ii) for the bound $F(p)$ we have, thanks to [20], in comparison to the Cauchy root $\varrho(p)$ and the true maximum modulus $\mu(p)$, that for p of degree n

$$\varrho(p) \leq F(p) \leq 2 \cdot \varrho(p),$$

and

$$\mu(p) \leq F(p) \leq 2 \cdot n \cdot \mu(p).$$

To retain the nice overestimation estimates, we combine both functionals $L(\cdot)$ and $F(\cdot)$ in a new one

$$LF(p) := \min\{L(p), F(p)\}. \tag{2}$$

Thus, the relative overestimation of $LF(\cdot)$ cannot exceed $2n$. But, as $LF(\cdot)$ is normal and absolute, the relative overestimation is attained for a polynomial with an n-fold root [20, Theorem 3.8.(d)]. By homogeneity [20], we may place this root at 1. We obtain the following

Theorem 1. *The relative overestimation of $LF(\cdot)$ for a polynomial of degree n is at most $n + \sqrt{\frac{n(n-1)}{2}}$, no larger than $n(1 + 1/\sqrt{2}) \leq 1.72n$.*

From our own analysis [1], $LF(p)$ compares favorably to Kojima's bound [12]

$$2\max\left\{\left|\frac{a_0}{2a_1}\right|;\left|\frac{a_1}{a_2}\right|,\ldots,\left|\frac{a_{n-2}}{a_{n-1}}\right|,|a_{n-1}|\right\}.$$

As van der Sluis [20] proved, any absolute root-bound functional must realize a maximum relative overestimation of at least $1.4n$. This motivated van der Sluis to call $F(\cdot)$, which has relative overestimation bounded by $2n$, a nearly optimal root-bound functional. Thus, $LF(\cdot)$ is closer-to-optimal in a precise sense with bounded relative overestimation $1.72n < 2n$. Let us state a first improvement of Theorem 2.7 in [20] as

Proposition 1. *$LF(\cdot)$ is a nearly optimal root-bound functional, a good approximation from above can be easily obtained, and it is to be recommended among all absolute root-bound functionals.*

Of course, all bounds that may be derived from $F(\cdot)$, especially Kioustelidis' bound for positive real roots [10], and Hong's absolute positivity bound [9], and those considered in the generalized framework in [2], can be re-modelled on $LF(\cdot)$, computing a joint minimum as in (2). Especially, in combination with $F(\cdot)$ (as above in (2)), we retain the improved overestimation bounds.

Moreover, $F(p)$ as well as $LF(p)$ and the other mentioned bounds can be computed in complexity $\mathcal{O}(\deg p)$ - thanks to the remarkable result in [13].

II. Sophisticated methods for root-bounds are unwanted, this is common sense *viz.* Runge's remark in [16, Sect. 2, p. 408], or, *e.g.*, [15, Chap. 8, p. 270]. What can we do, when p denotes a monic complex polynomial with Taylor expansion

$$p(x) = x^n + \sum_{i=1}^{n} a_{n-i}x^{n-i} \in \mathbb{C}[x],$$

to improve the upper bound $LF(p)$ of $\mu(p)$ without spending, say, more than $\mathcal{O}(n)$ field operations (and approximate logarithms from the floating-point representation)? To this end, let us first consider the latest proof of Lagrange's bound in [4]. If $\max_{i=1,\ldots,n}\{|a_{n-1}|^{1/i}\} =: R$, and

$$k := \min\{i \in \mathbb{N} : \quad 1 \leq i \leq n, \quad |a_{n-1}|^{1/i} = R\},$$

and

$$r := \max_{\substack{i=1,\ldots,n \\ i \neq k}} \{|a_{n-i}|^{1/i}\},$$

then we may express Lagrange's bound $L(p)$ as

$$L(p) = R + r.$$

To establish the bound, we may consider as in [4], the polynomials A_i defined by

$$A_i(x) := x^n - R^i x^{n-i} - \sum_{\substack{j=1 \\ j \neq i}}^{n} r^j x^{n-j},$$

with

$$\max_{i=1,\ldots,n} \{\mu(A_i)\} \geq \mu(p).$$

As noted in [4], we have

$$A_{i+1}(x + R + r) - A_i(x + R + r) \in \mathbb{R}_{\geq 0}[x] \quad \text{for} \quad i = n-1, \ldots, 1.$$

Thus, if $\mu(A_1) \leq R + r$, then $\mu(p) \leq R + r = L(p)$.
Claim. $\mu(A_1) \leq R + r$.

Proof. It is easy to verify that

$$\text{for} \quad n = 1 : L(p) = R \geq \mu(p),$$
$$\text{and} \quad n = 2 : L(p) = R + r \geq \mu(p).$$

Let us assume then $n \geq 3$, and compute

$$A_1(x)(x - r) = x^{n-1}(x^2 - (r + R)x + r(R - r)) + r^{n+1}.$$

This is obviously positive for $x > R + r$. Thus,

$$\mu(A_1) \leq R + r, \quad \text{and in conclusion we have} \quad \mu(p) \leq R + r = L(p). \qquad \square$$

Our condensation of the proof facilitates numerical improvements of the bound

$$L(p) = R + r.$$

Let us give one indicative sketch:
Taking into account the smallest index k, where $\max_{i=1,\ldots,n} \{|a_{n-i}|^{1/i}\} = R$ is realized, we may compute improved bounds for the fewnomial

$$A_k(x)(x - r)$$

as well as for those shifts $\tau \in \mathbb{R}_{\geq 0}$ which yield polynomial differences

$$A_{i+1}(x + \tau) - A_i(x + \tau) \in \mathbb{R}_{\geq 0}[x], \quad i = k, \ldots, n-1,$$

with non-negative coefficients.

Several simple, definite improvements of $L(p)$ stemming from different generalizations of a proof-technique by Lagrange [11] have been considered by the author. Some of the new root-bound functionals turn out to be non-normal (in the sense of [20, Definition 3.7]), which leads to a new quest for overestimation bounds.

References

1. Batra, P.: A property of the nearly optimal root-bound. J. Comput. Appl. Math. **167**(2), 489–491 (2004)
2. Batra, P., Sharma, V.: Bounds on absolute positiveness of multivariate polynomials. J. Symb. Comput. **45**(6), 617–628 (2010)
3. Burr, M.A., Krahmer, F.: SqFreeEVAL: an (almost) optimal real-root isolation algorithm. J. Symb. Comput. **47**(2), 153–166 (2012)
4. Collins, G.E.: Krandick's proof of Lagrange's real root bound claim. J. Symb. Comput. **70**, 106–111 (2015)
5. Collins, G.E.: Continued fraction real root isolation using the Hong bound. J. Symb. Comput. **72**, 21–54 (2016)
6. Collins, G.E., Krandick, W.: On the computing time of the continued fractions method. J. Symb. Comput. **47**(11), 1372–1412 (2012)
7. Dieudonné, J.: La théorie analytique des polynômes d'une variable (à coefficients quelconques). Gauthier-Villars, Paris (1938)
8. Fujiwara, M.: Über die obere Schranke des absoluten Betrages der Wurzeln einer algebraischen Gleichung. Tôhoku Math. J. **10**, 167–171 (1916)
9. Hong, H.: Bounds for absolute positiveness of multivariate polynomials. J. Symb. Comput. **25**(5), 571–585 (1998)
10. Kioustelidis, J.B.: Bounds for positive roots of polynomials. J. Comput. Appl. Math. **16**(2), 241–244 (1986)
11. Lagrange, J.-L.: Sur la résolution des équations numériques. In: Mémoires de l'Académie royale des Sciences et Belles-lettres de Berlin, t. XXIII, pp. 539–578 (1769)
12. Marden, M.: Geometry of Polynomials. AMS Mathematical Surveys 3, 2nd edn. AMS, Providence, Rhode Island (1966)
13. Mehlhorn, K., Ray, S.: Faster algorithms for computing Hong's bound on absolute positiveness. J. Symb. Comput. **45**(6), 677–683 (2010)
14. Ostrowski, A.: Solution of Equations in Euclidean and Banach Spaces, 3rd edn. Academic Press, New York (1973)
15. Rahman, Q.I., Schmeisser, G.: Analytic Theory of Polynomials. Oxford University Press, Oxford (2002)
16. Runge, C.: Separation und Approximation der Wurzeln. In: Encyklopädie der mathematischen Wissenschaften, vol. 1, pp. 404–448. Verlag Teubner, Leipzig (1899)
17. Sagraloff, M.: On the complexity of the Descartes method when using approximate arithmetic. J. Symb. Comput. **65**, 79–110 (2014)
18. Sharma, V.: Complexity of real root isolation using continued fractions. Theor. Comput. Sci. **409**(2), 292–310 (2008)
19. Specht, W.: Algebraische Gleichungen mit reellen oder komplexen Koeffizienten. In: Enzyklopädie der mathematischen Wissenschaften, Band I, Heft 3, Teil II. B.G. Teubner Verlagsgesellschaft, Stuttgart. Zweite, völlig neubearbeitete Auflage (1958)
20. van der Sluis, A.: Upperbounds for roots of polynomials. Numer. Math. **15**, 250–262 (1970)

Relative Hilbert-Post Completeness
for Exceptions

Jean-Guillaume Dumas[1]([⊠]), Dominique Duval[1], Burak Ekici[1], Damien Pous[2],
and Jean-Claude Reynaud[3]

[1] Laboratoire J. Kuntzmann, Université Grenoble Alpes, 51, rue des Mathématiques,
UMR CNRS 5224, bp 53X, 38041 Grenoble, France
{Jean-Guillaume.Dumas,Dominique.Duval,Burak.Ekici}@imag.fr
[2] Plume Team, CNRS, ENS Lyon, Université de Lyon, INRIA,
UMR 5668, Lyon, France
Damien.Pous@ens-lyon.fr
[3] Reynaud Consulting (RC), Claix, France
Jean-Claude.Reynaud@imag.fr

Abstract. A theory is complete if it does not contain a contradiction,
while all of its proper extensions do. In this paper, first we introduce
a relative notion of syntactic completeness; then we prove that adding
exceptions to a programming language can be done in such a way that
the completeness of the language is not made worse. These proofs are
formalized in a logical system which is close to the usual syntax for
exceptions, and they have been checked with the proof assistant Coq.

1 Introduction

In computer science, an exception is an abnormal event occurring during the execution of a program. A mechanism for handling exceptions consists of two parts:
an exception is *raised* when an abnormal event occurs, and it can be *handled*
later, by switching the execution to a specific subprogram. Such a mechanism is
very helpful, but it is difficult for programmers to reason about it. A difficulty for
reasoning about programs involving exceptions is that they are *computational
effects*, in the sense that their syntax does not look like their interpretation:
typically, a piece of program with arguments in X that returns a value in Y is
interpreted as a function from $X + E$ to $Y + E$ where E is the set of exceptions. On the one hand, reasoning with $f : X \to Y$ is close to the syntax, but
it is error-prone because it is not sound with respect to the semantics. On the
other hand, reasoning with $f : X + E \to Y + E$ is sound but it loses most of
the interest of the exception mechanism, where the propagation of exceptions is
implicit: syntactically, $f : X \to Y$ may be followed by any $g : Y \to Z$, since the
mechanism of exceptions will take care of propagating the exceptions raised by
f, if any. Another difficulty for reasoning about programs involving exceptions
is that the handling mechanism is encapsulated in a `try-catch` block, while the
behaviour of this mechanism is easier to explain in two parts (see for instance [11,
Chap. 14] for Java or [3, Sect. 15] for C++): the `catch` part may recover from

I.S. Kotsireas et al. (Eds.): MACIS 2015, LNCS 9582, pp. 596–610, 2016.
DOI: 10.1007/978-3-319-32859-1_51

exceptions, so that its interpretation may be any $f : X + E \rightarrow Y + E$, but the `try-catch` block must propagate exceptions, so that its interpretation is determined by some $f : X \rightarrow Y + E$.

In [8] we defined a logical system for reasoning about states and exceptions and we used it for getting certified proofs of properties of programs in computer algebra, with an application to exact linear algebra. This logical system is called the *decorated logic* for states and exceptions. Here we focus on exceptions. The decorated logic for exceptions deals with $f : X \rightarrow Y$, without any mention of E, however it is sound thanks to a classification of the terms and the equations. Terms are classified, as in a programming language, according to the way they may interact with exceptions: a term either has no interaction with exceptions (it is "pure"), or it may raise exceptions and must propagate them, or it is allowed to catch exceptions (which may occur only inside the `catch` part of a `try-catch` block). The classification of equations follows a line that was introduced in [4]: besides the usual "strong" equations, interpreted as equalities of functions, in the decorated logic for exceptions there are also "weak" equations, interpreted as equalities of functions on non-exceptional arguments. This logic has been built so as to be sound, but little was known about its completeness. In this paper we prove a novel completeness result: the decorated logic for exceptions is *relatively Hilbert-Post complete*, which means that adding exceptions to a programming language can be done in such a way that the completeness of the language is not made worse. For this purpose, we first define and study the novel notion of *relative* Hilbert-Post completeness, which seems to be a relevant notion for the completeness of various computational effects: indeed, we prove that this notion is preserved when combining effects. Practically, this means that we have defined a decorated framework where reasoning about programs with and without exceptions are equivalent, in the following sense: if there exists an unprovable equation not contradicting the given decorated rules, then this equation is equivalent to a set of unprovable equations of the pure sublogic not contradicting its rules.

Informally, in classical logic, a consistent theory is one that does not contain a contradiction and a theory is complete if it is consistent, and none of its proper extensions is consistent. Now, the usual (*"absolute"*) Hilbert-Post completeness, also called Post completeness, is a syntactic notion of completeness which does not use any notion of negation, so that it is well-suited for equational logic. In a given logic L, we call *theory* a set of sentences which is deductively closed: everything you can derive from it (using the rules of L) is already in it. Then, more formally, a theory is *(Hilbert-Post) consistent* if it does not contain all sentences, and it is *(Hilbert-Post) complete* if it is consistent and if any sentence which is added to it generates an inconsistent theory [21, Definition 4].

All our completeness proofs have been verified with the Coq proof assistant. First, this shows that it is possible to formally prove that programs involving exceptions comply to their specifications. Second, this is of help for improving the confidence in the results. Indeed, for a human prover, proofs in a decorated logic require some care: they look very much like familiar equational proofs, but the application of a rule may be subject to restrictions on the decoration of the premises of the rule. The use of a proof assistant in order to check that these unusual restrictions were never violated has thus proven to be quite useful. Then, many of the proofs we give in this paper require a structural induction. There, the correspondence between our proofs and their Coq counterpart was eased, as structural induction is also at the core of the design of Coq.

A major difficulty for reasoning about programs involving exceptions, and more generally computational effects, is that their syntax does not look like their interpretation: typically, a piece of program from X to Y is not interpreted as a function from X to Y, because of the effects. The best-known algebraic approach for dealing with this problem has been initiated by Moggi: an effect is associated to a monad T, in such a way that the interpretation of a program from X to Y is a function from X to $T(Y)$ [14]: typically, for exceptions, $T(Y) = Y + E$. Other algebraic approaches include effect systems [13], Lawvere theories [18], algebraic handlers [19], comonads [16,22], dynamic logic [15], among others. Some completeness results have been obtained, for instance for (global) states [17] and for local states [20]. The aim of these approaches is to extend functional languages with tools for programming and proving side-effecting programs; implementations include Haskell [2], Idris [12], Eff [1], while Ynot [23] is a Coq library for writing and verifying imperative programs.

Differently, our aim is to build a logical system for proving properties of some families of programs written in widely used non-functional languages like Java or C++[1]. The salient features of our approach are that:

(1) The syntax of our logic is kept close to the syntax of programming languages. This is made possible by starting from a simple syntax without effect and by adding decorations, which often correspond to keywords of the languages, for taking the effects into account [5,7].
(2) We consider exceptions in two settings, the programming language and the core language. This enables for instance to separate the treatment, in proofs, of the matching between normal or exceptional behavior from the actual recovery after an exceptional behavior.

In Sect. 2 we introduce a *relative* notion of Hilbert-Post completeness in a logic L with respect to a sublogic L_0. Then in Sect. 3 we prove the relative Hilbert-Post completeness of a theory of exceptions based on the usual `throw` and `try-catch` statement constructors. We go further in Sect. 4 by establishing the relative Hilbert-Post completeness of a *core* theory for exceptions with individualized `TRY` and `CATCH` statement constructors, which is useful for expressing the behaviour of the `try-catch` blocks. All our completeness proofs have been verified with the Coq proof assistant and we therefore give the main ingredients of the framework used for this verification in Sect. 5 and the correspondence between our Coq package and the theorems and propositions of this paper in [10].

---

[1] For instance, a denotational semantics of our framework for exceptions, which relies on the common semantics of exceptions in these languages, was given in [8, Sect. 4].

2   Relative Hilbert-Post Completeness

Each logic in this paper comes with a *language*, which is a set of *formulas*, and with *deduction rules*. Deduction rules are used for deriving (or generating) *theorems*, which are some formulas, from some chosen formulas called *axioms*. A *theory* T is a set of theorems which is *deductively closed*, in the sense that every theorem which can be derived from T using the rules of the logic is already in T. We describe a set-theoretic *intended model* for each logic we introduce; the rules of the logic are designed so as to be *sound* with respect to this intended model. Given a logic L, the theories of L are partially ordered by inclusion. There is a maximal theory T_{max}, where all formulas are theorems. There is a minimal theory T_{min}, which is generated by the empty set of axioms. For all theories T and T', we denote by $T + T'$ the theory generated from T and T'.

Example 1. With this point of view there are many different *equational logics*, with the same deduction rules but with different languages, depending on the definition of *terms*. In an equational logic, formulas are *pairs of parallel terms* $(f, g) : X \rightarrow Y$ and theorems are *equations* $f \equiv g : X \rightarrow Y$. Typically, the language of an equational logic may be defined from a *signature* (made of sorts and operations). The deduction rules are such that the equations in a theory form a *congruence*, i.e., an equivalence relation compatible with the structure of the terms. For instance, we may consider the logic "of naturals" L_{nat}, with its language generated from the signature made of a sort N, a constant $0 : \mathbb{1} \rightarrow N$ and an operation $s : N \rightarrow N$. For this logic, the minimal theory is the theory "of naturals" T_{nat}, the maximal theory is such that $s^k \equiv s^\ell$ and $s^k \circ 0 \equiv s^\ell \circ 0$ for all natural numbers k and ℓ, and (for instance) the theory "of naturals modulo 6" T_{mod6} can be generated from the equation $s^6 \equiv id_N$. We consider models of equational logics in sets: each type X is interpreted as a set (still denoted X), which is a singleton when X is $\mathbb{1}$, each term $f : X \rightarrow Y$ as a function from X to Y (still denoted $f : X \rightarrow Y$), and each equation as an equality of functions.

Definition 2. *Given a logic L and its maximal theory T_{max}, a theory T is consistent if $T \neq T_{max}$, and it is* Hilbert-Post complete *if it is consistent and if any theory containing T coincides with T_{max} or with T.*

Example 3. In Example 1 we considered two theories for the logic L_{nat}: the theory "of naturals" T_{nat} and the theory "of naturals modulo 6" T_{mod6}. Since both are consistent and T_{mod6} contains T_{nat}, the theory T_{nat} is not Hilbert-Post complete. A Hilbert-Post complete theory for L_{nat} is made of all equations but $s \equiv id_N$, it can be generated from the axioms $s \circ 0 \equiv 0$ and $s \circ s \equiv s$.

If a logic L is an extension of a sublogic L_0, each theory T_0 of L_0 generates a theory $F(T_0)$ of L. Conversely, each theory T of L determines a theory $G(T)$ of L_0, made of the theorems of T which are formulas of L_0, so that $G(T_{max}) = T_{max,0}$. The functions F and G are monotone and they form a *Galois connection*, denoted $F \dashv G$: for each theory T of L and each theory T_0 of L_0 we have $F(T_0) \subseteq T$ if and only if $T_0 \subseteq G(T)$. It follows that $T_0 \subseteq G(F(T_0))$ and $F(G(T)) \subseteq T$.

Until the end of Sect. 2, we consider: *a logic L_0, an extension L of L_0, and the associated Galois connection $F \dashv G$.*

Definition 4. *A theory T' of L is L_0-derivable from a theory T of L if $T' = T + F(T'_0)$ for some theory T'_0 of L_0. A theory T of L is* (relatively) Hilbert-Post complete with respect to L_0 *if it is consistent and if any theory of L containing T is L_0-derivable from T.*

Each theory T is L_0-derivable from itself, as $T = T + F(T_{min,0})$, where $T_{min,0}$ is the minimal theory of L_0. In addition, Theorem 6 shows that relative completeness lifts the usual "absolute" completeness from L_0 to L, and Proposition 7 proves that relative completeness is well-suited to the combination of effects.

Lemma 5. *For each theory T of L, a theory T' of L is L_0-derivable from T if and only if $T' = T + F(G(T'))$. As a special case, T_{max} is L_0-derivable from T if and only if $T_{max} = T + F(T_{max,0})$. A theory T of L is Hilbert-Post complete with respect to L_0 if and only if it is consistent and every theory T' of L containing T is such that $T' = T + F(G(T'))$.*

Proof. Clearly, if $T' = T + F(G(T'))$ then T' is L_0-derivable from T. So, let T'_0 be a theory of L_0 such that $T' = T + F(T'_0)$, and let us prove that $T' = T + F(G(T'))$. For each theory T' we know that $F(G(T')) \subseteq T'$; since here $T \subseteq T'$ we get $T + F(G(T')) \subseteq T'$. Conversely, for each theory T'_0 we know that $T'_0 \subseteq G(F(T'_0))$ and that $G(F(T'_0)) \subseteq G(T) + G(F(T'_0)) \subseteq G(T + F(T'_0))$, so that $T'_0 \subseteq G(T + F(T'_0))$; since here $T' = T + F(T'_0)$ we get first $T'_0 \subseteq G(T')$ and then $T' \subseteq T + F(G(T'))$. Then, the result for T_{max} comes from the fact that $G(T_{max}) = T_{max,0}$. The last point follows immediately.

Theorem 6. *Let T_0 be a theory of L_0 and $T = F(T_0)$. If T_0 is Hilbert-Post complete (in L_0) and T is Hilbert-Post complete with respect to L_0, then T is Hilbert-Post complete (in L).*

Proof. Since T is complete with respect to L_0, it is consistent. Since $T = F(T_0)$ we have $T_0 \subseteq G(T)$. Let T' be a theory such that $T \subseteq T'$. Since T is complete with respect to L_0, by Lemma 5 we have $T' = T + F(T'_0)$ where $T'_0 = G(T')$. Since $T \subseteq T', T_0 \subseteq G(T)$ and $T'_0 = G(T')$, we get $T_0 \subseteq T'_0$. Thus, since T_0 is complete, either $T'_0 = T_0$ or $T'_0 = T_{max,0}$; let us check that then either $T' = T$ or $T' = T_{max}$. If $T'_0 = T_0$ then $F(T'_0) = F(T_0) = T$, so that $T' = T + F(T'_0) = T$. If $T'_0 = T_{max,0}$ then $F(T'_0) = F(T_{max,0})$; since T is complete with respect to L_0, the theory T_{max} is L_0-derivable from T, which implies (by Lemma 5) that $T_{max} = T + F(T_{max,0}) = T'$.

Proposition 7. *Let L_1 be an intermediate logic between L_0 and L, let $F_1 \dashv G_1$ and $F_2 \dashv G_2$ be the Galois connections associated to the extensions L_1 of L_0 and L of L_1, respectively. Let $T_1 = F_1(T_0)$ and let $T = F_2(T_1)$. If T_1 is Hilbert-Post complete with respect to L_0 and T is Hilbert-Post complete with respect to L_1 then T is Hilbert-Post complete with respect to L_0.*

Proof. This is an easy consequence of the fact that $F = F_2 \circ F_1$.

Corollary 10 provides a characterization of relative Hilbert-Post completeness which is used in the next Sections and in the Coq implementation.

Definition 8. *For each set E of formulas let $Th(E)$ be the theory generated by E; and when $E = \{e\}$ let $Th(e) = Th(\{e\})$. Then two sets E_1, E_2 of formulas are T-equivalent if $T + Th(E_1) = T + Th(E_2)$; and a formula e of L is L_0-derivable from a theory T of L if $\{e\}$ is T-equivalent to E_0 for some set E_0 of formulas of L_0.*

Proposition 9. *Let T be a theory of L. Each theory T' of L containing T is L_0-derivable from T if and only if each formula e in L is L_0-derivable from T.*

Proof. Let us assume that each theory T' of L containing T is L_0-derivable from T. Let e be a formula in L, let $T' = T + Th(e)$, and let T'_0 be a theory of L_0 such that $T' = T + F(T'_0)$. The definition of $Th(-)$ is such that $Th(T'_0) = F(T'_0)$, so that we get $T + Th(e) = T + Th(E_0)$ where $E_0 = T'_0$. Conversely, let us assume that each formula e in L is L_0-derivable from T. Let T' be a theory containing T. Let $T'' = T + F(G(T'))$, so that $T \subseteq T'' \subseteq T'$ (because $F(G(T')) \subseteq T'$ for any T'). Let us consider an arbitrary formula e in T', by assumption there is a set E_0 of formulas of L_0 such that $T + Th(e) = T + Th(E_0)$. Since e is in T' and $T \subseteq T'$ we have $T + Th(e) \subseteq T'$, so that $T + Th(E_0) \subseteq T'$. It follows that E_0 is a set of theorems of T' which are formulas of L_0, which means that $E_0 \subseteq G(T')$, and consequently $Th(E_0) \subseteq F(G(T'))$, so that $T + Th(E_0) \subseteq T''$. Since $T + Th(e) = T + Th(E_0)$ we get $e \in T''$. We have proved that $T' = T''$, so that T' is L_0-derivable from T.

Corollary 10. *A theory T of L is Hilbert-Post complete with respect to L_0 if and only if it is consistent and for each formula e of L there is a set E_0 of formulas of L_0 such that $\{e\}$ is T-equivalent to E_0.*

3   Completeness for Exceptions

Exception handling is provided by most modern programming languages. It allows to deal with anomalous or exceptional events which require special processing. E.g., one can easily and simultaneously compute dynamic evaluation in exact linear algebra using exceptions [8]. There, we proposed to deal with exceptions as a decorated effect: a term $f : X \to Y$ is not interpreted as a function $f : X \to Y$ unless it is pure. A term which may raise an exception is instead interpreted as a function $f : X \to Y + E$ where "+" is the disjoint union operator and E is the set of exceptions. In this section, we prove the relative Hilbert-Post completeness of the decorated theory of exceptions in Theorem 15.

As in [8], decorated logics for exceptions are obtained from equational logics by classifying terms. Terms are classified as *pure* terms or *propagators*, which is expressed by adding a *decoration* or superscript, respectively (0) or (1); decoration and type information about terms may be omitted when they are clear from the context or when they do not matter. All terms must propagate exceptions, and propagators are allowed to raise an exception while pure terms are not. The fact of catching exceptions is hidden: it is embedded into the `try-catch` construction, as explained below. In Sect. 4 we consider a translation of the `try-catch` construction in a more elementary language where some terms are *catchers*, which means that they may recover from an exception, i.e., they do not have to propagate exceptions.

Let us describe informally a decorated theory for exceptions and its intended model. Each type X is interpreted as a set, still denoted X. The intended model is described with respect to a set E called the *set of exceptions*, which does not appear in the syntax. A pure term $u^{(0)} : X \to Y$ is interpreted as a function $u : X \to Y$ and a propagator $a^{(1)} : X \to Y$ as a function $a : X \to Y + E$; equations are interpreted as equalities of functions. There is an obvious conversion from pure terms to propagators, which allows to consider all terms as propagators whenever needed; if a propagator $a^{(1)} : X \to Y$ "is" a pure term, in the sense that it has been obtained by conversion from a pure term, then the function $a : X \to Y + E$ is such that $a(x) \in Y$ for each $x \in X$. This means that exceptions are always propagated: the interpretation of $(b \circ a)^{(1)} : X \to Z$ where $a^{(1)} : X \to Y$ and $b^{(1)} : Y \to Z$ is such that $(b \circ a)(x) = b(a(x))$ when $a(x)$ is not an exception and $(b \circ a)(x) = e$ when $a(x)$ is the exception e (more precisely, the composition of propagators is the Kleisli composition associated to the monad $X + E$ [14, Sect. 1]). Then, exceptions may be classified according to their *name*, as in [8]. Here, in order to focus on the main features of the proof of completeness, we assume that there is only one exception name. Each exception is built by *encapsulating* a parameter. Let P denote the type of parameters for exceptions. The fundamental operations for raising exceptions are the propagators $\mathtt{throw}_Y^{(1)} : P \to Y$ for each type Y: this operation throws an exception with a parameter p of type P and pretends that this exception has type Y. The interpretation of the term $\mathtt{throw}_Y^{(1)} : P \to Y$ is a function $\mathtt{throw}_Y : P \to Y + E$ such that $\mathtt{throw}_Y(p) \in E$ for each $p \in P$. The fundamental operations for handling exceptions are the propagators $(\mathtt{try}(a)\mathtt{catch}(b))^{(1)} : X \to Y$ for each terms $a : X \to Y$ and $b : P \to Y$: this operation first runs a until an exception with parameter p is raised (if any), then, if such an exception has been raised, it runs $b(p)$. The interpretation of the term $(\mathtt{try}(a)\mathtt{catch}(b))^{(1)} : X \to Y$ is a function $\mathtt{try}(a)\mathtt{catch}(b) : X \to Y + E$ such that $(\mathtt{try}(a)\mathtt{catch}(b))(x) = a(x)$ when a is pure and $(\mathtt{try}(a)\mathtt{catch}(b))(x) = b(p)$ when $a(x)$ throws an exception with parameter p.

More precisely, first the definition of the *monadic equational logic* L_{eq} is recalled in Fig. 1, (as in [14], this terminology might be misleading: the logic is called *monadic* because all its operations are have exactly one argument, this is unrelated to the use of the *monad* of exceptions).

A monadic equational logic is made of types, terms and operations, where all operations are unary, so that terms are simply paths. This constraint on arity will make it easier to focus on the completeness issue. For the same reason, we also assume that there is an *empty type* $\mathbb{0}$, which is defined as an *initial object*:

Terms are closed under composition:

$u_k \circ \cdots \circ u_1 : X_0 \to X_k$ for each $(u_i : X_{i-1} \to X_i)_{1 \le i \le k}$, and $id_X : X \to X$ when $k = 0$

Rules: (equiv) $\dfrac{u}{u \equiv u}$ $\dfrac{u \equiv v}{v \equiv u}$ $\dfrac{u \equiv v \quad v \equiv w}{u \equiv w}$

(subs) $\dfrac{u : X \to Y \quad v_1 \equiv v_2 : Y \to Z}{v_1 \circ u \equiv v_2 \circ u}$ (repl) $\dfrac{v_1 \equiv v_2 : X \to Y \quad w : Y \to Z}{w \circ v_1 \equiv w \circ v_2}$

Empty type $\mathbb{0}$ with terms $[\,]_Y : \mathbb{0} \to Y$ and rule: (initial) $\dfrac{u : \mathbb{0} \to Y}{u \equiv [\,]_Y}$

Fig. 1. Monadic equational logic L_{eq} (with empty type)

for each Y there is a unique term $[\,]_Y : \mathbb{0} \to Y$ and each term $u^{(0)} : Y \to \mathbb{0}$ is the inverse of $[\,]_Y^{(0)}$. In the intended model, $\mathbb{0}$ is interpreted as the empty set.

Then, the monadic equational logic L_{eq} is extended to form the *decorated logic for exceptions* L_{exc} by applying the rules in Fig. 2, with the following intended meaning:

– (initial$_1$): the term $[\,]_Y$ is unique as a propagator, not only as a pure term.
– (propagate): exceptions are always propagated.
– (recover): the parameter used for throwing an exception may be recovered.
– (try): equations are preserved by the exceptions mechanism.
– (try$_0$): pure code inside **try** never triggers the code inside **catch**.
– (try$_1$): code inside **catch** is executed when an exception is thrown inside **try**.

Pure part: the logic L_{eq} with a distinguished type P

Decorated terms: $\mathtt{throw}_Y^{(1)} : P \to Y$ for each type Y,

$(\mathtt{try}(a)\mathtt{catch}(b))^{(1)} : X \to Y$ for each $a^{(1)} : X \to Y$ and $b^{(1)} : P \to Y$, and

$(a_k \circ \cdots \circ a_1)^{(\max(d_1,\ldots,d_k))} : X_0 \to X_k$ for each $(a_i^{(d_i)} : X_{i-1} \to X_i)_{1 \le i \le k}$

with conversion from $u^{(0)} : X \to Y$ to $u^{(1)} : X \to Y$

Rules:

(equiv), (subs), (repl) for all decorations   (initial$_1$) $\dfrac{a^{(1)} : \mathbb{0} \to Y}{a \equiv [\,]_Y}$

(recover) $\dfrac{u_1^{(0)}, u_2^{(0)} : X \to P \quad \mathtt{throw}_Y \circ u_1 \equiv \mathtt{throw}_Y \circ u_2}{u_1 \equiv u_2}$

(propagate) $\dfrac{a^{(1)} : X \to Y}{a \circ \mathtt{throw}_X \equiv \mathtt{throw}_Y}$ (try) $\dfrac{a_1^{(1)} \equiv a_2^{(1)} : X \to Y \quad b^{(1)} : P \to Y}{\mathtt{try}(a_1)\mathtt{catch}(b) \equiv \mathtt{try}(a_2)\mathtt{catch}(b)}$

(try$_0$) $\dfrac{u^{(0)} : X \to Y \quad b^{(1)} : P \to Y}{\mathtt{try}(u)\mathtt{catch}(b) \equiv u}$ (try$_1$) $\dfrac{u^{(0)} : X \to P \quad b^{(1)} : P \to Y}{\mathtt{try}(\mathtt{throw}_Y \circ u)\mathtt{catch}(b) \equiv b \circ u}$

Fig. 2. Decorated logic for exceptions L_{exc}

The *theory of exceptions* T_{exc} is the theory of L_{exc} generated from some arbitrary consistent theory T_{eq} of L_{eq}; with the notations of Sect. 2, $T_{exc} = F(T_{eq})$. The soundness of the intended model follows: see [8, Sect. 5.1] and [6], which are based on the description of exceptions in Java [11, Chap. 14] or in C++ [3, Sect. 15].

Example 11. Using the naturals for P and the successor and predecessor functions (resp. denoted s and p) we can prove, e.g., that try(s(throw 3))catch(p) is equivalent to 2. Indeed, first the rule (propagate) shows that s(throw 3)) $\equiv$ throw 3, then the rules (try) and (try$_1$) rewrite the given term into p(3).

Now, in order to prove the completeness of the decorated theory for exceptions, we follow a classical method (see, e.g., [17, Propositions 2.37 and 2.40]): we first determine canonical forms in Proposition 12, then we study the equations between terms in canonical form in Proposition 13.

Proposition 12. *For each $a^{(1)} : X \to Y$, either there is a pure term $u^{(0)} : X \to Y$ such that $a \equiv u$ or there is a pure term $u^{(0)} : X \to P$ such that $a \equiv \text{throw}_Y \circ u$.*

Proof. The proof proceeds by structural induction. If a is pure the result is obvious, otherwise a can be written in a unique way as $a = b \circ \text{op} \circ v$ where v is pure, op is either throw_Z for some Z or $\text{try}(c)\text{catch}(d)$ for some c and d, and b is the remaining part of a. If $a = b^{(1)} \circ \text{throw}_Z \circ v^{(0)}$, then by (propagate) $a \equiv \text{throw}_Y \circ v^{(0)}$. Otherwise, $a = b^{(1)} \circ (\text{try}(c^{(1)})\text{catch}(d^{(1)})) \circ v^{(0)}$, then by induction we consider two cases.

- If $c \equiv w^{(0)}$ then by (try$_0$) $a \equiv b^{(1)} \circ w^{(0)} \circ v^{(0)}$ and by induction we consider two subcases: if $b \equiv t^{(0)}$ then $a \equiv (t \circ w \circ v)^{(0)}$ and if $b \equiv \text{throw}_Y \circ t^{(0)}$ then $a \equiv \text{throw}_Y \circ (t \circ w \circ v)^{(0)}$.
- If $c \equiv \text{throw}_Z \circ w^{(0)}$ then by (try$_1$) $a \equiv b^{(1)} \circ d^{(1)} \circ w^{(0)} \circ v^{(0)}$ and by induction we consider two subcases: if $b \circ d \equiv t^{(0)}$ then $a \equiv (t \circ w \circ v)^{(0)}$ and if $b \circ d \equiv \text{throw}_Y \circ t^{(0)}$ then $a \equiv \text{throw}_Y \circ (t \circ w \circ v)^{(0)}$.

Thanks to Proposition 12, the study of equations in the logic L_{exc} can be restricted to pure terms and to propagators of the form $\text{throw}_Y \circ v$ where v is pure.

Proposition 13. *For all $v_1^{(0)}, v_2^{(0)} : X \to P$ let $a_1^{(1)} = \text{throw}_Y \circ v_1 : X \to Y$ and $a_2^{(1)} = \text{throw}_Y \circ v_2 : X \to Y$. Then $a_1^{(1)} \equiv a_2^{(1)}$ is T_{exc}-equivalent to $v_1^{(0)} \equiv v_2^{(0)}$.*

Proof. Clearly, if $v_1 \equiv v_2$ then $a_1 \equiv a_2$. Conversely, if $a_1 \equiv a_2$, i.e., if $\text{throw}_Y \circ v_1 \equiv \text{throw}_Y \circ v_2$, then by rule (recover) it follows that $v_1 \equiv v_2$.

In the intended model, for all $v_1^{(0)} : X \to P$ and $v_2^{(0)} : X \to Y$, it is impossible to have $\text{throw}_Y(v_1(x)) = v_2(x)$ for some $x \in X$, because $\text{throw}_Y(v_1(x))$ is in the E summand and $v_2(x)$ in the Y summand of the disjoint union $Y + E$. This means that the functions $\text{throw}_Y \circ v_1$ and v_2 are distinct, as soon as their domain X is a non-empty set. For this reason, it is sound to make the following Assumption 14.

Assumption 14. *In the logic L_{exc}, the type of parameters P is non-empty, and for all $v_1^{(0)} : X \to P$ and $v_2^{(0)} : X \to Y$ with X non-empty, let $a_1^{(1)} = \mathtt{throw}_Y \circ v_1 : X \to Y$. Then $a_1^{(1)} \equiv v_2^{(0)}$ is T_{exc}-equivalent to $T_{max,0}$.*

Theorem 15. *Under Assumption 14, the theory of exceptions T_{exc} is Hilbert-Post complete with respect to the pure sublogic L_{eq} of L_{exc}.*

Proof. Using Corollary 10, the proof relies upon Propositions 12 and 13. The theory T_{exc} is consistent, because (by soundness) it cannot be proved that $\mathtt{throw}_P^{(1)} \equiv id_P^{(0)}$. Now, let us consider an equation between terms with domain X and let us prove that it is T_{exc}-equivalent to a set of pure equations. When X is non-empty, Propositions 12 and 13, together with Assumption 14, prove that the given equation is T_{exc}-equivalent to a set of pure equations. When X is empty, then all terms from X to Y are equivalent to $[\ ]_Y$ so that the given equation is T_{exc}-equivalent to the empty set of pure equations.

4 Completeness of the Core Language for Exceptions

In this section, following [8], we describe a translation of the language for exceptions from Sect. 3 in a *core* language with *catchers*. Thereafter, in Theorem 21, we state the relative Hilbert-Post completeness of this core language. Let us call the usual language for exceptions with $\mathtt{throw}$ and $\mathtt{try\text{-}catch}$, as described in Sect. 3, the *programmers' language* for exceptions. The documentation on the behaviour of exceptions in many languages (for instance in Java [11]) makes use of a *core language* for exceptions which is studied in [8]. In this language, the empty type plays an important role and the fundamental operations for dealing with exceptions are $\mathtt{tag}^{(1)} : P \to \mathbb{0}$ for encapsulating a parameter inside an exception and $\mathtt{untag}^{(2)} : \mathbb{0} \to P$ for recovering its parameter from any given exception. The new decoration (2) corresponds to *catchers*: a catcher may recover from an exception, it does not have to propagate it. Moreover, the equations also are decorated: in addition to the equations '$\equiv$' as in Sect. 3, now called *strong equations*, there are *weak equations* denoted '$\sim$'.

As in Sect. 3, a set E of exceptions is chosen; the interpretation is extended as follows: each catcher $f^{(2)} : X \to Y$ is interpreted as a function $f : X + E \to Y + E$, and there is an obvious conversion from propagators to catchers; the interpretation of the composition of catchers is straightforward, and it is compatible with the Kleisli composition for propagators. Weak and strong equations coincide on propagators, where they are interpreted as equalities, but they differ on catchers: $f^{(2)} \sim g^{(2)} : X \to Y$ means that the functions $f, g : X + E \to Y + E$ coincide on X, but maybe not on E. The interpretation of $\mathtt{tag}^{(1)} : P \to \mathbb{0}$ is an injective function $\mathtt{tag} : P \to E$ and the interpretation of $\mathtt{untag}^{(2)} : \mathbb{0} \to P$ is a function $\mathtt{untag} : E \to P + E$ such that $\mathtt{untag}(\mathtt{tag}(p)) = p$ for each parameter p. Thus, the fundamental axiom relating $\mathtt{tag}^{(1)}$ and $\mathtt{untag}^{(2)}$ is the weak equation $\mathtt{untag} \circ \mathtt{tag} \sim id_P$.

More precisely, the *decorated logic for the core language for exceptions L_{excore}* is defined in Fig. 3 as an extension of the monadic equational logic L_{eq}. There is

---

Pure part: the logic L_{eq} with a distinguished type P

Decorated terms: $\mathtt{tag}^{(1)} \colon P \to \mathbb{0}$, $\mathtt{untag}^{(2)} \colon \mathbb{0} \to P$, and

$(f_k \circ \cdots \circ f_1)^{(\max(d_1,\ldots,d_k))} \colon X_0 \to X_k$ for each $(f_i^{(d_i)} \colon X_{i-1} \to X_i)_{1 \le i \le k}$

with conversions from $f^{(0)}$ to $f^{(1)}$ and from $f^{(1)}$ to $f^{(2)}$

Rules:

(equiv$\equiv$), (subs$\equiv$), (repl$\equiv$) for all decorations

(equiv$\sim$), (repl$\sim$) for all decorations, (subs$\sim$) only when h is pure

(empty$\sim$) $\dfrac{f \colon \mathbb{0} \to Y}{f \sim [\,]_Y}$ ($\equiv$-to-$\sim$) $\dfrac{f \equiv g}{f \sim g}$ (ax) $\dfrac{}{\mathtt{untag} \circ \mathtt{tag} \sim id_P}$

(eq$_1$) $\dfrac{f_1^{(d_1)} \sim f_2^{(d_2)}}{f_1 \equiv f_2}$ only when $d_1 \le 1$ and $d_2 \le 1$

(eq$_2$) $\dfrac{f_1, f_2 \colon X \to Y \quad f_1 \sim f_2 \quad f_1 \circ [\,]_X \equiv f_2 \circ [\,]_X}{f_1 \equiv f_2}$

(eq$_3$) $\dfrac{f_1, f_2 \colon \mathbb{0} \to X \quad f_1 \circ \mathtt{tag} \sim f_2 \circ \mathtt{tag}}{f_1 \equiv f_2}$

Fig. 3. Decorated logic for the core language for exceptions L_{excore}

an obvious conversion from strong to weak equations ($\equiv$-to-$\sim$), and in addition strong and weak equations coincide on propagators by rule (eq$_1$). Two catchers $f_1^{(2)}, f_2^{(2)} \colon X \to Y$ behave in the same way on exceptions if and only if $f_1 \circ [\,]_X \equiv f_2 \circ [\,]_X \colon \mathbb{0} \to Y$, where $[\,]_X \colon \mathbb{0} \to X$ builds a term of type X from any exception. Then rule (eq$_2$) expresses the fact that weak and strong equations are related by the property that $f_1 \equiv f_2$ if and only if $f_1 \sim f_2$ and $f_1 \circ [\,]_X \equiv f_2 \circ [\,]_X$. This can also be expressed as a pair of weak equations: $f_1 \equiv f_2$ if and only if $f_1 \sim f_2$ and $f_1 \circ [\,]_X \circ \mathtt{tag} \sim f_2 \circ [\,]_X \circ \mathtt{tag}$ by rule (eq$_3$). The *core theory of exceptions* T_{excore} is the theory of L_{excore} generated from the theory T_{eq} of L_{eq}.

The operation $\mathtt{untag}$ in the core language can be used for decomposing the $\mathtt{try\text{-}catch}$ construction in the programmer's language in two steps: a step for catching the exception, which is nested into a second step inside the $\mathtt{try\text{-}catch}$ block: this corresponds to a translation of the programmer's language in the core language, as in [8], which is reminded below; then Proposition 16 proves the correctness of this translation. In view of this translation we extend the core language with:

- for each $b^{(1)} \colon P \to Y$, a catcher $(\mathtt{CATCH}(b))^{(2)} \colon Y \to Y$ such that $\mathtt{CATCH}(b) \sim id_Y$ and $\mathtt{CATCH}(b) \circ [\,]_Y \equiv b \circ \mathtt{untag}$: if the argument of $\mathtt{CATCH}(b)$ is non-exceptional then nothing is done, otherwise the parameter p of the exception is recovered and $b(p)$ is ran.
- for each $a^{(1)} \colon X \to Y$ and $k^{(2)} \colon Y \to Y$, a propagator $(\mathtt{TRY}(a, k))^{(1)} \colon X \to Y$ such that $\mathtt{TRY}(a, k) \sim k \circ a$: thus $\mathtt{TRY}(a, k)$ behaves as $k \circ a$ on non-exceptional arguments, but it does always propagate exceptions.

Then, a translation of the programmer's language of exceptions in the core language is easily obtained: for each type Y, $\mathtt{throw}_Y^{(1)} = [\ ]_Y \circ \mathtt{tag} : P \to Y$. and for each $a^{(1)} : X \to Y$, $b^{(1)} : P \to Y$, $(\mathtt{try}(a)\mathtt{catch}(b))^{(1)} = \mathtt{TRY}(a, \mathtt{CATCH}(b)) : X \to Y$. This translation is correct: see [10] for a proof of Proposition 16.

Proposition 16. *If the pure term* $[\ ]_Y : 0 \to Y$ *is a monomorphism with respect to propagators for each type* Y, *the above translation of the programmers' language for exceptions in the core language is correct.*

Example 17 (Continuation of Example 11). We here show that it is possible to separate the matching between normal or exceptional behavior from the recovery after an exceptional behavior: to prove that $\mathtt{try}(\mathtt{s}(\mathtt{throw}\ 3))\mathtt{catch}(\mathtt{p})$ is equivalent to 2 in the core language, we first use the translation to get: $\mathtt{TRY}(\mathtt{s}\circ[\ ]\circ\mathtt{tag}\circ 3, \mathtt{CATCH}(\mathtt{p}))$. Then $(\mathtt{empty}_\sim)$ shows that $\mathtt{s}\circ[\ ]\mathtt{tag}\circ 3 \sim [\ ]\circ\mathtt{tag}\circ 3$. Now, the $\mathtt{TRY}$ and $\mathtt{CATCH}$ translations show that $\mathtt{TRY}([\ ] \circ \mathtt{tag} \circ 3, \mathtt{CATCH}(\mathtt{p})) \sim \mathtt{CATCH}(\mathtt{p}) \circ [\ ] \circ \mathtt{tag} \circ 3 \sim \mathtt{p} \circ \mathtt{untag} \circ \mathtt{tag} \circ 3$. Finally the axiom (ax) and $(\mathtt{eq}_1)$ give $\mathtt{p} \circ 3 \equiv 2$.

In order to prove the completeness of the core decorated theory for exceptions, as for the proof of Theorem 15, we first determine canonical forms in Proposition 18, then we study the equations between terms in canonical form in Proposition 19. Let us begin by proving the *fundamental strong equation for exceptions* (1): by replacement in the axiom (ax) we get $\mathtt{tag}\circ\mathtt{untag}\circ\mathtt{tag} \sim \mathtt{tag}$, then by rule $(\mathtt{eq}_3)$:

$$\mathtt{tag} \circ \mathtt{untag} \equiv id_0. \tag{1}$$

Proposition 18. *1. For each propagator* $a^{(1)} : X \to Y$, *either a is pure or there is a pure term* $v^{(0)} : X \to P$ *such that* $a^{(1)} \equiv [\ ]_Y^{(0)} \circ \mathtt{tag}^{(1)} \circ v^{(0)}$. *And for each propagator* $a^{(1)} : X \to 0$ *(either pure or not), there is a pure term* $v^{(0)} : X \to P$ *such that* $a^{(1)} \equiv \mathtt{tag}^{(1)} \circ v^{(0)}$.
2. For each catcher $f^{(2)} : X \to Y$, *either f is a propagator or there is an propagator* $a^{(1)} : P \to Y$ *and a pure term* $u^{(0)} : X \to P$ *such that* $f^{(2)} \equiv a^{(1)} \circ \mathtt{untag}^{(2)} \circ \mathtt{tag}^{(1)} \circ u^{(0)}$.

Proof. 1. If the propagator $a^{(1)} : X \to Y$ is not pure then it contains at least one occurrence of $\mathtt{tag}^{(1)}$. Thus, it can be written in a unique way as $a = b \circ \mathtt{tag} \circ v$ for some propagator $b^{(1)} : 0 \to Y$ and some pure term $v^{(0)} : X \to P$. Since $b^{(1)} : 0 \to Y$ we have $b^{(1)} \equiv [\ ]_Y^{(0)}$, and the first result follows. When $X = 0$, it follows that $a^{(1)} \equiv \mathtt{tag}^{(1)} \circ v^{(0)}$. When $a : X \to 0$ is pure, one has $a \equiv \mathtt{tag}^{(1)} \circ ([\ ]_P \circ a)^{(0)}$.
2. The proof proceeds by structural induction. If f is pure the result is obvious, otherwise f can be written in a unique way as $f = g \circ \mathtt{op} \circ u$ where u is pure, $\mathtt{op}$ is either $\mathtt{tag}$ or $\mathtt{untag}$ and g is the remaining part of f. By induction, either g is a propagator or $g \equiv b \circ \mathtt{untag} \circ \mathtt{tag} \circ v$ for some pure term v and some propagator b. So, there are four cases to consider. (1) If $\mathtt{op} = \mathtt{tag}$ and g is a propagator then f is a propagator. (2) If $\mathtt{op} = \mathtt{untag}$ and g is a propagator then by Point 1 there is a pure term w such that $u \equiv \mathtt{tag} \circ w$, so that

$f \equiv g^{(1)} \circ \mathsf{untag} \circ \mathsf{tag} \circ w^{(0)}$. (3) If $op = \mathsf{tag}$ and $g \equiv b^{(1)} \circ \mathsf{untag} \circ \mathsf{tag} \circ v^{(0)}$ then $f \equiv b \circ \mathsf{untag} \circ \mathsf{tag} \circ v \circ \mathsf{tag} \circ u$. Since $v : \mathbb{0} \to P$ is pure we have $\mathsf{tag} \circ v \equiv id_{\mathbb{0}}$, so that $f \equiv b^{(1)} \circ \mathsf{untag} \circ \mathsf{tag} \circ u^{(0)}$. (4) If $op = \mathsf{untag}$ and $g \equiv b^{(1)} \circ \mathsf{untag} \circ \mathsf{tag} \circ v^{(0)}$ then $f \equiv b \circ \mathsf{untag} \circ \mathsf{tag} \circ v \circ \mathsf{untag} \circ u$. Since v is pure, by (ax) and (subs$_\sim$) we have $\mathsf{untag} \circ \mathsf{tag} \circ v \sim v$. Besides, by (ax) and (repl$_\sim$) we have $v \circ \mathsf{untag} \circ \mathsf{tag} \sim v$ and $\mathsf{untag} \circ \mathsf{tag} \circ v \circ \mathsf{untag} \circ \mathsf{tag} \sim \mathsf{untag} \circ \mathsf{tag} \circ v$. Since $\sim$ is an equivalence relation these three weak equations imply $\mathsf{untag} \circ \mathsf{tag} \circ v \circ \mathsf{untag} \circ \mathsf{tag} \sim v \circ \mathsf{untag} \circ \mathsf{tag}$. By rule (eq$_3$) we get $\mathsf{untag} \circ \mathsf{tag} \circ v \circ \mathsf{untag} \equiv v \circ \mathsf{untag}$, and by Point 1 there is a pure term w such that $u \equiv \mathsf{tag} \circ w$, so that $f \equiv (b \circ v)^{(1)} \circ \mathsf{untag} \circ \mathsf{tag} \circ w^{(0)}$.

Thanks to Proposition 18, in order to study equations in the logic L_{excore} we may restrict our study to pure terms, propagators of the form $[\ ]_Y^{(0)} \circ \mathsf{tag}^{(1)} \circ v^{(0)}$ and catchers of the form $a^{(1)} \circ \mathsf{untag}^{(2)} \circ \mathsf{tag}^{(1)} \circ u^{(0)}$. The proof of Proposition 19 is given in [10].

Proposition 19. *1. For all $a_1^{(1)}, a_2^{(1)} : P \to Y$ and $u_1^{(0)}, u_2^{(0)} : X \to P$, let $f_1^{(2)} = a_1 \circ \mathsf{untag} \circ \mathsf{tag} \circ u_1 : X \to Y$ and $f_2^{(2)} = a_2 \circ \mathsf{untag} \circ \mathsf{tag} \circ u_2 : X \to Y$, then $f_1 \sim f_2$ is T_{excore}-equivalent to $a_1 \circ u_1 \equiv a_2 \circ u_2$ and $f_1 \equiv f_2$ is T_{excore}-equivalent to $\{a_1 \equiv a_2,\ a_1 \circ u_1 \equiv a_2 \circ u_2\}$.*
2. For all $a_1^{(1)} : P \to Y$, $u_1^{(0)} : X \to P$ and $a_2^{(1)} : X \to Y$, let $f_1^{(2)} = a_1 \circ \mathsf{untag} \circ \mathsf{tag} \circ u_1 : X \to Y$, then $f_1 \sim a_2$ is T_{excore}-equivalent to $a_1 \circ u_1 \equiv a_2$ and $f_1 \equiv a_2$ is T_{excore}-equivalent to $\{a_1 \circ u_1 \equiv a_2,\ a_1 \equiv [\ ]_Y \circ \mathsf{tag}\}$.
3. Let us assume that $[\ ]_Y^{(0)}$ is a monomorphism with respect to propagators. For all $v_1^{(0)}, v_2^{(0)} : X \to P$, let $a_1^{(1)} = [\ ]_Y \circ \mathsf{tag} \circ v_1 : X \to Y$ and $a_2^{(1)} = [\ ]_Y \circ \mathsf{tag} \circ v_2 : X \to Y$. Then $a_1 \equiv a_2$ is T_{excore}-equivalent to $v_1 \equiv v_2$.

Assumption 20 is the image of Assumption 14 by the above translation.

Assumption 20. *In the logic L_{excore}, the type of parameters P is non-empty, and for all $v_1^{(0)} : X \to P$ and $v_2^{(0)} : X \to Y$ with X non-empty, let $a_1^{(1)} = [\ ]_Y \circ \mathsf{tag} \circ v_1 : X \to Y$. Then $a_1^{(1)} \equiv v_2^{(0)}$ is T_{exc}-equivalent to $T_{max,0}$.*

Theorem 21. *Under Assumption 20, the theory of exceptions T_{excore} is Hilbert-Post complete with respect to the pure sublogic L_{eq} of L_{excore}.*

Proof. Using Corollary 10, the proof is based upon Propositions 18 and 19. It follows the same lines as the proof of Theorem 15, except when X is empty: because of catchers the proof here is slightly more subtle. First, the theory T_{excore} is consistent, because (by soundness) it cannot be proved that $\mathsf{untag}^{(2)} \equiv [\ ]_P^{(0)}$. Now, let us consider an equation between terms $f_1, f_2 : X \to Y$, and let us prove that it is T_{excore}-equivalent to a set of pure equations. When X is non-empty, Propositions 18 and 19, together with Assumption 20, prove that the given equation is T_{excore}-equivalent to a set of pure equations. When X is empty, then $f_1 \sim [\ ]_Y$ and $f_2 \sim [\ ]_Y$, so that if the equation is weak or if both f_1 and f_2 are propagators then the given equation is T_{excore}-equivalent to the empty set of equations between pure terms. When X is empty and the equation is $f_1 \equiv f_2$ with at least one of f_1 and f_2 a catcher, then by Point 1 or 2 of Proposition 19, the given equation is T_{excore}-equivalent to a set of equations between propagators; but we have seen that each equation between propagators (whether X is empty or not) is T_{excore}-equivalent to a set of equations between pure terms, so that $f_1 \equiv f_2$ is T_{excore}-equivalent to the union of these sets of pure equations.

5   Verification of Hilbert-Post Completeness in Coq

All the statements of Sects. 3 and 4 have been checked in Coq. The proofs can be found in http://forge.imag.fr/frs/download.php/680/hp-0.7.tar.gz, as well as an almost dual proof for the completeness of the state. They share the same framework, defined in [9]:

1. the terms of each logic are inductively defined through the dependent type named **term** which builds a new **Type** out of two input **Types**. For instance, **term Y X** is the **Type** of all terms of the form **f: X → Y**;
2. the decorations are enumerated: **pure** and **propagator** for both languages, and **catcher** for the core language;
3. decorations are inductively assigned to the terms via the dependent type called **is**. The latter builds a proposition (a **Prop** instance in Coq) out of a **term** and a decoration. Accordingly, **is pure (id X)** is a **Prop** instance;
4. for the core language, we state the rules with respect to weak and strong equalities by defining them in a mutually inductive way.

The completeness proof for the exceptions core language is 950 SLOC in Coq where it is 460 SLOC in LaTeX. Full certification runs in 6.745 s on a Intel i7-3630QM @2.40 GHz using the Coq Proof Assistant, v. 8.4pl3. Below table details the proof lengths and timings for each library.

Proof lengths and benchmarks				
Package	Source	Length in Coq	Length in LaTeX	Execution time in Coq
exc_cl-hp	HPCompleteCoq.v	40 KB	15 KB	6.745 s
exc_pl-hp	HPCompleteCoq.v	8 KB	6 KB	1.704 s
exc_trans	Translation.v	4 KB	2 KB	1.696 s
st-hp	HPCompleteCoq.v	48 KB	15 KB	7.183 s

6   Conclusion and Future Work

This paper is a first step towards the proof of completeness of decorated logics for computer languages. It has to be extended in several directions: adding basic features to the language (arity, conditionals, loops, ...), proving completeness of the decorated approach for other effects (not only states and exceptions); the combination of effects should easily follow, thanks to Proposition 7.

References

1. Bauer, A., Pretnar, M.: Programming with algebraic effects and handlers. J. Log. Algebr. Methods Program. **84**, 108–123 (2015)
2. Benton, N., Hughes, J., Moggi, E.: Monads and effects. In: Barthe, G., Dybjer, P., Pinto, L., Saraiva, J. (eds.) APPSEM 2000. LNCS, vol. 2395, pp. 42–122. Springer, Heidelberg (2002)
3. C++ Working Draft: Standard for Programming Language C++. ISO/IEC JTC1/SC22/WG21 standard 14882:2011
4. Domínguez, C., Duval, D.: Diagrammatic logic applied to a parameterisation process. Math. Struct. Comput. Sci. **20**, 639–654 (2010)
5. Dumas, J.-G., Duval, D., Fousse, L., Reynaud, J.-C.: Decorated proofs for computational effects: states. In: ACCAT 2012, Electronic Proceedings in Theoretical Computer Science 93, pp. 45–59 (2012)
6. Dumas, J.-G., Duval, D., Fousse, L., Reynaud, J.-C.: A duality between exceptions and states. Math. Struct. Comput. Sci. **22**, 719–722 (2012)
7. Dumas, J.-G., Duval, D., Reynaud, J.-C.: Cartesian effect categories are Freyd-categories. J. Symb. Comput. **46**, 272–293 (2011)
8. Dumas, J.-G., Duval, D., Ekici, B., Reynaud, J.-C.: Certified proofs in programs involving exceptions. In: CICM 2014, CEUR Workshop Proceedings 1186 (2014)
9. Dumas, J.-G., Duval, D., Ekici, B., Pous, D.: Formal verification in Coq of program properties involving the global state. In: JFLA 2014, pp. 1–15 (2014)
10. Dumas, J.-G., Duval, D., Ekici, B., Pous, D., Reynaud, J.-C.: Hilbert-post completeness for the state and the exception effects (2015). arXiv:1503.00948 [v3]
11. Gosling, J., Joy, B., Steele, G., Bracha, G.: The Java Language Specification, 3rd edn. Addison-Wesley (2005)
12. Idris. The Effects Tutorial
13. Lucassen, J.M., Gifford, D.K.: Polymorphic effect systems. In: POPL. ACM Press, pp. 47–57 (1988)
14. Moggi, E.: Notions of computation and monads. Inf. Comput. **93**(1), 55–92 (1991)
15. Mossakowski, T., Schröder, L., Goncharov, S.: A generic complete dynamic logic for reasoning about purity and effects. Form. Asp. Comput. **22**, 363–384 (2010)
16. Petricek, T., Orchard, D.A., Mycroft, A.: Coeffects: a calculus of context-dependent computation. In: Proceedings of the 19th ACM SIGPLAN International Conference on Functional Programming, pp. 123–135 (2014)
17. Pretnar, M.: The logic and handling of algebraic effects. Ph.D. University of Edinburgh (2010)
18. Plotkin, G., Power, J.: Notions of computation determine monads. In: Engberg, U., Nielsen, M. (eds.) FOSSACS 2002. LNCS, vol. 2303, pp. 342–356. Springer, Heidelberg (2002)
19. Plotkin, G., Pretnar, M.: Handlers of algebraic effects. In: Castagna, G. (ed.) ESOP 2009. LNCS, vol. 5502, pp. 80–94. Springer, Heidelberg (2009)
20. Staton, S.: Completeness for algebraic theories of local state. In: Ong, L. (ed.) FOSSACS 2010. LNCS, vol. 6014, pp. 48–63. Springer, Heidelberg (2010)
21. Tarski, A.: III On some fundamental concepts in mathematics. In: Logic, Semantics, Metamathematics: Papers from 1923 to 1938 by Alfred Tarski, pp. 30–37. Oxford University Press (1956)
22. Uustalu, T., Vene, V.: Comonadic notions of computation. Electr. Notes Theor. Comput. Sci. **203**, 263–284 (2008)
23. Ynot. The Ynot Project

Optimal Coverage in Automotive Configuration

Rouven Walter[1,2]([✉]), Thore Kübart[1,2], and Wolfgang Küchlin[1,2]

[1] Symbolic Computation Group, WSI Informatics, Eberhard-Karls-Universität,
Tübingen, Germany
rouven.walter@uni-tuebingen.de
[2] Steinbeis Technology Transfer Centre STZ OIT, Tübingen, Germany
tkuebart@gmail.com
http://www-sr.informatik.uni-tuebingen.de

Abstract. It is a problem in automotive configuration to determine the minimum number of test vehicles which are needed for testing a given set of equipment options. This problem is related to the minimum set cover problem, but with the additional restriction that we can not enumerate all vehicle variants since in practice their number is far too large for each model type. In this work we illustrate different use cases of minimum set cover computations in the context of automotive configuration. We give formal problem definitions and we develop different approximate (greedy) and exact algorithms. Based on benchmarks of a German premium car manufacturer we evaluate our different approaches to compare their time and quality and to determine tradeoffs.

1 Introduction

In automotive configuration we face the minimum set cover problem in several use cases. For example, it is necessary to determine the minimum number of vehicles needed to cover all the equipment options of a set of tests in order to avoid the unnecessary construction of (very expensive) test vehicles. Since the size of the set of configurable vehicles can grow up to approximately 10^{100} for a model type [9], an enumeration of this set is not possible in practice. This problem can be solved by an implicit representation of this set as a Boolean formula, where each satisfying variable assignment represents a valid vehicle configuration [10,16]. Now we face the problem how to perform optimization tasks within this implicitly represented set.

In this work, we make the following contributions:

1. We illustrate use cases of optimal coverage computations in the automotive configuration domain.
2. We give formal problem definitions for each use case.
3. We present greedy and exact algorithms for the optimal coverage problem with an implicit representation of the models.

R. Walter and T. Kübart—Contributed equally to this work.

I.S. Kotsireas et al. (Eds.): MACIS 2015, LNCS 9582, pp. 611–626, 2016.
DOI: 10.1007/978-3-319-32859-1_52

4. We evaluate the performance and quality tradeoff of our algorithms on benchmarks based on real automotive configuration data from a premium German car manufacturer.

This work is organized as follows: In Sect. 2 we introduce our formal background and notation. In Sect. 3 we point out related work. We illustrate different use cases in Sect. 4 and give formal problem descriptions in Sect. 5. In Sect. 6 (resp. Sect. 7) we develop greedy (resp. exact) algorithms to address the optimal coverage problem. Section 8 contains a detailed evaluation of the different approaches and a comparison in terms of time and quality. Section 9 contains a conclusion.

2   Preliminaries

Our mathematical foundation is Propositional Logic over an infinite set of Boolean variables $\mathcal{V}$ and standard operators $\neg, \wedge$ and $\vee$ with constants $\top$ and $\bot$. The set of Boolean variables occurring in a Boolean formula φ is denoted by $\mathtt{vars}(\varphi)$. A *variable assignment* β is a mapping from $\mathtt{vars}(\varphi)$ to $\mathbb{B} = \{0, 1\}$. The evaluation of formula φ under β is denoted by $\beta(\varphi)$. If $\beta(\varphi) = 1$, then β is called a *model* of φ. A literal is a variable or its negation. A clause is a disjunction of literals. A formula is in *Conjunctive Normal Form* (CNF) if it is a conjunction of clauses. The NP-complete *SAT Problem* [5] is the question whether a propositional formula is satisfiable. For a more detailed description, see Chap. 2 of [4].

Furthermore, we consider *Pseudo-Boolean Constraints* (PBC) [12] of the form $\sum_{i=1}^{n} a_i l_i \geq b$ for $a_i \in \mathbb{Z}$, literals l_i and $b \in \mathbb{Z}$. A PBC is a generalization of a clause. The decision problem, called *Pseudo-Boolean Solving* (PBS), for a set of PBCs is the question whether an assignment β exists which simultaneously satisfies all constraints. The optimization problem for a target function $\sum_{i=1} c_i l_i$ for $c_i \in \mathbb{Z}$ and literals l_i is called *Pseudo-Boolean Optimization* (PBO).

3   Related Work

Singh and Rangaraj [15] describe a model to calculate vehicle configurations aimed at providing the production planning with a set of test variants before actual sales orders are received. The underlying ILP is solved by Branch & Price so that configurations are calculated by the so-called *column generation* method [7].

If a vehicle is not constructible after the selection of some options, a customer would like to receive an optimal repair suggestion, i.e., the maximum number of (prioritized) selections which are simultaneously constructible. Walter et al. [17] use partial weighted MaxSAT to formulate and solve such re-configuration use cases in the context of automotive configuration.

4 Use Cases in the Context of Automotive Configuration

A *product overview formula* φ_{POF} [10] is a Boolean formula which aggregates all configuration constraints of a vehicle series, such that each model β of φ_{POF} represents a configurable (and therefore constructible) vehicle. Such formulas have been developed in our Technology Transfer Centre for 5 German car manufacturers. Furthermore, all parts (materials) from which cars of a vehicle series may be constructed are listed in a bill-of-materials (BOM). A BOM is structured into *positions*, each with alternative position *variants* (e.g., the Engine position with Engine 1, Engine 2, etc., as variants). Each variant has an attached selection condition, which is a Boolean formula φ_i. Given any vehicle as a model β of φ_{POF}, each φ_i must evaluate to **true** if, and only if, the material of this variant belongs in the car β.

Next we present two use cases from automotive configuration involving φ_{POF}.

1. **Optimal Test Vehicle Coverage.** When testing a new type series of vehicles, the manufacturer builds test vehicles to validate the correct behavior of all vehicle configurations. For cost effectiveness, test vehicles are packed with a maximum number of equipment options. However, not all options are compatible with each other (e.g., different gear boxes). Therefore, the problem is to find the *minimum* number of test vehicles which contain all given equipment options.

2. **Optimal Verification Explanation.** In [10], different kinds of verification tests for the BOM in relation to the POF were pioneered. One verification consists of checking for each BOM position that the variant selection formulas do not overlap, i.e., no constructible vehicle will select two alternative materials within a position. This is done by solving the formula $\varphi_{POF} \wedge \varphi_i \wedge \varphi_j$ for all position variant combinations i and j with $i \neq j$. If the result is **true**, then there exists a constructible vehicle which selects two alternative materials at the same position. Furthermore, a position may cause multiple overlap errors. In practice it is important to give the user a comprehensive, but at the same time short, error description, and therefore we need to find a *minimum* number of models covering all overlap errors.

5 Formal Problem Descriptions

We will now reduce the solution of our use cases to the solution of a minimization version of the NP-complete Set Cover Problem [8], which is defined as follows:

Definition 1 *(Minimum Set Cover Problem). Let $U = \{e_1, \ldots, e_m\}$ be a set of elements called the* universe. *Let $\mathcal{S} = \{E_1, \ldots, E_n\} \subseteq \mathcal{P}(U)$ be a set of subsets of the universe U whose union $\bigcup_{E \in \mathcal{S}} E = U$ is the universe. A* cover *is a subset $\mathcal{C} \subseteq \mathcal{S}$ whose union $\bigcup_{E \in \mathcal{C}} E = U$ is the universe.*

The problem of finding a cover of minimum cardinality is the minimum set cover problem *which can be defined as a 0–1 ILP [14]:*

$$\min \sum_{E \in \mathcal{S}} x_E$$

$$\text{s.t.} \sum_{E:e \in E} x_E \geq 1 \qquad \forall e \in U$$

$$x_E \in \{0,1\} \quad \forall E \in \mathcal{S}$$

Next, we formulate both use cases of Sect. 4 in terms of a 0–1 ILP.

5.1   Encoding of a Variable Target Set

For the first use case, **Optimal Test Vehicle Coverage**, we consider a target set $\mathcal{T} = \{c_1, \ldots, c_m\} \subseteq \text{vars}(\varphi_{\text{POF}})$ of configurable options, i.e., $\varphi_{\text{POF}} \wedge c$ is satisfiable for each $c \in \mathcal{T}$. Otherwise, we have to remove the non-configurable options from $\mathcal{T}$ first. Then we consider the universe $U = \mathcal{T}$ and the (practically huge) set of all configurable vehicles $\mathcal{S} = \{\beta \mid \beta(\varphi_{\text{POF}}) = 1\} = \{\beta_1, \ldots, \beta_n\}$. We define the matrix $A(\varphi_{\text{POF}}, \mathcal{T})$ as follows:

$$A(\varphi_{\text{POF}}, \mathcal{T}) := \begin{pmatrix} \beta_1(c_1) & \cdots & \beta_n(c_1) \\ \vdots & & \vdots \\ \beta_1(c_m) & \cdots & \beta_n(c_m) \end{pmatrix}$$

Each column represents the projection of a model of φ_{POF} (i.e., of a constructible vehicle) onto the options in $\mathcal{T}$. Each 0/1 entry indicates whether the model covers the target variable $c_i \in \mathcal{T}$ or not. Then the problem of finding a minimum number of vehicles covering $\mathcal{T}$ can be defined by a 0–1 ILP as follows:

$$\min \sum_{i=1}^{n} x_i$$

$$\text{s.t.} \quad A(\varphi_{\text{POF}}, \mathcal{T}) \cdot \begin{pmatrix} x_1 \\ \vdots \\ x_n \end{pmatrix} \geq 1, \tag{1}$$

$$x_i \in \{0,1\}, \quad i = 1, \ldots, n$$

In other words, we want to find a minimum number of constructible vehicles whose chosen options (the `true` assigned variables) cover $\mathcal{T}$. Variable vector $(x_1 \ldots x_n)$ describes which vehicles are chosen.

5.2   Encoding of a Boolean Formula Target Set

Target set $\mathcal{T}$, as defined above, consists of variables but this is no restriction for the general case if we want to cover a set of Boolean formulas $\{\psi_1, \ldots, \psi_m\}$: For each ψ_i we introduce a new selector variable s_i and add the implication $s_i \rightarrow \psi_i$

to the set of constraints. The resulting target is $\mathcal{T} = \{s_1, \ldots, s_m\}$, consisting only of variables. If a selector variable s_i is covered by a model, then the model also satisfies the corresponding formula ψ_i.

In the second use case **Optimal Verification Explanation**, we consider a BOM position with k variants resulting in a set of overlap errors OE $\subseteq$ $\{\{i, j\} \mid i, j = 1, \ldots, k \text{ and } i \neq j\}$, i.e. formula $\varphi_{\text{POF}} \wedge \psi_i \wedge \psi_j$ is satisfiable for every $\{i, j\} \in$ OE. We can encode this use case by introducing a new selector variable s_i for every formula $\psi_i \wedge \psi_j$ with $\{i, j\} \in$ OE and following the steps described in the previous paragraph.

5.3   Implicit Vehicle Representation

In the context of automotive configuration we face the problem that enumerating all variants is not possible in practice, since the number of models for a model type, implicitly described by φ_{POF}, can grow up to an order of 10^{100} [9]. Thus, we cannot explicitly construct matrix $A\left(\varphi_{\text{POF}}, \mathcal{T}\right)$ and solve the corresponding 0–1 ILP. Instead, we solve the problem by using the implicit representation φ_{POF} of all vehicle configurations. In the following sections we will present greedy and exact algorithms to address the problem of implicit representation.

6   Greedy Algorithms

We present two greedy algorithms in this section. We assume that the target set $\mathcal{T}$ only contains configurable variables w.r.t. the constraints in φ, i.e., $\varphi \wedge v$ is satisfiable for all $v \in \mathcal{T}$.

Algorithm 1. SAT-Based Greedy

Input: Boolean formula φ, target $\mathcal{T} \subseteq \mathbf{vars}(\varphi)$
Output: $\{\beta_1, \ldots, \beta_l\}$

1 $B \leftarrow \emptyset$
2 **while** $\mathcal{T} \neq \emptyset$ **do**
3 $\quad$ $\beta \leftarrow$ SAT $\left(\varphi \wedge \bigvee_{v \in \mathcal{T}} v\right)$
4 $\quad$ $\mathcal{T} \leftarrow \mathcal{T} \setminus \{v \in \mathcal{T} \mid \beta(v) = 1\}$
5 $\quad$ $B \leftarrow B \cup \{\beta\}$
6 **return** B

Algorithm 1 shows a simple greedy algorithm based on iterative SAT calls. In each iteration we solve the formula φ with the additional condition that at least one target variable has to be covered, forced by the constraint $\bigvee_{v \in \mathcal{T}} v$. Thus, the target set $\mathcal{T}$ will be completely covered in some iteration and the algorithm terminates. No optimization computation at all is done, we only solve a decision problem in each iteration. To reduce the number of iterations, we exploit the model by removing its covered codes from the target. In the worst case, only one code is covered in each iteration, yielding a total of $|\mathcal{T}|$ SAT calls.

Here we use the SAT solver as a black box, i.e., any SAT solver can be used. Since the number of iterations depends on the model quality, we can modify the heuristics of the SAT solver. E.g., when deciding over a variable, we may choose a variable $v \in \mathcal{T}$ and branch on $\beta(v) = 1$ first.

Algorithm 2. PBO-Based Greedy

Input: Boolean formula φ, target $\mathcal{T} \subseteq \mathbf{vars}(\varphi)$, number of duplicates $k \in \mathbb{N}$
Output: $\{\beta_1, \ldots, \beta_l\}$

1 $B \leftarrow \emptyset$
2 **while** $\mathcal{T} \neq \emptyset$ **do**
3 $\quad$ CoverCondition $\leftarrow \bigwedge_{i=1}^{k} \varphi^{(i)} \wedge \bigwedge_{v \in \mathcal{T}} \left(s_v \rightarrow \bigvee_{i=1}^{k} v^{(i)} \right)$
4 $\quad$ TargetFuction $\leftarrow \max \sum_{v \in \mathcal{T}} s_v$
5 $\quad$ $\beta \leftarrow$ PBO (TargetFunction, CoverCondition)
6 $\quad$ $\mathcal{T} \leftarrow \mathcal{T} \setminus \left\{ v \in \mathcal{T} \mid \exists i \in \{1, \ldots, k\} : \beta\left(v^{(i)}\right) = 1 \right\}$
7 $\quad$ $B \leftarrow B \cup \text{Extract}(\beta)$
8 **return** B

We can improve Algorithm 1 by optimizing over the target set, i.e., by maximizing the target function $\sum_{v \in \mathcal{T}} v$. Optimization over a target function can be done by a PBO or an ILP solver. We then cover the maximum number of target variables for the next model. Furthermore, we can compute multiple models simultaneously by creating duplicates. Let $\varphi^{(i)}$ be a duplicate of φ by replacing each variable $v \in \mathbf{vars}(\varphi)$ by a fresh variable $v^{(i)}$. We then consider k duplicates $\varphi^{(1)}, \ldots, \varphi^{(k)}$ at the same time. In order to maximize over the target variables, we introduce fresh selector variables s_v for each $v \in \mathcal{T}$ and add the constraints $\bigwedge_{v \in \mathcal{T}} \left(s_v \rightarrow \bigvee_{i=1}^{k} v^{(i)} \right)$. The new target function is $\sum_{v \in \mathcal{T}} s_v$. If a variable s_v is assigned to 1, then at least one of the variables $v^{(1)}, \ldots, v^{(k)}$ is assigned to 1. Algorithm 2 shows this approach of simultaneously optimizing k duplicates of the input formula φ. In the set of models B we gather all models by extracting from the current model β models with original variable names. Since we only solve a local optimization problem this approach is not optimal in general.

Using k duplicates, the number of variables is $k \cdot |\mathbf{vars}(\varphi)|$. Since all duplicates represent the same formula, except for the variable names, we add plenty of symmetry. Symmetries slow down the search process because identified conflicts within a subset of duplicates hold for all combinations of duplicates but have to be re-identified again. Symmetry breaking techniques try to avoid this problem. For example, we could add a lexicographical order of the variables by additional constraints [6]. However, our experiments have shown that this technique does not improve the performance on our instances from automotive configuration, and therefore we discarded this technique in this application.

7   Exact Algorithms

Next we will present exact algorithms. We start by reusing the idea of duplicates of the input formula φ from greedy Algorithm 2. We have to choose the number of duplicates k large enough to simultaneously cover all target codes. To find the optimum number of k, we start by $k = 1$ and increase k by 1 in each iteration until k is large enough. In each iteration we want to ensure that all target variables are covered by at least one duplicate. Thus, we add the cover condition $\bigwedge_{v \in \mathcal{T}} \bigvee_{i=1}^{k} v^{(i)}$. Then we have to check if all duplicate constraints plus the cover condition can be satisfied. If satisfiable, k is large enough and we can extract the optimal cover from the delivered model β. If unsatisfiable, we increase k by 1. Algorithm 3 shows this approach.

Algorithm 3. SAT-Based Incremental Linear Search

Input: Boolean formula φ, target $\mathcal{T} \subseteq \text{vars}(\varphi)$
Output: $\{\beta_1, \ldots, \beta_l\}$
1 $k \leftarrow \text{EstimateLB}(\varphi, \mathcal{T})$
2 **while** true **do**
3   CoverCondition $\leftarrow \bigwedge_{v \in \mathcal{T}} \bigvee_{i=1}^{k} v^{(i)}$
4      $(st, \beta) \leftarrow \text{SAT}\left(\bigwedge_{i=1}^{k} \varphi^{(i)} \wedge \text{CoverCondition}\right)$
5   **if** $st = $ true **then** **return** $\{\beta_1, \ldots, \beta_l\} = \text{Extract}(\beta)$
6   **else** $k \leftarrow k + 1$

We can reduce iterations by estimating a good lower bound for k (subroutine EstimateLB in Algorithm 3). In automotive configuration there are structures which we can exploit. There are *regular* and *optional* groups of variables which are constrained to ensure that *exactly one* or *at most one* of a group of options is assigned to 1. For example, a constructible vehicle has exactly one engine from the regular group of engines. For optional groups like radio, navigation system or CD player, at most one element can be selected. Let $G_{\max}$ be the group of regular and optional groups such that $|G_{\max} \cap \mathcal{T}| \geq |G \cap \mathcal{T}|$ for all regular and optional groups G of φ. Then $|G_{\max} \cap \mathcal{T}|$ is a lower bound since the variables in $G_{\max} \cap \mathcal{T}$ have to be covered by separate models.

Algorithm 3 can also be used in a decremental mode. We start with k duplicates and decrease the value of k by 1 in each iteration until the formula becomes unsatisfiable. A decremental mode has the advantage that the SAT solver has to prove satisfiability in each iteration (except for the last). This is generally faster than proving unsatisfiability. Especially the instances in automotive configuration are not too restrictive, and a model can often be found quickly. To make decremental linear search competitive we have to estimate a good upper bound first. A trivial upper bound is $|\mathcal{T}|$, but we can also use any of the greedy algorithms presented in Sect. 6.

Algorithm 4. SAT-Based Binary Search

Input: Boolean formula φ, target $\mathcal{T} \subseteq \text{vars}(\varphi)$
Output: $\{\beta_1, \ldots, \beta_l\}$

1 $B \leftarrow \emptyset$
2 $lb \leftarrow \text{EstimateLB}(\varphi, \mathcal{T})$
3 $ub \leftarrow \text{EstimateUB}(\varphi, \mathcal{T})$
4 $mid \leftarrow \frac{ub-lb}{2} + lb$
5 **while** $lb \leq ub$ **do**
6 $\quad$ CoverCondition $\leftarrow \bigwedge_{v \in \mathcal{T}} \bigvee_{i=1}^{mid} v^{(i)}$
7 $\quad (st, \beta) \leftarrow \text{SAT}\left(\bigwedge_{i=1}^{mid} \varphi^{(i)} \wedge \text{CoverCondition}\right)$
8 $\quad$ **if** $st = \text{true}$ **then**
9 $\quad\quad B \leftarrow \text{Extract}(\beta)$
10 $\quad\quad ub \leftarrow mid - 1$
11 $\quad$ **else**
12 $\quad\quad lb \leftarrow mid + 1$
13 $\quad mid \leftarrow \frac{ub-lb}{2} + lb$
14 **return** B

Furthermore, we can conduct binary search with a trivial range between 1 and $|\mathcal{T}|$, or with improved ranges between $|G_{\max} \cap \mathcal{T}|$ and the result of a greedy algorithm as an upper bound. Algorithm 4 illustrates this approach.

A substantial disadvantage of the formula $X_k = \bigwedge_{i=1}^{k} \varphi^{(i)} \wedge$ CoverCondition in the previously presented linear and binary search are the contained symmetries. For example, if formula φ implies the constraint $\text{exact}(1, \{c_1, \ldots, c_{k+1}\})$ (e.g., a regular group of engines), then formula X_k contains the constraints of an unsatisfiable pigeon hole instance: It is impossible to distribute the $k+1$ options over the k models. Unsatisfiable pigeon hole instances are known to be very difficult for a SAT solver. Thus, Algorithms 3 and 4 are only suited for instances with a small optimum.

Linear programming combined with Branch & Bound provides an approach for calculating the models of an optimal coverage by iterative rather than simultaneous computation. Algorithm 5 illustrates the so called *Branch & Price* (B&P) approach [3] which calculates a solution x of the relaxation of the 0–1 ILP problem (1) ($x_i \geq 0$, $x_i \in \mathbb{R}$ for all $i = 1, \ldots, n$) by Column Generation [7], and which takes a non-integer x_i of the solution to preferably branch with $x_i = 1$.

The *Master Program* (MP) of Column Generation corresponds to the relaxation of the 0–1 ILP Problem (1). To calculate an optimal solution of MP, a feasible solution is required first. A feasible, qualitatively good starting solution for MP can be determined, for example, with the help of a greedy algorithm. The set $C = \{\beta_1, \ldots, \beta_t\}$ of models of such a starting solution leads to a *Restricted*

Algorithm 5. Branch & Price

Input: Boolean formula φ, target $\mathcal{T} \subseteq \text{vars}(\varphi)$
Output: $\{\beta_1, \ldots, \beta_l\}$

1 $C \leftarrow \text{InitialCover}(\varphi, \mathcal{T})$
2 **return** $\text{Solve}(\varphi, \mathcal{T}, C)$

3 **func** $\text{Solve}(\varphi, \mathcal{T}, C) : \{\beta_1, \ldots, \beta_l\}$
4 $(x = (x_1, \ldots, x_d), D) \leftarrow \text{ColumnGen}(\varphi, \mathcal{T}, C)$ // Real number solution x
5 **if** $\forall i : x_i \in \{0, 1\}$ **then return** $\{\beta_i \in D \mid x_i = 1\}$
6 $lb \leftarrow \sum x_i$
7 $B \leftarrow \text{Solve}(D, \mathcal{T})$ // Integer min. set cover with explicit models
8 $ub \leftarrow |B|$
9 **if** $ub - lb < 1$ **then return** B
10 **else**
11 | $\beta \leftarrow \text{Select}(x, D)$ // Pick a model to branch
12 | $\mathcal{T}_\beta \leftarrow \{v \in \mathcal{T} \mid \beta(v) = 0\}$
13 | $B \leftarrow \{\beta\} \cup \text{Solve}(\varphi, \mathcal{T}_\beta, D)$ // Including beta
14 | $ub = \min(ub, |B|)$
15 | **if** $ub - lb < 1$ **then return** B
16 | **else**
17 | | $\varphi_{\neg\beta} \leftarrow \varphi \wedge \bigvee_{v \in \mathcal{T}_\beta} v$
18 | | $D_{\neg\beta} \leftarrow \{\beta \in D \mid \beta(\varphi_{\neg\beta}) = 1\}$
19 | | $D_{\neg\beta} \leftarrow \text{ExtendToCompleteCover}(D_{\neg\beta})$
20 | | **return** $\text{Solve}(\varphi_{\neg\beta}, \mathcal{T}, D_{\neg\beta})$ // Excluding beta and neighbours

---

Master Program (RMP):

$$A(C, \mathcal{T}) := \begin{pmatrix} \beta_1(c_1) & \ldots & \beta_t(c_1) \\ \vdots & & \vdots \\ \beta_1(c_m) & \ldots & \beta_t(c_m) \end{pmatrix},$$

$$\min \sum_{i=1}^{t} x_i$$

$$\text{s.t.} \quad A(C, \mathcal{T}) \cdot \begin{pmatrix} x_1 \\ \vdots \\ x_t \end{pmatrix} \geq 1, \tag{RMP}$$

$$x_i \in \mathbb{R}_{\geq 0}, i = 1, \ldots, t$$

Derived from the strong duality theorem [14, 7.4] of linear programming, an optimal dual solution $y \in \mathbb{R}^m$, $y \geq \mathbf{0}$, of RMP provides a criterion that might indicate if an optimal solution x of RMP is an optimal solution of MP: If there is no model β of φ with $\sum y_i \beta(c_i) > 1$, then the corresponding basis to x is also optimal for MP. If the criterion does not hold, C is extended by a model β of φ that fulfills $\sum y_i \beta(c_i) > 1$. Determining such a model β is called the *pricing step*, which results in solving a PBO subproblem. In this manner, C is extended until the criterion verifies that the optimal values of RMP and MP are identical.

Method Solve($\varphi, \mathcal{T}, C$) of Algorithm 5 realizes our Branch & Price approach. It calls method ColumnGen($\varphi, \mathcal{T}, C$) first, which returns an optimal solution x of the last RMP solved in Column Generation. In addition, it returns the corresponding set D of models, which contains an optimal basis of MP. The optimal solution x of the relaxation provides a lower bound $lb = \sum x_i$. If x consists only of integers, Column Generation has already created an optimal cover. Otherwise an upper bound $ub = |B|$ can be derived from the models of D by calculating a minimum cover B consisting only of models of D. If $ub - lb < 1$ holds, B is optimal for the integer program. Otherwise, if the gap between upper bound and lower bound is too wide, we branch.

Since one model normally goes hand in hand with many similar models of low Hamming distance, a branch with $x_i = 1$ is examined first (subroutine Select(x, D)). A possible selection heuristic could be $\beta = \beta_j$ with $x_j = \max\{x_i \mid x_i < 1\}$. For that, we execute Solve($\varphi, \mathcal{T}_\beta, D$) with an updated target $\mathcal{T}_\beta$: Only those options have to be covered that are not yet covered by $\beta = \beta_i$. Also, all models of D are made available to the next RMP, so that it is not necessary to calculate them again. If the gap between lower bound and upper bound cannot be reduced below the value of 1, then $x_i = 0$ will be examined.

When applying $x_i = 0$, β is rejected as a possible model in the cover. We can extend the formula φ by an additional constraint so that from now on every model covers at least one option that is not covered by β. This is allowed, because we cut off only models weaker than β. This constraint is especially effective if we ensure that models created by column generation have the following property: There is no model γ of φ such that $\{c_i \mid \beta(c_i) = 1\} \subsetneq \{c_i \mid \gamma(c_i) = 1\}$. We ensure this property as follows: Instead of adding a model β with $\sum y_i \beta(c_i) > 1$ directly to C, we calculate a prime implicant p of φ with $\beta \models p$. Then we generate the model β' of φ by setting all Don't Cares of p to 1. The inequality $\sum y_i \beta'(c_i) > 1$ follows from $y \geq 0$. Finally we extend the RMP by adding β' to C.

Example 1 illustrates the execution of Algorithm 5.

Example 1. Let $\mathcal{T} = \{c_1, \ldots, c_9\}$ and $\varphi = \bigvee_{i=1}^{8} M_i$ such that:

$$A(\varphi, \mathcal{T}) = \begin{array}{c} \begin{array}{cccccccc} M_1 & M_2 & M_3 & M_4 & M_5 & M_6 & M_7 & M_8 \end{array} \\ \left(\begin{array}{cccccccc} 1 & 1 & 0 & 0 & 0 & 0 & 0 & 0 \\ 1 & 0 & 1 & 1 & 1 & 0 & 0 & 0 \\ 0 & 1 & 1 & 1 & 1 & 0 & 0 & 0 \\ 0 & 0 & 0 & 0 & 0 & 1 & 1 & 0 \\ 0 & 0 & 0 & 0 & 0 & 1 & 0 & 1 \\ 0 & 0 & 0 & 0 & 0 & 0 & 1 & 1 \\ 0 & 0 & 1 & 1 & 1 & 1 & 1 & 0 \\ 0 & 0 & 0 & 1 & 1 & 1 & 0 & 1 \\ 0 & 0 & 0 & 0 & 1 & 0 & 1 & 1 \end{array} \right) \begin{array}{c} c_1 \\ c_2 \\ c_3 \\ c_4 \\ c_5 \\ c_6 \\ c_7 \\ c_8 \\ c_9 \end{array} \end{array}$$

We assume that the initial cover is $C = \{M_1, M_2, M_6, M_7\}$. The execution of $B\&P(\varphi, \mathcal{T})$ leads to the following steps:

- ColumnGen$(\varphi, \mathcal{T}, C) \rightsquigarrow D^{(0)} = \{M_1, M_2, M_5, M_6, M_7, M_8\}$,
  $x^{(0)} = (0.5 \ \ 0.5 \ \ 0.5 \ \ 0.5 \ \ 0.5 \ \ 0.5)^T \notin \mathbb{Z}^6 \Rightarrow lb^{(0)} = 3.0$
- Solve$(D^{(0)}, \mathcal{T}) \rightsquigarrow B^{(0)} = \{M_1, M_2, M_6, M_7\} \subseteq D^{(0)} \Rightarrow ub^{(0)} = 4$
- Gap is too wide: $ub^{(0)} - lb^{(0)} = 1$
- Select$(x^{(0)}, D^{(0)}) \rightsquigarrow \beta = M_5$, $\mathcal{T}_\beta = \{c_1, c_4, c_5, c_6\}$
- Call Solve$(\varphi, \mathcal{T}_\beta, D^{(0)})$
  - ColumnGen$(\varphi, \mathcal{T}_\beta, D^{(0)}) \rightsquigarrow D^{(1)} = \{M_1, M_2, M_5, M_6, M_7, M_8\}$,
    $x^{(1)} = (1.0 \ \ 0.0 \ \ 0.0 \ \ 0.5 \ \ 0.5 \ \ 0.5)^T \notin \mathbb{Z}^6 \Rightarrow lb^{(1)} = 2.5$
  - Solve$(D^{(1)}, \mathcal{T}_\beta) \rightsquigarrow B^{(1)} = \{M_1, M_6, M_7\} \subseteq D^{(1)} \Rightarrow ub^{(1)} = 3$
  - $ub^{(1)} - lb^{(1)} = 0.5 < 1$, Return $B^{(1)}$
- $B^{(0)} = \{\beta\} \cup B^{(1)}$, $ub^{(0)}$ remains 4, gap is still too wide
- $\varphi_{\neg\beta} = \varphi \wedge (c_1 \vee c_4 \vee c_5 \vee c_6) = \varphi \wedge \neg M_3 \wedge \neg M_4 \wedge \neg M_5$
- $D^{(0)}_{\neg\beta} = \{M_1, M_2, M_6, M_7, M_8\}$
- Call Solve$(\varphi_{\neg\beta}, \mathcal{T}, D^{(0)}_{\neg\beta})$
  - ColumnGen$(\varphi_{\neg\beta}, \mathcal{T}, D^{(0)}_{\neg\beta}) \rightsquigarrow D^{(2)} = \{M_1, M_2, M_6, M_7, M_8\}$,
    $x^{(2)} = (1.0 \ \ 1.0 \ \ 0.5 \ \ 0.5 \ \ 0.5)^T \notin \mathbb{Z}^5 \Rightarrow lb^{(2)} = 3.5$
  - Solve$(D^{(2)}, \mathcal{T}) \rightsquigarrow B^{(2)} = \{M_1, M_2, M_6, M_7\} \Rightarrow ub^{(2)} = 4.0$
  - $ub^{(2)} - lb^{(2)} = 0.5 < 1$, Return $B^{(2)}$
- Return $B^{(2)}$.

8   Experimental Evaluation

Our tests were run with the following setup: Intel(R) Core(TM) i7-5600U CPU with 2.60 GHz and 4 GB main memory running 64 Bit Windows 7 Professional.

We used six product overview formulas from a German premium car manufacturer. Each product overview formula φ_{POF} represents the constructible vehicles on the level of a model type. Due to the sensitive kind of data, we cannot publish the formulas, but Table 1 shows some of their characteristics.

Table 1. Product overview formula characteristics

	φ_{POF}					
	1	2	3	4	5	6
$|\mathbf{vars}(\varphi_{POF})|$	1778	2252	2561	1928	2263	1886
$|\mathbf{vars}(\mathrm{CNF}(\varphi_{POF}))|$	4133	4687	5018	3547	4059	3971
$|\mathrm{CNF}(\varphi_{POF})|$	70986	82281	88133	60628	64200	72893

8.1   Use Case 1: Optimal Test Vehicle Coverage

In order to choose a realistic target set, we set the country option to the market Germany, which provides a huge variant space. Typically, we are not interested in finding an optimal coverage for worldwide constructible vehicles but only for a specific market. Further, we excluded regular groups (exactly one element has to be selected) which are not relevant when testing vehicle features, i.e., air bag warning label, user manual, paint, upholstery, etc. Thus, we have target sizes $|T|$ of $488, 622, 618, 334, 496$, and 340, for instances $1, 2, 3, 4, 5$, and 6, respectively.

In our evaluations, we used Java 1.8 with the two external solvers SAT4J [11] (with the default solver Glucose 2.1 [1]) and CPLEX [2].

Table 2 shows the greedy solver configurations we used, where k in parentheses is the number of duplicates used. The solver configuration PBO-based greedy algorithm using SAT4J-PBO uses the greedy variant of SAT4J-PBO, since exact PBO solving by SAT4J proved to be too inefficient for our test instances.

Table 2. Greedy solver configurations

Abbreviation	Algorithm	SAT solver	Decision heuristic
ASAT1	SAT-Based Greedy	CPLEX	Default
ASAT2	SAT-Based Greedy	SAT4J	Default
ASAT3	SAT-Based Greedy	SAT4J	Positive first
APBO1(k)	PBO-Based Greedy	CPLEX	Default
APBO2(k)	PBO-Based Greedy	SAT4J-PBO	Default

Table 3 shows the evaluation results of greedy solver settings for Use Case 1. Entries in bold are the best ones among the greedy solvers. Column 'Distance to Opt.' shows the difference $|cover| - |optimal\ cover|$, i.e., the distance to the optimal cover. A distance of 0 is optimal. The PBO-based greedy approach with configuration APBO2(k) is one of the fastest but the distance to the optimum increases for $k > 1$. The PBO-based greedy approach with configuration APBO1(k) is slower by more than a factor 10 but delivers better upper bounds.

Table 4 shows the exact solver configurations we used for the evaluation. Columns LB and UB show the method used to compute a lower bound and upper bound, respectively. Number k in parentheses is the number of duplicates used by the greedy solver for the computation of an upper bound. Linear search with incremental mode and binary search either exceeded the timeout limit, or an out-of-memory exception was thrown on most of the instances. Therefore, we left these two solver settings completely out of the evaluations. The reason behind this could be that these two solver settings perform a great number of satisfiability checks where the result is `false`, which amounts to exploring the whole search space with all of its symmetries to prove that there is no solution.

Table 3. Results of Use Case 1 with greedy algorithms

Solver	Time (s)						Distance to Opt.					
	1	2	3	4	5	6	1	2	3	4	5	6
ASAT1	45.14	92.87	96.74	45.31	47.71	59.24	96	146	144	98	99	110
ASAT2	5.97	9.48	10.99	3.47	7.48	4.17	352	437	449	223	366	210
ASAT3	0.73	2.41	2.70	0.79	1.06	1.33	29	60	73	36	44	44
APBO1(1)	6.44	21.66	19.48	8.35	9.30	10.85	0	6	5	3	3	0
APBO1(2)	7.23	34.13	35.21	16.28	14.88	15.74	1	4	4	4	3	0
APBO1(4)	10.18	150.52	869.43	157.92	75.57	37.85	3	2	4	2	1	0
APBO2(1)	**0.39**	**0.93**	**0.98**	**0.44**	**0.51**	**0.61**	0	8	10	7	7	0
APBO2(2)	0.61	1.27	1.28	0.50	0.67	0.67	1	10	10	8	7	0
APBO2(4)	0.51	1.71	1.54	0.66	0.91	0.88	3	10	16	10	13	0
APBO2(10)	1.69	5.50	4.15	1.52	2.30	2.23	27	42	34	20	29	20

Table 4. Exact solver configurations

Abbreviation	Algorithm	SAT solver	Mode	LB	UB
ELS1(k)	Linear search	CPLEX	decremental	$|G_{max} \cap \mathcal{T}|$	APBO2(k)
EBP1(k)	B&P	–	–	–	APBO1(k)
EBP2(k)	B&P	–	–	–	APBO2(k)

Table 5 shows the results of Use Case 1 using exact algorithms. For all Branch & Price settings we used a PBO-based greedy approach since it delivers good upper bounds, respectively initial covers, within a reasonable time. The best running times, except for one instance, are exhibited by the EBP2(1) configuration.

Table 5. Results of Use Case 1 with exact algorithms

Solver	Time (s)					
	1	2	3	4	5	6
ELS1(1)	21.51	t/o	t/o	748.08	717.99	63.43
ELS1(2)	32.80	t/o	t/o	526.33	597.19	76.39
EBP1(1)	7.38	58.79	171.93	25.79	30.77	12.31
EBP1(2)	10.01	95.32	234.6	34.22	42.12	17.87
EBP1(4)	13.29	232.16	1366.38	228.73	114.87	56.47
EBP2(1)	2.31	**46.35**	**167.01**	**13.91**	**12.17**	**1.55**
EBP2(2)	**1.83**	50.09	217.61	20.05	24.84	2.09
EBP2(4)	2.61	55.53	176.02	32.16	14.38	2.07
EBP2(10)	7.99	72.69	273.52	18.11	21.79	3.72

8.2   Use Case 2: Optimal Verification Explanation

For Use Case 2 we created random pairs $a \wedge b$ of options $a, b \in \mathtt{vars}(\varphi)$ to simulate overlap errors. The target set consists of all created pairs. To investigate the limits of the B&P algorithm we created different target set sizes. Figure 1 shows the increasing running time. For instance 3, we could increase the target size to 400. For all other instances we could increase the target size to 600. For greater target set sizes the algorithm aborted with an out-of-memory exception.

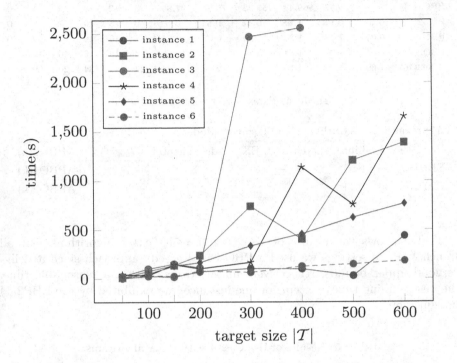

Fig. 1. Use Case 2: Running times for exact solver EBP1(1) (Color figure online)

Table 6 shows a comparison of the optimum results and the upper bounds computed by APBO1(1) for Use Case 1 and Use Case 2. We created this table for the same target size. Use Case 2 has a higher optimum in most cases and the upper bound is often worse.

In summary, it became clear that covering the target set is more difficult for Use Case 2: (i) The upper bounds are worse, (ii) the PBO instances are more complex in the pricing step, and (iii) the improvement in the MP is less.

However, we observed that in both use cases B&P solved the instances by branching with $x_i = 1$ and never had to revise this decision.

Table 6. Comparison of upper bounds of Use Case 1 and Use Case 2 for APBO(1)

		Instances					
		1	2	3	4	5	6
Use Case 1	Optimum	13	18	16	10	11	20
	Upper bound	13	24	21	13	14	20
	Distance	0	6	5	3	3	0
Use Case 2	Optimum	15	18	–	12	12	20
	Upper bound	18	25	–	15	17	23
	Distance	3	7	–	3	5	3

9 Conclusion and Future Work

We presented greedy and exact approaches to the minimum set cover problem with an implicit representation of the models. We evaluated our approaches on real data of a German premium car manufacturer. The exact Branch&Price approach with an upper bound computed by a PBO-based greedy approach was able to solve all instances of Use Case 1, and it was able to solve 1 instance of Use Case 2 up to a target size of 400, and 5 instances up to a target size of 600.

We plan to investigate further improvements: (i) Heuristics for the choice of k for the greedy Algorithm 2 (ii) different computations of upper bounds during Branch and Price, and (iii) a portfolio approach, where we analyze the instance and the target set first and afterwards select an appropriate algorithm.

Even though our first attempts in using symmetry breaking techniques did not help to improve the speed of linear search (cf. the description of Algorithm 3), a deeper investigation is necessary. The duplication of the input formula introduces plenty of symmetry and there may be a way to exploit these symmetries by further symmetry breaking techniques [13] to reduce the search space. This may help to make the linear and binary search algorithms competitive.

References

1. Glucose 2.1. http://www.labri.fr/perso/lsimon/glucose/. Accessed Sept 2015
2. IBM ILOG CPLEX Optimizer. http://www-01.ibm.com/software/commerce/optimization/cplex-optimizer/index.html. Accessed Sept 2015
3. Barnhart, C., Johnson, E.L., Nemhauser, G.L., Savelsbergh, M.W.P., Vance, P.H.: Branch-and-price: column generation for solving huge integer programs. Oper. Res. **46**, 316–329 (1996)
4. Ben-Ari, M.: Mathematical Logic for Computer Science, 3rd edn. Springer, Heidelberg (2012)
5. Cook, S.A.: The complexity of theorem-proving procedures. In: Proceedings of the 3rd Annual ACM Symposium on Theory of Computing, STOC 1971, pp. 151–158. ACM, New York, NY, USA (1971)

6. Crawford, J., Ginsberg, M., Luks, E., Roy, A.: Symmetry-breaking predicates for search problems. In: Aiello, L.C., Doyle, J., Shapiro, S.C. (eds.) Proceedings of the 5th International Conference on Principles of Knowledge Representation and Reasoning, pp. 148–159. Morgan Kaufmann (1996)

7. Desaulniers, G., Desrosiers, J., Solomon, M.M. (eds.): Column Generation. Springer, Heidelberg (2005)

8. Karp, R.M.: Reducibility among combinatorial problems. In: Proceedings of a Symposium on the Complexity of Computer Computations. The IBM Research Symposia Series, pp. 85–103. Plenum Press, New York (1972)

9. Kübler, A., Zengler, C., Küchlin, W.: Model counting in product configuration. In: Lynce, I., Treinen, R. (eds.) Proceedings 1st International Workshop on Logics for Component Configuration, LoCoCo 2010, vol. 29, pp. 44–53. EPTCS (2010)

10. Küchlin, W., Sinz, C.: Proving consistency assertions for automotive product data management. J. Autom. Reasoning **24**(1–2), 145–163 (2000)

11. Le Berre, D., Parrain, A.: The Sat4j library, release 2.2. J. Satisfiability Boolean Model. Comput. **7**(2–3), 59–66 (2010)

12. Roussel, O., Manquinho, V.M.: Pseudo-Boolean and cardinality constraints. In: Handbook of Satisfiability. Frontiers Artificial Intelligence and Applications, vol. 185, pp. 695–733. IOS Press, Amsterdam (2009). Chap. 22

13. Sakallah, K.A.: Symmetry and satisfiability. In: Biere, A., Heule, M., van Maaren, H., Walsh, T. (eds.) Handbook of Satisfiability. Frontiers Artificial Intelligence and Applications, vol. 185, pp. 289–338. IOS Press, Amsterdam (2009)

14. Schrijver, A.: Theory of Linear and Integer Programming. Wiley, Hoboken (1998)

15. Singh, T.R., Rangaraj, N.: Generation of predictive configurations for production planning. In: Aldanondo, M., Falkner, A. (eds.) Proceedings of the 15th International Configuration Workshop, pp. 79–86. Vienna, Austria, August 2013

16. Sinz, C., Kaiser, A., Küchlin, W.: Formal methods for the validation of automotive product configuration data. Artif. Intell. Eng. Des. Anal. Manuf. **17**(1), 75–97 (2003). special issue on configuration

17. Walter, R., Zengler, C., Küchlin, W.: Applications of MaxSAT in automotive configuration. In: Aldanondo, M., Falkner, A. (eds.) Proceedings of the 15th International Configuration Workshop, pp. 21–28. Vienna, Austria, August 2013

Author Index